			13 Group IIIA	14 Group IVA	15 Group VA	16 Group VIA	17 Group VIIA	18 Group VIIIA
								2 He 4.00
			5 B 10.81	6 C 12.01	7 N 14.01	8 O 16.00	9 F 19.00	10 Ne 20.18
10 Group	11 Group IB	12 Group IIB	13 Al 26.98	14 Si 28.09	15 P 30.97	16 S 32.07	17 Cl 35.45	18 Ar 39.95
28 Ni 58.69	29 Cu 63.55	30 Zn 65.38	31 Ga 69.72	32 Ge 72.59	33 As 74.92	34 Se 78.96	35 Br 79.90	36 Kr 83.80
46 Pd 106.42	47 Ag 107.87	48 Cd 112.41	49 In 114.82	50 Sn 118.71	51 Sb 121.75	52 Te 127.60	53 I 126.90	54 Xe 131.29
78 Pt 195.08	79 Au 196.97	80 Hg 200.59	81 Tl 204.38	82 Pb 207.2	83 Bi 208.98	84 Po (209)	85 At (210)	86 Rn (222)

Metals ← → Nonmetals

63 Eu 151.96	64 Gd 157.25	65 Tb 158.93	66 Dy 162.50	67 Ho 164.93	68 Er 167.26			
95 Am (243)	96 Cm (247)	97 Bk (247)	98 Cf (251)	99 Es (252)	100 Fm (257)	101 Md (258)	102 No (259)	103 Lr (260)

7

CREDITS FOR PART AND CHAPTER OPENING PHOTOS

Pt. I · Ira Kirschenbaum/Stock, Boston; Ch. 1 · U.S. Geological Survey; Ch. 3 · Enrico Fermi Institute, Chicago; Ch. 4 · Forsyth from Monkmeyer Press Photo Service; Ch. 5 · Visuals Unlimited/H. Oscar; Ch. 6 · Fred Slavin; Ch. 7 · Mettler Instrument Company; Pt. II · Monkmeyer Press Photo Service; Ch. 8 · Robert J. Ouellette; Ch. 9 · PAR/NYC; Ch. 10 · Dennis Tasa; Pt. III · AP Newsfeatures; Ch. 11 · Visuals Unlimited/Hank Andrews; Ch. 12 · Visuals Unlimited/Isis; Ch. 13 · Barry L. Runk from Grant Heilman; Ch. 14 · United Nations; Ch. 15 · Atomic Industrial Forum; Ch. 16 · Nancy Hays from Monkmeyer Press Photo Service; Ch. 17 · Ros Herion Freese; Ch. 18 · Grant Heilman; Ch. 19 · Union Pacific Railroad Co.; Ch. 20 · Dave Cupp; Ch. 21 · Visuals Unlimited/H. Oscar; Ch. 22 · Visuals Unlimited/Nick Noyes; Ch. 23 · David S. Strikler from Monkmeyer Press Photo Service.

A number of the photographs in the text were taken for Macmillan Publishing Company by Ken Lax.

Cover Photos by:
© Gerhard Gscheidle–Peter Arnold, Inc.
© Werner Müller–Peter Arnold, Inc.
© Michel Viard–Peter Arnold, Inc.

Macmillan Publishing Company
866 Third Avenue, New York, New York 10022

Collier Macmillan Canada, Inc.

LIBRARY OF CONGRESS CATALOGING IN PUBLICATION DATA

Stoker, H. Stephen (Howard Stephen), (date)
    Chemistry: a science for today.

    Includes index.
    1. Chemistry. I. Title.
QD31.2.S757 1988   540   87–26879
ISBN 0-02-417740-7

Printing: 1 2 3 4 5 6 7 8   Year: 9 0 1 2 3 4 5 6 7 8

H. STEPHEN STOKER

WEBER STATE COLLEGE, OGDEN, UTAH

CHEMISTRY
A Science
for Today

Preface

Chemistry: A Science for Today is primarily designed for a one-term course in college chemistry for nonscience majors who plan to take only one chemistry course. It may, however, also be used to provide a foundation for further study in chemistry.

The text has been written under the assumption that students using it have little or no background in chemistry. Early chapters in the text focus on fundamental chemical principles. The mathematics used in considering these principles is purposefully minimized although not entirely avoided. Some chemical principles cannot be divorced from mathematics. The later chapters in the book focus almost entirely on applications of chemistry that students encounter in everyday life. Enough material is included to permit the instructor and students to exercise a reasonable amount of choice concerning course content.

The author strongly feels that educated people, regardless of their areas of concentration, should know something about their physical selves and their environment. In addition, they should be able to relate to and function in that environment in a way that is satisfactory (to themselves). Most of us find ourselves in an environment consisting of other living organisms (including many people) as well as nonliving materials and objects. This textbook offers knowledge about the substances, living and nonliving, of which that environment is composed. Particular emphasis is given to chemicals that are regularly mentioned in the news media and those substances present in widely used consumer products. Such "chemical insights" are flagged in the margins.

Also featured throughout the text are Historical Profiles, brief biographical sketches of scientists who helped develop the theoretical foundations of modern chemistry and important practical applications. These Historical Profiles are designed to remind students that science is "made" by people. If it were not for these people, among others, chemistry would not be the significant science it is today and we would lack many of our important products and processes.

In today's society, it is extremely beneficial for a person to be "chemically literate." It is the goal of this text to cause this transformation in those students who seriously study it. The author hopes to impart to students some of the feelings of interest, appreciation, and (still) amazement that he experiences as he learns more about the human body, the drugs and chemicals used to keep it running properly, the substances used to raise our standard of living (which we take for granted), and the many diverse materials found in the physical environment in which we live.

To help both instructor and student, ancillary materials are available: a Study Guide, an Instructor's Manual, and a Test Bank. The Study Guide provides chapter summaries in outline form, worked-out solutions with explanations for selected problems, and student practice tests. The Instructor's Manual contains complete solutions to all text exercises. The Test Bank contains almost 600 multiple-choice examination questions.

The author appreciates the helpful comments and suggestions from those who reviewed the manuscript: Edward Cain of Rochester Institute of Technology, Irvin M. Citron of Fairleigh Dickinson University, Keith Harper of North Texas State University, and Stanley Johnson of Orange Coast College.

H. S. S.

Contents

I

The Basic Fundamentals of Chemistry

CHAPTER **1**

The Science of Chemistry
3

CHAPTER **2**

Classifications of Matter
19

CHAPTER **3**

Structure of the Atom
46

CHAPTER 4
Electronic Structure of Atoms and the Periodic Table   73

CHAPTER 5
Chemical Bonding   105

CHAPTER 6
Measurement   148

CHAPTER 7
Chemical Calculations: Formula Weights, Moles, and Chemical Equations   176

II
Chemical Principles and Their Applications

CHAPTER **8**
Gases, Liquids, and Solids   **211**

CHAPTER **9**
Solutions and Colloids   **250**

CHAPTER **10**
Chemical Reactions   **279**

III
Chemistry in the Modern World

CHAPTER 16
Introduction to Organic Chemistry—Hydrocarbons   488

CHAPTER 17
Hydrocarbon Derivatives   526

CHAPTER 18
Polymers: Macromolecules   558

CHAPTER 19
Fossil Fuels   590

CHAPTER **20**

Air Pollution   **622**

CHAPTER **21**

Water Pollution   **664**

CHAPTER **22**

Chemistry and Medicine: Drugs   **691**

CHAPTER **23**

Personal Care Products   **729**

I

The Basic Fundamentals of Chemistry

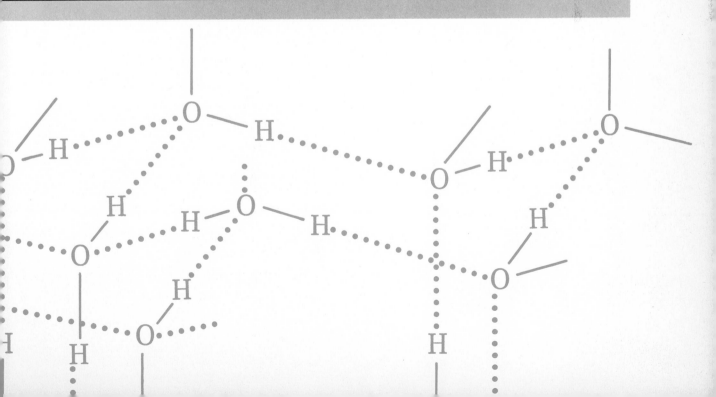

CHAPTER

1

The Science of Chemistry

CHAPTER HIGHLIGHTS

3

1.1 The Scope of Chemistry

Students who are required to take one semester or one quarter of college chemistry often ask the question "Why must I study chemistry? It has nothing to do with my major, which is art, music, accounting, history, English"—or whatever. This question is easily answered: educated people, regardless of their areas of concentration (expertise), should know something about themselves and their environment (the world in which we live). It is the science of chemistry, more than perhaps any other science, that addresses this area of knowledge. Whether we realize it or not, the rise in the standard of living we enjoy and appreciate, both from a health and a materials viewpoint, has paralleled the development and growth of chemistry.

Countless chemical substances we use every day to clothe, house, and transport ourselves are the products obtained from the chemical processing of natural resources such as coal, natural gas, petroleum, and minerals that are present within the earth. Products obtained from these natural resources include textiles, rubber, plastics, metals, fuels, and building materials.

Our bodies are a complex mixture of chemicals. Principles of chemistry are fundamental to an understanding of all processes of the living state. Chemical secretions (hormones) produced within our bodies help determine our outward physical characteristics such as height, weight, and appearance. Digestion of food involves a complex series of chemical reactions. Food itself is an extremely complicated array of chemical substances. Chemical reactions govern our thought processes and how knowledge is stored in and retrieved from our brains. In short, chemistry runs our lives.

Many people, including some now reading this printed page, are alive today because of chemicals used as medicines, drugs, anesthetics, antiseptics, and even artificial organs. Chemicals are used routinely for relief of fever, pain, and nervous tension and for protection from invading organisms (bacteria, germs) in our bodies. You expect doctors to be able to supply you with such chemicals. Central to the development of vaccines for protection from dreaded diseases such as smallpox and polio was the field of chemistry.

Chemical fertilizers and chemical pesticides have led to increased food production. Without the use of such chemicals, the food needs of the world's ever-increasing population could not be met.

The paper on which this textbook is printed was produced from naturally occurring materials (wood pulp) through a chemical process, and the ink used in printing the words on each page is a mixture of many chemicals. Thus, the dissemination of knowledge from one generation to the succeeding one involves chemistry.

Almost all of our recreational pursuits involve objects produced through chemical technology. Skis, boats, basketballs, bowling balls, fishing rods, musical instruments, television sets, and personal computers all contain materials that result from chemical technology. The movies we watch are possible because of synthetic materials called film. The images on film are produced through interaction of selected chemicals.

A formal course in chemistry can be a fascinating experience because it helps us understand ourselves and our surroundings. One cannot truly understand or even know very much about the world we live in or about our own bodies without being conversant with the fundamental ideas of chemistry. Chemistry is often referred to as the "central science" because of the many areas of our lives it touches.

1.2   Chemistry—A Scientific Discipline

The field of chemistry is part of a larger body of knowledge called science. **Science** *is the study in which humans attempt to organize and explain, in a systematic and logical manner, knowledge about themselves and their surroundings.*

The enormous range of types of information covered by science, the sheer amount of accumulated knowledge, and the limitations of human mental capacity to master such a large and diverse body of knowledge have led to the division of science into smaller subdivisions called scientific disciplines. **Scientific disciplines** *are branches of scientific knowledge limited in their size and scope to make them more manageable.* Chemistry is one of these disciplines. Astronomy, botany, geology, physics, and zoology are some of the other disciplines that have resulted from this substructuring of science.

The body of knowledge found within each scientific discipline is itself vast. No one can hope to master completely all knowledge found within a discipline, even in a lifetime of study. However, the fundamental concepts of a given discipline can be learned in a relatively short period of time.

The boundaries between scientific disciplines are not rigid boundaries. All scientific disciplines borrow information and methods from each other. No scientific discipline is a totally independent entity. Problems scientists have encountered in the last decade have particularly pointed out the interdependence of disciplines. For example, chemists attempting to solve the problem of chemical contamination of the environment find that they need some knowledge of geology, zoology, and botany. Because of this overlap, it is now common to talk not only of chemists, but also of geochemists, biochemists,

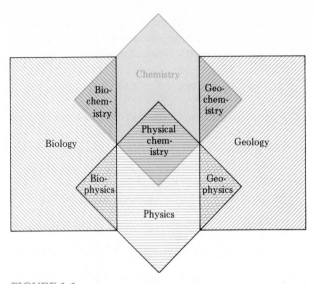

FIGURE 1.1
The overlap of the field of chemistry with selected other
scientific disciplines.

chemical physicists, and so on. Figure 1.1 shows how chemistry merges with
other selected scientific disciplines.

1.3   Chemistry—The Study of Matter

A definition for the scientific discipline of chemistry, given in terms of the areas of
major concern to chemists, will serve as our formal entry point into the realm of
chemistry. **Chemistry** *is the science concerned with (1) the composition and
structure of matter, (2) the properties of matter, and (3) the changes matter
undergoes and the conditions necessary to cause or prevent these changes.*

This definition has meaning only when the key scientific words used within it
are understood. The term *matter* occurs three times in the definition. What is
matter? Most people intuitively understand the meaning of this word. They
consider matter to be the materials of the physical universe—that is, the "stuff"
from which the universe is made. This interpretation is a proper one.

Matter *is anything that has mass and occupies space.* Mass, discussed in
detail in Section 6.2, is a measure of the total quantity of matter present in an

object. A substance need not be visible to the naked eye to be labeled "matter" as long as it meets the two qualifications of having mass and occupying space. Examples of matter include wood, paper, stone, the food we eat, the air we breathe, the fluids of our bodies, our bodies themselves, our clothing, and our shelter. What, then, does matter not include? Matter does not include various forms of energy, such as heat, light, and electricity; nor are wisdom, courage, ideas, thoughts, anger, and love included.

Our definition of chemistry mentions three aspects of matter of particular concern to chemists. These areas constitute the subject matter for the next three sections of this chapter.

1.4 Composition and Structure of Matter

The first of the three "concern areas" for chemists is the composition and structure of matter. The **composition** *of a substance is known when the identity and amount of each of its components have been determined.* It is not enough to know what is present; the amount of each component present must be specified before the composition can be known. **Structure** *is the manner in which the constituent parts of a substance are put together—that is, the order in which the constituent parts are arranged relative to each other.*

Knowledge of composition and structure can often be put to practical use. The chemical composition of certain body fluids, for example, is often used by a physician to pinpoint the cause of illness in a patient. In particular, the absence or excessive amount of certain substances can provide vital information to the physician.

Structural information about a substance often provides insights into its chemical behavior. In some cases these insights allow this behavior to be built into other substances. Detergents, substances with cleaning properties similar to those of soap, have structures patterned after that of soap. Synthetic rubbers, the predominant materials in automobile tires (Figure 1.2), have structures similar to that of natural rubber.

Developments in the field of genetic engineering (Section 25.11) now make it possible to use bacteria to produce human insulin for use by diabetic persons. Before these developments, diabetics had to use animal insulin obtained from slaughterhouse animals. Knowledge of the structure of insulin was a prerequisite for this genetic engineering breakthrough.

FIGURE 1.2
Much of the rubber in automobile tires is synthetic rather than natural. Knowledge
of the structure of natural rubber was the starting point for the development of
synthetic rubbers.
[Tom Kelly.]

1.5 Physical and Chemical Properties of Matter

Properties constitute the second aspect of matter that is of particular concern to
chemists (Section 1.3). **Properties** *are the distinguishing characteristics of a
substance that are used in its identification and description.* Just as we recognize
a friend by characteristics such as hair color, walk, tone of voice, or shape of
nose, we recognize various chemical substances by their properties. Each
chemical substance has a unique set of properties that distinguishes it from all
other substances. If two samples of pure materials have identical properties, they
are necessarily the same substance.

The properties of substances can be used in a number of practical ways, such
as the following.

1. *Identifying an unknown substance.*   Identifying a confiscated drug sample as marijuana involves comparing the properties of the drug with those of known marijuana samples.
2. *Distinguishing between different substances.*   A dentist can quickly tell the difference between a real tooth and a false tooth because of property differences.
3. *Characterizing a newly discovered substance.*   Any new substance must have a unique set of properties; they must be different from those of any previously characterized substance.
4. *Predicting the usefulness of a substance for specific applications.*   Water-soluble substances obviously should not be used in the manufacture of bathing suits. A substance that causes hair to fall out is not a proper ingredient for a new antidandruff formulation.

Two general categories of properties of matter exist: physical and chemical. **Physical properties** *are properties that are observable without changing a substance into another substance.* Color, odor, taste, size, physical state (solid, liquid, or gas), boiling point, melting point, and density are all examples of physical properties.

During the process of determining a physical property, a substance's physical appearance may change but not its identity. For example, the melting point of ice cannot be measured without changing the ice to liquid water. Although the liquid's appearance is much different from that of the solid ice, the substance is still water; its chemical identity has not changed. Hence, melting point is a physical property.

Chemical properties *are properties that matter exhibits as it undergoes changes in chemical composition.* Often these composition changes result from the interaction (reaction) of the matter with other substances. When copper objects are exposed to moist air for long periods of time, they turn green; this is a chemical property of copper. The green coating formed on the copper is a new substance; it results from the reaction of copper metal with the oxygen, carbon dioxide, and water in air. The properties of this new substance are very different from those of metallic copper.

Sometimes, under proper conditions, a single substance undergoes chemical change in the absence of any other substance in a process called *decomposition.* For example, hydrogen peroxide, in the presence of either heat or light, breaks down into the substances water and oxygen.

The *failure* of a substance to undergo change in the presence of another substance is also considered a chemical property. Both flammability and nonflammability are chemical properties.

Table 1.1 contrasts selected physical and chemical properties of the metal gold. Note that chemical properties cannot be described without reference to other substances. It does not make sense to say that a substance reacts. The

TABLE 1.1

Selected Physical and Chemical Properties of Gold

Physical Properties	*Chemical Properties*
Bright yellow in the solid state	Does not react with oxygen at common temperatures and pressures.
Soft and malleable in the solid state	Reacts with chlorine at elevated temperatures to give a red solid.
Melting point of 1065°C	Does not react with water at common temperatures.
Boiling point of 2808°C	Slowly dissolves in a mixture of nitric and hydrochloric acids to give a yellow solution

substance that it reacts with must be specified because a given substance usually reacts with many substances.

1.6 Physical and Chemical Changes

The third aspect of matter that is of particular interest to chemists (Section 1.3) is *changes* in matter. Changes in matter are common and familiar occurrences. Changes take place, for example, when snow melts, paper is burned, and iron rusts.

Like properties (Section 1.5), changes in matter are classified as physical or chemical. A **physical change** *is a process that does not alter the basic nature (chemical composition) of the substance undergoing change.* No new substances are ever formed as a result of a physical change. A **chemical change** *is a process that involves a change in the basic nature (chemical composition) of the substance.* These changes always involve conversion of the material or materials under consideration into one or more new substances with distinctly different properties and composition from those of the original materials.

A change in physical state is the most common type of physical change. Melting, freezing, evaporation, and condensation all represent changes of state. In any of these processes, the composition of the substance undergoing change remains the same even though its physical state and appearance change. The melting of ice does not produce a new substance; the substance is water before and after the change. Similarly, the steam produced from boiling water is still water.

Changes in size, shape, and state of subdivision are examples of physical changes that are not changes of state. Pulverizing an aspirin tablet into a fine

FIGURE 1.3
As the result of chemical change, bright shiny metal objects acquire a dull "rusty" covering when left exposed to moist air. [PAR/NUC.]

powder and cutting a piece of adhesive tape into small pieces are examples of physical changes that involve only the solid state.

The appearance of one or more new substances is always a characteristic of a chemical change. Consider, for example, the rusting of iron objects left exposed to moist air. The reddish brown substance formed (the rust) is a new substance with obviously different chemical properties from those of the original iron (see Figure 1.3).

Some chemical changes that take place in matter are beneficial because the resulting products are more useful than the starting materials. Other changes, such as the rusting of iron, are undesirable because the resulting products are not useful. By studying the nature of changes in matter, chemists learn how to bring about favorable changes and prevent undesirable ones.

The control of chemical change has been a major factor in reaching our modern standard of living. Two hundred years ago most of the familiar materials used by humans were only physically changed from the way they occurred in nature. Today, most of the familiar materials we use are the product of chemical change. Plastics, synthetic fibers, prescription drugs, and latex paints are all the result of controlled chemical change.

Table 1.2 classifies a number of changes for matter as being either physical or chemical.

Most changes in matter are easily classified as either physical or chemical. However, not all changes are "black" or "white"; there are some "gray" areas. Common salt dissolves easily in water to form a solution of salt water. The salt can easily be recovered by the physical process of evaporating the water. When

TABLE 1.2
Classification of Changes as Physical
or Chemical

Change	Classification
Rotting of a tree stump	Chemical
Sharpening of a pencil	Physical
Taking a bite of food	Physical
Digesting food	Chemical
Burning gasoline	Chemical
Detonation of dynamite	Chemical
Souring of milk	Chemical
Breaking of glass	Physical
Slicing a zucchini	Physical
Cracking a nut	Physical

gaseous hydrogen chloride is dissolved in water, again a solution results; in this case, however, the starting materials cannot be recovered by evaporation. The formation of salt water is considered a physical change because the original components can be recovered in an unchanged form. The second solution presents classification problems because of the possibility that a chemical reaction took place.

The changes involved in cooking an egg also present classification problems. The cooked egg contains the same structural units as the uncooked egg. However, since some changes in structural arrangement have taken place, is the change physical or chemical? Despite the existence of "gray" areas, we will continue to use the concepts of physical and chemical change because their usefulness far outweighs the problems created by a few exceptions.

1.7   The Terms Physical and Chemical

The terms *physical* and *chemical* are commonly used to qualify the meaning of scientific terms. For example, techniques used to accomplish physical change are called physical techniques. Similarly, chemical techniques are used to bring about chemical change. A physical separation is one in which none of the components experiences composition changes. Composition changes are part of a chemical separation process. The messages of the modifiers *physical* and *chemical* are constant: physical denotes no change in composition and chemical denotes change in composition. Table 1.3 summarizes how these terms are used to modify other terms.

TABLE 1.3

Use of the Terms *Physical* and *Chemical* as Modifiers

Physical	*Chemical*
Physical property: a property that can be observed without changing the composition of a substance	*Chemical property*: a property that can be observed only by attempting to change the substance into a new substance
Physical change: a change that does not alter the chemical composition of the substance present	*Chemical change*: a change that alters the chemical composition of one or more substances present
Physical technique: a technique that does not alter the chemical composition of the substances present	*Chemical technique*: a technique that alters the chemical composition of one or more substances present
Physical separation: a separation process that does not cause change in chemical composition	*Chemical separation*: a separation process that results in a change in chemical composition

1.8 How Chemists Discover Things— The Scientific Method

Chemistry is an experimental science; that is, chemical discoveries are made as the result of experimentation. This feature, experimentation, is what distinguishes chemistry (and other sciences) from other types of intellectual activity.

A majority of the scientific and technological advances of the twentieth century are the result of systematic experimentation using a method of problem solving known as the scientific method. The **scientific method** *is a set of specific procedures for acquiring knowledge and explaining phenomena.* Procedural steps in this method are as follows.

1. Identifying a problem and carefully planning procedures to obtain information about all aspects of this problem.
2. Collecting data concerning the problem through observation and experimentation.
3. Analyzing and organizing the data in terms of general statements (generalizations) that summarize the experimental observations.
4. Suggesting probable explanations for the generalizations.
5. Experimenting further to prove or disprove the proposed explanations.

Occasionally a great discovery is made by accident, but the majority of scientific discoveries are the result of the application of the preceding steps over long periods of time. There are no instantaneous steps in the scientific method; applying them requires a considerable amount of time. Even when "luck" is involved, it must be remembered that "chance favors the prepared mind." To take full advantage of an accidental discovery, a person must be well trained in the procedures of the scientific method.

There is a vocabulary associated with the scientific method and its use. This vocabulary includes the terms experiment, fact, law, hypothesis, and theory. Understanding the relationships among these terms is the key to a real comprehension of how chemical knowledge has been and still is obtained.

The beginning step in the search for chemical knowledge is to identify a problem concerning some chemical system that needs to be studied. After determining what other chemists have already learned concerning the selected problem, one sets up procedures for obtaining more information about the problem; that is, one designs experiments to be performed. An **experiment** *is a well-defined, controlled procedure for obtaining information about a system under study.* The exact conditions under which an experiment is carried out must always be noted because conditions (temperature, pressure, etc.) affect results.

New firsthand information is collected about the chosen chemical system by actually carrying out the experimental procedures; that is, new facts about the system are obtained. A **fact** *is a valid observation about some natural phenomenon.* Facts are reproducible pieces of information. If a given experiment is repeated, under exactly the same conditions, the same facts should be obtained. All facts, to be acceptable, must be verifiable by anyone who has the time, means, and knowledge needed to repeat the experiments that led to their discovery. It is important that scientific data be published so that other scientists have the opportunity to evaluate and double check both the data and the experimental design.

Next, an effort is made to determine how the facts about a given chemical system relate to each other and to facts about similar chemical systems. Quite often, repeating patterns become apparent among the collected facts. These patterns lead to generalizations, called laws, about how the chemical systems of concern behave under specific conditions. A **law** *is a generalization that concisely summarizes facts about natural phenomena.*

It should not be assumed that laws are easy to discover. Often, years and years of work and thousands upon thousands of facts are needed before the true relationships among variables in the area under study become apparent.

A law is a description of what happens in a given type of experiment. A law makes no mention of why what happens does happen; it simply summarizes experimental observations without attempting to clarify their causes.

A law may be expressed as a verbal statement or as a mathematical equation. An example of a verbally stated law is "If hot and cold pieces of metal are placed

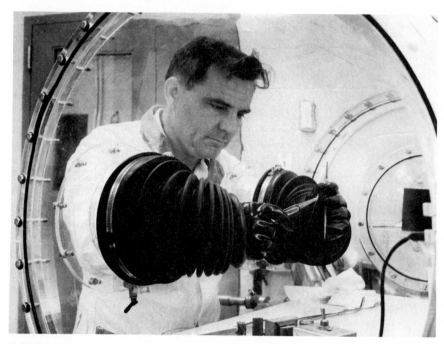

FIGURE 1.4

Experimentation and observation—the basis for discovering scientific laws. This research chemist is working with mildly radioactive materials. Such materials can be handled directly and safely in a glove box. [Brookhaven National Laboratory.]

in contact with each other, the hot piece always cools off and the cold piece always warms up.''

It is important to distinguish between the use of the word *law* in science and its use in a societal context. Scientific laws are *discovered* by research (see Figure 1.4). Researchers have *no control* over what the law turns out to be. Societal laws, which are designed to control aspects of human behavior, are *arbitrary conventions* agreed on (in a democracy) by the majority of those to whom the laws apply. Such laws *can be* and are *changed* when necessary. For example, the speed limit on a particular highway (a societal law) may be decreased or increased for various safety or political reasons.

Chemists, and scientists in general, are not content with just knowing about natural laws. They want to know why a certain type of observation is always made. Thus, after a law is discovered, plausible tentative explanations of the behavior encompassed by the law are worked out by scientists. Such explanations are called hypotheses. A **hypothesis** *is a tentative model or picture that offers an explanation for a law.*

Once a hypothesis has been proposed, experimentation begins again. Many more experiments, under varied conditions, are run to test the reliability of the proposed explanation. The hypothesis must be able to predict the outcome of as yet untried experiments. The validity of the hypothesis depends on the accuracy of its predictions.

It is much easier to disprove a hypothesis than to prove it. A negative result from an experiment indicates that the hypothesis is not valid as formulated and must be modified. Obtaining positive results supports the hypothesis, but does not definitely prove it. There is always the chance that someone will carry out a new type of experiment, never before thought of, that will disprove the hypothesis.

As further experimentation continues to validate a hypothesis, its acceptance in scientific circles increases. If, after extensive testing, the reliability of a hypothesis is still very high, confidence in it increases to the extent that it is accepted by the scientific community at large. After more time has elapsed and additional positive support has accumulated, the hypothesis assumes the status of a theory. A **theory** *is a hypothesis that has been tested and validated over a*

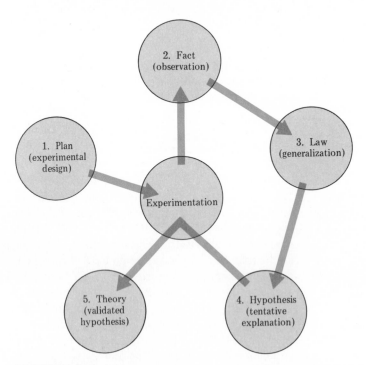

FIGURE 1.5

The central role of experimentation in carrying out the procedures of the scientific method.

long period of time. The dividing line between a hypothesis and a theory is arbitrary and cannot be precisely defined. No set number of supportive experiments must be performed to give theory status to a hypothesis.

Theories serve two important purposes: (1) they allow scientists to predict what will happen in experiments that have not yet been run, and (2) they simplify the very real problem of remembering all the scientific facts that have already been discovered.

Even theories often undergo modification. As scientific tools, particularly instrumentation, become more sophisticated, there is an increasing probability that some experimental observation will not be consistent with all aspects of a given theory. A theory affected in this manner must undergo modification to accommodate the new results or be restated in such a way that scientists know where it is useful and where it is not. All "truths" (theories) in science are provisional and subject to change in light of new experimental observations. Any science is like a living organism. It continues to grow and change. Science develops through a constant interplay between theory and experimentation.

The term *theory* is often misused by nonscientists in everyday contexts. "I have a theory that such and such is the case" is a comment often heard. Here theory is used to mean a "speculative guess," which is not what a theory is. A more appropriate term to describe speculation is hypothesis.

Figure 1.5 summarizes the sequence of steps scientists use when applying the scientific method to a given research problem. It also shows the central role that experimentation plays in the scientific method.

Practice Questions and Problems

CHEMISTRY—THE STUDY OF MATTER

1-1 Define the term *matter.*

1-2 Classify each of the following as matter or nonmatter.
  a. Light   b. Silver metal
  c. T-bone steak   d. Bacteria
  e. Heat   f. Water

1-3 List the three aspects of matter that are of particular concern to chemists.

COMPOSITION AND STRUCTURE OF MATTER

1-4 Define the terms *composition* and *structure.*

1-5 What two types of information are necessary to specify the composition of a substance?

1-6 What type of information is central to specifying the structure of a substance?

PHYSICAL AND CHEMICAL PROPERTIES OF MATTER

1-7 How does a physical property differ from a chemical property?

1-8 The following are properties of the metal lithium. Classify them as physical or chemical properties.
  a. Light enough to float on water
  b. Silvery gray in color
  c. Changes from silvery gray to black when placed in moist air
  d. In the liquid state boils at 1317°C
  e. Can be cut with a sharp knife
  f. Reacts violently with chlorine to form a white solid
  g. In the liquid state reacts spontaneously with its glass container, producing a hole in the container
  h. Burns in oxygen with a bright red flame

1-9 Indicate whether each of the following statements describes a physical property or a chemical property.
  a. An iron nail is attracted to a magnet.
  b. Beryllium metal vapor is extremely toxic to humans.
  c. Water boils at 78°C on top of a 18,500-ft mountain.
  d. Aspirin tablets can be pulverized with a hammer.
  e. Diamond is a very hard substance.
  f. A silver spoon tarnishes in air.
  g. Mercury is a liquid at room temperature.
  h. Chlorine gas has a yellowish green color.

PHYSICAL AND CHEMICAL CHANGES

1-10 What is the difference between a physical change and a chemical change?

1-11 Classify each of the following changes as physical or chemical.
  a. Rusting of an iron nail
  b. Melting of a block of ice
  c. Burning of your chemistry book
  d. Crushing of a dry leaf
  e. Cutting grass in your front yard
  f. Fashioning a table leg from a piece of wood
  g. "Scabbing over" of a skin cut
  h. Hammering of metal into a thin sheet

1-12 Does each of the following statements describe a physical change or a chemical change?
  a. Frozen orange juice is reconstituted by adding water to it.
  b. A board is sawed in two.
  c. A cake is baked in the oven.
  d. A flashlight beam slowly dims and finally goes out.
  e. A powdered drink mix is sweetened by adding sugar to it.
  f. A balloon is inflated using helium gas.
  g. Light is emitted by a flashbulb.
  h. In a blast furnace, iron ore is smelted with coke to produce cast iron.

THE SCIENTIFIC METHOD

1-13 Describe the five steps of the scientific method.

1-14 Match the following four definitions to the terms fact, law, hypothesis, and theory.
  a. A generalization that concisely summarizes information about natural phenomena.

  b. A tentative model or picture that offers an explanation for the behavior of a system.
  c. A tentative model that has been tested and validated over long periods of time.
  d. A valid observation about some natural phenomenon.

1-15 What are the differences between a scientific law and a societal law?

1-16 What are two important purposes that theories serve?

1-17 Why is it important that scientific data be published?

1-18 Why is it useless to conduct an experiment under uncontrolled conditions?

1-19 Indicate whether each of the following statements is true or false.
  a. A theory is a hypothesis that has not yet been subjected to experimental testing.
  b. A law is an explanation of why a particular natural phenomenon occurs.
  c. The validity of a hypothesis depends on its correctly predicting the outcome of yet untried experiments.
  d. A law is a general statement that summarizes a number of experimental facts.
  e. A hypothesis is a tentative law.
  f. A theory is an explanation for which no experimental basis has yet been found.
  g. A hypothesis is a summary of experimental facts.
  h. It is much easier to disprove than to prove a hypothesis.

1-20 Classify each of the following statements as a fact, a law, or a hypothesis.
  a. The ashes from a campfire weigh less than the wood from which they were produced because fire causes atoms to weigh less.
  b. All samples of a pure substance melt at the same temperature under a given set of conditions.
  c. A certain piece of glass, when heated in an open flame to a temperature near its melting point, imparts a yellow color to the flame.
  d. Metal objects, when thrown out a window, always fall to the ground.

2

Classification of Matter

CHAPTER HIGHLIGHTS

2.1 Pure Substances and Mixtures

Chemists are interested in matter. This was a major theme in Chapter 1. In this chapter, we learn more about matter by considering various classifications of matter that exist.

All specimens of matter can be divided into two categories—pure substances and mixtures. What are the distinctions between these two classifications of matter?

A **pure substance** *is a form of matter that always has a definite and constant composition.* This constancy of composition is reflected in the properties of the pure substance. All samples of a given pure substance, regardless of source, have the same properties under the same conditions. Collectively, these definite and constant physical and chemical properties of a pure substance form a set of properties not duplicated by any other pure substance. This unique set of properties provides the identification for the pure substance (Section 1.5).

A pure substance is exactly what the term implies—a single uncontaminated type of matter. All samples of a pure substance contain only that pure substance;

FIGURE 2.1

The individual properties of salt and pepper are easily recognizable in a mixture of the two substances because a mixture is a physical rather than a chemical combination of substances.

thus, pure water is water and nothing else, and pure gold is gold and nothing else.

It is important to note that there is a significant difference between the terms *substance* and *pure substance*. Substance is a general term used to denote any variety of matter. Pure substance is a specific term that applies only to matter with those characteristics just noted.

A **mixture** *is a physical combination of two or more pure substances in which the pure substances retain their identity.* Chemical identity of individual components is retained in a mixture because the components are physically rather than chemically combined. Consider, for example, a mixture of salt and pepper (see Figure 2.1). Close examination of such a mixture shows distinct particles of salt and pepper with no obvious interaction between them. The salt particles in the mixture are identical in properties and composition to the salt particles in the salt container, and the pepper particles in the mixture are no different from those in the pepper bottle.

Once a mixture is made up, its composition is constant. However, mixtures of the same components with different compositions can also be made up; thus, mixtures are characterized as having variable compositions. Consider the large number of salt–pepper mixtures that could be produced by varying the amounts of the two substances present.

An additional characteristic of any mixture is that its components can be

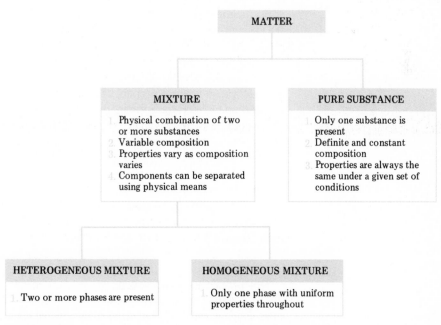

FIGURE 2.2

A comparison of the characteristics of mixtures and pure substances.

retrieved intact from the mixture by physical means, that is, without a chemical change. In many cases, the differences in properties of the various components make the separation relatively easy. For example, in our salt–pepper mixture, the two components could be separated manually, by picking out all of the pepper grains. Alternatively, the separation could be carried out by dissolving the salt particles in water, removing the insoluble pepper particles, and then evaporating the water to recover the salt. In theory, physical separation is possible for the components of all mixtures. In practice, however, separation is sometimes very difficult or nearly impossible. Consider the logistics involved in trying to separate out the components of a ready-made uncooked pizza.

Figure 2.2 summarizes the differences between mixtures and pure substances. Further discussion of these two categories of matter is presented in Sections 2.2 through 2.5.

2.2   Heterogeneous and Homogeneous Mixtures

Sometimes the fact that a sample of matter is a mixture can be determined just by looking at it; that is the case with a salt–pepper mixture. The individual components can be visually distinguished as separate entities. However, visual identification of a mixture's components is not always possible. Sometimes the separate ingredients of a mixture cannot be seen, even with the aid of a microscope. A mixture prepared by dissolving some sugar in water falls into this category. This mixture appears the same as pure water.

Mixtures can be classified as either heterogeneous or homogeneous based on the visual recognizability of the components present. A **heterogeneous mixture** *contains visibly different parts, or phases, each of which has different properties.* The phases present in a heterogeneous mixture have distinct boundaries and are usually easily observed. A pepperoni and cheese pizza is obviously a heterogeneous mixture that has numerous identifiable components. Even a piece of pepperoni contains a number of phases. Common materials such as rocks and wood are also heterogeneous mixtures; different parts of these materials clearly have different properties, such as hardness and color (see Figure 2.3).

The phases in a heterogeneous mixture may or may not be in the same physical state. Set concrete contains a number of phases, all of which are in the solid state. A mixture of sand and water contains two phases, and each is in a different state (solid and liquid). It is possible to have heterogeneous mixtures in which all the components are liquids. For these mixtures to occur, the mixed

FIGURE 2.3

Common materials such as rocks and wood are heterogeneous mixtures.
[Grant Heilman.]

liquids must have limited or no solubility in each other. When this is the case, the mixed liquids form separate layers with the least dense liquid on top. An oil-and-vinegar salad dressing is an example of such a liquid–liquid mixture (see Figure 2.4a). Oil-and-vinegar dressing consists of two phases (oil and vinegar), regardless of whether the mixture consists of two separate layers or of oil droplets dispersed throughout the vinegar, a condition caused by shaking the mixture (see Figure 2.4b). All of the oil droplets together are considered to be a single phase.

Oil

Vinegar

(a) Before shaking

(b) After shaking

FIGURE 2.4

Oil and vinegar salad dressing—a two-phase heterogeneous mixture. Shaking the mixture does not change the number of phases present.

A **homogeneous mixture** *contains only one phase, which has uniform properties throughout it.* This type of mixture can have only one set of properties, which are associated with the single phase present. A spoonful of sugar–water taken from the surface of a homogeneous sugar–water mixture tastes just as sweet as one taken from the bottom of the container. If this were not the case, the mixture would not be truly homogeneous.

All components present in a homogeneous mixture must be in the same state; otherwise heterogeneity would result. Homogeneous mixtures for all three states are common. Air is a homogeneous mixture of gases; motor oil and gasoline are multicomponent homogeneous mixtures of liquids; and metal alloys such as 14-karat gold (a mixture of copper and gold) are examples of solid homogeneous mixtures.

A thorough intermingling of the components in a homogeneous mixture is required for only a single phase to exist. Sometimes this occurs almost instantaneously during the preparation of the mixture, as when alcohol is added to water. At other times, an extended period of mixing or stirring is required. For example, when a hard sugar cube is added to a container of water, it does not dissolve instantaneously to give a homogeneous solution. Only after much stirring does the sugar completely dissolve. Before that point is reached, the mixture is heterogeneous because a solid phase (the undissolved sugar cube) is present.

A useful summary of the major concepts developed in both this section and Section 2.1 is presented in Figure 2.5. This summary is based on the interplay between the terms *heterogeneous* and *homogeneous* and the terms *chemical* and *physical*. From this interplay come the new expressions *chemically homogeneous, chemically heterogeneous, physically homogeneous*, and *physically heterogeneous.*

All pure substances are *chemically homogeneous.* Only one substance can be present in a chemically homogeneous material. Mixtures, which by definition must contain two or more substances, are always *chemically heterogeneous.* The term *physically homogeneous* describes materials consisting of only one phase. If two or more phases are present, then the term *physically heterogeneous* applies.

Pure water (Figure 2.5*a*) is both chemically and physically homogeneous because only one substance and only one phase are present. Sugar–water (Figure 2.5*b*) is physically homogeneous with only one phase present but is chemically heterogeneous because two substances, sugar and water, are present. A mixture of oil and water (Figure 2.5*c*) is both chemically and physically heterogeneous because it contains two substances and two phases. Ice cubes in liquid water (Figure 2.5*d*) represent the somewhat unusual situation of chemical homogeneity and physical heterogenity. This type of combination occurs only when two or more phases of the same substance are present, as in the case of ice and water.

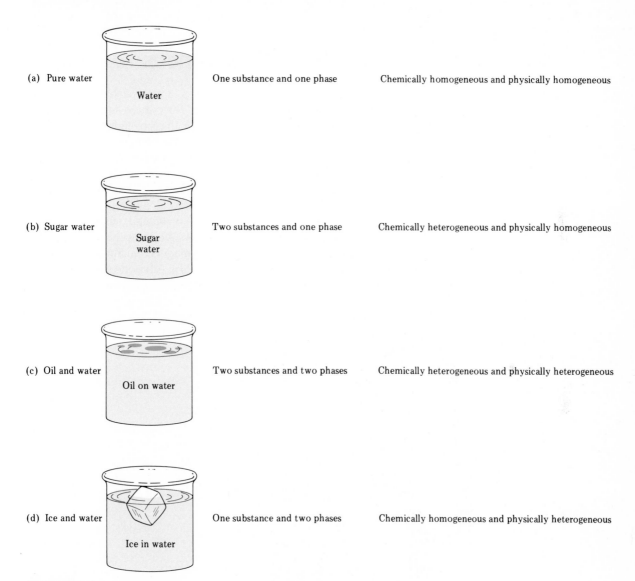

(a) Pure water — One substance and one phase — Chemically homogeneous and physically homogeneous

(b) Sugar water — Two substances and one phase — Chemically heterogeneous and physically homogeneous

(c) Oil and water — Two substances and two phases — Chemically heterogeneous and physically heterogeneous

(d) Ice and water — One substance and two phases — Chemically homogeneous and physically heterogeneous

FIGURE 2.5

Samples of matter described using the terms *chemically heterogeneous, physically heterogeneous, chemically homogeneous,* and *physically homogeneous.*

2.3   Characterization of Pure Substances

Today, many sophisticated physical techniques are available to separate mixtures into their component pure substances. After these techniques have been used, the question still arises as to whether the isolated substances are really pure, that is, how complete the separation into components really is. A substance is called *pure* when its properties, under a given set of conditions, cannot be changed by further attempts at purification. As many properties as possible must be examined, and numerous separation techniques must be used before one can conclude that a substance is pure.

Suppose, for example, that iron filings and powdered sulfur are mixed together. The iron is gray, and the sulfur is yellow. We can physically separate the components of the mixture by stirring it with a magnet; the iron filings cling to the magnet, leaving the sulfur behind (see Figure 2.6). After one stirring with the magnet, the remaining sulfur is still not pure; this is indicated by a gray-yellow

FIGURE 2.6
Iron fillings can be separated from powdered sulfur in a mixture of the two substances with a magnet. The iron, but not the sulfur, is attracted to the magnet.

color. Repeated stirrings with more powerful magnets eventually produce bright yellow sulfur whose color does not change with further treatments. The fact that the color no longer changes is significant, but not absolute, evidence that the sulfur is pure. The property of color should not be used as the sole basis for determining purity; other properties also have to be checked.

A mixture of table salt and fine white sand appears to be a pure substance if only color is used as a criterion for purity. However, a check of other properties quickly indicates a mixture. One part (salt) melts at a lower temperature than the other part (sand). One part (salt) dissolves in water, whereas the other part (sand) does not dissolve.

2.4   Subdivision of Pure Substances

In this section, we consider the processes of both physically and chemically subdividing a sample of a pure substance into smaller and smaller amounts. Throughout this discussion, it is important to remember the distinction between the terms *physical* and *chemical* when they modify other words. Recall, from Section 1.7, that the term *physical* indicates no change in the composition of the substances involved, and the term *chemical* is associated with composition change. Thus, physical subdivision is a subdivision process in which no composition change occurs; during chemical subdivision, on the other hand, composition change does occur.

Physical Subdivision of Pure Substances

Suppose a 1-gram sample of table sugar is divided into $\frac{1}{2}$-gram portions. Then the $\frac{1}{2}$-gram portions are cut in half again to give $\frac{1}{4}$-gram portions, and the subdivision process continues further. Is there a limit to the number of times this physical subdivision process can be carried out on the sugar sample? The answer is yes. There is a limit beyond which you cannot go and still have sugar. That limit is reached when only one unit of the substance called sugar remains. Correspondingly, a limit of physical subdivision exists for all pure substances.

The smallest unit of a pure substance that maintains all of its properties, the limit of physical subdivision, is called a *molecule*. A **molecule** *is the smallest unit of a pure substance obtainable through physical subdivision of the pure substance.* The smallest characteristic unit of every pure substance is a unique molecule.* Two substances with the same molecule as a basic unit would have

---

* The concept of a molecule, as described here, is modified when certain solids, called ionic solids, are considered (Section 5.6).

the same properties and would, thus, have to be classified as the same pure substances.

Chemical Subdivision of Pure Substances

Molecules, the limit of physical subdivision for pure substances, can be broken down into simpler units if chemical, rather than physical, subdivision methods are used. These techniques destroy the identity of the pure substance because chemical change occurs. The ultimate limit of chemical subdivision for a pure substance is the unit of matter called an *atom*. An **atom** *is the smallest unit of a pure substance obtainable through chemical subdivision of the pure substance.*

All molecules, except the few that contain only one atom, can be broken down into their smallest units if appropriate chemical methods are used. In practice, the breakdown of molecules often occurs in steps, with smaller molecules resulting from the intermediate steps. Ultimately, however, the limit of atoms is reached (see Figure 2.7).

It is important to remember that, by definition, chemical subdivision of a molecule destroys the identity of the molecule. A sugar molecule can be broken up into atoms, but the resulting atoms do not have the properties characteristic of sugar.

For the trivial case of a molecule that contains only one atom, the word molecule is seldom used. Such an entity is almost always called simply an atom. Use of the term molecule begins with the presence of two atoms in a molecular unit.

The atoms present in a molecule may all be the same kind, or two or more kinds may be present. On the basis of this observation, molecules are classified into two categories—homoatomic and heteroatomic molecules. **Homoatomic molecules** *are molecules in which all atoms present are the same kind.* Only one kind of atom can be produced from the chemical breakup of a homoatomic molecule. **Heteroatomic molecules** *are molecules in which two or more*

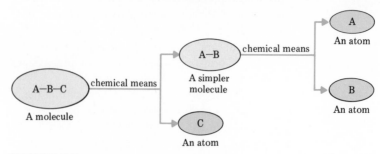

FIGURE 2.7

The breakdown of a molecule containing three atoms (A, B, and C) to yield the constituent atoms.

different kinds of atoms are present. Chemical breakup of heteroatomic molecules always produces two or more kinds of atoms. Classification of molecules as either homoatomic or heteroatomic provides the basis for a further classification of pure substances into the categories of elements and compounds; this is the topic of Section 2.5.

2.5   Elements and Compounds

There are two kinds of pure substances: elements and compounds. An **element** *is a pure substance whose basic unit is a homoatomic molecule.* Elements cannot be broken down into simpler substances by chemical means because they contain only one kind of atom. A **compound** *is a pure substance whose basic unit is a heteroatomic molecule.* Compounds can be broken down into two or more simpler substances by chemical means because at least two different kinds of atoms are present.

Presently 109 pure substances are classified as elements. These elements, which are the simplest known substances, are considered the building blocks of all other types of matter. Every object, regardless of its complexity, is a collection of substances made up from these 109 elements.

Compared to the total number of compounds characterized by chemists, the number of known elements is extremely small. More than 6 million different compounds are known. Each is a definite combination of two or more of the known elements. Figure 2.8 summarizes the differences between elements and compounds.

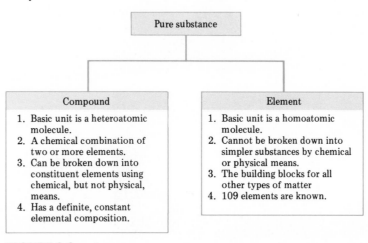

FIGURE 2.8

A comparison of the characteristics of elements and compounds.

Sulfur
1. Solid
2. Melts at 113°C
3. Boils at 444°C
4. Yellow

Zinc
1. Solid
2. Melts at 419°C
3. Boils at 907°C
4. Bluish white

Zinc sulfide
1. Solid
2. Melts at 1850°C
3. White

FIGURE 2.9
Compounds, chemical combinations of elements, always have properties that are
distinctly different from those of the elements of which they are composed.

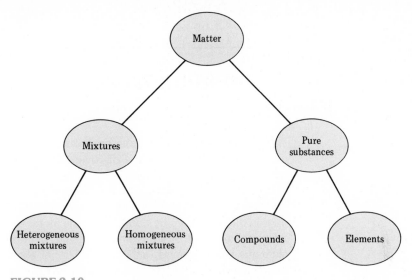

FIGURE 2.10
Categories of classification for matter.

The difference between compounds and mixtures is very important. Compounds are not mixtures, even though two or more simpler substances can be obtained from them. A compound is the result of the *chemical* combination of two or more elements; a mixture is formed by the *physical* mixing of two or more substances (elements or compounds). Compounds always have properties that are distinctly different from those of the elements used to produce them; mixtures retain the properties of their individual components. Figure 2.9 contrasts the properties of zinc sulfide, a compound that contains zinc and sulfur, with the properties of the elements zinc and sulfur. The properties of the compound are distinctly different from those of the two elements used to make the compound.

There are other differences between compounds and mixtures. Compounds have a definite composition, which is a property of all pure substances, and mixtures have variable compositions. Also, because mixtures are only physical combinations of substances, they can be separated by physical means. The separation of a compound into its constituent elements always requires chemical means; they never yield their constituent elements by means of physical separation techniques only.

Figure 2.10 summarizes the overall scheme developed in Sections 2.1 through 2.5 for classifying matter.

2.6 Discovery and Abundance of the Elements

The 109 known elements have been discovered and isolated over a period of several centuries. Most of the discoveries have occurred since 1700, with the 1800s being the most active discovery period. Table 2.1 shows the number of known elements at different times between 1700 and the present.

Eighty-eight of the 109 elements occur naturally, and 21 have been synthesized in the laboratory by bombarding the atoms of naturally occurring elements with small particles. Scientists generally accept that no more naturally occurring elements will be found, although additional elements may be prepared synthetically. The last of the naturally occurring elements was discovered in 1925. The elements synthesized by scientists are all unstable (radioactive) and usually change rapidly into stable elements as the result of radioactive emissions (Section 15.10).

The naturally occurring elements are not evenly distributed on earth or throughout the universe. This nonuniformity of distribution is startling. A small number of elements account for the majority of atoms. One must define the area to be considered before answering the question of which elements are the most abundant. The abundances of various elements found in the earth's crust differ considerably from the abundances found in the earth as a whole; they differ even more from the abundances of the elements in the universe as a whole. When living entities, such as vegetation or the human body, are considered, an

TABLE 2.1

Number of Elements Discovered During Various 50-Year Time Periods

Time Period	Number of Elements Discovered During Time Period	Total Number of Elements Known at End of Time Period
Ancient–1700	13	13
1701–1750	3	16
1751–1800	18	34
1801–1850	25	59
1851–1900	23	82
1901–1950	16	98
1951–present	11	109

TABLE 2.2
Abundance of the Elements from Various Viewpoints

Element	Abundance (atom %)	Element	Abundance (atom %)
Universe		Atmosphere	
Hydrogen	91	Nitrogen	78.3
Helium	9	Oxygen	21.0
Earth (including core)		Hydrosphere	
Oxygen	49.3	Hydrogen	66.4
Iron	16.5	Oxygen	33
Silicon	14.5	Human body	
Magnesium	14.2	Hydrogen	63
Earth's crust		Oxygen	25.5
Oxygen	60.1	Carbon	9.5
Silicon	20.1	Nitrogen	1.4
Aluminum	6.1	Vegetation	
Hydrogen	2.9	Hydrogen	49.8
Calcium	2.6	Oxygen	24.9
Magnesium	2.4	Carbon	24.9
Iron	2.2		
Sodium	2.1		

altogether different abundance perspective emerges. Table 2.2 lists the most abundant elements from a number of different perspectives. The numbers in the table are atom percents, that is, the percent of total atoms in the defined area that are of a given type. For example, Table 2.2 shows that 60.1% of the total atoms in the earth's.crust are oxygen atoms. Only element abundances greater than 1% are listed.

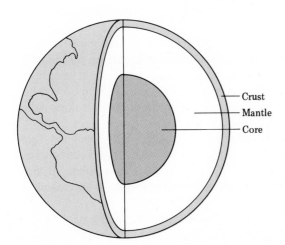

Crust
Mantle
Core

FIGURE 2.11
Structure of the earth. The earth's core, with a 2200-mile radius, is believed to be composed mainly of iron. Surrounding the core is fluid material, called the mantle, that is 1800 miles thick and is thought to be composed primarily of silicon, oxygen, iron, and magnesium. The very thin solid outer layer is the crust, which is 8–26 miles thick.

Note from Table 2.2 that the most abundant elements in the universe are not the most abundant ones on earth. The cosmic figures reflect the composition of stars, which are made up almost entirely of hydrogen and helium. The composition of the earth as a whole and that of its crust differ significantly. These differences reflect the fact that the figures for the earth's crust—its waters, atmosphere, and outer covering—do not take into account the composition of the earth's core or mantle (see Figure 2.11). Note that in both the atmosphere and the hydrosphere only two elements occur in large amounts (greater than 1% of atoms). Also note that the relative number of carbon atoms is greater and that of hydrogen atoms is less in vegetation than in the human body.

2.7   Names and Symbols of the Elements

Each element has a unique name, selected in most cases, by its discoverer. A wide variety of rationales for choosing names are found in studying the origins of element names. Some bear geographical names—germanium, for example, was named after the native country of its German discoverer and the elements francium and polonium acquired their names in a similar manner. The elements mercury, uranium, neptunium, and plutonium are all named for planets. Helium gets its name from the Greek word *helios,* for sun, because it was first observed spectroscopically in the sun's corona during an eclipse. Some elements carry names related to specific properties of the element or the compounds that contain it. Chlorine's name is derived from the Greek *chloros,* denoting greenish yellow, the color of chlorine gas. Iridium gets its name from the Greek *iris* meaning rainbow; this alludes to the varying colors of the compounds from which it was isolated.

In the early 1800s chemists adopted the practice of assigning chemical symbols to the elements. A **chemical symbol** *is an abbreviation for the name of an element.* These chemical symbols are used more frequently in referring to the elements than are the names themselves. The system of chemical symbols now in use was first proposed in 1814 by the Swedish chemist Jöns Jakob Berzelius (1779–1848). (See Historical Profile 1.)

A complete list of all the known elements and their symbols is given in Table 2.3. The symbols and names of the more frequently encountered elements are shown in color in this table. You would do well to learn the symbols of these more common elements. Learning them is a key to having a successful experience in studying chemistry.

Fourteen elements have one-letter symbols, 89 have two-letter symbols, and 6 have three-letter symbols. If a symbol consists of a single letter, it is capitalized.

TABLE 2.3

The Elements and Their Chemical Symbols

Ac	actinium	He	helium	Ra	radium
Ag	silver*	Hf	hafnium	Rb	rubidium
Al	aluminum	Hg	mercury*	Re	rhenium
Am	americium	Ho	holmium	Rh	rhodium
Ar	argon	I	iodine	Rn	radon
As	arsenic	In	indium	Ru	ruthenium
At	astatine	Ir	iridium	S	sulfur
Au	gold*	K	potassium*	Sb	antimony*
B	boron	Kr	krypton	Sc	scandium
Ba	barium	La	lanthanum	Se	selenium
Be	beryllium	Li	lithium	Si	silicon
Bi	bismuth	Lu	lutetium	Sm	samarium
Bk	berkelium	Lr	lawrencium	Sn	tin*
Br	bromine	Md	mendelevium	Sr	strontium
C	carbon	Mg	magnesium	Ta	tantalum
Ca	calcium	Mn	manganese	Tb	terbium
Cd	cadmium	Mo	molybdenum	Tc	technetium
Ce	cerium	N	nitrogen	Te	tellurium
Cf	californium	Na	sodium*	Th	thorium
Cl	chlorine	Nb	niobium	Ti	titanium
Cm	curium	Nd	neodymium	Tl	thallium
Co	cobalt	Ne	neon	Tm	thulium
Cr	chromium	Ni	nickel	U	uranium
Cs	cesium	No	nobelium	Une	unnilennium*
Cu	copper*	Np	neptunium	Unh	unnilhexium*
Dy	dysprosium	O	oxygen	Uno	unniloctium*
Er	erbium	Os	osmium	Unp	unnilpentium*
Es	einsteinium	P	phosphorus	Unq	unnilquadium*
Eu	europium	Pa	protactinium	Uns	unnilseptium*
F	fluorine	Pb	lead*	V	vanadium
Fe	iron*	Pd	palladium	W	tungsten*
Fm	fermium	Pm	promethium	Xe	xenon
Fr	francium	Po	polonium	Y	yttrium
Ga	gallium	Pr	praseodymium	Yb	ytterbium
Gd	gadolinium	Pt	platinum	Zn	zinc
Ge	germanium	Pu	plutonium	Zr	zirconium
H	hydrogen				

* These elements have symbols that were derived from non-English sources.

HISTORICAL PROFILE 1

Jöns Jakob Berzelius   1779–1848

Jöns Jakob Berzelius (1779–1848), a Swedish chemist, was a professor at Stockholm for 50 years. From 1815 to 1835, he was the "dominant" chemist in all of Europe, despite being isolated from the major centers of chemical activity in England and France.

In 1814, his first paper on chemical symbols was published. His symbols were not immediately popular among chemists. One contemporary wrote: "Berzelius' symbols are horrifying. A young student in chemistry might as well learn Hebrew as make himself acquainted with them." Today, his symbols are one of the "fundamentals" of chemistry.

Berzelius was a prolific writer, and one of his textbooks, which first appeared in 1808, was translated into many languages in many editions.

He improved chemical analysis and determined the composition of many substances. He was the first to isolate in pure form the elements silicon, titanium, and zirconium and is credited with the discovery of the elements thorium and selenium. Many laboratory innovations, among them filter paper and rubber tubing, were the work of Berzelius.

He was elevated to the peerage (as a Baron) by the King of Sweden in 1835 in recognition of the scientific labors to which his life had been devoted.

In all double-letter symbols, the first letter is capitalized, but the second letter is not. Double-letter symbols usually include the first letter of the element's name. The second letter of the symbol is frequently, but not always, the second letter of the name. Consider the elements terbium, technetium, and tellurium, whose symbols are respectively Tb, Tc, and Te. Obviously, a variety of choices of second letters is necessary because all of the elements' names begin with the same first two letters. In triple-letter symbols, the first letter of the symbol is capitalized and the other two letters are not. The six elements with triple-letter symbols are all synthetic elements and were the last six elements to be produced. These symbols come from a recently introduced systematic method for naming elements.

Eleven elements have "strange" symbols, that is, their symbols bear no relationship to the elements' English-language name. In 10 of these cases, the symbol is derived from the Latin name of the element; in the case of the element tungsten, a German name is the symbol source. Most of the elements with strange symbols have been known for hundreds of years and date back to the time when Latin was the language of scientists. Elements whose symbols are derived from non-English names are marked with an asterisk in Table 2.3.

2.8 Chemical Formulas

A most important piece of information about a compound is its composition. Chemical formulas represent a concise means of specifying compound composition. A **chemical formula** *is a notation made up of the symbols of the elements present in a compound and numerical subscripts (located to the right of each symbol) that indicate the number of atoms of each element present in a unit of the compound.*

The chemical formula for the compound we call aspirin is $C_9H_8O_4$. This formula provides us with the following information about an aspirin molecule: three elements are present—carbon (C), hydrogen (H), and oxygen (O); and 21 atoms are present—9 carbon atoms, 8 hydrogen atoms, and 4 oxygen atoms.

When only one atom of a particular element is present in a molecule of a compound, the element's symbol is written without a numerical subscript in the formula of the compound. In the formula for rubbing alcohol, C_3H_6O, for example, the subscript 1 for the element oxygen is not written.

To write formulas correctly, it is necessary to follow the capitalization rules for elemental symbols (Section 2.7). Making the error of capitalizing the second letter of an element's symbol dramatically alters the meaning of a chemical formula. The formulas $CoCl_2$ and $COCl_2$ illustrate this point; the symbol Co stands for the element cobalt, whereas CO stands for one atom of carbon and one atom of oxygen. The properties of the compounds $CoCl_2$ and $COCl_2$ are dramatically different. The compound $CoCl_2$ is a blue crystalline solid with a melting point of 724°C and a boiling point of 1029°C. The compound $COCl_2$ is a highly toxic colorless gas with a melting point of −118°C and a boiling point of 8°C.

Parentheses are sometimes used in chemical formulas, as in the example of $Ca(NO_3)_2$. The interpretation of this formula is straightforward—the grouping of atoms within the parentheses, NO_3, is present twice. The subscript following the parentheses always indicates the number of units of the multiatom entity inside the parentheses. As another example of the use of parentheses, consider the compound $Pb(C_2H_5)_4$; in this compound, four units of C_2H_5 are present. The formula $Pb(C_2H_5)_4$ represents 29 atoms—1 lead (Pb) atom, 8 (4 × 2) carbon (C) atoms, and 20 (4 × 5) hydrogen (H) atoms. The formula $Pb(C_2H_5)_4$ could be (but is not) written as PbC_8H_{20}. Both versions of the formula convey the same information in terms of atoms present; however, the former gives some information about structure (C_2H_5 units are present) that the latter does not. For this reason, the former is the preferred form for the formula. Further information concerning the use of parentheses is presented in Section 5.15. The important concern now is to be able to determine from formulas that contain parentheses the total number of atoms present. Example 2.1 further refines this skill.

EXAMPLE 2.1

Interpret each of the following formulas in terms of how many atoms of each element are present in one structural unit of the substance.
(a) $C_8H_{10}N_4O_2$—caffeine, a central nervous system stimulant
(b) $(NH_4)_3PO_4$—ammonium phosphate, an ingredient in some lawn fertilizers
(c) $Ca(NO_3)_2$—calcium nitrate, used in fireworks to give a red color

SOLUTION
(a) We simply look at the subscripts following the symbols for the elements. This formula indicates that 8 carbon atoms, 10 hydrogen atoms, 4 nitrogen atoms, and 2 oxygen atoms are present in one molecule of the compound.
(b) The subscript following the parentheses, a 3, indicates that 3 NH_4 units are present. We therefore have 3 nitrogen atoms and 12 (4×3) hydrogen atoms present in addition to 1 phosphorus atom and 4 oxygen atoms.
(c) The amounts of nitrogen and oxygen are affected by the subscript 2 outside the parentheses. In one unit of compound we have 1 calcium atom, 2 (2×1) nitrogen atoms, and 6 (2×3) oxygen atoms.

In addition to formulas, compounds have names. Naming compounds is not as simple as naming elements. Although the nomenclature of elements (Section 2.7) has been largely left up to the imagination of their discoverers, extensive sets of systematic rules exist for naming compounds. Rules must be used because of the large number of compounds that exist. Parts of Chapters 5 and 16 are devoted to rules for naming compounds, and we will not worry about this procedure until those chapters. For the time being, our focus is on the meaning and significance of chemical formulas; knowing how to name the compounds that the formulas represent is not a prerequisite for understanding the meaning of formulas.

Figure 2.12 pictorially relates chemical formulas to the classifications of matter previously discussed in this chapter. Note in the top third of the diagram that the formulas for molecules of an element contain only one elemental symbol. In the middle third of the diagram we see that formulas for compounds contain two or more elemental symbols. Finally, in the bottom third of the diagram, we see that in mixtures molecules with different formulas must be present.

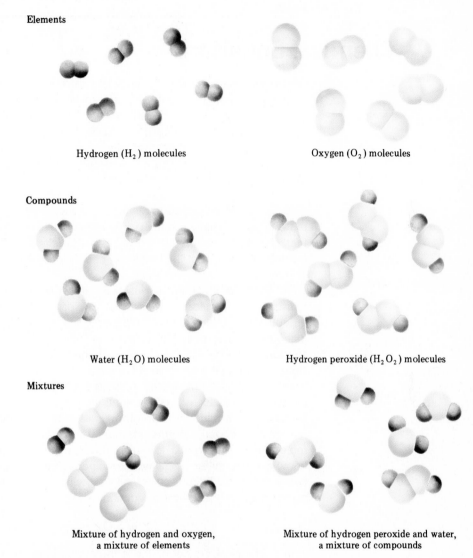

Elements

Hydrogen (H_2) molecules

Oxygen (O_2) molecules

Compounds

Water (H_2O) molecules

Hydrogen peroxide (H_2O_2) molecules

Mixtures

Mixture of hydrogen and oxygen,
a mixture of elements

Mixture of hydrogen peroxide and water,
a mixture of compounds

FIGURE 2.12

A contrast between formulas for elements and formulas for compounds. A formula for an element contains only one elemental symbol, whereas formulas for compounds must contain two or more elemental symbols.

2.9 Chemical Equations

During chemical change (Section 1.6) pure substances (elements and compounds) are changed into other pure substances. Carbon dioxide and water are two new substances produced in the chemical change associated with the burning of gasoline. Ashes are among the new products produced when wood is burned. Chemical changes are often called chemical reactions. A **chemical reaction** *is a process in which at least one new pure substance is produced as a result of chemical change.* The production of one or more new substances is always a characteristic of a chemical reaction.

A **chemical equation** *is a written statement that uses symbols and formulas to describe the changes that occur during a chemical reaction.* In the same way that chemical symbols are considered the *letters* of chemical language and formulas the *words* of the language, chemical equations can be considered the *sentences* of chemical language.

To illustrate the conventions used in writing chemical equations, let us consider the chemical change in which solid sulfur (an element) reacts with gaseous oxygen (an element) to produce sulfur trioxide (a compound). To write a chemical equation for this process, we first need to know the formulas of the substances involved in the chemical change. Formulas for elements in the solid state are simply the symbols of the elements; hence, we use S to denote sulfur. Formulas for elements in the gaseous state take into account the number of atoms in a molecule of that element; we use the formula O_2 to denote oxygen because it is a diatomic (two-atom) molecule. (The elemental gases oxygen, nitrogen, hydrogen, fluorine, chlorine, bromine,* and iodine* are all diatomic molecules.) The formula for sulfur trioxide is SO_3. Furthermore, we need to know that two sulfur atoms will combine with three oxygen molecules to form two sulfur trioxide molecules. All of the preceding information is concisely summarized in the chemical equation

$$2\,S + 3\,O_2 \longrightarrow 2\,SO_3$$

In a chemical equation, the numbers *in front* of the formulas are called coefficients. A **coefficient** *is a number placed to the left of the formula of a substance, which changes the amount, but not the identity, of the substance.* In the notation $2\,H_2O$, the 2 on the left is a coefficient; $2\,H_2O$ means two

---

* The elements bromine and iodine are not gases at room temperature, but vaporize at slightly higher temperatures. The resultant vapors are diatomic molecules. Even when not vaporized, these two elements are represented as diatomic molecules.

molecules of H_2O and $3\ H_2O$ means three molecules of H_2O. Thus, coefficients tell how many formula units of a given substance are present.

A coefficient placed in front of a formula applies to the *whole* formula. Subscripts, in contrast, affect only parts of a formula.

$$\text{coefficient (affects both H and O)}$$
$$2\ H_2O$$
$$\text{subscript (affects only H)}$$

The preceding notation denotes two molecules of H_2O; it also denotes a total of four H atoms and two O atoms.

Another example of a chemical equation is

$$2\ FeI_2 + 3\ Cl_2 \longrightarrow 2\ FeCl_3 + 2\ I_2$$

We will use this equation as our example in formalizing the conventions used in writing chemical equations.

1. The formulas of the *reactants* (starting materials) are always written on the *left* side of the equation.

$$2\ FeI_2 + 3\ Cl_2 \longrightarrow 2\ FeCl_3 + 2\ I_2$$

2. The formulas of the *products* (substances produced) are always written on the *right* side of the equation.

$$2\ FeI_2 + 3\ Cl_2 \longrightarrow 2\ FeCl_3 + 2\ I_2$$

3. The reactants and products are separated by an arrow pointing toward the products.

$$2\ FeI_2 + 3\ Cl_2 \longrightarrow 2\ FeCl_3 + 2\ I_2$$

4. Plus signs are used to separate different reactants or different products from each other.

$$2\ FeI_2 + 3\ Cl_2 \longrightarrow 2\ FeCl_3 + 2\ I_2$$

In chemical equations, the plus signs on the reactant side of the equation mean "reacts with"; the arrow means "to produce"; and plus signs on the product side mean "and."

5. Coefficients are used to indicate the amounts (number of formula units) of reactants needed and products produced.

$$2\ FeI_2 + 3\ Cl_2 \longrightarrow 2\ FeCl_3 + 2\ I_2$$

The procedure, for determining the coefficients in an equation when they are not given is called balancing an equation. This procedure is the subject of

Section 7.7. At present, we assume that the coefficients in an equation are given quantities.

EXAMPLE 2.2

Four ammonia (NH_3) molecules combine with five oxygen (O_2) molecules to form four nitric oxide (NO) molecules and six water (H_2O) molecules. Write a chemical equation that describes this chemical change.

SOLUTION

In this chemical reaction there are two reactants (NH_3 and O_2) and two products (NO and H_2O). The reactants are written on the left side of the equation and the products on the right side. An arrow separates reactants from products. Furthermore, coefficients (which are given in the problem statement) are used to indicate the number of molecules of each substance (reactant and product) needed. The chemical equation for the reaction is

$$4\,NH_3 + 5\,O_2 \longrightarrow 4\,NO + 6\,H_2O$$

The coefficient associated with a formula in a chemical equation can have the value of 1. When this is the case, the coefficient is not formally written; it is "understood" to be 1. We encountered this same concept in Section 2.8 in considering formula writing. The subscript 1 is not written. Following are examples of chemical equations in which one or more coefficients have the value 1.

$$CH_4 + 2\,O_2 \longrightarrow CO_2 + 2\,H_2O$$

$$2\,Al + 3\,S \longrightarrow Al_2S_3$$

$$Cr_2O_3 + 3\,C \longrightarrow 2\,Cr + 3\,CO$$

In addition to the essential plus sign and arrow notation used in chemical equations, a number of optional symbols can be used to convey more information about a chemical reaction than just the chemical species involved. In particular, it is often useful to know the physical state of the substances involved in a chemical reaction. These optional symbols, which are given in Table 2.4, enable physical state to be specified. The three equations given in the preceding paragraph appear as follows when the optional state symbols are included.

$$CH_4(g) + 2\,O_2(g) \longrightarrow CO_2(g) + 2\,H_2O(l)$$

$$2\,Al(s) + 3\,S(s) \longrightarrow Al_2S_3(s)$$

$$Cr_2O_3(s) + 3\,C(s) \longrightarrow 2\,Cr(s) + 3\,CO(g)$$

TABLE 2.4

Symbols Used in Equations

Symbol	Meaning
Essential	
$\rightarrow$	"to produce"
+	"reacts with" or "and"
Optional	
(s)	solid
(l)	liquid
(g)	gas
(aq)	aqueous solution (a substance dissolved in water)

Practice Questions and Problems

PURE SUBSTANCES AND MIXTURES

2-1 Consider the following classes of matter: heterogeneous mixtures, homogeneous mixtures, and pure substances.
  a. In which of these classes must two or more substances be present?
  b. Which of these classes could not possibly have a variable composition?
  c. For which of these classes is separation into simpler substances by physical means possible?

2-2 Assign each of the following descriptions of matter to one of the following categories: heterogeneous mixture, homogeneous mixture, or pure substance.
  a. Two substances present, two phases present.
  b. Two substances present, one phase present.
  c. One substance present, one phase present.
  d. One substance present, two phases present.
  e. Three substances present, four phases present.

2-3 Classify each of the following as a heterogeneous mixture, a homogeneous mixture, or a pure substance. Also indicate how many phases are present. (All substances are assumed to be present in the same container.)
  a. Water
  b. Water and dissolved table salt
  c. Water and white sand
  d. Water and oil
  e. Water and ice
  f. Water, ice, and oil
  g. Carbonated water (soda water) and ice

  h. Oil, ice, saltwater solution, sugar–water solution, and pieces of copper metal

2-4 The phrases chemically homogeneous, chemically heterogeneous, physically homogeneous, and physically heterogeneous are often used in describing samples of matter. In each of the following situations two of these phrases apply. Select the two correct phrases for each case.
  a. Pure water
  b. Tap water
  c. River water
  d. Pure water and ice
  e. Oil-and-vinegar salad dressing with spices
  f. Oil-and-vinegar salad dressing without spices
  g. Carbonated beverage
  h. Blood

2-5 Explain how the operational definition of a pure substance allows for the possibility that it is not actually pure.

2-6 Every sample taken from a homogeneous mixture has the same composition as every other sample. The same can be said about samples taken from a pure substance. Why, then, are homogeneous mixtures not considered to be pure substances?

2-7 How do the terms *substance* and *pure substance* differ in meaning?

ELEMENTS AND COMPOUNDS

2-8 Explain the difference between an element and a compound.

2-9 Based on the information given, classify each of the pure substances A through K as elements or compounds, or indicate that no such classification is possible because of insufficient information.

a. Analysis with an elaborate instrument indicates that *substance A* contains two elements.

b. *Substance B* and *substance C* react to give a new *substance D* (give three answers—one for B, one for C, and one for D).

c. *Substance E* decomposes upon heating to give *substance F* and *substance G* (give three answers).

d. Heating *substance H* to 1000°C causes no change.

e. *Substance I* cannot be broken down into simpler substances by chemical means.

f. *Substance J* cannot be broken down into simpler substances by physical means.

g. Heating *substance K* to 500°C causes it to change from a solid to a liquid.

2-10 Indicate whether each of the following statements is true or false.

a. Both elements and compounds are pure substances.

b. A compound results from the physical combination of two or more elements.

c. For matter to be heterogeneous, at least two compounds must be present.

d. Pure substances cannot have a variable composition.

e. Compounds, but not elements, can have a variable composition.

f. Compounds can be separated into their constituent elements using chemical means.

g. A compound must contain at least two elements.

2-11 Is it possible to have a mixture of two elements and also to have a compound of the same two elements? Explain. Can you think of an example?

2-12 Compounds are classified as pure substances even though two or more substances (elements) are present. Explain.

2-13 What is the difference between heteroatomic molecules and homoatomic molecules?

2-14 Assign each of the following descriptions of matter to one of the following categories: mixture, element, or compound.

a. One substance present, heteroatomic molecules present.

b. One phase present, both heteroatomic and homoatomic molecules present.

c. One substance present, two phases present, homoatomic molecules present.

d. Two substances present, one phase present, heteroatomic molecules present.

2-15 Indicate whether each of the following statements is true or false.

a. Molecules must contain three or more atoms if they are heteroatomic.

b. The smallest characteristic unit of a pure substance is an atom.

c. A molecule of a compound must be heteroatomic.

d. A molecule of an element can be homoatomic or heteroatomic, depending on which element is involved.

e. The limit of physical subdivision for a pure substance is a molecule.

f. There is only one kind of molecule for any given compound.

g. Heteroatomic molecules do not maintain the properties of their constituent elements.

h. The main difference between the molecules of elements and the molecules of compounds is the number of atoms they contain.

DISCOVERY AND ABUNDANCE OF THE ELEMENTS

2-16 How many elements were discovered in each of the following time periods?

a. ancient–1700   b. 1701–1800
c. 1801–1900   d. 1901–1950
e. 1951–present

2-17 What are the two most abundant elements in each of the following realms of matter?

a. Universe   b. Earth (including core)
c. Earth's crust   d. Atmosphere
e. Hydrosphere   f. Human body
g. Vegetation

NAMES AND SYMBOLS OF THE ELEMENTS

2-18 The symbols for elements found in the human body include C, H, O, N, P, S, I, Na, Cl, K, Ca, Fe, Br, Cu, Zn, Co, and Mg. What elements do these symbols represent?

2-19 There are 12 elements that have an abundance of 1000 parts per million or greater (by mass) in the earth's crust: hydrogen, oxygen, sodium, magnesium, aluminum, silicon, phosphorus, potassium, calcium, titanium, manganese, and iron. What are the chemical symbols for these elements?

2-20 What elements are represented by the following chemical symbols?
a. Ne   b. Hg   c. Pb   d. Ar
e. Au   f. Ba   g. Sn   h. Cd

2-21 What are the chemical symbols of the following elements?
a. Lithium   b. Helium   c. Silver
d. Boron   e. Fluorine   f. Beryllium
g. Nickel   h. Arsenic

2-22 Make a list of the elements that have chemical symbols with the following characteristics.
a. Begin with a letter other than the first letter of the English name for the element.
b. Consist of the first two letters of the element's English name.
c. Contain only one letter.

2-23 For which letters of the alphabet are there at least five elements whose chemical symbol begins with them?

CHEMICAL FORMULAS

2-24 The notation HF stands for a compound (hydrogen fluoride) rather than an element. How can we tell this from the notation?

2-25 On the basis of its formula, classify each of the following substances as an element or compound.
a. $NaClO_2$ b. CO   c. Co   d. $COCl_2$
e. $CoCl_2$ f. AlN   g. S_8 h. Hf

2-26 From the information given about one molecule of each of these compounds, write the chemical formula for the compound.
a. Nitroglycerin contains 3 atoms of carbon, 5 atoms of hydrogen, 3 atoms of nitrogen, and 9 atoms of oxygen.
b. Vitamin A contains 20 atoms of carbon, 30 atoms of hydrogen, and 1 atom of oxygen.
c. Nicotine contains 10 atoms of carbon, 14 atoms of hydrogen, and 2 atoms of nitrogen.

2-27 How many atoms of oxygen are represented in each of the following formulas?
a. H_2SO_4 b. $HClO_3$ c. O_2F_2
d. C_2H_6O e. $(NH_4)_2SO_4$ f. $Ca(NO_3)_2$
g. $Al(OH)_3$ h. $Mg(C_2H_3O_2)_2$

2-28 How many atoms of each kind are represented by the following formulas?
a. SO_2 b. NaOH
c. NH_4ClO_3 d. $CaSO_4$
e. K_3PO_4 f. $Mg_3(PO_4)_2$
g. $Be(CN)_2$ h. $Ca(C_6H_{12}NSO_3)_2$

CHEMICAL EQUATIONS

2-29 What symbol is used in a chemical equation to represent the following phrases?
a. To produce   b. Reacts with

2-30 Write a chemical equation that summarizes the following chemical information: "Three oxygen molecules containing two oxygen atoms per molecule react with one carbon disulfide molecule containing one carbon atom and two sulfur atoms per molecule to produce one carbon dioxide molecule containing one carbon atom and two oxygen atoms per molecule and two sulfur dioxide molecules containing one sulfur atom and two oxygen atoms per molecule.

2-31 Describe in words the chemical change described by the following equation.

$$2 H_2 + O_2 \longrightarrow 2 H_2O$$

2-32 Use the information contained in the chemical equation

$$2 H_2SO_4 + Cu \longrightarrow SO_2 + 2 H_2O + CuSO_4$$

to answer the following questions.
a. How many products are formed in this reaction?
b. How many of the substances involved in this reaction are compounds?
c. List the coefficients found in the equation.
d. Which of the products are triatomic molecules?
e. How many total atoms are represented by the notation $2 H_2SO_4$?

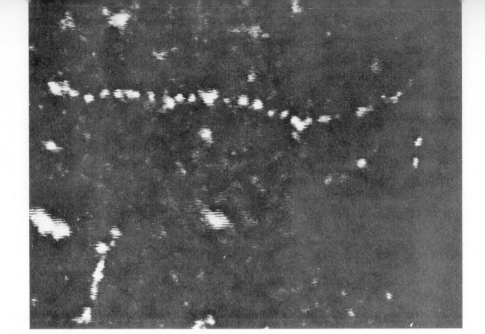

CHAPTER

3

Structure of the Atom

3.1   The Atomic Theory of Matter

In Chapter 2 we defined a number of terms that dealt with the classification of matter. One term was *atom*, which is the limit of chemical subdivision for a pure substance. We noted that 109 different kinds of atoms are known, and all matter is made from them. In this chapter, we consider the topic of atoms in greater detail.

The concept of an atom is an old one, dating back to ancient Greece. Records indicate that around 460 B.C., the Greek philosopher Democritus suggested that continued subdivision of matter would ultimately yield small, indivisible particles, which he called atoms (from the Greek word *atomos* meaning "uncut or indivisible"). However, Democritus' ideas about matter were forgotten during the Middle Ages.

It was not until the beginning of the nineteenth century that the concept of the atom was "rediscovered." In a series of papers published in the period 1803–1807, the English chemist John Dalton (1766–1844) (see Historical Profile 2) again proposed that the fundamental building block for all kinds of matter was an atom. This time, however, there was a firm basis for the proposal. Dalton's proposal had as its basis experimental observations. This is in marked contrast to the early Greek concept of atoms, which was based solely on philosophical speculation. Because of its experimental basis, Dalton's idea received wide attention and stimulated new research and thought about the ultimate building blocks of matter.

Additional research carried out by many scientists has now validated Dalton's basic conclusion that the building blocks for all types of matter are atomic. Some of the details of Dalton's original proposals have had to be modified in the light of more sophisticated experiments, but the basic concept of atoms remains.

Among today's scientists, the concept that atoms are the building blocks for matter is a foregone conclusion. The large amount of supporting evidence for atoms is impressive. Key concepts about atoms, in terms of current knowledge, are found in what is known as the atomic theory of matter. The **atomic theory of matter** *is a set of five statements that summarize modern-day scientific thought about atoms.* These five statements are

1. All matter is made up of small particles called atoms, of which 109 different "types" are known, with each "type" corresponding to atoms of a different element.
2. All atoms of a given type are similar to one another and significantly different from all other types.

47

[Smithsonian Institution,
Washington DC.]

HISTORICAL PROFILE 2

John Dalton   1766–1844

John Dalton (1766–1844), an English chemist, spent most of his life in Manchester, England. Largely self-taught because of the poverty of his family, he is considered the father of chemical theory.

Throughout his life, Dalton had a particular interest in the study of weather. In 1787 he made his first entry in a notebook entitled "Observations on the Weather." He continued to record temperature, barometric pressure, rainfall, dew point, and so on, for the next 57 years; the last of over 200,000 observations was made the evening before his death.

His interest in meteorology was responsible for his greatest contribution to chemistry, the atomic theory. From "weather" he turned his attention to the nature of the atmosphere and then to the study of gases in general. Dalton's atomic theory, first published in 1808, was based on his observations of the behavior of gases. He is also the formulator of the gas law now called Dalton's law of partial pressures.

A devout Quaker, he began teaching at a Quaker school when he was 12 years old. During most of his life he supported himself by private tutoring. He shunned glory and never found time for marriage. Honors both from his country and from abroad poured in upon Dalton in later life. In 1826 the Royal Society of London awarded him the first of the Royal Medals given by the king. A public funeral was held in Manchester upon his death; over 40,000 persons passed by his casket.

3. The relative number and arrangement of the different types of atoms found within a pure substance determine its identity.
4. Chemical change is a union, separation, or rearrangement of atoms to give new substances.
5. Only whole atoms can participate in or result from any chemical change, because atoms are considered to be indestructible during these changes.

Atoms are incredibly small particles. No one has seen or ever will see an atom with the naked eye. The question may thus be asked: "How can you be absolutely sure that something as minute as an atom really exists?" The achievements of twentieth-century scientific instrumentation have gone a long way toward removing any doubt about the existence of atoms. Electron microscopes, which are capable of producing magnification factors in the

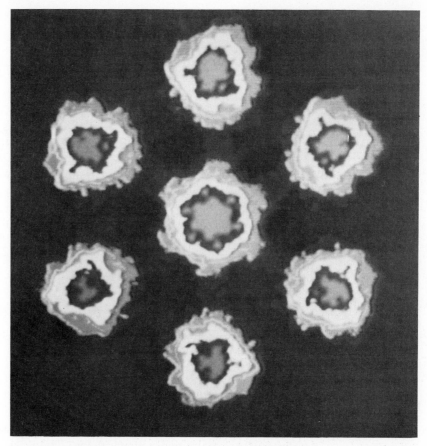

FIGURE 3.1

An electron microscope photograph of a uranyl acetate cluster on a very thin carbon substrate. The individual uranium atoms are the roundish spots with dark gray centers. [Courtesy of M. Isaacson (Cornell University) and M. Ohtsuki (University of Chicago).]

millions, have made it possible to photograph "images" of individual atoms. An example of such an "atom picture" is shown in Figure 3.1.

Just how small is an atom? Although atomic dimensions and masses are not directly measurable, they are known quantities that are obtained by calculation. The data used for the calculations come from measurements made on macroscopic amounts of pure substances. Because such calculated numbers for atomic dimensions and masses are extremely small, they are expressed in exponential notation. A detailed discussion of exponential notation is given in Appendix A for use by those unfamiliar with this notation or those who need to be "refreshed in memory" concerning this notation.

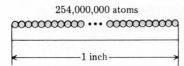

FIGURE 3.2
Comparison of atomic diameters and the common
measuring unit 1 inch.

The diameter of an atom is approximately 1×10^{-8} centimeter. If you arranged atoms of diameter 1×10^{-8} centimeter in a straight line, it would take 254 million of them to extend the length of 1 inch (see Figure 3.2).

The mass of an atom is obviously also very small. The mass of a uranium atom, which is one of the heaviest of the known kinds of atoms, is 4×10^{-22} gram. To produce a mass of 1 gram would require 2×10^{21} atoms of uranium. The number 2×10^{21} is so large it is difficult to comprehend. The following comparison gives some idea of its magnitude. If each of the 5 billion people on earth were made a millionaire (receiving 1 million dollar bills), we would still need 400,000 other worlds, each inhabited by the same number of millionaires, to have 2×10^{21} dollar bills in circulation.

3.2   Chemical Evidence Supporting the Atomic Theory

The law of conservation of mass and the law of definite proportions are typical of the evidence that supports atomic theory. These are two of a number of known laws that summarize the results of thousands of different investigations into the chemical behavior of matter.

Both of these laws were formulated in the late 1700s. It was during that century that scientists developed experimental methods for accurately determining masses and volumes of chemical substances. These methods enabled chemists to study chemical change *quantitatively* for the first time.

The Law of Conservation of Mass

Countless experimental studies of chemical reactions have shown that the substances produced in a chemical reaction have the same total mass as the starting materials. The law of conservation of mass is based on this fact. The **law of conservation of mass** *states that there is no experimentally detectable gain or loss in total mass for the substances involved in a chemical reaction.* The French chemist Antoine Laurent Lavoisier (1743–1794) (see Historical Profile

3) is credited as the first to state this important relationship between the reactants and products of a chemical reaction.

The validity of the law of conservation of mass can be checked by carefully determining the masses of all reactants and all products in a chemical reaction. To illustrate, let us consider the reaction of known masses of the elements copper (Cu) and sulfur (S) to form the compound copper sulfide (CuS). As shown in Figure 3.3, 66.5 grams of copper reacts with *exactly* 33.5 grams of sulfur. After that reaction, no copper or sulfur remains in elemental form; the only substance present is the product copper sulfide, the chemically combined copper and sulfur. The mass of the product is 100.0 grams, which is the same as the combined masses of the reactants. Thus, the law of conservation of mass has been obeyed.

In addition, when the 100.0 grams of the product copper sulfide is heated to a high temperature in the absence of air, it decomposes into copper and sulfur, producing 66.5 grams of copper and 33.5 grams of sulfur. Once again, no detectable mass change is observed; the mass of the reactant is equal to the masses of the products.

$$\underbrace{100.0 \text{ g CuS}}_{\text{100.0 g of reactant}} \longrightarrow \underbrace{66.5 \text{ g Cu} + 33.5 \text{ g S}}_{\text{100.0 g of products}}$$

The law of conservation of mass is consistent with the statements of atomic theory (Section 3.1). Because all reacting chemical substances are made up of atoms (statement 1), each with its unique identity (statement 2), and these atoms can be neither created nor destroyed in a chemical reaction, but merely

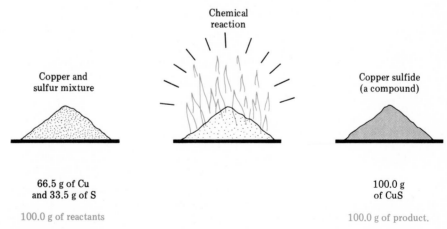

Copper and sulfur mixture

66.5 g of Cu
and 33.5 g of S

100.0 g of reactants

Chemical reaction

Copper sulfide (a compound)

100.0 g
of CuS

100.0 g of product.

FIGURE 3.3

Experimental verification of the law of conservation of mass. The combined masses of the reactants (copper and sulfur) are equal to the mass of the product (copper sulfide).

[Burndy Library, Norwalk CT.]

HISTORICAL PROFILE 3

Antoine Laurent Lavoisier 1743–1794

Antoine Laurent Lavoisier (1743–1794), the son of a wealthy Parisian lawyer, is often called the "father of 'modern' chemistry." It is he who first appreciated the importance of carrying out very accurate (quantitative) measurements of chemical change. His work was performed on balances he had specially designed that were more accurate than any then known. From his "balance work" came the discovery of the law of conservation of mass.

Originally trained as a lawyer, Lavoisier entered the field of science through geology and from there became interested in chemistry. This change in focus took place during his early twenties.

Lavoisier's studies on combustion are considered his major work. He was the first to realize that combustion involves the reaction of oxygen from air with the substance that is burned. He gave the name oxygen to the gas then known as "dephlogisticated air." In later years he became interested in physiological chemistry.

Lavoisier's *Elementary Treatise on Chemistry*, published in 1789, was the first textbook based on quantitative experiments. He was also one of the first to use systematic nomenclature for elements and a few compounds.

Early in life he invested money in, and for some time was involved in the management of, a private company commissioned by the French government to collect taxes. He used his earnings from this endeavour (about 100,000 francs a year) to support his scientific work. During the French Revolution all associated with this type of activity, known as "tax-farming," were denounced, arrested, and guillotined. Lavoisier's death came just two months before the end of the Revolution. On the day of his death, one of Lavoisier's scientific colleagues gave this tribute: "It took but a moment to cut off that head; perhaps a hundred years will be required to produce another like it."

rearranged (statement 4), it follows that the total mass after the reaction must equal the total mass before the reaction. We have the same number of each kind of atom after the reaction as we had at the start of the reaction.

The Law of Definite Proportions

We can determine the composition of a compound by decomposing a weighed amount of the compound into its elements and then determining the masses of the individual elements present. Alternatively, we can also determine composi-

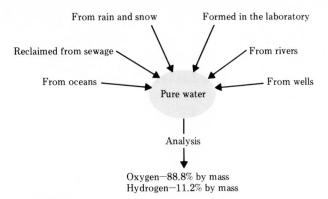

From rain and snow

Formed in the laboratory

Reclaimed from sewage

From rivers

From oceans

Pure water

From wells

Analysis

Oxygen—88.8% by mass
Hydrogen—11.2% by mass

FIGURE 3.4

In accordance with the law of definite proportions, samples of pure water from different sources will have the same composition by mass.

tion by weighing a compound that is formed by the combination of known masses of elements.

Studies of composition data for many compounds show that the percentage by mass of each element present in a given compound does not vary. The law of definite proportions is a formal statement of this conclusion. The **law of definite proportions** *states that in a pure compound the proportion by mass of the elements present is always the same.* This law of definite proportions is sometimes referred to as the *law of constant composition*. The French chemist Joseph Louis Proust (1754–1826) (see Historical Profile 4) is responsible for the work that established this law as one of the "fundamentals" of chemistry.

In accordance with this law, once we determine the composition of a pure compound, we know that the composition of all other pure samples of the compound will be the same. For example, water has a composition by mass of 88.8% oxygen and 11.2% hydrogen. All samples of pure water, regardless of their sources, will have this composition (see Figure 3.4).

An alternative way of examining the constancy of composition of compounds involves looking at the mass ratios in which elements combine to form compounds. From this viewpoint, constancy in composition can easily be demonstrated experimentally.

Suppose we attempt to combine various masses of sulfur with a fixed mass of calcium to make calcium sulfide (CaS). Possible experimental data for this attempt are given in the first four lines of Table 3.1. Notice that, regardless of the mass of S present, only a certain amount, 44.4 grams, reacts with 55.6 grams of Ca. The excess S is left over in an unreacted form. The data illustrate that Ca and S will react in only one fixed-mass ratio (55.6/44.4 = 1.25) to form CaS. Note also that if the amount of Ca is doubled (as in line 5 of Table 3.1), the amount of S with which it reacts also doubles (compare lines 1 and 5 of

HISTORICAL PROFILE 4

Joseph Louis Proust   1754–1826

[E. F. Smith Memorial Collection, Center for the History of Chemistry, University of Pennsylvania.]

Joseph Louis Proust (1754–1826), born in Angers, France, began his chemistry studies at home in his father's apothecary shop. Later he studied in Paris. It is Proust's researches on the law of definite proportions, his chief contribution to chemistry, which brought this fundamental law general recognition.

Almost all of Proust's work, including that on the law of definite proportions, was carried out in Spain rather than France. At the invitation of the Spanish government, in 1791, he became director of the Royal Laboratory at Madrid. This laboratory, financed by the government, was elegantly equipped. Almost all of the vessels, even those in common use, were made of platinum.

While in Spain, Proust devoted most of his time to the analysis of Spanish minerals and natural products. His studies included work on starch, camphor, preparation of mercury from cinnabar, and isolation of grape sugar (glucose) from grapes.

His research work in Madrid stopped abruptly in 1808 when his laboratory was destroyed by French troops occupying Spain. Reduced to poverty, he still refused an offer of 100,000 francs from Napoleon to come back to France and supervise the manufacture of glucose from grapes (a most abundant crop in both Spain and France).

Later, in 1817, after Louis XVIII had come to power, he returned to France where he remained until his death.

Table 3.1). Nevertheless, the ratio in which the substances react (111.2/88.8) remains at 1.25.

The data in Table 3.1 are consistent with the statements of atomic theory according to the following line of reasoning: (1) A 55.6-gram sample of calcium contains a certain number of Ca atoms. (2) A 44.0-gram sample of sulfur contains a number of S atoms just sufficient to react with the number of Ca atoms present. (3) Because the atoms of Ca and S always react in the same ratio, according to the law of definite proportions, any amount of sulfur greater than 44.4 grams contains more atoms than are needed and some of the atoms will go unreacted.

TABLE 3.1
Data Illustrating the Law of Definite Proportions

Mass of Ca Used (g)	Mass of S Used (g)	Mass of CaS Formed (g)	Mass of Excess Unreacted Sulfur (g)	Ratio in Which Substances React
55.6	44.4	100.0	none	1.25
55.6	50.0	100.0	5.6	1.25
55.6	100.0	100.0	55.6	1.25
55.6	200.0	100.0	155.6	1.25
111.2	88.8	200.0	none	1.25

3.3 Subatomic Particles: Electrons, Protons, and Neutrons

Until the closing decades of the nineteenth century, scientists believed that atoms were solid, indivisible spheres that did not have substructures. Today, this concept is known to be incorrect. Evidence from a variety of sources indicates that atoms are made up of smaller, more fundamental particles called subatomic particles. **Subatomic particles** *are very small particles that are the building blocks from which atoms are made.*

Three major types of subatomic particles exist: electrons, protons, and neutrons. These subatomic particles can be distinguished from each other on the basis of their charge. **Electrons** *are subatomic particles that possess a negative (−) charge.* Electrons are the smallest of the subatomic particles in terms of mass. **Protons** *are subatomic particles that possess a positive (+) charge.* The amount of charge associated with a proton is the same as that for an electron; however, the character of the charge is opposite (positive versus negative). The fact that electrons and protons bear opposite charges is extremely important because of the way in which charged particles interact. It is a natural scientific law that particles of opposite (or unlike) charge attract each other and particles of the same (or like) charge repel each other. The third type of subatomic particle, the neutron, has no charge. **Neutrons** *are subatomic particles that have no charge associated with them; that is, they are neutral.*

All of the subatomic particles are extremely small. The lightest of the three types, the electron, has a mass of 9.11×10^{-28} gram. Both protons and neutrons are massive particles compared with the electrons. Protons are 1836 times heavier (1.672×10^{-24} gram) and neutrons are 1839 times heavier (1.675×10^{-24} gram). For most purposes, the masses of protons and neutrons

TABLE 3.2
Charges and Masses of the Principal Subatomic Particles

	Electron	Proton	Neutron
Charge	−1	+1	0
Actual mass (g)	9.109×10^{-28}	1.672×10^{-24}	1.675×10^{-24}
Relative mass (based on the electron being one unit)	1	1836	1839
Relative mass (based on the neutron being one unit)	0 (1/1839)	1	1

can be considered equal. The mass of an electron is almost negligible compared with the masses of the other heavier subatomic particles. The properties of the three types of subatomic particles are summarized in Table 3.2.

The arrangement of subatomic particles within an atom is not haphazard. As is shown in Figure 3.5, an atom is composed of two regions: (1) a nuclear region and (2) an extranuclear region, the space outside the nuclear region.

Located at the center of an atom, the nuclear region is most often simply called the nucleus. The **nucleus** is *the very small, dense, positively charged core of an atom.* All protons and neutrons present in the atom are found within the nucleus. A nucleus always carries a positive charge because of the presence of the positively charged protons within it. Almost all (over 99.9%) of the mass of an atom is concentrated in its nucleus; all of the heavy subatomic particles (protons and neutrons) are there. The smallness of the nucleus, coupled with the large mass there, makes nuclear material extremely dense material. Protons and neutrons, because of their presence in the nucleus, are often called *nucleons.*

The extranuclear region of an atom contains all of the electrons. It is an

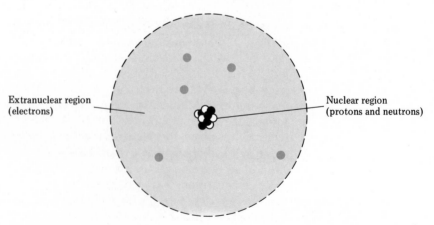

Extranuclear region (electrons)

Nuclear region (protons and neutrons)

FIGURE 3.5
Arrangement of subatomic particles in an atom.

extremely large region, comprising mostly empty space, in which the electrons randomly move about the nucleus. The motion of the electrons in this extranuclear region determines the volume (size) of the atom in the same way that the blade of a fan determines a volume by its motion. The volume occupied by the electrons is sometimes referred to as the *electron cloud*. Because electrons are negatively charged, the electron cloud is also negatively charged.

An atom, as a whole, is neutral; it has no net charge. How can it be that an entity possessing positive charge (protons) and negative charge (electrons) can end up neutral? For this to occur, the same amounts of positive and negative charge must be present in the atom. Equal amounts of positive and negative charge cancel each other. Thus, equal numbers of protons and electrons are always present in an atom. This relationship between protons and electrons is a key relationship.

Number of protons = Number of electrons

It is important that the size relationships between the parts of an atom be correctly visualized. The nucleus is extremely small compared with the total atom size. A nonmathematical conceptual model will help you visualize this relationship. Visualize a large spherical or nearly spherical fluffy cloud in the sky. Let the cloud represent the negatively charged extranuclear region of the atom. Buried deep within the cloud is the positively charged nucleus, which is the size of a small pebble. A more quantitative example involves the enlargement (magnification) of the nucleus until it would be the size of a baseball (7.4 centimeters in diameter). If the nucleus were this large, the whole atom would have a diameter of approximately 4 kilometers (2.5 miles). Within the large extranuclear region (4 kilometers diameter), the electrons (smaller than the periods used to end sentences in this text) move about at random. The volume of the nucleus is only 1/100,000 of the atom's total volume; that is, almost all (more than 99.99%) of the volume of an atom consists of the electron cloud.

The concentration of almost all of the mass of an atom in the nucleus can best be illustrated with an example. If a copper penny contained copper nuclei (copper atoms stripped of their electrons) rather than copper atoms (which are mostly empty space), the penny would weigh 170,000,000,000 kilograms (190,000,000 tons). Nuclei are indeed very dense matter.

Despite the existence of subatomic particles, we will continue to use the concept of atoms as the fundamental building blocks for all types of matter (Section 2.4). Subatomic particles do not lead an independent existence for any appreciable length of time; they gain stability by joining together to form atoms.

3.4   Evidence Supporting the Existence and Arrangement of Subatomic Particles

A significant collection of evidence is consistent with and supports the existence, nature, and arrangement of subatomic particles as described in Section 3.3. Two historically important types of experiments illustrate some of the sources of this evidence. *Discharge-tube experiments* resulted in the original concept that the atom contained negatively and positively charged particles. *Metal-foil experiments* provided evidence for the existence of a nucleus within the atom.

Discharge-Tube Experiments

Neon signs, fluorescent lights, and television tubes are all basic ingredients of our modern technological society. The forerunner for all three of these developments was the *gas discharge tube*. Gas discharge tubes also provided some of the first evidence that an atom consisted of still smaller particles (subatomic particles).

The principle behind the operation of a gas discharge tube—that gases at low pressure conduct electricity—was discovered in 1821 by the English chemist Humphry Davy (1778–1829). Subsequently, studies with gas discharge tubes were carried out by many scientists.

A simplified diagram of a gas discharge tube is shown in Figure 3.6. The apparatus consists of a sealed glass tube containing two metal disks called *electrodes*. The glass tube also has a side arm for attachment to a vacuum pump. During operation, the electrodes are connected to a source of electrical power. (The positive electrode, the electrode attached to the positive side of the

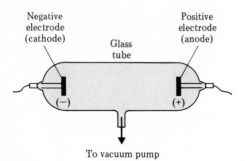

To vacuum pump

FIGURE 3.6
A simplified version of a gas discharge tube.

electrical power source is called the *anode*; the negative electrode, the one attached to the negative side is known as the *cathode*.) Use of the vacuum pump allows the amount of gas within the tube to be varied. The smaller the amount of gas present, the lower the pressure is within the tube.

Early studies with gas discharge tubes showed that when the tube was almost evacuated (low pressure) electricity flowed from one electrode to the other and the residual gas became luminous (it glowed). Different gases in the tube gave the ''glow'' different colors. After the pressure in the tube was reduced to still lower levels (very little gas remaining), it was found that the luminosity disappeared but the electrical conductance continued—as shown by a ''greenish glow'' given off by the tube's glass walls. This ''glow'' was the initial discovery of what became known as *cathode rays*. Their discovery marked the beginning of nearly 40 years of discharge-tube experimentation that ultimately led to the discovery of both the electron and the proton.

The term cathode rays comes from the observation that when an obstacle is placed between the negative electrode (cathode) and the opposite glass wall, a sharp shadow the shape of the obstacle is cast on that wall. This indicates that the rays are coming from the cathode.

It was further found that these cathode rays caused certain minerals such as sphalerite (zinc sulfide) to glow. Detailed observation of sphalerite-coated glass plates being bombarded with cathode rays showed that the light emitted by the sphalerite coating consisted of many pinpoint flashes. This observation suggested that cathode rays were, in reality, a stream of extremely small particles.

Joseph John Thomson (1856–1940), an English physicist, provided many facts related to the nature of cathode rays. Using a variety of materials as cathodes, he showed that cathode ray production was a general property of matter. Using a specially designed cathode ray tube (see Figure 3.7), he also found that cathode rays could be deflected by charged plates or a magnetic field. The rays were repelled by the north pole, or negative plate, and attracted to the south pole, or positive plate, thus indicating that they were negatively charged. In 1897, Thomson concluded that cathode rays were streams of negatively charged particles. Further experiments by others proved that his conclusions were correct. Today we call these negatively charged particles electrons.

In 1886, a German physicist, Eugene Goldstein (1850–1930), showed that positive particles were also present in discharge tubes. He used a discharge tube in which the cathode was a metal plate with a large number of holes drilled in it. The usual cathode rays were observed to stream from cathode to anode. In addition, rays of light appeared to stream from each of the holes in the cathode in a direction opposite to that of the cathode rays (see Figure 3.8). Because these rays were observed streaming through the holes or channels in the cathode, Goldstein called them *canal rays*.

Further research showed that canal rays were of many types, in contrast with cathode rays which had only one type, and that the particles making up canal rays were much heavier than those of cathode rays. The type of canal rays

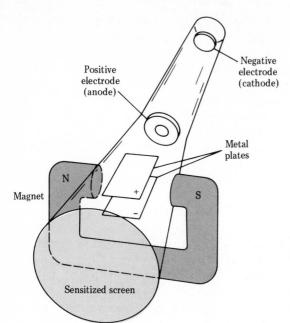

Positive
electrode
(anode)

Negative
electrode
(cathode)

Metal
plates

Magnet

N

S

+

−

Sensitized screen

FIGURE 3.7
J. J. Thomson's cathode ray tube involved the use of
both electrical and magnetic fields.

produced depended on the gas in the tube. The simplest canal rays were
eventually identified as the particles now called protons.

Canal rays are now known to be gas atoms that have lost one or more
electrons. Their origin and behavior in a discharge tube can be understood as
follows: electrons (cathode rays) emitted from the cathode collide with residual
gas molecules (air) on the way to the anode. Some of these electrons have
enough energy to knock electrons away from the gas molecules, leaving behind
a positive particle (the remainder of the gas molecule). These positive particles
are attracted to the cathode, and some of them pass through the holes or
channels. The fact that atoms, under certain conditions, can lose electrons is
discussed further in Section 5.4.

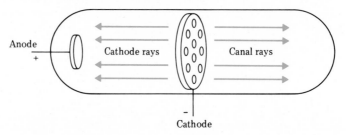

Anode
+

Cathode rays

Canal rays

Cathode

FIGURE 3.8
A gas discharge tube with a perforated cathode through which canal rays stream.

In 1898, on the basis of discharge-tube experiment results, Thomson proposed that the atom was composed of a sphere of positive electricity containing most of the mass, and that small negative electrons were attached to the surface of the positive sphere. He postulated that a high voltage could pull off surface electrons to produce cathode rays. Thomson's model of the atom, sometimes referred to as the "plum pudding" or "raisin muffin" model—with the electrons as the plums or raisins—is now known to be incorrect. Its significance is that it set the stage for an experiment, commonly called the gold-foil experiment, that led to the currently accepted model of the atom.

Metal-Foil Experiments

In 1911 Ernest Rutherford (1871–1937) (see Historical Profile 5) designed an experiment to test the Thomson model of the atom. In this experiment thin sheets of metal foil were bombarded by alpha particles from a radioactive source. Alpha particles, which are positively charged, are ejected at high speed from some radioactive materials. The phenomenon of radioactivity had been discovered in 1896 and gave further evidence that electrical charges existed within the atom. (Further details concerning alpha particles and radioactivity in general are found in Chapter 15.) Gold was chosen as the target metal because it is easily hammered into very thin sheets. The experimental setup for Rutherford's experiment is shown in Figure 3.9. Because alpha particles do not

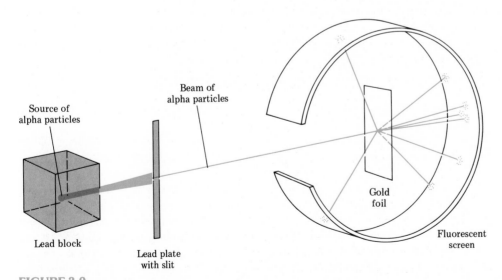

FIGURE 3.9

Rutherford's gold-foil alpha-particle experiment. Most of the alpha articles went straight through the foil, but a few were deflected at large angles.

Ernest Rutherford   1871–1937

[The Bettmann Archive.]

Ernest Rutherford (1871–1937) was born in New Zealand, but did most of his work in England at Manchester and Cambridge Universities. Considered the father of nuclear science, Rutherford contributed much more than the "gold-foil experiment." Indeed, 3 years before he carried out this famous experiment, in 1908, he received the Nobel prize in chemistry for his investigations into atomic structure.

His first work in England, with Joseph John Thomson, involved gas discharge tubes. He then spent 9 years in Canada working in the field of radioactivity. During this time he was the first to characterize what are now known as alpha and beta particles.

Then it was back to England (Manchester University), where further studies on alpha particles led to his famous gold-foil experiment, considered to be Rutherford's greatest and most fruitful contribution to atomic theory. Many years later Rutherford described the unexpected results of this experiment as follows: "It was about as credible as if you had fired a 15-inch shell at a piece of tissue paper and it came back and hit you."

World War I brought an abrupt change in direction, a switch from atoms to submarines. He studied underwater acoustics, supplying his government with much information needed to advance the technology of submarine detection.

Following the war, he spent the rest of his career at Cambridge University. Here, he was again at the "forefront" of scientific advances, this time discovering nuclear transformation, the process in which an atom of an element changes its identity.

appreciably penetrate lead, this metal was used to produce a narrow alpha-particle beam. Each time an alpha particle hit the fluorescent screen, a flash of light was produced.

Rutherford expected that all the alpha particles, since they were so energetic, would pass *straight* through the thin gold foil. His reasoning was based on the Thomson model, in which the mass and positive charge of the gold atoms were distributed uniformly through the atom. As each positive particle neared the foil, Rutherford assumed that it would be confronted by a uniform positive charge and thus there would be no preferred direction of deflection from positive–positive replusions. All particles would be affected the same way (no deflection) as they went through the foil, which would support the Thomson model of the atom.

The results from the experiment were very surprising. Most of the particles—more than 99%—went straight through, as expected. A few, however, were appreciably deflected by something that had to be difficult to hit and much heavier than the alpha particles themselves (see Figure 3.9). A very few particles were deflected almost directly back toward the alpha-particle source. Similar results were obtained when elements other than gold were used as targets.

Extensive study of the results of his experiments led Rutherford to propose the following explanation.

1. A very dense, small nucleus exists in the center of the atom. This nucleus contains most of the mass of the atom and all of the positive charge.
2. Electrons occupy most of the total volume of the atom and are located outside the nucleus.
3. When an alpha particle scores a "direct hit" on a nucleus, it is deflected back along the incoming path.
4. A near miss of a nucleus by an alpha particle results in repulsion and deflection.
5. Most of the alpha particles pass through without any interference because most of the atomic volume is empty space.
6. Electrons have so little mass that they do not deflect the much larger alpha particles (almost 8000 times heavier than the electron).

Many other experiments have since verified Rutherford's conclusion that at the center of an atom is a nucleus that is positively charged, very heavy, very small, and very dense.

3.5 Atomic Number and Mass Number

The number of protons, neutrons, and electrons in a given atom is specified by two numbers: the atomic number and the mass number.

The **atomic number** *is equal to the number of protons in the nucleus of an atom.* Atomic numbers are always integers (whole numbers). All atoms of a given element must contain the same number of protons. Thus, all atoms of a given element have the same atomic number. Because an atom has the same number of electrons as protons, the atomic number also specifies the number of electrons present.

Atomic number = Number of protons = Number of electrons

A second necessary quantity in specifying atomic identities is the mass number. The **mass number** *is equal to the number of protons plus the* number of neutrons in the nucleus of the atom—that is, the total number of nucleons present. The mass of an atom is almost totally accounted for by the protons and neutrons (Section 3.3); hence the term *mass* number. Like the atomic number, the mass number is always an integer.

$$\text{Mass number} = \text{Number of protons} + \text{Number of neutrons}$$

The atomic number and mass number of an atom uniquely specify the atom's makeup in terms of subatomic particles. The following equations show the relationship between subatomic particle makeup and the two numbers.

$$\text{Number of protons} = \text{Atomic number}$$

$$\text{Number of electrons} = \text{Atomic number}$$

$$\text{Number of neutrons} = \text{Mass number} - \text{Atomic number}$$

EXAMPLE 3.1

An atom has an atomic number of 12 and a mass number of 26. Determine
(a) the number of protons present
(b) the number of neutrons present
(c) the number of electrons present

SOLUTION

(a) There are 12 protons because the atomic number is always equal to the number of protons present.
(b) There are 14 neutrons because the number of neutrons is always obtained by subtracting the atomic number from the mass number.

$$\underbrace{(\text{Protons} + \text{Neutrons})}_{\text{Mass number}} - \underbrace{\text{Protons}}_{\substack{\text{Atomic} \\ \text{number}}} = \text{Neutrons}$$

$$26 - 12 = 14$$

(c) There are 12 electrons because the number of protons and the number of electrons are always the same in a neutral atom.

An alphabetical listing of the 109 known elements, along with selected information about each element, is printed on the inside back cover of this text. One of the pieces of information given for each element is its atomic number. If you carefully checked the atomic number data, you would find that an element exists for each atomic number in the numerical sequence 1 through 109; no numbers in the sequence are missing. The existence of an element that corresponds to each of these numbers is an indication of the order existing in nature. Thus, elements exist with any number of protons (or electrons) up to and including 109.

Mass numbers are not tabulated in a manner similar to atomic numbers because, as we will see in the next section, most elements lack a unique mass number.

3.6 Isotopes

The identity of an atom is determined by its atomic number, which is the number of protons it contains. All atoms of an element must contain the same number of protons.

The chemical properties of an atom, which are the basis for its identification, are determined by the number and arrangement of electrons about the nucleus. All atoms that have the same atomic number will have the same number of electrons and the same chemical properties.

All of the atoms of an element do not have to be identical. They can differ in the number of neutrons they have. The presence of one or more extra neutrons in the tiny nucleus of an atom has essentially no effect on the way the atom behaves chemically. For example, all carbon atoms have six protons and six electrons. Most carbon atoms also contain six neutrons. However, some carbon atoms contain seven neutrons, and others contain eight. Thus, three different kinds of carbon atoms exist, each with the same chemical properties—that is, three carbon *isotopes* exist. **Isotopes** *are atoms that have the same number of protons and electrons but different numbers of neutrons.* The isotopes of an element always have the same atomic number and different mass numbers.

When it is necessary to distinguish between isotopes, the following notation is used.

$$^{\text{Mass number}}_{\text{Atomic number}}\text{Symbol}$$

The atomic number, whose general symbol is Z, is written as a subscript to the left of the elemental symbol for the atom. The mass number, whose general symbol is A, is also written to the left of the elemental symbol, but as a superscript. Using this symbolism, the three previously mentioned carbon isotopes would be designated, respectively, as

$$^{12}_{6}\text{C}, \quad ^{13}_{6}\text{C}, \quad \text{and} \quad ^{14}_{6}\text{C}$$

The existence of isotopes clarifies, to some degree, the statements of the atomic theory made earlier (Section 3.1). Statement 1 reads: "All matter is made up of small particles called atoms, of which 109 different 'types' are known." It should now be apparent why the word types was put in quotation marks. Because of the existence of isotopes, types mean "similar, but not identical."

Atoms of a given element are similar because they have the same atomic number, but are not identical because they may have different mass numbers.

Statement 2 reads: "All atoms of a given type are similar to one another and significantly different from all other types." All atoms of an element are similar in chemical properties, and they differ significantly from atoms of other elements, which have different chemical properties.

Most naturally occurring elements are mixtures of isotopes. The various isotopes of a given element vary in abundance; usually one isotope is predominant. Typical of this situation is the element magnesium, which exists in nature in three isotopic forms: $^{24}_{12}Mg$, $^{25}_{12}Mg$, and $^{26}_{12}Mg$. The percentage abundances for these three isotopes are 78.70%, 10.13%, and 11.17%, respectively. Percentage abundances are number percentages (number of atoms) rather than mass percentages. A sample of 10,000 magnesium atoms contains 7870 $^{24}_{12}Mg$ atoms, 1013 $^{25}_{12}Mg$, and 1117 $^{26}_{12}Mg$ atoms. Table 3.3 gives natural isotopic abundances and isotopic masses for selected elements. The unit used for specifying the mass of the various isotopes, the amu, is discussed in Section 3.7. Twenty elements exist in nature in only one form; these elements are listed in the footnote to Table 3.3.

EXAMPLE 3.2

Indicate whether the members of each of the following pairs of atoms are isotopes.

(a) $^{28}_{13}X$ and $^{28}_{14}Y$ (b) $^{40}_{20}X$ and $^{42}_{20}Y$

(c) An atom X with eight protons and nine neutrons and an atom Y with nine protons and nine neutrons.

SOLUTION

(a) These atoms are not isotopes. Isotopes must have the same atomic number. The first atom of this pair, X, has an atomic number of 13 and the second atom, Y, has an atomic number of 14. The mass numbers of the two atoms are the same, but that does not make them isotopes.

(b) These atoms are isotopes. Both atoms have the same atomic number of 20. Isotopes differ from each other only in neutron content, which is the case here. Atom X has 20 neutrons and atom Y has 22 neutrons.

(c) These atoms are not isotopes. Because they differ in the number of protons present, they differ in atomic number. The number of protons and the atomic number are the same.

It is because of the existence of isotopes that mass numbers are not unique for elements. Quite often atoms of two different elements have the same mass number. For example, the element iron, whose atomic number is 26, exists in

TABLE 3.3
Naturally Occurring Isotopic Abundances for Selected Common Elements*

Element	Isotope	Percent Natural Abundance	Isotopic Mass (amu)
Hydrogen	$^{1}_{1}H$	99.98	1.01
	$^{2}_{1}H$	0.02	2.01
Carbon	$^{12}_{6}C$	98.89	12.00
	$^{13}_{6}C$	1.11	13.00
	$^{14}_{6}C$	trace	14.00
Nitrogen	$^{14}_{7}N$	99.63	14.00
	$^{15}_{7}N$	0.37	15.00
Oxygen	$^{16}_{8}O$	99.76	15.99
	$^{17}_{8}O$	0.04	17.00
	$^{18}_{8}O$	0.20	18.00
Sulfur	$^{32}_{16}S$	95.00	31.97
	$^{33}_{16}S$	0.76	32.97
	$^{34}_{16}S$	4.22	33.97
	$^{36}_{16}S$	0.01	35.97
Chlorine	$^{35}_{17}Cl$	75.53	34.97
	$^{37}_{17}Cl$	24.47	36.97
Copper	$^{63}_{29}Cu$	69.09	62.93
	$^{65}_{29}Cu$	30.91	64.93
Titanium	$^{46}_{22}Ti$	7.93	45.95
	$^{47}_{22}Ti$	7.28	46.95
	$^{48}_{22}Ti$	73.94	47.95
	$^{49}_{22}Ti$	5.51	48.95
	$^{50}_{22}Ti$	5.34	49.94
Uranium	$^{234}_{92}U$	0.01	234.04
	$^{235}_{92}U$	0.72	235.04
	$^{238}_{92}U$	99.27	238.05

* Twenty elements have only one naturally occurring form: $^{9}_{4}Be$, $^{19}_{9}F$, $^{23}_{11}Na$, $^{27}_{13}Al$, $^{31}_{15}P$, $^{45}_{21}Sc$, $^{55}_{25}Mn$, $^{59}_{27}Co$, $^{75}_{33}As$, $^{89}_{41}Nb$, $^{103}_{45}Rh$, $^{127}_{53}Cs$, $^{141}_{59}Pr$, $^{159}_{65}Tb$, $^{165}_{67}Ho$, $^{169}_{69}Tm$, $^{197}_{79}Au$, $^{209}_{83}Bi$.

nature in four isotopic forms, one of which is $^{58}_{26}Fe$. The element nickel, whose atomic number is 28, exists in nature in five isotopic forms, one of which is $^{58}_{28}Ni$. Thus, atoms of both iron and nickel exist with a mass number of 58.

3.7 Atomic Weights

An atom of a specific element can have one of several different masses if the element exists in isotopic forms. Magnesium atoms, for example, can have any one of three masses because there are three isotopes of magnesium. It might seem necessary to specify isotopic identity every time masses of atoms are encountered; however, this is not the case. In practice, isotopes are seldom mentioned in discussions about the masses of atoms because atoms of an element are treated as if they all had a single mass. The mass used is the "average mass," a number that takes into account the existence of isotopes. The use of this average mass reduces the number of masses needed for calculations from many hundreds (one for each isotope) to 109 (one for each element).

The term *atomic weight* is used to describe the average mass of the isotopes of an element. An **atomic weight** *is the average mass for the isotopes of an element expressed on a scale using atoms of* $^{12}_{6}C$ *as the reference.*

Three pieces of information are needed to calculate an atomic weight

1. The number of isotopes that exist for the element.
2. The masses of the isotopes on the $^{12}_{6}C$ scale.
3. The percentage abundance of each isotope.

Before we actually calculate an atomic weight, let us first consider the $^{12}_{6}C$ scale that is mentioned in the definition of atomic weight.

The usual standards of mass, such as grams or pounds, are not convenient for use with atoms because very small numbers are always encountered. For example, the mass in grams of a $^{238}_{92}U$ atom, one of the heaviest atoms known, is 3.95×10^{-22}. To avoid repeatedly encountering such small numbers, scientists have chosen to work with relative rather than actual mass values.

A relative mass value of an atom is the mass of that atom relative to some standard rather than the actual mass value of the atom in grams. The term *relative* means "as compared to." The choice of the standard is arbitrary; this gives scientists control over the magnitude of the numbers on the relative scale and avoids very small numbers.

The standard for specifying atomic mass involves a particular isotope of carbon, $^{12}_{6}C$. The mass of this isotope has a defined value of 12.00 *atomic mass units (amu)*. The masses of all other atoms are then determined relative to that of $^{12}_{6}C$. For example, if an atom is twice as heavy as $^{12}_{6}C$, its mass is 24.00 amu on the scale, and if an atom has a mass one-half that of $^{12}_{6}C$, its mass is 6.00 amu. Mass values, on the amu scale, for isotopes of selected elements are found in Table 3.3.

On the basis of the amu values given in Table 3.3, it is possible to make statements such as $^{238}_{92}U$ is 13.22 times as heavy as $^{18}_{8}O$ (238.05 amu/18.00 amu = 13.22), or $^{32}_{16}S$ is 2.46 times as heavy as $^{13}_{6}C$ (31.97 amu/13.00 amu = 2.46).

Example 3.3 shows how information from the amu ($^{12}_{6}C$) scale, percentage abundances, and number of isotopes are combined to calculate an atomic weight.

SECTION 3.7—*Atomic Weights*

EXAMPLE 3.3

Naturally occuring chlorine exists in two isotopic forms, $^{35}_{17}Cl$ and $^{37}_{17}Cl$. The relative mass of $^{35}_{17}Cl$ is 34.97 amu, and its abundance is 75.53%; the relative mass of $^{37}_{17}Cl$ is 36.97 amu and it has a 24.47% abundance. Calculate the atomic weight of chlorine.

SOLUTION

The atomic weight is calculated by multiplying the relative mass of each isotope by its fractional abundance and then totaling the products.

The fractional abundance is the percentage abundance converted to decimal form (divided by 100).

$$^{35}_{17}Cl: \quad \frac{75.53}{100} \times 34.97 \text{ amu} = (0.7553) \times 34.97 \text{ amu} = 26.41 \text{ amu}$$

$$^{37}_{17}Cl: \quad \frac{24.47}{100} \times 36.97 \text{ amu} = (0.2447) \times 36.97 \text{ amu} = \underline{9.04 \text{ amu}}$$

$$\text{Atomic weight of chlorine} = 35.45 \text{ amu}$$

This calculation involves an element containing just two isotopes. A similar calculation for an element with three isotopes would be carried out the same way, but it would have three terms in the final sum; an element possessing four isotopes would have four terms in the final sum.

The validity of the average mass (atomic weight) concept rests on two points. First, extensive studies of naturally occurring elements have shown that the percentage abundance of the isotopes for a given element is nearly constant. Element samples contain nearly the same percentage of each isotope, regardless of their source. Because of these relatively constant isotopic ratios, the mass of an "average atom" is considered to be constant. Second, chemical operations are always carried out with very large numbers of atoms. The tiniest piece of matter visible to the eye contains more atoms than can be counted in a lifetime. Any collection of atoms a chemist encounters is very unlikely to contain an isotopic ratio different from that of the naturally occurring element.

The alphabetical listing of the known elements printed inside the back cover of this text gives, in the last column, the calculated atomic weight for each of the elements.

Isotope masses, although not whole numbers, have values that are very close to whole numbers. This fact may be verified by looking at the numbers in the last column of Table 3.3. If an isotopic mass is rounded off to the closest whole number, this value is the same as the mass number of the isotope—compare the second and fourth columns in Table 3.3.

Practice Questions and Problems

THE ATOMIC THEORY OF MATTER

3-1 List the five statements of modern-day atomic theory.

3-2 Which of the following concepts are not consistent with the statements of modern-day atomic theory?
a. Atoms are the basic building blocks for all kinds of matter.
b. Different "types" of atoms exist.
c. All atoms of a given "type" are identical.
d. Only whole atoms can participate in chemical reactions.
e. Atoms change identity during chemical change processes.
f. 113 different "types" of atoms are known.

3-3 What is the approximate diameter of an atom (in centimeters)? What is the approximate mass of an atom (in grams)?

LAWS OF CHEMICAL CHANGE

3-4 Consider a hypothetical reaction in which A and B are reactants and C and D are products. If 15 grams of A completely reacts with 12 grams of B to produce 14 grams of C, how many grams of D will be produced?

3-5 A 4.2 gram sample of sodium hydrogen carbonate is added to a solution of acetic acid weighing 10.2 grams. The two substances react, releasing carbon dioxide gas to the atmosphere. After the reaction, the contents of the reaction vessel weighs 12.2 grams. What is the mass of carbon dioxide given off during the reaction?

3-6 When a log burns, the ashes weigh less than the log. When sodium reacts with sulfur, the sodium–sulfur compound formed weighs the same as the sodium and sulfur. When phosphorus burns in air, however, the compound formed weighs more than the original phosphorus. Explain these observations in terms of the law of conservation of mass.

3-7 Two different samples of a pure compound containing elements A and B were analyzed with the following results.
Sample I: 15.8 grams of compound yielded 9.8 grams of A and 6.0 grams of B.

Sample II: 25.0 grams of compound yielded 15.5 grams of A and 9.5 grams of B.
Show that these data are consistent with the law of definite proportions.

SUBATOMIC PARTICLES: PROTONS, NEUTRONS, AND ELECTRONS

3-8 List the two kinds of subatomic particles found in the nucleus. Compare their properties with each other and with those of an electron.

3-9 What is meant by the statement: "Atoms as a whole are neutral"?

3-10 Indicate which subatomic particle (proton, neutron, or electron) correctly matches each of the following statements. More than one particle can be used as an answer.
a. Possesses a negative charge.
b. Has no charge.
c. Has a mass slightly less than that of a neutron.
d. Has a charge equal to but opposite in sign to that of an electron.
e. Is not found in the nucleus.
f. Has a positive charge.
g. Can be called a nucleon.
h. Is the heaviest of the three particles.
i. Has a relative mass of 1836 if the mass of an electron is 1
j. Has a relative mass of 1839 if the mass of an electron is 1.

3-11 Indicate whether each of the following statements about the nucleus of an atom is true or false.
a. The nucleus of an atom is neutral.
b. The nucleus of an atom contains only neutrons.
c. The number of nucleons present in the nucleus is equal to the number of electrons present outside the nucleus.
d. The nucleus accounts for almost all of the mass of an atom.
e. The nucleus accounts for almost all of the volume of an atom.
f. The nucleus can be positively or negatively charged, depending on the identity of the atom.

3-12 Explain why atoms are considered to be the fundamental building blocks of matter even though smaller particles (subatomic particles) exist.

3-13 What subatomic particles were identified as a result of discharge-tube experiments?

3-14 Contrast the properties of cathode rays with those of canal rays.

3-15 How did the results of Rutherford's gold-foil experiment suggest that an atom has a nucleus?

ATOMIC NUMBER AND MASS NUMBER

3-16 What information about the subatomic makeup of an atom is given by the following?
a. Atomic number
b. Mass number
c. Mass number − Atomic number
d. Mass number + Atomic number

3-17 Determine the number of protons, neutrons, and electrons present in each of the following atoms.
a. $^{53}_{24}Cr$ b. $^{103}_{44}Ru$ c. $^{256}_{101}Md$ d. $^{34}_{16}S$
e. $^{67}_{30}Zn$ f. $^{9}_{4}Be$ g. $^{40}_{20}Ca$ h. $^{3}_{1}H$

3-18 Write complete symbols, with the help of the information listed inside the back cover, for atoms with the following characteristics.
a. Contains 28 protons and 30 neutrons.
b. Contains 22 protons and 21 neutrons.
c. Contains 15 electrons and 19 neutrons.
d. Oxygen atom with 10 neutrons.
e. Chromium atom with a mass number of 54.
f. Has an atomic number of 11 and a mass number of 25.
g. Gold atom that contains 276 subatomic particles.
h. Beryllium atom that contains 10 nucleons.

3-19 The notation $^{7}_{3}Li$ and ^{7}Li may be used interchangeably. Why is this allowable? Could the notations $^{7}_{3}Li$ and $_{3}Li$ be used interchangeably without loss of meaning? Why or why not?

3-20 Can the atomic number of an element be larger than its mass number?

ISOTOPES

3-21 Identify the atoms that are isotopes in each of the following sets of four atoms.
a. $^{14}_{6}X$, $^{14}_{7}X$, $^{15}_{7}X$, $^{15}_{8}X$
b. $^{20}_{10}X$, $^{21}_{10}X$, $^{22}_{10}X$, $^{23}_{10}X$

c. $^{41}_{19}X$, $^{41}_{20}X$, $^{41}_{21}X$, $^{41}_{22}X$

3-22 What restrictions on the numbers of protons, neutrons, and electrons apply to each of the following?
a. Atoms of different isotopes of the same element.
b. Atoms of a particular isotope of an element
c. Atoms of different elements with the same mass number
d. Atoms of the same element with different mass numbers

3-23 How do you write the symbol for an atom of an element if you wish to specify a particular isotope of that element?

3-24 Account for the fact that there are only 109 known elements and yet nearly 2000 different kinds of atoms.

3-25 The following are selected properties for the most abundant isotope of a particular element. Which of these properties would be the same for the second most abundant isotope of the element?
a. Mass number is 70.
b. Atomic number is 31.
c. Melting point is 29.8°C.
d. Isotopic mass is 69.92 amu.
e. Reacts with chlorine to give the compound $GaCl_3$.

3-26 Write complete symbols for atoms with the following characteristics.
a. An isotope of boron that contains two fewer neutrons than $^{10}_{5}B$.
b. An isotope of boron that contains three more subatomic particles than $^{9}_{5}B$.
c. An isotope of boron that contains the same number of neutrons as $^{14}_{7}N$.
d. An isotope of boron that contains the same number of subatomic particles as $^{11}_{6}C$.

ATOMIC WEIGHTS

3-27 A certain isotope of silver is 8.91 times heavier than $^{12}_{6}C$. What is the mass of this silver isotope on the amu scale?

3-28 What three types of information are needed to calculate an atomic weight?

3-29 A sample of lithium of natural origin is found to consist of two isotopes: $^{6}_{3}Li$ (abundance 7.42%) and $^{7}_{3}Li$ (abundance 92.58%). The masses of these isotopes on the amu scale are 6.01 and 7.02, respectively. Calculate the atomic weight of lithium.

3-30 A sample of silicon of natural origin is found to consist of three isotopes: $^{28}_{14}Si$ (abundance 92.21%), $^{29}_{14}Si$ (abundance 4.70%), and $^{30}_{14}Si$ (abundance 3.09%). Their relative masses are 27.98, 28.98, and 29.97 amu, respectively. Calculate the atomic weight of silicon.

3-31 How many times heavier, on the average, is an atom of gold than an atom of beryllium?

3-32 The atomic weight of fluorine is 18.998 amu and the atomic weight of iron is 55.847 amu. All fluorine atoms have a mass of 18.998 amu and not a single iron atom has a mass of 55.847 amu. Explain?

3-33 An isotope has a mass of 33.98 amu on the $^{12}_{6}C$ relative mass scale. What is the mass number of this isotope?

3-34 Three naturally occurring isotopes of potassium exist: $^{39}_{19}K$, $^{40}_{19}K$, and $^{41}_{19}K$. The atomic weight of potassium is 39.102 amu. Which of the three potassium isotopes is most abundant. Explain your answer.

Electronic Structure of Atoms and the Periodic Table

HISTORICAL PROFILE 6

Dmitri Ivanovich Mendeleev   1834–1907

Dmitri Ivanovich Mendeleev (1834–1907), born in Siberia, the youngest of 17 children, is Russia's most famous chemist. His periodic classification of the elements is heralded as his greatest achievement. Other areas of research for him included petroleum chemistry, properties and behavior of gases at high and low pressure, and the critical temperature for gases.

Mendeleev's talents as a prolific writer and a popular teacher directly influenced his discovery of the periodic law. In 1867, two years prior to his great discovery, Mendeleev was appointed professor of general chemistry at the University of St. Petersburg (now Leningrad). To help his students, he immediately began writing a new textbook. In preparation for writing some material about the elements and their properties, he wrote down the properties of each known element on separate cards. While sorting through the cards trying to "organize things," he "discovered" the periodic repetition of properties for the elements. Within a month he had written a paper on the subject that was delivered before the Russian Chemical Society. The textbook he was writing, *General Chemistry*, became a classic, going through eight Russian editions and numerous English, French, and German editions.

Mendeleev remained a professor at St. Petersburg until 1890, when he resigned because of a controversy with the Ministry of Education concerning student political rights. He took the side of the students. In 1893, he became director of the Russian Bureau of Weights and Measures, a post he held until his death.

It should be pointed out that the German chemist Julius Lothar Meyer (1830–1895) discovered the periodic law simultaneously and independently of Mendeleev. Unknown to each other, both published papers on the subject in 1869. Because Mendeleev did more with the periodic law, once it was published, he is generally given more credit for its development.

In 1882, when selected to receive the Davy medal of the British Royal Society for his work concerning systematic classification of the elements, Mendeleev insisted that Lothar Meyer be corecipient of the award, formally recognizing Meyer's independent conception of the periodic law.

4.1 The Periodic Law

During the early part of the nineteenth century, an abundance of chemical facts became available from detailed studies of the then known elements. With the hope of providing a systematic approach to the study of chemistry, scientists began to look for some form of order in the increasing amount of chemical information. They were encouraged in their search by the unexplained, but well-known fact that certain elements had properties that were very similar to those of other elements. Numerous attempts were made to explain these similarities and to use them as a means for arranging or classifying the elements.

In 1869, these efforts culminated in the discovery of what is now called the *periodic law.* Proposed independently by both the Russian chemist Dmitri Ivanovich Mendeleev (1834–1907) (see Historical Profile 6) and the German chemist Julius Lothar Meyer (1830–1895), this law is one of the most important chemical laws. Given in its modern form, the **periodic law** *states that when elements are arranged in order of increasing atomic numbers, elements with similar properties occur at periodic (regularly recurring) intervals.* Figure 4.1 illustrates this regularly repeating pattern for chemical properties for the sequence of elements with atomic numbers 3 through 20. The elements within similar geometric symbols (circles and squares) have similar chemical properties. For the sake of simplicity, only two of the periodic relationships (repeating patterns) are shown; for the elements listed, similar properties are found in every eighth element.

4.2 The Periodic Table

A **periodic table** *is a graphical representation of the behavior described by the periodic law.* In this table, the elements are arranged according to

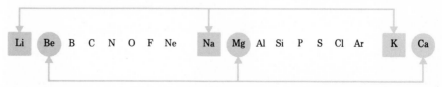

FIGURE 4.1
Periodicity in properties when elements are arranged in order of increasing atomic number.

1 Group IA												13 Group IIIA	14 Group IVA	15 Group VA	16 Group VIA	17 Group VIIA	18 Group VIIIA
1 H 1.01	2 Group IIA																2 He 4.00
3 Li 6.94	4 Be 9.01											5 B 10.81	6 C 12.01	7 N 14.01	8 O 16.00	9 F 19.00	10 Ne 20.18
11 Na 22.99	12 Mg 24.30	3 Group IIIB	4 Group IVB	5 Group VB	6 Group VIB	7 Group VIIB	8 Group	9 Group ←VIIIB→	10 Group	11 Group IB	12 Group IIB	13 Al 26.98	14 Si 28.09	15 P 30.97	16 S 32.07	17 Cl 35.45	18 Ar 39.95
19 K 39.10	20 Ca 40.08	21 Sc 44.96	22 Ti 47.88	23 V 50.94	24 Cr 52.00	25 Mn 54.94	26 Fe 55.85	27 Co 58.93	28 Ni 58.69	29 Cu 63.55	30 Zn 65.38	31 Ga 69.72	32 Ge 72.59	33 As 74.92	34 Se 78.96	35 Br 79.90	36 Kr 83.80
37 Rb 85.47	38 Sr 87.62	39 Y 88.91	40 Zr 91.22	41 Nb 92.91	42 Mo 95.94	43 Tc (98)	44 Ru 101.07	45 Rh 102.91	46 Pd 106.42	47 Ag 107.87	48 Cd 112.41	49 In 114.82	50 Sn 118.71	51 Sb 121.75	52 Te 127.60	53 I 126.90	54 Xe 131.29
55 Cs 132.91	56 Ba 137.33	57 La 138.91	72 Hf 178.49	73 Ta 180.95	74 W 183.85	75 Re 186.21	76 Os 190.2	77 Ir 192.22	78 Pt 195.08	79 Au 196.97	80 Hg 200.59	81 Tl 204.38	82 Pb 207.2	83 Bi 208.98	84 Po (209)	85 At (210)	86 Rn (222)
87 Fr (223)	88 Ra (226)	89 Ac (227)	104 Unq (261)	105 Unp (262)	106 Unh (263)	107 Uns (262)	108 Uno (265)	109 Une (266)									

Metals ← → Nonmetals

59 Ce 140.12	60 Pr 140.91	61 Nd 144.24	62 Pm (145)	63 Sm 150.36	64 Eu 151.96	65 Gd 157.25	Tb 158.93	66 Dy 162.50	67 Ho 164.93	68 Er 167.26	69 Tm 168.93	70 Yb 173.04	71 Lu 174.97
90 Th (232)	91 Pa (231)	92 U (238)	93 Np (237)	94 Pu (244)	95 Am (243)	96 Cm (247)	97 Bk (247)	98 Cf (251)	99 Es (252)	100 Fm (257)	101 Md (258)	102 No (259)	103 Lr (260)

FIGURE 4.2

The modern periodic table.

increasing atomic number in such a way that the similarities predicted by the periodic law become readily apparent.

The most commonly used form of the periodic table is shown in Figure 4.2 (and also inside the front cover of the text). In this periodic table, the vertical columns contain elements with similar chemical properties. Each element is represented by a rectangular box, which contains the symbol, atomic number, and atomic weight of the element, as shown in Figure 4.3.

Special chemical terminology exists for specifying the position (location) of an element within the periodic table. This terminology involves the use of the words *group* and *period*. A **group** *in the periodic table is a vertical column of elements*. There are two notations in use for designating individual periodic table

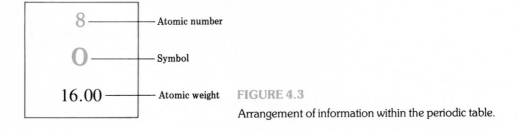

8 —— Atomic number

O —— Symbol

16.00 —— Atomic weight

FIGURE 4.3

Arrangement of information within the periodic table.

groups. In the first notation, which has been in use for many years, groups are designated using Roman numerals and the letters A and B. In the second notation, which has been recently recommended for use by an international scientific commission, the Arabic numbers 1 through 18 are used. Notice that both group notations are given at the top of each group in the periodic table (Figure 4.2). The elements with atomic numbers 8, 16, 34, 52, and 84 (O, S, Se, Te, and Po) constitute group VIA (old notation) or group 16 (new notation). Because it may be some time before the new group numbering system is widely accepted, we will not use it in succeeding chapters of the text.

A **period** *in the periodic table is a horizontal row of elements.* For identification purposes, the periods are numbered sequentially, with Arabic numbers, starting at the top of the periodic table. (These period numbers are not explicitly shown on the periodic table.) The elements Na, Mg, Al, Si, P, S, Cl, and Ar are all members of period 3. Period 3 is the third row of elements, period 4 the fourth row of elements, and so forth. Period 1 has only two elements—H and He.

The location of any element in the periodic table is specified by giving its group number and its period number. The element gold, with an atomic number of 79, belongs to group IB (or 11) and is in period 6. The element nitrogen, with an atomic number of 7, belongs to group VA (or 15) and is in period 2.

A careful study of the periodic table shows that there is one area in the table where the practice of arranging the elements according to increasing atomic numbers seems to be violated. This is the area where elements 57 and 89, both located in group IIIB (or 3) of the table, are found. Element 72 follows element 57 and element 104 follows element 89. The missing elements, 58 through 71 and 90 through 103, are located in two rows at the bottom of the periodic table. Technically, the elements at the bottom of the table should be included in the body of the table, as shown in Figure 4.4*a*. However, to have a more compact table, they are placed in the position shown in Figure 4.4*b*. This arrangement should present no problems to the user of the periodic table as long as it is recognized for what it is—a space-saving device.

The mass number, which is one of the fundamental identifying characteristics of an atom (Section 3.5), is not part of the information given on a periodic table. The reason for this is that mass number is not a unique quantity for most elements; different isotopes of an element have different mass numbers (Section 3.6). Thus, using the information on a periodic table, you can quickly determine the number of protons and electrons for atoms of an element. However, you cannot derive the actual number of neutrons present in a given isotope from the table.

For many years, the periodic law and periodic table were considered to be empirical; the law worked and the table was useful, but there was no explanation available either for the law or for why the periodic table had the shape it had. We now know that the theoretical basis for both the periodic law and periodic table is found in electronic theory. The properties of the elements repeat themselves in a periodic manner because the arrangement of electrons about the nucleus of an

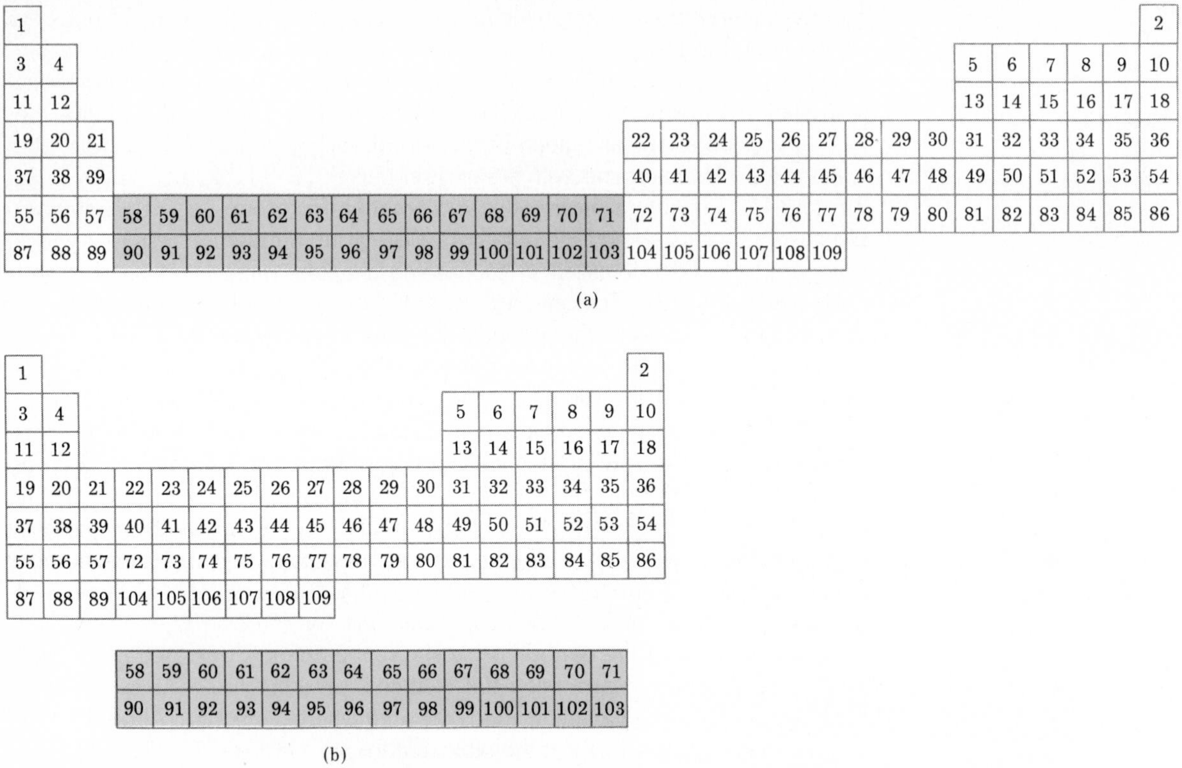

1																															2
3	4																									5	6	7	8	9	10
11	12																									13	14	15	16	17	18
19	20	21															22	23	24	25	26	27	28	29	30	31	32	33	34	35	36
37	38	39															40	41	42	43	44	45	46	47	48	49	50	51	52	53	54
55	56	57	58	59	60	61	62	63	64	65	66	67	68	69	70	71	72	73	74	75	76	77	78	79	80	81	82	83	84	85	86
87	88	89	90	91	92	93	94	95	96	97	98	99	100	101	102	103	104	105	106	107	108	109									

(a)

1																	2
3	4											5	6	7	8	9	10
11	12											13	14	15	16	17	18
19	20	21	22	23	24	25	26	27	28	29	30	31	32	33	34	35	36
37	38	39	40	41	42	43	44	45	46	47	48	49	50	51	52	53	54
55	56	57	72	73	74	75	76	77	78	79	80	81	82	83	84	85	86
87	88	89	104	105	106	107	108	109									

58	59	60	61	62	63	64	65	66	67	68	69	70	71
90	91	92	93	94	95	96	97	98	99	100	101	102	103

(b)

FIGURE 4.4

"Long" and "short" forms of the periodic table. (a) Periodic table with elements 58–71 and 90–103 (in color) in their proper positions. (b) Periodic table modified to conserve space with elements 58–71 and 90–103 placed at the bottom of the table.

atom follows a periodic pattern. This arrangement of electrons and an explanation of the periodic law and periodic table in terms of electronic theory are the subject for the remainder of this chapter.

4.3   Important Properties of an Electron

In Section 3.3 we learned the following facts about electrons.

1. They are one of the three fundamental subatomic particles.
2. They have very little mass in comparison to protons and neutrons.
3. They are found outside the nucleus of an atom.

4. They move rapidly about the nucleus in a volume that defines the size of the atom.

Much more must be known about electrons to understand the chemical behavior of the various elements. It is the arrangement of electrons about the nucleus that determines an element's chemical properties, that is, its behavior toward other substances.

Information about the behavior of electrons as they move about the nucleus comes from a very complex mathematical theory called *quantum mechanics*. The theory of quantum mechanics, which deals with the laws that govern the motion of very small particles like electrons, is beyond the scope of an introductory course. However, some of the concepts obtained from this theory are simple enough for us to easily understand. These concepts will enable us to understand the basis of the periodic law and periodic table, as well as the basic rules governing compound formation (Chapter 5).

A major force in the development of quantum mechanics was the Austrian physicist Erwin Schrödinger (1887–1961) (see Historical Profile 7). In 1926 he showed that the laws of quantum mechanics could be used to characterize the motion of electrons. Most of the concepts of this chapter come from solutions to equations developed by Schrödinger.

Quantum-mechanical theory describes the arrangement of an atom's electrons in terms of their energies. Indeed, when considering an electron's behavior about the nucleus, its energy is its most important property.

A most significant characteristic of an electron's energy is that it is a quantized property. A **quantized property** *is a property that can have only certain values; that is, not all values are allowed.* Because an electron's energy is quantized, it can have only certain specific energies.

Quantization is a phenomenon that is not commonly encountered in the macroscopic world. A somewhat analogous condition to quantization is encountered as a person climbs a flight of stairs. In Figure 4.5a you can see six steps between ground level and the level of the entrance door. Note that as a person

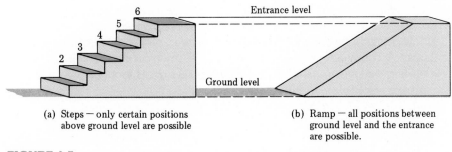

(a) Steps — only certain positions above ground level are possible

(b) Ramp — all positions between ground level and the entrance are possible.

FIGURE 4.5
A stairway with quantized position levels versus a ramp with continuous position levels.

HISTORICAL PROFILE 7

Erwin Schrödinger   1887–1961

[The Nobel Institute, Stockholm.]

Erwin Schrödinger, an only child, was born in 1887 in Vienna, Austria. He was taught at home until age 11, and science was always part of his environment—his father was a botanist, and his maternal grandfather a professor of chemistry. His formal education began in 1898 when he entered a gymnasium, where he received a sound classical education. This early schooling in the classics influenced Schrödinger's actions throughout his life. The culmination of his formal education was a doctorate in theoretical physics, in 1910, from the University of Vienna.

After service as an artillery officer during World War 1, Schrödinger held professorships in Stuttgart and Zurich. During his years in Switzerland his pioneering papers on the equations of quantum theory were formulated and published.

To have more contact with notable physicists (including Albert Einstein) he moved, in 1927, to Berlin. All went well until Hitler's new regime (which he could not accept) came into power. With the rise of Nazism, he accepted an invitation to be a guest professor at Oxford. In 1933, only a few weeks after leaving Germany for England, he received the Nobel prize in physics for his quantum theory work.

Hitler's activities continued to affect his career. After a short stay in England he returned to his native Austria despite an uncertain political situation there. When the Germans occupied Austria in 1938, Schrödinger fled to Italy and from there to the United States. Following a short stay at Princeton University, he became director of the School of Theoretical Physics in Dublin, Ireland. For the next 17 years he worked in Dublin. In 1956, he again returned to Austria, where he remained until his death.

In addition to his pioneering work in quantum theory, Schrödinger's research touched the areas of thermodynamics, specific heats of solids, theory of color, and physical aspects of the living cell.

Schrödinger was "many-sided." He loved poetry and wrote verse of his own. He was a sculptor and also widely acquainted with modern and classical art. He quoted Greek philosophers as he lectured on problems of theoretical physics. A colleague gave the following tribute upon his death: "The breadth of his knowledge was as marvelous as the power of his thought."

climbs these stairs, there are only six permanent positions he or she can occupy (with both feet together). Thus, the person's position (height above ground level) is quantized; only certain heights are allowed. The opposite of quantization is a *continuum*. A person climbing a ramp up to the entrance (Figure 4.5b) could assume a continuous set of heights above ground level; in this case all values are allowed.

The energy of an electron determines its behavior about the nucleus. Because electron energies are quantized, only certain behavior patterns are allowed. Descriptions of these behavior patterns involve the use of the terms *shell*, *subshell*, and *orbital*. Sections 4.4 through 4.6 consider the meaning of these terms and the interrelationships between them.

4.4 Electron Shells

Electrons can be grouped into *shells* or *main energy levels*. A **shell** *contains electrons that have approximately the same energy and spend most of their time approximately the same distance from the nucleus.*

Electron shells are identified by a number, n, which can have the values 1, 2, 3, ... The lowest-energy shell is assigned an n value of 1, the next higher a 2, then a 3, and so on. Only values of $n = 1$–7 are used at present in designating electron shells. No known atom has electrons that are farther from the nucleus than the seventh main energy level ($n = 7$).

The maximum number of electrons that can be found in an electron shell varies; the higher the energy of the shell, the more electrons the shell can accommodate. The farther away electrons are from the nucleus (a high-energy shell), the greater the volume of space available for electrons is—hence, the larger the number of electrons in the shell.

The lowest-energy shell ($n = 1$) accommodates a maximum of two electrons. In the second, third, and fourth shells, 8, 18, and 32 electrons are allowed, respectively. A very simple mathematical equation can be used to calculate the maximum number of electrons allowed in any given shell.

$$\text{Shell electron capacity} = 2n^2 \qquad (\text{where } n = \text{shell number})$$

For example, when n is 4, the value is

$$2n^2 = 2(4)^2 = 32$$

This is the number previously given for the number of electrons allowed in the fourth shell. Although a maximum electron-occupancy level exists for each shell (group of electrons), these main energy levels can hold less than the allowable number of electrons in a given situation.

Table 4.1 summarizes the information about electron shells presented in this section.

TABLE 4.1

Important Characteristics of Electron Shells

Shell	Number Designation (n)	Electron Capacity ($2n^2$)
1st	1	$2 \times 1^2 = 2$
2nd	2	$2 \times 2^2 = 8$
3rd	3	$2 \times 3^2 = 18$
4th	4	$2 \times 4^2 = 32$
5th	5	$2 \times 5^2 = 50*$
6th	6	$2 \times 6^2 = 72*$
7th	7	$2 \times 7^2 = 98*$

* The maximum number of electrons in this shell has never been attained in any element now known.

4.5　Electron Subshells

Each electron shell (main energy level) is divided into energy sublevels called *subshells*. A **subshell** *contains electrons that all have the same energy.* The number of subshells within a shell depends on the n value for the shell; in each shell there are n subshells.

Subshells in a shell = n　　(where n = shell number)

Each successive shell has one more subshell than the previous one; shell 1 contains one subshell, shell 2 contains two subshells, shell 3 contains three subshells, and shell 4 contains four subshells.

Subshells are identified with both a number and a lowercase letter. The notations 3s, 2p, 4d, and 5f are subshell notations. The number indicates the shell to which the subshell belongs. Electrons are found in four types of subshells, which are denoted by the lowercase letters *s, p, d,* and *f*. These four letters, in the order listed, denote subshells of increasing energy within a shell. The lowest energy subshell within a shell is always the *s* subshell, the next highest is the *p* subshell, then the *d*, and finally the *f*.

Not all types of subshells are found in all shells. The lowest-energy shell ($n = 1$) has only one subshell, the 1s. The second shell ($n = 2$) has two subshells: 2s and 2p. The 2p subshell is of higher energy than the 2s. The third

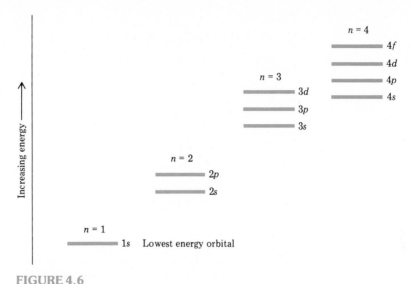

FIGURE 4.6

The energy of electron shells increases with increasing *n*, and the energy of subshells within a shell increases in the order *s, p, d,* and *f.*

subshell ($n = 3$) has three subshells: 3s, 3p, and 3d. The 3d subshell is of higher energy than the 3p, which in turn is of higher energy than the 3s. It is not until the fourth shell ($n = 4$) that we encounter all four types of subshells. In order of increasing energy, the 4s, 4p, 4d, and 4f subshells are found in the fourth shell. Figure 4.6 summarizes the shell–subshell relationships for the first four shells.

A close look at the energy relationships depicted in Figure 4.6 shows that there is an "energy overlap" between shells 3 and 4. Although the *average* energy of the electrons in the third shell is lower than that of electrons in shell 4, the energies of the different subshells are such that there is one subshell in shell 4, the 4s, that is lower in energy than the highest-energy subshell of shell 3—the 3d. The ramifications of this "energy overlap" are discussed in Section 4.7.

In Section 4.4 we noted that the seventh shell is the highest-numbered shell we need to describe the electron arrangements of the 109 known elements. Table 4.2 gives information about the subshells found in the first seven shells. Note that number–letter designations are not given for all of the subshells in shells 5, 6, and 7. This is because some subshells in these shells (the dashes) are not needed, as we show in Section 4.7.

The maximum number of electrons that a subshell can hold varies from 2 to 14, depending on whether the subshell is an *s, p, d,* or *f* subshell. An *s* subshell can accommodate only two electrons. The shell in which the *s* subshell is located does not affect the maximum electron-occupancy figure of 2. Subshells of the *p, d,* and *f* types can accommodate maxima of 6, 10, and 14 electrons, respectively. Again, the maximum number of electrons in these subshell types

TABLE 4.2

Subshell Arrangements Within Shells

Shell number (n)	Subshells
1	1s
2	2s 2p
3	3s 3p 3d
4	4s 4p 4d 4f
5	5s 5p 5d 5f —
6	6s 6p 6d — — —
7	7s — — — — — —

TABLE 4.3

Distribution of Electrons Within Subshells

Shell	Number of Subshells Within Shell	Maximum Number of Electrons Within Each Subshell				Maximum Number of Electrons Within Shell ($2n^2$)
		s	p	d	f	
$n = 1$	1	2				2
$n = 2$	2	2	6			8
$n = 3$	3	2	6	10		18
$n = 4$	4	2	6	10	14	32

depends only on the type of subshell and is independent of shell number. Table 4.3 summarizes information about distribution of electrons in subshells for shells 1 through 4. Notice the consistency between the numbers in columns 3 and 4 of Table 4.3. Within a shell, the sum of the subshell electron occupancies is the same as the shell electron occupancy ($2n^2$). For example, in shell 4, an s subshell containing 2 electrons, a p subshell containing 6 electrons, a d subshell containing 10 electrons, and an f subshell containing 14 electrons add up to a total of 32 electrons, which is the maximum occupancy of shell 4 as calculated by the $2n^2$ formula.

4.6   Electron Orbitals

The term *orbital* is the last and most basic of the three terms used to describe electron arrangements about nuclei. An **orbital** *is the region of space around a nucleus where an electron with a specific energy is most likely to be found.*

An analogy for the relationship between shells, subshells, and orbitals is the physical layout of a high-rise condominium complex. In our analogy, a shell is the counterpart of a floor of the condominium. Just as each floor contains apartments (of varying size), a shell contains subshells (of varying size). In addition, just as apartments contain rooms, subshells contain orbitals. An apartment is a collection of rooms; a subshell is a collection of orbitals.

The following are characteristics of orbitals, the rooms in our "electron apartment house."

1. The number of orbitals in a subshell varies; there is one for an *s* subshell, three for a *p* subshell, five for a *d* subshell, and seven for an *f* subshell.
2. The maximum number of electrons found in an orbital does not vary; it is always 2.
3. The notation used to designated orbitals is the same as that used for subshells. Thus, the seven orbitals in the 4*f* subshell are called 4*f* orbitals.

We have already noted (Section 4.5) that all electrons in a subshell have the same energy. Thus, all electrons in orbitals of the same subshell also have the same energy. This means shell and subshell locations are sufficient to specify the energy of an electron. This statement is of great importance in the discussions of Section 4.7.

Orbitals have a definite size and shape that is related to the type of subshell in which they are found. (Remember, an orbital is a region of space. We are not talking about the size and shape of an electron, but rather the size and shape of a region of space in which an electron is found.) An electron can be at only one point in an orbital, at any given time, but because of its rapid movement throughout the orbital, it "occupies" the entire orbital. An analogy would be the definite volume that is occupied by a rotating fan blade. The fan blade "occupies" all of that volume even though it cannot be "everywhere" at once. Typical *s*, *p*, *d*, and *f* orbital shapes are shown in Figure 4.7. Notice

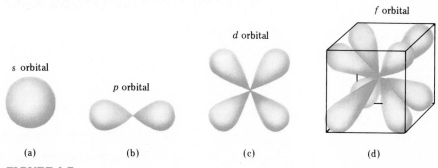

FIGURE 4.7

The shapes of atomic orbitals. The *f* orbital is shown within a cube to illustrate that its lobes are directed toward the corners of a cube.

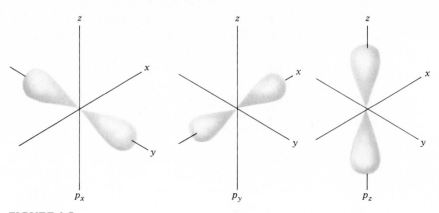

FIGURE 4.8
The orientation of the three orbitals in a *p* subshell.

1s   2s   3s

FIGURE 4.9
Size relationships among *s* orbitals located in different subshells.

that the shapes increase in complexity in the order *s, p, d,* and *f.* Some of the more complex *d* and *f* orbitals have shapes related to, but not identical to, those shown in Figure 4.7.

Orbitals that are in the same subshell differ mainly in their orientation. For example, as shown in Figure 4.8, the three 2*p* orbitals look the same, but they are aligned in different directions—along the *x, y,* and *z* axes in a Cartesian coordinate system.

When orbitals of the same type are found in different subshells, such as 1*s*, 2*s*, and 3*s*, they have the same general shape but differ in size (volume), as shown in Figure 4.9.

EXAMPLE 4.1

Determine the following for the third electron shell (third main energy level) of an atom.
(a) The number of subshells it contains.
(b) The designation used to describe each subshell.
(c) The number of orbitals in each subshell.

(d) The maximum number of electrons that could be contained in each subshell.

(e) The maximum number of electrons that could be contained in the shell.

SOLUTION

(a) The number of subshells in a shell is the same as the value of n for the shell. Therefore, the third shell ($n = 3$) contains three subshells.

(b) The lowest-energy of the three subshells is designated $3s$, the next highest is $3p$, and the final subshell, with the highest energy, is $3d$.

(c) The number of orbitals in a given type (s, p, d, or f) of subshell is independent of the shell number. All s subshells ($1s$, $2s$, or $3s$) contain one orbital, all p subshells contain three orbitals, and all d subshells contain five orbitals.

(d) Regardless of type, an orbital can contain a maximum of only two electrons. Therefore, the $3s$ subshell (one orbital) contains 2 electrons; the $3p$ subshell (three orbitals) contains 6 electrons; and the $3d$ subshell (five orbitals) contains 10 electrons.

(e) The maximum number of electrons in a shell is found by the formula $2n^2$, where n is the shell number. Because $n = 3$ in this problem, $2n^2 = 2 \times 3^2 = 18$. Alternatively, from part c, we note that shell 3 contains nine orbitals ($1 + 3 + 5$). Because each orbital can hold two electrons, the maximum number of electrons is 18 (2×9).

EXAMPLE 4.2

Characterize the similarities and differences between a $3s$ and a $3p$ orbital.

SOLUTION

Similarities: Both orbitals are located in shell 3 and therefore extend approximately the same distance from the nucleus. Each orbital accommodates a maximum of two electrons.

Differences: The orbitals belong to different subshells (s and p) and therefore have different shapes. Electrons in each orbital possess different energies.

4.7   Electron Configurations

An **electron configuration** *is a statement of how many electrons an atom has in each of its subshells.* Because subshells group electrons according to energy (Section 4.5), electron configurations indicate how many electrons an atom has of various energies.

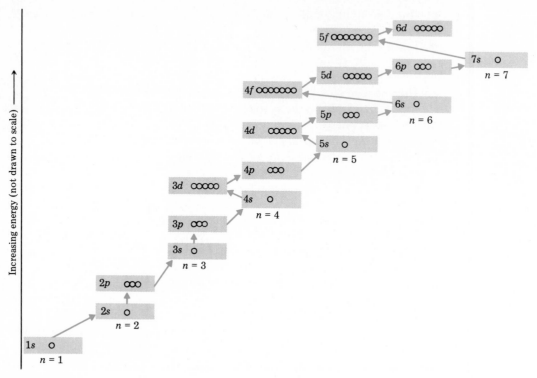

FIGURE 4.10
Relative energies and filling order for the electron subshells.

Electron configurations are not written out in words; a shorthand system with symbols is used. Subshells containing electrons, listed in order of increasing energy, are designated using number–letter combinations (1s, 2s, 2p, and so on). A superscript following each subshell designation indicates the number of electrons in that subshell. The electron configuration for nitrogen using this shorthand notation is

$$1s^2 2s^2 2p^3 \quad \text{(read ``one-s-two, two-s-two, two-p-three'')}$$

Thus, a nitrogen atom has an electron arrangement of two electrons in the 1s subshell, two electrons in the 2s subshell, and three electrons in the 2p subshell.

To find the electron configuration for an atom, a procedure called the *Aufbau principle* (German *aufbauen*, "to build") is used. The **Aufbau principle** *states that electrons normally occupy the lowest-energy subshell available.* This guideline brings order to what could be a very disorganized situation. There are many orbitals about the nucleus of any given atom. Electrons do not occupy these orbitals in a random, haphazard fashion; electron orbital occupancy has a very predictable pattern, governed by the Aufbau principle. Orbitals are filled in order of increasing energy.

Use of the Aufbau principle requires knowledge about the electron capacities of orbitals and subshells and their relative energies.

Figure 4.10 shows the order in which electron subshells about a nucleus acquire electrons. Note that the sequence of subshell filling is not as simple a pattern as you would probably predict. All subshells within a given shell do not necessarily have lower energies than all subshells of higher numbered shells. Because of "energy overlaps," beginning with shell 4, one or more lower-energy subshells of a specific shell have energies that are lower than those of the upper subshells of the preceding shell; thus, they acquire electrons first. For example, the 4s subshell acquires electrons before the 3d subshell (see Figure 4.10). As another example, the s subshell of the sixth energy level fills before the d subshell of the fifth energy level or the f subshell of the fourth energy level (Figure 4.10).

The sequence in which subshells acquire electrons must be learned before electron configurations can be written. A useful mnemonic (memory) device called an *Aufbau diagram* helps considerably in this learning process. An **Aufbau diagram** *is a listing of subshells in the order in which electrons occupy them.* In this diagram, illustrated in Figure 4.11, all s subshells are located in column 1, all p subshells in column 2, and so on. Subshells belonging to the same shell are found in the same row. The order of subshell filling is given by following the diagonal arrows, starting with the top one. The 1s subshell fills first. The second arrow points to (goes through) the 2s subshell, which fills next. The third arrow points to both the 2p and 3s subshells. The 2p fills first and the 3s next. Any time a single arrow points to more than one subshell, you should start at the tail of the arrow and work toward its head to determine the proper filling

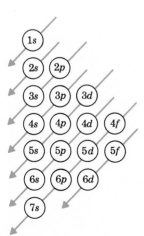

FIGURE 4.11

The Aufbau diagram, which illustrates subshell filling order.

sequence. An Aufbau diagram is an easy means of cataloging the information given in Figure 4.10.

We are now ready to write electron configurations. Let us systematically consider electron configurations for the first few elements in the periodic table.

Hydrogen ($Z = 1$) has only one electron, which goes into the 1s subshell; this subshell has the lowest energy of all subshells. Hydrogen's electron configuration is written as

$$1s^1$$

Helium ($Z = 2$) has two electrons, both of which occupy the 1s subshell. (Remember, an s subshell contains one orbital and an orbital can accommodate two electrons.) Helium's electron configuration is

$$1s^2$$

Lithium ($Z = 3$) has three electrons, and the third electron cannot enter the 1s subshell because its maximum capacity is two electrons. (All s subshells are completely filled with two electrons; see Section 4.5.) The third electron is placed in the next highest energy subshell, 2s. The electron configuration for lithium is

$$1s^2 2s^1$$

For beryllium ($Z = 4$), the additional electron is placed in the 2s subshell, which is now completely filled, giving beryllium the electron configuration.

$$1s^2 2s^2$$

For boron ($Z = 5$), the 2p subshell, which is the subshell of next highest energy, becomes occupied for the first time. Boron's electron configuration is

$$1s^2 2s^2 2p^1$$

A p subshell can accommodate six electrons because it contains three orbitals (Section 4.5). The 2p subshell can thus accommodate the additional electrons found in C, N, O, F, and Ne. The electron configurations of these elements are

$$\text{C } (Z = 6): \quad 1s^2 2s^2 2p^2$$
$$\text{N } (Z = 7): \quad 1s^2 2s^2 2p^3$$
$$\text{O } (Z = 8): \quad 1s^2 2s^2 2p^4$$
$$\text{F } (Z = 9): \quad 1s^2 2s^2 2p^5$$
$$\text{Ne } (Z = 10): \quad 1s^2 2s^2 2p^6$$

With sodium ($Z = 11$) the 3s subshell acquires an electron for the first time.

$$\text{Na } (Z = 11): \quad 1s^2 2s^2 2p^6 3s^1$$

Note the pattern that is developing in the electron configurations we have written so far. Each element has an electron configuration that is the same as the one just before it except for the addition of one electron.

Electron configurations for other elements are obtained by simply extending the principles we have just illustrated. A subshell of lower energy is always filled before electrons are added to the next highest subshell; this continues until the correct number of electrons have been accommodated.

EXAMPLE 4.3

Write out the electron configuration for
(a) Rubidium ($Z = 37$)   (b) Bismuth ($Z = 83$)

SOLUTION

(a) The number of electrons in a rubidium atom is 37. Remember, the atomic number (Z) gives the number of electrons (Section 3.5). We need to fill subshells, in order of increasing energy, until 37 electrons have been accommodated.

The $1s$, $2s$, and $2p$ subshells fill first, accommodating a total of 10 electrons among them.

$$1s^2 2s^2 2p^6 \ldots$$

Next, according to Figure 4.10 or 4.11, the $3s$ subshell fills and then the $3p$ subshell.

$$1s^2 2s^2 2p^6 3s^2 3p^6 \ldots$$

We have accommodated 18 electrons at this point. We still need to add 19 more electrons to get our desired number of 37.

The $4s$ subshell fills next, followed by the $3d$ subshell, giving us 30 electrons at this point.

$$1s^2 2s^2 2p^6 3s^2 3p^6 4s^2 3d^{10} \ldots$$

Note that the maximum electron population for a d subshell is 10 electrons.

Seven more electrons are needed, which are added to the next two higher subshells, the $4p$ and the $5s$. The $4p$ subshell can accommodate six electrons, and the final electron is then added to the $5s$ subshell.

$$1s^2 2s^2 2p^6 3s^2 3p^6 4s^2 3d^{10} 4p^6 5s^1$$

To double-check that we have the correct number of electrons, 37, we add the superscripts in our final electron configuration.

$$2 + 2 + 6 + 2 + 6 + 2 + 10 + 6 + 1 = 37$$

The sum of the superscripts in any electron configuration should add up to the atomic number if the configuration is for a neutral atom.

(b) To write this configuration, we continue along the same lines as in part a, remembering that the maximum electron subshell populations are $s = 2$, $p = 6$, $d = 10$, and $f = 14$.

Bismuth, with an atomic number of 83, contains 83 electrons that are distributed among the various subshells. We add 83 electrons to subshells in the following order.

$$1s^2 2s^2 2p^6 3s^2 3p^6 4s^2 3d^{10} 4p^6 5s^2 4d^{10} 5p^6 6s^2 4f^{14} 5d^{10} 6p^3$$

$$2 \quad 4 \quad 10 \quad 12 \quad 18 \quad 20 \quad 30 \quad 36 \quad 38 \quad 48 \quad 54 \quad 56 \quad 70 \quad\quad 80 \quad 83$$

The numbers below the electron configuration give a running total of added electrons, which is obtained by adding the superscripts up to that point. We stop when we hit 83 electrons.

Notice, in the electron configuration, that the $6p$ subshell contains only three electrons, even though it can hold a maximum of six. We put only three electrons in this subshell because that is sufficient to give 83 total electrons. If we had completely filled this subshell, we would have had 86 electrons, which is too many.

We should note that for a few elements in the middle of the periodic table, the actual distribution of electrons within subshells differs slightly from that obtained by using the Aufbau principle and Aufbau diagram. These exceptions are caused by very small energy differences between some subshells and are not important in the uses we will make of electron configurations.

4.8　Abbreviated Electron Configurations: Shell Notation

The electron configuration for an element is often written in an abbreviated form in which only electron shell occupancy is shown. In this notation a circle with the value of the nuclear charge written inside it is used to denote the nucleus and then arcs are used to denote the shells. The number of electrons in the shell is written within each arc. The shell notation for the element chlorine, which has the electron configuration

$$\underbrace{1s^2}_{\text{1st shell}} \quad \underbrace{2s^2 2p^6}_{\text{2nd shell}} \quad \underbrace{3s^2 3p^5}_{\text{3rd shell}}$$

is as follows.

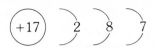

EXAMPLE 4.4

The electron configuration for the element calcium, which has an atomic number of 20, is

$$1s^2 2s^2 2p^6 3s^2 3p^6 4s^2$$

What is the abbreviated electron configuration (shell notation) for this element?

SOLUTION

We first group calcium's electrons into shells. All subshells with a particular n value belong to the same shell.

$1s^2$	$2s^2 2p^6$	$3s^2 3p^6$	$4s^2$
2 electrons	8 electrons	8 electrons	2 electrons

The total number of electrons in a shell is given by the sum of the superscripts of the involved subshells.

With the total number of electrons in each shell determined we are ready to write the abbreviated electron configuration.

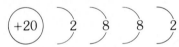

4.9 Electron Configurations and the Periodic Law

Knowledge of the electron configurations of the elements provides an explanation for the periodic law. You will recall, from Section 4.1, that the periodic law points out that the properties of the elements repeat themselves in a regular manner when the elements are ordered in sequence of increasing atomic number. The elements with similar chemical properties are placed under one another, in vertical columns (groups), in the periodic table.

Groups of elements have similar chemical properties because of similarities in their electron configurations. Chemical properties repeat themselves regularly among the elements because electron configurations among elements repeat themselves in a regular manner.

To illustrate this similar-chemical-property–similar-electron-configuration correlation, let us look at the electron configurations of two groups of elements known to have similar chemical properties.

We begin with the elements lithium, sodium, potassium, and rubidium, all members of group IA of the periodic table. The electron configurations for these elements are

$$_3\text{Li:}\quad 1s^2 2s^1$$

$$_{11}\text{Na:}\quad 1s^2 2s^2 2p^6 3s^1$$

$$_{19}\text{K:}\quad 1s^2 2s^2 2p^6 3s^2 3p^6 4s^1$$

$$_{37}\text{Rb:}\quad 1s^2 2s^2 2p^6 3s^2 3p^6 4s^2 3d^{10} 4p^6 5s^1$$

Note that each of these elements has one outer electron (shown in color) in an *s* subshell, the last electron added by the Aufbau principle. It is this similarity in outer-shell electron arrangements that causes these elements to have similar chemical properties. In general, elements with similar outer-shell electron configurations have similar chemical properties.

Let us consider another group of elements known to have similar chemical properties: fluorine, chlorine, bromine, and iodine of group VIIA of the periodic table. The electron configurations for these four elements are

$$_9\text{F:}\quad 1s^2 2s^2 2p^5$$

$$_{17}\text{Cl:}\quad 1s^2 2s^2 2p^6 3s^2 3p^5$$

$$_{35}\text{Br:}\quad 1s^2 2s^2 2p^6 3s^2 3p^6 4s^2 3d^{10} 4p^5$$

$$_{53}\text{I:}\quad 1s^2 2s^2 2p^6 3s^2 3p^6 4s^2 3d^{10} 4p^6 5s^2 4d^{10} 5p^5$$

Once again similarities in electron configurations are readily apparent. This time, the repeating pattern involves an outermost shell containing seven electrons (shown in color).

Section 5.2 considers in depth the fact that the most important electrons in determining chemical properties are found in the outermost shell of an atom.

4.10   Electron Configurations and the Periodic Table

One of the strongest pieces of supporting evidence for the assignment of electrons to shells, subshells, and orbitals is the periodic table itself. The basic shape and structure of the table, which was determined many years before

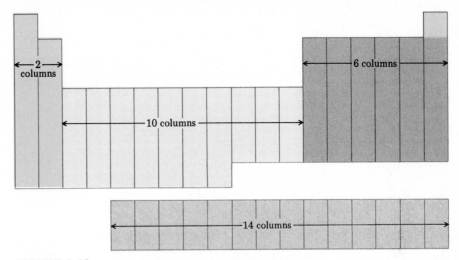

FIGURE 4.12

Structure of the periodic table in terms of columns of elements.

electrons were even discovered, is consistent with and can be explained by electron configurations. Indeed, the specific location of an element in the periodic table can be used to obtain information about its electron configuration.

The concept of *distinguishing* electrons is the key to obtaining electron configuration information from the periodic table. The **distinguishing electron** *for an element is the last electron that is added to its electron configuration when the configuration is written using the Aufbau principle.* This last electron is the one that causes an element's electron configuration to differ from that of the element immediately preceding it in the periodic table.

As the first step in linking electron configurations to the periodic table, let us analyze the general shape of the periodic table in terms of columns of elements. As shown in Figure 4.12, on the extreme left of the table there are 2 columns of elements; in the center is an area containing 10 columns of elements; to the right there is a block of 6 columns of elements; and in two rows at the bottom of the table are 14 columns of elements.

The number of columns of elements in the various regions of the periodic table (Figure 4.12)—2, 6, 10, 14—is the same as the maximum number of electrons that the various types of subshells can accommodate. We will see shortly that this is a very significant observation; the number match-up is no coincidence. The various columnar regions of the periodic table are called the *s* area (2 columns), the *p* area (6 columns), the *d* area (10 columns), and the *f* area (14 columns), as shown in Figure 4.13.

For all elements located in the *s* area of the periodic table, the distinguishing electron is always found in an *s* subshell. All *p*-area elements have distinguishing

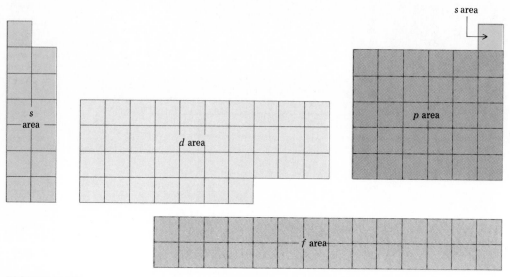

FIGURE 4.13
Areas of the periodic table.

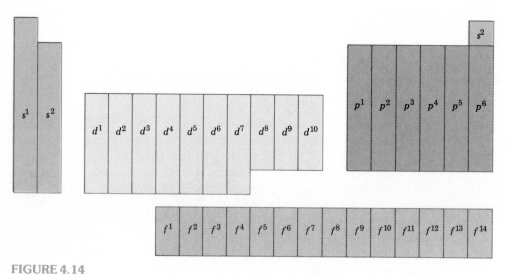

FIGURE 4.14
The extent of subshell filling as a function of periodic table position.

electrons in p subshells. Similarly, elements in the d and f areas of the periodic table have distinguishing electrons located in d and f subshells, respectively. Thus, the area location of an element in the periodic table can be used to determine the type of subshell that contains the distinguishing electron. Note that the element helium belongs to the s rather than the p area of the periodic table, even though its table position is on the right side. (The reason for this placement of helium is explained in Section 5.3.)

The extent to which the subshell containing an element's distinguishing electron is filled can also be determined from the element's position in the periodic table. All elements in the first column of a specific area contain only one electron in the subshell; all elements in the second column contain two electrons in the subshell, and so on. Thus, all elements in the first column of the p area (group IIIA) have an electron configuration ending in p^1. Elements in the second column of the p area (group IVA) have electron configurations ending in p^2 and so on. Similar relationships hold in other areas of the table, as shown in Figure 4.14.

4.11 Classification Systems for the Elements

The elements can be classified in several ways. The two most common classification systems are

1. A system based on the electron configurations of the elements, in which elements are described as *noble-gas, representative, transition,* or *inner-transition elements.*
2. A system based on selected physical properties of the elements, in which elements are described as *metals* or *nonmetals.*

The classification scheme based on electron configurations of the elements is depicted in Figure 4.15.

The **noble-gas elements** *are found in the far-right column of the periodic table.* They are all gases at room temperature, and they have little tendency to form chemical compounds. With one exception, the distinguishing electron for a noble gas completes the p subshell; therefore, they have electron configurations ending in p^6. The exception is helium, in which the distinguishing electron completes the first shell—a shell that has only two electrons. Helium's electron configuration is $1s^2$.

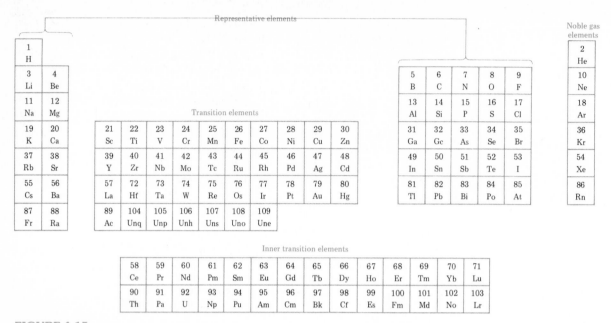

FIGURE 4.15

Elemental classification scheme based on the electron configurations of the elements.

The **representative elements** *are all of the elements of the s and p areas of the periodic table, with the exception of the noble gases.* The distinguishing electron in these elements partially or completely fills an *s* subshell or partially fills a *p* subshell. The representative elements include most of the more common elements.

The **transition elements** *are all of the elements of the d area of the periodic table.* The common feature in the electron configurations of the transition elements is the presence of the distinguishing electron in a *d* subshell.

The **inner-transition elements** *are all of the elements of the f area of the periodic table.* The characteristic feature of their electron configurations is the presence of the distinguishing electron in an *f* subshell. There is very little variance in the properties of either the 4*f* or 5*f* series of inner-transition elements.

On the basis of selected physical properties of the elements, the second of the two classification schemes divides the elements into the categories of metals and nonmetals. A **metal** *is an element that has the characteristic properties of luster, thermal conductivity, electrical conductivity, and malleability.* With the exception of mercury, all metals are solids at room temperature (25°C). Metals are good conductors of heat and electricity. Most metals are ductile (can be drawn into wires) and malleable (can be rolled into sheets). Most metals have

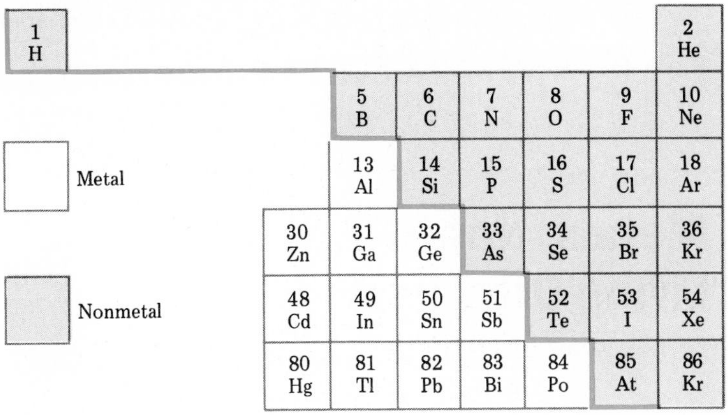

							2 He
1 H							
	5 B	6 C	7 N	8 O	9 F	10 Ne
☐ Metal	13 Al	14 Si	15 P	16 S	17 Cl	18 Ar
30 Zn	31 Ga	32 Ge	33 As	34 Se	35 Br	36 Kr
☐ Nonmetal	48 Cd	49 In	50 Sn	51 Sb	52 Te	53 I	54 Xe
80 Hg	81 Tl	82 Pb	83 Bi	84 Po	85 At	86 Kr

FIGURE 4.16

A portion of the periodic table showing the dividing line between metals and nonmetals. All elements that are not shown are metals.

high luster (shine), high density, and high melting points. Among the more familiar metals are the elements iron, aluminum, copper, silver, and gold.

A **nonmetal** *is an element characterized by the absence of the properties of luster, thermal conductivity, electrical conductivity, and malleability.* Many of the nonmetals such as hydrogen, oxygen, nitrogen, and the noble gases are gases at room temperature (25°C). The only nonmetal that is a liquid at room temperature is bromine. Solid nonmetals include carbon, sulfur, and phosphorus. In general, the nonmetals have lower densities and melting points than metals.

The majority of the elements are metals—only 22 elements are nonmetals; the rest (87) are metals. It is not necessary to memorize which elements are nonmetals and which are metals, as this information is obtainable from a periodic table. As can be seen from Figure 4.16, the location of an element in the periodic table correlates directly with its classification as a metal or nonmetal. Note the steplike heavy line that runs through the *p* area of the periodic table. This line separates the metals from the nonmetals; metals are on the left and nonmetals on the right. Note also that the element hydrogen is a nonmetal.

The fact that the vast majority of elements are metals in no way indicates that metals are more important than nonmetals. Most nonmetals are relatively common and are found in many important compounds. For example, water (H_2O) is a compound involving two nonmetals. An analysis of the previously given abundance of the elements in the earth's crust (Section 2.6) in terms of metals and nonmetals shows that the two most abundant elements, which account for 80.2% of all atoms, are nonmetals—oxygen and silicon. The four most abundant elements in the human body (Section 2.6), which constitute more than 99% of all atoms in the body, are nonmetals—hydrogen, oxygen, carbon, and nitrogen.

4.12   Metals, Nonmetals, and Living Organisms

Twenty-six elements are known to be essential for the proper functioning of living organisms. As shown in Figure 4.17, 12 of these elements are nonmetals and 14 are metals. The 26 "essential" elements can be divided into three categories based on the amounts needed. The *macronutrients* are needed in large amounts, the *micronutrients* are needed in small amounts, and the *trace elements* are needed in minute amounts.

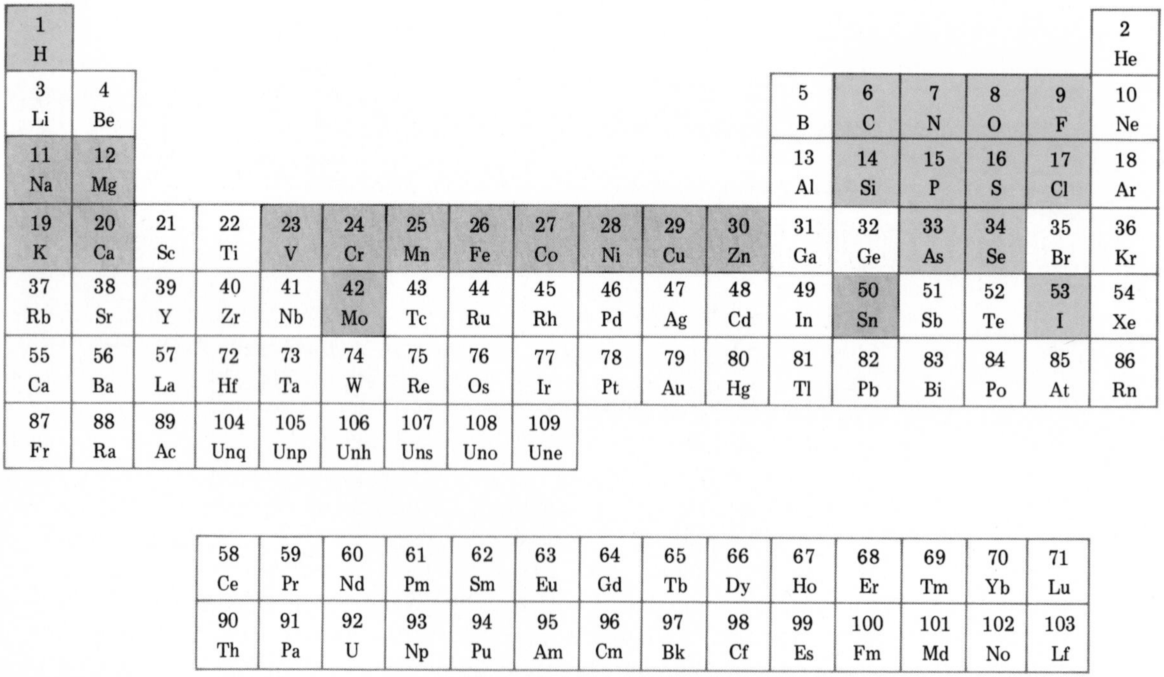

FIGURE 4.17

The twelve nonmetals (colored areas) and the fourteen metals (grey areas) known to be essential for the proper functioning of living organisms.

TABLE 4.4

The Most Abundant Elements Present in the Human Body
(*nonmetals shown in color*)

Element	Percent of Total Number of Atoms in the Human Body	Number of Grams in a 70-kg Man
Macronutrients		
Hydrogen	63	6,580
Oxygen	25.5	43,550
Carbon	9.5	12,590
Nitrogen	1.4	1,815
Micronutrients		
Calcium	0.31	1,700
Phosphorus	0.22	680
Potassium	0.06	250
Sulfur	0.05	100
Chlorine	0.03	115
Sodium	0.03	70
Magnesium	0.01	42

We have noted previously that the macronutrients are carbon, hydrogen, oxygen, and nitrogen (Section 2.6). These nonmetals are found in most of the compounds present within the human body. Table 4.4 gives further information about the macronutrients.

Table 4.4 also gives information about the micronutrients, a group of four metals and three nonmetals. The amounts of these elements in the body vary from 0.31 atom percent for calcium to 0.01 atom percent for magnesium. Collectively, the micronutrients constitute less than 1% of the total atoms in the human body. Despite these small numbers, life could not go on without these elements. Body fluid chemistry is an area in which the micronutrients are particularly important. Sulfur and phosphorus are also important in the makeup of molecular building blocks such as proteins and nucleic acids (Chapter 25).

Table 4.5 lists the trace elements known to be essential in humans and animals. Ten of the 15 trace elements are metals. Although they are needed in only minute amounts (much less than 0.01 atom percent) by living organisms, they are still of vital importance for the proper functioning of an organism. Table 4.5 also includes information about the sources and functions of the trace elements.

TABLE 4.5

Trace Elements Known to Be Essential in Humans and Animals
(*nonmetals shown in color*)

Element	Date Discovered to Be an Essential Element	Dietary Sources
Need in Humans Established and Quantified		
Iron	17th century	Meat, liver, fish, poultry, beans, peas, raisins, prunes
Iodine	1850	Iodized table salt, shellfish, kelp
Zinc	1934	Meat, liver, eggs, shellfish
Need in Humans Established But Not Yet Quantified		
Copper	1928	Nuts, liver, shellfish
Manganese	1931	Nuts, fruits, vegetables, whole-grain cereals
Cobalt	1935	Meat, dairy products
Molybdenum	1953	Organ meats, green leafy vegetables, legumes
Selenium	1957	Meat, seafood
Chromium	1959	Meat, beer, unrefined wheat flour
Need Established in Animals But Not Yet in Humans		
Tin	1970	*
Vanadium	1971	*
Fluorine	1972	Fluoridated water
Silicon	1972	*
Nickel	1974	*
Arsenic	1975	*

* Present in trace quantities in many foods and in the environment.

Practice Questions and Problems

PERIODIC LAW AND PERIODIC TABLE

4-1 Give a modern-day statement of the periodic law.

4-2 How do the terms *group* and *period* relate to the periodic table?

4-3 Give the symbol of the element that occupies each of the following positions in the periodic table.
 a. Period 4, group IIIA
 b. Period 5, group IVB
 c. Group IA, period 2
 d. Group VIIA, period 3
 e. Period 1, group IA
 f. Period 6, group IB
 g. Group IIIB, period 4
 h. Group IVA, period 5

4-4 The following statements either define or are closely related to the terms *periodic law, period,* and *group*. Match the terms to the appropriate statements.
 a. This is a vertical arrangement of elements in the periodic table.

b. This is a horizontal arrangement of elements in the periodic table.

c. The properties of the elements repeat in a regular way as the atomic numbers increase.

d. Element 19 begins this arrangement in the periodic table.

e. The chemical properties of elements 12, 20, and 38 demonstrate this principle.

f. The element carbon is the first member of this arrangement.

g. Elements 24 and 33 belong to this arrangement.

h. Elements 10, 18, and 36 belong to this arrangement.

TERMINOLOGY ASSOCIATED WITH ELECTRON ARRANGEMENTS

4-5 What does the term *quantization* mean, and what property of an electron is quantized?

4-6 The following statements define or are closely related to the terms *shell, subshell,* and *orbital.* Match the terms to the appropriate statements.

a. In terms of electron capacity, this unit is the smallest of the three.

b. This unit can contain a maximum of two electrons.

c. This unit can contain as many electrons as or more electrons than either of the other two.

d. The term *subenergy level* is closely associated with this unit.

e. This unit is designated by just a number.

f. The formula $2n^2$ gives the maximum number of electrons that can occupy this unit.

g. This unit is designated in the same way as the orbitals contained within it.

4-7 Determine the following for the fifth electron shell (the fifth main energy level) of an atom.

a. The number of subshells it contains.

b. The designation used to describe each of the first four subshells.

c. The number of orbitals in each of the first four subshells.

d. The maximum number of electrons that can occupy this shell.

e. The maximum number of electrons that can occupy each of the first four subshells.

4-8 Fill in the numerical value(s) that correctly complete(s) each of the following statements.

a. The maximum number of electrons in the second electron shell is _____.

b. The maximum number of electrons in the sixth electron main energy level is _____.

c. A *4f* subshell holds a maximum of _____ electrons.

d. A *3d* orbital holds a maximum of _____ electrons.

e. A *2s* orbital holds a maximum of _____ electrons.

f. The fifth shell contains _____ subshells, _____ orbitals, and a maximum of _____ electrons.

4-9 Describe the general shape of each of the following orbitals.

a. *2s* orbital   b. *3s* orbital

c. *4p* orbital   d. *3d* orbital

4-10 Characterize the similarities and differences between the following.

a. A *3s* orbital and *3d* orbital

b. A *2p* orbital and *3p* orbital

c. The third shell and fourth shell

4-11 Give the maximum number of electrons that can occupy each of the following units.

a. *3d* subshell   b. *2s* orbital

c. Second shell   d. Shell with $n = 4$

e. *4p* subshell   f. *4p* orbital

WRITING ELECTRON CONFIGURATIONS

4-12 Explain the difference between the Aufbau principle and an Aufbau diagram.

4-13 Within any given shell, how do the energies of the *s, p, d,* and *f* subshells compare?

4-14 For each of the following sets of subshells, determine which has the lowest energy and which has the highest energy.

a. *2s, 3s, 4s*   b. *4p, 4d, 4f*

c. *3d, 4d, 4f*   d. *6s, 3d, 5f*

4-15 Explain what each number and letter mean in the notations $6s^2$ and $4f^{14}$.

4-16 With the help of an Aufbau diagram, write the complete electron configuration for each of the following atoms.

a. $_9$F   b. $_{17}$Cl   c. $_{50}$Sn   d. $_{20}$Ca

e. $_{44}$Ru   f. $_{24}$Cr   g. $_{86}$Rn   h. $_{36}$Kr

4-17 What is wrong with each of the following attempts to write electron configurations?

a. $1s^2 1p^6 2s^2 2p^6$ b. $1s^2 2s^2 3s^2$

c. $1s^2 2s^2 2p^6 3s^2 3p^6 3d^{10}$ d. $1s^2 2s^2 2p^6 2d^{10}$

THE PERIODIC LAW AND ELECTRON CONFIGURATIONS

4-18 Group the following six electron configurations of elements into pairs that you would expect to show similar chemical properties.
a. $1s^2 2s^2 2p^6 3s^1$
b. $1s^2 2s^2 2p^6 3s^2 3p^5$
c. $1s^2 2s^2 2p^6 3s^2 3p^6 4s^2 3d^{10} 4p^5$
d. $1s^2 2s^2 2p^6 3s^2 3p^6 4s^2 3d^{10} 4p^3$
e. $1s^2 2s^2 2p^6 3s^2 3p^6 4s^2 3d^{10} 4p^6 5s^1$
f. $1s^2 2s^2 2p^6 3s^2 3p^3$

THE PERIODIC TABLE AND ELECTRON CONFIGURATIONS

4-19 Indicate the position in the periodic table of each of the following elements in terms of *s, p, d,* or *f* area.
a. $_{13}$Al   b. $_{29}$Cu   c. $_{20}$Ca
d. $_{10}$Ne   e. $_{92}$U   f. $_{21}$Sc

4-20 For each of the following elements, indicate whether the distinguishing electron is in an *s, p, d,* or *f* subshell.
a. $_8$O   b. $_{11}$Na   c. $_{53}$I
d. $_{47}$Ag   e. $_{96}$Cm   f. $_3$Li

4-21 For each of the following elements, indicate the extent that the subshell containing the distinguishing electron is filled.
a. $_7$N   b. $_{25}$Mn   c. $_{17}$Cl

d. $_{56}$Ba   e. $_{16}$S   f. $_{50}$Sn

CLASSIFICATION SYSTEMS FOR THE ELEMENTS

4-22 Identify each of the following as a noble gas, representative element, transition element, or inner-transition element.
a. $_{54}$Xe   b. $_{45}$Rh   c. $_2$He   d. $_{106}$Unh
e. $_{16}$S   f. $_3$Li   g. $_{19}$K   h. $_{95}$Am

4-23 Classify each of the following elements as a metal or nonmetal.
a. $_{15}$P   b. $_4$Be   c. $_{78}$Pt   d. $_{10}$Ne
e. $_1$H   f. $_{47}$Ag   g. $_{33}$As   h. $_{13}$Al

4-24 Twenty-six elements are known to be essential for living organisms. List, by elemental symbols, the essential elements that are
a. Macronutrients
b. Micronutrients
c. Trace elements
d. Metals
e. Nonmetals
f. Representative elements
g. Representative nonmetals
h. Representative metals
i. Transition elements
j. Transition metals

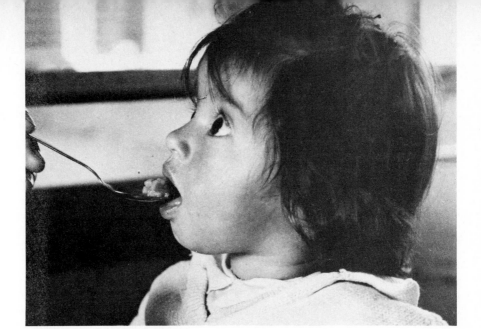

5

Chemical Bonding

CHAPTER HIGHLIGHTS

5.1 Types of Chemical Bonds

As scientists study living organisms and the world in which we live, they rarely encounter free isolated atoms. Instead, under normal conditions of temperature and pressure, they almost always find atoms associated in aggregates or clusters ranging in size from two atoms to numbers too large to count. In this chapter, we discuss reasons why atoms tend to collect together into larger units and provide information about the binding forces (chemical bonds) involved in these units.

As we examine the nature of the attractive forces between atoms, we will discover that both the tendency and capacity of an atom to be attracted to other atoms are dictated by its electron configuration. Thus, the concepts introduced in Chapter 4 are applied extensively in our discussion of chemical bonding.

Chemical bonds *are the attractive forces that hold atoms together in more complex units.* Current chemical theory describes chemical bonds in terms of two major types: ionic bonds and covalent bonds. The major difference between these types of bonds is the perceived mechanism through which bond formation occurs.

An **ionic bond** *results from the TRANSFER of one or more electrons from one atom or group of atoms to another.* This electron-transfer model is used to describe the bonding in compounds that contain both metallic and nonmetallic elements.

A **covalent bond** *results from the SHARING of one or more pairs of electrons between atoms.* This electron-sharing model is used for bonding situations in which all of the atoms are nonmetals.

Consideration of the properties of numerous compounds suggests the existence of two types of chemical bonds. Some compounds, such as sodium chloride (table salt), ammonium nitrate, and calcium sulfate, have rigid crystalline structures, are brittle, and have relatively high melting points (above 300°C). Additionally, such compounds conduct an electric current in the liquid state or in aqueous solution. The bonding in compounds with the preceding general properties is now known to involve ionic bonds.

In contrast, many other compounds have general properties quite different from those previously mentioned. In this group are water, carbon dioxide, ammonia, ethyl alcohol, sucrose (table sugar), and aspirin. Such compounds have much lower melting points and tend to be gases, liquids, or low-melting solids. They do not conduct an electric current when liquid or in aqueous solution, as do compounds containing ionic bonds. The bonding in this second type of compound is now known to involve covalent bonding.

In Sections 5.4 and 5.8 we describe the characteristics of ionic bonds and covalent bonds. First, however, we discuss two fundamental concepts that are common to and necessary for understanding both bonding models.

1. Certain electrons, called valence electrons, are the electrons involved in bonding.
2. Certain arrangements of electrons are more stable than other arrangements of electrons. This concept is known as the *octet rule*.

Section 5.2 deals with valence electrons and Section 5.3 with the octet rule.

5.2 Valence Electrons and Electron-Dot Structures

Certain electrons, which are called valence electrons, are particularly important in determining the bonding characteristics of a given atom. For representative and noble-gas elements (Section 4.10), **valence electrons** *are the electrons in the outermost electron shell, which is the shell with the highest shell number* (n). Valence electrons are always found in either s or p subshells. Note the restriction on the use of this definition; it applies only to representative and noble-gas elements. Most of the common elements are representative elements; thus, this definition is quite useful. We will not consider the more complicated valence electron definitions for transition or inner-transition elements (Section 4.10) in this text; the presence of incompletely filled *inner d* or *f* subshells is the complicating factor in the definitions for these elements.

EXAMPLE 5.1

Determine the number of valence electrons present in atoms of each of the following elements.

(a) $_{20}Ca$ (b) $_{16}S$ (c) $_{35}Br$

SOLUTION

(a) The element calcium has two valence electrons, as can be seen by examining its electron configuration.

Number of valence electrons

$$1s^2 2s^2 2p^6 3s^2 3p^6 4s^2$$

Highest value of the electron shell number

The highest value of the electron shell number is $n = 4$. Only two electrons are found in shell 4—the two electrons in the 4s subshell.

(b) The element sulfur has six valence electrons.

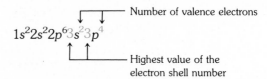

Electrons in two different subshells can simultaneously be valence electrons. The highest shell number is 3, and both the 3*s* and 3*p* subshells belong to shell 3. Hence, all of the electrons in both of these subshells are valence electrons.

(c) The element bromine has seven valence electrons.

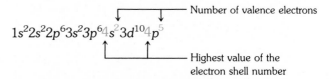

The 3*d* electrons are not counted as valence electrons because the 3*d* subshell is in shell 3 and this shell does not have the maximum *n* value. Shell 4, the outermost shell, has the maximum *n* value.

It seems reasonable that the outermost electrons of atoms are involved in bonding when you remember that the outermost electrons are the first to come into close proximity when atoms collide — an event that is necessary before atoms can combine. Also, these electrons are located the greatest distance away from the nucleus, and so they are the least tightly bound (attraction to the nucleus decreases with distance) and most susceptible to change (transfer or sharing).

Scientists have developed a shorthand system for designating numbers of valence electrons that uses electron-dot structures. This system makes it easier to picture the role that valence electrons play in chemical bonding. An **electron-dot structure** *consists of an element's symbol, with one dot for each valence electron placed around the elemental symbol.* Electron-dot structures for the first 20 elements (all representative or noble-gas elements), arranged as in the periodic table, are given in Figure 5.1.

Note that the location of the dots around the elemental symbols is not critical. The following notations all have the same meaning

$$\text{Mg·} \quad \text{Mg·} \quad \text{·Mg} \quad \text{·Mg} \quad \text{·Mg·}$$

Electron-dot structures are also often called Lewis dot structures. The American chemist Gilbert Newton Lewis (1875–1946), an early contributor to chemical bonding theory, was the first to use such structures (see Historical Profile 8).

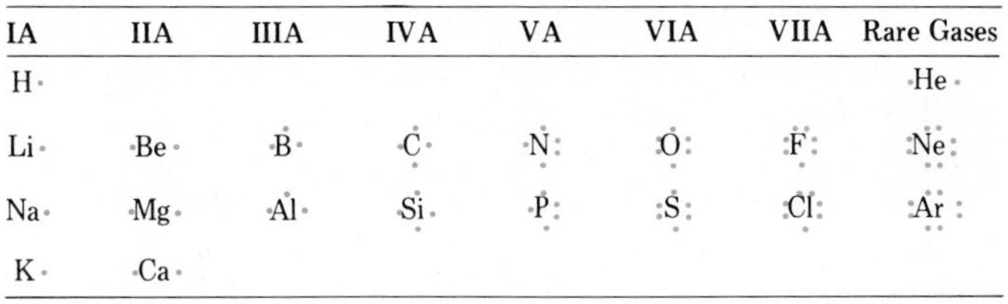

IA	IIA	IIIA	IVA	VA	VIA	VIIA	Rare Gases
H·							·He·
Li·	·Be·	·Ḃ·	·Ċ·	·N̈:	:Ö:	:F̈:	:N̈e:
Na·	·Mg·	·Al·	·Si·	·P̈:	:S̈:	:Cl̈:	:Är:
K·	·Ca·						

FIGURE 5.1
Electron-dot structures for selected elements.

Three important generalizations about valence electrons can be drawn from a study of the structures shown in Figure 5.1.

1. *Representative elements in the same group of the periodic table have the same number of valence electrons.* This should not be surprising. Elements in the same group in the periodic table have similar chemical properties as a result of their similar outer-shell electron configurations (Section 4.8). The electrons in the outermost shell are the valence electrons.

2. *The number of valence electrons for representative elements in a group is the same as the Roman numeral periodic table group number.* For example, the electron-dot structures for oxygen and sulfur, which are both members of group VIA, show six dots. Similarly, the electron-dot structures of hydrogen, lithium, sodium, and potassium, which are all members of group IA show one dot.

3. *The maximum number of valence electrons for any element is eight.* Only the noble gases (Section 4.10), beginning with neon, have the maximum number of eight electrons. Helium, which has only two valence electrons, is the exception in the noble gas family. Obviously, an element with a grand total of two electrons cannot have eight valence electrons. Although shells with n greater than 2 are capable of holding more than eight electrons, they do so only when they are no longer the outermost shell and are thus not the valence shell. For example, bromine, whose electron configuration was given in Example 5.1c, has 18 electrons in the third shell; however, shell 4 is the valence shell for bromine.

5.3 The Octet Rule

A key concept in modern elementary bonding theory is that certain arrangements of valence electrons are more stable than others. The term *stable* as used here refers to the idea that a stable system, which in this case is an arrangement of electrons, does not easily undergo spontaneous change.

HISTORICAL PROFILE 8

Gilbert Newton Lewis   1875–1946

Gilbert Newton Lewis, an American chemist, is recognized as one of the fore-most chemists of this century. The son of a lawyer, he received his primary education at home. He was very precocious and was reading at age 3.

His early years as a chemist were spent at Harvard and MIT, but most of his career, from 1912 to 1946, was spent at the University of California, Berkeley.

In 1916 he published a paper proposing that the chemical bond was a pair of electrons shared or held jointly by two atoms. This proved to be one of the most fruitful ideas in the history of chemistry. Within a few years of his 1916 paper, with contributions from other scientists, a formal theory of chemical bonding based on sharing of electron pairs had been developed. Lewis is also the one who developed the symbolism used in electron-dot structures. His memoirs indicate that he first had the idea of shared electrons while he was lecturing to an introductory chemistry class in 1902.

Lewis also made significant contributions in many other areas of chemistry besides bonding theory. Thermodynamic studies were a major interest during most of his career. In 1938 a generalized theory for acids and bases (now called Lewis acid–base theory) came from his mind and pen. Still later, he was the first to isolate heavy hydrogen ($_{1}^{2}H$).

[E. F. Smith Memorial Collection, Center for the History of Chemistry, University of Pennsylvania.]

The valence-electron configurations of the noble gases are considered to be the most stable of all valence–electron configurations. All of the noble gases except helium possess eight valence electrons, which is the maximum number possible. Helium's valence–electron configuration is $1s^2$. All of the other noble gases possess ns^2np^6 valence-electron configurations, where n has the maximum value found in the atom.

He: $1s^2$

Ne: $1s^22s^22p^6$

Ar: $1s^22s^22p^63s^23p^6$

Kr: $1s^22s^22p^63s^23p^64s^23d^{10}4p^6$

Xe: $1s^22s^22p^63s^23p^64s^23d^{10}4p^65s^24d^{10}5p^6$

Rn: $1s^22s^22p^63s^23p^64s^23d^{10}4p^65s^24d^{10}5p^66s^24f^{14}5d^{10}6p^6$

Except for helium, each of the noble-gas valence-electron configurations has the common characteristic of having the outermost *s* and *p* subshells *completely filled.*

The conclusion that an ns^2np^6 ($1s^2$ for helium) configuration is the most stable of all valence-electron configurations is based on the chemical properties of the noble gases. The noble gases are the *most unreactive* of all the elements. They are the only elemental gases found in nature in the form of individual uncombined atoms. There are no known compounds of helium, neon, and argon and only a few compounds of krypton, xenon, and radon. The noble gases appear to be "happy" the way they are; they have little or no desire to form bonds to other atoms.

Atoms of many elements that lack this very stable, noble-gas valence-electron configuration tend to attain this configuration through chemical reactions that result in compound formation. This observation is known as the octet rule because of the eight valence electrons possessed by atoms having a noble-gas electron configuration. A formal statement of the **octet rule** is: *In compound formation, atoms of elements lose, gain, or share electrons in such a way as to produce a noble-gas electron configuration for each of the atoms involved.*

Applications of the octet rule to many different systems have shown that it has value in correctly predicting the observed combining ratios of atoms. For example, it explains why two hydrogen atoms, instead of some other number, are bonded to one oxygen atom in the compound water. It explains why the formula of the ionic compound sodium chloride is $NaCl$ rather than $NaCl_2$, $NaCl_3$ or Na_2Cl.

There are exceptions to the octet rule, but it is still used because of the large amount of information that it is able to correlate. It is particularly effective in explaining compound formation involving only representative elements.

5.4 The Ionic-Bond Model

The concept of transferring one or more electrons between two or more atoms is central to the ionic-bond model. This electron-transfer process produces charged particles called ions. An **ion** *is an atom (or group of atoms) that is electrically charged as the result of the loss or gain of electrons.* Neutrality of atoms is the result of the number of protons (positive charges) being equal to the number of electrons (negative charges). Loss or gain of electrons destroys this proton–electron balance and leaves a net charge on the atom.

If one or more electrons is gained by an atom, a negatively charged ion is produced; excess negative charge is present because electrons now outnumber protons. The loss of one or more electrons by an atom results in the formation of a positively charged ion; more protons are now present than electrons, resulting in excess positive charge. Note that the excess positive charge associated with a positive ion is never caused by proton gain but always by electron loss. If the number of protons remains constant, and the number of electrons decreases, the result is net positive charge. The number of protons, which determines the identity of an element, never changes during ion formation.

The charge on an ion is directly correlated with the number of electrons lost or gained. Loss of one, two, or three electrons gives ions with +1, +2, or +3 charges, respectively. Similarly, a gain of one, two, or three electrons gives ions with −1, −2, or −3 charges, respectively. (Ions that have lost or gained more than three electrons are very seldom encountered).

The notation for charges on ions is a superscript placed to the right of the elemental symbol. Some examples of ion symbols are the following:

$$\text{Positive ions:} \quad Na^+, K^+, Be^{2+}, Mg^{2+}, Al^{3+}$$

$$\text{Negative ions:} \quad Cl^-, Br^-, O^{2-}, S^{2-}, N^{3-}$$

Note that a single plus or minus sign is used to denote a charge of one, instead of using the notation $^{1+}$ or $^{1-}$. Also note that in multicharged ions the number precedes the charge sign; that is, the notation for a charge of plus two is $^{2+}$ rather than $^{+2}$. (Some older textbooks have the charge sign preceding the number.)

EXAMPLE 5.2

Give the symbol for each of the following ions.
(a) The ion formed when a calcium atom ($Z = 20$) loses two electrons.
(b) The ion formed when a phosphorus atom ($Z = 15$) gains three electrons.

SOLUTION

(a) A neutral calcium atom contains 20 protons and 20 electrons. The calcium ion formed by the loss of two electrons would still contain 20 protons, but would have only 18 electrons because two electrons were lost.

$$\begin{aligned} 20 \text{ protons} \ &= 20 + \text{charges} \\ 18 \text{ electrons} &= \underline{18 - \text{charges}} \\ \text{Net charge} &= \ \ 2 - \end{aligned}$$

The symbol of the calcium ion is thus Ca^{2+}.

(b) The atomic number of phosphorus is 15. Thus, 15 protons and 15 electrons are present in a neutral phosphorus atom. A gain of three electrons raises the electron count to 18.

$$\begin{aligned} 15 \text{ protons} \ &= 15 + \text{charges} \\ 18 \text{ electrons} &= \underline{18 - \text{charges}} \\ \text{Net charge} &= \ \ 3 - \end{aligned}$$

The symbol for the ion is P^{3-}.

So far our discussion of electron transfer and ion formation has focused on the loss or gain of electrons by isolated individual atoms. In reality, loss and gain of electrons are partner processes; that is, one does not occur without the other also occurring. Ion formation occurs only when atoms of two elements are present—an element to donate electrons (electron loss) and an element to accept electrons (electron gain). The electrons lost by the one element are the same ones gained by the other element. Thus, positive and negative ion formations always occur together.

The mutual attraction between positive and negative ions that results from electron transfer constitutes the force that holds the ions together as an ionic compound. This force is referred to as an ionic bond. An **ionic bond** *is the attractive force between positive and negative ions that causes them to remain together as a group.*

5.5 Formulas for Ionic Compounds

A simple example of ionic bonding occurs between the elements sodium and chlorine in the compound NaCl. A sodium atom loses (transfers) one electron to a chlorine atom, producing Na^+ and Cl^- ions. The ions combine in a one-to-one ratio to form the compound NaCl.

Why do sodium atoms form Na^+ ions and not Na^{2+} or Na^- ions? Why do chlorine atoms form Cl^- ions rather than Cl^{2-} or Cl^+ ions? In general, what determines the specific number of electrons lost or gained in electron transfer processes?

The octet rule (Section 5.3) provides very simple and straightforward answers to these questions. *Atoms tend to gain or lose electrons until they have obtained an electron configuration that is the same as that of a noble gas.*

Consider the element sodium, which has the electron configuration

$$1s^2 2s^2 2p^6 3s^1$$

Written in shell notation this configuration becomes

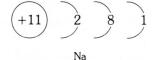

Na

The element sodium can attain a noble-gas configuration by losing one electron (to give it the electron configuration of neon) or by gaining seven electrons (to give it the electron configuration of argon).

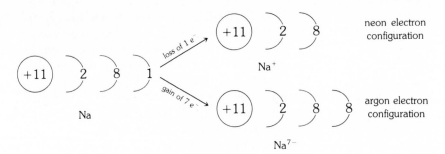

The first process, the loss of one electron, is more energetically favorable than the gain of seven electrons and is the process that occurs. The process that involves the least number of electrons is always the more energetically favorable process and thus is the process that occurs.

Consider the element chlorine, which has the electron configuration

$$1s^2 2s^2 2p^6 3s^2 3p^5$$

Written in shell notation this configuration is

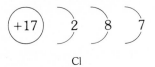

Chlorine can attain a noble-gas configuration by losing seven electrons, which gives it a neon electron configuration, or by gaining one electron, which gives it an argon electron configuration. The latter occurs for the reason just given.

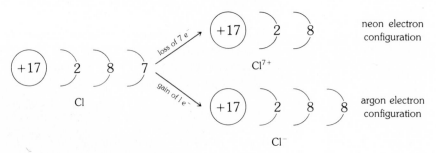

The preceding considerations, for the elements sodium and chlorine, lead to the following generalizations.

1. Metal atoms containing one, two, or three valence electrons (the metals in groups IA, IIA, and IIIA of the periodic table) tend to lose electrons to acquire a noble-gas electron configuration. The noble gas involved is the one preceding the metal in the periodic table.

Group IA metals form $+1$ ions

Group IIA metals form $+2$ ions

Group IIIA metals form $+3$ ions

2. Nonmetal atoms containing five, six, or seven valence electrons (the non-metals in groups VA, VIA, and VIIA of the periodic table) tend to gain electrons to acquire a noble-gas configuration.

Group VIIA nonmetals form -1 ions

Group VIA nonmetals form -2 ions

Group VA nonmetals form -3 ions

The use of electron-dot structures will help you further visualize the relationships between noble-gas electron configurations, electron transfer, and ionic compound formation. Let us consider, again, the reaction between sodium, which has one valence electron, and chlorine, which has seven valence electrons, to form NaCl. This reaction can be represented as follows with electron-dot structures.

$$\text{Na}\cdot + \cdot\ddot{\text{Cl}}\text{:} \longrightarrow \text{Na}^+ \left[:\ddot{\text{Cl}}\text{:} \right]^- \longrightarrow \text{NaCl}$$

The loss of an electron by sodium empties its valence shell. The next inner shell, which contains eight electrons (a noble-gas configuration), then becomes the valence shell. After the valence shell of chlorine gains one electron, it has the desired eight valence electrons.

When sodium, which has one valence electron, combines with oxygen, which has six valence electrons, the oxygen atom requires two sodium atoms to meet its need of two additional electrons.

$$\begin{array}{c} \text{Na}\cdot \\ + \cdot\ddot{\text{O}}\text{:} \\ \text{Na}\cdot \end{array} \longrightarrow \begin{array}{c} \text{Na}^+ \\ \text{Na}^+ \end{array} \left[:\ddot{\text{O}}\text{:} \right]^{2-} \longrightarrow \text{Na}_2\text{O}$$

Note how the need of oxygen for two additional electrons dictates that two sodium atoms are required per oxygen atom—hence the formula Na_2O.

An opposite situation occurs in the reaction between calcium, which has two valence electrons, and chlorine, which has seven valence electrons. Here, two chlorine atoms are required to accommodate electrons transferred from one calcium atom because a chlorine atom can accept only one electron. (It has seven valence electrons and needs only eight.)

$$\text{Ca}\cdot + \begin{array}{c} \cdot\ddot{\text{Cl}}\text{:} \\ \cdot\ddot{\text{Cl}}\text{:} \end{array} \longrightarrow \text{Ca}^{2+} \begin{array}{c} \left[:\ddot{\text{Cl}}\text{:} \right]^- \\ \left[:\ddot{\text{Cl}}\text{:} \right]^- \end{array} \longrightarrow \text{CaCl}_2$$

It is not always necessary or convenient to write out electron-dot structures to determine the formula for an ionic compound. Formulas for ionic compounds can be written directly by using the charges associated with the ions being combined and the fact that the total amount of positive and negative charge must add up to zero. *Electron loss always equals electron gain in an electron-transfer process.* Consequently, ionic compounds are always neutral; no net charge is present. The total positive charge present on the ions that have lost electrons is always exactly counterbalanced by the total negative charge on the ions that have gained electrons. Thus, *the ratio in which positive and negative ions combine is the ratio that achieves charge neutrality for the resulting compound.*

The correct combining ratio when K^+ ions and S^{2-} ions combine is two to one. Two K^+ ions (each of $+1$ charge) are required to balance the charge on a single S^{2-} ion.

$$
\begin{aligned}
2(K^+): \quad & (2 \text{ ions}) \times (\text{charge of } +1) = +2 \\
S^{2-}: \quad & \underline{(1 \text{ ion}) \ \times (\text{charge of } -2) = -2} \\
& \qquad\qquad\qquad\quad \text{Net charge} = \quad 0
\end{aligned}
$$

Example 5.3 further illustrates the procedures needed to determine correct combining ratios between ions and write correct ionic formulas from the combining ratios. Three items to note about all ionic formulas are the following

1. The symbol for the positive ion is always written first.
2. The charges on the ions that are present are *not* shown in the formula. Knowledge of charges is necessary to determine the formula, but once it is determined, the charges are not explicitly written.
3. The numbers in the formula (the subscripts) give the combining ratio for the ions.

EXAMPLE 5.3

Determine the formula for the compound that is formed when each of the following pairs of elements interact.
(a) Na and P   (b) Be and P   (c) Al and P

SOLUTION

(a) Sodium (a group IA element) has one valence electron, which it would like to lose to give a Na^+ ion. Phosphorus (a group VA element) has five valence electrons and would thus like to acquire three more to give a P^{3-} ion. Na^+ and P^{3-} ions will combine in a three-to-one ratio because this combination causes the total charge to add up to zero. Three Na^+ ions give a total positive charge of 3. One P^{3-} ion results in a total negative charge of 3. Thus, the formula for the compound is Na_3P.

(b) Beryllium (a group IIA element) has two valence electrons that it would like to lose to form a Be^{2+} ion. Phosphorus will again form a P^{3-} ion. The numbers in the charges for the two ions are 2 and 3. The lowest common multiple of 2 and 3 is 6 ($2 \times 3 = 6$). Thus, we will need six units of positive charge and six units of negative charge. Three Be^{2+} ions are needed to give the six units of positive charge, and two P^{3-} ions are needed to give the six units of negative charge. The combining ratio of ions is three to two and the formula is Be_3P_2.

　　The strategy of finding the lowest common multiple of the numbers in the charges of the ions always works.

(c) Aluminum (a group IIIA element) has three valence electrons that it would like to lose to give a Al^{3+} ion. This ion combines with P^{3-} ion in a one to one ratio. One Al^{3+} ion contributes three units of positive charge, and that is counterbalanced by three units of negative charge from one P^{3-} ion. The formula of the compound is simply AlP.

Before leaving the subject of ions, ionic bonds, and formulas for ionic compounds, let us quickly review the key principles of ionic bonding.

1. Ionic compounds usually contain both a metallic and nonmetallic element.
2. The metallic element atoms lose electrons to produce positive ions and the nonmetallic element atoms gain electrons to produce negative ions.
3. The electrons lost by the metal atoms are the same ones that are gained by the nonmetal atoms. Electron loss must always equal electron gain.
4. The ratio in which positive metal ions and negative nonmetal ions combine is the ratio that achieves charge neutrality for the resulting compound.
5. Metals from groups IA, IIA, and IIIA of the periodic table form ions with charges of +1, +2, and +3, respectively. Nonmetals of groups VIIA, VIA, and VA of the periodic table form ions with charges of −1, −2, and −3, respectively. The number of electrons that are lost or gained by a particular type of atom is governed by the octet rule. Electron loss or gain is such that the resulting ion has an electron configuration like that of a noble-gas element.

5.6　Structure of Ionic Compounds

The term *molecule* is not appropriate for describing the smallest unit of an ionic compound. In the solid state, ionic compounds consist of an extended array of alternating positive and negative ions. **Ionic solids** *consist of positive and*

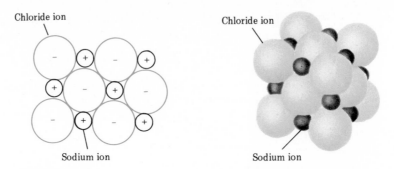

FIGURE 5.2
Two-dimensional cross section and three-dimensional view of the ionic solid NaCl.

negative ions arranged in such a way that each ion is surrounded by nearest neighbors of the opposite charge. Any given ion is bonded by electrostatic (positive-negative) attractions to all of the other ions of opposite charge immediately surrounding it. Figure 5.2 shows a two-dimensional cross section and a three-dimensional view of the arrangement of ions for the ionic compound NaCl (table salt).

We can see in Figure 5.2 that discrete molecules do not exist in an ionic solid. Therefore, the formulas for these solids (Section 5.5) cannot represent the composition of a molecule of the substance. Instead, formulas for this type of solid represent the simplest ratio in which the atoms combine. For example, in NaCl, there is no single partner for a sodium ion; six immediate neighbors (chloride ions) are equidistant from it. A chloride ion in turn has six immediate sodium neighbors. The formula NaCl represents the fact that in this solid, sodium and chloride ions are present in a 1 to 1 ratio.

Although the formulas for ionic solids only represent ratios, they are used in equations and chemical calculations in the same way as the formulas for molecular species. Remember, however, that they cannot be interpreted as indicating that molecules exist for these substances. Thus, the molecule is not the smallest unit capable of a stable existence for *all* pure substances (Section 2.4)—ionic solids are exceptions.

5.7 Nomenclature for Ionic Compounds

All of the examples of ionic compounds in Sections 5.4 and 5.5 were examples of *binary* ionic compounds—that is, ionic compounds in which only two elements are present. Names for this type of ionic compound are assigned by using the following rule.

TABLE 5.1

Names of Some Common Nonmetal Ions

Element	Stem	Name of Ion	Formula
Bromine	brom-	bromide	Br^-
Carbon	carb-	carbide	C^{4-}
Chlorine	chlor-	chloride	Cl^-
Fluorine	fluor-	fluoride	F^-
Hydrogen	hydr-	hydride	H^-
Iodine	iod-	iodide	I^-
Nitrogen	nitr-	nitride	N^{3-}
Oxygen	ox-	oxide	O^{2-}
Phosphorus	phosph-	phosphide	P^{3-}
Sulfur	sulf-	sulfide	S^{2-}

The full name of the metallic element is given first, followed by a separate word containing the stem of the nonmetallic element name and the suffix -ide.

Thus, to name the compound NaF, we start with the name of the metal (sodium), follow it with the stem of the name of the nonmetal (fluor-), and then add the suffix -ide. The name becomes *sodium fluoride.*

The stem of the name of the nonmetal is always the first few letters of the nonmetal's name, that is, the name of the nonmetal with its ending chopped off. Table 5.1 gives the stem part of the name for the most common nonmetallic elements. The name of the metal ion present is always exactly the same as the name of the metal itself; the metal's name is never shortened. Example 5.4 illustrates the use of the rule for naming binary ionic compounds.

EXAMPLE 5.4

Name the following binary ionic compounds.
(a) CaO   (b) AlF_3 (c) K_3N (d) Na_2S

SOLUTION

The general pattern for naming binary ionic compounds is

Name of metal + stem of name of nonmetal + -ide

(a) The metal is calcium and the nonmetal is oxygen. Thus, the compound name is calcium <u>ox</u>ide (the stem of the nonmetal name is underlined).

(b) The metal is aluminum and the nonmetal is fluorine; the compound name is aluminum <u>fluor</u>ide. Note that no mention is made of the subscript present in the formula—the 3. The name of an ionic compound never contains any reference to formula subscripts. There is only one ratio in which aluminum and fluorine atoms combine. Thus, just giving the names of the elements present in the compound is adequate nomenclature.

(c) Potassium (K) and nitrogen (N) are present in the compound, and its name is potassium <u>nitr</u>ide.

(d) The compound name is sodium <u>sulf</u>ide.

So far in our discussion of ionic compounds, it has been assumed that the only behavior allowable for an element is predicted by the octet rule. This is a good assumption for nonmetals and most representative metals. However, many other metals exhibit a less predictable behavior because they are able to form more than one type of ion. For example, iron forms both Fe^{2+} ions and Fe^{3+} ions depending on chemical circumstances. All of the inner-transition elements, most of the transition elements, and a few representative elements (some metals in the p area of the periodic table) exhibit this variable ionic charge behavior. A detailed discussion of why this phenomenon exists and prediction of what the charges on those ions will be are beyond the scope of this text. We mention this phenomenon here, however, because it must be taken into account when we name the ionic compounds that contain such metals.

When naming compounds that contain metals with variable ionic charges, the charge on the metal ion must be incorporated in the name. The magnitude of the charge on the metal ion is indicated with a Roman numeral, inside parentheses, placed immediately after the name of the metal. This Roman numeral is considered to be part of the metal's name, as shown by the following examples.

$$Fe^{3+}: \quad \text{iron(III) ion}$$

$$Fe^{2+}: \quad \text{iron(II) ion}$$

$$Au^{+}: \quad \text{gold(I) ion}$$

The chlorides of Fe^{2+} and Fe^{3+} ($FeCl_2$ and $FeCl_3$, respectively) are named iron(II) chloride and iron(III) chloride. Likewise, CuO is named copper(II) oxide. If you are uncertain about the charge on the metal ion in an ionic compound, use the charge on the nonmetal ion (which does not vary) to calculate it. For example, to determine the charge on the copper ion in CuO, you can note that the oxide ion carries a -2 charge because oxygen is in group VIA. This means that the copper ion must have a $+2$ charge to counterbalance the -2 charge.

EXAMPLE 5.5

Name the following binary ionic compounds, each of which contains a metal whose ionic charge can vary.

(a) $AuCl_3$ (b) Cu_2O

SOLUTION

We will need to indicate the magnitude of the charge on the metal ion in the name of each of these compounds by means of a Roman numeral.

(a) To calculate the charge on the gold ion, use the fact that total ionic charge (both positive and negative) must add up to zero.

$$\text{Gold charge} + 3(\text{Chlorine charge}) = 0$$

Note in the preceding equation that we have to take into account the presence of three chlorine atoms.

The chloride ion has a -1 charge (Section 5.5). Therefore,

$$\text{Gold charge} + 3(-1) = 0$$

Solving this equation shows that the gold charge must be a $+3$. Therefore, the gold ion present is Au^{3+}, and the name of the compound is gold(III) chloride.

(b) For charge balance in this compound we have the equation

$$2(\text{Copper charge}) + \text{Oxygen charge} = 0$$

Oxide ions carry a -2 charge (Section 5.5). Therefore,

$$2(\text{Copper charge}) + (-2) = 0$$

$$2(\text{Copper charge}) = +2$$

$$\text{Copper charge} = +1$$

Here we are interested in the charge on a single copper ion $(+1)$ and not in the total positive charge present $(+2)$. The compound is named copper(I) oxide because Cu^+ ions are present. As is the case for all ionic compounds, the name does not contain any reference to the numerical subscripts in the compound's formula.

To name binary ionic compounds correctly, we must know which metals form only one type of ion and which ones form more than one type of ion. In the former case, we do not use a Roman numeral in the compound's name, but in the latter case we do. It is easy to know when to use Roman numerals in the names of ionic compounds if you learn the metals that form only one type of ion; relatively few such metals exist. Figure 5.3 shows the metals that always form a single type of ion in ionic compound formation. Ionic compounds that contain these metals are the only ones without Roman numerals in their names.

5.8 The Covalent-Bond Model

In binary ionic compounds, the two atoms involved in a given ionic bond are a metal and a nonmetal. These atoms are quite *dissimilar* but are complementary to each other; one atom (the metal) likes to lose electrons and the other (the

1	2	3	4	5	6	7	8	9	10	11	12	13	14	15	16	17	18
H																	He
Li	Be											B	C	N	O	F	Ne
Na	Mg											Al	Si	P	S	Cl	Ar
K	Ca	Sc	Ti	V	Cr	Mn	Fe	Co	Ni	Cu	Zn	Ga	Ge	As	Se	Br	Kr
Rb	Sr	Y	Zr	Nb	Mo	Tc	Ru	Rh	Pd	Ag	Cd	In	Sn	Sb	Te	I	Xe
Ca	Ba	La	Hf	Ta	W	Re	Os	Ir	Pt	Au	Hg	Tl	Pb	Bi	Po	At	Rn
Fr	Ra	Ac	Unq	Unp	Unh	Uns	Uno	Une									

Ce	Pr	Nd	Pm	Sm	Eu	Gd	Tb	Dy	Ho	Er	Tm	Yb	Lu
Th	Pa	U	Np	Pu	Am	Cm	Bk	Cf	Es	Fm	Md	No	Lr

FIGURE 5.3

Periodic table showing the metals (in color) that form only one type of ion. The names for ionic compounds containing these metals do not need a Roman numeral.

nonmetal) likes to gain electrons. The net result is the transfer of one or more electrons from metal to nonmetal.

Covalent bonds, resulting from electron sharing (Section 5.1), are formed between *similar* or even *identical* atoms. Most often two nonmetal atoms are involved. It is not reasonable to suppose that one atom would give up electrons to another atom when the atoms are identical or very similar. The concept of *electron sharing* explains bonding between similar or identical atoms. In electron sharing, two nuclei attract the same electrons, and the resulting attractive forces hold the two nuclei together. The formation of a covalent bond always involves this process of electron sharing.

The hydrogen molecule (H_2) is the simplest covalent bonding situation that exists. Hydrogen, with just one 1s electron, needs one more electron to obtain the noble-gas configuration of helium ($1s^2$). A hydrogen atom accomplishes this by sharing its lone electron with another hydrogen atom, which reciprocates by sharing its *electron* with the first hydrogen atom. The net result is the formation of an H_2 molecule. The two shared electrons in an H_2 molecules do "double duty," helping each of the hydrogen atoms achieve a noble-gas configuration.

Covalent bonds are represented by electron-dot structures in much the same way that we used them for ionic bonds. A pair of dots placed between the

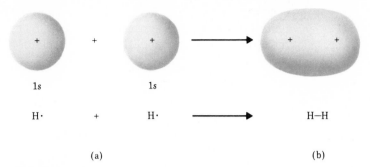

(a)　　　　　　　　　　　　　　　　　　(b)

FIGURE 5.4
Overlap of orbitals creates a situation where electron sharing can take place.

symbols of the bonded atoms indicates the shared pair of electrons. The electron-dot notation for H_2 is

Shared electron pair

H: ⟷ :H ⟶ (H:H)

An alternative way of representing the shared electron pair in a covalent bond is to draw a dash between the symbols of the bonded atoms.

$$H—H$$

Both of the atoms in H_2 have access to the two electrons of the shared electron pair. The concept of overlap of orbitals helps you visualize this. Suppose two hydrogen atoms are moving toward each other, as shown in Figure 5.4a, to eventually form H_2. As long as the atoms are well separated, the 1s electrons on each of the two atoms are independent of each other. As the atoms get closer together, the orbitals containing the electrons eventually overlap and create a common orbital (Figure 5.4b). When this happens, the two electrons move throughout the overlap region between the nuclei and are *shared* by both nuclei.

A situation in which two hydrogen atoms in an H_2 molecule are sharing two electrons between them is more stable than one in which two separate hydrogen atoms exist, each with one electron. Thus, hydrogen atoms are always found in pairs (as H_2 molecules) in samples of elemental hydrogen.

Using the octet rule and electron-dot structures, let us consider some other simple molecules in which covalent bonding is present. The element chlorine, which is located in group VIIA of the periodic table, has seven valence electrons. Its electron-dot structure is

$$\cdot \overset{..}{\underset{..}{Cl}} :$$

Chlorine needs only one electron to achieve the octet of electrons that enables it to have a noble-gas electron configuration. When chlorine bonds to other nonmetals, the octet of electrons is completed by means of electron sharing. The molecules HCl, Cl_2, and BrCl, whose electron-dot structures follow, are representative of this situation.

Note that in each of these molecules, chlorine atoms have four pairs of electrons around them (an octet) and only one of the four pairs is involved in electron sharing. The three pairs of electrons on each chlorine atom that are not taking part in the bonding are called *nonbonding electron pairs* or *unshared electron pairs*.

The number of covalent bonds that an atom forms is equal to the number of electrons it needs to achieve a noble-gas configuration. Note that the chlorine atoms in HCl, Cl_2, and BrCl all formed one covalent bond. For chlorine, seven valence electrons plus one electron acquired by electron sharing (one bond) gives the eight valence electrons needed for a noble-gas electron configuration.

The elements oxygen, nitrogen, and carbon have six, five, and four valence electrons, respectively. Therefore, these elements form two, three, and four covalent bonds, respectively. The number of covalent bonds that these elements form is reflected in the formulas of their simplest hydrogen compounds—H_2O, NH_3, and CH_4. Electron-dot structures for these molecule are

We see here that, just as the octet rule was useful in determining the ratio of ions in ionic compounds, it can also be used to predict formulas in molecular compounds. Example 5.6 gives additional illustrations of the use of the octet rule to determine formulas for molecular compounds.

EXAMPLE 5.6

Write electron-dot structures for the simplest binary compound that can be formed from the following pairs of elements.
(a) Phosphorus and bromine   (b) Sulfur and hydrogen

SOLUTION

(a) Phosphorus is in group VA of the periodic table and has five valence electrons. It will want to form three covalent bonds. Bromine, in group VIIA of the periodic table, has seven valence electrons and will want to form only one covalent bond. Therefore, three bromine atoms are needed to meet the needs of one phosphorus atom. The electron-dot structure for this molecule is

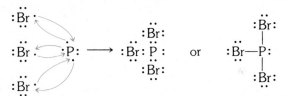

Each atom of PBr_3 has an octet of electrons, which is circled in color in the following diagram.

(b) Sulfur has six valence electrons, and hydrogen has one. Thus, sulfur forms two covalent bonds ($6 + 2 = 8$) and hydrogen forms one covalent bond ($1 + 1 = 2$). Remember that for hydrogen an "octet" of electrons is two electrons; the noble gas that hydrogen mimics is helium, which has only two valence electrons. The formula of the compound is H_2S.

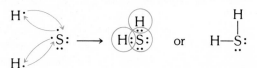

5.9 Multiple Covalent Bonds

In our discussion of covalent bonding in Section 5.8, all of the molecules that were chosen as examples to illustrate various aspect of bonding contained only *single covalent bonds*. A **single covalent bond** *is a bond in which a single pair of electrons is shared between two atoms.*

In many molecules, two atoms share two or three pairs of electrons to provide a complete octet of electrons for each atom involved in the bonding. These bonds are called double covalent bonds and triple covalent bonds. A **double covalent bond** *is a bond in which two atoms share two pairs of electrons.* A double covalent bond is stronger than a single covalent bond, but not twice as strong, because the two electron pairs repel each other and cannot become fully concentrated between the two atoms. (Bond strength is a measure of how much energy it takes to break a bond.) A **triple covalent bond** *is a bond in which two atoms share three pairs of electrons.* A triple covalent bond is stronger than a single or a double covalent bond, but not three times as strong as a single bond for the reason previously mentioned. Now let us consider some molecules in which double or triple bonds are present.

A diatomic N_2 molecule, the form in which nitrogen occurs in the atmosphere, contains a triple covalent bond. It is the simplest known triple covalent bond. A nitrogen atom has five valence electrons and needs three additional electrons to complete its octet.

$$\cdot \overset{\displaystyle \cdot}{\underset{\displaystyle \cdot \cdot}{N}} \cdot$$

In a N_2 molecule, the only sharing that can take place is between two nitrogen atoms because they are the only atoms present. Thus, to acquire a noble-gas electron configuration each nitrogen atom must share three of its electrons with the other nitrogen atom.

$$:\!N\cdot \quad \cdot N\!: \longrightarrow \ :N\!:\!:\!:\!N\!: \quad \text{or} \quad :N\!\equiv\!N\!:$$

Notice how all three shared electron pairs are placed in the area between the two nitrogen atoms in this bonding diagram. Note also that three lines are used to denote a triple covalent bond, paralleling the use of one line to denote a single covalent bond.

In "bookkeeping" electrons in an electron-dot structure, to make sure that all atoms in the molecule have achieved their octet of electrons, *all* electrons in a double or triple bond are considered to belong to *both* of the atoms involved in that bond. The bookkeeping for the N_2 molecule would be

$$:\!N\!:\!:\!:\!N\!:$$

Each of the circles around a nitrogen atom contains eight valence electrons. Again, all of the electrons in a double or triple bond are considered to belong to each of the atoms in the bond. Circles are never drawn to include just some of the electrons in a double or triple bond.

A slightly more complicated molecule that contains a triple covalent bond is the molecule C_2H_2. A carbon–carbon triple bond is present, as well as two carbon–hydrogen single bonds. The arrangement of valence electrons in C_2H_2 is

$$H \cdot \quad \cdot \ddot{C} \cdot \quad \cdot \ddot{C} \cdot \quad \cdot H \longrightarrow H \colon C \vdots \vdots C \colon H \quad \text{or} \quad H—C\equiv C—H$$

The two atoms in a triple covalent bond are usually the same element. They do not, however, have to be the same element. The molecule HCN contains a heteroatomic triple bond.

$$H \colon C \vdots \vdots N \colon \quad \text{or} \quad H—C\equiv N \colon$$

Double covalent bonds are found in numerous molecules. A common molecule that contains bonding of this type is carbon dioxide (CO_2). In fact, two carbon–oxygen double covalent bonds are present in CO_2.

$$\colon \ddot{O} \cdot \quad \cdot \ddot{C} \cdot \quad \cdot \ddot{O} \colon \longrightarrow \colon \ddot{O} \colon \colon C \colon \colon \ddot{O} \colon \quad \text{or} \quad \colon \ddot{O}=C=\ddot{O} \colon$$

Note in the following diagram how the circles are drawn for the octet of electrons around each of the atoms in CO_2.

$$\colon \ddot{O} \colon \colon C \colon \colon \ddot{O} \colon$$

Not all elements can form double or triple bonds. There must be at least two vacancies in an atom's valence electron shell before bond formation if it is to participate in a double bond and at least three vacancies are necessary for triple bond formation. This requirement eliminates Group VIIA elements (F, Cl, Br, I) and hydrogen as participants in such bonds. The Group VIIA elements have seven valence electrons and one vacancy, and hydrogen has one valence electron and one vacancy. All bonds formed by these elements are single covalent bonds.

EXAMPLE 5.7

Write electron-dot structures to describe the bonding in each of the following covalent compounds, given the formula and the arrangement of atoms within a molecule for each compound.

(a) C_2H_4 (b) CH_2O

arrangement of atoms:
$$\begin{matrix} H & & & H \\ & C & C & \\ H & & & H \end{matrix}$$

arrangement of atoms:
$$\begin{matrix} & & H & \\ & C & & O \\ & H & & \end{matrix}$$

SOLUTION

We will follow these four steps to determine the electron-dot structures.

STEP 1   Determine the number of valence electrons each atom in the molecule possesses.

STEP 2   Determine the "needs" of each atom—that is, the number of electrons each atom must obtain from other atoms through sharing to obtain a noble-gas electron configuration.

STEP 3   Determine how the "needs" of each atom will be met. When doing this, we must take into account the arrangement of the atoms within the molecules. We need to know which atoms are bonded to each other.

STEP 4   Write an electron-dot structure that is consistent with our "needs analysis."

(a) 1. Each carbon atom (group IVA) has four valence electrons. Each hydrogen atom (group IA) has one valence electron.

2. Each hydrogen atom needs one more electron to have an "octet." Each carbon atom needs four more electrons to acquire an octet of electrons.

3. The one electron each hydrogen atom needs must be obtained from the carbon atom to which it is bonded. Each carbon atom is bonded to three other atoms, each of which must share electrons with the carbon. Two hydrogen atoms each share one electron with the carbon, and the other carbon atom shares two electrons with it.

4. The electron-dot structure consistent with the information in step 3 is

$$
\begin{array}{cc}
\text{H} & \text{H} \\
& \\
\ddot{\text{C}} :: \ddot{\text{C}} \\
& \\
\text{H} & \text{H}
\end{array}
$$

Each carbon atom has eight electrons, and each hydrogen atom has two electrons.

Using lines to denote bonds, we can also write the bonding structure for C_2H_4 as

$$
\begin{array}{cc}
\text{H} & \text{H} \\
\ \ \diagdown & \diagup \\
\text{C} = \text{C} \\
\ \ \diagup & \diagdown \\
\text{H} & \text{H}
\end{array}
$$

(b) 1. Each hydrogen atom (group IA) has one valence electron, the carbon atom (group IVA) has four valence electrons, and the oxygen atom (group VIA) has six valence electrons.

2. Each hydrogen atom needs to gain one electron through sharing; the carbon atom needs to gain four electrons; and the oxygen atom needs to gain two electrons.

3. For hydrogen, the needed electron comes from carbon, the only atom to which it is bonded. Similarly, the two electrons that oxygen needs must come from the carbon, the only atom to which it is bonded.

   The carbon atom is bonded to three other atoms, each of which contributes to its needs. One electron comes from each hydrogen atom and two electrons come from the oxygen atom.

4. The electron-dot structure consistent with the information in step 3 is

Both the carbon atom and the oxygen atom have eight electrons and the two hydrogen atoms each have two electrons.

Using lines to denote bonds, we can also write the bonding structure for CH_2O as

5.10   Coordinate Covalent Bonds

In our examples of single-, double-, and triple-covalent bonding encountered so far, each of the bonded atoms has contributed an equal number of electrons to the bond—one each for a single covalent bond, two each for a double covalent bond, and three each for a triple covalent bond. A few molecules exist in which all of the covalent bonding within the molecule cannot be explained in this manner; instead, the concept of coordinate covalency must be invoked.

A **coordinate covalent bond** *is a bond in which both electrons of a shared pair come from one of the two atoms involved in the bond.* Coordinate covalent bonding allows an atom that has two or more vacancies in its valence shell to share a pair of nonbonding electrons that are located on another atom.

Once a coordinate covalent bond is formed, there is no way to distinguish it from any of the other covalent bonds in a molecule; all electrons are identical, regardless of their source. The main use of the concept of coordinate covalency is to help rationalize the existence of certain molecules and ions whose electron-bonding arrangement would otherwise present problems. Again, once a coordinate covalent bond is formed, it is no different from any other covalent bond. Electrons are electrons; they are all identical.

The molecule N_2O contains a single coordinate covalent bond.

$$\text{Coordinate covalent bond}$$
$$\overset{\circ}{_\circ}N \overset{\circ\circ\circ}{_{\circ\circ\circ}} N \overset{xx}{\underset{xx}{:}} \overset{x}{\underset{x}{O}}$$

In this bonding diagram different symbols are used for electrons originating from each of the three atoms. The nitrogen–nitrogen triple bond in N_2O is a "normal" covalent bond; the nitrogen–oxygen bond is a coordinate covalent bond, in which both electrons are supplied by the nitrogen atom. Again, once the bonds form, original "ownership" of the electrons is immaterial.

The triple covalent bond joining carbon and oxygen in carbon monoxide (CO) is a coordinate covalent bond.

$$\overset{x}{\underset{x}{:}}C\overset{x}{\underset{x}{::}}O:$$

Four of the six electrons in the triple bond can be considered to have come from the oxygen atom. Because carbon has only four valence electrons before it bonds, it must share four electrons (from oxygen) to achieve an octet of electrons. On the other hand, oxygen has six valence electrons before it bonds, and thus needs to share only two electrons (from carbon).

5.11   Electronegativity and Bond Polarity

At first glance, the ionic and covalent models for bonding seem to represent two very distinct forms of bonding. Actually, the two models are closely related to each other; they are the extremes of a broad continuum of bonding patterns. The close relationship between the two bonding models is apparent when the concept of *electronegativity* is considered.

The electronegativity concept, which was developed by the American chemist Linus Pauling (1901——), has its origins in the fact that the atoms of various elements differ in their abilities to attract shared electrons (in a bond) to themselves. Some elements are better electron-attractors than others. **Elec-**

H
2.1

IA	IIA
Li	Be
1.0	1.5
Na	Mg
0.9	1.2
K	Ca
0.8	1.0

IIIA	IVA	VA	VIA	VIIA
B	C	N	O	F
2.0	2.5	3.0	3.5	4.0
Al	Si	P	S	Cl
1.5	1.8	2.1	2.5	3.0
				Br
				2.8
				I
				2.5

FIGURE 5.5

Relative electronegativity values for selected representative elements.

tronegativity *is a measure of the relative attraction that an atom has for the shared electrons in a bond.*

The element fluorine, which is located in the upper right-hand corner of the periodic table, has the highest electronegativity of all the elements. Other very good electron-attractors are the elements whose periodic table positions are close to that of fluorine; these include oxygen, nitrogen, chlorine, and bromine.

The actual numerical values of electronegativity for the more common representative elements are presented in Figure 5.5. Note that electronegativity values are unitless numbers on a relative scale that runs from 0 to 4.0. Fluorine, the most electronegative of all the elements, has a value of 4.0 on the scale (the maximum value possible). *The higher the electronegativity value for an element is, the greater the electron-attracting ability of atoms of that element for shared electrons in bonds is.*

Note the following trends in electronegativity values that are present in the data of Figure 5.5. Electronegativity increases from left to right across a period of the periodic table and decreases from top to bottom within a group of the periodic table. These two trends result in nonmetals generally having higher electronegativities than metals. This fact is consistent with our previous generalization (Section 5.5) that metals tend to lose electrons and nonmetals tend to gain electrons when an ionic bond is formed. Metals (low electronegativities, poor electron attractors) will give up electrons to nonmetals (high electronegativities, good electron attractors).

It is important to note for future consideration how the electronegativity of the element hydrogen (located to the far left in the periodic table) compares with that of the period 2 elements. Hydrogen's value of electronegativity is between that of boron and carbon.

Li	Be	B	H	C	N	O	F
1.0	1.5	2.0	2.1	2.5	3.0	3.5	4.0

Increasing electronegativity →

Electronegativity for an element is not a directly measurable quantity. Rather, electronegativity values are calculated from bond energy information and other related experimental data. Values differ from element to element because of differences in atom size, nuclear charge, and number of inner-shell (nonvalence) electrons.

The difference in the electronegativity values of the atoms in a bond is the key to predicting the *polarity* of that bond. **Polarity** *is a measure of the inequality in the sharing of bonding electrons.*

When two identical atoms (atoms of equal electronegativity) share one or more pairs of electrons, each atom exerts the same attraction for the electrons, which results in the electrons being shared *equally*. This type of bond is called a *nonpolar covalent bond*. A **nonpolar covalent bond** *is one in which the sharing of electrons is equal.*

When two atoms involved in a covalent bond are not identical, and thus have different electronegativities, the electron-sharing situation is quite different. The atom that has highest electronegativity atttracts the electrons more strongly than the other atom; this results in an *unequal* sharing of electrons. This type of covalent bond is called a *polar covalent bond*. A **polar covalent bond** *is one in which the sharing of bonding electrons is unequal.*

The significance of a polar covalent bond is that it creates partial positive and negative charges. Although both atoms involved in the bond were initially uncharged, the uneven sharing of electrons produces a partial negative charge, a charge of less than one unit, on one bonded atom (the more electronegative one) and an equivalent partial positive charge on the other bonded atom. This means that one end of the bond is negative with respect to the other end.

The partial positive and negative charges associated with a polar covalent bond are often indicated by a notation that involves the lowercase Greek letter δ (delta). A $\delta-$ symbol, meaning a "partial negative charge," is placed above the relatively negative atom of the bond and a $\delta+$ symbol, meaning a "partial positive charge," is placed above the relatively positive atom. For example, the bond in hydrogen fluoride (HF) would be depicted as

$$\overset{\delta+ \quad \delta-}{\text{H—F}}$$

Fluorine is the more electronegative of the two elements; it dominates the electron-sharing process and draws the electrons closer to itself. Hence, the fluorine end of the bond has the $\delta-$ designation; the more electronegative element always has the $\delta-$ designation.

We now see that most chemical bonds are neither 100% covalent (equal sharing) nor 100% ionic (no sharing); instead, they fall somewhere in between (unequal sharing). Figure 5.6 provides pictorial representation of the continuum of bonding types that are possible because of the occurrence of unequal sharing. Figure 5.6a shows the equal sharing situation that results when both atoms are

(a) Equal sharing

(b) Slightly unequal sharing

(c) Very unequal sharing

(d) Electron transfer

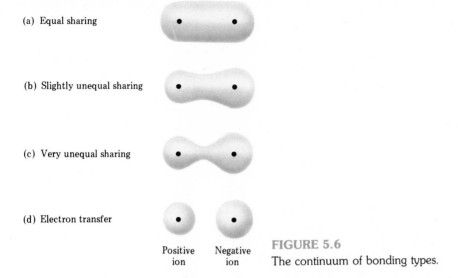

Positive   Negative
ion   ion

FIGURE 5.6

The continuum of bonding types.

identical. The sharing must be equal because identical nuclei must affect the bonding electrons in the same way.

Whenever two nuclei differ in their abilities to attract a pair of bonding electrons, unequal sharing results. Figure 5.6b shows the situation in which the electronegativity difference is small. Electron sharing is "close" to but not exactly equal. Note, in Figure 5.6b, that the electron distribution is no longer symmetrical. The bonding electrons spend more time associated with the nucleus that has the greater electronegativity; this is the nucleus on the right side in Figure 5.6b.

Figure 5.6c depicts electron density distribution when a relatively large difference in electron-attracting ability exists between nuclei. The sharing of electrons here can be described as "very unequal."

Finally, when the electron-attracting ability difference is very large, one atom "wins the battle," and electron transfer is said to occur. This situation is depicted in Figure 5.6d. It corresponds to what we previously called an ionic bond.

We can obtain an estimate of the "degree of unequality" of electron sharing in a chemical bond from the electronegativity values of the bonded atoms. The greater the difference in electronegativity is, the greater the inequality in the electron-sharing process is.

It is still convenient to use the terms ionic and covalent in describing chemical bonds, based on the following guidelines, which take into account electronegativity differences and the resulting unequal sharing of electrons.

1. When there is no difference in electronegativity between bonded atoms, the bond is called a *nonpolar covalent bond*.

Electronegativity Difference	Bond Type	Degree of Covalent Character	Degree of Ionic Character
Zero	Nonpolar covalent		
Greater than zero but less than 1.7	Polar covalent	Decreases	Increases
1.7 or greater	Ionic		

FIGURE 5.7

The relationships between bonding terminology and electronegativity differences.

2. When the electronegativity difference between bonded atoms is greater than zero but less than 1.7, the bond is called a *polar covalent bond*.

3. When the difference in electronegativity between bonded atoms is 1.7 or greater, the bond is called an *ionic bond*.

The preceding guidelines are the basis for deciding whether to formulate the bonding description for a compound in terms of an ionic electron-dot structure or a covalent electron-dot structure. Figure 5.7 summarizes the relationships developed in this section among electronegativity differences, bond types, and bond terminology.

EXAMPLE 5.8

Indicate whether each of the following bonds should be designated as ionic, polar covalent, or nonpolar covalent. Also designate the direction of polarity using delta notation for any polar covalent bonds.
(a) N—O   (b) Mg—O

SOLUTION

(a) The electronegativities of nitrogen and oxygen are 3.0 and 3.5, respectively. The electronegativity difference, which is obtained by subtracting the smaller electronegativity value from the larger value, is

$$\text{Electronegativity difference} = 3.5 - 3.0 = 0.5$$

The bond is *polar covalent* because the electronegativity difference is greater than zero but less than 1.7.

The oxygen end of the bond will be negative relative to the nitrogen end, because oxygen is the more electronegative of the two elements. Thus, the direction of the polarity is

$$\overset{\delta+}{N}—\overset{\delta-}{O}$$

(b) The electronegativity of oxygen is 3.5 and that of magnesium is 1.2. The electronegativity difference is

$$\text{Electronegativity difference} = 3.5 - 1.2 = 2.3$$

The term *ionic* is used to describe bonds when the electronegativity difference is 1.7 or greater. The polarity of the Mg—O bond can best be described in terms of ions—complete transfer of electrons

$$[Mg^{2+}][O^{2-}]$$

Bond polarity plays an important role in many of the discussions in later chapters of this text. Many times bond polarities are factors that influence the products formed in a reaction. The collective effects of bond polarities within a molecule can cause the molecule as a whole to be polar (Section 5.12). Molecular polarity influences many physical properties of a substance, such as its boiling and melting points and solubility in various solvents.

5.12   Molecular Polarity

Molecules, as well as bonds, can have polarity. As we will see shortly, just because a molecule contains polar bonds does not mean the molecule as a whole is polar. Molecular polarity depends on the polarity of the bonds within a molecule and the geometry of the molecule (when three or more atoms are present).

Molecular geometry *describes the way in which atoms in a molecule are arranged in space relative to each other.* All molecules containing three or more atoms have characteristic three-dimensional shapes. For example, the triatomic CO_2 molecule is linear; that is, its three atoms lie in a straight line (Figure 5.8a). On the other hand, an H_2O molecule, which is also a triatomic molecule, has a nonlinear, or bent, geometry (Figure 5.8b). An NH_3 molecule has a trigonal pyramidal geometry, with the nitrogen atom at the apex and the hydrogen atoms at the base of the pyramid (Figure 5.8c).

Determining the molecular polarity of a *diatomic* molecule is simple because only one bond is present. If that bond is nonpolar, the molecule is nonpolar; if the bond is polar, the molecule is polar.

The collective effect of individual bond polarities must be considered to determine molecular polarity for molecules containing more than one bond (triatomic molecules, tetraatomic molecules, and so on). Molecular geometry plays an important role in determining this collective effect. In some instances, because of the symmetrical nature of the geometry of the molecule, the effects of

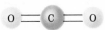

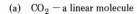

(a) CO_2 — a linear molecule   (b) H_2O — a nonlinear   (c) NH_3 — a trigonal
                                                       or angular molecule   pyramidal molecule

FIGURE 5.8
Molecular geometries of selected molecules.

polar bonds are canceled and a nonpolar molecule results. Let us consider the polarities of three specific triatomic molecules—CO_2, H_2O, and HCN.

In the linear CO_2 molecule (Figure 5.8a), both bonds are polar (oxygen is more electronegative than carbon). Despite the presence of these polar bonds, CO_2 molecules are *nonpolar*. The effects of the two polar bonds are canceled out as a result of the oxygen atoms being arranged symmetrically around the carbon atom. The shift of electronic charge toward one oxygen atom is exactly compensated by the shift of electronic charge toward the other oxygen atom. Thus, one end of the molecule is not negatively charged relative to the other end (a requirement for polarity) and the molecule is nonpolar. This cancellation of individual bond polarities is diagrammed as

$$\overset{\longleftarrow \quad \longrightarrow}{O=C=O}$$

The two individual bond polarities (denoted by arrows) are of equal magnitude (each oxygen affects the carbon atom in the same way), but because they are opposite in direction, they cancel each other.

The nonlinear (bent) triatomic H_2O molecule (Figure 5.8b) is polar. The bond polarities associated with the two hydrogen–oxygen bonds do not cancel each other because of the nonlinearity of the molecule.

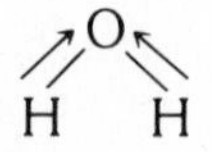

$$\overset{O}{\underset{H \quad\; H}{\diagup\!\diagup \;\; \diagdown\!\diagdown}}$$

As a result of their orientation, both bonds contribute to an accumulation of negative charge on the oxygen atom. The two bond polarities are equal in magnitude but are not opposite in their direction.

The generalization that linear triatomic molecules are nonpolar and nonlinear triatomic molecules are polar, which you might be tempted to make on the basis of our discussion of CO_2 and H_2O molecular polarities, is not valid. The linear molecule HCN, which is polar, invalidates this statement. Both bond polarities contribute to nitrogen acquiring a partial negative charge relative to hydrogen in HCN.

$$\overset{\longrightarrow \qquad \longrightarrow}{H-C\equiv N}$$

(Note that the two polarity arrows point in the same direction because nitrogen is more electronegative than carbon and carbon is more electronegative than hydrogen.)

Tetraatomic and pentatomic molecules commonly have trigonal planar and tetrahedral geometries, respectively. The arrangement of atoms associated with these two geometries is

Trigonal planar   Tetrahedral

Trigonal planar and tetrahedral molecules in which all of the atoms attached to the central atom are identical, such as BF_3 (trigonal planar) and CH_4 (tetrahedral), are *nonpolar*. The individual bond polarities cancel as the result of the highly symmetrical arrangement of atoms around the central atom. (Proof that cancellation does occur involves some trigonometric considerations; it is not an obvious situation.)

If two or more kinds of atoms are attached to the central atom in a trigonal planar or tetrahedral molecule, the molecule is polar. The high symmetry required for cancellation of the individual bond polarities is no longer present. For example, if one of the hydrogen atoms in CH_4 (a nonpolar molecule) is replaced by a chlorine atom, a polar molecule results, even though the resulting CH_3Cl molecule is still a tetrahedral molecule. A carbon–chlorine bond has a greater polarity than a carbon–hydrogen bond; chlorine has an electronegativity of 3.0 and hydrogen has an electronegativity of only 2.1. Figure 5.9 contrasts the polar CH_3Cl and nonpolar CH_4 molecules. Note that the direction of polarity of the carbon–chlorine bond is opposite to that of the carbon–hydrogen bonds.

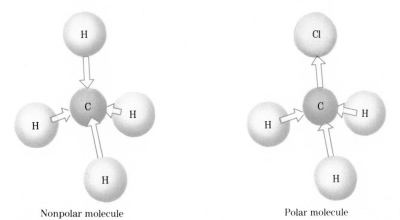

Nonpolar molecule   Polar molecule

FIGURE 5.9

The nonpolar and polar tetrahedral molecules CH_4 and CH_3Cl.

5.13 Nomenclature for Covalent Compounds

The simplest and most common type of covalent compound is one that contains two nonmetals (Section 5.8). The names of such compounds are derived by using a rule very similar to the one used for naming ionic compounds (Section 5.6); however, one major difference exists. Names for covalent compounds always contain Greek numerical prefixes that give the number of each type of atom present, in addition to the names of the elements present. This is in direct contrast to ionic nomenclature in which formula subscripts are never mentioned when naming compounds.

The basic rule used when constructing the name of a covalent compound is as follows.

The full name of the nonmetal with the lower electronegativity preceded with a Greek numerical prefix is given first, followed by a separate word containing a Greek numerical prefix, the stem of the name of the more electronegative nonmetal, and the suffix -ide.

Prefix use is necessary because most pairs of nonmetals form many different compounds. For example, all of the following nitrogen–oxygen compounds are known: NO, NO_2, N_2O, N_2O_3, N_2O_4, and N_2O_5. Such diverse behavior between two elements exists because single, double, and triple covalent bonds are possible. The prefixes used are the standard Greek numerical prefixes, which are given for the numbers 1 through 10 in Table 5.2. Example 5.9 shows how these prefixes are used in binary covalent compound nomenclature.

TABLE 5.2

Greek Numerical Prefixes for the Numbers 1 through 10

Greek Prefix	Number
Mono-	1
Di-	2
Tri-	3
Tetra-	4
Penta-	5
Hexa-	6
Hepta-	7
Octa-	8
Ennea-	9
Deca-	10

EXAMPLE 5.9

Name the following binary covalent compounds.

(a) N_2O (b) N_2O_3 (c) PCl_5 (d) P_4S_{10} (e) CCl_4

SOLUTION

The names of each of these compounds consist of two words, which have the following general formats.

First word: (prefix) + (full name of least electronegative nonmetal)

Second word: (prefix) + (stem of name of more electronegative nonmetal) + (ide)

(a) The elements present are nitrogen and oxygen. The two portions of the name, before the Greek numerical prefixes are added, are *nitrogen* and *oxide*. Adding the prefixes gives *dinitrogen* (two nitrogen atoms are present) and *monoxide* (one oxygen atom is present). (When an element name begins with an *a* or *o*, the a or o at the end of the Greek prefix is dropped for ease of pronunciation—monoxide instead of monooxide.) Thus, the name of this compound is *dinitrogen monoxide*.

(b) The elements present are again nitrogen and oxygen. This time the two portions of the name are *dinitrogen* and *trioxide*, which are combined to give the name *dinitrogen trioxide*.

(c) When there is only one atom of the first nonmetal present, it is common to omit the prefix *mono-* for that element. Following this guideline, we have *phosphorus pentachloride* as the name of this compound.

(d) The prefix is *tetra-* for 4 atoms and *deca-* for 10 atoms. This compound thus has the name *tetraphosphorus decasulfide*.

(e) Omitting the initial *mono-* (see part c), we name this compound *carbon tetrachloride*.

There is one standard exception to the use of Greek numerical prefixes in naming covalent compounds. Covalent compounds with hydrogen as the first element in the formula are named without any prefix. Thus, the compounds H_2S and HC1 are named hydrogen sulfide and hydrogen chloride, respectively.

A few covalent compounds have names that are completely unrelated to the rules we have been discussing. These have common names that were coined before the systematic nomenclature rules were developed. At one time, in the early history of chemistry, all compounds had common names. With the advent of systematic nomenclature, most common names were discontinued. A few, however, have persisted and are now officially accepted. The most famous example of this is the compound H_2O, which has the systematic name hydrogen

TABLE 5.3

Selected Covalent Compounds That Have Common
Names

Compound Formula	Accepted Common Name
H_2O	water
H_2O_2	hydrogen peroxide
NH_3	ammonia
N_2H_4	hydrazine
CH_4	methane
C_2H_6	ethane

oxide, but this name is never used. H_2O is *water*—a name that is not going to change. Another well-known example of common nomenclature is the name *ammonia* for the compound NH_3. Table 5.3 lists additional examples of compounds for which common names are used in preference to systematic names. The common name exceptions are actually very few in relation to the total number of compounds that are named by systematic rules.

5.14   Polyatomic Ions

A **polyatomic ion** *is a group of covalently bonded atoms that have acquired a charge through the loss or gain of electrons.* Numerous ionic compounds exist in which the positive or negative ion (sometimes both) is polyatomic. Compounds that contain polyatomic ions offer an interesting combination of both ionic and covalent bonding; covalent bonding occurs *within* the polyatomic ion and ionic bonding occurs *between* it and the other ion.

Polyatomic ions are very stable species that generally maintain their identity during chemical reactions. However, they are not molecules; they never occur alone as molecules do. Instead, they always associate with other ions that have opposite charges. Polyatomic ions are *pieces* of compounds, not compounds. Ionic compounds require the presence of both positive and negative ions and are neutral overall. Polyatomic ions are always charged species.

Table 5.4 lists the names and formulas of some of the more common polyatomic ions. The table is organized around key elements (other than oxygen) that are present in the polyatomic ions. Note from the table that almost all the polyatomic ions listed contain oxygen atoms. The names of some of these common polyatomic ions should be familiar even if you have had no previous chemistry, because many of them are found in commercial products— for example, fertilizers (phosphates, sulfates, nitrates), baking powder and soda (bicarbonates), and building materials (carbonates, sulfates).

TABLE 5.4

Formulas and Names of Selected Polyatomic Ions (*the most common ions are in color*)

Key Element Present	Formula	Name of ion
Nitrogen	NO_3^-	nitrate
	NO_2^-	nitrite
	NH_4^+	ammonium
Sulfur	SO_4^{2-}	sulfate
	HSO_4^-	bisulfate or hydrogen sulfate
	SO_3^{2-}	sulfite
	HSO_3^-	bisulfite or hydrogen sulfite
Phosphorus	PO_4^{3-}	phosphate
	HPO_4^{2-}	hydrogen phosphate
	$H_2PO_4^-$	dihydrogen phosphate
	PO_3^{3-}	phosphite
Carbon	CO_3^{2-}	carbonate
	HCO_3^-	bicarbonate or hydrogen carbonate
	$C_2O_4^{2-}$	oxalate
	$C_2H_3O_2^-$	acetate
	CN^-	cyanide
Chlorine	ClO_4^-	perchlorate
	ClO_3^-	chlorate
	ClO_2^-	chlorite
	ClO^-	hypochlorite
Boron	BO_3^{3-}	borate
Hydrogen	H_3O^+	hydronium
	OH^-	hydroxide
Metals	MnO_4^-	permanganate
	CrO_4^{2-}	chromate
	$Cr_2O_7^{2-}$	dichromate

There is no easy way to learn the common polyatomic ions; memorization is required. The charges and formulas for the various polyatomic ions cannot be easily related to the periodic table, as was the case for many of the monoatomic ions (Section 5.5). The inability to recognize the presence of polyatomic ions in a compound is a major stumbling block for many chemistry students. It requires effort to learn the names and formulas of the more common polyatomic ions and thus avoid this obstacle.

Note from Table 5.4 the following relationships among polyatomic ions.

1. Most of the ions have a negative charge, which can vary from -1 to -3.

Only two positive ions are listed in the table: NH_4^+ (ammonium) and H_3O^+ (hydronium).

2. Two of the polyatomic ions, OH^- (hydroxide) and CN^- (cyanide), have names ending in −ide. These names represent exceptions to the rule that the suffix −ide is reserved for binary ionic compounds (Section 5.6).

3. A number of -ate/-ite pairs of ions exist, as in SO_4^{2-} (sulfate) and SO_3^{2-} (sulfite). The -ate ion always has one more oxygen atom than the -ite ion. Both the -ate and -ite ions carry the same charge.

4. In a number of pairs of ions one member of the pair differs from the other by having a hydrogen atom present, as in CO_3^{2-} (carbonate) and HCO_3^- (hydrogen carbonate or bicarbonate). In such pairs, the charge on the ion containing hydrogen is always one less than that on the other ion.

5.15 Formulas and Names for Compounds Containing Polyatomic Ions

Formulas for ionic compounds containing polyatomic ions are determined in the same way as they are for ionic compounds containing monoatomic ions (Section 5.5). The total positive and negative charge present must add up to zero.

Two conventions not encountered previously in formula writing often arise in writing formulas for compounds containing polyatomic ions.

1. When more than one polyatomic ion of a given kind is required in a formula, the polyatomic ion is enclosed in parentheses and a subscript placed outside the parentheses is used to indicate the number of polyatomic ions needed.

2. To preserve the identity of polyatomic ions, the same elemental symbol may be used more than once in a formula. In the formula NH_4NO_3, which represents an ionic compound containing ammonium ions (NH_4^+) and nitrate ions (NO_3^-), the symbol for the element nitrogen (N) appears in two locations.

EXAMPLE 5.10

Determine the formulas for the ionic compounds containing the following pairs of ions.

(a) Na^+ and PO_4^{3-} (b) Ca^{2+} and NO_3^-

SOLUTION

(a) To equalize the total positive and negative charge, we need three sodium ions (+1 charge) for each phosphate ion (−3 charge). We indicate the presence of three Na^+ ions with the subscript 3 following the symbol of this ion. The formula of the compound is Na_3PO_4. The convention that the positive ion is always written first in the formula still holds when polyatomic ions are present.

(b) Two nitrate ions (−1 charge) are required to balance the charge on one calcium ion (+2 charge). Because more than one polyatomic ion is needed, the formula contains parentheses, $Ca(NO_3)_2$. The subscript 2 outside the parentheses indicates two of what is inside the parentheses. If parentheses were not used, the formula would appear to be $CaNO_{32}$, which conveys false information. The formula $Ca(NO_3)_2$ indicates a formula unit containing one Ca atom, two N atoms, and six O atoms; the formula $CaNO_{32}$ indicates a formula unit containing one Ca atom, one N atom, and 32 O atoms. The correct formula, $Ca(NO_3)_2$, is read as "C-A" (pause) "N-O-three-taken-twice."

The names of ionic compounds containing polyatomic ions are derived in a manner that is similar to that for binary ionic compounds (Section 5.7). The rule for naming binary ionic compounds is to give the name of the metallic element first (including, when needed, a Roman numeral indicating ion charge), followed by a term containing the stem of the nonmetallic name to which the suffix −ide is appended.

For our present situation, if the polyatomic ion is positive, its name is substituted for that of the metal. If the polyatomic ion is negative, its name is substituted for the nonmetal stem plus -ide. When both positive and negative ions are polyatomic, dual substitution occurs and the resulting term includes just the names of the polyatomic ions. Example 5.11 illustrates the use of these rules.

EXAMPLE 5.11

Name the following compounds, which contain one or more polyatomic ions:
(a) $Ca(OH)_2$ (b) $Fe_2(CO_3)_3$ (c) $(NH_4)_2SO_4$

SOLUTION

(a) The positive ion present is the calcium ion (Ca^{2+}). We will not need a Roman numeral to specify the charge on a Ca^{2+} ion because it is always a +2. The negative ion is the polyatomic hydroxide ion (OH^-). The name of the compound is *calcium hydroxide*. As in naming binary ionic compounds, subscripts in the formula are not incorporated into the name.

(b) The positive ion present is iron(III). The negative ion is the polyatomic carbonate ion (CO_3^{2-}). The name of the compound is *iron(III) carbonate*.

The determination that iron is present as iron(III) involves the following calculation dealing with charge balance.

$$2(\text{Iron charge}) + 3(\text{Carbonate charge}) = 0$$

The carbonate charge is -2. (You had to memorize that.) Therefore,

$$2(\text{Iron charge}) + 3(-2) = 0$$
$$2(\text{Iron charge}) = +6$$
$$\text{Iron charge} = +3$$

(c) Both the positive and negative ions in this compound are polyatomic—the ammonium ion (NH_4^+) and the sulfate ion (SO_4^{2-}). The name of the compound is simply the combination of the names of the two polyatomic ions: *ammonium sulfate*.

Practice Questions and Problems

VALENCE ELECTRONS

5-1 How many valence electrons do atoms with the following electron configurations have?
a. $1s^2 2s^1$
b. $1s^2 2s^2 2p^3$
c. $1s^2 2s^2 2p^6$
d. $1s^2 2s^2 2p^6 3s^2$
e. $1s^2 2s^2 2p^6 3s^2 3p^6 4s^2 3d^{10} 4p^5$
f. $1s^2 2s^2 2p^6 3s^2 3p^6 4s^2 3d^{10} 4p^6 5s^1$

5-2 How many valence electrons do atoms of each of the following elements have?
a. $_4Be$ b. $_9F$ c. $_{11}Na$
d. $_{19}K$ e. $_{34}Se$ f. $_{50}Sn$

ELECTRON-DOT STRUCTURES FOR ATOMS

5-3 Draw electron-dot structures for atoms of the following elements.
a. $_{20}Ca$ b. $_7N$ c. $_{35}Br$
d. $_{10}Ne$ e. $_1H$ f. $_{52}Te$

5-4 Each of the following electron-dot structures represents a period 3 element. In each case what is the element's identity?
a. $\cdot \overset{\cdot}{X}$ b. $\cdot X \cdot$ c. $\cdot \overset{\cdot}{X} \cdot$ d. $: \overset{\cdot}{X} :$

OCTET RULE

5-5 What is unique about the electron configurations of the noble gases?

5-6 State the octet rule.

NOTATION FOR IONS

5-7 Why does an atom that loses electrons become positively charged?

5-8 What would be the charge, if any, on particles with the following subatomic makeups?
a. 13 protons, 14 neutrons, 10 electrons
b. 8 protons, 10 neutrons, 8 electrons
c. 15 protons, 18 neutrons, 18 electrons
d. 3 protons, 3 neutrons, 2 electrons

5-9 What is the difference in meaning associated with the notations S and S^{2-}?

5-10 Calculate the number of protons and electrons in each of the following ions.
a. $_{12}Mg^{2+}$ b. $_9F^-$ c. $_7N^{3-}$
d. $_3Li^+$ e. $_{19}K^+$ f. $_8O^{2-}$

IONIC CHARGE

5-11 Predict the charge on the ion formed by each of the following elements.

a. $_{20}Ca$ b. $_{15}P$ c. $_{17}Cl$
d. $_{11}Na$ e. $_{38}Sr$ f. $_{13}Al$

5-12 Indicate the number of electrons lost or gained when each of the following atoms forms an ion.

a. $_4Be$ b. $_{34}Se$ c. $_{35}Br$
d. $_{37}Rb$ e. $_{52}Te$ f. $_{53}I$

IONIC COMPOUND FORMATION

5-13 Show the formation of the following ionic compounds using electron-dot structures.

a. Na_2O b. MgO c. K_3N d. $AlBr_3$

5-14 Using electron-dot structures, show how ionic compounds are formed by atoms of the following elements.

a. Be and S   b. Ca and O
c. Na and P   d. Mg and Cl

FORMULAS FOR BINARY IONIC COMPOUNDS

5-15 Write the formula for an ionic compound formed from K^+ ion and each of the following ions.

a. Cl^- b. O^{2-} c. Br^-
d. S^{2-} e. N^{3-} f. P^{3-}

5-16 Write the formula for an ionic compound formed from O^{2-} ion and each of the following ions.

a. Magnesium ion   b. Beryllium ion
c. Sodium ion   d. Lithium ion
e. Aluminum ion   f. Calcium ion

5-17 Write the formulas for the ionic compounds formed from the following elements.

a. Sodium and sulfur
b. Beryllium and fluorine
c. Potassium and oxygen
d. Aluminum and phosphorus
e. Chlorine and lithium
f. Bromine and magnesium

STRUCTURE OF IONIC COMPOUNDS

5-18 Describe in words the general structure of an ionic solid.

5-19 What does the formula for an ionic compound actually represent?

5-20 Explain why it is inappropriate to talk about molecules of an ionic compound.

NOMENCLATURE FOR BINARY IONIC COMPOUNDS

5-21 Name the following binary ionic compounds.

a. K_3P b. NaI c. $BeCl_2$
d. CaO e. $AlBr_3$ f. Ca_2C

5-22 Indicate whether or not a Roman numeral is required in the name of each of the following binary ionic compounds.

a. $AuCl$ b. Fe_2O_3 c. $LiCl$
d. $AgBr$ e. ZnO f. Cu_2S

5-23 Calculate the charge on the metal ion in each of the following binary ionic compounds.

a. $AuCl_3$ b. FeO c. $CuCl$
d. $SnCl_4$ e. PbO_2 f. CoI_3

5-24 Name each compound in the following pairs of binary ionic compounds.

a. CuO and Cu_2O b. MnO and Mn_2O_3
c. PbO and PbO_2 d. $SnBr_2$ and $SnBr_4$
e. $FeCl_2$ and $FeCl_3$ f. Au_2O and Au_2O_3

5-25 Name the following binary ionic compounds.

a. $AuCl$ b. $AgCl$ c. KCl
d. $AlCl_3$ e. $CrCl_3$ f. $MnCl_3$

5-26 Write formulas for the following binary ionic compounds.

a. Potassium bromide   b. Silver oxide
c. Magnesium bromide   d. Beryllium phosphide
e. Gallium nitride   f. Copper(II) iodide

5-27 Write formulas for the following binary ionic compounds.

a. Nickel(II) sulfide   b. Cobalt(II) sulfide
c. Tin(IV) sulfide   d. Gold(III) sulfide
e. Aluminum sulfide   f. Zinc sulfide

SINGLE COVALENT BONDS

5-28 Draw electron-dot structures to illustrate the covalent bonding found in each of the following molecules.

a. Br_2 b. $BrCl$ c. HBr
d. NCl_3 e. OF_2 f. PH_3

5-29 Write electron-dot structures for the simplest compound formed between the following pairs of elements.

a. Phosphorus and chlorine
b. Hydrogen and fluorine
c. Iodine and chlorine
d. Carbon and bromine

e. Silicon and hydrogen
f. Fluorine and sulfur

MULTIPLE COVALENT BONDS

5-30 What is the difference between a single covalent bond, a double covalent bond, and a triple covalent bond?

5-31 Draw the electron-dot structures for the following molecules, each of which contains at least one double bond or triple bond. (The skeletal arrangement of atoms in each molecule is shown under its formula.)

a. N_2F_2 b. C_3H_4

    F N   N F   H   H
                           C   C   C
                       H   H

c. C_2H_3N d. C_2N_2

       H   N C   C N
    H C   C N
       H

COORDINATE COVALENT BONDS

5-32 What is a coordinate covalent bond?

5-33 Once formed, how (if at all) does a coordinate covalent bond differ from an ordinary covalent bond?

5-34 Draw an electron-dot structure for the molecule S_2O (S S O), a molecule that contains a coordinate covalent bond. Indicate the location of the coordinate covalent bond.

ELECTRONEGATIVITY AND POLARITY OF BONDS

5-35 In each of the following pairs of elements, indicate which element is the more electronegative element.

a. H and F   b. N and O
c. O and S   d. Na and Mg
e. Be and N   f. Cl and Br

5-36 Arrange each of the following sets of bonds in order of increasing polarity.

a. H—Cl, H—O, H—F
b. N—O, P—O, Al—O
c. H—Cl, Br—Br, B—N
d. P—N, S—O, Be—F

5-37 Place a $\delta+$ above the atom that is relatively positive and a $\delta-$ above the atom that is relatively negative in

each of the following bonds.

a. B—N   b. F—C   c. O—Cl
d. Al—N   e. S—Cl   f. H—Si

5-38 Classify each of the following bonds as nonpolar covalent, polar covalent, or ionic.

a. Carbon–oxygen   b. Beryllium–chlorine
c. Nitrogen–phosphorus   d. Iodine–iodine
e. Potassium–oxygen   f. Sodium–fluorine

MOLECULAR POLARITY

5-39 It is possible for a molecule to be nonpolar when it contains polar bonds? Explain.

5-40 For each of the following hypothetical triatomic molecules, indicate whether the bonds are polar or nonpolar and whether the molecule is polar or nonpolar. Assume A, X, and Y have different electronegativities.

a. X—A—X   b. A—X—X   c. Y—A—X

d.   e. X—X—X   f.

5-41 Indicate whether each of the following molecules is polar or nonpolar. The geometry of each molecule is given in parentheses.

a. H_2S (bent)—the sulfur atom is in the middle.
b. CS_2 (linear)—the carbon atom is in the middle.
c. $CHCl_3$ (tetrahedral)—the carbon atom is in the center.
d. NCl_3 (trigonal pyramid)—the nitrogen atom is the apex.

NOMENCLATURE FOR BINARY COVALENT COMPOUNDS

5-42 Name the following binary covalent compounds.

a. SF_6 b. P_4O_6 c. CO
d. ClO_2 e. S_4N_2 f. N_2O

5-43 Write formulas for the following binary covalent compounds.

a. Iodine monochloride   b. Nitrogen trichloride
c. Silicon tetrabromide   d. Dinitrogen trioxide
e. Ammonia   f. Hydrogen peroxide

COMPOUNDS CONTAINING POLYATOMIC IONS

5-44 Write formulas (including charge) for the following polyatomic ions.

a. Nitrate   b. Sulfate   c. Ammonium
d. Hydroxide   e. Carbonate   f. Phosphate

5-45 Indicate which of the following compounds contain polyatomic ions and identify the polyatomic ion by name if it is present.
a. Al_2S_3 b. $ZnSO_4$ c. KCN
d. NaClO   e. NH_4Cl f. Co_2O_3

5-46 Write formulas for the compounds formed by the following positive and negative ions.
a. Ba^{2+} and NO_3^- b. Al^{3+} and CO_3^{2-}
c. K^+ and PO_4^{3-} d. Au^+ and sulfate
e. Be^{2+} and hydroxide   f. Fe^{3+} and cyanide

5-47 Name the following compounds that contain one or more polyatomic ions.

a. Na_2CO_3 b. NH_4NO_3 c. $CuSO_4$
d. $Ca(OH)_2$ e. AgCN   f. $Mg_3(PO_4)_2$

5-48 Write formulas for the following compounds containing polyatomic ions.
a. Potassium bicarbonate
b. Silver carbonate
c. Aluminum phosphate
d. Copper(II) hydroxide
e. Gold(III) nitrate
f. Ammonium sulfate

6

Measurement

CHAPTER HIGHLIGHTS

6.1   Observations and Measurements

Chemists, as well as other scientists, rely on observations to explain the nature of the substances they study. Two general types of observations exist: qualitative and quantitative. A **qualitative observation** *is an observation made with the senses and is usually expressed using words instead of numbers.* Qualitative observations about a person sick in the hospital might include that the person is breathing rapidly, has a high temperature, and is very thin.

A **quantitative observation** *is an observation that requires a numerical measurement and describes something in terms of "how much."* The quantitative observation that a person has a body temperature of 103.6°F is much more useful information than just knowing that the person has a high fever. Quantitative observations are preferred by scientists.

One or more measurements is always a part of any quantitative observation. A **measurement** *determines the dimensions, capacity, quantity, or extent of something.* The most common types of measurements made in chemical laboratories are those of mass, volume, length, temperature, pressure, and concentration.

Measurements always consist of two parts: a *number*, which tells the amount of the quantity measured, and a *unit*, which tells the nature of the quantity measured.

6.2   Systems of Measurement Units

A unit is a label that describes something that is being measured or counted. It can be almost anything: 4 quarts, 4 dimes, 4 dozen frogs, 4 bushels, 4 inches, or 4 pages. Two formal systems of measurement are in use in the United States today. Common measurements of commerce, such as those used in a grocery store, are made in the *English system*. The units of this system include the familiar inch, foot, pound, quart, and gallon. A second system, the *Metric system*, is used in scientific work. Units in this system include the gram, meter, and liter. The United States is one of a few countries that use different unit systems for commerce and scientific work. The metric system is used in most countries for both commercial and scientific work.

The United States is in the process of a voluntary conversion to the metric system. Many metric system units now appear on consumer products (see

FIGURE 6.1
Metric units are becoming increasingly evident on many highway signs and consumer products. [(a) Dr. Georg Gerster/Photo Researchers, Inc.]

Figure 6.1). Soft drinks can now be bought in 2-liter containers. Road signs in some states display distances in both miles and kilometers. Canned and packaged goods such as cereals and mixes on grocery store shelves now have their content masses listed in grams as well as in pounds and ounces.

Why should the United States convert to the metric system? The answer is simple. The metric system is superior to the English system. Its superiority lies in the area of interrelationships between units of the same type such as volume or length. Metric unit interrelationships are less complicated than English unit interrelationships because the metric system is a decimal unit system. In the metric system, conversion from one unit size to another can be accomplished simply by moving the decimal point to the right or left an appropriate number of places. The metric system is no more precise than the English system; it is simply more convenient.

The metric system was "updated" in 1960. By international agreement, a revision of the traditional metric system, called the *International System of Units* (abbreviated *SI* after the French name *Le Système International d'Unités*) was adopted by scientists. Acceptance of the SI units has varied within the scientific community. In many sciences, movement toward use of SI units has been very slow.

Because the differences between the two unit systems are not of major importance for the subjects to be covered in this text, we will continue to use the more familar traditional metric system units. Switching to SI units sometime in the future, when their use is more extensive, will present no major problems to the student who properly understands the traditional metric system because the two systems are so so similar.

Metric System Prefixes

In the metric system there is one base unit for each type of measurement—length, volume, mass, and so on. These base units are multiplied by appropriate powers of 10 to form smaller or larger units. The names of the larger and smaller units are constructed from the base unit name by attaching prefixes that tell which power of 10 is involved. These prefixes are given in Table 6.1, along with their symbols or abbreviations and mathematical meanings. The prefixes in color are the most frequently used.

The use of numerical prefixes should not be new to you. Consider the use of the prefix *tri-* in the words *triangle*, *tricycle*, *trio*, *trinity*, and *triple*. Each of these words conveys the idea of three of something. The metric system prefixes are used in the same way.

The meaning of a metric system prefix remains constant; it is independent of the base it modifies. For example, a kilosecond is 1000 seconds; a kilowatt is 1000 watts; and a kilocalorie is 1000 calories. The prefix *kilo* always means 1000.

TABLE 6.1

Metric System Prefixes and Their Mathematical Meanings

		Mathematical Meaning		
Prefix	Symbol	Exponential Number	Common Number	
Giga-	G	10^9	1,000,000,000	
Mega-	M	10^6	1,000,000	Prefixes for
Kilo-	k	10^3	1,000	multiple units
Hecto-	h	10^2	100	
Deca-	da	10^1	10	
Deci-	d	10^{-1}	0.1	
Centi-	c	10^{-2}	0.01	Prefixes for
Milli-	m	10^{-3}	0.001	subunits
Micro-	μ	10^{-6}	0.000001	
Nano-	n	10^{-9}	0.000000001	

Metric Units of Length

The **meter** *is the basic unit of length in the metric system.* Other units of length in the metric system are derived from the meter by using the prefixes listed in Table 6-1. The kilometer (km) is 1000 times larger than the meter (m), whereas the centimeter (cm) and millimeter (mm) are 100 and 1000 times smaller than the meter, respectively.

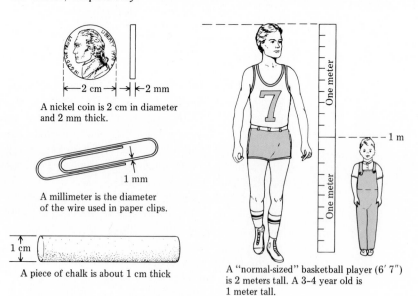

A nickel coin is 2 cm in diameter and 2 mm thick.

A millimeter is the diameter of the wire used in paper clips.

A piece of chalk is about 1 cm thick

A "normal-sized" basketball player (6′ 7″) is 2 meters tall. A 3-4 year old is 1 meter tall.

FIGURE 6.2

Metric units of length in the "everyday realm."

Figure 6.2 relates the metric units of meters, centimeters, and millimeters to objects and situations we encounter in everyday life.

A comparison of metric lengths with the commonly used English system lengths of mile, yard, and inch reveals that a meter is slightly larger than a yard, a kilometer is approximately five-eighths of a mile, and a centimeter is slightly less than one-half of an inch.

Metric Units of Mass

The **gram** *is the basic unit of mass in the traditional metric system.* It is a very small unit compared with the commonly used English mass units of ounce and pound. It takes approximately 28 grams to equal 1 ounce and nearly 454 grams to equal 1 pound. Because of the small size of the gram (g) the kilogram (kg) is a very commonly used unit. A kilogram is equivalent to an English mass of slightly more than 2 pounds.

Figure 6.3 relates the mass units of milligram (mg), gram, and kilogram to everyday objects.

The terms mass and weight are frequently used interchangeably. Although in most cases this practice does no harm, technically it is incorrect to interchange the terms. Mass and weight refer to different properties of matter, and the difference in their meaning should be understood.

Mass *is a measure of the total quantity of matter in an object.* **Weight** *is a measure of the force exerted on an object by the pull of gravity.* The mass of a substance is a constant; the weight of an object is a variable dependent on its geographical location.

Matter at the equator weighs less than it would at the North Pole because the earth is not a perfect sphere, but bulges at the equator. It weighs less

A single staple has a mass of about 3 mg.

2 thumbtacks have a mass of 1 g.

A nickel has a mass of about 5 g.

1 quart of milk in a cardboard container has a mass of 1 kg.

A 220-lb football player is equivalent to a mass of 100 kg.

FIGURE 6.3

Metric units of mass in the "everyday realm."

FIGURE 6.4
An astronaut on a "space walk" is weightless but not massless. He or she will have exactly the same mass as on earth. [NASA.]

because the magnitude of gravitational attraction (the measure of weight) is less at the equator. An object would weigh less on the moon than on earth because of the smaller size of the moon and the correspondingly lower gravitational attraction. Quantitatively, a mass weighing 22.0 pounds at the earth's North Pole would weigh 21.9 pounds at the earth's equator and only 3.7 pounds on the moon. In outer space an astronaut may be weightless but never massless. In fact, he or she has the same mass in space as on earth (see Figure 6.4).

Metric Units of Volume

It is frequently faster and more convenient to measure a substance's volume rather than its mass. This is particularly true for liquids and gases. For example, volume is used instead of mass to determine the amount of gasoline put into an

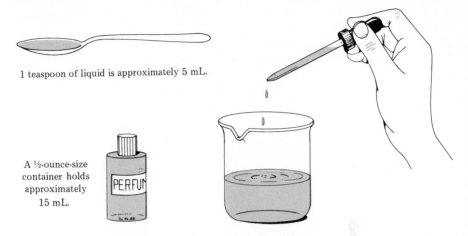

1 teaspoon of liquid is approximately 5 mL.

A ½-ounce-size container holds approximately 15 mL.

20 drops from an eyedropper is about 1 mL.

FIGURE 6.5
The metric volume unit of milliliter and the "everyday realm."

automobile fuel tank. The **liter** *is the basic unit of volume in the metric system.* A liter is abbreviated with a capital L; a lowercase l might be confused with the number 1.

As with the units of length and mass, the basic unit of volume is modified with prefixes to represent smaller or larger units. The most commonly used volume unit using a prefix is the milliliter (mL), which is 1/1000 of a liter. A milliliter is much smaller than a fluid ounce. A quart and a liter are almost the same volume, but a quart is slightly smaller. Figure 6.5 relates the unit of milliliter to everyday situations.

In some laboratory work, the volume unit cubic centimeters (abbreviated cm^3 or cc) is used. What is its relationship to the units of liter and milliliter? Cubic centimeter is an alternative term for milliliter, that is

$$1 \text{ milliliter} = 1 \text{ cubic centimeter}$$

Consequently, the units milliliter and cubic centimeter are interchangeable.

6.3 Conversion Factors and Dimensional Analysis

Often the need arises to change the units of a quantity or measurement. The new units needed can be in either the same measurement system or a different system from the old units. With two unit systems in common use in the United

States, it is frequently necessary to change measurements from one system to their equivalent in the other system.

The mathematical tool we use to accomplish this task is a general method of problem solving called *dimensional analysis*. Central to the use of the dimensional analysis problem-solving method is the concept of conversion factors. A **conversion factor** *is a ratio, with a value of unity, that specifies how one unit of measurement is related to another.*

Let us construct some conversion factors to see how they originate and why they always have values of unity.

The quantities 1 hour and 60 minutes both describe the same amount of time. We can thus write the following equation:

$$1 \text{ hour} = 60 \text{ minutes}$$

This equation, which describes a fixed relationship, can be used to construct a pair of conversion factors that relate hours and minutes. (Conversion factors always occur in pairs.)

Dividing both sides of our hour–minute equation by the quantity "1 hour" gives

$$\frac{1 \text{ hour}}{1 \text{ hour}} = \frac{60 \text{ minutes}}{1 \text{ hour}}$$

Because the numerator and denominator of the fraction on the left side of the equation are identical, this fraction has a value of unity.

$$1 = \frac{60 \text{ minutes}}{1 \text{ hour}}$$

The fraction on the right side of the equation is our conversion factor and its value is one. Note that the numerator and denominator of the conversion factor describe the same amount of time.

Two conversion factors are always obtainable from any given equality. For the equality we are considering (1 hour = 60 minutes), the second conversion factor is

$$\frac{1 \text{ hour}}{60 \text{ minutes}}$$

It is obtained by dividing both sides of the equality by 60 minutes instead of 1 hour. The two conversion factors in a pair are reciprocals.

$$\frac{60 \text{ minutes}}{1 \text{ hour}} \quad \text{and} \quad \frac{1 \text{ hour}}{60 \text{ minutes}}$$

In general, we can always construct a reciprocal pair of conversion factors, each with a value of unity, from any two terms that describe the same amount of whatever we are considering.

Metric–Metric Conversion Factors

Metric–metric conversion factors are used to change one metric unit into another metric unit. Both the numerator and the denominator of these conversion factors involve metric system units. Metric system prefix meanings are used to derive these conversion factors. For example, the set of conversion factors involving kilometer and meter is derived from the meaning of the prefix *kilo–*, which is 10^3. The two conversion factors are

$$\frac{10^3 \text{ m}}{1 \text{ km}} \quad \text{and} \quad \frac{1 \text{ km}}{10^3 \text{ m}}$$

Note the reciprocal relationship between the two conversion factors of the set.
The conversion factors relating microgram and gram involve the number 10^{-6}, the mathematical equivalent of *micro–*, and are

$$\frac{1 \text{ } \mu g}{10^{-6} \text{ g}} \quad \text{and} \quad \frac{10^{-6} \text{ g}}{1 \text{ } \mu g}$$

Note the placement of the number 10^{-6} within the conversion factors. The numerical equivalent of the prefix always goes with the base (unprefixed) unit.

Metric–English and English–Metric Conversion Factors

Conversion factors that relate metric units to English units and vice versa are not exactly defined quantities because they involve two different systems of measurement. The numbers associated with these conversion factors must be determined experimentally. Table 6.2 lists commonly encountered relationships between metric and English system units. These few factors are sufficient to solve most of the problems that we will encounter.

Dimensional Analysis

Dimensional analysis *is a general problem-solving method that uses the units associated with numbers as a guide in setting up calculations.* In this method, units are treated in the same way as numbers, that is, they can be multiplied, divided, canceled, and so on. For example, just as

$$5 \times 5 = 5^2 \quad \text{(5 squared)}$$

we have

$$\text{mL} \times \text{mL} = \text{mL}^2 \quad \text{(mL squared)}$$

TABLE 6.2

Equalities and Conversion Factors That Relate the English and Metric Systems of Measurement

	Metric to English	English to Metric
Length		
1 inch = 2.54 centimeters	$\dfrac{1.00 \text{ in.}}{2.54 \text{ cm}}$	$\dfrac{2.54 \text{ cm}}{1.00 \text{ in.}}$
1 meter = 39.4 inches	$\dfrac{39.4 \text{ in.}}{1.00 \text{ m}}$	$\dfrac{1.00 \text{ m}}{39.4 \text{ in.}}$
1 kilometer = 0.621 mile	$\dfrac{0.621 \text{ mi}}{1.00 \text{ km}}$	$\dfrac{1.00 \text{ km}}{0.621 \text{ mi}}$
Mass		
1 pound = 454 grams	$\dfrac{1.00 \text{ lb}}{454 \text{ g}}$	$\dfrac{454 \text{ g}}{1.00 \text{ lb}}$
1 kilogram = 2.20 pounds	$\dfrac{2.20 \text{ lb}}{1.00 \text{ kg}}$	$\dfrac{1.00 \text{ kg}}{2.20 \text{ lb}}$
1 ounce = 28.3 grams	$\dfrac{1.00 \text{ oz}}{28.3 \text{ g}}$	$\dfrac{28.3 \text{ g}}{1.00 \text{ oz}}$
Volume		
1 quart = 946 milliliters	$\dfrac{1.00 \text{ qt}}{946 \text{ mL}}$	$\dfrac{946 \text{ mL}}{1.00 \text{ qt}}$
1 liter = 1.06 quarts	$\dfrac{1.06 \text{ qt}}{1.00 \text{ L}}$	$\dfrac{1.00 \text{ L}}{1.06 \text{ qt}}$
1 milliliter = 0.034 fluid ounce	$\dfrac{0.034 \text{ fl oz}}{1.00 \text{ mL}}$	$\dfrac{1.00 \text{ mL}}{0.034 \text{ fl oz}}$

Also, just as the 3 s cancel in the expression

$$\frac{\cancel{3} \times 5 \times 7}{\cancel{3} \times 2}$$

the centimeters cancel in the expression

$$\frac{\cancel{(\text{cm})} \times (\text{inch})}{\cancel{(\text{cm})}}$$

Like units found in the numerator and denominator of a fraction always cancel, just as like numbers do.

The following steps show how to set up a problem using dimensional analysis.

STEP 1 *Identify the known or given quantity (both numerical value and units) and the units of the new quantity to be determined.*

This necessary information, which serves as the starting point for setting up the problem, is always found in the statement of the problem. Write an equation with the given quantity on the left and the units of the desired quantity on the right.

STEP 2 *Multiply the given quantity by one or more conversion factors in such a manner that the unwanted (original) units are canceled out, leaving only the desired units.*

The general format for the multiplication is

$$\left(\begin{array}{c}\text{Information}\\ \text{given}\end{array}\right) \times \left(\begin{array}{c}\text{Conversion}\\ \text{factors}\end{array}\right) = \left(\begin{array}{c}\text{Information}\\ \text{sought}\end{array}\right)$$

The number of conversion factors used depends on the individual problem. Except in the simplest problems, it is a good idea to predetermine the sequence of unit changes to be used formally. This sequence will be called the unit pathway.

STEP 3 *Perform the mathematical operations indicated by the conversion factor setup.*

When performing the calculation, double check to make sure that all units except the desired set have canceled out.

Now let us work a number of sample problems using dimensional analysis and the steps just outlined. Our first two examples involve only metric–metric conversion factors. The first problem (Example 6.1) is very simple. Even though you can do this problem "in your head" let us formally set it up using the steps of dimensional analysis. Much can be learned about dimensional analysis from this simple problem.

EXAMPLE 6.1

A vitamin C tablet is found to contain 0.00500 gram of vitamin C. How many milligrams of vitamin C does this tablet contain?

SOLUTION

STEP 1   The given quantity is 0.00500 gram, the mass of vitamin C in the tablet. The unit of the desired quantity is milligrams.

$$0.00500 \text{ g} = ? \text{ mg}$$

STEP 2   Only one conversion factor will be needed to convert from grams to milligrams, one that relates grams to milligrams. Two forms of this factor exist.

$$\frac{1 \text{ mg}}{10^{-3} \text{ g}} \quad \text{and} \quad \frac{10^{-3} \text{ g}}{1 \text{ mg}}$$

The first factor is used because it allows for the cancellation of the gram units, leaving us with milligrams as the new units.

$$0.00500 \ \cancel{g} \times \left(\frac{1 \ \text{mg}}{10^{-3} \ \cancel{g}} \right) = ? \ \text{mg}$$

For cancellation, a unit must appear in both the numerator and the denominator. Because the given quantity (0.00500 g) has grams in the numerator, the conversion factor used must be the one with grams in the denominator.

If the other conversion factor had been used, we would have

$$0.00500 \ \text{g} \times \left(\frac{10^{-3} \ \text{g}}{1 \ \text{mg}} \right)$$

No unit cancellation is possible in this setup. Multiplication gives g^2/mg as the final units, which is certainly not what we want. In all cases, only one of the two conversion factors of a reciprocal pair will correctly fit into a dimensional-analysis setup.

STEP 3   Step 2 takes care of the units. All that is left is to combine numerical terms to get a final answer; we still have to do the arithmetic. Collecting the numerical terms gives

$$\left(0.00500 \times \frac{1}{10^{-3}} \right) \text{mg} = 5.00 \ \text{mg}$$

Number from   Numbers from
first factor   second factor

EXAMPLE 6.2

One drop of water has a volume of 5×10^{-8} kiloliter. What is the volume of this drop in milliliters?

SOLUTION

STEP 1   From the problem statement, we identify 5×10^{-8} kL of water as the given quantity and milliliters of water as the units of the desired quantity.

$$5 \times 10^{-8} \ \text{kL} = ? \ \text{mL}$$

STEP 2   When dealing with metric–metric unit changes in which both the original and the desired units carry prefixes, it is recommended that you always channel unit changes through the basic unit (the unprefixed unit). If this is done, you do not need to deal with any conversion factors other than those resulting from prefix definitions. Following this recommendation, the unit "pathway" for this problem is

$$kL \longrightarrow L \longrightarrow mL$$

In the setup for this problem we will need two conversion factors, one for the kiloliter-to-liter change and one for the liter-to-milliliter change.

$$5 \times 10^{-8} \, \text{kL} \times \left(\frac{10^3 \, \text{L}}{1 \, \text{kL}} \right) \times \left(\frac{1 \, \text{mL}}{10^{-3} \, \text{L}} \right) = ? \, \text{mL}$$

This conversion factor converts kL to L

This conversion factor converts L to mL

Note how all the units cancel except for milliliters.

STEP 3 Carrying out the indicated numerical calculation gives

Number from first factor

$$\left(5 \times 10^{-8} \times \frac{10^3}{1} \times \frac{1}{10^{-3}} \right) \text{mL} = 5 \times 10^{-2} \, \text{mL}$$

$$= 0.05 \, \text{mL}$$

Numbers from second factor

Numbers from third factor

If you are mystified as to how the preceding answer was obtained from the given numbers, you should review the material in Appendix B—material that covers the topic of exponential notation and mathematical operations.

Our next two sample problems involve the use of both the English and metric systems of units. As mentioned previously, conversion factors between these two systems do not arise from definitions, but rather are determined experimentally. Some of these experimentally determine factors were given in Table 6.2.

EXAMPLE 6.3

The diameter of an aluminum atom is 2.86×10^{-8} cm. What is this diameter in inches?

SOLUTION

STEP 1 The given quantity is 2.86×10^{-8} cm and the units of the desired quantity are inches.

$$2.86 \times 10^{-8} \, \text{cm} = ? \, \text{in}.$$

STEP 2   The given units are metric units and the desired units are English units. Going to the information in Table 6.2, we note that one of the three conversion factor sets in the length area involves the units centimeters and inches

$$\frac{1.00 \text{ in.}}{2.54 \text{ cm}} \quad \text{and} \quad \frac{2.54 \text{ cm}}{1.00 \text{ in.}}$$

The first of these conversion factors is the one we need to convert from centimeters to inches.

$$(2.86 \times 10^{-8} \text{ cm}) \times \left(\frac{1.00 \text{ in.}}{2.54 \text{ cm}}\right) = ? \text{ in.}$$

STEP 3   Performing the indicated arithmetic gives

$$\left(2.86 \times 10^{-8} \times \frac{1.00}{2.54}\right) \text{ in.} = 1.13 \times 10^{-8} \text{ in.}$$

EXAMPLE 6.4

In Europe gasoline is sold by the liter rather than the gallon. How many liters of gasoline are needed to fill a 20.0-gallon gas tank?

SOLUTION

STEP 1   The given quantity is 20.0 gallons of gasoline and the units of the desired quantity are liters of gasoline.

$$20.0 \text{ gal} = ? \text{ L}$$

STEP 2   None of the volume conversion factors given in Table 6–2 directly effect the conversion we need. Consequently, we will need to use a two-step pathway. First, we convert gallons to quarts. Then one of the conversion factors of Table 6–2 (quarts to liters) can be used.

$$\text{gal} \longrightarrow \text{qt} \longrightarrow \text{L}$$

The correct conversion factor setup is

$$20.0 \text{ gal} \times \left(\frac{4 \text{ qt}}{1 \text{ gal}}\right) \times \left(\frac{1.00 \text{ L}}{1.06 \text{ qt}}\right) = ? \text{ L}$$

All of the units except for liters cancel, which is what is needed. The information for the second conversion factor was obtained from your "everyday knowledge."

This setup illustrates the fact that sometimes the given units must be changed to intermediate units before common conversion factors (see Table 6.2) are applicable.

STEP 3   Collecting the numerical factors and performing the indicated math gives

$$\left(\frac{20.0 \times 4 \times 1.00}{1 \times 1.06}\right) L = 75.5 \text{ L}$$

6.4   Density and Specific Gravity

Density is the ratio of the mass of an object to the volume occupied by that object, that is

$$\text{Density} = \frac{\text{Mass}}{\text{Volume}}$$

People often speak of a substance as being heavier or lighter than another substance. What they actually mean is that the two substances have different densities; a specific volume of one substance is heavier or lighter than the same volume of the second substance.

A correct density expression includes a number, a mass unit, and a volume unit. Although any mass and volume units can be used, densities are usually expressed in grams per cubic centimeter (g/cm^3) for solids, grams per milliliter (g/mL) for liquids, and grams per liter (g/L) for gases. The use of these units avoids the problem of having density values that are extremely small or large. Table 6.3 gives density values for a number of substances. Note that temperature must be specified with density values because substances expand and

TABLE 6.3

Densities of Selected Substances

Solids (25°C)		Liquids (25°C)		Gases (25°C and 1 atmosphere pressure)	
Gold	19.3 g/cm^3	Mercury	13.55 g/mL	Chlorine	3.17 g/L
Lead	11.3 g/cm^3	Milk	1.028–1.035 g/mL	Carbon dioxide	1.96 g/L
Copper	8.93 g/cm^3	Blood plasma	1.027 g/mL	Oxygen	1.42 g/L
Aluminum	2.70 g/cm^3	Urine	1.003–1.030 g/mL	Air (dry)	1.29 g/L
Table salt	2.16 g/cm^3	Water	0.997 g/mL	Nitrogen	1.25 g/L
Bone	1.7–2.0 g/cm^3	Olive oil	0.92 g/mL	Methane	0.66 g/L
Table sugar	1.59 g/cm^3	Ethyl alcohol	0.79 g/mL	Hydrogen	0.08 g/L
Wood, pine	0.30–0.50 g/cm^3	Gasoline	0.56 g/mL		

contract with changes in temperature. Similarly, the pressure of gases is given with their density values.

In a mathematical sense, density can be thought of as a conversion factor that relates the volume of a substance to its mass. This use of density enables the volume of a substance to be calculated if its mass is known. Conversely, the mass can be calculated if the volume is known.

EXAMPLE 6.5

Blood plasma has a density of 1.027 g/mL at 25°C. What mass, in grams, would 125 mL of plasma occupy, at 25°C.

SOLUTION

STEP 1   The given quantity is 125 mL of blood plasma. The units of the desired quantity are grams. Thus, our starting point is

$$125 \text{ mL} = ? \text{ g}$$

STEP 2   The conversion from milliliters to grams can be accomplished in one step because density, used as a conversion factor, directly relates milliliters to grams.

$$125 \text{ mL} \times \left(\frac{1.027 \text{ g}}{1 \text{ mL}} \right) = ? \text{ g}$$

STEP 3   Doing the necessary arithmetic gives us our answer.

$$\left(\frac{125 \times 1.027}{1} \right) \text{g} = 128 \text{ g}$$

Specific Gravity

Specific gravity is a quantity that is closely related to density. The **specific gravity** of a solid or liquid is the ratio of the density of that substance to the density of water at 4°C.

$$\text{Specific gravity of solid or liquid} = \frac{\text{Density of substance}}{\text{Density of water at } 4°C}$$

At 4°C the density of water is 1.00 g/mL. For gases, specific gravity involves a density comparison with air rather than water. At 25°C, the density of dry air is 1.29 g/L.

When you are calculating the specific gravity of a substance, both densities must be expressed in the same units. Specific gravity is a *unitless* quantity because the identical sets of density units always cancel.

EXAMPLE 6.6

The density of a sample of copper metal is 8.93 g/cm^3. What is the specific gravity of this copper metal?

SOLUTION

The specific gravity of a sample in the solid state is equal to the density of the sample divided by the density of water at 4°C.

$$\text{Specific gravity of copper sample} = \frac{8.93 \text{ g/cm}^3}{1.00 \text{ g/cm}^3} = 8.93$$

Note how all the units cancel to make specific gravity a unitless quantity. The copper is 8.93 times as dense as water.

Although the determination of the specific gravity of a solid or gas usually requires both a mass and a volume measurement, it is possible to measure the specific gravity of a liquid directly by using a hydrometer. A hydrometer is a glass float that has a weighted bottom and calibrations on its stem (see Figure 6.6). The higher the density of the liquid, the higher the hydrometer tube floats.

Specific gravity has many uses in many occupations. The attendant at a service station can tell the extent to which a car battery (Section 10.12) is charged by measuring the specific gravity of the battery acid (sulfuric acid). The

STATE OF CHARGE IN CAR BATTERIES

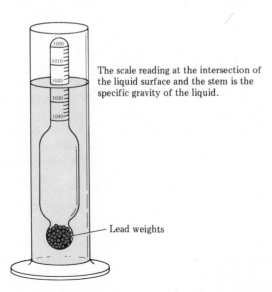

The scale reading at the intersection of the liquid surface and the stem is the specific gravity of the liquid.

Lead weights

FIGURE 6.6

Specific gravity measurement using a hydrometer.

strength of the antifreeze (Section 17.3) in a car radiator may also be checked with a hydrometer. In clinical and hospital laboratories, the specific gravity of urine samples is often determined; such information is helpful in diagnosing certain diseases.

6.5 Temperature Scales

The instrument most commonly used to measure temperature is the mercury thermometer, which consists of a glass bulb containing mercury sealed into a slender glass capillary tube. A small change in the volume of the mercury in the glass bulb caused by expansion or contraction of the liquid, leads to a large change in the height of the mercury in the capillary tube. Graduations on the capillary tube indicate the extent of mercury expansion in terms of defined units called *degrees*.

Three different degree scales (temperature scales) are in common use— Celsius, Kelvin, and Fahrenheit. Both the Celsius and Kelvin scales are part of the metric measurement system and the Fahrenheit scale belongs to the English measurement system. Different sized degrees and different reference points are what produce the various temperature scales.

The *Celsius scale* is the scale most commonly encountered in scientific work. The normal boiling and freezing points of water serve as reference points on this scale, with the former having a value of 100°C and the later, 0°C. Thus, there are 100 degree intervals between the two reference points.

The *Kelvin scale* is a close relative of the Celsius scale. They have the same size of unit, and the number of units between the freezing and boiling points of water is the same. The two scales differ only in the numerical unit values assigned to the reference points. On the Kelvin scale, the boiling point of water is 373 K and its freezing point is 273 K. The choice of these reference points, a shift upward by 273 units from the Celsius scale, makes all temperature readings on the Kelvin scale positive values. A temperature of zero on the Kelvin scale corresponds to the lowest temperature known to scientists. In calculations involving the properties of gases (Chapter 8), temperature must always be specified on the Kelvin scale. A temperature of zero on the Kelvin scale is often called *absolute zero* and the scale itself is referred to as the absolute scale. Kelvin scale units are called *kelvins* rather than degrees. The symbol for a kelvin is K rather than °K.

The *Fahrenheit scale* has a smaller degree size than the other two temperature scales. On this scale, there are 180 degrees between the freezing and boiling points of water. Thus, the Celsius (and Kelvin) degree size is 1.8 times the Fahrenheit degree. Reference points on the Fahrenheit scale are 32°F for the freezing point of water and 212°F for its boiling point. Figure 6.7 compares the three temperature scales.

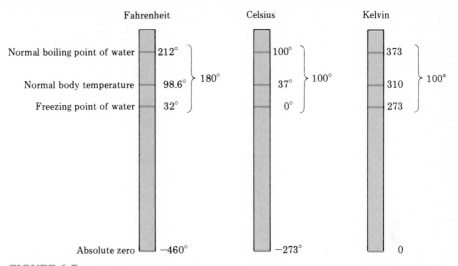

FIGURE 6.7

The relationships among the Fahrenheit, Celsius, and Kelvin temperature scales.

Because the size of the degree is the same, the relationship between the Kelvin and Celsius scales is very simple. No conversion factors are needed; all that is required is an adjustment for the different numerical scale values. The adjustment factor is 273, the number of degrees by which the two scales are offset from each other.

$$K = °C + 273$$

$$°C = K - 273$$

The relationship between the Fahrenheit and Celsius scales can also be stated in an equation format.

$$°F = \tfrac{9}{5}(°C) + 32$$

$$°C = \tfrac{5}{9}(°F - 32)$$

Example 6.7 illustrates the use of these equations.

EXAMPLE 6.7

Normal human body temperature on the Fahrenheit scale is 98.6°. What temperature is this equivalent to on (a) the Celsius scale and (b) the Kelvin scale?

SOLUTION

(a) Plugging the value 98.6 for the Fahrenheit temperature into the equation

$$°C = \tfrac{5}{9}(°F - 32)$$

and then solving for °C gives

$$°C = \tfrac{5}{9}(98.6 - 32)$$

$$= \tfrac{5}{9}(66.6)$$

$$= 37.0$$

(b) Using the answer from part (a) and the equation

$$K = °C + 273$$

we get, by substitution

$$K = 37.0 + 273$$

$$= 310$$

6.6   Types and Forms of Energy

On some days you wake up feeling very energetic. On these days you usually accomplish a great deal. By the end of the day you are tired, you have "no energy left," and you do not feel like doing any more "work." The scientific definition for energy closely parallels the concept just presented. **Energy** *is the capacity to do work.*

Energy can exist in any one of several forms. Common forms include radiant (light) energy, chemical energy, thermal (heat) energy, electrical energy, and mechanical energy. These forms of energy are interconvertible. The heating of a home is a process that illustrates energy interconversion. As the result of burning natural gas or some other fuel, chemical energy is converted into heat energy. In large conventional power plants that are used to produce electricity, the heat energy obtained from burning coal is used to change water into steam, which can then turn a turbine (mechanical energy) to produce electricity (electrical energy).

During energy interconversions, the law of conservation of energy is always obeyed. The **law of conservation of energy** *states that in any chemical or physical change, energy can be converted from one form to another, but it is neither created nor destroyed.* Studies of energy changes in numerous systems have shown that no system acquires energy except at the expense of energy possessed by another system.

Almost all of our energy on earth originates from the sun in the form of radiant or light energy. Green plants convert this radiant energy into chemical energy by means of a process called photosynthesis. The chemical energy is stored within the living plant. **Chemical energy** *is energy stored in a substance that can be* *released during a chemical change.* Energy for the human body is obtained, either directly or indirectly, from plants when they are consumed as food.

Chemical change can be used to produce other forms of energy. Chemical changes in an automobile battery produce the electrical energy needed to start a car. Chemical changes that occur in a magnesium flash bulb generate the light energy needed for photographic purposes. The burning of fuels releases both heat and light energy. The energy that "runs" our life processes—for example, breathing, muscle contraction, and blood circulation—is produced by chemical changes occurring within the cells of the body. The energy required for or generated by chemical changes is an important point of focus in many discussions in later chapters of this text.

In addition to the various *forms* of energy, there are two *types* of energy: potential energy and kinetic energy. The basis for determining energy types depends on whether the energy is available but not being used or is actually in use. **Potential energy** *is stored energy that results from an object's position, condition, and/or composition.* Water backed up behind a dam represents potential energy because of its position. When the water is released, it can be used to produce electrical energy at a hydroelectric plant. A compressed spring can spontaneously expand and do work as the result of potential energy associated with condition. Chemical energy, such as that stored in gasoline, is potential energy arising from composition. This stored energy is released when the gasoline is burned.

Kinetic energy *is energy that matter possesses because of its motion.* An object that is in motion has the capacity to do work. If it collides with another object, it will do work on that object. A hammer held in the air above an object possesses potential energy of position. As the hammer moves downward toward an object, this potential energy becomes kinetic energy that can be used to drive a nail into a board. When water behind a dam is released and allowed to flow, its potential energy of position becomes kinetic energy. During the operation of a hydroelectric plant, some of this kinetic energy becomes mechanical and electrical energy.

Figure 6.8 summarizes the forms and types of energy discussed in the preceding paragraphs.

The concepts of potential and kinetic energy play a part in many discussions in later chapters. For example, in Chapter 8, we explain the differences between

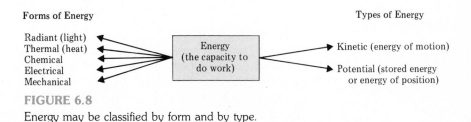

FIGURE 6.8
Energy may be classified by form and by type.

the solid, liquid, and gaseous states of matter in terms of relative amounts of potential and kinetic energy. The pressure that a gas exerts (Section 8.7) is related to kinetic energy.

6.7   Heat Energy and Specific Heat

The form of energy that is most often required for or released by the chemical and physical changes considered in this text is heat energy. For this reason, we now consider further details about this form of energy.

The Calorie

The most commonly used unit of measurement for heat energy is the calorie, which is abbreviated cal. A **calorie** *is the amount of heat energy needed to raise the temperature of 1 gram of water by 1 degree Celsius.* For large amounts of heat energy the measurement is usually expressed in kilocalories.

$$1 \text{ kilocalorie} = 1000 \text{ calories}$$

The SI unit (Section 6.2) of energy is the *joule*. The relationship between the joule and the calorie is

$$1 \text{ calorie} = 4.184 \text{ joules}$$

Heat energy values in calories can be converted to joules by multiplying the calorie value by 4.184.

In discussions involving the energy content of food (nutrition), the term *Calories* is also used. The dietetic Calorie (spelled with a capital C) is actually 1 kilocalorie (1000 calories). The statement that an oatmeal raisin cookie contains 60 Calories means that 60 kilocalories (60,000 calories) of energy are released when the cookie is metabolized (undergoes chemical change) within the body. Dieters who are aware that a banana split supplies 1500 Cal of energy might be less apt to succumb to temptation if they noted that the banana split contains 1,500,000 calories (with a lowercase c).

$$1500 \text{ Calories} = 1500 \text{ kilocalories} = 1,500,000 \text{ calories}$$

An active adult can easily "burn up" 2800 Calories (kcal) of energy in a day. Calories that are consumed in excess of amounts needed are used to maintain body heat or are deposited as fat. Approximately 3500 excess Calories produce 1 pound of fat.

Different types of foods have different caloric values. Both carbohydrates and proteins supply approximately 4.1 Calories per gram, whereas fats have a caloric

TABLE 6.4
Average Nutritional Values for Selected Menu Items at Fast-Food Restaurants

Food	*Calories*	*Protein (grams)*	*Fat (grams)*	*Carbohydrates (grams)*
Taco	193	15	8	14
French fries	243	3	12	29
Onion rings	280	3	16	29
Hamburger	314	16	13	31
Sliced roast beef	350	25	13	31
Cheeseburger	368	19	18	31
Milk shake	380	10	10	60
2 pieces fried chicken	401	28	26	11
Beef burrito	470	30	21	37
Fish sandwich	494	20	24	46
1/4-lb hamburger	504	26	26	38
Chicken sandwich	546	26	27	46
Double hamburger	577	34	33	32
One-half cheese pizza (13 in., thin crust)	705	38	22	84

FAST-FOOD
NUTRITIONAL VALUES

content of more than double that value (9.3 Calories per gram). Table 6.4 gives the caloric values, as well as the carbohydrate, protein, and fat content, for the most popular items available at fast-food restaurants. Contrary to common belief, fast foods have good nutritional value.

EXAMPLE 6.8

A student, after taking a "bear" of a chemistry exam, goes to a fast-food restaurant and consumes a taco, French fries, onion rings, a chicken sandwich, and a milk shake. Based on the average nutritional values given in Table 6.4, calculate the student's caloric intake in
(a) Calories   (b) kilocalories   (c) calories

SOLUTION

(a) The energy values in Table 6.4 are given in terms of dietetic Calories. Adding the caloric values together for the various foods consumed gives

$$(193 + 243 + 280 + 546 + 380) \text{ Calories} = 1642 \text{ Calories}$$

(b) A dietetic Calorie is equal to 1 kilocalorie. Thus,

$$1642 \text{ Cal} \times \left(\frac{1 \text{ kcal}}{1 \text{ Cal}} \right) = 1642 \text{ kcal}$$

(c) A dietetic Calorie is equal to 1000 calories. Thus,

$$1642 \; \cancel{\text{Cal}} \times \left(\frac{1000 \text{ cal}}{1 \; \cancel{\text{Cal}}} \right) = 1,642,000 \text{ cal}$$

Specific Heat

The **specific heat** *of a substance is the quantity of heat energy, in calories, necessary to raise the temperature of 1 gram of the substance by 1 degree Celsius.* Specific heats for a number of substances in various states are given in Table 6.5.

The higher the specific heat of a substance, the less its temperature will change when it absorbs a given amount of heat. For liquids, water has a relatively high specific heat; it is thus a very effective coolant. The moderate climates of geographical areas where large amounts of water are present—for example, the Hawaiian Islands—are related to water's ability to absorb large amounts of heat without undergoing drastic temperature changes. Desert areas, areas that lack water, are the places where the extremes of high temperature are encountered on the earth. The temperature of a living organism remains relatively constant because of the large amounts of water present in it.

TEMPERATURE
EXTREMES IN DESERT
AREAS

Specific heat is an important quantity because it can be used to calculate the number of calories required to heat a known mass of a substance from one temperature to another. It can also be used to calculate how much the temperature of a substance increases when it absorbs a known number of calories of heat. The equation used for such calculations is

Heat absorbed = Specific heat × Mass × Temperature change

If any three of the four quantities in this equation are known the fourth quantity can be calculated. If the units for specific heat are cal/g°C, the units for mass are

TABLE 6.5
Specific Heats of Selected Common
Substances

Substance	Specific Heat (cal/g °C)
Water, liquid	1.00
Ethyl alcohol	0.58
Olive oil	0.47
Wood	0.42
Aluminum	0.21
Glass	0.12
Silver	0.057
Gold	0.031

grams, and the units for temperature change are degrees Celsius, then the heat absorbed will have the units of calories.

EXAMPLE 6.9

Calculate the amount of heat in calories needed to increase the temperature of one cup of water (237 grams) from 81°C to 93°C.

SOLUTION

We substitute known quantities into the equation

Heat absorbed = Specific heat × Mass × Temperature change

From Table 6.5 we determine that the specific heat of liquid water is 1.00 cal/g °C. The mass of the water is given as 237 grams. The temperature change in going from 81°C to 93°C is 12 degrees Celsius. Substituting these values into the preceding equation gives

$$\text{Heat absorbed} = \left(1.00 \, \frac{\text{cal}}{g \, °C} \times 237 \, g \times 12 °C \right)$$

$$= 2844 \text{ calories}$$

EXAMPLE 6.10

A small serving (½ cup) of zucchini has a caloric value of approximately 12,000 calories (12 Calories). What would be the temperature change in a cup of water (237 grams) at 20°C if this amount of energy was added to it?

SOLUTION

The equation

Heat absorbed = Specific heat × Mass × Temperature change

is rearranged to isolate temperature change on the left side of the equation.

$$\text{Change in temperature (°C)} = \frac{\text{Heat absorbed (cal)}}{\text{Mass(g)} \times \text{Specific heat(cal/g °C)}}$$

Substituting the known values into the equation gives

$$\text{Change in temperature (°C)} = \frac{12,000 \, \text{cal}}{237 \, g \times 1.00 \, \text{cal/g} \, °C} = 51 °C$$

The temperature of the water would increase from 20°C to 71°C.

Practice Questions and Problems

OBSERVATIONS

6-1 Classify each of the following observations about a patient in the hospital as qualitative or quantitative.
a. Blood pressure is high.
b. Temperature is 100.2°F.
c. Pulse is normal.
d. Urine specific gravity is 1.020.
e. Face is very pale.
f. Blood sugar level is 150 mg/dL.

METRIC SYSTEM UNITS

6-2 What is the main advantage of the metric system of units over the English system of units?

6-3 Write the metric system prefixes associated with each of the following mathematical meanings.
a. 10^{-6} b. 10^{-3} c. 1/100
d. 0.001   e. 1000   f. 10^{-2}

6-4 Write out the names of the metric system units that have the following abbreviations.
a. cg   b. kL   c. mm
d. Mm   e. μL   f. ng

6-5 Arrange each of the following in sequence, from smallest to largest.
a. Milligram, centigram, decigram
b. Microliter, nanoliter, decaliter
c. Kilometer, millimeter, centimeter
d. Microliter, megaliter, milliliter
e. Centigram, kilogram, nanogram
f. Gigagram, hectogram, megagram

UNIT CONVERSIONS WITHIN THE METRIC SYSTEM

6-6 Give both forms of the conversion factor that you would use to relate the following sets of metric units to each other.
a. Gram and kilogram
b. Liter and milliliter
c. Meter and centimeter
d. Meter and micrometer
e. Gram and decagram
f. Liter and nanoliter

6-7 A nickel coin has a mass of 4832 mg. Express this mass in each of the following units.
a. Grams   b. Kilograms
c. Micrograms   d. Centigrams
e. Gigagrams   f. Nanograms

6-8 Convert each of the following measurements to kilometers.
a. 0.003 m   b. 1.6×10^3 cm
c. 3×10^8 mm   d. 24 nm
e. 0.67 m   f. 13,000 Mm

6-9 The diameter of a human hair is about 0.040 mm. What is this diameter in nanometers?

6-10 The human stomach produces approximately 2500 mL of gastric juice per day. What is the volume, in liters, of gastric juice produced?

6-11 Diamonds are weighed in terms of carats. One carat is equal to 200 mg. How much does a 5.0-carat diamond weigh in grams?

6-12 A drop of blood has a volume of 0.05 mL How many drops of blood are there in an adult body that has 5.3 L of blood?

METRIC–ENGLISH AND ENGLISH—METRIC UNIT CONVERSIONS

6-13 For each of the pairs of units listed, indicate which quantity is larger.
a. 1 cm, 1 in.   b. 1 km, 1 yd
c. 1 km, 1 mile   d. 1 g, 1 lb
e. 1 kg, 1 lb   f. 1 L, 1 qt

6-14 A standard unsharpened pencil with eraser is 7.5 in. long. What is this length in centimeters?

6-15 The smallest bone in the human body, which is in the ear, has a mass of approximately 3.0 mg. What is the approximate mass of this bone in pounds.

6-16 A standard basketball has a volume of 7.47 L. What is the volume of the basketball in gallons?

6-17 A certain chemical process requires 55 gal of pure water per day. Express this daily water requirement in milliliters.

6-18 The speed limit on many United States highways

is 55 miles per hour. What is this speed limit in kilometers per hour?

DENSITY AND SPECIFIC GRAVITY

6-19 A sample of sand is found to have a mass of 51.3 g and a volume of 20.2 cm^3. What is its density in grams per cm^3?

6-20 Acetone, the solvent in many nail polish removers, has a density of 0.791 g/mL. What is the volume, in mL, of 12.0 g of acetone?

6-21 The density of homogenized milk is 1.03 g/mL. How much does 1 cup (236 mL) of homogenized milk weigh, in pounds?

6-22 An automobile gasoline tank holds 24.0 gal when full. How many pounds of gasoline will it hold, if the gasoline has a density of 0.560 g/mL?

6-23 A copper penny has a mass of 3.21 g. If eight pennies of this mass occupy 2.83 cm^3, what is the density of copper?

6-24 The density of silver is 10.40 g/cm^3. What is the specific gravity of silver?

6-25 What is the specific gravity of lithium, the least dense of all metals, if 5.34 g of lithium occupies a volume of 10.0 cm^3?

TEMPERATURE SCALE CONVERSIONS

6-26 On a hot summer day the temperature outside may reach 105°F. What is this temperature in degrees Celsius?

6-27 The element gold has a melting point of 1063°C and a boiling point of 2966°C. What are its melting point and freezing point in degrees Fahrenheit?

6-28 An oven for baking pizza operates at approximately 525°F. What is this temperature in degrees Kelvin?

6-29 The temperature difference between the boiling and freezing points for the metal aluminum is 1807 degrees on the Celsius scale. What is this temperature difference on the Kelvin scale?

6-30 A comfortable temperature for bathtub water is 95°F. What temperature is this in degrees Kelvin?

TYPES AND FORMS OF ENERGY

6-31 What is the scientific definition for energy?

6-32 State the law of conservation of energy.

6-33 List the predominant *forms* of energy produced when each of the following processes occur.
a. An electric light bulb is turned on.
b. A log is burned in a fireplace.
c. A green plant "grows."
d. A bicycle is pedaled.
e. A flashlight is turned on.
f. A photographer's flash bulb "goes off."

6-34 What is the difference between potential energy and kinetic energy?

6-35 Identify the principal *type* of energy (kinetic or potential) exhibited by each of the following.
a. A car parked on a hill.
b. A car traveling at 60 miles per hour.
c. Chemical energy.
d. Water behind a dam.
e. A falling rock.
f. A compressed metal spring.

HEAT ENERGY AND SPECIFIC HEAT

6-36 What is the relationship between a calorie and the following?
a. A kilocalorie   b. A Calorie

6-37 How many calories of heat energy are required to raise the temperature of 30.0 g of each of the following substances from 50°C to 60°C?
a. Liquid water   b. Olive oil
c. Aluminum   d. Gold

6-38 If it takes 18.6 cal of heat to raise the temperature of 12.0 g of a substance by 10.0°C, what is the specific heat of the substance?

6-39 What is the temperature change, in degrees Celsius, if 75 cal of heat energy is added to 40.0 g of the following?
a. Water   b. Aluminum   c. Silver

6-40 The body contains approximately 5.7 L of blood. Assuming that the density of blood is 1.06 g/mL and that the specific heats of blood and water are the same, how many calories of heat are required to raise the temperature of this amount of blood by 1.0°C?

7

Chemical Calculations:
Formula Weights, Moles, and Chemical Equations

CHAPTER HIGHLIGHTS

176   Practice Questions and Problems

7.1 Formula Weights

In this chapter, we consider "chemical arithmetic," the quantitative relationships between elements and compounds. Anyone dealing with chemical processes needs to understand the simpler aspects of this topic. All chemical processes, regardless of where they occur—in the human body, at a steel mill, on top of the kitchen stove, or in a clinical laboratory setting—are governed by the same "mathematical rules."

Our entry point into the realm of "chemical arithmetic" is a discussion of the quantity called *formula weight*, an entity that plays a role in almost all chemical calculations. Once the formula of a substance has been established, it is a simple matter to calculate its formula weight. The **formula weight** *of a substance is the sum of the atomic weights of the atoms in its formula*. Formula weights, like the atomic weights from which they are calculated, are relative masses based on the $^{12}_{6}C$ relative mass scale (see Section 3.7). Example 7.1 illustrates how formula weights are calculated.

EXAMPLE 7.1

Calculate the formula weight for each of the following substances.
(a) Ag_2S silver sulfide (silverware tarnish)
(b) $C_{10}H_{14}N_2$ nicotine (a common stimulant)
(c) $(NH_4)_2SO_4$ ammonium sulfate (a lawn fertilizer)

SOLUTION

Formula weights are obtained simply by adding the atomic weights of the constituent elements, counting each atomic weight as many times as the symbol for the element occurs in the formula.

(a) A formula unit of Ag_2S contains three atoms: two atoms of Ag and one atom of S. The formula weight, the collective mass of these three atoms, is calculated as follows:

$$2 \text{ atoms Ag} \times \left(\frac{107.9 \text{ amu}}{1 \text{ atom Ag}}\right) = 215.8 \text{ amu}$$

$$1 \text{ atom S} \times \left(\frac{32.1 \text{ amu}}{1 \text{ atom S}}\right) = \underline{\ 32.1 \text{ amu}}$$

$$\text{Formula weight} = 247.9 \text{ amu}$$

We derive the conversion factors in the calculation from the atomic weights listed on the inside back cover of the text. Our rules for the use of conversion factors are the same as those discussed in Section 6.3.

Conversion factors are not usually explicitly shown in a formula-weight calculation, as they are in our example; the calculation is simplified as follows.

$$
\begin{aligned}
\text{Ag:} \quad & 2 \times 107.9 \text{ amu} = 215.8 \text{ amu} \\
\text{S:} \quad & 1 \times 32.1 \text{ amu} = \underline{32.1 \text{ amu}} \\
& \text{Formula weight} = 247.9 \text{ amu}
\end{aligned}
$$

(b) Similarly, for nicotine ($C_{10}H_{14}N_4$) we calculate the formula weight as

$$
\begin{aligned}
\text{C:} \quad & 10 \times 12.0 \text{ amu} = 120.0 \text{ amu} \\
\text{H:} \quad & 14 \times 1.0 \text{ amu} = 14.0 \text{ amu} \\
\text{N:} \quad & 2 \times 14.0 \text{ amu} = \underline{28.0 \text{ amu}} \\
& \text{Formula weight} = 162.0 \text{ amu}
\end{aligned}
$$

(c) The formula for this compound contains parentheses. Improper interpretation of parentheses (see Section 2.8) is a common error made by students doing formula-weight calculations. In the formula $(NH_4)_2SO_4$, the subscript 2 outside the parentheses affects all of the symbols inside the parentheses. Thus, we have

$$
\begin{aligned}
\text{N:} \quad & 2 \times 14.0 \text{ amu} = 28.0 \text{ amu} \\
\text{H:} \quad & 8 \times 1.0 \text{ amu} = 8.0 \text{ amu} \\
\text{S:} \quad & 1 \times 32.1 \text{ amu} = 32.1 \text{ amu} \\
\text{O:} \quad & 4 \times 16.0 \text{ amu} = \underline{64.0 \text{ amu}} \\
& \text{Formula weight} = 132.1 \text{ amu}
\end{aligned}
$$

7.2   Percent Composition

The **percent composition** *of a compound specifies the percentage by mass of each element present in the compound.* For example, the percent composition of water is 88.81% oxygen and 11.19% hydrogen.

Percent compositions are frequently used to compare compound compositions. For example, the gold compounds AuI_3, $Au(NO_3)_3$, and $AuCN$ contain 34.1, 51.4, and 88.3% gold by mass, respectively. If you were given the choice of receiving a gift of 1 pound of one of these three gold compounds, which would you choose?. Percent composition is a very important factor in your decision.

We can calculate the percent composition for a compound from its chemical formula, Example 7.2 illustrates how this is done.

EXAMPLE 7.2

Cortisone, a drug used in the treatment of rheumatoid arthritis, has the formula $C_{24}H_{28}O_5$. Determine the percent composition by mass for this compound.

SOLUTION

First, we calculate the formula weight for the compound from the atomic weights of the constituent elements using the method shown in Example 7.1.

$$
\begin{aligned}
\text{C:} &\quad 24 \times 12.0 \text{ amu} = 288.0 \text{ amu} \\
\text{H:} &\quad 28 \times \ 1.0 \text{ amu} = \ \ 28.0 \text{ amu} \\
\text{O:} &\quad \ \ 5 \times 16.0 \text{ amu} = \ \ \underline{80.0 \text{ amu}} \\
&\quad \text{Formula weight} = 396.0 \text{ amu}
\end{aligned}
$$

We determine the mass percent of each element in the compound by dividing the mass contribution of each element, in amu, by the total mass (formula weight), in amu, and multiplying by 100.

$$
\text{Mass percent of element} = \frac{\text{Mass of element in one formula unit}}{\text{Formula weight}} \times 100
$$

Using this equation, the percent by mass of each element present is calculated as follows.

$$
\% \text{ C:} \quad \frac{288.0 \text{ amu}}{396.0 \text{ amu}} \times 100 = 72.7\%
$$

$$
\% \text{ H:} \quad \frac{28.0 \text{ amu}}{396.0 \text{ amu}} \times 100 = \ 7.1\%
$$

$$
\% \text{ O:} \quad \frac{80.0 \text{ amu}}{396.0 \text{ amu}} \times 100 = 20.2\%
$$

To check our work, we can add the percentages of all the parts; of course, they have to total 100. (On occasion, round-off errors can cause the total percent to be 99.9% or 100.1%. This is acceptable.)

$$
72.7\% + 7.1\% + 20.2\% = 100.0\%
$$

7.3 The Mole: The Chemist's Counting Unit

The quantity of material in a sample of a substance can be specified either in terms of units of *mass* or units of *amount*. Mass is specified in terms of units such as grams, kilograms, or pounds. The amount of a substance is specified by indicating the number of objects present—4, 19, or 362, for example.

HISTORICAL PROFILE 9

Lorenzo Romano Amedeo Carlo Avogadro   1776–1856

[Burndy Library, Norwalk CT.]

Amedeo Avogadro, born in Turin, Italy, followed in the footsteps of his father, obtained a law degree (at age 16), and for some years engaged in the practice of law. At age 24, he began to study privately mathematics and physics. Abandoning the law profession at age 30, he spent most of his later life as a professor of mathematical physics at the University of Turin.

Avogadro was the first scientist to distinguish between atoms and molecules. His most important paper, published in 1811, suggested that all gases (at a given temperature) contained the same number of particles per unit volume and that these particles need not be individual atoms but might be combinations of atoms. The thrust of this paper, known now as Avogadro's law, received little attention from his contemporaries. Turin, where Avogadro was professor, was outside the mainstream of scientific activity in 1811.

It was not until after his death that his genius was recognized. In 1858, two years after his death, Stanislao Cannizzaro (1826–1910), one of his countrymen, pointed out the full significance of Avogadro's law at an international gathering of scientists. The "time was right," and acceptance was rapid. One chemist, after reading Cannizzaro's paper, wrote "The scales seemed to fall from my eyes. Doubts disappeared and feeling of quiet certainty took their place." In 1911, in Turin, in commemoration of the 100th anniversary of the first publication of his law, a monument honoring Avogadro was unveiled by the king in one of the greatest posthumous tributes to a scientist in history.

Avogadro's name is attached to a number he never determined. During the early 1900s, chemical theory developed to the point that estimates could be made of the number of molecules in a given volume of gas. That equal volumes of all gases really contain the same number of molecules (Avogadro's law) was a provable reality. Jean Baptiste Perrin (1870–1942), in 1909, was the first to use the phrase "Avogadro's number" to denote the calculated number of molecules in a specific volume of gas. The honor is appropriate since Avogadro's law was the springboard for Perrin's work as well as that of many other scientists working in the same research area.

Both units of mass and units of amount are used on a daily basis by all of us. We work well with this dual system. Sometimes it does not matter which type of unit is used; at other times one system is preferred over the other. When buying potatoes at the grocery store we can decide on quantity in either mass units (10-lb bag or 20-lb bag) or amount units (7 potatoes or 13 potatoes). When buying eggs, amount units are used almost exclusively—12 eggs (1 dozen) or 24 eggs (2 dozen). We do not ordinarily buy 2 pounds of eggs. On the other hand, grapes are almost always purchased in weighed quantities. It is impractical to count the number of grapes in a bunch. Very few people go to the store with the idea that they will buy 189 grapes.

In chemistry, as in everyday life, both the mass and amount methods of specifying quantity are used. Again, the specific situation dictates the method used. In laboratory work, practicality dictates working with quantities of known mass. (Counting out a given number of atoms for a laboratory experiment is somewhat impractical, because we cannot see individual atoms.)

When performing chemical calculations after the laboratory work has been done, it is often useful and even necessary to think of the quantities of substances present in terms of amounts such as numbers of atoms or molecules instead of masses. The problem in doing this is that very large numbers are always encountered. Any macroscopic-sized sample of a chemical substance contains many trillions of atoms or molecules.

To cope with this "large number problem" chemists have found it convenient to use a special unit when counting atoms and molecules. The employment of such a unit should not surprise you because specialized counting units are used in many areas. The two most common counting units are *dozen* and *pair*. Other more specialized counting units exist; for example, at an office supply store paper is sold by the *ream* (500 sheets), pencils by the *gross* (144), and stencils by the *quire* (24) (see Figure 7.1a–d).

The chemist's counting unit is called a *mole*. What is unusual about the mole is its magnitude. A **mole** *is 6.02 × 10²³ objects*. The extremely large size of the mole unit is necessitated by the extremely small size of atoms and molecules (see Figure 7.1e).

To the chemist, *1 mole* always means 6.02×10^{23} objects just as *one dozen* always means 12 objects. Two moles of objects would be 2 times 6.02×10^{23} objects and 5 moles of objects would be 5 times 6.02×10^{23} objects.

The number of objects in a mole, 6.02×10^{23}, has a special name. **Avogadro's number** *is the name given to the numerical value 6.02×10^{23}*. This designation honors the Italian physicist Lorenzo Romano Amedeo Carlo Avogadro (1776–1856) (see Historical Profile 9), whose pioneering work on gases later proved to be valuable in determining the number of particles present in a given volume of a substance.

When solving problems dealing with the number of objects (atoms or

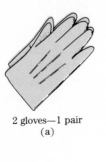

2 gloves—1 pair
(a)

12 rolls—1 dozen
(b)

144 pencils—1 gross
(c)

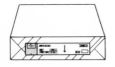

500 sheets of paper—1 ream
(d)

6.02×10^{23} iron atoms—1 mole
(e)

FIGURE 7.1

Examples of counting units used to denote quantities in terms of groups of various sizes.

molecules) present in a given number of moles of a substance, Avogadro's number becomes part of the conversion factor used to relate the number of objects present to the number of moles present.

From the definition

$$1 \text{ mole} = 6.02 \times 10^{23} \text{ objects}$$

two conversion factors can be derived.

$$\left(\frac{6.02 \times 10^{23} \text{ objects}}{1 \text{ mole}} \right) \quad \text{and} \quad \left(\frac{1 \text{ mole}}{6.02 \times 10^{23} \text{ objects}} \right)$$

Example 7.3 illustrates the use of these conversion factors in solving problems.

EXAMPLE 7.3

How many objects are there in each of the following quantities?
(a) 1.78 moles of sulfur atoms
(b) 0.345 mole of carbon dioxide molecules
(c) 0.75 mole of watermelon seeds

SOLUTION

We use dimensional analysis (see Section 6.3) to solve each of these problems. All of the problems are similar in that we are given a certain number of moles of substance and want to find the number of objects present in the given number of moles. We need Avogadro's number to solve each of these moles-to-particles problems.

(a) The objects of concern are atoms of sulfur. The given quantity is 1.78 moles of sulfur atoms and the desired quantity is the number of sulfur atoms.

$$1.78 \text{ moles sulfur atoms} = ? \text{ sulfur atoms}$$

The setup, using dimensional analysis, involves the use of a single conversion factor, one that relates moles and atoms.

$$1.78 \text{ moles S atoms} \times \left(\frac{6.02 \times 10^{23} \text{ S atoms}}{1 \text{ mole S atoms}}\right) = 1.07 \times 10^{24} \text{ S atoms}$$

(b) This time we are dealing with molecules instead of atoms. This switch does not change the way we work the problem. We need the same conversion factor.

The given quantity is 0.345 mole of carbon dioxide (CO_2) molecules and the desired quantity is the actual number of carbon dioxide molecules.

$$0.345 \text{ mole } CO_2 \text{ molecules} = ? \text{ } CO_2 \text{ molecules}$$

The setup is

$$0.340 \text{ mole } CO_2 \text{ molecules} \times \left(\frac{6.02 \times 10^{23} \text{ } CO_2 \text{ molecules}}{1 \text{ mole } CO_2 \text{ molecules}}\right)$$
$$= 2.08 \times 10^{23} \text{ } CO_2 \text{ molecules}$$

(c) Use of the mole as a counting unit is usually found only in a chemical context. Technically, however, any type of object can be counted in units of moles, even watermelon seeds. One mole denotes 6.02×10^{23} objects; it does not matter what the objects are. Just as we can talk about three dozen watermelon seeds, we can talk about 3 moles, or any other number of moles, of watermelon seeds.

The given quantity is 0.75 mole of watermelon seeds and the desired quantity is the number of watermelon seeds

$$0.75 \text{ mole watermelon seeds} = ? \text{ watermelon seeds}$$

The setup for the problem, using the same conversion factor as in previous parts, is

$$0.75 \text{ mole watermelon seeds} \times \left(\frac{6.02 \times 10^{23} \text{ watermelon seeds}}{1 \text{ mole watermelon seeds}} \right)$$

$$= 4.5 \times 10^{23} \text{ watermelon seeds}$$

This number of seeds would certainly be sufficient to plant an above-average-size patch of watermelons.

It is somewhat unfortunate that the name mole was selected as the name for the chemist's counting unit because of its similarity to the word molecule. Students often think of the word mole as being an abbreviation for molecule. That is not the case. The word mole comes from the Latin *moles*, which means "heap or pile." The word molecule is a diminutive form of moles meaning "the smallest piece" of that heap or pile. A mole is a macroscopic amount, a heap or pile of objects, that can easily be seen. A molecule is an invisible particle too small to be seen with the naked eye.

In Example 7.3 we calculated the number of objects present in molar-sized samples ranging from 0.345 to 1.78 moles. Our answers were numbers carrying the exponents 10^{23} or 10^{24}. Numbers with these exponents are inconceivably large. The magnitude of Avogadro's number itself is so large that it is almost incomprehensible. There is nothing in our experience to relate it to. (When chemists count, they really count in "big jumps.") Many attempts have been made to create word pictures of the vast size of Avogadro's number. Such pictures, however, really only hint at its magnitude because other large numbers must be used in the word pictures. Two such word pictures are as follows.

1. Suppose a fraternity decided they would throw a large pizza party. So as not to run out of food, 1 mole of pizza pies was ordered. Two thousand students showed up at the party. It would take these 2000 students, each "snarfing" one pizza every 3 minutes, 2×10^{15} years to finish the stack of pizzas. If everyone living on this planet attended the party (5 billion pizza eaters), it would still take 1×10^9 years to eat the pizzas, just 1 billion years of nonstop pizza eating. (You have to wonder who the caterer was who supplied all the pizzas. How many years ahead of time did the caterer have to start baking pizza to supply the fraternity's order?)
2. Avogadro's number (1 mole) of chemistry textbooks the approximate thickness of this one (3 cm) piled on top of each other would reach to the moon and back 1.3×10^{10} times (13 billion times). Outer space would no longer be empty.

7.4   The Mass of a Mole

How much does a mole weigh? Are you uncertain about the answer to that question? Let us, then, consider a similar (but more familar) question first: "How much does a dozen weigh? Your response is now immediate. "A dozen what?" you reply. The mass of a dozen identical objects obviously depends on the identity of the object. For example, the mass of a dozen elephants will be somewhat greater than the mass of a dozen peanuts. The mole, like the dozen, is a counting unit. Similarly, the mass of a mole, like the mass of a dozen, depends on the identity of the object. Thus, the mass of a mole, or *molar mass*, is not a set number; it varies and is different for each chemical substance. This is in direct contrast to the *molar number*, Avogadro's number, which is the same for all chemical substances.

The **molar mass** *of a substance is the mass, in grams, that is numerically equal to the substance's formula weight.* For example, the formula weight (atomic weight) of the element sodium is 23.0 amu; therefore, 1 mole of sodium would weigh 23.0 g. In Example 7.1, we calculated that the formula weight of nicotine is 162.0 amu; therefore, 1 mole of nicotine molecules would have a mass of 162.0 g. We can obtain the actual mass, in grams, of 1 mole of any substance by computing its formula weight and writing "grams" after it. Thus, when we add atomic weights to get the formula weight (in amu) of a compound, we are simultaneously finding the mass of 1 mole of compound (in grams).

Figure 7.2 shows molar quantities of a number of common substances. Note how the numerical value of the mass of a mole varies from substance to substance.

It is not a coincidence that the molar mass of a substance and its formula weight or atomic weight match numerically. Avogadro's number has the value that it has in order to cause this relationship to exist. Avogadro's number represents the experimentally determined number of atoms, molecules, or formula units contained in a sample of a pure substance with a mass in grams numerically equal to the atomic weight or formula weight of the pure substance.

The numerical match between molar mass and atomic or formula weights makes calculating the mass of any given number of moles of a substance a very simple procedure. When you solve problems of this type, the numerical value of the molar mass becomes part of the conversion factor used to convert from moles to grams.

For example, for the compound CO_2, which has a formula weight of 44.0 amu, we can write the equality

$$44.0 \text{ g } CO_2 = 1 \text{ mole } CO_2$$

FIGURE 7.2
One mole of a substance is that amount of substance with a mass, in grams,
numerically equal to its atomic weight or formula weight.

From this statement (equality), two conversion factors can be written.

$$\frac{44.0 \text{ g } CO_2}{1 \text{ mole } CO_2} \quad \text{and} \quad \frac{1 \text{ mole } CO_2}{44.0 \text{ g } CO_2}$$

Example 7.4 illustrates the use of gram–mole conversion factors like these in problems.

EXAMPLE 7.4

What is the mass, in grams, of each of the following molar quantities of chemical substances?
(a) 4.63 moles of water (H_2O)
(b) 0.72 mole of table sugar ($C_{12}H_{22}O_{11}$)
(c) 2.07 moles of copper (Cu)

SOLUTION

We use dimensional analysis to solve each of these problems. The relationship between molar mass and atomic or formula weight serves as a conversion factor in the setup of each problem.

(a) The given quantity is 4.63 moles of H_2O, and the desired quantity is grams of the same substance.

$$4.63 \text{ moles } H_2O = ? \text{ grams } H_2O$$

The formula weight of H_2O is calculated to be 18.0 amu

$$
\begin{array}{llll}
H: & 2 \times & 1.0 = & 2.0 \text{ amu} \\
O: & 1 \times & 16.0 = & \underline{16.0 \text{ amu}} \\
& & & 18.0 \text{ amu}
\end{array}
$$

Knowing this value, we can write the equality

$$18.0 \text{ g } H_2O = 1 \text{ mole } H_2O$$

The dimensional analysis setup for the problem, with the gram–mole equation serving as the basis for the conversion factor, is

$$4.63 \text{ moles } H_2O \times \left(\frac{18.0 \text{ g } H_2O}{1 \text{ mole } H_2O} \right) = 83.3 \text{ g } H_2O$$

(b) The given quantity is 0.72 mole of $C_{12}H_{22}O_{11}$, and the desired quantity is grams of $C_{12}H_{22}O_{11}$.

$$0.72 \text{ mole } C_{12}H_{22}O_{11} = ? \text{ g } C_{12}H_{22}O_{11}$$

The calculated formula weight of $C_{12}H_{22}O_{11}$ is 342 amu. Thus,

$$342 \text{ g } C_{12}H_{22}O_{11} = 1 \text{ mole } C_{12}H_{22}O_{11}$$

With this relationship as a conversion factor, the setup for the problem becomes

$$0.72 \text{ mole } C_{12}H_{22}O_{11} \times \left(\frac{342 \text{ g } C_{12}H_{22}O_{11}}{1 \text{ mole } C_{12}H_{22}O_{11}} \right) = 246 \text{ g } C_{12}H_{22}O_{11}$$

(c) The given quantity is 2.07 moles of copper and the desired quantity is grams of copper.

$$2.07 \text{ moles } Cu = ? \text{ g } Cu$$

Here, we are dealing with an element rather than a compound. Thus,

63.5 amu, the atomic weight of copper, is used in the mole–gram equality statement.

$$63.5 \text{ g Cu} = 1 \text{ mole Cu}$$

With this relationship as a conversion factor, the setup becomes

$$2.07 \text{ moles Cu} \times \left(\frac{63.5 \text{ g Cu}}{1 \text{ mole Cu}}\right) = 131 \text{ g Cu}$$

7.5   The Mole and Chemical Formulas

A chemical formula has two meanings or interpretations: a microscopic level interpretation and a macroscopic level interpretation. The first interpretation was discussed in Section 2.8. At a microscopic level, a chemical formula indicates the number of atoms of each element present in one molecule or formula unit of a substance. *The numerical subscripts in the formula give the number of atoms of the various elements present in one formula unit of the substance.* The formula C_2H_6, interpreted at the microscopic level, conveys the information that two atoms of carbon and six atoms of hydrogen are present in one molecule of C_2H_6.

Now that the mole concept has been introduced, a macroscopic interpretation of chemical formulas is possible. At the macroscopic level, a chemical formula indicates the number of moles of atoms of each element present in one mole of a substance. *The numerical subscripts in the formula give the number of moles of atoms of the various elements present in 1 mole of the substance.* The designation macroscopic is given to this molar interpretation because moles are "laboratory-sized" quantities of atoms. The formula C_2H_6, interpreted at the macroscopic level, conveys the information that 2 moles of carbon atoms and 6 moles of hydrogen atoms are present in 1 mole of C_2H_6. Thus, the subscripts in a formula always carry a dual meaning: "atoms" at the microscopic level and "moles of atoms" at the macroscopic level.

The validity of the molar interpretation for subscripts in a formula derives from the following line of reasoning. In x molecules of C_2H_6, where x is any number, there are $2x$ atoms of carbon and $6x$ atoms of hydrogen. Regardless of the value of x, there are always twice as many carbon atoms as molecules and six times as many hydrogen atoms as molecules; that is

$$\text{Number of } C_2H_6 \text{ molecules} = x$$

$$\text{Number of C atoms} = 2x$$

$$\text{Number of H atoms} = 6x$$

Now let x equal 6.02×10^{23}, the value of Avogadro's number. With this x value, the following statements are true.

$$\text{Number of } C_2H_6 \text{ molecules} = 6.02 \times 10^{23}$$

$$\text{Number of C atoms} = 2 \times (6.02 \times 10^{23})$$

$$\text{Number of H atoms} = 6 \times (6.02 \times 10^{23})$$

Because 6.02×10^{23} is equal to 1 mole, the preceding statements can be changed, by substitution, to read

$$\text{Number of } C_2H_6 \text{ molecules} = 1 \text{ mole}$$

$$\text{Number of C atoms} = 2 \text{ moles}$$

$$\text{Number of H atoms} = 6 \text{ moles}$$

Thus, the mole ratio is the same as the subscript ratio: 2 to 6.

When it is necessary to know the number of moles of a particular element *within* a compound, the subscript of that element in the chemical formula becomes part of the conversion factor used to convert from moles of compound to moles of element *within* the compound. Using C_2H_6 as our chemical formula, we can write the following as conversion factors.

For C: $\left(\dfrac{2 \text{ moles C atoms}}{1 \text{ mole } C_2H_6 \text{ molecules}} \right)$ and $\left(\dfrac{1 \text{ mole } C_2H_6 \text{ molecules}}{2 \text{ moles C atoms}} \right)$

For H: $\left(\dfrac{6 \text{ moles H atoms}}{1 \text{ mole } C_2H_6 \text{ molecules}} \right)$ and $\left(\dfrac{1 \text{ mole } C_2H_6 \text{ molecules}}{6 \text{ moles H atoms}} \right)$

Example 7.5 illustrates the use of this type of conversion factor in a problem-solving context.

EXAMPLE 7.5

How many moles of each type of atom are present in each of the following molar quantities?
(a) 0.753 mole of CO_2 molecules
(b) 1.31 moles of P_4O_{10} molecules

SOLUTION

(a) One mole of CO_2 contains 1 mole of carbon atoms and 2 moles of oxygen atoms. We obtain the following conversion factors from this statement.

$$\frac{1 \text{ mole C atoms}}{1 \text{ mole } CO_2 \text{ molecules}} \quad \text{and} \quad \frac{2 \text{ moles O atoms}}{1 \text{ mole } CO_2 \text{ molecules}}$$

Using the first conversion factor, the moles of carbon atoms present are calculated as follows:

$$0.753 \text{ mole } CO_2 \text{ molecules} \times \left(\frac{1 \text{ mole C atoms}}{1 \text{ mole } CO_2 \text{ molecules}} \right)$$

$$= 0.753 \text{ mole C atoms}$$

Similarly, using the second conversion factor, the moles of oxygen atoms present are calculated as follows:

$$0.753 \text{ mole } CO_2 \text{ molecules} \times \left(\frac{2 \text{ moles O atoms}}{1 \text{ mole } CO_2 \text{ molecules}} \right)$$

$$= 1.51 \text{ moles O atoms}$$

(b) Interpreting the formula P_4O_{10} in terms of moles, we obtain the following conversion factors:

$$\frac{4 \text{ moles P atoms}}{1 \text{ mole } P_4O_{10} \text{ molecules}} \quad \text{and} \quad \frac{10 \text{ moles O atoms}}{1 \text{ mole } P_4O_{10} \text{ molecules}}$$

The setup for calculating the moles of P atoms present is

$$1.31 \text{ moles } P_4O_{10} \text{ molecules} \times \left(\frac{4 \text{ moles P atoms}}{1 \text{ mole } P_4O_{10} \text{ molecules}} \right)$$

$$= 5.24 \text{ moles P atoms}$$

The setup for calculating the moles of O atoms present is

$$1.31 \text{ moles } P_4O_{10} \text{ molecules} \times \left(\frac{10 \text{ moles O atoms}}{1 \text{ mole } P_4O_{10} \text{ molecules}} \right)$$

$$= 13.1 \text{ moles O atoms}$$

7.6   The Mole and Chemical Calculations

In this section we combine the major points we have learned about moles to produce a general approach to problem solving that is applicable to a variety of chemical situations.

In Section 7.3, we learned that Avogadro's number provides a relationship between the number of particles of a substance and the number of moles of that

same substance

In Section 7.4 we learned that molar mass provides a relationship between the number of grams of a substance and the number of moles of that substance.

In Section 7.5, we learned that molar interpretation of chemical formula subscripts provides a relationship between the number of moles of a substance and the number of moles of its component parts.

The preceding three concepts can be combined into a single diagram that is very useful in problem solving. This diagram (Figure 7.3) can be viewed as a "road map" from which conversion factor sequences (pathways) can be obtained. It gives all of the needed relationships for solving two general types of problems:

1. Calculations in which information (moles, grams, particles) is given about a particular substance and additional information (moles, grams, particles) is needed concerning the *same* substance.
2. Calculations in which information (moles, grams, particles) is given about a particular substance and information (moles, grams, particles) is needed concerning a *component* of that same substance.

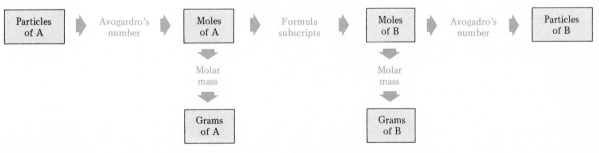

FIGURE 7.3
Useful relationships in chemical-formula-based problem-solving situations.

For the first type of problem, only the left side of Figure 7.3 (the A boxes) is needed. For problems of the second type, both sides of the diagram (A and B boxes) are used.

The thinking pattern needed to use Figure 7.3 is very simple.

1. Determine which box in the diagram represents the *given* quantity in the problem.
2. Next, locate the box that represents the *desired* quantity.
3. Finally, follow the indicated pathway that takes you from the given quantity to the desired quantity. This involves simply "following the arrows." There is always only one pathway possible for the needed transition.

Examples 7.6 through 7.8 indicate some of the types of problems that can be solved using the relationships shown in Figure 7.3.

EXAMPLE 7.6

Vitamin A has the formula $C_{20}H_{30}O$. Calculate the number of vitamin A molecules present in a 0.250-g (250-mg) tablet of vitamin A.

SOLUTION

We will solve this problem by using the three steps of dimensional analysis (see Section 6.3) and Figure 7.3.

STEP 1   The given quantity is 0.250 g of $C_{20}H_{30}O$ and the desired quantity is molecules of $C_{20}H_{30}O$.

$$0.250 \text{ g } C_{20}H_{30}O = ? \text{ molecules } C_{20}H_{30}O$$

In terms of Figure 7.3, this is a "grams of A" to "particles of A" problem. We are given grams of a substance, A, and desire to find molecules (particles) of that same substance.

STEP 2   Figure 7.3 gives us the pathway needed to solve the problem. Starting with "grams of A," we convert to "moles of A" and finally reach "particles of A." The arrows between the boxes along the path give the type of conversion factor needed for each step.

Using dimensional analysis, the setup for this sequence of conversion factors is

a. $Na_2S_2O_3$ (a photographic chemical)
b. $C_{18}H_{21}NO_3$ (codeine, a pain-killing drug)
c. H_2SO_4 (an industrial acid)
d. $C_6H_8O_6$ (vitamin C)
e. $C_{10}H_{19}O_6PS_2$ (malathion, an insecticide)
f. $Ca(NO_3)_2$ (a chemical that gives red color to fireworks)

a. NaCl (table salt)
b. $C_{12}H_{22}O_{11}$ (table sugar)
c. $Pb(OH)_2$ [lead(II) hydroxide]
d. PbI_2 [lead(II) iodide]
e. NaCN (sodium cyanide)
f. $Ca_3(PO_4)_2$ (calcium phosphate)

7-10 What is the mass, in grams, of each of the following quantities of matter?
a. 5.21 moles of nitrogen atoms
b. 7.31 moles of beryllium atoms
c. 0.755 mole of sodium atoms
d. 3.00 moles of H_2O molecules
e. 3.00 moles of CH_4 molecules
f. 0.100 mole of Cl_2 molecules

PERCENTAGE COMPOSITION

7-3 Calculate the percent composition for each of the following compounds.
a. H_2O_2 (hydrogen peroxide, a bleach)
b. C_7H_{16} (a component of gasoline)
c. NaCN (a chemical that extracts silver and gold ores)
d. C_4H_8S (a chemical responsible in part for the odor of skunks)
e. $C_7H_{14}O_2$ (a chemical responsible in part for the odor and taste of bananas)
f. $C_{16}H_{18}N_2O_5S$ (Penicillin V, an important antibiotic)

7-4 Decomposition of a sample of a compound yields 20.67 g of C, 4.62 g of H, and 24.12 g of N. What is the percent composition for this compound?

7-11 How many moles are present in a sample of each of the following substances if each sample weighs 2.00 g?
a. Br atoms
b. S atoms
c. U atoms
d. B_4H_{10} molecules
e. SF_6 molecules
f. Si_2H_6 molecules

7-12 Explain why an easier number to work with, such as the numbers 1.0×10^{23} or 1.0×10^{24}, was not selected as Avogadro's number.

THE MOLE AS A COUNTING UNIT

7-5 What is the relationship between Avogadro's number and the mole counting unit?

7-6 How many molecules are present in each of the following amounts of substance?
a. 3.00 moles of SO_2 molecules
b. 1.25 moles of N_2O_3 molecules
c. 0.444 mole of CO molecules
d. Avogadro's number of NH_3 molecules

7-7 You are given a sample containing 0.750 mole of a substance.
a. How many atoms are present if the sample is nickel metal?
b. How many atoms are present if the sample is gold metal?
c. How many molecules are present if the sample is nitric acid, HNO_3?
d. How many molecules are present if the sample is aspirin, $C_9H_8O_4$?

MOLAR INTERPRETATION OF CHEMICAL FORMULAS

7-13 Construct conversion factors that relate the numbers of moles of atoms of hydrogen, phosphorus, and oxygen to 1 mole of H_3PO_4 molecules.

7-14 Calculate the number of moles of each type of atom that are present in each of the following molar quantities.
a. 3.00 moles of N_2O molecules
b. 3.00 moles of NO_2 molecules
c. 3.00 moles of NO molecules
d. 2.00 moles of O_2 molecules
e. 2.00 moles of O_3 molecules
f. 4.00 moles of B_6H_{14} molecules

7-15 How many *total* moles of atoms are present in 2.00 moles of each of the following substances?
a. NH_3 b. HN_3 c. CaS
d. $CaSO_4$ e. P_4 f. $Al(NO_3)_3$

MOLAR MASS

7-8 Define the term *molar mass*.

7-9 How much does 1.00 mole of each of the following substances weigh in grams?

CHEMICAL-FORMULA-BASED CALCULATIONS

7-16 Calculate the number of atoms in each of the following quantities of an element.
a. 4.7 g of Be   b. 10.0 g of S
c. 2.0 g He   d. 243 g U

7-17 Calculate the mass, in grams, of each of the following quantities of copper.
  a. 6.02×10^{23} atoms   b. 1.32×10^{19} atoms
  c. 43,500 atoms   d. 1 atom

7-18 Calculate the number of moles of substance present in each of the following quantities.
  a. 20.0 g Ag   b. 30.0 g Au
  c. 4.0×10^{20} atoms P   d. 4.0×10^{22} atoms S

7-19 Calculate the number of atoms of phosphorus present in each of the following quantities.
  a. 10.0 g H_3PO_4 b. 30.0 g PH_3
  c. 50.0 g Ca_3P_2 d. 70.0 g $Sn_3(PO_4)_2$

7-20 The active ingredient in the illegal drug marijuana is the compound tetrahydrocannabinol, commonly called THC. The formula of THC is $C_{21}H_{36}O_2$. Calculate the number of grams of THC that you need to obtain 1 million (1.0×10^6) of the following.
  a. Molecules   b. Atoms
  c. C atoms   d. H atoms

BALANCING CHEMICAL EQUATIONS

7-21 Show that the law of conservation of mass is observed for the equation

$$2 HNO_3 + Ca(OH)_2 \longrightarrow Ca(NO_3)_2 + 2 H_2O$$

7-22 Balance the following equations.
  a. $N_2 + O_2 \longrightarrow NO$
  b. $SO_2 + O_2 \longrightarrow SO_3$
  c. $TiCl_4 + H_2O \longrightarrow TiO_2 + HCl$
  d. $Na + H_2O \longrightarrow NaOH + H_2$

7-23 Balance the following equations.
  a. $CH_4 + O_2 \longrightarrow CO_2 + H_2O$
  b. $C_2H_6 + O_2 \longrightarrow CO_2 + H_2O$
  c. $C_3H_8 + O_2 \longrightarrow CO_2 + H_2O$
  d. $C_4H_{10} + O_2 \longrightarrow CO_2 + H_2O$

7-24 Balance the following equations.
  a. $Fe_2O_3 + C \longrightarrow Fe + CO_2$
  b. $CaC_2 + H_2O \longrightarrow C_2H_2 + Ca(OH)_2$
  c. $Fe(OH)_3 + HCl \longrightarrow FeCl_3 + H_2O$
  d. $Al(NO_3)_3 + H_2SO_4 \longrightarrow Al_2(SO_4)_3 + HNO_3$

CHEMICAL EQUATIONS AND THE
MOLE CONCEPT

7-25 Write the 12 mole-to-mole conversion factors that can be derived from the equation.

$$N_2H_4 + 2 H_2O_2 \longrightarrow N_2 + 4 H_2O$$

7-26 How many moles of the first listed product in each of the following equations could be obtained by reacting 2.00 moles of the first listed reactant with an excess of the other reactant?
  a. $FeO + CO \longrightarrow Fe + CO_2$
  b. $3 O_2 + CS_2 \longrightarrow CO_2 + 2 SO_2$
  c. $6 HCl + 2 Al \longrightarrow 3 H_2 + 2 AlCl_3$
  d. $4 Fe_3O_4 + O_2 \longrightarrow 6 Fe_2O_3$

7-27 Using each of the following equations, calculate the number of moles of N_2 gas that can be obtained from 2.00 moles of the first listed reactant and an excess of any other reactants.
  a. $4 NH_3 + 3 O_2 \longrightarrow 2 N_2 + 6 H_2O$
  b. $(NH_4)_2Cr_2O_7 \longrightarrow Cr_2O_3 + N_2 + 4 H_2O$
  c. $N_2H_4 + 2 H_2O_2 \longrightarrow N_2 + 4 H_2$
  d. $2 NH_3 \longrightarrow N_2 + 3 H_2$

CHEMICAL-EQUATION-BASED CALCULATIONS

7-28 How many grams of the first reactant in each of the following equations would be needed to produce 30.0 g of CO_2 gas?
  a. $C_7H_{16} + 11 O_2 \longrightarrow 7 CO_2 + 8 H_2O$
  b. $2 HCl + CaCO_3 \longrightarrow CaCl_2 + CO_2 + H_2O$
  c. $Na_2SO_4 + 2 C \longrightarrow Na_2S + 2 CO_2$
  d. $Fe_3O_4 + CO \longrightarrow 3 FeO + CO_2$

7-29 The catalytic converter, now required equipment on American automobiles, converts carbon monoxide (CO) to carbon dioxide (CO_2) by the following reaction.

$$2 CO + O_2 \longrightarrow 2 CO_2$$

  a. What mass of O_2, in grams, is needed to completely react with 50.0 g of CO?
  b. What mass of CO_2, in grams, is produced when 20.0 g of CO undergo reaction?

7-30 A common laboratory method for preparing oxygen gas (O_2) involves decomposing potassium chlorate ($KClO_3$), as shown by the following equation.

$$2 KClO_3 \longrightarrow 2 KCl + 3 O_2$$

Based on this equation, how many grams of $KClO_3$ must be decomposed to produce the following?
  a. 6.25 moles KCl
  b. 3.0×10^{23} molecules O_2
  c. 225 formula units KCl
  d. 23.4 g O_2

II

Chemical Principles and Their Applications

2 MgO

$\rightarrow$ 2 MgO

$\longrightarrow$ 2 MgO

CHAPTER HIGHLIGHTS

8.1   Properties of the Physical States of Matter

In Chapters 2 and 3, we considered the structure of matter from a submicroscopic point of view—in terms of molecules, atoms, protons, neutrons, and electrons. In this chapter we are concerned with the macroscopic characteristics of matter, as represented by the three physical states of matter: solid, liquid, and gas.

The differences among solids, liquids, and gases are so great that only a few easily observable distinguishing features need to be mentioned to differentiate them clearly. For example, everyone knows that ice cubes retain their shape when placed in a glass and that, when these ice cubes melt, the resultant liquid water takes the shape of the glass to the extent that it fills it (see Figure 8.1). Definite shape is a characteristics of the solid state; indefinite shape is a characteristic of the liquid state.

A comparison of four macroscopically observable properties of each of the physical states of matter will serve as our starting point for our discussion of these states. The four properties are (1) volume and shape, (2) density, (3) compressibility, and (4) thermal expansion. We discussed the property of density in Section 6.4. *Compressibility* involves the change in volume resulting from a pressure change. *Thermal expansion* involves the volume change resulting from a temperature change. Table 8.1 contrasts the previously listed properties for the three states of matter.

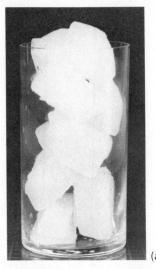 

(a)   (b)

FIGURE 8.1
Solids such as ice cubes have a definite shape, but liquids take the shape of their container.

8.2 The Kinetic Molecular Theory of Matter

The physical characteristics of the solid, liquid, and gaseous states listed in Table 8.1 can be explained by *kinetic molecular theory*, which is one of the fundamental theories of chemistry. The **kinetic molecular theory of matter** *is a set of five statements that are used to explain the physical behavior of the*

TABLE 8.1
Distinguishing Properties of Solids, Liquids, and Gases

Property	Solid State	Liquid State	Gaseous State
Volume and shape	definite volume and definite shape	definite volume and indefinite shape: takes the shape of container to the extent it is filled	indefinite volume and indefinite shape; takes the volume and shape of container that it fills
Density	high	high, but usually lower than corresponding solid	low
Compressibility	small	small, but usually greater than corresponding solid	large
Thermal expansion	very small: about 0.01% per °C	small: about 0.10% per °C	moderate: about 0.30% per °C

three states of matter (solids, liquids, and gases). The foundation for this theory is the concept that the particles (atoms, molecules, or ions) present in a substance, independent of the physical state of the substance, are *always* in *motion.* The word *kinetic* comes from the Greek word *kinesis*, which means movement; hence, the term *kinetic molecular theory.*

The following specific statements constitute the kinetic molecular theory of matter.

1. Matter is ultimately composed of tiny particles (atoms, molecules, or ions) with definite and characteristic sizes that do not change.
2. The particles are in constant random motion and therefore possess kinetic energy.
3. The particles interact with each other through attractions and repulsions and therefore possess potential energy.
4. The velocity of the particles increases as the temperature is increased. The average kinetic energy of all particles in a system depends on the temperature, and it increases as the temperature increases.
5. The particles in a system transfer energy to each other during collisions in which no net energy is lost from the system. The energy of any given particle is thus continually changing.

Both kinetic energy and potential energy are mentioned in the kinetic molecular theory of matter. We considered definitions for these two forms of energy in Section 6.6. Recall that kinetic energy is energy associated with motion, and potential energy is stored energy that results from an object's position, condition, and/or composition. We need to consider these two forms of energy in more detail if we are to make full use of kinetic molecular theory.

The amount of kinetic energy a particle possesses depends on both its mass and its velocity. The exact mathematical relationship between kinetic energy and the mass and velocity of a particle is

$$\text{Kinetic energy} = \tfrac{1}{2}mv^2$$

where m is the mass of the particle and v is its velocity. We can see from this expression that any differences in kinetic energy between particles of the same mass must be caused by differences in their velocities. Similarly, differences in kinetic energy between particles moving at the same velocity must be caused by mass differences.

Because of the diverse origins of potential energy (Section 6.6), no single, simple equation can be used to calculate the amount of potential energy an object has. The potential energy of greatest importance when considering the three states of matter is that related to composition—specifically, that potential energy which originates from electrostatic interactions between particles. **Electrostatic interactions** *are attractions and repulsions that occur between charged particles; particles of opposite charge (one positive and one negative)* *attract each other, and particles of identical charge (both positive or both negative) repel each other.* The magnitude of electrostatic interactions depends on the sizes of the charges associated with the particles ($+1$, $+2$, -1, -2, and so on) and the distance of separation between particles.

When we use kinetic molecular theory to explain the differences among the solid, liquid, and gaseous states of matter, the relative magnitudes of kinetic energy and potential energy (electrostatic interactions) for the system under consideration are key factors. Kinetic energy can be considered a *disruptive force* within the chemical system that tends to make the particles of the system increasingly independent of each other. As a result of the energy of motion, the particles tend to move away from each other. Potential energy of attraction can be considered to be a *cohesive force* that tends to cause order and stability among the particles of the system.

The kinetic energy of a chemical system depends on its temperature. Kinetic energy increases as temperature increases (statement 4 of the kinetic molecular theory of matter). Thus, the higher the temperature is, the greater the magnitude of disruptive influences within a chemical system is. Potential energy magnitude, that is, cohesive force magnitude, is essentially independent of temperature. The fact that one of the types of forces (disruptive forces) depends on temperature and the other (cohesive forces) does not makes temperature the factor that determines in which of the three physical states a given sample is found.

Sections 8.3, 8.4, and 8.5 explain the general properties of the solid, liquid, and gaseous states in terms of disruptive forces, cohesive forces, and temperature magnitude—that is, in terms of kinetic molecular theory.

8.3   The Solid State

The solid state is characterized by a dominance of potential energy (cohesive forces) over kinetic energy (disruptive forces). The particles in a solid are drawn close together in a regular pattern by the strong cohesive forces present. Each particle occupies a fixed position about which it vibrates because of disruptive kinetic energy. Using this model, we can explain the characteristic properties of solids (see Table 8.1) as follows.

1. *Definite volume and definite shape.*   The strong, cohesive forces hold the particles in essentially fixed positions, resulting in a definite volume and definite shape.
2. *High density.*   The constituent particles of solids are located as close together as possible (essentially touching each other). Therefore, a given volume contains a large number of particles, resulting in a high density.

3. *Small compressibility.*   Because there is very little space between particles, increased pressure cannot push the particles any closer together;
   therefore, it has little effect on the solid's volume.
4. *Very small thermal expansion.*   An increased temperature increases the
   kinetic energy (disruptive forces), thereby causing more vibrational motion
   of the particles. Each particle "occupies" a slightly larger volume, and
   the result is a slight expansion of the solid. The strong, cohesive forces
   prevent this effect from becoming large.

8.4   The Liquid State

The liquid state consists of particles that are randomly packed but relatively near
to each other. The molecules are in constant, random motion, freely sliding over
one another, but they do not have sufficient energy to separate from each other.
Thus, *the liquid state is one in which neither potential energy (cohesive forces)
nor kinetic energy (disruptive forces) dominates.* The particles freely sliding over
each other indicates the influence of disruptive forces; however, the inability to
separate indicates fairly strong influences from cohesive forces. Using this model,
we can explain the characteristic properties of liquids (see Table 8.1) as follows.

1. *Definite volume and indefinite shape.*   The attractive forces are strong
   enough to restrict particles to movement within a definite volume. They
   are not strong enough, however, to prevent the particles from moving
   over each other in a random manner that is limited only by the container
   walls. Thus, liquids have no definite shape, with the exception that they
   maintain a horizontal upper surface in containers that are not completely
   filled.
2. *High density.*   The particles in a liquid are not widely separated; they
   are still touching each other. Therefore, a large number of particles are
   present in a given volume, resulting in high density.
3. *Small compressibility.*   Because the particles in a liquid are still touching
   each other, there is very little empty space. Therefore, an increase in
   pressure cannot squeeze the particles much closer together.
4. *Small thermal expansion.*   Most of the particle movement in a liquid is
   vibrational because a particle can move only a short distance before it
   collides with a neighbor. The increased particle velocity that accompanies
   a temperature increase results only in increased vibrational amplitudes.
   The net effect is an increase in the effective volume a particle "occupies,"
   which causes a slight volume increase in the liquid.

8.5 The Gaseous State

The gaseous state is characterized by a complete dominance of kinetic energy (disruptive forces) over potential energy (cohesive forces). As a result, the particles of a gas move essentially independently of each other, in a totally random manner. Under ordinary pressure, the particles are relatively far apart except when they collide with each other. Between collisions with each other or with the container walls, gas particles travel in straight lines. The particle velocities and resultant collision frequencies are extremely high; at room temperature, the collisions experienced by one molecule in one second at one atmosphere pressure is of the order of 10^{10}.

The kinetic theory explanation of gaseous-state properties follows the same pattern we saw earlier for solids and liquids.

1. *Indefinite volume and indefinite shape.*   The attractive (cohesive) forces between particles have been overcome by high kinetic energy and the particles are free to travel in all directions. Therefore, gas particles completely fill their container and the shape of the gas is that of the container.
2. *Low density.*   The particles of a gas are widely separated. There are relatively few particles in a given volume (compared with liquids and solids), which means little mass per volume (a low density).
3. *Large compressibility.* Particles in a gas are widely separated; essentially a gas is mostly empty space. When pressure is applied, the particles are easily pushed closer together, decreasing the amount of empty space and the volume of the gas (see Figure 8.2).
4. *Moderate thermal expansion.*   An increase in temperature means an increase in particle velocity. The increased kinetic energy of the particles enables them to push back whatever barrier is confining them within a given volume, and the volume increases.

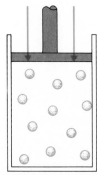

Gas at low pressure

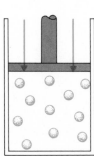

Gas at higher pressure

FIGURE 8.2
The compression of a gas—decreasing the amount of empty space in the container.

Note that the particles do not change size as a gas, liquid, or solid is expanded or compressed; they merely move either farther apart or closer together; it is the space between the particles that changes.

8.6 A Comparison of Solids, Liquids and Gases

Two obvious conclusions about the similarities and differences among the various states of matter can be drawn by comparing the descriptive materials in Sections 8.3 through 8.5.

1. One of the states of matter, the gaseous state, is markedly different from the other two states.
2. Two of the states of matter, the solid and liquid states, have many similar characteristics.

These two conclusions are illustrated diagrammatically in Figure 8.3.

The average distance between particles is only slightly different in the solid and liquid states, but it is markedly different in the gaseous state. Roughly speaking, at ordinary temperatures and pressures, particles in a liquid are about 10% farther apart than those in the solid state. However, particles in a gas are about 1000% farther apart than solid-state particles. The distance ratio between particles in the three states (solid to liquid to gas) is thus 1 to 1.1 to 10.

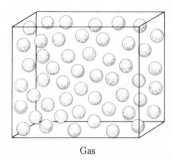

Gas

Molecules far apart and disordered
Negligible interactions between molecules

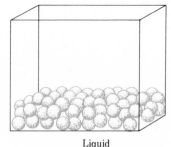

Liquid

Intermediate situation

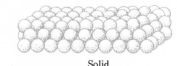

Solid

Molecules close together and ordered
Strong interactions between molecules

FIGURE 8.3
Similarities and differences in the states of matter.

8.7 Gas Laws and Gas Law Variables

Thus far in this chapter, we have discussed solids, liquids, and gases in a very qualitative manner. We have used few numbers in our consideration of the general properties of the three states of matter.

We now consider the gaseous state from a quantitative (numerical) viewpoint. Behavior of matter in the gaseous state can be described by some simple quantitative relationships called *gas laws*. **Gas laws** *are generalizations that describe the relationships among the pressure, temperature, and volume of a specific quantity of gas in mathematical terms.*

Only the gaseous state can be described by simple mathematical relationships. Laws describing liquid-state and solid-state behaviors are mathematically extremely complex. Consequently, quantitative treatments of liquid-state and solid-state behavior are not given in this text.

Before we discuss the mathematical form of the various gas laws, some comments concerning the major variables involved in these laws—volume, temperature, and pressure—are in order. Two of these variables, volume and temperature, were discussed previously in Sections 6.2 and 6.5. The units *liter* and *milliliter* are generally used to specify gas volume. Only one of the three temperature scales discussed in Section 6.5, the *Kelvin scale*, can be used in gas law calculations if the results are to be valid. Therefore, you should review the conversion of Celsius and Fahrenheit scale readings to the Kelvin scale (Section 6.5).

Pressure *is the force applied per unit area—that is, the total force on a surface divided by the area of that surface.*

$$P \text{ (pressure)} = \frac{F \text{ (force)}}{A \text{ (area)}}$$

For a gas, the force involved in pressure is that which is exerted by the gas molecules or atoms as they constantly collide with the walls of their container. Barometers, manometers, and gauges are the instruments most commonly used to measure gas pressures.

Millimeters of mercury, atmospheres, and *pounds-per-square inch* are the most commonly used pressure units. Barometers and manometers measure pressure in terms of the height of a column of mercury in millimeters. Gauges are usually calibrated in terms of force per area, as in pounds per square inch.

The air that surrounds the earth exerts a pressure on all the objects it is in contact with. A **barometer** *is a device used to measure atmospheric pressure.* The essential components of a simple barometer are shown in Figure 8.4. A

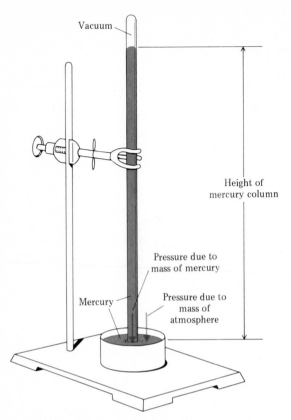

Vacuum

Height of
mercury column

Pressure due to
mass of mercury

Mercury

Pressure due to
mass of
atmosphere

FIGURE 8.4

The essential components of a mercury barometer.

barometer can be constructed by taking a long glass tube that is sealed at one end and filling it to the top with mercury; then the tube is inverted in a dish of mercury without letting any air into it. The mercury in the tube falls until the pressure from the mass of the mercury in the tube is just balanced by the pressure of the atmosphere on the mercury in the dish. The pressure of the atmosphere is expressed in terms of the height of the supported column of mercury. Mercury is the liquid of choice in a barometer for two reasons: (1) because it is a very dense liquid, only a short tube is needed, and (2) because it does not readily vaporize (evaporate), the pressure reading does not have to be corrected for the presence of mercury vapor.

The pressure unit *millimeters of mercury* is usually written in the abbreviated form mm Hg. An alternative term for this unit is *torr*; this name honors the Italian physicist Evangelista Torricelli (1608–1647), the inventor of the barometer.

$$1 \text{ mm Hg} = 1 \text{ torr}$$

The pressure of the atmosphere varies with altitude, decreasing at the rate of approximately 25 mm Hg per 1000 feet increase in altitude. It also fluctuates with weather conditions. Think about the terminology used in a weather report—high pressure ridge, low pressure front, and so on. At sea level barometric pressure fluctuates between 740 and 770 mm Hg and averages about 760 mm Hg, depending on weather conditions. The pressure unit *atmospheres*, abbreviated *atm*, is defined in terms of this average sea level pressure. By definition, we have

$$1 \text{ atm} = 760 \text{ mm Hg} = 760 \text{ torr}$$

Because of its size, which is 760 times larger than 1 mm Hg, the atmosphere unit is used whenever high pressures are encountered.

Television weather forecasters give barometric pressure readings in terms of inches rather than millimeters of mercury (torr). Barometric pressure readings usually fall between 29 and 30 inches of mercury. The relationship, among atmospheres, millimeters of mercury, and inches of mercury are

$$1 \text{ atm} = 760 \text{ mm Hg} = 29.92 \text{ in. Hg}$$

The relationship between the pressure units pounds per square inch and atmospheres is

$$1 \text{ atmosphere} = 14.7 \text{ pounds per square inch}$$

Two abbreviations exist for the pounds per square inch pressure unit—lb/in.^{2} and psi.

8.8 Boyle's Law: A Pressure–Volume Relationship

Of the several relationships that exist between gas-law variables, the first to be discovered was that between gas pressure and gas volume. Known as Boyle's law, it was formulated over 300 years ago, in 1662, by the British chemist and physicist Robert Boyle (1627–1691) (see Historical Profile 10). **Boyle's law states that** *the volume of a sample of a gas is* INVERSELY PROPORTIONAL *to the pressure applied to the gas if the temperature is kept constant.* This means that if the pressure on the gas increases, the volume decreases proportionally; conversely, if the pressure is decreased, the volume will increase. Doubling the pressure cuts the volume in half; tripling the pressure cuts the volume to one-third its original value; quadrupling the pressure cuts the volume to one-fourth; and so on. Any time two quantities are *inversely proportional*, which is the case with pressure and volume of a gas (Boyle's law), one increases as the other decreases. Figure 8.5 illustrates Boyle's law.

[Smithsonian Institution, Washington DC.]

HISTORICAL PROFILE 10

Robert Boyle   1627–1691

Robert Boyle, one of the most prominent of 17th century scientists, spent most of his life in England although he was born in Ireland. Born to wealth and nobility, the fourteenth of 15 children of the Earl of Cork, he was an infant prodigy. At the age of 14 he was in Italy studying the works of the recently deceased Galileo.

Like most men of the 17th century who devoted themselves to science, Boyle was self-taught. Primarily known as a chemist and natural philosopher, he is best known for the gas law that bears his name. However, he made many other significant contributions in chemistry and physics.

During his lifetime his name was known throughout the learned world as a leading advocate of the "experimental philosophy." Through his efforts the true value of experimental investigation was first realized.

Historically, before chemistry could become a flourishing science a period of "housecleaning" was needed to get rid of accumulated false concepts of previous centuries. Perhaps Boyle's greatest work in chemistry was in this area. He was instrumental in demolishing the false doctrine of only four elements—earth, air, water, and fire.

His book *The Sceptical Chymist* (1661), which attacked many of the false concepts of previous years, influenced generations of chemists. In this book he developed the concept of "primary particles," a forerunner of our present concept of atoms and elements. He attempted to explain different kinds of matter in terms of the organization and motion of these primary particles.

Boyle was a prolific writer not only on scientific but also on philosophical and religious topics. In later years he became increasingly interested in religion. Upon his death, Boyle left a sum of money to fund the Boyle lectures, which were not on science, but on the defense of Christianity against unbelievers.

Boyle's law may be stated mathematically as

$$P_1 \times V_1 = P_2 \times V_2$$

P_1 and V_1 are the initial pressure and volume of a gas sample, and P_2 and V_2 are the new pressure and volume of the gas sample. This equation is valid only if the temperature remains constant.

When we know any three of the four quantities in the Boyle's law equation, we can calculate the fourth, which is usually the new pressure, P_2, or the new volume, V_2.

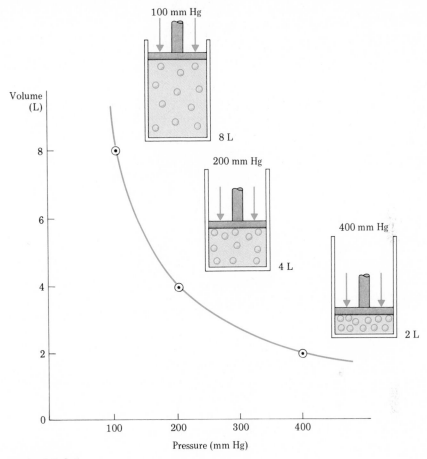

FIGURE 8.5

A pressure–volume graph illustrating Boyle's law data.

EXAMPLE 8.1

An air bubble forms at the bottom of a lake, where the total pressure is 2.45 atm. At this pressure, the bubble has a volume of 3.1 mL. What volume will it have when it rises to the surface, where the pressure is that of the atmosphere, 0.93 atm? Assume that the temperature remains constant, as does the amount of gas within the bubble.

SOLUTION

A suggested first step in working gas-law problems involving two sets of conditions is to analyze the given data in terms of initial and final conditions.

Doing this, we find that

$$P_1 = 2.45 \text{ atm} \qquad P_2 = 0.93 \text{ atm}$$

$$V_1 = 3.1 \text{ mL} \qquad V_2 = \text{? mL}$$

We know three of the four variables in the Boyle's law equation, so we can calculate the fourth, V_2. We rearrange Boyle's law to isolate V_2 (the quantity to be calculated) on one side of the equation. This is accomplished by dividing both sides of the Boyle's law equation by P_2

$$P_1V_1 = P_2V_2 \qquad \text{(Boyle's law)}$$

$$\frac{P_1V_1}{P_2} = \frac{P_2V_2}{P_2} \qquad \left(\begin{array}{l}\text{Divide each side of} \\ \text{the equation by } P_2.\end{array}\right)$$

$$V_2 = \frac{P_1V_1}{P_2}$$

Substituting the given data into the rearranged equation and doing the arithmetic gives

$$V_2 = \frac{(2.45 \text{ atm}) \times (3.1 \text{ mL})}{0.93 \text{ atm}} = 8.2 \text{ mL}$$

Our answer is reasonable. Decreasing the pressure should increase the volume; the volume went from 3.1 mL to 8.2 mL.

Boyle's law is consistent with kinetic molecular theory (Section 8.2). The pressure exerted by a gas results from collisions of the gas molecules with the sides of the container. The number of collisions within a given area on the container wall during a given time is proportional to the pressure of the gas at a given temperature. If the volume of a container holding a specific number of gas molecules is increased, the total wall area of the container will also increase and the number of collisions in a given area (the pressure) will decrease because of the greater wall area. Conversely, if the volume of the container is decreased, the wall area will be smaller and more collisions will occur within a given wall area. Figure 8.6 illustrates the situation that occurs when the volume is decreased.

HELIUM-FILLED
RESEARCH BALLOONS

The phenomenon described by Boyle's law has practical importance. Helium-filled research balloons used to study the upper atmosphere are only half-filled with helium when launched. As the balloon ascends, it encounters increasingly low pressures. As the pressure decreases, the balloon expands until it reaches full inflation. If the balloon were fully inflated when launched, it would burst in the upper atmosphere because of the reduced external pressure.

OPERATION OF A
MEDICAL SYRINGE

The process of filling a medical syringe with liquid also involves Boyle's law. As the plunger is drawn out of the syringe (see Figure 8.7), the increase in

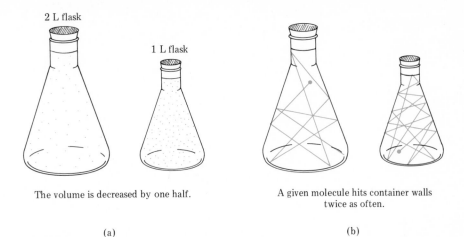

2 L flask

1 L flask

The volume is decreased by one half.

A given molecule hits container walls twice as often.

(a)

(b)

FIGURE 8.6

The average number of times a molecule hits container walls is doubled when the volume of the gas, at constant temperature, decreases by one half.

volume inside the syringe chamber results in decreased pressure there. The liquid, which is under the influence of atmospheric pressure, flows into this reduced-pressure area. The liquid is then expelled from the chamber by pushing the plunger back in. This ejection of the liquid does not involve Boyle's law; a liquid is incompressible and mechanical force pushes it out.

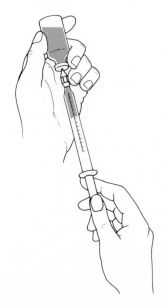

FIGURE 8.7

The filling of a syringe with liquid is an application of Boyle's law.

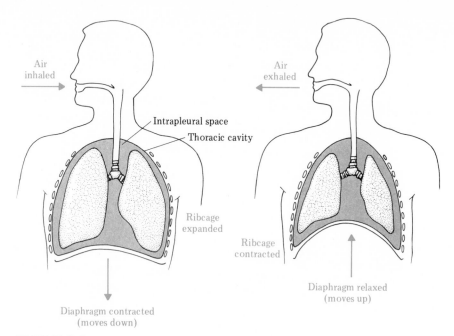

Air
inhaled

Air
exhaled

Intrapleural space
Thoracic cavity

Ribcage
expanded

Ribcage
contracted

Diaphragm relaxed
(moves up)

Diaphragm contracted
(moves down)

FIGURE 8.8
The mechanics of breathing is based on Boyle's law.

**HUMAN LUNGS AND
BREATHING**

A direct application of Boyle's law is found in the process of breathing, illustrated in Figure 8.8. Human lungs, which are elastic (expandable) structures open to the atmosphere, are located within an airtight region of the body called the thoracic cavity. The diaphragm, which is a muscle that is able to move up and down (contract and relax), forms the floor of the thoracic cavity. Breathing in (inhaling or inspiration) occurs when the diaphragm flattens out (contracts). This contraction causes the volume of the thoracic cavity to increase. At the same time, the pressure within the cavity drops (Boyle's law) below atmospheric (external) pressure. Air flows into the lungs, and expands them because the pressure is greater outside than within the lungs. Breathing out (exhaling or expiration) occurs when the diaphragm relaxes (moves up). This action decreases the volume of the thoracic cavity and simultaneously increases the pressure (Boyle's law) within the cavity to a value greater than the external pressure. Air flows out of the lungs. The direction of air flow is always from a high-pressure to a low-pressure region.

**THE HEIMLICH
MANEUVER**

The Heimlich maneuver, which is an emergency procedure used to help someone who is choking, also works on the principle of Boyle's law. The procedure involves wrapping one's arms below the rib cage of the choking person and giving the person an "upward bearhug." The thoracic cavity volume decreases, the pressure within the cavity increases (Boyle's law), and material that is lodged in a person's trachea is often expelled.

8.9 Charles' Law: A Temperature–Volume Relationship

The relationship between the temperature and volume of a gas at constant pressure is called Charles' law, after the French scientist Jacques Alexander Cesar Charles (1746–1823) (see Historical Profile 11). This law was discovered in 1787, more than 100 years after the discovery of Boyle's law. **Charles' law** states that the *volume of a sample of gas is* DIRECTLY PROPORTIONAL *to its Kelvin temperature if the pressure is kept constant.* Notice the phrase *directly proportional,* and recall the contrasting phrase *inversely proportional* in Boyle's law. Whenever two quantities are in *direct* proportion, one increases when the other increases, and one decreases when the other decreases. Thus, direct proportion and inverse proportion are opposites. The directly proportional relationship of Charles' law means that if the temperature increases, the volume will also increase and if the temperature decreases, the volume will also decrease. Data given in Figure 8.9 illustrate Charles' law.

A balloon filled with air provides a qualitative illustration of Charles' law. If the balloon is placed near a heat source, such as a light bulb that has been on for some time, the heat will cause the balloon to increase in size (volume). The change in volume is visually noticeable. Putting the same balloon in the refrigerator will cause it to shrink.

Charles' law, stated in mathematical form, is

$$\frac{V_1}{T_1} = \frac{V_2}{T_2}$$

In this expression, V_1 is the initial volume of a gas sample and T_1 is the initial *Kelvin* temperature of the gas. V_2 and T_2 are the new volume and *Kelvin* temperature of the gas sample. This equation is valid only if the pressure remains constant. When you use the Charles' law equation, temperatures must always be expressed in Kelvin units.

When we know any three of the four quantities in the Charles' law equation, we can calculate the fourth, which is usually the final volume, V_2, or the final temperature, T_2.

Jacques Alexandre Cesar Charles

1746–1823

[Burndy, Library, Norwalk CT.]

Jacques Alexandre Cesar Charles, a French physicist, had an early career completely removed from science. He worked at a routine job in the Bureau of Finances in Paris. All was well until, during a period of government austerity, he was discharged from his job. Unemployed, he decided to study physics.

During this time of transition in Charles' life, hot air ballooning, both unmanned and manned, was in its first stages of development. Charles was best known to his colleagues for his contributions to the developing science of ballooning. He was the first to design a hydrogen-filled, rather than air-filled, balloon for manned flight. In constructing the balloon Charles showed much innovation in the design of scientific apparatus.

The first hot air balloon flight with people aboard was made on November 17, 1783. A few days later Charles made an ascent in his hydrogen-filled balloon. His landing put him in a rural area. The peasants in the area, terrified at their first exposure to "manned flight" attacked and destroyed the balloon using pitchforks. Burning straw was the source of heat for these original balloons.

In the process of working with balloons, about 1787, Charles observed that several different gases expanded in the same way when heated. From further observations he was able to estimate the degree of thermal expansion for gases as a function of temperature. This led to what is now known as Charles' law.

Charles never published his work. He did, however, tell another scientist, Joseph Louis Gay-Lussac (Historical Profile 12), about it and it was Gay-Lussac who made it public. In an article on the expansion of gases by heat, published in 1802, Gay-Lussac described, criticized, and considerably improved on Charles' experimental procedures.

Charles' contributions to science were limited to those associated with his ballooning.

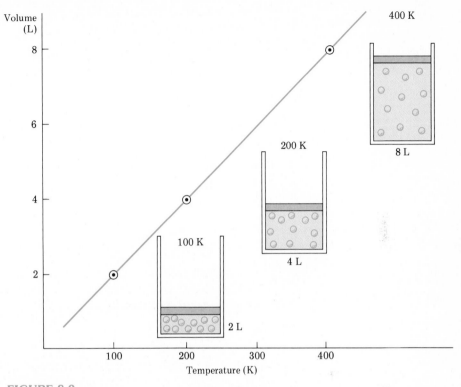

FIGURE 8.9

A temperature–volume curve illustrating Charles' law data.

EXAMPLE 8.2

A certain quantity of gas in a balloon has a volume of 223 mL at a temperature of 42°C. What is the volume of the balloon if the temperature is lowered to −13°C? Assume that the pressure remains constant.

SOLUTION

First, we analyze the data in terms of initial and final conditions. We find that

$$V_1 = 223 \text{ mL} \qquad\qquad V_2 = ? \text{ mL}$$

$$T_1 = 42°C + 273 = 315 \text{ K} \qquad T_2 = -13°C + 273 = 260 \text{ K}$$

Note that both of the given temperatures have been converted to Kelvin scale readings. This change is accomplished by simply adding 273 to the Celsius scale value (see Section 6.5).

We know three of the four variables in the Charles' law equation, so we can calculate the fourth, V_2. We rearrange Charles' law to isolate V_2 (the quantity desired) by multiplying each side of the equation by T_2.

$$\frac{V_1}{T_1} = \frac{V_2}{T_2} \quad \text{(Charles' law)}$$

$$\frac{V_1 T_2}{T_1} = \frac{V_2 T_2}{T_2}$$

$$V_2 = \frac{V_1 T_2}{T_1}$$

Substituting the given data into the equation and doing the arithmetic gives

$$V_2 = \frac{(223 \text{ mL}) \times (260 \text{ K})}{(315 \text{ K})} = 184 \text{ mL}$$

Our answer is consistent with what reasoning says it should be. A decrease in the temperature of a gas at constant pressure should result in a volume decrease; the volume decreases from 223 mL to 184 mL.

Charles' law is easy to understand in terms of the kinetic molecular theory. The theory states that when the temperature of a gas increases, the velocity (kinetic energy) of the gas molecules increases. The speedier particles hit the container walls harder and more often. For the pressure of the gas to remain constant, it is necessary for the container volume to increase. In a larger volume, the particles hit the container walls less often and the pressure can remain the same. Similarly if the temperature of the gas is lowered, the velocity of the molecules decreases and the wall area (volume) must decrease to increase the number of collisions in a given area in a given time.

Charles' law indicates that, at constant pressure, heating a gas will cause it to expand (increase in volume). As a result of this expansion, the density of the gas decreases; the same amount of gas now fills a larger volume.

HOT AIR BALLOONS

The density changes associated with Charles' law behavior are important in a number of situations. Operation of hot air balloons (see Figure 8.10) is based on density change. The hot air of the balloon is lower in density than the surrounding air. Therefore, the balloon rises in the dense, cooler air.

The phenomenon called *temperature inversion*, associated with air pollution episodes, results from Charles' law behavior of gases. In temperature inversions, denser, cool air containing air pollutants is trapped in a valley by an overlying layer of warm air. Section 20.11 considers the subject of temperature inversions in detail.

FIGURE 8.10
Hot air balloon operation is based on the gas density change that occurs when a gas is heated under constant pressure (Charles' law). [Courtesy of Robert J. Ouellette, The Ohio State University.]

8.10 Gay-Lussac's Law: A Temperature–Pressure Relationship

If a gas is placed in a rigid container—one that cannot expand—the volume of the gas must remain constant. What is the relationship between the pressure and temperature of a fixed amount of gas in such a situation? The question was answered in 1802 by the French scientist Joseph Louis Gay-Lussac (1778–1850) (*see* Historical Profile 12). The relationship between these two variables is a direct proportion. **Gay-Lussac's law** states that *the pressure of a sample of*

HISTORICAL PROFILE 12

Joseph Louis Gay-Lussac   1778–1850

[Burndy Library, Norwalk CT.]

Joseph Louis Gay-Lussac, the eldest of five children, is considered one of the greatest scholars France has produced. Because of the French Revolution and the uncertainties created by it, Gay-Lussac's formal education had to be delayed until he was 16. He quickly distinguished himself as a student.

His grandfather was a physician and his father a lawyer. The family surname was actually Gay, the surname by which he was baptized. His father added the Lussac (which has geographical connotations) to the family name to distinguish his family from many others in the area with the same surname.

Gay-Lussac devoted nearly all of his life to pure and applied science. Much of it, from 1809 on, was spent as a professor at the École Polytechnique in Paris. He did have a brief political career (1831–1838) as an elected French legislator.

Gay-Lussac's first major research (1801–1802) involved the thermal expansion of gases. Here he followed up on work performed earlier by Charles. His greatest achievement, which came in 1808, was work involving the ratio, by volume, in which gases combine during chemical reactions. From this work came the law of combining volumes that now bears his name.

He was the first to isolate the elements potassium and boron. With a co-worker, he made the first thorough studies of iodine and cyanogen. He was a pioneer in the area of volumetric analysis. The "Gay-Lussac tower," an essential part of sulfuric acid plants for many years, is evidence of his interest in chemical technology.

gas, at constant volume, is DIRECTLY PROPORTIONAL *to its Kelvin temperature.* Thus, at constant volume, as the temperature of a gas increases, the pressure increases, and as temperature decreases pressure decreases.

Gay-Lussac's law, stated in mathematical form, is

$$\frac{P_1}{T_1} = \frac{P_2}{T_2}$$

Again, as with Charles' law (Section 8.9), the temperature must be expressed on the Kelvin scale. Any pressure unit is acceptable; P_1 and P_2 must, however, be in the same units.

The kinetic-molecular-theory explanation for Gay-Lussac's law is very simple. As the temperature increases, the velocity of the gas molecules increases. This increases the frequency with which the molecules hit the container walls, which translates into an increase in pressure.

Gay-Lussac's law explains why one should not heat a closed container. Heating causes the pressure within the container to build up; an explosion may result if the container cannot withstand the increased pressure. This is why aerosal spray cans of deodorant, paint, and other substances have warning labels stating that they should not be heated and that they should be stored in locations where the temperature does not exceed certain levels.

INCINERATION OF
AEROSOL SPRAY CANS

HOME CANNING

While canning fruits and vegetables, one should leave the lids loose enough for steam to escape so that pressure buildup does not crack the jar.

AUTOMOBILE TIRE
PRESSURE

To obtain a "true" pressure reading for an automobile tire, the pressure should always be measured before a drive rather than during or immediately after a drive. As a car travels, the constant friction between tires and road generates sufficient heat to cause significant increases in tire pressure. (You need not be alarmed at the increased pressure in a "hot" tire because tires are designed to withstand the higher pressure that develops.)

8.11 The Combined Gas Law

The **combined gas law** *is an expression obtained by mathematically combining Boyle's, Charles' and Gay-Lussac's laws.* Its mathematical form is

$$\frac{P_1 V_1}{T_1} = \frac{P_2 V_2}{T_2}$$

This combined gas law is a much more versatile equation than the individual gas laws. With it, we can calculate a change in any one of the three gas-law variables, brought about by changes in *both* of the other two variables. Each of the individual gas laws requires that one of the three variables be held constant.

The three most used forms of the combined gas law are those that isolate V_2, P_2, and T_2, respectively, on the left side of the equation.

$$V_2 = V_1 \times \frac{P_1}{P_2} \times \frac{T_2}{T_1}$$

$$P_2 = P_1 \times \frac{V_1}{V_2} \times \frac{T_2}{T_1}$$

$$T_2 = T_1 \times \frac{P_2}{P_1} \times \frac{V_2}{V_1}$$

EXAMPLE 8.3

A sample of O_2 gas occupies a volume of 1.62 L at 755 mm Hg pressure and at a temperature of 0°C. What volume will this gas sample occupy at 725 mm Hg pressure and 50°C temperature?

SOLUTION

The form of the combined gas law that we need is the one in which V_2 is isolated on the left side of the equation.

$$V_2 = V_1 \times \frac{P_1}{P_2} \times \frac{T_2}{T_1}$$

Before we insert values into this equation, let us analyze the given data in terms of initial and final conditions.

$P_1 = 755$ mm Hg $P_2 = 725$ mm Hg

$V_1 = 1.62$ L $V_2 = ?$ L

$T_1 = 0°C + 273 = 273$ K $T_2 = 50°C + 273 = 323$ K

Substituting these numerical values into the above form of the combined gas law gives

$$V_2 = 1.62 \text{ L} \times \left(\frac{755 \text{ mm Hg}}{725 \text{ mm Hg}}\right) \times \left(\frac{323 \text{ K}}{273 \text{ K}}\right)$$

$$= 2.00 \text{ L}$$

8.12   Changes of State

A **change of state** is a process in which a substance is transformed from one physical state to another. Changes of state are usually accomplished by heating or cooling a substance. Pressure change is also a factor in some systems. As noted previously (in Section 1.6), changes of state are examples of physical change—that is, change in which chemical composition remains constant. No new substances are ever formed as a result of state change.

There are six possible changes of state. Figure 8.11 identifies each of these changes and gives the terminology used to describe the changes. Four of the six terms used in describing state changes are familiar: freezing, melting, evaporation, and condensation. The other two terms—sublimation and deposition—describe changes between the solid and gaseous states; these changes are not as common as the other types.

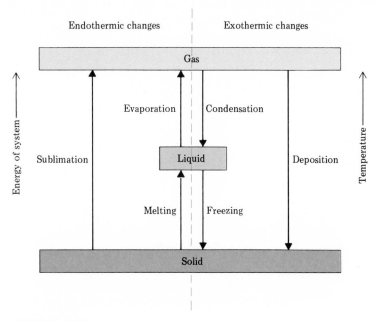

FIGURE 8.11
The six changes of state and the terms used to describe them.

Changes of state can be classified into two categories, based on whether heat (thermal energy) is given up or absorbed during the change process. An **endothermic change of state** *is a change that requires the input (absorption) of heat.* The endothermic changes of state are melting, sublimation, and evaporation (the processes in Figure 8.11 in which the arrows point up). An **exothermic change of state** *is a change that requires heat to be given up (released).* Exothermic changes of state are the reverse of endothermic changes and include freezing, condensation, and deposition (the processes in Figure 8.11 in which the arrows point down.)

8.13   Evaporation of Liquids

Evaporation *is the process by which molecules escape from the liquid phase to the gas phase.* It is a familiar process to us. We are all aware that water left in an open container at room temperature slowly "disappears" by evaporation.

The phenomenon of evaporation is readily explained using kinetic molecular theory. Statement 5 of this theory (Section 8.2) indicates that the molecules in a liquid (or a solid or gas) do not all possess the same kinetic energy. At any given instant, some molecules have above-average kinetic energies and others

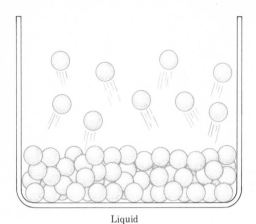

Liquid

FIGURE 8.12

Evaporation is a surface phenomenon in which molecules with above-average kinetic energy are able to escape from the bulk of the liquid.

below-average kinetic energies as a result of collisions between molecules. A given molecule's energy constantly changes as a result of collisions with neighboring molecules. Molecules that happen to be considerably above average in kinetic energy at any given moment can overcome the attractive forces (potential energy) holding them in the liquid and escape if they are on the liquid surface and are moving in a favorable direction relative to the surface (see Figure 8.12).

Evaporation is a surface phenomenon. Molecules within the interior of a liquid are surrounded on all sides by other molecules, which makes escape very improbable. Surface molecules are subject to fewer attractive forces because they are not completely surrounded by other molecules; thus escape is much more probable. Liquid surface area is an important factor in determining the rate at which evaporation occurs. Increased surface area results in an increased evaporation rate because a greater fraction of molecules occupy "surface" locations.

Water evaporates faster from a glass of hot water than from a glass of cold water. Why is this so? A certain minimum kinetic energy is required for a molecule to escape from the attractions of its neighboring molecules. As the temperature of a liquid increases, a larger fraction of the molecules present acquire this needed minimum kinetic energy. Consequently, the rate of evaporation always increases as liquid temperature increases. Figure 8.13 contrasts the fraction of molecules possessing the needed minimum kinetic energy for escape at two temperatures. Note that at both the lower and higher temperatures, a broad distribution of kinetic energies is present, and some molecules possess the needed minimum kinetic energy at each temperature. At the higher temperature, however, a larger fraction of the molecules present have the requisite kinetic energy.

The escape of high-energy molecules from a liquid during evaporation affects the liquid in two ways: the amount of liquid decreases and the liquid temperature is lowered. The lower temperature reflects the fact that the average kinetic

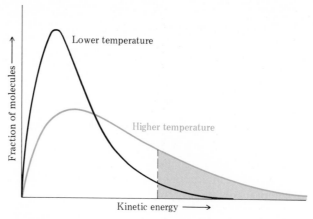

FIGURE 8.13

The distribution of kinetic energies for molecules in a liquid at two different temperatures. The dashed line represents the minimum kinetic energy required for molecules within the liquid to overcome attractive forces and escape into the gas phase. Molecules in the shaded area have that necessary energy.

energy of the remaining molecules is lower than the preevaporation value because of the loss of the most energetic molecules. (Analogously, if all the tall people are removed from a classroom of students, the average height of the remaining students decreases.) A lower average kinetic energy corresponds to a lower temperature (statement 4 of kinetic molecular theory); hence, a cooling effect is produced.

Evaporative cooling is important in many processes. Our own bodies use evaporation to maintain a constant body temperature. We perspire in hot weather because evaporation of the perspiration cools our skin. The cooling effect of evaporation is quite noticeable when someone first emerges from an outdoor swimming pool on a breezy day. In a humid atmosphere—an atmosphere that already contains sizable amounts of water vapor—the rate of evaporation of perspiration is slower than in dry air; this explains why the same temperature feels hotter in humid (muggy) climates than it does in dry ones.

In medicine, the local skin anesthetic ethyl chloride (C_2H_5Cl) is used to reduce pain near the surface of the skin. It is sprayed on the skin and exerts its effect through evaporative cooling. The evaporation rate for ethyl chloride at human body temperature is so fast that the cooling effect from the evaporation "numbs" the tissue near the surface of the skin and results in temporary loss of feeling in the region of application.

The molecules that escape from an evaporating liquid are collectively referred to as vapor, rather than gas. The term **vapor** *describes gaseous molecules of a substance at a temperature and pressure at which we ordinarily would think of the substance as a liquid or solid.* For example, at room temperature and atmospheric pressure, the normal state for water is the liquid state. Molecules that escape (evaporate) from liquid water at these conditions are called *water vapor.*

8.14   Vapor Pressure of Liquids

The evaporative behavior of a liquid in a closed container is quite different from its behavior in an open container. We observe that in a closed container some liquid evaporation occurs; this is indicated by a drop in liquid level. However, unlike the open container system, with time the liquid level ceases to drop (becomes constant); thus not all of the liquid will evaporate.

Kinetic molecular theory explains these observations in the following way. The molecules that evaporate in the closed container are unable to move completely away from the liquid, as they did in the open container. They find themselves confined in a fixed space immediately above the liquid (see Figure 8.14a). These "trapped" molecules of vapor undergo many random collisions with the container walls, other vapor molecules, and the liquid surface. Molecules that collide with the liquid surface are recaptured by the liquid. Thus, two processes—evaporation (escape) and condensation (recapture)—take place in a closed container.

For a short time, the rate of evaporation in a closed container exceeds the rate of condensation and the liquid level drops. As more of the liquid evaporates, however, the number of vapor molecules increases and the chance of their striking the liquid surface and being recaptured also increases. Eventually, the rate of condensation becomes equal to the rate of evaporation and the liquid level stops dropping (see Figure 8.14c). At this point, the number of molecules that escape in a time is the same as the number recaptured; a steady-state situation has been reached. The amount of liquid and vapor in the container

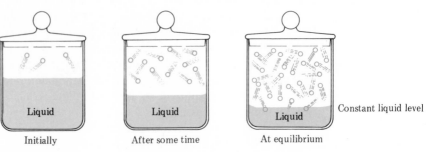

FIGURE 8.14
Evaporation of a liquid in a closed container.

TABLE 8.2
Vapor Pressure of Water at Various Temperatures

Temperature (°C)	Vapor Pressure (mm Hg)	Temperature (°C)	Vapor Pressure (mm Hg)
0	4.6	60	149.4
10	9.2	70	233.7
20	17.5	80	355.1
30	31.8	90	525.8
40	55.3	100	760.0
50	92.5		

does not change, even though both evaporation and condensation are still occurring.

This steady-state situation, which will continue as long as the temperature of the system remains constant, is an example of a *state of equilibrium*. A **state of equilibrium** *is a situation in which two opposite processes take place at equal rates*. For systems in a state of equilibrium, no net macroscopic changes can be detected. However, the system is dynamic; the forward and reverse processes are occurring at equal rates.

In a liquid–vapor equilibrium situation in a closed container, the vapor in the fixed space immediately above the liquid exerts a constant pressure on both the liquid surface and the walls of the container. This pressure is called the vapor pressure of the liquid. **Vapor pressure** *is the pressure exerted by a vapor above a liquid when the liquid and vapor are in equilibrium.*

The magnitude of a vapor pressure depends on the nature and temperature of a liquid. Liquids that have strong attractive forces between molecules will have lower vapor pressures than liquids that have weak attractive forces between particles. Substances that have high vapor pressures at room temperature are said to be *volatile*.

The vapor pressure of all liquids increases with temperature. Why? An increase in temperature results in more molecules having the minimum kinetic energy required for evaporation. Hence, the pressure of the vapor is greater at equilibrium. Table 8.2 shows the variation in vapor pressure with increasing temperature for the compound water.

The size (volume) of space that the vapor occupies does not affect the magnitude of the vapor pressure. A larger fixed space enables more molecules to be present in the vapor at equilibrium. However, the larger number of molecules spread over a larger volume results in the same pressure as a small number of molecules in a small volume.

Vapor pressures for liquids are commonly measured using an apparatus similar to that shown in Figure 8.15. At the moment that liquid is added to the vessel, the space above the liquid is filled only with air and the levels of mercury in the U-tube are equal. With time, some liquid evaporates and equilibrium is

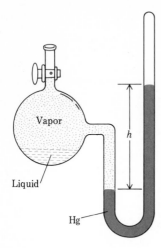

FIGURE 8.15
Apparatus for determining the vapor pressure of a
liquid.

established. The vapor pressure of the liquid is proportional to the difference
between the heights of the mercury columns. The larger the vapor pressure, the
greater is the extent to which the mercury column is pushed up.

8.15 Boiling and Boiling Point

If a molecule is to escape from the liquid state, it usually must be on the surface
of the liquid. **Boiling** *is a special form of evaporation in which conversion
from the liquid state to the vapor state occurs within the body of a liquid
through bubble formation.* Boiling begins when the vapor pressure of a liquid,
which steadily increases as the liquid is heated, reaches a value equal to that
of the prevailing external pressure on the liquid; for liquids in open containers,
this value is atmospheric pressure. When these two pressures become equal,
bubbles of vapor form around any speck of dust or around any irregularity
associated with the container surface. These vapor bubbles quickly rise to the
surface and escape because they are less dense than the liquid itself. The quick
ascent of the bubbles causes the agitation associated with a boiling liquid.

Let us consider this "bubble phenomenon" in more detail. The first small
bubbles that form on the bottom of the container being heated are bubbles of
dissolved air (oxygen and nitrogen) that have been driven out of solution by the
rising liquid temperature. (The solubility of oxygen and nitrogen in water decreases with increasing temperature.)

As the liquid is heated further, larger bubbles form and begin to rise. These
bubbles, which usually form on the bottom of the container where the liquid is

BUBBLE FORMATION
AND BOILING

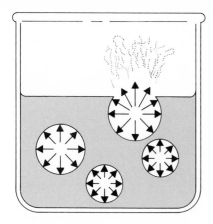

FIGURE 8.16
Bubble formation associated with a liquid that is boiling.

hottest, are liquid vapor bubbles rather than air bubbles. Initially, these vapor bubbles "disappear" as they rise and never reach the liquid surface. In the hotter lower portions of the liquid, the vapor pressure is high enough to sustain bubble formation. For the bubble to exist, the pressure within it (vapor pressure) must equal external pressure (atmospheric pressure). In the cooler, higher portions of the liquid, where the vapor pressure is lower, the bubbles are collapsed by external pressure. With further heating, however, the temperature throughout the liquid becomes high enough to sustain bubbles. The bubbles rise all the way to the surface and escape (see Figure 8.16). At this point we say the liquid is boiling.

The **boiling point** of a liquid is *the temperature at which the vapor pressure of the liquid becomes equal to the external (atmospheric) pressure exerted on it.* Because atmospheric pressure fluctuates from day to day, the boiling point of a liquid does also. Thus, to compare the boiling points of different liquids, the external pressure must be the same. The boiling point of a liquid that is most often used to compare and tabulate is called the *normal* boiling point. A liquid's **normal boiling point** is *the temperature at which the liquid boils under a pressure of 760 mm Hg.*

At any given location, the changes in the boiling point of a liquid caused by *natural* variations in atmospheric pressure seldom exceed a few degrees; in the case of water, the maximum is about 2°C. However, variations in boiling points *between* locations at different elevations can be quite striking, as is shown in the data in Table 8.3.

BOILING POINT AND
ELEVATION

PRESSURE COOKERS

The boiling point of a liquid can be raised by increasing the external pressure. This principle is used in the operation of a pressure cooker. Foods cook faster in pressure cookers because the elevated pressure causes water to boil above 100°C. With an increase in temperature of only 10°C, food cooks in approximately one-half the normal time. (Cooking food involves chemical reactions, and the rate of a chemical reaction generally doubles with every 10°C increase in temperature.) Table 8.4 gives the boiling temperatures reached by water under

TABLE 8.3

Variation of the Boiling Point of Water with Elevation

Location	Elevation (feet above sea level)	Boiling Point of Water (°C)
San Francisco, California	sea level	100.0
Salt Lake City, Utah	4,390	95.6
Denver, Colorado	5,280	95.0
La Paz, Bolivia	12,795	91.4
Mount Everest	28,028	76.5

TABLE 8.4

Boiling Point of Water at Various Pressure-Cooker Pressures

Pressure Above Atmospheric		Boiling Point of Water (°C)
$lb/in.^2$	mm Hg	
5	259	108
10	517	115
15	776	120

FROZEN JUICE
CONCENTRATES

normal household pressure-cooker conditions. Hospitals use this same principle (boiling-point elevation) to sterilize instruments; there sufficiently high temperatures are reached to destroy bacteria.

Liquids can be made to boil at low temperatures by reducing the external pressure. This principle is used in preparing numerous food products, including frozen fruit juice concentrates. Some of the water in a fruit juice is boiled away at a reduced pressure, thus concentrating the juice without having to heat it to a high temperature. Heating juices to high temperatures causes changes that spoil the taste of the juice and reduce its nutritional value.

8.16   Intermolecular Forces in Liquids

Boiling points of substances vary greatly. Some are well below 0°C; for example, oxygen has a boiling point of −183°C. As we know, numerous other substances do not boil until the temperature is much higher. An explanation of this variation in boiling points involves a consideration of the nature of the *intermolecular forces* that molecules must overcome to escape from the liquid state into the

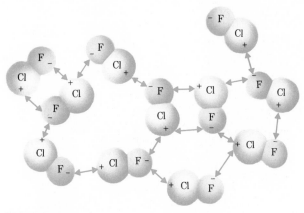

FIGURE 8.17

Dipole–dipole interactions between randomly arranged ClF molecules. The positive end of one molecule is attracted to the negative end of another, neighboring molecule.

vapor state. **Intermolecular forces** *are forces that act* BETWEEN *one molecule or ion and another molecule or ion.*

Intermolecular forces are similar in one way to the previously discussed *intramolecular* (*within* molecules) forces involved in covalent bonding (see Sections 5.8 and 5.9); they are electrostatic in origin. A major difference between inter- and intramolecular forces is their magnitude; the former are much weaker. However, despite their relative weakness, intermolecular forces exert "strong" effects because of the large numbers of such interactions present. The properties of many liquids are affected in dramatic ways by intermolecular forces.

A major type of intermolecular force is the dipole–dipole interaction. **Dipole–dipole interactions** *are electrostatic attractions between polar molecules.* Polar molecules (Section 5.12), often called *dipoles*, are electrically unsymmetrical. Therefore, when polar molecules approach each other, they tend to line up so that the relatively positive end of one molecule is directed toward the relatively negative end of the other molecule. As a result, an electrostatic attraction occurs between the molecules. The greater the polarity of the molecules, the greater the strength of the dipole–dipole interactions; the greater the strength of the dipole–dipole interactions, the higher the boiling point of the liquid is. Figure 8.17 shows selected dipole–dipole interactions that are possible for a random arrangement of polar ClF molecules.

Particularly strong dipole–dipole interactions occur between polar molecules that contain a hydrogen atom bonded to a highly electronegative element such as fluorine, oxygen, and nitrogen. Two factors account for the extra strength of

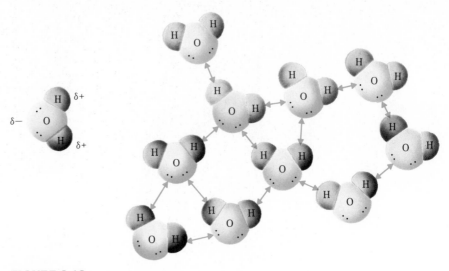

FIGURE 8.18

Hydrogen bonds (double-headed arrows) among water molecules.

these dipole–dipole interactions: (1) the great polarity of the bond between hydrogen and a highly electronegative element and (2) the close approach of the dipoles, which is allowed by the very small size of the hydrogen atom. Dipole–dipole interactions of this type are given a special name—hydrogen bonds. A **hydrogen bond** *is a stronger than usual dipole–dipole interaction that occurs between polar molecules when one or both molecules contain a hydrogen atom bonded to a very electronegative element.*

Water (H_2O) is the most commonly encountered substance that shows hydrogen bonding. Figure 8.18 depicts the process of hydrogen bonding among water molecules. Because of the polarity of water molecules (see Section 5.12), each hydrogen atom carries a partial positive charge (denoted as $\delta+$), and each oxygen atom carries a partial negative charge (denoted as $2\delta-$). Hydrogen bonds result from the close approach of the very small, partially charged hydrogen atoms to nonbonding pairs of electrons on the partially charged oxygen atoms of other water molecules.

Hydrogen bonds are much weaker than ionic or covalent bonds. The strongest hydrogen bonds are about one-tenth as strong as a covalent bond. Nevertheless, hydrogen bonds profoundly affect the properties of substances in which they occur. The effects of hydrogen bonding on the properties of water are considered in Chapter 11.

Many molecules of biological importance, such as deoxyribonucleic acid (DNA) and proteins, contain O—H and N—H bonds, and hydrogen bonding plays a role in the behavior of these substances. Certain bonds in these compounds must be capable of breaking and reforming with relative ease, and only hydrogen bonds have the right energies to permit this. It is not an overstatement to say that hydrogen bonding makes life possible. The importance of hydrogen bonding in molecules of biological significance is explored further in Chapters 24 and 25.

8.17   Sublimation and Melting

Solids, like liquids, have vapor pressures. The odors associated with mothballs, spices, and instant coffee are evidence of this. The kinetic molecular theory applies to solids in much the same way as it does to liquids. Particles in the solid state have a range of energies, and particles at a solid's surface can escape into the gas phase if they acquire sufficiently high energies. **Sublimation** *is the process whereby a solid changes directly to a gas (vapor).*

The vapor pressures for solids are generally much lower than those of liquids because the cohesive forces in solids are not overcome by opposing disruptive forces to the extent that they are in liquids (see Section 8.3). As expected, the vapor pressure of a solid increases with increasing temperature. At room temperature and pressure, there are very few solids for which the vapor pressure is sufficiently high that the transition from solid to vapor is noticeable.

Two sublimation processes that most people have encountered in everyday life involve solid carbon dioxide (dry ice) and naphthalene (mothballs). Sublimation of solid water (ice) is also a common process during winter months. Ice disappears (sublimes) from sidewalks and driveways in the coldest part of the winter even though the temperature does not rise above freezing. Wet laundry that is hung out to dry in freezing weather eventually dries as the frozen water sublimes. A commercial sublimation process called *freeze-drying* is used to dry biological materials or foodstuffs that would be damaged by heating (see Figure 8.19). Reduced pressure is usually a necessary part of such commercial freeze-drying processes. Freeze-drying can also be used to preserve tissue cultures and bacteria. These freeze-dried specimens can be stored for an extended period of time.

Although all solids have vapor pressures and the potential to sublime, most of them do not sublime. Instead, most solids melt when heated. **Melting** *is the process in which a solid changes to a liquid.* As the temperature of a solid is increased, the motion of the particles about their fixed positions becomes more vigorous and the particles are forced farther apart. Eventually, the particles gain sufficient kinetic energy to break down (collapse) the rigid structure associated with the solid state. When this happens, we say that the substance has melted.

ICE AND WINTER
WEATHER

DEHYDRATED
FOODSTUFFS

FIGURE 8.19
Commercial packages of freeze-dried foods. Water is removed from these foods through sublimation.

8.18 Decomposition

Nearly every gas becomes a liquid and then a solid if its constituent particles are slowed (cooled) enough to allow the cohesive forces that are present in all matter to dominate the disruptive forces. It is not true, however, that heating changes all solids into liquids or all liquids into gases. In some cases, the atoms that make up the molecules of a solid or liquid can acquire enough energy and vibrate so violently that bonds within the molecules are broken before the solid or liquid can change state. When this occurs, we say that the substance has decomposed. **Decomposition** *is the process in which chemical bonds (intramolecular forces) are overcome (broken) before intermolecular forces are overcome.* For example, paper and cotton do not melt when heated, but char or decompose instead.

Practice Questions and Problems

STATES OF MATTER

8-1 Contrast the following properties for the solid, liquid, and gaseous states.
a. Shape
b. Volume
c. Density
d. Compressibility
e. Thermal expansion

8-2 The following statements relate to *solid state, liquid state*, and *gaseous state*. Match the appropriate term to each statement.
a. This state is characterized by the lowest density of the three states.
b. This state is characterized by an indefinite shape and a high density.
c. Temperature changes significantly influence the volume of this state.
d. Pressure changes influence the volume of this state more than the other two states.
e. In this state, constituent particles are less free to move around than in other states.

KINETIC MOLECULAR THEORY OF MATTER

8-3 Answer the following questions about the kinetic molecular theory of matter.
a. What two types of energy do particles possess?
b. How do molecules transfer energy to each other?
c. What is the relationship between temperature and the average velocity with which the particles move?

8-4 Answer each of the following questions about *cohesive forces* and *disruptive forces*.
a. What type of energy is associated with each of these types of forces?
b. What effect does temperature have on the magnitude of each of these types of forces?
c. What are the effects of each of these types of forces on a system of particles?

8-5 Distinguish among the gaseous, liquid, and solid states of a substance from the point of view of the relative magnitude of the kinetic and potential energies of the constituent particles.

8-6 Explain each of the following observations using the kinetic molecular theory of matter.
a. Gases have a low density.
b. Solids maintain characteristic shapes.
c. Liquids show little change in volume with changes in temperature.
d. Both liquids and solids are practically incompressible.
e. A gas always exerts a pressure on the object or container with which it is in contact.

8-7 Using kinetic molecular theory, explain why it is dangerous to throw an aerosol can in an open fire.

8-8 Illustrate the difference among the three states of matter on the microscopic (molecular) level by drawing three different arrangements of 15 to 20 small spheres.

GAS-LAW VARIABLES

8-9 What volume units are most often used in gas law calculations?

8-10 What temperature scale must always be used to specify temperature when it is used in a gas law equation?

8-11 What are the three most common units for specifying the pressure a gas exerts and what are the interrelationships among these three units?

8-12 Carry out the following pressure-unit conversions using the dimensional analysis method of problem solving.
a. 725 mm Hg to atmospheres
b. 1.20 atm to mm Hg
c. 1.20 atm to torr
d. 12.0 psi to atmospheres
e. 652 mm Hg to pounds per square inch

BOYLE'S LAW

8-13 Give a written statement and a mathematical equation for Boyle's law.

8-14 When we say that two quantities are inversely proportional, what do we mean?

8-15 A balloon is inflated to a volume of 8.3 L on a day when the atmospheric pressure is 652 mm Hg. The next day, as a storm front arrives, the atmospheric pressure drops to 620 mm Hg. Assuming the temperature remains constant, what is the new volume for the balloon, in liters?

8-16 A sample of carbon monoxide gas (CO) in an expandable container exerts a pressure of 525 mm Hg when the volume is set at 3.00 L. Determine the volume settings that would produce the following pressures, assuming the temperature remains constant.
a. 750 mm Hg   b. 900 mm Hg
c. 425 mm Hg   d. 200 mm Hg

8-17 A sample of N_2 gas occupies a volume of 3.00 L at 27°C and 1.2 atm. Determine the pressure that this sample exerts at the same temperature if the volume of the gas is changed to each of the following values.
a. 1.00 L   b. 1.76 L
c. 3.25 L   d. 7.20 L

8-18 Explain Boyle's law using the kinetic molecular theory of matter.

CHARLES' LAW

8-19 Give a written statement and a mathematical equation for Charles' law.

8-20 When we say that two quantities are directly proportional, what do we mean?

8-21 An adult human breathes in approximately 0.500 L of air at 36°C and 1.00 atm pressure with every breath. What would this "breath volume" be at the same pressure if the temperature drops to 24°C?

8-22 A sample of H_2 gas has a volume of 2.71 mL at 27°C and 1.2 atm pressure. What volume will the H_2 gas occupy at each of the following temperatures if the pressure is held constant?
a. 327°C   b. 227°C   c. − 73°C   d. 1500°C

8-23 A sample of O_2 gas occupies a volume of 325 mL at 25°C and a pressure of 725 torr. At the same pressure, determine the temperature, in degrees Celsius, at which the volume of the gas would be equal to each of the following.
a. 375 mL   b. 75 mL
c. 775 mL   d. 2200 mL

8-24 Explain Charles' law using the kinetic molecular theory of matter.

GAY-LUSSAC'S LAW

8-25 Give a written statement and a mathematical equation for Gay-Lussac's law.

8-26 An aerosol spray can is empty, except for the propellant gas, which exerts a pressure of 1.2 atm at 24°C. If the can is thrown into a fire ($T = 485°C$), what will be the pressure inside the hot can?

8-27 A sample of F_2 gas at 55°C in a constant volume container exerts a pressure of 1.31 atm. To what temperatures, in degrees Celsius, would the gas have to be heated for it to exert each of the following pressures?
a. 2.00 atm   b. 10.0 atm
c. 2.63 atm   d. 6.33 atm

8-28 At room temperature, 27°C, the pressure exerted by some SO_2 gas stored in a 20.0-L steel cylinder is 1.50 atm. What will be the pressure of the SO_2 in the cylinder if the temperature is allowed to reach the following levels?
a. 86°C   b. 250°C   c. 450°C   d. 773°C

8-29 Explain Gay-Lussac's law using the kinetic molecular theory of matter.

COMBINED GAS LAW

8-30 A sample of Cl_2 gas has a volume of 25.0 L at a pressure of 4.00 atm and a temperature of 27°C. What volume, in liters, will the gas occupy at 25.0 atm and 327°C.

8-31 A sample of N_2O gas has a volume of 4.32 L at a pressure of 0.34 atm and a temperature of 127°C. What pressure, in atmospheres, will the gas exert at a temperature of 53°C and a volume of 2.25 L?

8.32 A sample of C_2H_2 gas has a volume of 5.00 L at a pressure of 1.00 atm and a temperature of 100°C. What will be the temperature of the gas, in degrees Celsius, if the volume is decreased to 1.00 L and the pressure is increased to 5.00 atm?

CHANGES OF STATE

8-33 Which two changes of state involve each pairing?
a. Gaseous and liquid states
b. Gaseous and solid states
c. Liquid and solid states

8-34 Indicate whether each of the following is an exothermic or endothermic change of state.

a. Sublimation   b. Freezing
c. Melting   d. Deposition
e. Evaporation   f. Condensation

8-35 Match the following statements to the appropriate term: evaporation, vapor pressure, boiling point, deposition, and decomposition.

a. This is a temperature at which the liquid vapor pressure is equal to the external pressure on the liquid.

b. This process corresponds to a direct change from the gaseous state to the solid state.

c. This process takes place when a liquid changes to a vapor.

d. In this process, intramolecular forces rather than intermolecular forces are "overcome."

e. This property can be measured by allowing a liquid to evaporate in a closed container.

f. At this temperature, bubbles of vapor form within a liquid.

g. This temperature is affected by changes in atmospheric pressure.

8-36 Clearly explain each of the following observations.

a. All liquids do not have the same vapor pressure at a given temperature.

b. Changing the volume of a container in which there is a liquid–vapor equilibrium does not change the magnitude of the vapor pressure.

c. Increasing the temperature of a liquid–vapor equilibrium system causes an increase in the magnitude of the vapor pressure.

8-37 Clearly explain each of the following observations.

a. The boiling point of a liquid varies with atmospheric pressure.

b. It takes more time to cook an egg in boiling water on a mountain top than at sea level.

c. Food will cook just as fast in boiling water with the stove set at "low heat" as in boiling water at "high heat."

d. Foods cook faster in a pressure cooker than in an open pan.

8-38 Clearly explain each of the following observations.

a. A person feels more uncomfortable in humid air at 90°F than in dry air at 90°F.

b. A person emerging from an outdoor swimming pool on a breezy day gets the "shivers."

c. During the cold winter months, snow often slowly disappears without melting.

8-39 Distinguish between the terms *boiling point* and *normal boiling point*.

8-40 What two factors affect the rate at which a substance evaporates?

8-41 What factors affect the magnitude of the vapor pressure a liquid exerts?

8-42 Criticize the statement "All solids will melt if heated to a high enough temperature."

INTERMOLECULAR FORCES IN LIQUIDS

8-43 What is the difference in meaning between the terms *intermolecular force* and *intramolecular force*?

8-44 In liquids, what is the relationship between boiling point and the strength of intermolecular forces?

8-45 What is a dipole–dipole interaction?

8-46 What is the difference between a hydrogen bond and a dipole–dipole interaction?

8-47 Which is harder to break, an ordinary covalent bond or a hydrogen bond?

Solutions and Colloids

9.1 Characteristics of Solutions

All samples of matter can be classified into two categories: pure substances and mixtures. We discussed this concept of matter classification in Section 2.1, but have tended to limit our discussion to pure substances so far. Now we turn our attention to mixtures, with emphasis in this chapter on homogeneous mixtures (Section 2.2).

Another name for a homogenous mixture is a solution; we will use this terminology in this chapter. A **solution** *is a homogenous (uniform) mixture of two or more substances (components).* Solutions are common in nature and they represent an abundant form of matter. Solutions carry nutrients to the cells of our bodies and carry away waste products. The ocean is a solution of water, sodium chloride, and many other substances (even gold). A large percentage of chemical reactions take place in solution, including most of those discussed in later chapters in this text.

When discussing solutions, it is often convenient to call one component of the solution the *solvent* and other components present *solutes.* The **solvent** *is the component of a solution that is present in the greatest amount.* The solvent can be thought of as the medium in which the other substances present are dissolved. A **solute** *is a solution component that is present in a small amount compared to the solvent.* More than one solute can be present in a solution. For example, both sugar and salt (two solutes) can be dissolved in a container of water to give a salty sugar water solution.

In most of the situations we encounter, the solutes present in a solution will be of more interest to us than the solvent. The solutes are the ''active ingredients'' in the solution. These are the substances that undergo reaction when solutions are mixed.

The solutions used in laboratories and clinical settings are most often liquids, and the solvent is almost always water. However, as we shall see shortly, gaseous solutions and solid solutions of numerous types do exist.

Because a solution is homogenous, it will have the same properties throughout. No matter where we take a sample from a solution, we will obtain material with the same composition as that of any other sample from the same solution. The composition of a solution can be varied, usually within certain limits, by changing the relative amounts of solvent and solute present. If the composition limits are transgressed, a hetereogeneous mixture is formed.

Nine types of two-component solutions exist, according to a classification scheme based on the physical states of the solvent and solute before mixing.

TABLE 9.1

Examples of Various Types of Solutions

Solution Type *(solute listed first)*	*Example*
Gaseous Solutions	
Gas dissolved in gas	Dry air (oxygen and other gases dissolved in nitrogen)
Liquid dissolved in gas*	Wet air (water vapor in air)
Solid dissolved in gas*	Moth repellent (or moth balls) sublimed into air
Liquid Solutions	
Gas dissolved in liquid	Carbonated beverage (CO_2 in water)
Liquid dissolved in liquid	Vinegar (acetic acid dissolved in water
Solid dissolved in liquid	Salt water
Solid Solutions	
Gas dissolved in solid	Hydrogen in platinum
Liquid dissolved in solid	Dental filling (mercury dissolved in silver)
Solid dissolved in solid	Sterling silver (copper dissolved in silver)

* An alternative viewpoint is that liquid-in-gas and solid-in-gas solutions do not actually exist as true solutions. From this viewpoint water vapor or moth repellent in air is considered to be a gas-in-gas solution since the water or moth repellent must evaporate or sublime first, to enter the air.

Table 9.1 lists these types of solutions and gives an example of each type. The type of solution that is emphasized in this book is that in which the final state of the solution components is liquid.

The physical state of a solute becomes that of the solvent when a solution is formed. For example, solid naphthalene (moth repellent) must be sublimed in order to dissolve in air. Finely pulverizing a solid and dispersing it in air does not produce a solution. (Dust particles in air are an example of this.) The particles of the solid must be subdivided to the molecular level; that is, the solid must be sublimed. Similarly, fog is a suspension of water droplets in air; the droplets are large enough to reflect light, a fact that becomes evident when we drive an automobile on a foggy night. Thus, fog is not a solution. However, water vapor is present in solution form in air. When hydrogen gas dissolves in platinum metal (a gas in solid solution), the gas molecules take up fixed positions in the structure of the metal; the gas becomes "solidified."

9.2 Terminology for Describing Solutions

In addition to solvent and solute, several other terms are used to describe characteristics of solutions. The **solubility** *of a solute is the amount of solute that will dissolve in a given amount of solvent.* Many factors affect the numerical value of a solute's solubility in a given solvent, including the nature of the solvent itself, the temperature, and in some cases, the pressure and the presence of other solutes.

Common units for expressing solubility are grams of solute per 100 g of solvent. The temperature of the solvent must also be specified. Table 9.2 gives the solubilities of selected solutes in the solvent water at three different temperatures.

A **saturated solution** *is a solution that contains the maximum amount of solute that can be dissolved under the conditions at which the solution exists.* When additional solute is added to a saturated solution it will not dissolve; instead, it forms a second phase, resulting in a heterogeneous mixture. A saturated solution containing excess undissolved solute is an equilibrium situation in which the rate of dissolution of undissolved solute is equal to the rate of crystallization of dissolved solute. Consider the process of adding table sugar (sucrose) to a container of water. Initially, the added sugar dissolves as the solution is stirred. Finally, as we add more sugar, we reach a point where no amount of stirring will cause the added sugar to dissolve. The last-added sugar remains as a solid on the bottom of the container; the solution is saturated. Although it appears to the eye that nothing is happening once the saturation point is reached, this is not the case on the molecular level. Solid sugar from the

TABLE 9.2

Solubilities of Various Compounds in Water at 0°C, 50°C, and 100°C

Solute	Solubility (g solute/100 g H_2O)		
	0°C	*50°C*	*100°C*
Lead(II) bromide ($PbBr_2$)	0.455	1.94	4.75
Silver sulfate (Ag_2SO_4)	0.573	1.08	1.41
Copper(II) sulfate ($CuSO_4$)	14.3	33.3	75.4
Sodium chloride (NaCl)	35.7	37.0	39.8
Silver nitrate ($AgNO_3$)	122	455	952
Cesium chloride (CsCl)	161.4	218.5	270.5

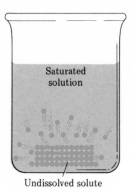

Saturated
solution

Undissolved solute

FIGURE 9.1
The dynamic equilibrium process occuring in a
saturated solution that contains undissolved *excess*
solute.

bottom of the container is continuously dissolving in the water and an equal
amount of sugar is coming out of solution. Accordingly, the net number of sugar
molecules in the liquid remains the same so that, outwardly, the dissolution
process appears to have stopped. The equilibrium situation in the saturated
solution is somewhat similar to the previously discussed evaporation of a liquid
in a closed container (Section 8.14). Figure 9.1 illustrates the dynamic equilibrium process occurring in a saturated solution that contains undissolved excess
solute.

An **unsaturated solution** *is a solution in which less solute than the*
maximum amount possible is dissolved in the solution. Most solutions we
encounter fall into this category.

The terms *dilute* and *concentrated* are also used to convey qualitative information about the degree of saturation of a solution. A **dilute solution** *is a*
solution that contains a small amount of solute relative to the amount that could
dissolve. On the other hand, a **concentrated solution** *is a solution that*
contains a large amount of solute relative to the amount that could dissolve. A
concentrated solution does not have to be a saturated solution.

When dealing with liquid-in-liquid solutions, the terms *miscible, partially*
miscible, and *immiscible* are frequently used to describe solubility characteristics
associated with the liquids. **Miscible** *substances dissolve in any amount in*
each other. For example, methyl alcohol (CH_3OH) is completely miscible with
water; that is, they completely mix with each other in any and all proportions. After
these two liquids are mixed, only one phase is present. **Partially miscible**
substances have limited solubility in each other. Benzene (C_6H_6) and water are
partially miscible. If benzene is slowly added to water, a small amount of
benzene initially dissolves, and a single phase results. However, as soon as the
benzene solubility limit is reached, the excess benzene forms a separate layer on
top of the water; it is on top because it is less dense than water. **Immiscible**
substances do not dissolve in each other. When these substances are mixed,
two layers (phases) immediately form. Very few liquids are totally immiscible

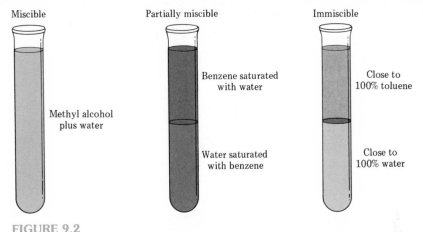

Miscible — Methyl alcohol plus water

Partially miscible — Benzene saturated with water / Water saturated with benzene

Immiscible — Close to 100% toluene / Close to 100% water

FIGURE 9.2

Miscibility of selected liquids with each other.

in each other; however, toluene (C_7H_8) and water approach this limiting case. Figure 9.2 illustrates the results obtained by mixing liquids of various miscibilities with each other.

9.3 Solution Formation

In a solution, solute particles are uniformly dispersed throughout the solvent. Considering what happens at the molecular level during the solution process helps us understand how this is achieved.

For a solute to dissolve in a solvent, two types of interparticle attractions must be overcome: (1) attractions between solute particles (solute–solute attractions) and (2) attractions between solvent particles (solvent–solvent attractions). Only when these attractions are overcome can particles in both pure solute and pure solvent separate from each other and begin to intermingle. A new type of interaction, which does not exist before solution forms, arises as the result of the mixing of solute and solvent. This new interaction is the attraction between solute and solvent particles (solvent–solute attractions). These attractions are the primary driving force for solution formation. The extent to which a substance dissolves depends on the degree to which the newly formed solvent–solute attractions are able to compensate for the energy needed to overcome the solute–solute and solvent–solvent attractions. A solute will not dissolve in a solvent if either solute–solute or solvent–solvent attractions are too strong to be compensated for by the formation of the new solvent–solute attractions.

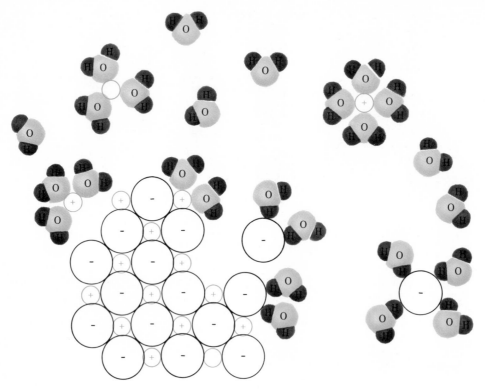

FIGURE 9.3
The solution process for an ionic solid in water.

An important type of solution process is one in which an ionic solid dissolves in water. Let us consider in detail the process of dissolving sodium chloride, a typical ionic solid, in water. We will consider this process in steps. The fact that water molecules are polar (Section 5.12) is very important in our considerations.

Figure 9.3 shows what is thought to happen when sodium chloride is placed in water. The polar water molecules become oriented in such a way that the negative oxygen portion points toward positive sodium ions and the positive hydrogen portion points toward negative chloride ions. As the polar water molecules begin to surround ions on the crystal surface, they exert sufficient attraction to cause these ions to break away from the crystal surface. After leaving the crystal, an ion retains its surrounding group of water molecules; it has become a *hydrated ion*. As each hydrated ion leaves the surface, other ions are exposed to the water, and the crystal is picked apart ion by ion. Once in solution, the hydrated ions are uniformly distributed either by stirring or through random collisions with other molecules or ions.

The random motion of solute ions in a solution causes them to collide with each other, with solvent molecules, and occasionally with the surface of any undissolved solute. Ions undergoing the latter type of collision occasionally stick to the solid surface and thus leave the solution. When the number of ions in solution is low, the chances for collision with the undissolved solute are low. However, as the number of ions in solution increases, so do the chances for collisions, and more ions are recaptured by the undissolved solute. Eventually the number of ions in solution reaches a level at which ions return to the undissolved solute at the same rate as other ions leave. At this point, the solution is saturated, and the equilibrium process discussed in the last section is in operation.

Care must be taken to avoid confusing the solubility of a solute with the *rate* at which it dissolves. It is true that some very soluble solutes dissolve rapidly, but others do not. A number of factors contribute to the rate at which a solute will dissolve.

1. *Particle size of solute.* The smaller the state of subdivision of the solute (powder versus bulk solid), the more surface area there is for solvent attack and therefore the greater the rate of dissolution.
2. *Solvent temperature.* At high temperatures solvent molecules are moving faster (Section 8.2) and have more frequent interactions with solute.
3. *Agitation or stirring of solution.* Stirring removes locally saturated, or at least concentrated, solution from the vicinity of the solute and allows less concentrated solution to take its place.

9.4 Solubility Rules

In this section, we present some rules for qualitatively predicting solute solubilities. These rules concisely summarize the results of thousands of experimental solute–solvent solubility determinations.

A very useful generalization that relates polarity to solubility is that *substances of like polarity tend to be more soluble in each other than substances that differ in polarity.* This conclusion is often expressed as the simple phrase *likes dissolve likes.* Polar substances, in general, are good solvents for other polar substances, but not for nonpolar substances. Similarly, nonpolar substances exhibit greater solubility in nonpolar solvents than they do in polar solvents.

Polarity plays an important role in the solubility of many substances in the fluids and tissues of the human body. For example, let us briefly consider vitamin solubilities. Some vitamins are water soluble and others are fat soluble.

Vitamins that are water soluble have polar molecular structures as water does. On the other hand, vitamins that are fat soluble have nonpolar molecular structures that are compatible with the nonpolar nature of fats.

Vitamin C is an example of a water-soluble vitamin. Because of its water-solubility, vitamin C is not stored in the body but must be taken in as part of our daily diet. Unneeded (excess) vitamin C is eliminated rapidly from the body through body fluids. Vitamin A is a fat-soluble vitamin. It can be stored by the body in fat tissue for later use. If the quantities of vitamin A consumed are too large (excessive vitamin supplements), illness may result. Because of its limited water solubility, vitamin A cannot be rapidly eliminated from the body by means of body fluids.

The generalization "likes dissolve likes" is a useful tool for predicting solubility behavior in many, but not all, solute–solvent situations. Results confirming this generalization are almost always obtained in gas-in-liquid and liquid-in-liquid solutions and for solid-in-liquid solutions in which the solute is not an ionic compound. In the case of a solution in which the solute is an ionic solid, a very common situation, this rule is not adequate. One would predict, because of their polar nature, that all ionic compounds are soluble in a polar solvent such as water. This is not the case. The failure of the previous generalization for ionic compounds is related to the complexity of the factors involved in determining the magnitude of the solute–solute (ion–ion) and solvent–solute (solvent–ion) interactions. Among other things, both the charge and size of the ions in the solute must be considered. Changes in these factors affect both types of interactions, but not to the same extent.

Some guidelines concerning the solubility of ionic compounds in water, which should be used in place of "likes dissolve likes," are given in Table 9.3.

TABLE 9.3

Solubility Guidelines for Ionic Compounds in Water

Ion Contained in the Compound	Solubility	Exceptions
Group IA (Li^+, Na^+, K^+, etc.)	soluble	
Ammonium (NH_4^+)	soluble	
Acetates ($C_2H_3O_2^-$)	soluble	
Nitrates (NO_3^-)	soluble	
Chlorides (Cl^-), bromides (Br^-), and iodides (I^-)	soluble	Ag^+, Pb^{2+}, Hg_2^{2+}
Sulfates (SO_4^{2-})	soluble	Ca^{2+}, Sr^{2+}, Ba^{2+}, Pb^{2+}
Carbonates (CO_3^{2-})	insoluble	group IA and NH_4^+
Phosphates (PO_4^{3-})	insoluble	group IA and NH_4^+
Sulfides (S^{2-})	insoluble	groups IA and IIA and NH_4^+
Hydroxides (OH^-)	insoluble	group IA, Ba^{2+}, Sr^{2+}

> EXAMPLE 9.1
>
> Predict the solubility of the following solutes in the solvent indicated.
> (a) NH_3 (a polar gas) in water
> (b) O_2 (a nonpolar gas) in water
> (c) AgCl (an ionic solid) in water
> (d) Na_2SO_4 (an ionic solid) in water
> (e) $AgNO_3$ (an ionic solid) in water
>
> SOLUTION
>
> (a) Soluble. Both substances are polar, so they should be relatively soluble in each other—likes dissolve likes.
> (b) Insoluble. The two substances are of unlike polarity because water is polar. The actual solubilities of NH_3 (part a) and O_2 (part b) in water at 20°C are 51.8 g/100 g H_2O and 0.0043 g/100 g H_2O, respectively.
> (c) Insoluble. Table 9.3 indicates that all chlorides except those of silver, lead, and mercury(I) are soluble. Thus, AgCl is one of the exceptions.
> (d) Soluble. Table 9.3 indicates that all ionic sodium-containing compounds are soluble. Sodium is a group IA element.
> (e) Soluble. Table 9.3 indicates that all compounds containing the nitrate ion (NO_3^-) are soluble.

All ionic compounds, even the most insoluble ones, dissolve to some slight extent in water. Thus, the "insoluble" classification used in Table 9.3 and Example 9.1 really refers to ionic compounds that have very limited solubility in water.

9.5 Solution Concentrations

By using solubility values such as those in Table 9.2, we can determine the amount of solute contained in a specified quantity of a saturated solution. Most solutions of interest to us, however are not saturated. Therefore, we need a method to express the amount of solute present in unsaturated solutions over and beyond the qualitative terms *dilute* and *concentrated* (Section 9.2). Our methods will need to be quantitative.

In this section, we discuss two methods for quantitatively representing solution concentration. For the concentration units we will consider, the **concentration** *of a solution is the amount of solute present in a specified amount of solution.* Thus, concentration is a ratio of two quantities.

$$\text{Concentration} = \frac{\text{Amount of solute}}{\text{Amount of solution}}$$

Percent Concentration

The concentration of a solution is often specified in terms of the percent of solute in the total amount of solution. Different types of percent units exist because the amounts of solute and solution present can be stated in terms of either mass or volume. The three most common types of percent units are

1. Mass–mass percent (or percent by mass).
2. Volume–volume percent (or percent by volume).
3. Mass–volume percent.

Mass–mass percent (or percent by mass) is the percentage unit most often used in chemical laboratories. **Percent by mass** *is equal to the mass of solute divided by the total mass of solution multiplied by 100 (to put the value in terms of percentage).*

$$\text{Percent by mass} = \frac{\text{Mass of solute}}{\text{Mass of solution}} \times 100$$

The solute and solution masses must be measured in the same units, which is usually grams. The mass of the solution is equal to the mass of the solute plus the mass of the solvent.

$$\text{Mass of solution} = \text{Mass of solute} + \text{Mass of solvent}$$

A solution whose mass percent concentration is 5.0% would contain 5.0 g of solute per 100.0 g of solution (5.0 g of solute and 95.0 g of solvent). Thus, percent by mass directly gives the number of grams of solute in 100 g of solution. The percent by mass concentration unit is often abbreviated as % (m/m).

EXAMPLE 9.2

What is the percent by mass, % (m/m), concentration of sucrose (table sugar) in a solution made by dissolving 3.2 g of sucrose in 61.9 g of water?

SOLUTION

Both the mass of solute and mass of solvent are known. Substituting these numbers into the equation

$$\% \ (m/m) = \frac{\text{Mass of solute}}{\text{Mass of solution}} \times 100$$

gives

$$\% \ (m/m) = \frac{3.2 \text{ g sucrose}}{3.2 \text{ g sucrose} + 61.9 \text{ g water}} \times 100$$

Note that the denominator of the preceding equation is mass of solution, which is the combined mass of solute and solvent.

Doing the mathematics gives

$$\% \ (m/m) = \frac{3.2 \text{ g}}{65.1 \text{ g}} \times 100 = 4.9\%$$

Volume–volume percent (or percent by volume), which is abbreviated % (v/v), is used as a concentration unit when the solute and solvent are both liquids or both gases. In these cases, it is often more convenient to measure volumes than masses. **Percent by volume** *is equal to the volume of solute divided by the total volume of solution multiplied by 100.*

$$\text{Percent by volume} = \frac{\text{Volume of solute}}{\text{Volume of solution}} \times 100$$

Solute and solution volumes must always be expressed in the same units when percent by volume is used.

When the numerical value of a concentration is expressed as a percent by volume, it directly gives the number of milliliters of solute in 100 mL of solution. Thus, a 100-mL sample of a 5.0% (v/v) alcohol-in-water solution contains 5.0 mL of alcohol dissolved in enough water to give 100 mL of solution. Note that such a 5.0% (v/v) solution could not be made by adding 5 mL of alcohol to 95 mL of water because the volumes of different liquids are not usually additive. Differences in the way molecules are packed, as well as differences in distances between molecules, almost always result in the volume of the solution being less than the sum of the volumes of solute and solvent. For example, the final volume resulting from the addition of 50.0 mL of ethyl alcohol to 50.0 mL of water is 96.5 mL of solution.

EXAMPLE 9.3

When 80.0 mL of methyl alcohol and 80.0 mL of water are mixed, they make a solution that has a final volume of 154 mL. What is the concentration of the solution, expressed as percent by volume of methyl alcohol?

SOLUTION

To calculate the percent by volume, the volumes of solute and solution are needed. Both are given in the problem statement.

$$\text{Solute volume} = 80.0 \text{ mL}$$

$$\text{Solution volume} = 154 \text{ mL}$$

Note that the solution volume is not the sum of the solute and solvent volumes. As mentioned previously, the volumes of two different liquids are not usually additive.

Substituting the given quantities into the equation

$$\text{Percent by volume} = \frac{\text{Volume solute}}{\text{Volume solution}} \times 100$$

gives

$$\% \ (v/v) = \left(\frac{80.0 \text{ mL}}{154 \text{ mL}}\right) \times 100 = 51.9\%$$

The third type of percent unit in common use is mass–volume percent. This unit, which is often encountered in clinical and hospital settings, is particularly convenient to use in working with a solid solute, which is easily weighed, and a liquid solvent. Solutions of drugs for external and internal use, intravenous and intramuscular injectables, and solutions used for testing for the presence of particular substances are usually labeled in percent mass–volume. This unit is abbreviated as % (m/v).

Mass–volume percent *is equal to the mass of solute in grams divided by the total volume of solution in milliliters multiplied by 100:*

$$\text{Mass–volume percent} = \frac{\text{Mass of solute (g)}}{\text{Volume of solution (mL)}} \times 100$$

Note that in defining mass–volume percent, specific mass and volume units are given. This is necessary because the units do not cancel as was the case with mass percent and volume percent.

Mass–volume percent indicates the number of grams of solute dissolved in each 100 mL of solution. Thus, a 2.3% (m/v) solution of any solute contains 2.3 g of solute in each 100 mL of solution and a 7.9% (m/v) solution contains 7.9 g of solute in each 100 mL of solution.

The steps in preparing a % (m/v) solution are shown in Figure 9.4. First, the amount of solute needed is weighed out (Figure 9.4a). This solute is then added to a 100-mL flask (Figure 9.4b). A small amount of solvent is added to the flask,

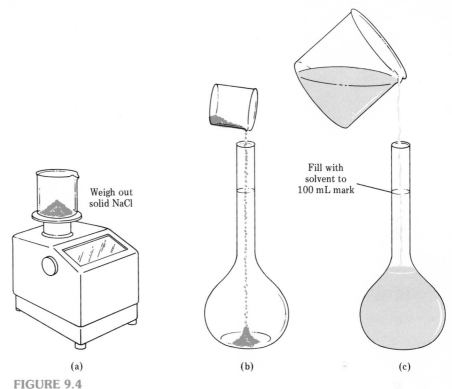

FIGURE 9.4

Steps in the preparation of a mass–volume percent NaCl solution.

and the solute is dissolved in the solvent. The flask is then filled up to the 100-mL mark (Figure 9.4c). Note that one does not add 100 mL of solvent, but enough solvent to make 100 mL of solution. If weighed solute was added to 100 mL of solvent, the final volume of the solution would be greater than 100 mL because the solute occupies some volume in the final solution.

EXAMPLE 9.4

Normal saline solution that is used to dissolve drugs for intravenous use is a 0.92% (m/v) NaCl solution. How many grams of NaCl are required to prepare 35.0 mL of normal saline solution?

SOLUTION

The given quantity is 35.0 mL of solution and the desired quantity is grams of NaCl.

$$35.0 \text{ mL solution} = ? \text{ g NaCl}$$

The given concentration, 0.92% (m/v), which means 0.92 g NaCl per 100 mL of solution, is used as a conversion factor to go from milliliters of solution to grams of NaCl. The setup for the conversion is

$$35.0 \text{ mL solution} \times \left(\frac{0.92 \text{ g NaCl}}{100 \text{ mL solution}} \right)$$

Doing the arithmetic, after canceling the units, gives

$$\left(\frac{35.0 \times 0.92}{100} \right) \text{ g NaCl} = 0.32 \text{ g NaCl}$$

Molarity

Because chemical reactions occur between molecules and atoms, it is often convenient to express the concentration of a given solute (element or compound) in terms of the number of particles (moles) present. The mole, it should be recalled (Section 7.3), is the unit used for counting large numbers of atoms or molecules. *Molarity* is a concentration unit in which the amount of solute is specified in terms of moles. The **molarity** *of a solution is a ratio giving the number of moles of solute per liter of solution*:

$$\text{Molarity} = \frac{\text{Moles of solute}}{\text{Liters of solution}}$$

The abbreviation for the molarity concentration unit is a capital *M*.

Two solutions of the same molarity contain the same number of solute molecules. In contrast, two solutions of equal mass percent do not necessarily contain the same number of solute molecules. Such solutions contain equal masses of solute. If the formula weights of the two solutes differ, which is usually the case, the number of moles (and molecules) differ.

The use of solution molarities allows us to measure an exact number of moles of solute simply by measuring a solution volume. For this reason, molarity is the concentration unit most often used in a chemical laboratory.

To find the molarity of a solution, we need to know the solution volume in liters and the number of moles of solute present. An alternative to knowing the number of moles of solute is knowing the number of grams of solute present and the solute's formula weight. The number of moles can be calculated from these two quantities.

EXAMPLE 9.5

Determine the molarities of the following solutions.

(a) 2.63 moles of Na_2SO_4 is dissolved in enough water to give 545 mL of solution.

(b) 20.0 g of NaOH is dissolved in enough water to give 2.25 L of solution.

SOLUTION

(a) The number of moles of solute is given in the problem statement.

$$\text{Moles of } Na_2SO_4 = 2.63$$

The volume of the solution is also given in the problem statement, but not in the right units. Molarity requires liters for the volume units and we are given milliliters of solution. Making the unit change gives

$$545 \text{ mL} \times \left(\frac{1 \text{ L}}{1000 \text{ mL}} \right) = 0.545 \text{ L}$$

The molarity of the solution is obtained by substituting the known quantities into the equation

$$M = \frac{\text{Moles of solute}}{\text{Liters of solution}}$$

which gives

$$M = \frac{2.63 \text{ moles } Na_2SO_4}{0.545 \text{ L solution}}$$

$$= 4.83 \frac{\text{moles } Na_2SO_4}{\text{L solution}}$$

Note that the units for molarity are always moles per liter.

(b) This time, the volume of solution is given in the correct units.

$$\text{Volume of solution} = 2.25 \text{ L}$$

The moles of solute must be calculated from the grams of solute (given) and the solute's formula weight, which is 40.0 amu (calculated from a table of atomic weights).

$$20.0 \text{ g NaOH} \times \left(\frac{1 \text{ mole NaOH}}{40.0 \text{ g NaOH}} \right) = 0.500 \text{ mole NaOH}$$

Substituting the known quantities into the defining equation for molarity gives

$$M = \frac{0.500 \text{ mole NaOH}}{2.25 \text{ L solution}} = 0.222 \frac{\text{mole NaOH}}{\text{L solution}}$$

9.6   Colligative Properties of Solutions

The physical properties that a solvent possesses when it is pure undergo change when a solute is added. Some of these physical property changes depend on the chemical nature of the solute added to the solvent. Solution color is an example of such a physical property. If $KMnO_4$ (potassium permanganate) is added to water, a purple solution results; a water solution of $CuSO_4$ [copper(II) sulfate] gives a blue solution, and a water solution of $Ni(NO_3)_2$ [nickel(II) nitrate] has a green color. Other physical properties of a solvent undergo changes that do not depend on the chemical nature of the solute but only on its amount (number of solute particles present). Such physical properties are called *colligative properties*. The **colligative properties** *of a solution are those physical properties whose extent of change depend only on the number (concentration) of solute particles (molecules or ions) in a given quantity of solvent and not on their chemical identities.* Four important colligative properties of a solution are vapor pressure, boiling point, freezing point, and osmotic pressure. In this section we briefly consider each of these four colligative properties.

Vapor Pressure

The vapor pressure of a solution containing a nonvolatile solute is always less than that of pure solvent at the same temperature. (A nonvolatile solute does not have a measurable vapor pressure of its own; it has a low tendency to vaporize.) Vapor pressure depends on the ability of molecules to escape from the surface of a liquid (Section 8.14). In a solution, solute molecules literally get in the way of solvent molecules and interfere with their escape. Since some of the solute molecules occupy positions on the surface of the liquid, their presence decreases the probability of solvent molecules escaping. Figure 9.5 illustrates the decrease in surface concentration of solvent molecules when a solute is added. As the *number* of solute particles increases, the extent of vapor pressure reduction also increases; thus, vapor pressure is a colligative property. What is important is not the identity of the solute molecules, but the fact that they ''take up room'' on the surface of the liquid.

Boiling Point

The addition of a nonvolatile solute to a solvent *raises* the boiling point of the resulting solution above that of the pure solvent; that is, it causes boiling-point

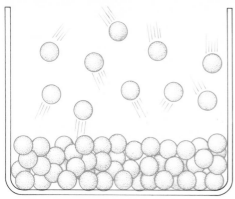

(a) Pure solvent
All surface positions are
occupied by solvent molecules.

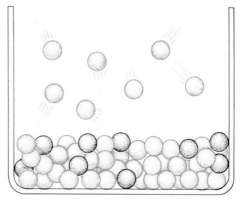

(b) Solution
Some surface positions
are occupied by solute
molecules.

FIGURE 9.5

Close-up comparison of the surface of pure solvent and a solution formed from
adding solute to the solvent.

elevation. This is logical when we remember that the vapor pressure of the
solution is lower than that of pure solvent, and that the boiling point depends on
vapor pressure (Section 8.15). A higher temperature will be required to raise the
depressed vapor pressure of the solution to atmospheric pressure (external
pressure); this is the condition necessary for boiling to occur.

AUTOMOBILE ENGINE
COOLANT

The coolant ethylene glycol (a nonvolatile solute) is added to car radiators to
prevent boil over in hot weather. The engine may not run any cooler, but the
coolant–water mixture does not boil until a temperature well above the normal
boiling point of water is reached.

CANDY MAKING

Boiling point elevation is a key factor in many types of candy making. Most
candy is made by dissolving sugar and other ingredients in water and cooking
the mixture at the boiling point of the resulting solution. As the mixture boils,
water evaporates, the concentration of sugar increases, and the boiling point of
the mixture also increases (boiling point elevation). Heating of the mixture
continues until the sugar concentration, as indicated by boiling point, reaches a
specific level. Thus, the "candy thermometer" is used to determine sugar
concentration. Perfect candy results if the cook also remembers to correct the
boiling point for any difference in atmospheric pressure due to altitude (Section 8.15).

Freezing Point

The addition of nonvolatile solute to a solvent lowers the freezing point of the
resulting solution. This freezing point depression, like boiling point elevation, is a

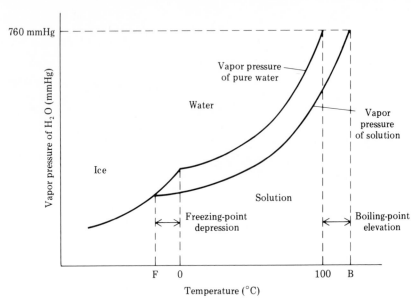

FIGURE 9.6

The effect of solute on the freezing and boiling points of a solvent.

direct consequence of the decrease in vapor pressure associated with the presence of solute. A liquid freezes at the temperature at which the vapor pressure of the liquid becomes equal to the vapor pressure of the solid state of the substance. Vapor-pressure lowering for the liquid causes this equality point to be reached at a lower temperature, as shown in the vapor pressure graph (Figure 9.6). (The same graph can also be used to show boiling-point elevation.)

Applications of freezing-point depression are many. In climates where temperatures drop below 0°C in the winter, it is necessary to protect water-cooled automobile engines from freezing. This is done by adding antifreeze (usually ethylene glycol) to the radiator (see Figure 9.7). The addition of this nonvolatile material causes the vapor pressure and freezing point of the resulting solution to be much lower than that of pure water. Also in winter, salt, usually NaCl or $CaCl_2$, is spread on roads and sidewalks to melt ice or prevent it from forming. The salt dissolves in the water to form a solution that will not freeze until the temperature drops lower than 0°C, the normal freezing point of water. This method is ineffective, however, if the temperature of the ice is below that to which the addition of salt can depress the freezing point of water.

Freezing-point depression gives fruit farmers a little safety margin against crop failure caused by springtime freezes. The blossoms and fruit of trees contain solutions that freeze at approximately −2.3°C (28°F). Smudge pots and other devices are used to keep orchard temperatures higher than the critical level.

The making of homemade ice cream demonstrates freezing-point depression in two different locations—inside the container holding the ice cream mix and in

SALT AND WINTER ICE

SPRINGTIME FRUIT
BLOSSOM FREEZE

HOME-MADE ICE CREAM

FIGURE 9.7

A woman pours antifreeze into the radiator of her car in preparation for cold-weather driving. The scientific principle involved in the use of antifreeze in automobile radiators is freezing-point depression.

the salt–ice–water mixture surrounding the container. The ice cream mix (a mixture of milk, sugar, and other materials) has a freezing point lower than that of pure water (or an ice–water mixture), since it is actually a solution. To become frozen, the ice cream mix must be surrounded by a solution that will freeze at an even lower temperature. The salt–ice–water mixture is used for this purpose, and it must contain enough solute (salt) to cause a freezing point depression in the ice–water mixture greater than that which has occurred in the ice cream mix.

Osmosis and Osmotic Pressure

The process of osmosis and the colligative property of osmotic pressure are extremely important phenomena when considering biological solutions. These phenomena govern many of the processes that occur in a functioning human body.

Osmosis *involves the passage of a solvent from a dilute solution (or pure solvent), through a semipermeable membrane, into a more concentrated solution.* The simple apparatus shown in Figure 9.8a is helpful in explaining, at the molecular level, what actually occurs during the osmotic process. The apparatus consists of a tube containing a concentrated saltwater solution that has been immersed in a dilute saltwater solution. The immersed end of the tube is

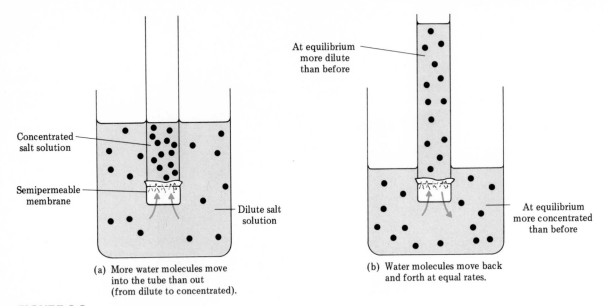

Concentrated salt solution

Semipermeable membrane

Dilute salt solution

At equilibrium more dilute than before

At equilibrium more concentrated than before

(a) More water molecules move into the tube than out (from dilute to concentrated).

(b) Water molecules move back and forth at equal rates.

FIGURE 9.8

An apparatus in which osmosis, the flow of solvent through a semipermeable membrane from a dilute to a concentrated solution, can be observed.

covered with a semipermeable membrane. A **semipermeable membrane** *is a thin layer of material that allows certain types of molecules to pass through, but prohibits the passage of others.* The particles that are allowed to pass through (usually just solvent molecules such as water) are relatively small. Thus, the membrane functions somewhat like a sieve. Using the experimental setup of Figure 9.8a, we can observe a net flow of solvent from the dilute to the concentrated solution over the course of time. This is indicated by a rise in the level of the solution in the tube and a drop in the level of the dilute solution, as shown in Figure 9.8b.

What is actually happening on a molecular level as the process of osmosis occurs? Water is flowing in both directions through the membrane. However, the rate of flow into the concentrated solution is greater than the rate of flow in the other direction (see Figure 9.9). Why? The presence of solute molecules diminishes the ability of water molecules to cross the membrane. The solute molecules literally "get in the way"; they occupy some of the surface positions next to the membrane. As there are more solute molecules on one side of the membrane than on the other, the flow rates differ.

The net transfer of solvent across the membrane continues, at a diminishing rate, until the pressure of the solution on the concentrated side of the membrane becomes sufficient to counterbalance the greater escaping tendency of molecules from the dilute side. (This pressure results from the increased mass and volume of the concentrated solution.)

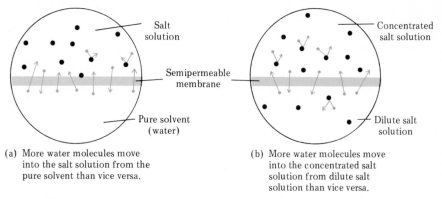

(a) More water molecules move into the salt solution from the pure solvent than vice versa.

(b) More water molecules move into the concentrated salt solution from dilute salt solution than vice versa.

FIGURE 9.9

An enlarged view of a semipermeable membrane separating (a) pure water and a salt solution and (b) a dilute saltwater solution and a concentrated saltwater solution. Because the solute molecules (large dots) interfere with the movement of water molecules (small dots), water moves more readily from an area of lower solute concentration to an area of greater solute concentration.

Osmotic pressure is the amount of pressure that must be applied to prevent the flow of solvent through a semipermeable membrane from a solution of lower solute concentration to a solution of higher solute concentration. In terms of Figure 9.8, osmotic pressure is the pressure required to prevent water from rising in the tube. Figure 9.10 shows how this pressure can be measured.

The greater the original concentration difference between the separated solutions, the greater the magnitude of the osmotic pressure is. Note that the concentrations of the two solutions involved in osmosis never become equal, even at equilibrium. However, the more concentrated solution is less concentrated than it originally was because of its increased volume. Of course, the amount of solute stays the same.

Cell membranes in both plants and animals are semipermeable in nature. The selective passage of fluid materials through these membranes governs the balance of fluids in living systems. Thus, osmotic-type phenomena are of prime importance for life. We say "osmotic-type phenomena" instead of "osmosis" because the semipermeable membranes found in living cells usually permit the passage of small solute molecules (nutrients and waste products) in addition to solvent. The term *osmosis* implies the passage of solvent only. The substances prohibited from passing through the membrane in osmotic-type processes are large molecules and insoluble suspended materials.

PLANTS AND SALTWATER

Plants die if they are watered with saltwater because of an osmotic-type process. Because the salt solution outside the root membranes is more concentrated than the solution in the root, water flows out of the roots; the plant dehydrates and dies. This same principle is behind the reason for not drinking

P (osmotic pressure)

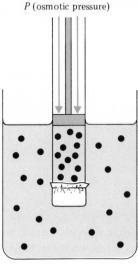

No net flow into the tube
because of the applied pressure

FIGURE 9.10

Osmotic pressure is the amount of pressure needed to prevent the solution in the tube from rising as the result of the process of osmosis.

saltwater, even if you are stranded on a raft in the middle of the ocean. When saltwater is taken into the stomach, water flows out of the stomach wall and into the stomach; then the tissues become dehydrated. Drinking seawater increases your thirst because the body loses water rather than absorbs it.

The manufacture of pickles involves osmotic-type behavior. The skin of a cucumber placed in a strong salt (brine) solution acts as a semipermeable membrane. Because the solution inside the cucumber is more dilute, water diffuses out of the cucumber, and the cucumber shrinks in size.

Solutions of 0.92% (m/v) sodium chloride are found to exert an osmotic pressure about equal to that exerted by the fluids in a red blood cell. Red blood cells placed in this solution (which is called physiological saline) neither swell nor shrink. Two solutions such as these, for which the osmotic pressure is the same, are said to be *isotonic*. A 5.5% (m/v) glucose solution is also isotonic with the fluid inside red blood cells. The processes of replacing body fluids intravenously and supplying nutrients to the body by intravenous drip both require the use of isotonic solutions.

9.7 Colloidal Dispersions

Colloidal dispersions (or colloids) are heterogeneous mixtures that have many properties similar to those of solutions, although they are not true solutions. In a broad sense, colloids may be thought of as "solutions" in which a material is *suspended* rather than *dissolved*. A **colloidal dispersion** *is a dispersion*

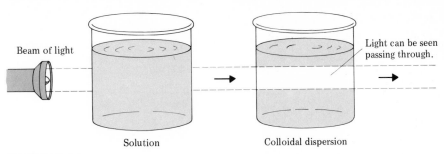

Solution   Colloidal dispersion

FIGURE 9.11

A beam of light travels through a true solution without being scattered. This is not true for a colloidal dispersion.

(*suspension*) *of small particles of one substance in another substance.* The terms *solute* and *solvent* are not used to indicate the components of a colloidal dispersion. Instead, the particles suspended in a colloidal dispersion are called the *dispersed phase*, and the material in which they are suspended is called the *dispersing medium*.

Suspended colloidal material is small enough in size that it (1) is usually not discernible by the naked eye, (2) does not settle out under the influence of gravity, and (3) cannot be filtered out with filter paper that has relatively large pores. In these aspects, the suspended material behaves similarly to the solute of a solution. However, the suspended material is sufficiently large to make the dispersion nonhomogeneous.

A beam of light detects the nonhomogeneity of a colloidal dispersion. If we shine a beam of light through a true solution, we cannot see the "track" of the light. However, a beam of light passing through a colloid can be observed because the light is scattered by the colloidal particles. This scattered light is reflected in our eyes. Figure 9.11 contrasts the different behaviors of a true solution and a colloidal dispersion with a beam of light.

The diameters of the suspended particles in a colloidal dispersion are in the range of 10^{-7} to 10^{-5} cm. This compares with diameters of less than 10^{-7} cm for particles such as ions, atoms, and molecules. Thus, colloidal particles are up to 1000 times larger than those present in a true solution. The suspended particles are usually aggregates of molecules, but this is not always the case. Some protein molecules are sufficiently large to form colloidal dispersions that contain single molecules in suspension. Colloidal dispersions containing particles with diameters larger than 10^{-5} cm are usually not encountered. Suspended particles of this size usually settle out under the influence of gravity.

Types of Colloidal Dispersions

Colloids can be classified into categories by using the system that was previously employed for classifying solutions (Section 9.1). This system is based on the

TABLE 9.4

Types of Colloidal Dispersions

Type			
Dispersing Medium	*Dispersed Phase*	*Name*	*Examples*
Gas	liquid	aerosol	fog, mist, aerosol sprays, some types of air pollutants
Gas	solid	aerosol	smoke, some types of air pollutants
Liquid	gas	foam	whipped cream, shaving cream
Liquid	liquid	emulsion	milk, mayonnaise, cosmetic creams
Liquid	solid	sol	paint, ink, blood, gelatin, hot chocolate
Solid	gas	solid foam	marshmallow, foam rubber
Solid	liquid	solid emulsion	butter, cheese
Solid	solid	solid sol	pearls, opals, colored glass

physical states of the pure constituents before they interact. Only eight types of colloidal dispersions are possible, in contrast with nine types of solutions. A gas-in-gas system cannot produce a colloid because gas molecules are not large enough to be colloidal. The eight types of colloidal systems are listed in Table 9.4, along with examples and the special names by which the general types are known.

Note the wide variety of substances that are colloidal in nature. They range from natural items such as blood and milk to manufactured food items (cheese and butter), shaving cream, paint, and ink (see Figure 9.12).

Gelatin desserts and jellies belong to a special subclass of sols called *gels*. In these colloids, the solid dispersed phase has a very high affinity for the dispersing medium. The gel sets by forming a three-dimensional network of particles and water. Another example of a gel is canned heat (jellied alcohol), which is made from a mixture of ethyl alcohol and a saturated solution of calcium acetate.

Many different biochemical colloidal dispersions occur within the human body. Foremost among them is blood, which has numerous components that are colloidal in size. The colloidal nature of bile salts and bile protein help keep slightly soluble cholesterol in suspension in the blood. Fat is transported in the blood and lymph systems as colloidal-sized particles.

Colloidal Stability

Colloidal dispersions vary in stability. Some dispersions in water have been kept for hundreds of years and others separate in a short time. Two factors that affect colloid stability are Brownian movement and electrical charges.

FIGURE 9.12

Examples of colloids found in the home: milk, mayonnaise, and butter (liquid in liquid); whipped cream (gas in liquid); gelatin dessert (liquid in solid); milk of magnesia (solid in liquid).

A colloidal dispersion that is illuminated from the side and viewed through a microscope appears to be a collection of tiny sparkling specks. Observations reveal that the suspended particles move around in a very erratic, zigzag manner. This erratic motion, which was first observed by British botanist Robert Brown in 1827, is known as *Brownian movement* (see Figure 9.13). This motion is attributed to constant collisions between the suspended particles and the molecules of the liquid dispersing medium. The fact that colloidal particles are small enough to be buffeted around by molecular collisions is one major reason why they do not settle out of the dispersing medium. Evidently, the collisions exert a greater force on the particles than gravity does.

Colloidal particles tend to adsorb (physically attract and hold on their surface) charged ions that are present in the dispersing medium. The sign (positive or negative) of the adsorbed ions depends on the nature of the colloid, but all particles within a particular system attract ions of the same charge. Thus, the colloidal particles, having acquired the same charge, repel each other. This repulsion helps prevent the particles from coalescing into aggregates that are large enough to settle out. This repulsion is another reason why colloidal particles can remain indefinitely suspended in their medium.

If the electrical charge is removed from colloidal particles, they tend to coalesce and precipitate (settle out). This is the principle behind the operation of electrostatic precipitators, which are finding increasing use in air pollution control procedures. In this equipment, the charged colloids (air pollutants), which are suspended in a gaseous medium (air), are exposed to metal plates of opposite

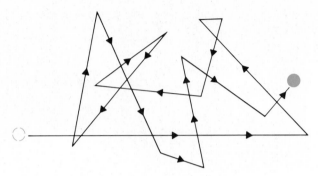

FIGURE 9.13

The erratic, zigzag path followed by a colloidal particle as a result of Brownian motion.

charge. They are attracted to these plates, give up their charge, coalesce into larger particles, and finally settle out of the air.

The formation of a river delta is an important natural example of coalescence and precipitation of colloidal particles. As a river flows toward the ocean, it picks up a lot of finely divided soil and mud that become colloidally dispersed in the river. When the river empties into the ocean, saltwater (a solution of Na^+ and Cl^- ions) is encountered. These ions neutralize the electrical charges on the surface of the tiny colloidal mud particles. This destabilizes the colloid, and mud settles out, forming the river delta.

Practice Questions and Problems

SOLUTION TERMINOLOGY

9-1 Indicate which substance is the *solvent* and which is the *solute* in each of the following solutions.
  a. A solution containing 10.0 g of NaCl and 100.0 g of water.
  b. A solution containing 20.0 mL of ethyl alcohol and 15.0 mL of water.
  c. A solution containing 0.50 g of $AgNO_3$ and 10.0 mL of water.

9-2 Give an example of each of the following types of solutions.
  a. Solute is a gas and solvent is a gas.
  b. Solute is a solid and solvent is a liquid.
  c. Solute is a liquid and solvent is a solid.
  d. Solute is a gas and solvent is a liquid.

9-3 Use Table 9.2 to determine whether each of the following solutions is saturated or unsaturated.
  a. 1.94 g of $PbBr_2$ in 100 g of H_2O at 50°C
  b. 34.0 g of NaCl in 100 g of H_2O at 0°C
  c. 75.4 g of $CuSO_4$ in 200 g of H_2O at 100°C
  d. 0.540 g of Ag_2SO_4 in 50 g of H_2O at 50°C

9-4 In a saturated solution containing undissolved solute, solute is continually dissolving, but the concentration of the solution remains the same. Explain.

9-5 Based on the solubilities given in Table 9.2, characterize each of the following solutions as being a dilute or concentrated solution.
  a. 0.20 g of $CuSO_4$ dissolved in 100 g of H_2O at 100°C
  b. 1.50 g of $PbBr_2$ dissolved in 100 g of H_2O at 50°C
  c. 60 g of $AgNO_3$ dissolved in 100 g of H_2O at 50°C
  d. 0.50 g of Ag_2SO_4 dissolved in 100 g of H_2O at 0°C

9-6 How are immiscible liquids distinguished from those that are miscible?

SOLUBILITY OF SOLUTES

9-7 What attractive forces must be broken down (overcome) in dissolving an ionic solid? What new attractive forces are formed in dissolving an ionic solid?

9-8 Predict whether the following solutes are very soluble or slightly soluble in water.
a. SO_2 (a polar gas)
b. C_2H_5OH (a polar liquid)
c. CCl_4 (a nonpolar liquid)
d. Na_2CO_3 (an ionic solid)
e. $AlPO_4$ (an ionic solid)

9-9 Using Table 9.3, predict whether each of the following ionic compounds is soluble or insoluble in water.
a. NaI
b. AgI
c. $KC_2H_3O_2$
d. $Ba(NO_3)_2$
e. $CaSO_4$
f. $CuCO_3$
g. K_3PO_4
h. Al_2S_3

PERCENT CONCENTRATION UNITS

9-10 Calculate the mass percent of solute in the following solutions.
a. 3.27 g of KCl in 45.0 g of H_2O
b. 2.13 g of Li_2SO_4 in 35.0 g of H_2O
c. 25.0 g of KNO_3 in 850 g of H_2O

9-11 Would the same quantity of solute be present, in grams, in 200 mL of 5% (m/m) formic acid and in 200 mL of 5% (m/m) acetic acid?

9-12 Calculate the mass, in grams, of solute needed to prepare 24.0 g of a 13.5% by mass sucrose ($C_{12}H_{22}O_{11}$) solution.

9-13 Calculate the volume percent of solute in the following solutions.
a. 20.0 mL of methyl alcohol in enough water to give 500 mL of solution.
b. 4.00 mL of bromine in enough carbon tetrachloride to produce 87.0 mL of solution.
c. 75.0 mL of ethylene glycol in enough water to give 200 mL of solution.

9-14 What is the percent by volume of isopropyl alcohol in an aqueous solution made by diluting 20 mL of pure isopropyl alcohol to a volume of 125 mL?

9-15 Calculate the mass-volume percent of magnesium chloride in each of the following solutions.
a. 6.0 g of $MgCl_2$ in enough water to give 330 mL of solution.
b. 25.0 g of $MgCl_2$ in enough water to give 5.0 L of solution.
c. 1.0 g of $MgCl_2$ in enough water to give 45 mL of solution.

MOLARITY CONCENTRATION UNIT

9-16 Calculate the molarity of the following solutions.
a. 3.0 moles of potassium nitrate (KNO_3) in 0.50 L of solution.
b. 25.0 g of sodium chloride (NaCl) in 1.25 L of solution.
c. 2.25 g of baking soda ($NaHCO_3$) in 225 mL of solution.

9-17 How many grams of solute are present in 375 mL of 7.5 M $CaCl_2$ solution?

9-18 What volume, in milliliters, of 0.20 M NaCl solution is needed to obtain 1.00 g of NaCl?

COLLIGATIVE PROPERTIES

9-19 What is a colligative property of a solution?

9-20 Explain why the vapor pressure of a solution containing a nonvolatile solute is always less than that of pure solvent.

9-21 What effect does increasing the concentration of dissolved solute have on each of the following solution properties?
a. Boiling point
b. Freezing point
c. Vapor pressure

9-22 Why does seawater evaporate more slowly than fresh water at the same temperature?

9-23 In cooking, what effect does adding salt to water have on the time required to bring the water to a boil?

9-24 Sketch an arrangement used to detect osmotic pressure. Indicate where pressure needs to be applied to balance this pressure.

9-25 Humans must drink fresh water, not seawater. Explain why this is so.

9-26 A 1% (m/m) glucose solution is separated from a 3% (m/m) glucose solution by an osmotic membrane. Initially the two solution levels are the same, but after a

short period of time the levels are different.
a. What process is taking place?
b. In which direction is there a net flow of water?
c. Which solution becomes more concentrated?
d. Which solution level rises?

9-27 Explain why cut flowers wilt quickly when they are placed in a sugar solution, and why they will quickly revive when transferred to pure water.

9-28 Explain why a red blood cell would burst if immersed in pure water.

COLLOIDAL DISPERSIONS

9-29 What is the relationship between a colloidal dispersion and a true solution?

9-30 What terms are used in place of solute and solvent to describe the makeup of a colloidal dispersion?

9-31 Give two reasons for the ability of colloidal particles to stay suspended in the dispersing medium.

9-32 Describe the difference in the passage of light through a colloidal dispersion and a solution.

9-33 Distinguish between the following pairs of terms.
a. Aerosol and foam
b. Sol and gel
c. Emulsion and sol

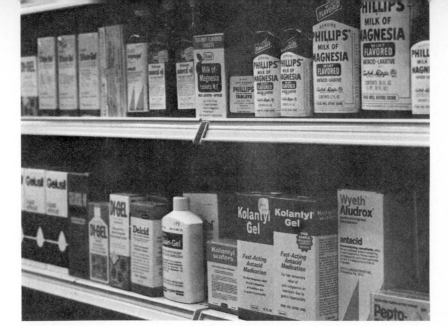

CHAPTER

10

Chemical Reactions

CHAPTER HIGHLIGHTS

10.1   Conditions Necessary for Chemical Reaction
10.2   Factors That Influence Chemical Reaction Rates
       *Chemical insights: Coal and grain dust explosions; Respiratory problems and
                           oxygen use; Refrigeration of food; Fevers and human
                           body processes; Open-heart surgery; Hibernation of
                           animals; Automobile emission control systems*
10.3   Hydrogen-Ion Transfer and Electron Transfer Reactions
10.4   Acid–Base Definitions
       Historical Profile 13: Svante August Arrhenius (1859–1927)
10.5   Strengths of Acids
10.6   Polyprotic Acids
10.7   Acid–Base Neutralization Reactions
10.8   The Dissociation of Water
10.9   The pH Scale
       Chemical insights: pH and human body fluids; pH of selected foods
10.10 Oxidation–Reduction Reactions
10.11 Oxidation Numbers

279

10.1   Conditions Necessary for Chemical Reaction

The starting point in a general discussion of chemical reactions is a consideration of the conditions necessary for a chemical reaction to take place. **Collision theory** *is a set of statements that give the conditions required for a chemical reaction to occur.* Collision theory contains three fundamental concepts developed from analysis of information from many different reactions.

1. Reactant particles must collide with each other for a reaction to occur.
2. Colliding particles must impact with a certain minimum energy for the collision to be effective—that is, to result in reaction.
3. In some cases, reactants must be oriented in a specific way when they collide for reaction to occur.

Let us consider each of these concepts separately.

Molecular Collisions

When reactions involve two or more reactants, collision theory assumes (statement 1) that the reactant molecules, ions, or atoms must come in contact (collide) with each other for any chemical change to occur. The validity of this statement is fairly obvious. Reactants cannot react with each other if they are miles apart.

Most reactions are carried out either in liquid solution or in the gaseous phase. The reason for this is directly related to the concept of molecular collisions. In both liquid solution and the gaseous phase, reacting particles are more free to move about; thus, it is easier for the reactants to come in contact with each other. Reactions involving solid reactants do occur; however, the conditions for molecular collisions are not as favorable as they are for liquids and gases. Reactions of solids usually take place only on the solid surface and thus involve only a small fraction of the total particles present in the solid. As the reaction proceeds and products dissolve, diffuse, or fall from the surface, fresh solid is

FIGURE 10.1

The heat energy generated by friction when a match head goes against the rough striking surface supplies the activation energy needed for the ignition of the match head.

exposed. In this way, the reaction eventually consumes all of the solid. The rusting of iron is an example of this type of process.

Activation Energy

The collisions between reactant particles do not always result in the formation of reaction products. Reactant particles sometimes rebound, unchanged, from a collision. Statement 2 of the collision theory indicates that, for a reaction to occur, colliding particles must impact with a certain minimum energy; that is, the sum of the kinetic energies of the colliding particles must add up to a certain mininum value. **Activation energy** *is the minimum combined kinetic energy reactant particles must possess for their collision to result in a reaction.* Every chemical reaction has a different activation energy.

In a slow reaction, the activation energy is far above the average energy content of the reacting particles. Only those few particles with above-average energy will undergo collisions that result in reaction—hence the slowness of the reaction.

Sometimes a reaction cannot start unless activation energy is provided. Once the reaction is started, enough energy is released to activate other molecules and keep the reaction going. The striking of a kitchen match is an example of such a situation (see Figure 10.1). Activation energy is initially provided by rubbing the

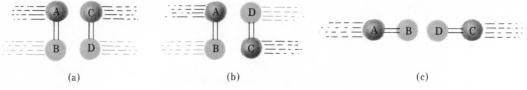

FIGURE 10.2

Different collision orientations for the reacting molecules AB and CD.

match head against a rough surface; heat is generated by friction. Once the reaction is started, the match continues to burn.

Collision Orientation

Reaction rates are sometimes very slow because reactant molecules must be oriented in a certain way for collisions to lead to products. How can this be? Statement 3 of the collision theory explains this situation. For nonspherical molecules and polyatomic ions, their orientation in relation to each other at the moment of collision affects whether or not a collision is effective.

Consider the following hypothetical reaction of molecules AB and CD.

$$A\text{—}B + C\text{—}D \longrightarrow A\text{—}C + B\text{—}D$$

In this reaction, B and C exchange places. The most favorable orientation during reactant molecule collisions would be one that simultaneously puts A and C in close proximity to each other (to form the molecule A—C) and B and D near each other (to form the molecule B—D). A possible orientation in which this situation exists is shown in Figure 10.2a. The possibility for a reaction resulting from this orientation is much greater than if the molecules were to collide while oriented as shown in Figure 10.2b or 10.2c. In Figure 10.2b, A is not near C and B is not near D. In Figure 10.2c, B is near D, but A and C are far removed from each other. Thus, certain collision orientations are preferred over others. The undesirable collision orientations of Figures 10.2b and 10.2c, however, could still result in a reaction if the molecules collided with abnormally high energies.

10.2 Factors That Influence Chemical Reaction Rates

The **rate of a chemical reaction** *is the rate or speed at which reactants are consumed or products are produced.* A number of variables affect the rate of a reaction; we encounter most of them routinely. One variable is temperature.

Food is stored in a refrigerator to reduce the rate of spoiling (chemical reactions). The state of subdivision of solids is another reaction rate variable. Sawdust and kindling wood burn (a chemical reaction) much faster than large logs do.

In this section we consider four factors that affect reaction rate: (1) the physical nature of the reactants, (2) reactant concentrations, (3) reactant temperature, and (4) the presence of catalysts.

Physical Nature of Reactants

The physical nature of reactants refers not only to the physical state of each reactant (solid, liquid, or gas), but also to the state of subdivision, or particle size. In reactions in which the reactants are all in the same physical state, the reaction rate is generally faster between liquid state reactants than between solid reactants, and fastest between gaseous reactants. Of the three states of matter, the gaseous state is the one with the most freedom of movement; a greater frequency of collision (reaction) between reactants occurs in this state.

In the solid state, reactions occur at the boundary surface between reactants. The reaction rate increases as the boundary surface area increases. Subdividing a solid into smaller particles increases surface area and thus increases reaction rate. A crushed aspirin tablet will exert its pain-relieving effect much faster than a whole tablet because its surface area is increased.

COAL AND GRAIN DUST EXPLOSIONS

When particle size is extremely small, reaction rates can be so fast that an explosion results. A lump of coal is difficult to ignite; coal dust ignites explosively. Spontaneous ignition of coal dust is a real threat to underground coal mining operations. Grain dust (very finely divided grain particles) is a problem in grain storage elevators; the possibility of an explosive ignition of the dust from an accidental spark is always present. Figure 10.3 shows the destruction that can result from such an accidental ignition.

Reactant Concentration

RESPIRATORY PROBLEMS AND OXYGEN USE

An increase in the concentration of a reactant causes an increase in the rate of the reaction. Combustible substances burn much more rapidly in pure oxygen than they do in air (21% by volume oxygen). A person with respiratory problems (such as pneumonia or emphysema) is often given air enriched with oxygen because an increased oxygen concentration facilitates the absorption of oxygen in the lungs and thus expedites all subsequent steps in respiration.

With increased concentration of a reactant, more molecules of that reactant are present in the reaction mixture; thus, the possibility is greater for collisions to occur between this reactant and other reactant particles. An analogy to the reaction rate–reactant concentration relationship can be drawn from the game

FIGURE 10.3
Extremely rapid combustion (reaction with oxygen) of grain dust produced the
explosive effect that destroyed this grain elevator. [AP/Wide World Photo.]

of billiards. The more billiard balls there are on the table, the greater the
probability that a moving cue ball will strike one of them.

When the concentration of reactants is increased, the actual quantitative
change in reaction rate varies with the specific reaction. The rate always
increases, but not to the same extent in all cases. You cannot determine how
changes in concentration will affect the reaction rate by simply looking at the
balanced equation for a reaction. This must be determined by actual experimentation. Sometimes the rate doubles with a doubling of concentration;
however, this is not always the case.

Reaction Temperature

The effect of temperature on reaction rates can also be explained by using the
molecular-collision concept. Increasing the temperature of a system increases
the average kinetic energy of the reacting molecules (Section 8.2). The

FIGURE 10.4

The chemical reactions associated with the spoiling of food are much slower at lower temperatures.

increased molecular speed causes more collisions to take place within a given time. Also, because the average energy of the colliding molecules is greater, a larger fraction of the collisions will result in reaction, from the point of view of activation energy (Section 10.1).

As a rough rule of thumb, chemists have found that the rate of a chemical reaction doubles for every 10°C increase (a difference of about 18°F) in temperature for the temperature ranges we normally encounter. The chemical reactions involved in cooking foods take place faster in a pressure cooker because the temperature is higher (Section 8.15). On the other hand, foods are cooled or frozen to slow down the chemical reactions that result in spoiling, souring, and ripening (see Figure 10.4).

When a person has a fever, many chemical reactions in the body proceed at a faster rate than normal. The fever's effect translates into increased pulse and breathing rates. For every 1°C increase in body temperature, body tissues require 13% more oxygen.

A decrease in body temperature has the opposite effect. Cells use less oxygen than they normally do. This knowledge is applied clinically in certain medical situations, such as in open-heart surgery. During open-heart surgery, a patient's body temperature is usually lowered 2–3°C; both the metabolism and oxygen requirements of the patient decrease correspondingly.

The reason many animals can withstand a long period of hibernation during

REFRIGERATION OF
FOOD

FEVERS AND HUMAN
BODY PROCESSES

OPEN-HEART SURGERY

HIBERNATION OF
ANIMALS
the winter is connected to decreased body temperature. For some animals, body temperature drops to within a few degrees of freezing. Chemical reactions are slowed down to the extent that the animals can live on stored body fat alone. Studies involving a hibernating woodchuck indicate the extent of the "slow-down." An active woodchuck's heart beats approximately 80 times a minute, while a hibernating woodchuck's heart beats only 4 times a minute.

Presence of Catalysts

A **catalyst** *is a substance that increases a reaction rate without being consumed in the reaction.* In most cases only extremely small amounts of catalyst are needed. Catalysts enhance reaction rates by providing alternate reaction pathways with lower activation energies (Section 10.1) than that of the origin uncatalyzed reaction.

A nonchemical analogy for the mechanism by which catalysts function involves a highway that goes up and over a mountain. A catalyst is analogous to a tunnel through the mountain; using the tunnel is much easier and faster than taking the highway over the mountain (Figure 10.5).

FIGURE 10.5

Going through a tunnel is easier and faster than driving up and over the mountain. A catalyst has the same effect on a chemical reaction by providing an alternate pathway by which the reaction can occur. [Jeff Albertson/Stock, Boston.]

Catalysts exert their effects in varying ways. Some catalysts provide a lower-energy pathway by entering into a reaction and forming an "intermediate," which then reacts further to produce the desired products and regenerate the catalyst. The following equations, where C is the catalyst, illustrate this concept.

Uncatalyzed Reaction $\qquad\qquad$ X + Y $\longrightarrow$ XY

Catalyzed Reaction $\qquad$ Step 1: $\ $ X + C $\longrightarrow$ XC

$\qquad\qquad\qquad\qquad$ Step 2: $\ $ XC + Y $\longrightarrow$ XY + C

In step 1 of the catalyzed reaction, the intermediate XC is formed from the interaction of the catalyst with one of the reactants (X). The activation energy for this reaction is different from (lower than) that of the uncatalyzed reaction. The intermediate from step 1, XC, serves as a reactant for step 2; this produces the desired product, XY, and regenerates the catalyst.

Solid-state catalysts often act by providing a surface to which impacting reactant molecules are physically attracted and held with a particular orientation. These reactants are sufficiently close and favorably oriented toward each other for the reaction to take place. The products of the reaction then leave the surface, making it available to catalyze other reactants.

AUTOMOBILE EMISSION CONTROL SYSTEMS

Catalysts are a key element in the functioning of automobile emission control systems. In these systems, solid catalysts speed up reactions that convert air pollutants present in the exhaust to less harmful products. For example, carbon monoxide is converted to carbon dioxide through reaction with oxygen.

Catalysts are extremely important to the proper functioning of the human body and other biological systems. In the human body special proteins called *enzymes* function as catalysts. They cause many reactions to take place rapidly and under mild conditions at body temperature; without these enzymes, the reactions would proceed very slowly and then only under harsher conditions. One of the dangers of an extremely high fever during illness is the possibility that some enzymes will be deactivated at the higher temperature. Enzymes are very sensitive to temperature changes.

10.3 Hydrogen-Ion Transfer and Electron Transfer Reactions

Two of the most important classes of chemical reactions are those of *hydrogen-ion transfer* and *electron transfer*. Both of these reaction types have far-reaching applications in our lives. An understanding of how hydrogen ions and electrons are transferred in chemical reactions is the key to understanding many important chemical phenomena.

The terms *acid* and *base* are familiar to almost everyone, even those with nonscience backgrounds. Acids and bases are particular types of compounds. Central to understanding the chemistry of these compounds is the concept of hydrogen-ion transfer. Indeed, these compounds are defined in terms of hydrogen-ion transfer. In Sections 10.4 through 10.8 we consider acids and bases, that is, hydrogen-ion transfer reactions.

The terms *oxidation* and *reduction* are also familiar to most people. These two terms are associated with chemical reactions involving electron transfer. The rusting of iron and the burning of gasoline in an automobile engine are examples of oxidation processes. Metallic copper is obtained from an ore like chalcocite (Cu_2S) by means of a reduction process. In Sections 10.9 through 10.11 we consider electron transfer reactions.

10.4   Acid–Base Definitions

Several definitions for acids and bases are now in use. We will consider two sets: the Arrhenius definitions and the Brønsted–Lowry definitions.

Arrhenius Definitions

In 1884, the Swedish chemist Svante August Arrhenius (1859–1927) (see Historical Profile 13) proposed that acids and bases be defined in terms of the species they form upon dissolution in water. His definitions are the simplest and most commonly used today. An **Arrhenius acid** *is a substance that releases hydrogen ions (H^+) in aqueous solution.* An **Arrhenius base** *is a substance that releases hydroxide ions (OH^-) in aqueous solution.*

Some common examples of acids, according to the Arrhenius definition, are the substances HNO_3 and HCl.

$$HNO_3(l) \xrightarrow{\text{H}_2\text{O}} H^+(aq) + NO_3^-(aq)$$

$$HCl(g) \xrightarrow{\text{H}_2\text{O}} H^+(aq) + Cl^-(aq)$$

When Arrhenius acids are in the pure state (not in solution), they are covalent compounds; that is, they do not contain H^+ ions. These ions are formed through a chemical reaction when the acid is mixed with water. The chemical reaction between water and the acid molecules results in the removal of H^+ ions from acid molecules.

Two common examples of Arrhenius bases are NaOH and KOH.

$$NaOH(s) \xrightarrow{H_2O} Na^+(aq) + OH^-(aq)$$

$$KOH(s) \xrightarrow{H_2O} K^+(aq) + OH^-(aq)$$

Arrhenius bases are usually ionic compounds in the pure state, in direct contrast to acids. When these compounds dissolve in water, the ions separate to yield the OH^- ions.

At an introductory level the Arrhenius definitions adequately explain the behaviors of acids and bases in aqueous solution and are the only ones found in some textbooks. However, these definitions have two drawbacks: (1) they are adequate only in the case of aqueous solutions, and (2) the identity of the acidic species present in aqueous solutions is oversimplified. Regarding drawback 2, unknown to Arrhenius, *free* hydrogen ions (H^+) cannot exist in water. The attraction between a hydrogen ion and polar water molecules is sufficiently strong to bond the hydrogen ion with a water molecule to form a hydronium ion (H_3O^+).

$$H^+ + :\overset{..}{O}{-}H \longrightarrow \left[H:\overset{..}{O}{-}H \right]^+$$

Hydronium ion

The bond holding the hydrogen ion to the water molecule is a coordinate covalent bond (Section 5.10) because both electrons are furnished by the oxygen atom.

In reality more than one water molecule is usually attracted to a H^+ ion in aqueous solution. A more accurate description of a hydrogen ion in water would be $H^+(H_2O)_n$, where n is a constantly changing number, with a value of 4 or 5 in dilute solutions at room temperature. For simplicity, however, only one attracted water molecule is written in the formula of the hydronium ion—H_3O^+ or $H^+(H_2O)$.

Brønsted–Lowry Definitions

In 1923, Johannes Nicholas Brønsted (1879–1947), a Danish scientist, and Thomas Martin Lowry (1874–1936), a British scientist, independently and almost simultaneously proposed definitions for acids and bases that expanded on the ideas of Arrhenius. Their definitions (1) extend the number of substances that can be considered to be acids and bases, (2) are not restricted to aqueous solutions, and (3) account for the fact that the acidic species in aqueous solution is the hydronium ion.

A **Brønsted–Lowry acid** *is any substance that can donate a hydrogen ion* (H^+) *to some other substance.* A **Brønsted–Lowry base** *is any substance that*

HISTORICAL PROFILE 13

Svante August Arrhenius   1859–1927

[Burndy Library, Norwalk CT.]

Svante August Arrhenius, born in 1859 near Upsala, Sweden, is considered one of the founders of modern physical chemistry. His roots are those of a Swedish farming family. An infant prodigy, on his own accord and against his parents wishes, he taught himself to read at age 3.

In 1884, as a university student, he proposed his now famous definitions for acids and bases and simultaneously presented a new theory about ionic dissociation, which was that ionic substances, when dissolved in water, dissociate into ions. This theory, which came directly from his university doctoral work in Stockholm, was given a hostile reception by many chemists including his mentors at the University of Stockholm. He was awarded his doctoral degree with reluctance and given the lowest possible passing grade. It was the opinion of his teachers that his theory was "too far-fetched."

Completely rebuffed by Swedish scientists he decided to approach the scientific world elsewhere. He did find acceptance in some places and building on this acceptance he further developed and refined his theory.

Acceptance in his homeland eventually came. In 1903, Arrhenius was awarded the Nobel prize in chemistry for his "far-fetched" ideas concerning ions in solution. In 1905, the king of Sweden founded the Nobel Institute for Physical Research at Stockholm and installed Arrhenius as director. This appointment came as a counter offer to that of a major professorship in Berlin. Arrhenius remained at the Nobel Institute until shortly before his death.

Arrhenius made other major contributions to chemistry besides those dealing with ions, acids, and bases. In 1889, while studying how rates of reactions increased with temperature, he worked out the concept of activation energy (Section 10.1). In later years he became interested in such diverse things as serum chemistry and astronomy. He spent considerable time speculating on the origin of life on earth.

can accept a hydrogen ion (H^+) from some other substance. In short, a Brønsted–Lowry acid is a hydrogen-*ion donor*, and a Brønsted–Lowry base is a *hydrogen-ion acceptor*. The Brønsted–Lowry definitions change the focus on acids and bases from the production of specific chemical species (H^+ and OH^-) to the chemical reactions involving hydrogen-ion exchange.

The Brønsted–Lowry acid–base definitions are best illustrated by example. Let us consider the formation reaction for hydrochloric acid, which involves the dissolving of hydrogen chloride gas in water

$$HCl(g) + H_2O(l) \longrightarrow H_3O^+(aq) + Cl^-(aq)$$

The hydrogen chloride behaves as a Brønsted–Lowry acid by donating a hydrogen ion to a water molecule. Note that a hydronium ion is formed as a result. The base in this reaction is water because it has accepted a hydrogen ion; no hydroxide ions are involved. The Brønsted–Lowry definition of a base includes any species capable of accepting a hydrogen ion; hydroxide ions can do this, but so can many other substances.

It is not necessary that a water molecule be one of the reactants in a Brønsted–Lowry acid–base reaction; the reaction does not have to take place in the liquid state. Brønsted–Lowry acid–base theory can be used to describe gas-phase reactions. The white solid haze that often covers glassware in a chemistry laboratory results from the gas-phase reaction between HCl and NH_3.

$$HCl(g) + NH_3(g) \longrightarrow NH_4^+(g) + Cl^-(g) \longrightarrow NH_4Cl(s)$$

This is a Brønsted–Lowry acid–base reaction because the HCl molecules donate protons to the NH_3, forming NH_4^+ and Cl^- ions. These ions combine instantaneously to form the white solid NH_4Cl.

All Arrhenius acids are Brønsted–Lowry acids, and all Arrhenius bases are Brønsted–Lowry bases. However, the converse of this statement is not true. Brønsted–Lowry theory includes Arrhenius theory and much more.

10.5   Strengths of Acids

Acids can be classified as strong or weak based on the percentage of acid molecules that dissociate (Arrhenius viewpoint) or transfer hydrogen ions (Brønsted–Lowry viewpoint) when dissolved in water. A **strong acid** *dissociates 100%, or very nearly 100%, in aqueous solution.* Thus, if an acid is strong, almost all of the acid molecules present dissociate into ions. Because of this extensive dissociation, many hydrogen ions are present in the solution of a strong acid. A **weak acid** *dissociates only slightly, usually less than 5%, in aqueous solution.* Most acid molecules are present in undissociated form in a solution of a weak acid. Relatively, only a few hydrogen ions are present in the solution of a weak acid.

The extent to which an acid dissociates in solution depends on the acid's molecular structure; molecular polarity and the strength and polarity of individual bonds are particularly important factors in determining whether an acid is strong or weak.

Only a few strong acids exist. The formulas and structures of the six most commonly encountered strong acids are listed in Table 10.1. The vast majority of acids are weak. The formulas for numerous weak acids are listed in Table 10.3.

TABLE 10.1

Commonly Encountered Strong Acids

Name	Molecular Formula	Molecular Structure
Nitric acid	HNO_3	$H-O-N-O$ with $\overset{\parallel}{O}$ below N
Sulfuric acid	H_2SO_4	$H-O-\overset{O}{\underset{O}{S}}-O-H$
Perchloric acid	$HClO_4$	$H-O-\overset{O}{\underset{O}{Cl}}-O$
Hydrochloric acid	HCl	H—Cl
Hydrobromic acid	HBr	H—Br
Hydroiodic acid	HI	H—I

An alternative approach to visualizing the difference between a strong acid and a weak acid involves a consideration of the reaction

$$HA \longrightarrow H^+ + A^-$$

where HA represents the acid and H^+ and A^- are the products of its dissociation. For solutions of strong acids the predominant species are H^+ and A^-, since the majority of HA molecules undergo dissociation. For solutions of weak acids, the opposite situation applies. The predominant species is HA, the undissociated acid. Only a few H^+ and A^- ions are present. A graphical representation of the differences between strong and weak acids, in terms of species present, is given in Figure 10.6.

Just as there are strong and weak acids, there are also strong and weak bases. As with acids, only a few strong bases exist. These are limited to the hydroxides of Groups IA and IIA listed in Table 10.2. Of the strong bases, only NaOH and KOH are commonly encountered in a chemical laboratory. The limited solubility in water of the listed Group IIA hydroxides restricts their use. Despite their limited solubility, however, these hydroxides are still considered to be strong bases; that which dissolves dissociates into ions 100%.

Only one of the many weak bases is fairly common—aqueous ammonia. In this solution of ammonia gas (NH_3) in water, small amounts of OH^- ions are produced through the reaction of NH_3 molecules with water. Water functions as the Brønsted–Lowry acid and NH_3 as the Brønsted–Lowry base.

$$H_2O(l) + NH_3(g) \longrightarrow NH_4^+(aq) + OH^-(aq)$$

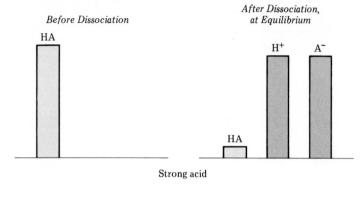

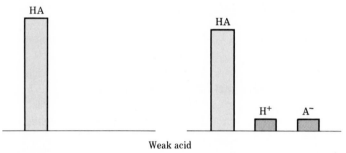

FIGURE 10.6

A strong acid is completely dissociated in solution; in a weak acid only a small fraction of the molecules is dissociated in solution.

A solution of aqueous ammonia is sometimes erroneously called ammonium hydroxide. Aqueous ammonia is the preferred designation because most of the NH_3 present has not reacted with water. Only a few ammonium ions (NH_4^+) and hydroxide ions (OH^-) are present.

It is important to remember that the terms *strong* and *weak* apply to the extent of dissociation and not to the concentrations of acid or base. For example,

TABLE 10.2

Common Strong Bases

Group IA Hydroxides	Group IIA Hydroxides
LiOH	
NaOH	
KOH	
RbOH	$Sr(OH)_2$
CsOH	$Ba(OH)_2$

stomach acid (gastric juice) is a dilute (not weak) solution of a strong acid; it is 5% by mass hydrochloric acid. On the other hand, a 36% by mass solution of hydrochloric acid is considered to be a concentrated (not strong) solution of a strong acid.

10.6   Polyprotic Acids

Acids can be classified according to the number of hydrogen ions they produce *per molecule* upon dissociation in solution. A **monoprotic acid** *is an acid that yields one hydrogen ion (H^+) per molecule upon dissociation.* The "protic" portion of the term monoprotic acid comes from the word *proton.* A H^+ ion is actually a proton. When a ^1_1H atom (99.98% abundance), a proton plus an electron, loses its electron to produce a H^+ ion, all that remains is a proton. Hydrochloric acid (HCl) and nitric acid (HNO_3) are both monoprotic acids.

A **diprotic acid** *is an acid that yields two hydrogen ions (H^+) per molecule upon dissociation.* Carbonic acid (H_2CO_3) is a diprotic acid. The dissociation process for a diprotic acid always occurs in steps. For H_2CO_3, the two steps are

$$H_2CO_3 \longrightarrow H^+ + HCO_3^-$$

$$HCO_3^- \longrightarrow H^+ + CO_3^{2-}$$

The second proton is not as easily removed as the first, because it must be pulled away from a negatively charged particle, HCO_3^-. (Remember, particles with opposite charges attract each other.) Accordingly, HCO_3^- is a weaker acid that H_2CO_3. In general, each successive step in the dissociation of an acid occurs to a lesser extent than the previous step.

A few triprotic acids exist. A **triprotic acid** *is an acid that yields three hydrogen ions (H^+) per molecule upon dissociation.* Phosphoric acid (H_3PO_4) is the most common triprotic acid. The three dissociation steps involved in the removal of H^+ ions from this molecule are

$$H_3PO_4 \longrightarrow H^+ + H_2PO_4^-$$

$$H_2PO_4^- \longrightarrow H^+ + HPO_4^{2-}$$

$$HPO_4^{2-} \longrightarrow H^+ + PO_4^{3-}$$

The general term **polyprotic acid** *describes acids that are capable of producing two or more hydrogen ions (H^+) per molecule upon dissociation.*

The number of hydrogen atoms present in one molecule of an acid cannot always be used to classify the acid as mono-, di-, or triprotic. For example, a molecule of acetic acid contains four hydrogen atoms, and yet it is monoprotic.

Only one of the hydrogen atoms in acetic acid is *acidic*; that is, only one of the hydrogen atoms leaves the molecule when it is in solution.

Whether or not a hydrogen atom is acidic is related to its location in a molecule, that is, to which other atom it is bonded. Let us consider our acetic acid example in more detail by looking at the structures of the species involved in the dissociation process. From a structural viewpoint, the dissociation of acetic acid can be represented by the equation

Acetic acid   Acetate ion

Note that one hydrogen atom is bonded to an oxygen atom; the other three hydrogen atoms are bonded to a carbon atom. The hydrogen atom bound to the oxygen atom is the acidic one. Why? Hydrogen atoms bonded to carbon

TABLE 10.3

Selected Common Mono-, Di- and Triprotic Acids

Name	Formula	Classification	Common Occurrence
Acetic	$HC_2H_3O_2$	monoprotic; weak	vinegar
Lactic	$HC_3H_5O_3$	monoprotic; weak	sour milk; cheese; produced during muscle contraction
Salicylic	$HC_7H_5O_3$	monoprotic; weak	present in chemically combined form in aspirin
Hydrochloric	HCl	monoprotic; strong	constituent of gastric juice, industrial cleaning agent
Nitric	HNO_3	monoprotic; strong	used in urinalysis test for protein; used in manufacture of dyes and explosives
Tartaric	$H_2C_4H_4O_6$	diprotic; weak	grapes
Carbonic	H_2CO_3	diprotic; weak	carbonated beverages; produced in the body from carbon dioxide
Sulfuric	H_2SO_4	diprotic; strong	storage batteries; manufacture of fertilizer
Citric	$H_3C_6H_5O_7$	triprotic; weak	citrus fruits
Boric	H_3BO_3	triprotic; weak	antiseptic eyewash
Phosphoric	H_3PO_4	triprotic; weak	found in dissociated form (HPO_4^{2-}, $H_2PO_4^-$) in intracellular fluid; component of DNA; fertilizer manufacture

atoms are too tightly held to be removed by reaction with water molecules. Water has very little effect on a carbon–hydrogen bond because that bond is only slightly polar. On the other hand, the hydrogen–oxygen bond is very polar because of oxygen's great electronegativity (see Section 5.11). Water, which is a polar molecule, readily attacks this polar bond.

The formula for acetic acid is written as $HC_2H_3O_2$ instead of $C_2H_4O_2$. We now understand why; one of the hydrogen atoms is acidic and the other three are not. In such situations it is accepted procedure to write the acidic hydrogens first, thus separating them from the other hydrogens in the formula. Citric acid, the principal acid in citrus fruits, it another example of an acid that contains both acidic and nonacidic hydrogens. Its formula, $H_3C_6H_5O_6$, indicates that three of the eight hydrogen atoms present in a molecule are acidic. Table 10.3 contains information about the use and occurrence of selected common mono-, di-, and triprotic acids, many of which contain nonacidic hydrogen atoms.

The preceding discussion has focused on acids. It should be noted that, in a similar manner, molecules of Arrhenius bases can be the source of more than one hydroxide ion. For example, calcium hydroxide, $Ca(OH)_2$, is a base that yields two hydroxide ions per molecule when it dissociates in solution. Base dissociation differs from acid dissociation in that dissociation occurs in one step rather than stepwise.

10.7   Acid–Base Neutralization Reactions

Many acid solutions have a sour taste. (Concentrated solutions of acids are very corrosive and should not be tasted.) Solutions of bases have a slippery feeling and a bitter taste. When equal amounts of acid (H_3O^+ or H^+ ions) and base (OH^- ions) are reacted together, the properties of both acid and base disappear—*neutralization* has occurred. **Neutralization** *is a chemical reaction in which an acid (H_3O^+ or H^+ ion) reacts with a base (OH^- ion) to form water.*

To obtain a better understanding of acid–base neutralization, we consider what happens, at the molecular level, when a solution of nitric acid (HNO_3) is reacted with a solution of the base potassium hydroxide (KOH). The nitric acid solution contains H^+ ions (H_3O^+ ions) and NO_3^- ions, and the potassium hydroxide solution contains K^+ and OH^- ions. When these two solutions are mixed, a reaction between the H_3O^+ ions (acidic species) and OH^- ions (basic species) occurs.

$$H_3O^+ + OH^- \longrightarrow H_2O + H_2O$$

The hydronium ions donate protons to hydroxide ions; as a result of this hydrogen-ion transfer, water molecules are produced. If equal numbers of

H_3O^+ and OH^- ions are present, the reaction between these two ions produces a neutral solution; both the acidic and basic species are converted to water.

The overall equation for the acid–base neutralization involving HNO_3 and KOH is

$$HNO_3 + KOH \longrightarrow H_2O + KNO_3$$

Note that a second product besides water has been formed. It contains the negative ion from the acid and the positive ion from the base. This product is a salt. A **salt** *is an ionic compound that contains any negative ion except hydroxide and any positive ion except hydrogen ion.* Salts, as a class of compounds, differ from acids and bases in that they contain no common ion or species that characterizes the class. Acids, bases, and salts are related in the following way: a salt is one of the products resulting from the reaction of an acid with a base.

$$Acid + Base \longrightarrow Salt + Water$$

Any time an acid and base neutralize each other, a salt is one of the products.

Many salts occur in nature, and numerous others have been prepared in the laboratory. The wide variety of uses found for salts can be seen in Table 10.4, which lists selected salts and their uses.

Much information concerning salts has been presented in previous chapters, although the term *salt* was not explicitly used in these discussions. Formula writing and nomenclature for binary ionic compounds (salts) was covered in Sections 5.5 and 5.7. As can be seen from examples in Table 10.4, many salts contain polyatomic ions such as nitrate and sulfate. These ions were discussed in

TABLE 10.4

Some Common Salts and Their Uses

Name	Formula	Uses
Ammonium nitrate	NH_4NO_3	fertilizer, explosive manufacture
Barium sulfate	$BaSO_4$	X-rays of gastrointestinal tract
Calcium carbonate	$CaCO_3$	chalk; limestone
Calcium chloride	$CaCl_2$	drying agent for removal of small amounts of water
Iron(II) sulfate	$FeSO_4$	treatment for anemia
Potassium chloride	KCl	"salt" substitute for low-sodium diets
Sodium chloride	NaCl	table salt; used as a deicer (to melt ice)
Sodium bicarbonate	$NaHCO_3$	ingredient in baking powder; stomach antacid
Sodium hypochlorite	NaClO	bleaching agent
Silver bromide	AgBr	light-sensitive material in photographic film
Tin(II) fluoride	SnF_2	toothpaste additive

Sections 5.14 and 5.15. The solubility of ionic compounds (salts) in water was the topic of Section 9.4.

All common soluble salts are dissociated into ions in solution (Section 9.3). Even if a salt is only slightly soluble, the small amount that does dissolve dissociates completely. Thus, the terms weak and strong, which are used to denote qualitatively the percent dissociation of acids and bases, are not applicable to salts. We do not use the terms *strong salt* and *weak salt*.

10.8   The Dissociation of Water

In Sections 10.4 through 10.6 we discussed the dissociation of acids in aqueous solution to produce H^+ ions and the dissociation of bases to produce OH^- ions. We now discuss another dissociation process, one that involves water molecules themselves.

Although we usually think of water as a covalent nondissociating substance, experiments show that, in reality, an *extremely small* percentage of water molecules in pure water dissociate to form ions. This dissociation reaction can be thought of as the transfer of a proton from one water molecule to another (Brønsted–Lowry theory, Section 10.4).

$$H_2O + H_2O \longrightarrow H_3O^+ + OH^-$$

or simply as the dissociation of a single water molecule (Arrhenius theory, Section 10.4)

$$H\!-\!OH \longrightarrow H^+ + OH^-$$

In this equation the formula of water is purposely written as H—OH so you can better visualize how the dissociation occurs. From either viewpoint, Brønsted–Lowry or Arrhenius, the net result is the formation of *equal amounts* of hydrogen (hydronium) ion and hydroxide ion.

At any given time, the number of H^+ and OH^- ions present in a sample of pure water is always extremely small. At 25°C, the H^+ and OH^- concentrations are $1.00 \times 10^{-7} \, M$, that is, 0.000000100 M. This concentration is equivalent to 1 out of every 500 million water molecules dissociating. The ramifications of this small amount of dissociation are very important.

The constant concentration of H^+ and OH^- ions present in pure water, at 25°C, can be used to calculate a very useful relationship called the *ion product* for water. The **ion product of water** is *the numerical value* 1.00×10^{-14}, *obtained by multiplying the molar concentrations of H^+ and OH^- ion present in pure water.* We have the following equation for the ion product of water.

$$\text{Ion product for water} = [H^+] \times [OH^-]$$
$$= (1.00 \times 10^{-7}) \times (1.00 \times 10^{-7})$$
$$= 1.00 \times 10^{-14}$$

Whenever square brackets are used in an equation, as in the equation for the ion product for water, they should be interpreted to mean *concentration in moles per liter* (molarity).

The ion-product expression for water is valid not only in pure water but also when solutes are present in water. At all times, the product of the hydrogen and hydroxide ion molarities in an aqueous solution at 25°C must equal 1.00×10^{-14}. Thus, if the $[H^+]$ is increased by the addition of an acidic solute, the $[OH^-]$ must decrease so that their product is still 1.00×10^{-14}. Similarly, if OH^- ions are added to the water, the $[H^+]$ must decrease correspondingly.

If we know the concentration of either H^+ or OH^-, we can easily calculate the concentration of the other ion present in an aqueous solution by simply rearranging the ion-product expression.

$$[H^+] = \frac{1.00 \times 10^{-14}}{[OH^-]}$$

$$[OH^-] = \frac{1.00 \times 10^{-14}}{[H^+]}$$

EXAMPLE 10.1

Sufficient acidic solute is added to a quantity of water to produce a $[H^+] = 4.0 \times 10^{-3}$ *M*. What is the $[OH^-]$ in this solution?

SOLUTION

The $[OH^-]$ can be calculated using the ion product expression for water, rearranged in the form

$$[OH^-] = \frac{1.00 \times 10^{-14}}{[H^+]}$$

Substituting into this expression the known $[H^+]$ and doing the arithmetic gives

$$[OH^-] = \frac{1.00 \times 10^{-14}}{4.0 \times 10^{-3}} = 2.5 \times 10^{-12} \, M$$

The interdependence of $[H^+]$ and $[OH^-]$ (if one increases, the other decreases) is illustrated diagramatically in Figure 10.7. Note that the increase–decrease relationship is a direct proportion. If the $[H^+]$ is increased by a factor of 10^2, then the $[OH^-]$ decreases by the same factor, 10^2.

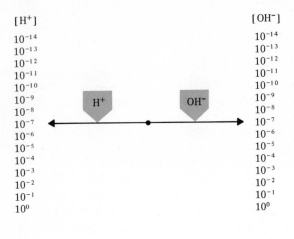

(a) In pure water the concentrations of H^+ and OH^- are equal. Each is $1.00 \times 10^{-7} M$ at 25°C.

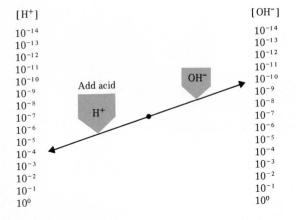

(b) If the H^+ ion concentration is increased by a factor of 10^3 (from 10^{-7} to $10^{-4} M$), the OH^- concentration is decreased by a factor of 10^3 (from 10^{-7} to $10^{-10} M$).

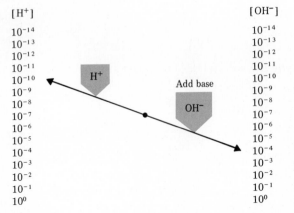

(c) If the OH^- concentration is increased by a factor of 10^3 (from 10^{-7} to $10^{-4} M$), the H^+ ion concentration is decreased by a factor of 10^3 (from 10^{-7} to $10^{-10} M$).

FIGURE 10.7

The relationship between H^+ and OH^- concentrations in an aqueous solution is fixed. The product of the two ion concentrations must always equal $1.00 \times 10^{-14} M$.

TABLE 10.5

Relationships Between [H⁺] and [OH⁻] in Aqueous Solution at 25°C

Neutral solution	$[H^+] = [OH^-] = 1.00 \times 10^{-7}\,M$
Acidic solution $[H^+] > [OH^-]$	$[H^+]$ is greater than $1.00 \times 10^{-7}\,M$ $[OH^-]$ is less than $1.00 \times 10^{-7}\,M$
Basic solution $[OH^-] > [H^+]$	$[H^+]$ is less than $1.00 \times 10^{-7}\,M$ $[OH^-]$ is greater than $1.00 \times 10^{-7}\,M$

Hydrogen ion and hydroxide ion are present in *all* aqueous solutions; the dissociation of water molecules ensures this. What, then, determines whether a solution is acidic or basic? It is the relative amounts of these two ions present. An **acidic solution** *is one in which the concentration of H⁺ ion is higher than that of OH⁻ ion.* Similarly, a **basic solution** *is one in which the concentration of the OH⁻ ion is higher than that of the H⁺ ion.* A basic solution is also often called an *alkaline solution*. It is possible to have an aqueous solution that is neither acidic or basic; such a solution is neutral. A **neutral solution** *is one in which the concentrations of H⁺ and OH⁻ are equal.* Table 10.5 summarizes the relationships between [H⁺] and [OH⁻] that we have just considered.

10.9 The pH Scale

It is convenient to have a simple way of specifying the acidity or basicity of a solution. The pH scale is used for this purpose. The **pH scale** *is a scale of small numbers that are used to specify molar hydrogen-ion concentration in an aqueous solution.* The relationship between the pH scale and molar hydrogen-ion concentration is given in Table 10.6.

From Table 10.6 we see that the pH value for a solution whose molar hydrogen-ion concentration is an exact power of 10, such as 1.0×10^{-4}, 1.0×10^{-8}, or 1.0×10^{-11}, is given directly by the negative of the exponent value on the power of 10.

$$[H^+] = 1.0 \times 10^{-x}\,M$$

$$pH = x$$

Thus, if the hydrogen ion concentration is $1.0 \times 10^{-9}\,M$, the pH is 9. Note that the preceding simple relationship between pH and [H⁺] is valid only when the coefficient in the exponential expression for the hydrogen-ion concentration is 1.0. We will comment on how to calculate the pH when the coefficient is not 1.0 later in this section.

TABLE 10.6

Relationships Between pH Values and
$[H^+]$ and $[OH^-]$ at 25°C

pH	$[H^+]$	$[OH^-]$	
0	1	10^{-14}	↑
1	10^{-1}	10^{-13}	
2	10^{-2}	10^{-12}	
3	10^{-3}	10^{-11}	Acidic
4	10^{-4}	10^{-10}	
5	10^{-5}	10^{-9}	
6	10^{-6}	10^{-8}	
7	10^{-7}	10^{-7}	NEUTRAL
8	10^{-8}	10^{-6}	
9	10^{-9}	10^{-5}	
10	10^{-10}	10^{-4}	
11	10^{-11}	10^{-3}	Basic
12	10^{-12}	10^{-2}	
13	10^{-13}	10^{-1}	
14	10^{-14}	1	↓

EXAMPLE 10.2

Calculate the pH of each of the following solutions.
(a) $[H^+] = 1.0 \times 10^{-6}\, M$
(b) $[H^+] = 0.001\, M$

SOLUTION

(a) The power of 10 associated with the molar hydrogen-ion concentration is −6. Thus, the pH is 6, which is the negative of the exponential power.
(b) We must first express the given molar hydrogen-ion concentration in scientific notation.

$$0.001 = 1 \times 10^{-3}$$

The power of 10 associated with the molar hydrogen-ion concentration is −3. Thus, the pH is 3, which is the negative of the exponential power.

From Table 10.6 we also note that acidic, neutral, and basic solutions can be identified by their pH values. At 25°C, a neutral solution has a pH value of 7. Values of pH that are less than 7 correspond to acidic solutions, and values of pH that are greater than 7 are associated with basic solutions.

The following are two other important concepts concerning pH values.

1. Increasing the hydrogen-ion concentration of a solution, which makes the solution more acidic, always lowers the pH value. The higher the concentration of hydrogen ion, the smaller the pH value.

2. A change of one unit in pH always corresponds to a 10-fold change in hydrogen-ion concentration. For example,

$$\text{when the pH} = 1, \text{ the } H^+ = 0.1\ M$$

Change of one unit | 10-fold change in concentration

$$\text{and when the pH} = 2, \text{ the } H^+ = 0.01\ M$$

In a laboratory, solutions corresponding to any pH can be prepared. The range of pH values that are displayed by natural solutions is more limited than that of prepared solutions, but solutions corresponding to most pH values can be found. Some natural solutions have nearly constant pH values. For example, human blood plasma has a pH value between 7.3 and 7.5. Any significant deviation from this range—even a few tenths of a pH unit—can cause severe physiological problems.

pH AND HUMAN BODY FLUIDS

The pH values of several human body fluids are given in Table 10.7. Most human body fluids, except for gastric juices, have pH values in a narrow range centered on neutrality. Blood plasma and spinal fluid are always slightly basic.

pH OF SELECTED FOODS

The pH values of selected foods are given in Table 10.8. Note the acidic nature of most fruits and vegetables. Tart or sour taste is associated with food of low pH.

Most of the pH values in Tables 10.7 and 10.8 are nonintegral, that is, they are not whole numbers. Obtaining such pH values from measured hydrogen-ion concentrations involves the use of logarithms, a mathematical concept we will not deal with in this text. However, when given nonintegral pH values, we can still interpret them in a qualitative way without resorting to logarithms. For example, a pH of 4.75 is between pH 4 and pH 5. We know that pH 4

TABLE 10.7

The Normal pH Range of Some Body Fluids

Type of Fluid	pH Value
Bile	6.8–7.0
Blood plasma	7.3–7.5
Gastric juices	1.0–3.0
Milk	6.6–7.6
Saliva	6.5–7.5
Spinal fluid	7.3–7.5
Urine	4.8–8.4

TABLE 10.8

Approximate pH Values for
Selected Foods

Food	pH Value
Apples	2.9–3.3
Butter	6.1–6.4
Carrots	4.9–5.3
Eggs, fresh white	7.6–8.0
Grapefruit	3.0–3.3
Limes	1.8–2.0
Milk	6.3–6.6
Peaches	3.4–3.6
Pears	3.6–4.0
Peas	5.8–6.4
Soft drinks	2.0–4.0
Strawberries	3.0–3.5
Tomatoes	4.0–4.4
Vinegar	2.4–3.4
Water, drinking	6.5–8.0

corresponds to $[H^+] = 1 \times 10^{-4}\,M$ and that pH 5 corresponds to $[H^+] = 1 \times 10^{-5}\,M$. Hence, a pH of 4.75 will correspond to a hydrogen-ion concentration that is between 1×10^{-4} and $1 \times 10^{-5}\,M$.

10.10 Oxidation–Reduction Reactions

In the same way that acid–base reactions (Sections 10.4 and 10.7) are regarded as hydrogen-ion-transfer reactions, oxidation–reduction reactions involve electron transfer between the reactants. Oxidation–reduction reactions occur all around us and even inside us. The bulk of the energy needed for the functioning of all living organisms is obtained from food through oxidation–reduction processes. Diverse phenomena such as the electricity obtained from a battery to start a car, the use of natural gas to heat a home, rusting of iron, illumination from a flashlight, and the action of antiseptic agents in killing or preventing the growth of bacteria all involve oxidation–reduction reactions. In short, knowledge of this type of reaction is fundamental to understanding many biological and technological processes.

Oxidation and *reduction*, like the terms *acid* and *base* (Section 10.4), have several definitions. The word *oxidation* was first used to describe the reaction of a substance with oxygen. According to this definition, each of the following reactions involves oxidation.

$$4 \, Fe + 3 \, O_2 \longrightarrow 2 \, Fe_2O_3$$

$$S + O_2 \longrightarrow SO_2$$

$$CH_4 + 2 \, O_2 \longrightarrow CO_2 + 2 \, H_2O$$

The reactant on the far left in each of these reactions is said to have been oxidized.

Originally, the term *reduction* referred to removal of oxygen from a compound. A common type of reduction reaction, according to this definition, is the production of a free metal from its oxide.

$$CuO + H_2 \longrightarrow Cu + H_2O$$

$$2 \, Fe_2O_3 + 3 \, C \longrightarrow 4 \, Fe + 3 \, CO_2$$

The term reduction comes from the reduction in mass of the metal-containing species; the metal has a mass less than that of the metal oxide.

Today the words oxidation and reduction are used in a much broader sense. Current definitions include the previous examples and much more. Scientists now recognize that the changes produced in a substance by reaction with oxygen can also be caused by reaction with numerous substances that do not contain oxygen. For example, consider the following reactions.

$$2 \, Mg + O_2 \longrightarrow 2 \, MgO$$

$$Mg + S \longrightarrow MgS$$

$$Mg + F_2 \longrightarrow MgF_2$$

$$3 \, Mg + N_2 \longrightarrow Mg_3N_2$$

In each of these reactions, magnesium metal is converted to a magnesium compound that contains Mg^{2+} ions. The process in each case is the same— magnesium atoms change to magnesium ions by losing two electrons; the only difference is in the identity of the substance that causes magnesium to undergo the change. All of these reactions involve oxidation according to the modern definition. **Oxidation** *is the process whereby a substance in a chemical reaction loses one or more electrons.* The modern definition for reduction is similar. **Reduction** *is the process whereby a substance in a chemical reaction gains one or more electrons.*

Oxidation and reduction are complementary processes, not isolated phenomena. They always occur together; you cannot have one without the other. If electrons are lost by one species, they cannot just disappear; they must be gained by another species. Electron transfer, then, is the basis for oxidation and reduction. The phrase **oxidation–reduction reaction** *is used to describe any*

TABLE 10.9

Oxidation–Reduction Terminology in Terms of Electron Transfer

Terms Associated with Loss of Electrons	*Terms Associated with Gain of Electrons*
Process of oxidation	Process of reduction
Substance oxidized	Substance reduced
Reducing agent	Oxidizing agent

reaction in which electrons are transferred from one reactant to another. This designation is often shortened to *redox reaction.*

There are two ways of looking at the reactants in a redox reaction. First, the reactants can be viewed as being "acted upon." From this viewpoint, one reactant is oxidized (the one that loses electrons) and one is reduced (the one that gains electrons). Second, the reactants can be looked at as "bringing about" the reaction. In this approach the terms *oxidizing agent* and *reducing agent* are used. An **oxidizing agent** *causes oxidation by accepting electrons from the other reactant.* This acceptance of electrons means that the oxidizing agent itself is reduced. Similarly, the **reducing agent** *causes reduction by providing electrons for the other reactant to accept.* Thus, the reducing agent is the substance that is oxidized.

Substance oxidized = Reducing agent

Substance reduced = Oxidizing agent

The terms oxidizing agent and reducing agent sometimes cause confusion because the oxidizing agent is not oxidized (it is reduced) and the reducing agent is not reduced (it is oxidized). A simple analogy is that a travel agent is not the one who takes a trip but the one who arranges for the trip to be taken.

Table 10.9 summarizes the terminology presented in this section.

10.11 Oxidation Numbers

Oxidation numbers are used to help determine whether oxidation and reduction have occurred in a reaction, and if they have, to identify the oxidizing and reducing agents. An **oxidation number** *is the charge that an atom appears to have when the electrons in each bond it is participating in are assigned to the more electronegative of the two atoms involved in the bond.*

Oxidation numbers are *calculated* charges on atoms rather than actual charges on atoms. This is why the phrase "appears to have" is used in the definition. When assigning oxidation numbers, we assume that each bond is ionic (complete transfer of electrons) when we assign the bonding electrons to the more electronegative element. We know that this is not always the case. Sometimes it is a good approximation and sometimes it is not. Why do we make such assignments when we know they do not always correspond to reality? Oxidation numbers, as we will see shortly, serve as a very convenient device for keeping track of electron transfer in redox reactions. Even though they do not always correspond to physical reality, they are very useful entities.

The following set of operational rules, which are derivable from the general definition for oxidation numbers, is used in determining oxidation numbers.

RULE 1   *The oxidation number of an atom in its element state is zero.*
For example, the oxidation number of Cu is zero, and the oxidation number of Cl in Cl_2 is zero.

RULE 2   *The oxidation number of any monoatomic ion is equal to the charge on the ion.*
For example, the Na^+ ion has an oxidation number of $+1$ and the S^{2-} ion has an oxidation number of -2.

RULE 3   *The oxidation numbers of groups IA and IIA elements are always +1 and +2, respectively.*

RULE 4   *The oxidation number of fluorine is always −1 and that of other group VIIA elements (Cl, Br, and I) is usually −1.*
The exception for these latter elements occurs when they are bonded to more electronegative elements. In this case, they exhibit positive oxidation numbers.

RULE 5   *The usual oxidation number of oxygen is −2.*
The exceptions occur when oxygen is bonded to the more electronegative fluorine (O then has a positive oxidation number) or is found in compounds containing oxygen–oxygen bonds (peroxides). In peroxides, the oxidation number for oxygen is -1. Peroxides form only between oxygen and hydrogen (H_2O_2), group IA metals (M_2O_2), and group IIA metals (MO_2).

RULE 6   *The usual oxidation number for hydrogen is +1.*
The exception occurs in hydrides, compounds in which hydrogen is bonded to a metal of lower electronegativity. In these compounds, hydrogen is assigned an oxidation number of -1. Examples of hydrides are NaH, CaH_2, and LiH.

RULE 7   *The algebraic sum of the oxidation numbers of all atoms in a neutral molecule must be zero.*

RULE 8   *The algebraic sum of the oxidation numbers of all atoms in a polyatomic ion is equal to the charge on the ion.*

Example 10.3 illustrates the use of these rules.

EXAMPLE 10.3

Assign oxidation numbers to each element in the following compounds or polyatomic ions.

(a) N_2O_5 (b) $KMnO_4$ (c) PO_4^{3-}

SOLUTION

(a) The sum of the oxidation numbers of all of the atoms must add up to zero (rule 7).

$$2(\text{oxid. no. N}) + 5(\text{oxid. no. O}) = 0$$

The oxidation number of oxygen is -2 (rule 5). Substituting this value into the previous equation enables us to calculate the oxidation number of nitrogen, an element for which no specific oxidation number rule exists.

$$2(\text{oxid. no. N}) + 5(-2) = 0$$

$$2(\text{oxid. no. N}) = +10$$

$$(\text{oxid. no. N}) = +5$$

Thus, the oxidation numbers for the elements involved in this compound are

$$\underset{+5\ \ -2}{N_2\,O_5}$$

Note that the oxidation number of nitrogen is not $+10$; this is the calculated charge associated with two nitrogen atoms. The oxidation number is always specified on a *per atom* basis.

(b) The sum of the oxidation numbers of all of the atoms present must be zero (rule 7).

$$(\text{oxid. no. K}) + (\text{oxid. no. Mn}) + 4(\text{oxid. no. O}) = 0$$

The oxidation number of potassium, a group IA element, is $+1$ (rule 3), and the oxidation number of oxygen is -2 (rule 5). Substituting these two values into the rule 7 equation enables us to calculate the oxidation number of manganese.

$$(+1) + (\text{oxid. no. Mn}) + 4(-2) = 0$$

$$(\text{oxid. no. Mn}) = 8 - 1 = 7$$

Thus, the oxidation numbers for the elements involved in this compound are

$$\underset{+1\ +7\ -2}{K\,Mn\,O_4}$$

Note that all of the oxidation numbers add up to zero when we take into account that there are four oxygen atoms.

$$(+1) + (+7) + 4(-2) = 0$$

$$0 = 0$$

(c) The species $PO_4{}^{3-}$ is a polyatomic ion rather than neutral compound. Thus, rule 8 rather than rule 7 applies; the oxidation number sum must add up to the charge on the ion rather than to zero.

$$(\text{oxid. no. P}) + 4(\text{oxid. no. O}) = -3$$

The oxidation number of oxygen is -2 (rule 5). Substituting this value into the sum equation gives

$$(\text{oxid. no. P}) + 4(-2) = -3$$

$$(\text{oxid. no. P}) = -3 + (+8) = +5$$

Thus, the oxidation numbers for the elements involved in this polyatomic ion are

$$\underset{+5\ -2}{P\,O_4{}^{3-}}$$

Many elements display a range of oxidation numbers in their various compounds. For example, nitrogen exhibits oxidation numbers ranging from -3 to $+5$ in various compounds. Selected examples are

NH_3	N_2O	NO	N_2O_3	NO_2	N_2O_5
-3	$+1$	$+2$	$+3$	$+4$	$+5$

As shown in this listing of nitrogen-containing compounds (and also in Example 10.3), the oxidation number of an atom is written *underneath* the atom in the formula. This convention is used to avoid confusion with the charges on ions.

Oxidizing and reducing agents can be defined in terms of changes in oxidation numbers. The **oxidizing agent** *in a redox reaction is the substance containing the atom that shows a decrease in oxidation number.* Because the oxidizing agent is the substance reduced in a reaction, reduction involves a decrease in oxidation number; the oxidation number is reduced (decreased) in a reduction. The **reducing agent** *in a redox reaction is the substance containing the atom*

TABLE 10.10

Oxidation–Reduction Terminology in Terms of Oxidation Number
Change

Terms Associated with an Increase in Oxidation Number	*Terms Associated with a Decrease in Oxidation Number*
Process of oxidation	Process of reduction
Substance oxidized	Substance reduced
Reducing agent	Oxidizing agent

that shows an increase in oxidation number. Oxidation involves an increase in oxidation number because the reducing agent is the substance oxidized in a reaction. Table 10.10 summarizes the relationships between oxidation–reduction terminology and the change in oxidation number. A comparison of Table 10.10 with Table 10.9 shows that the loss of electrons and increase in oxidation number are synonomous, as are gain of electrons and decrease in oxidation number. The fact that the oxidation number becomes more positive (increases) as electrons are lost is consistent with our understanding of the proton–electron charge relationships in an atom.

EXAMPLE 10.4

Determine oxidation numbers for each atom in the following reactions, and identify the oxidizing and reducing agents.

(a) $2 SO_2 + O_2 \longrightarrow 2 SO_3$

(b) $2 Fe_2O_3 + 3 C \longrightarrow 4 Fe + 3 CO_2$

(c) $CaCO_3 \longrightarrow CaO + CO_2$

SOLUTION

The oxidation numbers are calculated by the methods illustrated in Example 10.3.

(a) $2 SO_2 \quad + \quad O_2 \longrightarrow 2 SO_3$

$\quad\quad$ +4 −2 $\quad\quad\quad$ 0 $\quad\quad\quad\quad$ +6 −2

$\quad\quad$ Rules 5, 7 $\quad\quad$ Rule 1 $\quad\quad$ Rules 5, 7

The oxidation number of sulfur has increased from +4 to +6. Therefore, the substance that contains sulfur, SO_2, has been oxidized and is the reducing agent.

$\quad$ The oxidation number of the oxygen in O_2 has decreased from 0 to −2. Therefore, the O_2 has been reduced and is the oxidizing agent.

(b) $2 Fe_2O_3 + 3 C \longrightarrow 4 Fe \quad + \quad 3 CO_2$

$\quad\quad$ +3 −2 $\quad\quad$ 0 $\quad\quad\quad\quad$ 0 $\quad\quad\quad$ +4 −2

$\quad\quad$ Rules 5, 7 $\quad$ Rule 1 $\quad\quad\quad$ Rule 1 $\quad$ Rules 5, 7

The oxidation number of carbon has increased from 0 to +4. An increase in oxidation number is associated with oxidation. Therefore, because carbon has been oxidized, it is the reducing agent.

The oxidation number of iron has decreased from +3 to 0. Because a decrease in oxidation number is associated with reduction, the Fe_2O_3, the iron-containing compound, is the oxidizing agent.

(c) $CaCO_3 \longrightarrow CaO + CO_2$

   +2+4−2   +2−2   +4−2

   Rules 3, 5, 7   Rules 3, 5   Rules 5, 7

None of the elements has experienced a change in oxidation number. Calcium is +2 on both sides of the equation, carbon is +4 on both sides, and oxygen remains at −2 throughout the equation. Not all reactions are redox reactions. Electron transfer (oxidation and reduction) is not a prerequisite for a chemical reaction. This equation is not a redox equation, and so oxidizing and reducing agents are not present.

10.12 Some Important Oxidation–Reduction Reactions

In this section we consider two important applications of redox reactions.

1. A *spontaneous* oxidation–reduction reaction can be used to convert chemical energy into electrical energy. For this to occur, a redox reaction must be carried out in a specially designed apparatus, called a *galvanic cell*.
2. A *nonspontaneous* oxidation–reduction reaction can be caused to occur by using electrical energy to produce chemical energy. Such a process is called *electrolysis*, and the apparatus in which the reaction is carried out is called an *electrolysis cell*.

Galvanic Cells

When a strip of zinc metal is placed in a solution of copper(II) sulfate, a source of Cu^{2+} ion, a coating of copper metal forms on the zinc strip (see Figure 10.8). At the same time this occurs, some of the zinc dissolves to give Zn^{2+} ions in solution. The reaction occurring is

$$Zn(s) + Cu^{2+}(aq) \longrightarrow Zn^{2+}(aq) + Cu(s)$$

The sulfate ions (SO_4^{2-}) present in the solution remain unaffected by this change.

This reaction is an example of a spontaneous oxidation–reduction reaction. When Zn metal and a solution of Cu^{2+} ions come in contact with each other,

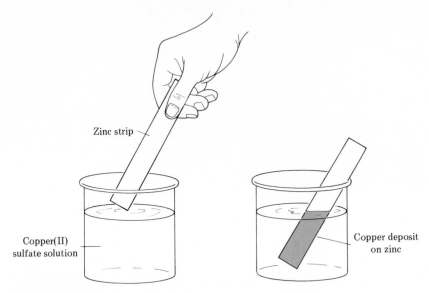

Zinc strip

Copper(II)
sulfate solution

Copper deposit
on zinc

FIGURE 10.8

A spontaneous redox reaction occurs when zinc metal is placed in a solution of Cu^{2+} ions. A coating of copper metal quickly deposits on the zinc metal.

spontaneously the Zn metal is oxidized and the Cu^{2+} ions are reduced; a direct transfer of electrons from the zinc atoms to the Cu^{2+} ions occurs. The products of this reaction are copper atoms (Cu^{2+} ions that have gained electrons) and Zn^{2+} ions (zinc atoms that have lost electrons). Heat is liberated as the reaction proceeds, as evidenced by a slight warming of the solution.

By rearranging the reactants for this spontaneous redox reaction, one can obtain energy in the form of electricity rather than heat. The desired arrangement, a simple galvanic cell, is shown in Figure 10.9. A **galvanic cell** *is an apparatus in which a spontaneous redox reaction is used to convert chemical energy to electrical energy.*

The galvanic cell of Figure 10.9 has two compartments separated by a porous partition. One compartment contains a strip of zinc metal immersed in a solution of zinc sulfate ($ZnSO_4$), and the other contains a strip of copper metal immersed in a solution of copper(II) sulfate ($CuSO_4$). The porous partition prevents the solutions from mixing completely; it does, however, allow for passage of ions from one compartment to the other, a necessity for proper operation of the cell. The two strips of metal, called *electrodes*, are connected by a wire. This wire allows electrons to be transferred from one electrode to the other. As the spontaneous reaction between Zn and Cu^{2+} ions occurs, the flow of electrons through the wire can be demonstrated by placing a lightbulb in the external circuit (see Figure 10.9). The light bulb glows.

What is actually happening in the cell to cause the lightbulb to glow?

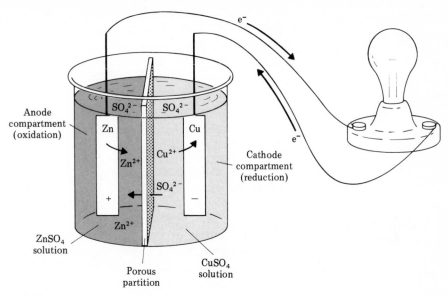

FIGURE 10.9

A zinc-copper galvanic cell. The electrical energy produced by this cell is generated by the spontaneous redox reaction $Zn(s) + Cu^{2+}(aq) \rightarrow Zn^{2+}(aq) + Cu(s)$.

1. Electrons are produced at the zinc electrode through the process of oxidation.

$$Zn(s) \longrightarrow Zn^{2+}(aq) + 2\, \bar{e}$$

2. The electrons pass from the zinc electrode to the copper electrode through the external circuit (the wire).

3. Electrons enter the copper electrode and are accepted by Cu^{2+} ions in solution adjacent to the electrode. This is a reduction reaction.

$$Cu^{2+}(aq) + 2\, e^{-} \longrightarrow Cu(s)$$

4. To complete the circuit, ions (both positive and negative) move through the solution, passing through the porous membrane as needed.

Special names are given to the two electrodes in a galvanic cell; one is called the *cathode* and the other is the *anode*. The **cathode** *is the electrode at which reduction takes place.* Here electrons enter the galvanic cell from the external circuit. The **anode** *is the electrode at which oxidation takes place.* Here electrons leave the cell for the external circuit. In the cell now under discussion, the copper electrode is the cathode and the zinc electrode is the anode.

Many students have a hard time remembering the relationship between cathode–anode and oxidation–reduction. A mnemonic device can be helpful here. The two words that begin with vowels (anode and oxidation) go together and the two words beginning with consonants (cathode and reduction) go together.

In principle, any spontaneous oxidation–reduction reaction can be used to build a galvanic cell, and many such cells have been studied in the laboratory. A selected few of such galvanic cells are now major commercial entities. We will discuss two that are commercially important: (1) the dry cell, and (2) the lead storage battery.

The Dry Cell

The *dry cell* is widely used in flashlights, portable radios and tape recorders, and battery-powered toys; it is often referred to as a flashlight battery. Two versions of the dry cell are marketed—an acidic version and an alkaline version.

The *acidic version* of the dry cell contains a zinc outer surface (covered with cardboard or paint for protection) that acts as the anode and a carbon (graphite) rod in contact with a moist paste that serves as the cathode (see Figure 10.10). The moist paste (the cell is not truly dry) is a mixture of solid MnO_2, solid NH_4Cl, and graphite powder. In the operation of the cell, zinc is oxidized to Zn^{2+} ion, and MnO_2 (a paste component) is reduced to Mn_2O_3. The reduction takes place at the interface between the carbon cathode and the paste; the inert carbon electrode conducts the electrons to the external circuit. The electrode reactions are

Anode (oxidation): $Zn(s) \longrightarrow Zn^{2+}(aq) + 2\ e^-$

Cathode (reduction): $2\ MnO_2(s) + 2\ NH_4^+(aq) + 2\ e^- \longrightarrow$

$$Mn_2O_3(s) + 2\ NH_3(aq) + H_2O$$

The useful life of acidic dry cells can be shortened by the slightly acidic paste

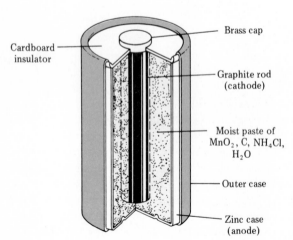

Cardboard insulator

Brass cap

Graphite rod (cathode)

Moist paste of MnO_2, C, NH_4Cl, H_2O

Outer case

Zinc case (anode)

FIGURE 10.10
The acidic Zn–MnO_2 dry cell is commonly known as a flashlight battery.

corroding the zinc can. A protective paper is inserted between the paste and zinc to help minimize this problem.

In the *alkaline version* of the dry cell, the solid NH_4Cl is replaced with KOH and a steel rod rather than a graphite one is the cathode. The anode reaction still involves oxidation of Zn, but the Zn is present as a powder in a gel formulation. The cathode reaction also still involves the reduction of MnO_2. Equations for the electrode reactions are

Anode (oxidation): $Zn(s) + 2 OH^-(aq) \longrightarrow ZnO(s) + H_2O(l) + 2 e^-$

Cathode (reduction):

$$2 MnO_2(s) + H_2O + 2 e^- \longrightarrow Mn_2O_3(s) + 2 OH^-(aq)$$

COST OF FLASHLIGHT BATTERIES

The alkaline dry cell costs roughly three times as much to produce as the acidic version. A major cost factor is the more elaborate internal construction needed to prevent leakage of the KOH solution. These cells provide up to 50% more total energy than the less expensive acidic model because they maintain usable voltage over a larger fraction of the lifetime of the cathode and anode materials. Miniature alkaline cells find extensive use in calculators, watches, and camera exposure controls.

Lead Storage Battery

The lead storage battery provides the starting power for automobiles. A 12-volt lead storage battery, the standard size, consists of six galvanic cells connected together (see Figure 10.11). Each cell generates 2 volts.

Both electrodes in a lead storage battery involve the metal lead. Lead serves as the anode and lead coated with lead dioxide serves as the cathode. Glass-fiber spacers separate electrodes to prevent them from touching each other. The electrodes are immersed in a 38% by mass sulfuric acid (H_2SO_4) solution. Details of a single cell of a lead storage battery are shown in Figure 10.12.

The electrode reactions when a lead storage battery is used to supply power (discharging) are

Anode (oxidation): $Pb(s) + SO_4{}^{2-}(aq) \longrightarrow PbSO_4(s) + 2 e^-$

Cathode (reduction):

$$PbO_2(s) + 4 H^+(aq) + SO_4{}^{2-}(aq) + 2 e^- \longrightarrow PbSO_4(s) + 2 H_2O(l)$$

The chemical results of discharge are a buildup of $PbSO_4$ on the electrodes and a decrease in the density of the sulfuric acid. [The state of charge of a lead storage battery can thus be checked by a service-station attendant by measuring the specific gravity (Section 6.4) of the sulfuric acid.]

Unlike dry cells, a lead storage battery may be recharged. This reverse process, which is nonspontaneous, must be started by an external source of

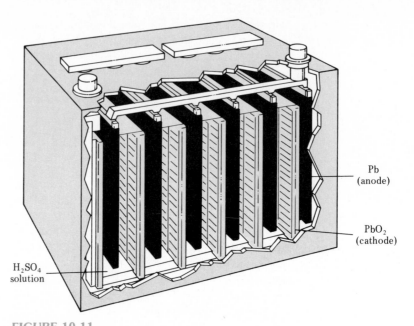

FIGURE 10.11

The six galvanic cells in a 12-volt automobile battery.

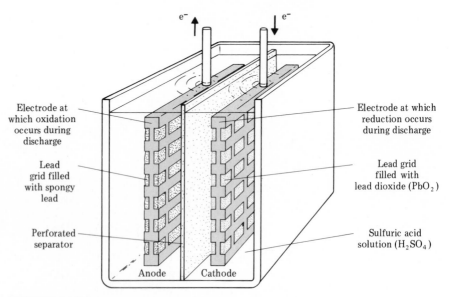

FIGURE 10.12

A single lead storage battery cell.

energy; in the automobile the external energy source is an alternator driven by the automobile engine.

In recharging, the $PbSO_4$ on the electrodes (formed during discharge) is converted back to Pb (at one electrode) and to PbO_2 (at the other). The electrode reactions are the reverse of what occurs during discharge.

RECHARGING AN
AUTOMOBILE BATTERY

Theoretically a lead storage battery should be rechargeable indefinitely. In practice, such batteries have a lifetime of 3 to 5 years because small amounts of lead sulfate continually fall from the electrodes (to the bottom of the cell) as a result of "roadshock" and chemical side reactions. Eventually the electrodes lose so much lead sulfate that the recharging process is no longer effective.

In "standard" lead storage batteries, water must be added to the individual cells on a regular basis. Recharging the battery, besides converting $PbSO_4$ back to Pb and PbO_2, also decomposes small amounts of water to give H_2 and O_2; hence, the H_2O must be replenished. Because of the possible presence of H_2 gas in a lead storage battery, a person should wear glasses (for eye protection) when releasing the cap of a battery since escaping gas can force sulfuric acid out. In addition, a person should not smoke while doing this since hydrogen gas is flammable and forms explosive mixtures with oxygen. Newer automobile batteries have electrodes made of an alloy of calcium and lead. The presence of the calcium minimizes the decomposition of water during recharging. Thus, batteries with these alloy electrodes can be sealed; there is no need to add water.

Electrolysis Cells

Application of electrical energy, from an external power source, can be used to cause a nonspontaneous redox reaction. The charging of the lead storage battery previously discussed is an example of this. The general term for such a process is electrolysis. **Electrolysis** *is the process in which electrical energy is used to produce a nonspontaneous redox reaction.* The lead storage battery during recharging is functioning as an electrolysis cell. An **electrolysis cell** *is an apparatus in which chemical change is caused to occur through the application of electrical energy.*

A number of important commercial applications of electrolysis exist. They include (1) production of important industrial chemicals, (2) electrorefining and purifying of metals, and (3) electroplating. Let us consider examples in all three of these areas.

Chlorine gas, hydrogen gas, and sodium hydroxide—three important industrial chemicals—can be produced simultaneously from the electrolysis of a concentrated aqueous sodium chloride solution (saltwater or brine solution). The type of electrolysis cell needed is shown in Figure 10.13.

In this cell the two electrodes are inert, that is, they do not participate in the redox reactions themselves but serve as surfaces on which the redox reactions

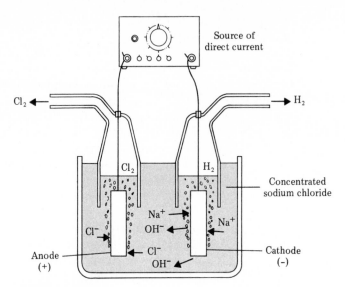

FIGURE 10.13

The electrolysis of aqueous sodium chloride solution (saltwater brine) produces hydrogen gas, chlorine gas, and sodium hydroxide solution.

occur. As with galvanic cells, reduction occurs at the cathode and oxidation at the anode.

The negative ions in the solution, the Cl^- ions, react at the anode, where they give up electrons and are reduced to produce Cl_2 gas.

$$\text{Anode (oxidation):} \quad 2\,Cl^-(aq) \longrightarrow Cl_2(g) + 2\,e^-$$

The positive ions in the solution, the Na^+ ions, you would expect to react at the cathode, where they could pick up electrons to produce Na metal atoms. However, this process does not occur. Instead, water, the solvent in the solution, reacts at the cathode; it is more easily reduced than Na^+ ion. The cathode reaction is

$$\text{Cathode (reduction):} \quad 2\,H_2O(l) + 2\,e^- \longrightarrow H_2(g) + 2\,OH^-(aq)$$

From this reduction H_2 gas is produced as well as OH^- ion. The OH^- ions remain in solution, producing a solution that now contains Na^+ ions and OH^- ions; our solution of sodium chloride has been changed to one of sodium hydroxide. Thus, three substances—Cl_2 gas, H_2 gas, and NaOH solution—result from the electrolysis of a concentrated NaCl solution.

Electrolysis of bauxite, an aluminum ore, is the primary method of producing aluminum. We consider this process in detail in Section 14.6.

Metal purification frequently depends on electrolysis. All copper used in electrical wire is electrolytically purified; impurities cut the electrical conductance of the wire. In an electrolysis cell used to purify copper, large slabs of

PURITY OF COPPER
WIRE

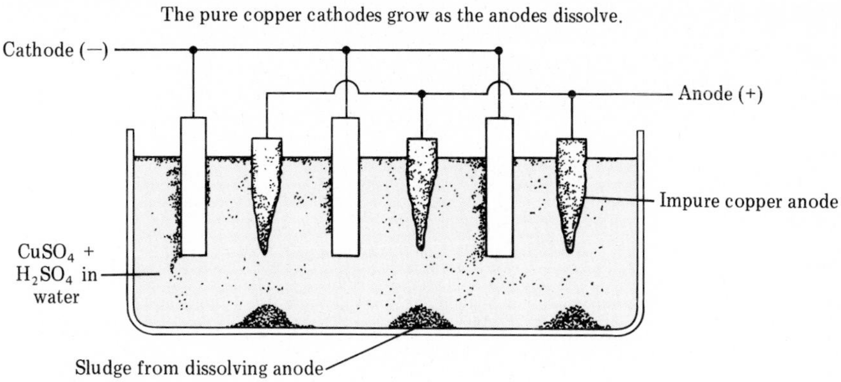

The pure copper cathodes grow as the anodes dissolve.

FIGURE 10.14

Cross section of an electrolysis cell for purifying copper.

impure copper (obtained from the reduction of copper ores) serve as anodes and thin sheets of very pure copper serve as cathodes. The solution in which the electrodes are immersed is an acidic copper(II) sulfate solution. As the cell is operated the anodes (the impure copper decrease in size and the cathodes (the pure copper) increase in size (see Figure 10.14). What is happening? At the anodes oxidation of Cu causes it to dissolve. The reaction is

$$\text{Anode (oxidation):} \quad Cu(s) \longrightarrow Cu^{2+}(aq) + 2\,e^-$$

As the anodic copper dissolves, the impurities present also go into solution (are also reduced) or fall to the bottom of the cell. At the cathode, reduction causes copper to come out of solution.

$$\text{Cathode (reduction):} \quad Cu^{2+}(aq) + 2\,e^- \longrightarrow 2\,Cu(s)$$

The net effect of this oxidation–reduction process is that copper is transferred via solution from one electrode (the impure one) to the other (the pure one). The electricity supplied to the cell is set at a value that allows copper ions to be reduced but is not sufficient to reduce any dissolved impurities. Thus, only copper, and not impurities, deposits on the cathode. More information concerning this process is presented in Section 14.3, where the element copper is discussed.

ELECTROPLATING AND "TIN CANS"

Electroplating *is the deposition of a thin layer of a metal on an object through the process of electrolysis.* Electroplated objects are common in our society. Jewelry is plated with silver and gold. Tableware is often plated with silver. Gold-plated electrical contacts are used extensively. "Tin cans" are actually steel cans with a thin coating of tin. Chromium-plated steel automobile bumpers have been used for many years. The thin metallic layer deposited during electroplating is generally only 0.001 to 0.002 inch thick.

Figure 10.15 shows a typical apparatus used for electroplating the metal silver. The object to be plated is made the cathode in a solution containing Ag^+ ions.

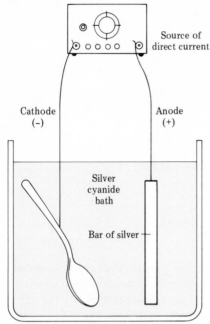

Source of direct current

Cathode (−)

Anode (+)

Silver cyanide bath

Bar of silver

FIGURE 10.15

An apparatus for electroplating silver.

The anode is a bar of the plating metal, silver in this case. At the cathode, Ag^+ ion from solution is deposited as metallic silver.

$$\text{Cathode (reduction):} \quad Ag^+(aq) + e^- \longrightarrow Ag(s)$$

At the anode, silver from that electrode is oxidized to give Ag^+ ion in solution; this replenishes the supply of Ag^+ ion in solution needed for the plating process.

$$\text{Anode (oxidation):} \quad Ag(s) \longrightarrow Ag^+(aq) + e^-$$

The electroplating bath, the solution around the electrodes, usually contains other chemicals besides the plating metal. For example, silver plating is usually done from a solution containing both AgCN and KCN.

Practice Questions and Problems

COLLISION THEORY

10-1 What is collision theory and what are the main concepts of this theory?

10-2 Define the term *activation energy*.

10-3 What two factors determine whether a collision between two reactant molecules will result in reaction?

FACTORS THAT INFLUENCE REACTION RATES

10-4 List four factors that influence the rate of a reaction.

10-5 Define the term *reaction rate*.

10-6 Substances burn more rapidly in pure oxygen than in air. Explain why, using colllision theory.

10-7 Milk will sour in a couple of days when left at room temperature; however, it will remain unspoiled for 2 weeks when refrigerated. Explain why this is so, using collision theory.

10-8 What is a catalyst?

10-9 Describe two ways by which catalysts can exert their effect.

ACID–BASE DEFINITIONS

10-10 What are the Arrhenius definitions for acids and bases?

10-11 What are the Brønsted–Lowry definitions for acids and bases, and what are their advantages over the Arrhenius definitions?

10-12 Write equations for the dissociation of the following Arrhenius acids and bases in water.
a. HI (hydroiodic acid)
b. HClO (hypochlorous acid)
c. LiOH (lithium hydroxide)
d. KOH (potassium hydroxide)

10-13 Identify the Brønsted–Lowry acid and base in the following reactions.
a. $HF + H_2O \longrightarrow H_3O^+ + F^-$
b. $H_2O + S^{2-} \longrightarrow HS^- + OH^-$
c. $H_2O + H_2CO_3 \longrightarrow H_3O^+ + HCO_3^-$
d. $HCO_3^- + H_2O \longrightarrow H_3O^+ + CO_3^{2-}$

10-14 Write equations to illustrate the acid–base reactions that can take place between the following Brønsted–Lowry acids and bases.
a. Acid: HClO, base: H_2O
b. Acid: $HClO_4$, base: NH_3
c. Acid: H_3O^+, base: NH_2^-
d. Acid: H_3O^+, base: OH^-

STRENGTHS OF ACIDS AND BASES

10-15 What is the principal distinction between strong and weak acids? Is the distinction between strong and weak bases similar?

10-16 Classify each of the following as a weak or strong acid.
a. H_2SO_4 b. HNO_3 c. H_3PO_4
d. HClO   e. H_3BO_3 f. HCl

10-17 Classify each of the following as a weak or strong base.
a. NaOH   b. $Ba(OH)_2$ c. NH_3 d. KOH

10-18 Make a listing of the six common strong acids and seven common strong bases.

10-19 Distinguish between the terms *weak acid* and *dilute acid*.

POLYPROTIC ACIDS

10-20 Identify the following acids as monoprotic, diprotic, or triprotic.
a. $HClO_4$ (perchloric acid)
b. $H_2C_2O_4$ (oxalic acid)
c. $HC_2H_3O_2$ (acetic acid)
d. $HC_4H_7O_2$ (butyric acid)
e. H_3PO_4 (phosphoric acid)
f. H_2SO_4 (sulfuric acid)

10-21 Write equations showing all steps in the dissociation of the following acids.
a. H_2CO_3 (carbonic acid)
b. $H_3C_6H_5O_7$ (citric acid)
c. $HC_4H_7O_2$ (butyric acid)

10-22 The formula for lactic acid is preferably written as $HC_3H_5O_3$, rather than as $C_3H_6O_3$. Explain why this is so.

10-23 Pyruvic acid, which is produced in metabolic reactions within the human body, has the following structure.

Would you predict that this acid is mono-, di-, tri-, or tetraprotic? Give your reasoning for your answer.

ACID–BASE NEUTRALIZATION REACTIONS

10-24 Describe, in general terms, the products produced when an acid neutralizes a base.

10-25 Write a balanced molecular equation to represent each of the following acid–base neutralizations.
a. Acid: HCl, base: NaOH
b. Acid: HNO_3, base: KOH
c. Acid: H_2SO_4, base: LiOH
d. Acid: H_3PO_4, base: $Ba(OH)_2$

10-26 What is a salt?

10-27 Classify each of the following substances as an acid, base, or salt.
a. HBr   b. NaI
c. NH_4NO_3 d. $AlPO_4$
e. $Ba(OH)_2$ f. KOH
g. HNO_3 h. $HC_2H_3O_2$

10-28 The term weak is used to describe certain acids and bases, but it is never used to describe salts. Why?

THE DISSOCIATION OF WATER

10-29 Pure water contains small amounts of H^+ and OH^- ions. What is the origin of these ions.

10-30 What is the concentration of hydrogen and hydroxide ions, in moles per liter, in pure water at 25°C.

10-31 Calculate the H^+ ion concentration of a solution given that the OH^- concentration is
  a. $1.0 \times 10^{-3} M$ b. $1.0 \times 10^{-11} M$
  c. $1.0 \times 10^{-7} M$ d. $2.0 \times 10^{-6} M$

10-32 How are the terms acidic solution, basic solution, and neutral solution defined in terms of $[H^+]$ and $[OH^-]$?

10-33 Indicate whether each solution is acidic, basic, or neutral.
  a. $[H^+] = 1.0 \times 10^{-4} M$
  b. $[H^+] = 1.0 \times 10^{-11} M$
  c. $[H^+] = 1.0 \times 10^{-9} M$
  d. $[OH^-] = 1.0 \times 10^{-6} M$

THE pH SCALE

10-34 Calculate the pH of the following solutions.
  a. $[H^+] = 1.0 \times 10^{-7} M$
  b. $[H^+] = 1.0 \times 10^{-12} M$
  c. $[H^+] = 0.0001 M$
  d. $[H^+] = 0.1 M$

10-35 How are the terms acidic solution, basic solution, and neutral solution defined in terms of pH values?

10-36 Determine the $[H^+]$ in a solution that has a pH of
  a. 3   b. 7   c. 10   d. 13

10-37 Solution A has a pH of 2 and solution B has a pH of 5. Which solution is more acidic? How many times more acidic is the one solution than the other?

10-38 Would you expect solutions of 0.1 M HCl and 0.1 M HCN to be of equal pH value? Explain your answer.

OXIDATION–REDUCTION TERMINOLOGY

10-39 Give definitions of *oxidation* and *reduction* in terms of the following.
  a. Loss and gain of electrons.
  b. Increase and decrease in oxidation number.

10-40 Give definitions of *oxidizing agent* and *reducing*

agent in terms of
  a. Loss and gain of electrons.
  b. Increase and decrease in oxidation number.
  c. Substance oxidized and substance reduced.

10-41 In each of the following statements, choose the word in parentheses that best completes the statement.
  a. An element that has lost electrons in a redox reaction is said to have been (oxidized, reduced).
  b. Reduction always results in an (increase, decrease) in the oxidation number.
  c. The substance oxidized in a redox reaction is the (oxidizing agent, reducing agent).
  d. The reducing agent (gains, loses) electrons during a redox reaction.
  e. The reducing agent causes an (increase, decrease) in the oxidation number of the oxidizing agent in a redox reaction.

OXIDATION NUMBERS

10-42 Determine the oxidation number of each of the following.
  a. S in SO_2 b. S in SO_3
  c. S in SO_4^{2-} d. Ba in Ba^{2+}
  e. Zn in $ZnSO_4$ f. Cl in $HClO_3$
  g. P in H_3PO_4 h. Cl in Cl_2

10-43 Determine the oxidation number of Cl in each of the following species.
  a. ClF_4^+ b. $BeCl_2$ c. $Ba(ClO_2)_2$
  d. Cl_2O_7 e. $AlCl_4^-$ f. NCl_3
  g. ClF h. ClO^-

10-44 Indicate whether each of the following chemical equations does or does not represent an oxidation–reduction reaction.
  a. $2\,CO + O_2 \longrightarrow 2\,CO_2$
  b. $2\,HBr + Mg \longrightarrow MgBr_2 + H_2$
  c. $HCl + NaOH \longrightarrow H_2O + NaCl$
  d. $SO_3 + H_2O \longrightarrow H_2SO_4$
  e. $2\,Al + 3\,Cl_2 \longrightarrow 2\,AlCl_3$
  f. $Zn + CuCl_2 \longrightarrow ZnCl_2 + Cu$

10-45 For each of the following reactions, identify the substance oxidized, the substance reduced, the oxidizing agent, and the reducing agent.
  a. $Fe_2O_3 + 2\,Al \longrightarrow Al_2O_3 + 2\,Fe$
  b. $4\,NH_3 + 3\,O_2 \longrightarrow 2\,N_2 + 6\,H_2O$
  c. $3\,H_2S + 2\,HNO_3 \longrightarrow 3\,S + 2\,NO + 4\,H_2O$
  d. $F_2 + 2\,NaCl \longrightarrow Cl_2 + 2\,NaF$

IMPORTANT COMMERCIAL
OXIDATION–REDUCTION PROCESSES

10-46 What is the difference between a galvanic and an electrolysis cell?

10-47 What processes occur at the anode and cathode in the following?
a. A galvanic cell.　　b. An electrolysis cell.

10-48 When a strip of copper is dipped into a solution of silver nitrate (Ag^+ ions and NO_3^- ions), the copper is silver-plated. Write a chemical equation to describe this process.

10-49 What are the anode and cathode reactions during the following operations?
a. The operation of an acidic dry cell.
b. The operation of an alkaline dry cell.
c. The discharging of a lead storage battery.
d. The charging of a lead storage battery.

10-50 Why does the specific gravity of the H_2SO_4 in a lead storage battery decrease as the cell discharges?

10-51 Why is it not possible to recharge a lead storage battery an infinite number of times?

10-52 Write equations for what happens at each electrode during the electrolysis of a concentrated aqueous NaCl solution.

10-53 Describe how copper metal is purified using electrolysis.

10-54 In electroplating, to which electrode is the object to be plated attached? Explain your answer.

III
Chemistry in the Modern World

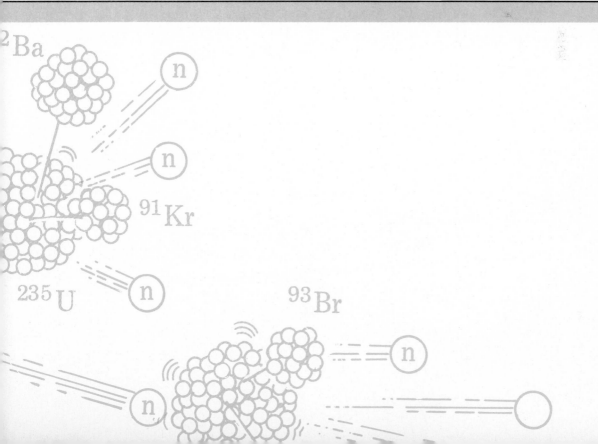

CHAPTER

11

Water—An Abundant and Vital Compound

CHAPTER HIGHLIGHTS

11.1   Abundance and Distribution of Water

Water is the most abundant compound on the face of the earth. We encounter it everywhere we go: as water vapor in the air, as a liquid in rivers, lakes, and oceans, and as a solid (ice and snow) both on land and in the oceans.

Approximately 75% of the earth's surface is covered by water. As indicated in Table 11.1, the oceans are the major repositories for water. Water is also the most abundant compound in the human body. It constitutes more than 65% of the total body mass. A human being can survive for only a few days without taking in water.

Water serves a variety of functions in the body. It is the fluid medium in which the chemical reactions of the cells take place. In the blood, water is the major transport medium for distributing oxygen and nutrients to the body cells, as well as for carrying away their waste products. Water plays an important role in the excretion of wastes through the kidneys, intestines, and sweat glands. The water that is eliminated through the sweat glands also plays a key role in the body's temperature-control mechanisms.

WATER CONTENT OF
FOODS

Water enters the body in three different ways—two well known and one not so well known. The two well-known sources are the liquids we ingest and the foods we eat. Table 11.2 gives the water content of selected foods. The not so well-known source is the cells of the human body. These cells produce water as a by-product of the reactions in which food is broken down into simpler components. This water enters the bloodstream. (Some desert insects are capable of using this metabolically generated water as their sole source of water.)

WATER BALANCE IN
THE HUMAN BODY

Water normally leaves the body by four exits: kidneys (urine), lungs (water in expired air), skin (by diffusion and sweat), and intestines (feces). Because a body must have a fluid balance to function normally, the total amount of water entering the body usually equals the total amount of water leaving the body. The body's water content is continuously monitored in the brain. Thirst is triggered if

TABLE 11.1

Distribution of Water on the Earth

Location	Percent
Oceans	97.2
Glaciers and polar caps	2.16
Subsurface water	0.62
Lakes and rivers	0.019
Atmospheric water	0.001

TABLE 11.2

Water Content of Selected Foods

Food	Mass Percent Water
Apples	85
Broccoli	90
Cheddar cheese	35
Eggs	75
Grapes	80
Milk	87
Potatoes	78
Steak	73
Tuna fish	60

the body needs more water. Too much water in the tissues triggers control mechanisms that restore fluid balance by increasing elimination of fluid.

Approximtely 2400 mL of water enters and leaves the body daily. Table 11.3 lists typical normal values for each portal of water entry and exit. These values can vary considerably and still be considered normal. Much of the variation relates to the fat content of the body. Fat people have a lower water content per kilogram of body mass than slender people. Also, the amount of water in the body decreases with age. A newborn infant's body contains about 77% water, and an older adult man's body contains only about 60% water.

TABLE 11.3

Typical Normal Values for Each Portal of Water Entry and Exit in the Human Body

	Amount (mL)
Intake	
Ingested liquids	1500
Water in foods	700
Water formed as a by-product in the metabolism of food	200
	2400
Output	
Kidneys (urine)	1400
Lungs (water in expired air)	350
Skin	
By diffusion	350
By sweat	100
Intestines (in feces)	200
	2400

11.2   Properties of Water

From a chemical viewpoint, the most fascinating thing about water is the "uniqueness" of many of its properties. Numerous properties of water have values that fall outside the normal range exhibited by compounds similar in structure. It is the unusual properties of water that make it the key substance for forms of life on this planet.

Most of water's unusual behavior is a consequence of the extensive hydrogen bonding (Section 8.16) that occurs between water molecules, in both its liquid state and its solid state. Three important effects of hydrogen bonding in water are (1) higher-than-expected boiling and freezing points, (2) higher-than-expected thermal properties, and (3) an unusual temperature-density relationship. The next three sections of the text consider these three effects.

11.3   Boiling and Freezing Points of Water

The vapor pressures (Section 8.14) of liquids in which extensive hydrogen bonding occurs are significantly lower than those of similar liquids in which little or no hydrogen bonding occurs. The presence of hydrogen bonds makes it more difficult for molecules to escape from the condensed state; additional energy is needed to overcome the hydrogen bonds. The greater the strength of the hydrogen bonds the lower the vapor pressure is at any given temperature.

The boiling point of a liquid depends on vapor pressure (Section 8.15), so liquids with low vapor pressures have to be heated to higher temperatures to bring their vapor pressures up to the boiling point (atmospheric pressure). As a result, boiling points are much higher for liquids in which hydrogen bonding occurs.

The effect that hydrogen bonding has on water's boiling point can be seen by comparing it with the boiling points of other hydrogen compounds of group VIA elements—H_2S, H_2Se, and H_2Te. In this series of compounds—H_2O, H_2S, H_2Se, and H_2Te—water is the only one in which significant hydrogen bonding occurs. Normally, the boiling points of a series of compounds that contain elements in the same periodic table group increase with increasing molecular weight. Thus, in the hydrogen–group VIA series, we would expect that H_2Te, the heaviest member of the series, would have the highest boiling point and that

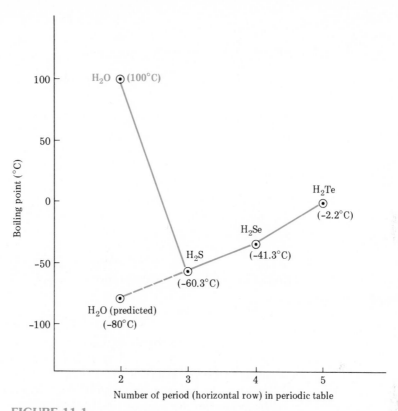

FIGURE 11.1

Boiling points of the hydrogren compounds of group VIA elements. Water is the only one of the four compounds in which there is significant hydrogen bonding.

water, the compound of lowest molecular weight, would have the lowest boiling point. Contrary to expectation, H_2O has a boiling point higher than that of any other member of the series, as can be seen from the boiling point data shown in Figure 11.1.

The data in Figure 11.1 indicate that water should have a boiling point of approximately −80°C; this value is obtained by extrapolation (extension) of the line connecting the three heavier compounds. The actual boiling point of water, 100°C, is nearly 200° higher than predicted. Indeed, in the absence of hydrogen bonding, water would be a gas at room temperature and life as we know it on earth would not be possible.

A higher-than-expected freezing point is also characteristic of water, and a plot of freezing points for similar compounds would have the same general shape as the boiling point plot of Figure 11.1.

11.4   Thermal Properties of Water

Three important thermal properties for any substance are specific heat, heat of vaporization, and heat of fusion. The property *specific heat* was defined and discussed in Section 6.7. Liquid water has a specific heat of 1.00 cal/g °C. This specific heat value is at least double that of most other substances (*see* Table 6.5).

Heat energy is absorbed or evolved in changes of state (*see* Section 8.12). The **heat of fusion** *is the amount of energy required to convert one gram of a solid to a liquid at its melting point.* This same amount of energy is released to the surroundings when 1 g of the liquid is changed back to solid during a freezing process. The **heat of vaporization** *is the amount of energy required to convert one gram of a liquid to a gas at its boiling point.* This same amount of energy is released to the surroundings when 1 g of the gas condenses to a liquid.

The numerical values of the heats of fusion and vaporization for water are 79.8 and 540 cal/g, respectively. Both of these values are higher than those for most other substances. By comparison, liquid CH_4 (a nonhydrogen-bonded liquid) has a heat of fusion of 14.0 cal/g and a heat of vaporization of 122 cal/g, values that are less than one-fourth those of water.

The presence of extensive hydrogen bonding among water molecules significantly increases the ability of water to absorb heat energy when evaporating (heat of vaporization) and to release heat energy when freezing (heat of fusion). Additional heat energy is required in the evaporation process to overcome (break) the hydrogen bonds present.

TEMPERATURE-
MODERATING EFFECT
OF WATER ON CLIMATE

Water's thermal properties and its abundance account for its widespread use as a coolant. Large bodies of water exert a temperature-moderating effect on their surroundings, primarily because of water's ability to absorb and release large amounts of heat energy. In the heat of a summer day, extensive water evaporation occurs, and in the process, energy is absorbed from the surroundings. The net effect of the evaporation is a lowering of the temperature of the surroundings. In the cool of the evening, some of this water vapor condenses back to the liquid state, releasing heat, which raises the temperature of the surroundings. In this manner, the temperature variation between night and day is reduced. A similar process occurs in the winter; water freezes on cold days and releases heat energy to the surroundings. The hottest and coldest regions on earth are all inland regions that are farthest away from the moderating effects of large bodies of water.

Water's ability to absorb large amounts of heat during the evaporation process

FIGURE 11.2
Water's high heat of vaporization reduces the amount of water (perspiration) needed to carry away body heat during strenuous exercise. [Jaye R. Phillips/The Picture Cube.]

WATER-BASED
TEMPERATURE
REGULATION IN THE
HUMAN BODY

is a factor in the cooling of the human body. We noted in Section 8.13 that evaporation of perspiration from the skin reduces body temperature. Water's high heat of vaporization minimizes the amount of water lost from the body; this makes it easier to maintain fluid balance (Section 11.1) within the body (see Figure 11.2).

Some organisms do not have a water-based temperature-regulating system like the one in humans. For example, many types of aquatic life can function only within a very narrow range of temperatures. These organisms are usually found in large bodies of water where the temperature normally varies only a few degrees. The addition of heated water from industrial complexes to natural waters, called thermal pollution (Section 21.8), can cause serious problems for these organisms.

11.5   Water's Temperature–Density Relationship

An unusual behavior pattern of water is the variation of its density with temperature. For most liquids, density increases with decreasing temperature and reaches a maximum for the liquid at its freezing point. The density pattern for water is different. Water's maximum density is reached at a temperature that is a few degrees higher than its freezing point. As shown in Figure 11.3, the maximum density for liquid water occurs at 4°C. This abnormality—that water at its freezing point is less dense than water at slightly higher temperatures—has tremendous ecological significance. Furthermore, at 0°C, solid water (ice) is significantly less dense than liquid water—0.9170 g/mL versus 0.9998 g/mL. Water's unusual density behavior is directly related to hydrogen bonding.

The fact that hydrogen bonding among water molecules is directional in nature is the prime factor in explaining water's peculiar density behavior. Hydrogen bonds can form only between molecules that have specific orientations to each other. These orientations are dictated by the location of the

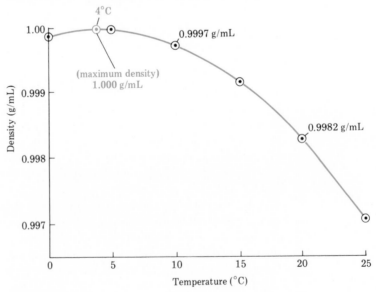

FIGURE 11.3
A plot of water density versus temperature for liquid water.

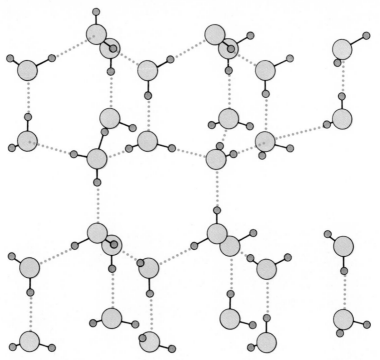

FIGURE 11.4

The crystalline structure of ice. Normal covalent bonds between oxygen and hydrogen, which hold the water molecules together, are shown by solid short lines. The weaker hydrogen bonds between molecules are shown by dotted longer lines.

nonbonding pairs of electrons of water's oxygen atom. The net result is that when water molecules are hydrogen bonded, they are farther apart than when they are not hydrogen bonded. A diagram in Chapter 8 (Figure 8.18) shows hydrogen-bonding interactions between water molecules in the liquid state. Figure 11.4 shows the hydrogen-bonding pattern characteristic of solid water (ice).

On a molecular scale, let us now consider what happens to water molecules when the temperature is lowered. At high temperatures such as 80°C, the kinetic energy of the water molecules is great enough to prevent hydrogen bonding from having much of an orientation effect on the molecules. Hydrogen bonds are rapidly and continually being formed and broken. As the temperature is lowered, the accompanying decrease in kinetic energy reduces molecular motion and the molecules move closer together. This results in an increase in density. The kinetic energy is still sufficient to negate most of the orientation effects of hydrogen bonding. When the temperature is lowered still further, the kinetic energy finally becomes insufficient to prevent hydrogen bonding from orienting molecules into definite patterns that require "open spaces" between molecules. At 4°C, the temperature at which water has its maximum density, the balance between random motion from kinetic energy and orientation from hydrogen bonding brings the molecules as close together as they will ever be. Cooling below 4°C decreases the density because hydrogen bonding produces more and more "open spaces" in the liquid. Density continues to decrease down to the freezing point, at which temperature the hydrogen bonding causes molecular orientation to the maximum degree; then the solid crystal lattice of ice is formed. This solid crystal lattice (Figure 11.4) is an extremely open structure; this results in solid ice having a lower density than liquid water. Water is one of only a few known substances in which the solid phase is less dense than the liquid phase.

FIGURE 11.5
Because ice has a lower density than liquid water, it floats in liquid water. Only 8% of floating ice is above the surface of the water. You see in this photograph only the "tip" of the iceberg. [Ira Kirschenbaum/Stock, Boston.]

ICEBERGS

Two important consequences of ice being less dense than water are the facts that ice floats in liquid water and that liquid water expands upon freezing.

When ice floats in liquid water, approximately 8% of the ice volume is above water. In the case of icebergs found in far nothern locations, 92% of the volume of the icebergs is located below the liquid surface (see Figure 11.5). The common expression "it is only the tip of the iceberg" is true in its literal sense. What is seen is only a very small part of what is actually there.

If a container is filled with water and sealed, the force generated from the expansion of the water upon freezing will break the container. This is the reason that antifreeze is used in car radiators in the winter. Water in the radiator will freeze and crack the engine block; the force generated from expansion when water freezes is sufficient to burst even iron or copper parts. During the winter season, the weathering of rocks and concrete and the formation of potholes in streets are hastened by the expansion of freezing water in cracks.

Water's density pattern also explains why lakes freeze from top to bottom and not vice versa, and why aquatic life can continue to exist for extended periods of time in bodies of water that are "frozen over."

FROZEN-OVER LAKES

Let us consider some details of the freezing over of a lake in the winter. In the fall of the year, surface water is cooled through contact with cold air. This water becomes denser than the warmer water underneath and sinks. In this way, cool water is circulated from the top of the lake to the bottom until the entire lake has reached the temperature of water's maximum density, 4°C. During the circulation process, oxygen and nutrients are distributed throughout the water. As the

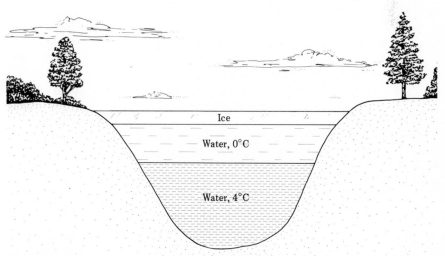

FIGURE 11.6

In the winter, because of water's unusual density pattern, bodies of water freeze over from the top. The ice formed insulates the water below from the colder atmosphere and the entire body does not freeze.

water finally cools below 4°C, a new behavior pattern emerges. When the surface water cools, it no longer sinks because it is less dense than the water underneath. Eventually a thin layer of surface water is cooled to the freezing point and is changed to ice, which floats because of its lower density (see Figure 11.6). Even in the coldest winters, lakes usually do not freeze to a depth of more than a few feet, because the ice forms an insulating layer over the water. Thus, aquatic life can live throughout the winter under the ice in water that is thermally insulated and contains nutrients. If water behaved as "normal" substances do, freezing would occur from the bottom up, and most, if not all, aquatic life would be destroyed.

Circulation occurs in a body of water during the melting and warming of the surface water for a few weeks in the early spring. This process stops when the surface water is warmed above 4°C. Surface water above this temperature is less dense and forms a layer over the colder water of higher density. This causes a thermal stratification that persists through the hot summer months. In the fall, circulation begins again and the cycle is repeated.

11.6 Natural Waters

All water as it occurs in nature is impure in a chemical sense. Impurities present include suspended matter, microbiological organisms, dissolved minerals, and dissolved gases. The diversity of dissolved substances present in natural waters illustrates the broad solvent properties of this compound. Water's "great" solvent ability stems from the highly polar nature of water molecules (Sections 5.12 and 9.3).

The solvent properties of water are both beneficial and detrimental, from the viewpoint of humans. They are beneficial because water is a useful medium for carrying nutrients and oxygen throughout the human body and a useful substance in washing and cleansing processes. They are detrimental because materials that are harmful to the health of humans tend to dissolve in water.

Natural waters can be classified into three categories on the basis of their dissolved mineral content: fresh water, inland brackish water, and seawater. Water containing less than 0.1% by mass dissolved solids is generally considered to be fresh, but the U.S. Public Health Service has set 0.05% by mass as the recommended upper limit for drinking water. The inland underground and surface supply of brackish water contains dissolved solids in the range of 0.1 to 3.5% by mass, with an average of 0.6% by mass. The dissolved mineral content of seawater varies from 3.3 to 3.7% by mass, with an average of 3.5% by mass. However, bodies of water such as the Great Salt Lake and the Dead Sea can contain much higher levels—up to 25% by mass dissolved solids.

TABLE 11.4

Ions That Are Present in Seawater in Concentrations of 1 ppm by Mass or Greater

Ion	Concentration (ppm by mass)	Concentration (molarity)
Chloride (Cl^-)	19,400	0.55
Sodium (Na^+)	10,600	0.46
Sulfate (SO_4^{2-})	2,700	0.028
Magnesium (Mg^{2+})	1,300	0.054
Calcium (Ca^{2+})	410	0.010
Potassium (K^+)	390	0.010
Hydrogen carbonate (HCO_3^-)	140	2.3×10^{-3}
Bromide (Br^-)	67	8.3×10^{-4}
Carbonate (CO_3^{2-})	18	3×10^{-4}
Strontium (Sr^{2+})	8	9×10^{-5}
Fluoride (F^-)	1	7×10^{-5}

Seawater is often referred to as *saline water*; the salinity of the water is the percent by mass of dissolved solids present in the water. More than 70 different elements contribute to the salinity of seawater, but most of these are present in only a very low concentration. Table 11.4 lists the 11 ionic species present in seawater in the largest amounts. The concentrations of these species are given in parts per million by mass and in molarity. (The parts per million concentration unit, abbreviated ppm, is the number of parts of solute per one million parts of solution.)

Of the fresh water on earth, 75% is in the form of ice in the polar regions and glaciers. Groundwater, the water found below land surfaces, represents 20% of the fresh water and the remaining 5% is found in lakes, rivers, the soil as moisture, and air. Fortunately, fresh water, unlike most natural resources, is

NATURE'S HYDROLOGIC CYCLE

renewable because it is part of a natural hydrologic cycle. Evaporation from oceans and lakes, as well as transpiration by plants (loss of water vapor from the leaves of plants) places water in the atmosphere. Eventually, fresh water finds its way back to the earth in the form of rain and snow. Figure 11.7 diagramatically shows nature's hydrologic cycle.

CHEMICAL COMPOSITION OF FRESH WATER AND SEAWATER

It is interesting to compare ion concentrations in fresh water and seawater. This is done in Table 11.5. Note the total concentration of ions in each type of water at the bottom of the table. On a relative scale, where the total concentration of ions in fresh water is assigned a value of 1, seawater would have a value of approximately 500 (actually 474). Seawater has a concentration of dissolved ions 500 times greater than that of fresh water. Note also that Na^+ ion is the dominant positive ion and Cl^- is the dominant negative ion in seawater. This contrasts with fresh water, in which Ca^{2+} and Mg^{2+} are the most abundant positive ions and HCO_3^- is the most abundant negative ion.

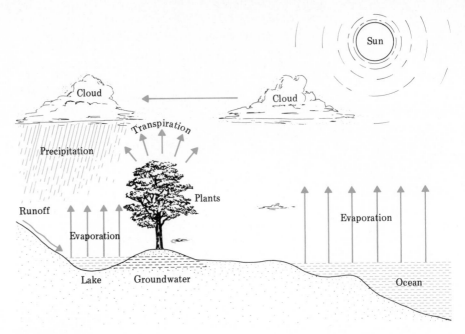

FIGURE 11.7

Natures's hydrologic cycle causes fresh water to be a renewable natural resource.

TABLE 11.5

Comparison of the Concentrations of the Major
Ionic Species in Fresh Water and Sea water
[*in millimolarity units (molarity/1000)*]

	Fresh water	Seawater
	Positive Ions	
Na^+	0.27 (11%)*	460 (41%)
K^+	0.06 (2.5%)	10 (0.90%)
Ca^{2+}	0.38 (16%)	10 (0.90%)
Mg^{2+}	0.34 (14%)	54 (4.9%)
	Negative Ions	
Cl^-	0.22 (8.5%)	550 (49%)
SO_4^{2-}	0.12 (5.1%)	28 (2.5%)
HCO_3^-	0.96 (41%)	2.3 (0.21%)
Total ions	2.35 (100%)	1114 (100%)

* Percent of the total ionic concentration of the solution.

11.7 Drinking Water Purification

The treatment of fresh water for domestic use involves processes that are designed to remove or destroy suspended particles, disease-causing agents, and objectionable colors and odors. Dissolved substances are not usually present in sufficient quantities to require their removal for health or aesthetic reasons.

A typical municipal water treatment program involves five steps: (1) coarse filtration, (2) sedimentation, (3) sand filtration, (4) aeration, and (5) disinfection.

Coarse filtration, which removes macroscopic-sized material from the water, involves letting the water flow through a screen. In the sedimentation step, finely divided materials are removed by treating the water with chemicals that react with each other to form a *flocculating agent*, that is, a spongy, gelatinous solid that traps small particles and some microorganisms. The flocculating agent settles to the bottom of sedimentation tanks and carries most finely divided suspended matter with it. Aluminum hydroxide and iron(II) hydroxide are the two most commonly used flocculating agents. They are formed in water by a reaction of the sulfates of these two metals with slaked lime, $Ca(OH)_2$.

$$3\,Ca(OH)_2 + Al_2(SO_4)_3 \longrightarrow 2\,Al(OH)_3 + 3\,CaSO_4$$

$$Ca(OH)_2 + FeSO_4 \longrightarrow Fe(OH)_2 + CaSO_4$$

The by-product $CaSO_4$, which is formed along with the hydroxide flocculating agents, also settles out.

After the flocculation treatment, the water is passed through filtering beds of gravel and sand, where any remaining flocculating agent, suspended particles, and some bacteria are removed.

In the aeration step, the water is sprayed into the air. This enables dissolved, odor-causing gases to escape more easily and hastens the oxidation of some dissolved substances. As an alterative to aeration, some treatment plants use finely divided carbon to absorb and remove odor-causing substances.

The final step in water purification processes involves treating the water with a chemical agent that destroys bacteria. Both ozone (O_3) and chlorine (Cl_2) are used for this purpose. Ozone has the disadvantage that it must be generated where it is used because of its reactivity. Chlorine, as a liquefied gas, can be shipped in tanks and then dispensed from the tanks (see Figure 11.8). Both ozone and chlorine function as oxidizing agents (Section 10.10). They destroy microorganisms by converting the carbon and hydrogen atoms present in the organisms to carbon dioxide and water.

Figure 11.9 summarizes the five steps that are used at water purification facilities to make fresh water fit to drink.

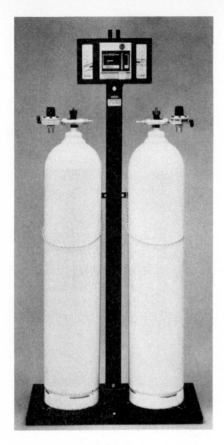

FIGURE 11.8

The chlorine used at water purification plants to disinfect water is dispensed from steel cylinders containing liquefied chlorine gas. These two chlorine cylinders are on a scale that continuously monitors the amount of Cl_2 left in the tanks. [Wallace & Tiernan Division, Pennwalt Corporation.]

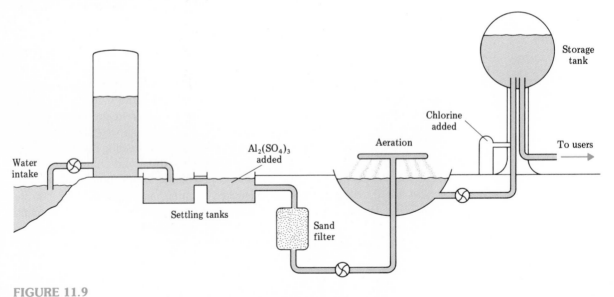

FIGURE 11.9

Steps used in municipal water-treatment plants to change natural fresh water to drinking water.

In Section 21.10 we look at a different type of water purification: general wastewater treatment. A community's wastewater comes from four important sources: (1) sewage, (2) commercial waste, (3) industrial waste, and (4) storm runoff.

11.8 Hard and Soft Water

Hard water *is water containing metal ions (principally Ca^{2+}, Mg^{2+}, and Fe^{2+}) that form insoluble compounds (precipitates) either with soap or upon heating.* Almost all natural fresh waters contain these metal ions; thus, almost all natural fresh waters are hard.

Conventional water purification steps like those discussed in Section 11.6 do not alter the hardness of water because they do not remove dissolved minerals. The focus in water purification is on removing substances that are harmful to human health. At the concentrations normally present in natural fresh waters, dissolved minerals are not harmful to health. Indeed, some of the taste of water is caused by these ions; water without metal ions would taste ''unpleasant'' to most people.

Although hardness does not affect the drinkability of water, it does affect other uses for water. Two very noticeable effects of water hardness are the formation of a sticky, curdy precipitate (scum) with soap and deposit of a hard scale in steam boilers, tea kettles, and hot water pipes. The scum, among other things, produces a ring on bathtubs and leaves washed clothes looking dull, and gray. The hard, scaly deposit inside boilers and pipes forms an effective insulating layer, and heat is not transferred efficiently to the water inside. Ultimately, pipes can become completely clogged with these deposits (see Figure 11.10). Both the ''scum'' and ''scale'' effects of hard water are related to the insolubility of certain calcium, magnesium, and iron salts.

Water hardness can be classified into two types, based on the identity of the predominant negative ion species that accompany the hard water ions (Ca^{2+}, Mg^{2+}, and Fe^{2+}). The three most common negative ions found in hard water are bicarbonate (HCO_3^-), sulfate (SO_4^{2-}), and chloride (Cl^-). Hard water in which bicarbonate ion is the predominant negative species displays *carbonate hardness* (or *temporary hardness*). Boiling this hard water will remove much of the hardness. Upon heating, bicarbonate ions produce carbon dioxide (CO_2) and carbonate ions (CO_3^{2-}).

$$2\,HCO_3^- \xrightarrow{\text{heat}} CO_2 + CO_3^{2-}$$

These carbonate ions react with the hard water ions that are present and produce insoluble carbonates that come out of solution.

FIGURE 11.10
Boiler scale deposits in a section of a hot water pipe over a two-year time period has
nearly closed off the pipe. [Nalco Chemical Company.]

$$M^{2+}(aq) + 2\ HCO_3^-(aq) \longrightarrow MCO_3(s) + CO_2(g) + H_2O(g)$$

$$(M = Ca^{2+}, Mg^{2+}, or\ Fe^{2+})$$

The metal ions are no longer in solution, so the hardness of the water has been
reduced. However, scum or scale (the insoluble carbonates) has been produced.
Hard water that contains predominantly Cl^- or SO_3^{2-} ions rather than HCO_3^-
ions possesses *noncarbonate hardness* (or *permanent hardness*). The hardness
of this water cannot be removed by boiling.

   Soft water *is water from which hardness-causing ions have been removed
(or tied up chemically).* Many different methods for accomplishing this are in use
today. One of the simplest ways to remove the offending metal ions from hard
water is to add chemicals that precipitate them—that is, cause them to form
insoluble compounds. Washing soda, $Na_2CO_3 \cdot 10H_2O$, is used in home laundry situations for this purpose. The washing soda is added before the addition of soap, and the offending ions react with it to form insoluble carbonates
instead of reacting with the more expensive soap, as they would if the chemical
softening agent had not been added.

   Detergents also soften water, but in a different manner than washing soda. All
detergents contain chemicals called *builders*, which exert a softening effect on

water by tying up the metal ions in the form of *soluble complexes*. The offending ions remain in solution, but they no longer have the ability, in their complexed state, to react with cleansing agents in the detergent. The complexing method of water softening has one big advantage over the precipitation method—no scum (insoluble material) is produced.

HOME WATER-SOFTENING EQUIPMENT

The most popular method of softening water in the home is a process called *ion exchange*. In this process, the offending Ca^{2+}, Mg^{2+}, and Fe^{2+} ions are exchanged for sodium ions (Na^+). Sodium ions do not form a precipitate with soap, and they do not form scale; recall from the solubility rules (Section 9.4) that all Na^+ ion salts are soluble.

Ion exchange is accomplished by letting hard water percolate through a container (commercially, the water softener tank) that holds a finely divided substance capable of performing ion exchange. Both naturally occurring materials called *zeolites*, which are complex sodium aluminum silicates, and synthetic materials are available to perform ion exchange.

The reaction that occurs in zeolite ion exchangers can be represented as follows (M^{2+} represents the hard water ions).

$$2\ NaAlSiO_4 + M^{2+}(aq) \longrightarrow M(AlSiO_4)_2 + 2\ Na^+(aq)$$

Zeolite	To be removed from solution		Released into the water

The preceding ion-exchange process is shown pictorially in Figure 11.11. Note that the hard water ions become attached to the ion-exchange material.

Over time, an ion-exchange material becomes saturated with hard water ions and can no longer effectively soften water; all of the Na^+ ions have been replaced. The ion-exchange material can be regenerated by washing it with a concentrated salt (NaCl) solution. This treatment, which "recharges" the water softener, reverses the original "discharging" process; the hard water ions are replaced by Na^+ ions.

$$M(AlSiO_4)_2 + 2\ Na^+(aq) \longrightarrow 2\ NaAlSiO_4 + M^{2+}(aq)$$

"Recharged" zeolite   Discarded in wastewater from recharging

Concentrated sodium chloride solutions are used in the recharging process simply because sodium chloride is the cheapest source of Na^+ ion available.

It should be obvious that soft water produced by the preceding zeolite ion exchange has a much higher than normal Na^+ ion concentration. People with high blood pressure (hypertension) or kidney problems are often advised to avoid drinking softened water. An increased intake of sodium can lead to water retention in the body's tissues and add to the work of the kidneys. A good rule is that hard water is for drinking and soft water is for washing and heating.

Ion-exchange materials that are more expensive than those used in standard home water-softening equipment can produce extremely pure water, called

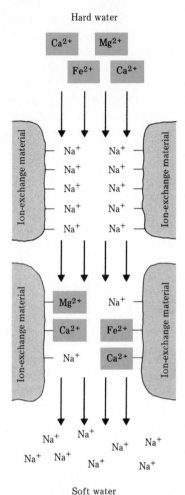

Hard water

FIGURE 11.11

Schematic representation of what happens when hard water is percolated through a zeolite ion-exchange material.

deionized water. **Deionized water** *is water from which all dissolved ions (both positive and negative) have been removed through a double ion-exchange process.* In the production of deionized water, the hard water is first passed through an "acidic" ion-exchange material, which replaces all positive ions with H^+ ions. Then the water is passed through a "basic" ion-exchange material that replaces all negative ions with OH^- ions. The H^+ and OH^- ions now present in the water immediately react with each other (neutralization; see Section 10.7) to produce additional water.

DEIONIZED WATER

$$H^+ + OH^- \longrightarrow H_2O$$

The net result is almost pure water. The gallon jugs of water that are available at grocery and variety stores for home use in steam irons are almost always deionized water.

The pure water used in most laboratory situations, where only small amounts of pure water are needed, is usually distilled water. **Distilled water** *is water purified through distillation, the process in which water is selectively vaporized and condensed to remove it from other substances (the impurities).* The water is boiled and the resulting steam, which is ion-free, is condensed back to water. At present, distillation is too expensive (because of the large amount of heat energy required) for large-scale use as a softening or purification process.

DISTILLED WATER

11.9 Desalination of Seawater

Not having adequate fresh water to meet the needs of an ever-increasing world population is a problem that is on the horizon. The demand for fresh water in the United States has increased over 10-fold since 1900. This increase is not all a result of population increase; per capita demand has increased fourfold, as is shown in Table 11.6.

A seemingly obvious answer to the sure-to-come freshwater supply problems is to reclaim some of the huge quantities of water found in the oceans, a process called desalination. **Desalination** *is the process of removing salts from seawater and brackish water to the extent that the water becomes usable (fresh).* Normal seawater is unfit for human consumption (Section 9.6) and for most of the other uses to which we put water, including irrigation and industry. In industrial processes, the high salt content causes severe corrosion problems in equipment.

Several methods of desalination are in use. We will discuss three of them: (1) distillation, (2) freezing, and (3) reverse osmosis. A major consideration for all desalination processes is cost. The technology in many cases exists but the cost for obtaining the fresh water is not socially acceptable.

TABLE 11.6

United States Freshwater Demands, by Total Use and Per Capita, 1900–1980

Year	Total Demand (billion gal/day)	Per Capita Demand (gal/day)
1900	40	526
1920	92	864
1940	136	1027
1960	275	1500
1980	450	2000

Distillation

Distillation is the oldest method of desalination and also the most-used method at present. This process accounts for about 90% of desalination capacity at present. Total worldwide desalination capacity is approximately 500 million gallons per day.

Distillation desalination is based on the fact that water is a volatile (Section 8.14) substance whereas salts are not volatile. Thus, salts are left behind when water evaporates. Economic benefits are obtained by using multistage distillation units such as the one diagrammed in Figure 11.12.

In a multistage distillation process, heat obtained from condensing vapor in one stage is used to heat water for the next stage. This technique takes maximum advantage of water's high heat vaporization (Section 11.4). Steam gives up 540 cal/g of heat as it condenses. This multistage process also maximizes the use made of the heat stored in the hot seawater. It is not energy efficient to heat new batches of cold seawater continually. Figure 11.13 shows a multistage distillation unit capable of producing about 2 million gallons of fresh water each day.

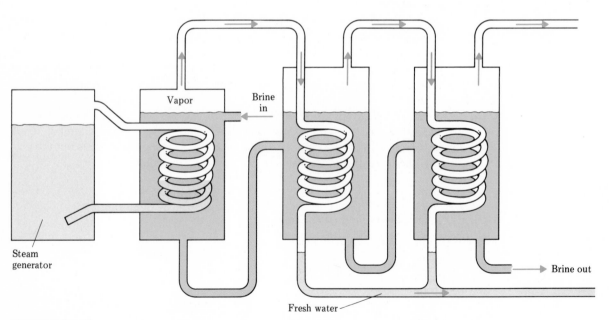

FIGURE 11.12
A multiple-stage distillation unit for desalination of seawater.

FIGURE 11.13

Multistage distillation unit capable of producing about 2 million gallons of fresh water from seawater each day. [© Dan McCoy/Rainbow.]

Freezing

When a stream of cold seawater is sprayed into a vacuum, the evaporation of some of the water cools the remainder, and ice crystals form in the brine. (Remember, from Section 8.13, that evaporation is a cooling process.) The ice crystals have a much lower salt concentration than the brine from which they were formed. "Purer" water can be obtained by collecting the crystals and then melting them. The process is repeated until the desired degree of purity is obtained. An advantage of freezing over distillation is its lower energy consumption. The heat of fusion of water is only 79.8 cal/g compared to 540 cal/g for the heat of vaporization (Section 11.4). Major disadvantages of freezing are the problems associated with the growing, handling, and washing of the ice crystals.

Reverse Osmosis

Both distillation and freezing involve changes of state that require considerable energy input. Reverse osmosis is a process that does not involve a phase change, a situation which is economically desirable.

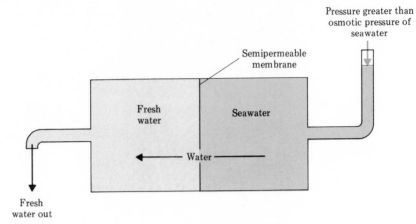

FIGURE 11.14
When a pressure greater than the osmotic pressure of seawater is exerted on the seawater side of the semipermeable membrane, water is forced through the membrane in the direction from right to left. This is the process of reverse osmosis.

FIGURE 11.15
A small-scale, manually operated reverse osmosis desalinator is now used by the U.S. Navy as equipment for life rafts. Slightly more than a gallon of drinkable water per hour can be produced by this device. [Robert Sherrow/People Weekly/© 1987 Time Inc.]

The process of osmosis was discussed previously (Section 9.6). In this process, water (solvent) moves through a semipermeable membrane from an area of low concentration to one of higher concentration. To stop this net flow of water through the membrane, a backpressure, equal to the osmotic pressure of the solution, must be applied. If a backpressure exceeding osmotic pressure is applied to the system, the water can be made to flow in the opposite direction—from concentrated solution to less concentrated solution; this is the process of reverse osmosis (see Figure 11.14).

Reverse osmosis desalination is considerably cheaper than distillation desalination. The main problem preventing more extensive use of this method involves development of suitable semipermeable membranes. The membranes must be permeable to water but not to dissolved substances (ions) and must also be able to withstand high pressures for prolonged periods of time. The osmotic pressure of seawater is about 30 atm. Thus, pressures greater than 30 atm are required to reverse the osmosis process.

A small-scale, manually operated reverse osmosis desalinator is now used by the United States Navy in life rafts. It replaces bulky cases of fresh water previously stored on life rafts. Its construction involves a cellophanlike membrane wrapped around a tube with holes in it. Seawater is forced through the membrane by a hand-operated pump at pressures of about 70 atm. Slightly more than a gallon of drinkable water can be produced per hour by this device (see Figure 11.15).

11.10   Fluoridation of Drinking Water

In the previous two sections we have considered the removal of various ions from water. Fluoridation is a process in which an ion, the fluoride ion (F^-), is purposely added to water. Many communities fluoridate their drinking water. In this process about 1 ppm by mass F^- ion is added to drinking water. Salts such as NaF serve as the source of F^- ion. The purpose of fluoridation is to reduce dental caries. What is the chemistry involved here? The compound hydroxyapatite, $Ca_{10}(PO_4)_6(OH)_2$, is the major constituent of teeth (Section 24.10). Ions associated with this compound, such as OH^- ion, are readily replaced by F^- ions. The resulting fluoride-containing material is less soluble and less reactive than the nonfluoride material is. This decreased solubility and decreased reactivity afford greater protection against caries. There is some opposition to fluoridation of water because many fluorine-containing compounds (at concentrations much greater than in fluoridated drinking water) exhibit toxic effects.

Figure 11.16 shows the extent of fluoridation of public water supplies in the United States by state. Fifty-four percent of the population uses fluoridated drinking water.

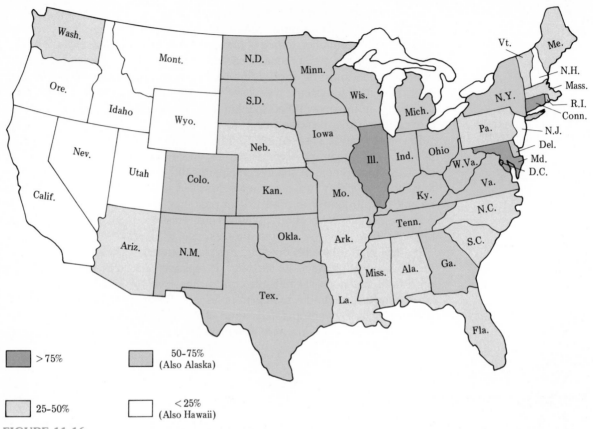

FIGURE 11.16

Fifty-four percent of the United States population uses fluoridated drinking water as of 1986. The percent of state population using fluoridated water and state ranking are given for each state plus Washington, DC, and Puerto Rico.

Fluoridated toothpaste (Section 24.10) is much less effective than fluoridated drinking water in reducing the incidence of tooth decay. In many areas in which drinking water is not fluoridated dentists prescribe fluoride tablets or drops to children.

Practice Questions and Problems

ABUNDANCE AND DISTRIBUTION OF WATER

11-1 What are the major reservoirs for water on earth, and how do they compare to each other in size?

11-2 Approximately what percentage of a human being's total body mass is water?

11-3 What are the major sources of fluid intake and output for the human body?

PROPERTIES OF WATER

11-4 If water were a "normal" compound, what would its boiling point be?

11-5 Explain, in terms of hydrogen bonding, why water has a higher-than-expected boiling point.

11-6 Explain, in terms of water's thermal properties, why large bodies of water have a moderating effect on the climate of the surrounding area.

11-7 Explain why the maximum density of water occurs at a temperature that is higher than its freezing point.

11-8 Explain why lakes freeze from the top to the bottom during the winter rather than from the bottom up.

11-9 Explain why the formation of potholes in streets is hastened during the winter.

NATURAL WATERS

11-10 Discuss the basis for the classification of natural waters as fresh, brackish, or seawater.

11-11 What are the three most abundant metallic ionic species and the three most abundant nonmetallic ionic species in seawater?

11-12 What are the three most abundant metallic ionic species and the three most abundant nonmetallic ionic species in fresh water?

11-13 Discuss the distribution of fresh water on earth.

11-14 Explain why fresh water is considered a renewable natural resource.

DRINKING WATER PURIFICATION

11-15 What five steps are typically involved in drinking water treatment programs?

11-16 Discuss the chemistry involved in the following drinking water treatment steps.
a. The sedimentation step
b. The chemical disinfection step

HARD AND SOFT WATER

11-17 What metal ions found in natural waters are responsible for the hardness of water?

11-18 What are the two major problems caused by water hardness?

11-19 Distinguish between temporary and permanent hardness of water.

11-20 From the point of view of what happens to the "hard-water ions" present, compare hard water and water softened by reaction with the following substances.
a. Washing soda
b. A detergent builder
c. A zeolite-type substance

11-21 What chemistry is involved in recharging a mechanical water softener unit by running concentrated saltwater solution through it?

11-22 What is the difference between distilled water and deionized water?

DESALINATION OF SEAWATER

11-23 What is the principal process in use today for desalination of seawater, and what percent of present desalination capacity does it account for?

11-24 In a multistage distillation process for desalination, what use is made of the heat obtained when water vapor is condensed?

11-25 Describe how desalination can be accomplished using freezing techniques.

11-26 What is the difference between the processes of osmosis and reverse osmosis?

11-27 What major advantage does reverse osmosis have over distillation and freezing as a method for desalination?

11-28 What is the major problem preventing more extensive use of reverse osmosis as a desalination technique?

FLUORIDATION OF DRINKING WATER

11-29 What chemistry is involved in using fluoridated water to help prevent dental caries?

Some Important Industrial Chemicals

CHAPTER HIGHLIGHTS

12.1 High-Volume Industrial Chemicals

Hundreds of thousands of different chemicals are used every day in the United States. Which are the most important? The answer depends on the criteria used to define the term *important*. One answer relates the importance of a chemical to the amount produced; the most important are those produced in the largest quantities. We use this definition of importance to discuss the most important chemicals produced and used in the United States.

Which chemical substance is produced in the greatest amount in the United States? The answer is not debatable. There is no doubt. It is sulfuric acid, a runaway winner, as can be seen in Table 12.1, which gives the average annual production figures (in billions of pounds) for the period since 1980.

Sulfuric acid has an almost 2 to 1 lead over any other chemical. The race for the number 2 position is closer, with elemental nitrogen not so far ahead of the other three contenders for this position. Small differences in production levels separate the chemicals in positions 3, 4, and 5. (No other chemicals have had an annual average production level exceeding 30 billion lb since 1980.)

Two elements, nitrogen and oxygen, are listed among the top five high-volume chemicals. These two elements are considered to be "chemicals" rather than naturally occurring substances because technological processes must be used to obtain them. Neither nitrogen nor oxygen is found in its pure state in the environment; they are always mixed with each other and other gases.

Let us now consider why such great amounts of these five substances are needed in the United States each year, and also how each of these five substances is obtained in its desired form. We will begin with sulfuric acid and then consider each of the other chemicals in the "top five."

TABLE 12.1

Chemicals Produced in the Greatest Amount in the United States

Rank	Chemical Name	Chemical Formula	Average Annual Production Since 1980 (billion pounds)
1	Sulfuric acid	H_2SO_4	77.85
2	Nitrogen	N_2	41.44
3	Ammonia	NH_3	33.13
4	Oxygen	O_2	32.19
5	Lime	CaO	32.13

12.2   Sulfuric Acid

Sulfuric acid is a colorless, corrosive, oily liquid in the pure state. It is usually marketed as a concentrated aqueous solution of concentration 96% by mass. The chemical has been known for hundred of years; it was the *oil of vitriol* used by ancient alchemists.

People rarely have direct contact with this substance because it is seldom part of finished consumer products. The closest encounter most people have with this acid (other than in a chemical laboratory) is in an automobile battery. The battery acid in the standard automobile battery is an aqueous solution of sulfuric acid (Section 10.12). However, battery acid is not a major use for sulfuric acid. Less than 1% of the annual sulfuric acid production ends up in car batteries.

Approximately 70% of sulfuric acid production is used in the manufacture of chemical fertilizers. These fertilizer compounds are an absolute necessity if the food needs of an ever-increasing population are to be met.

The six nutrients needed in the greatest amounts by plants are carbon, hydrogen, oxygen, nitrogen, phosphorus, and potassium. Carbon, hydrogen, and oxygen are obtained by plants from both soil and air; the soil is the only source of the other three nutrients. Chemical fertilizers are used to replenish the soil's nitrogen, phosphorus, and potassium.

PHOSPHORUS-
CONTAINING
FERTILIZERS

The sulfuric acid–chemical fertilizer connection centers on the element phosphorus. The raw material for phosphate fertilizers, the form in which phosphorus is supplied to plants, is phosphate rock. Phosphate rock, which is mainly $Ca_3(PO_4)_2$, is a highly insoluble material that is almost useless to plants as a source of phosphorus. The treatment of phosphate rock with H_2SO_4 results in the formation of phosphoric acid (H_3PO_4).

$$Ca_3(PO_4)_2 + 3\ H_2SO_4 \longrightarrow 3\ CaSO_4 + 2\ H_3PO_4$$

Phosphate
rock

Phosphoric
acid

The phosphoric acid is then used to produce soluble phosphate salts, which are used as a source of phosphorus for plants. The major phosphoric acid fertilizer derivative is diammonium phosphate (DAP) [$(NH_4)_2HPO_4$].

Petroleum refining and metallurgical processes constitute the next largest uses of sulfuric acid. Each of these use areas account for approximately 4% of sulfuric acid production. These are small amounts compared to the 70% of production used in fertilizer manufacture.

One remarkable property of sulfuric acid is its great affinity for water. This makes sulfuric acid an excellent dehydrating agent. For example, when this acid is mixed with sugar, it dehydrates the sugar (extracts hydrogen and oxygen

(a) (b) (c)

FIGURE 12.1
Table sugar (sucrose) before (left) and after (right) treatment with concentrated sulfuric acid. The acid extracts water from the sugar to form black carbon and steam. The steam mixes with the carbon and the mixture rises above the top of the container.

atoms to form water) and leaves behind a column of black carbon (see Figure 12.1). Because of this property, sulfuric acid finds industrial use as a dehydrating agent in the production of explosives, dyes, and detergents.

The raw materials needed for the production of the large amounts of sulfuric acid consumed in the United States are sulfur, air, and water. Elemental sulfur can be obtained in a relatively pure form from underground deposits of this element, using the *Frasch process*. In this process superheated water is pumped into the underground sulfur deposit to melt the sulfur (mp = 113°C). Compressed air, also pumped into the deposit, forces to the surface a bubbly froth of air, water, and molten sulfur. The molten sulfur is collected and allowed to solidify. Figure 12.2 shows a schematic view of the Frasch method for recovering sulfur from underground deposits.

About 60% of the sulfur produced in the United States is mined from

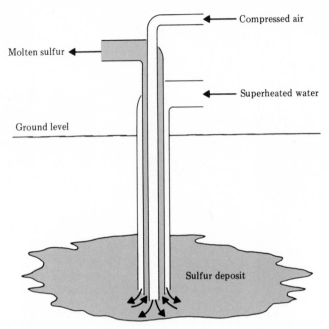

FIGURE 12.2

The Frasch method for recovering sulfur. The hot water melts the sulfur, which is then forced to the surface by compressed air.

underground deposits using the Frasch process. The other 40% of sulfur produced in the United States results from air pollution control procedures. It is either by-product sulfur removed from fossil fuels (particularly natural gas) before combustion to prevent pollution or sulfur obtained from removal of SO_2 from exhaust gases produced from burning fossil fuels (particularly coal). The topic of sulfur dioxide as an air pollutant is considered in Section 20.8.

The first step in the production of sulfuric acid is the reaction of sulfur with oxygen (from air) to produce sulfur dioxide.

$$S(s) + O_2(g) \longrightarrow SO_2(g)$$

Next the SO_2 gas is combined with additional O_2 (air) to produce sulfur trioxide (SO_3) gas. The oxidation of SO_2 to SO_3 is very slow at ordinary temperatures. To speed up the reaction, both elevated temperatures (400°C) and a catalyst [usually solid-state vanadium oxide (V_2O_5)] are used. Because the reaction between SO_2 and O_2 takes place when these two gases come in contact with the solid catalytic surface, this method for making sulfuric acid is known as the *contact process.*

$$2\,SO_2(g) + O_2(g) \xrightarrow[\text{V_2O_5 catalyst}]{400°C} 2\,SO_3(g)$$

The final step in the process is the combination of SO_3 with water to give H_2SO_4.

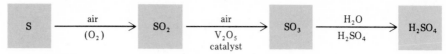

FIGURE 12.3

Steps in the contact process for producing sulfuric acid (H_2SO_4).

$$SO_3(g) + H_2O(l) \xrightarrow{H_2SO_4} H_2SO_4(l)$$

The SO_3 is not directly reacted with water, because such reaction produces a mist of H_2SO_4, which is extremely difficult to condense. Instead, gaseous SO_3 is added to a flowing solution of 98% H_2SO_4, to which water is constantly added to keep the sulfuric acid concentration constant.

Figure 12.3 summarizes the steps involved in producing sulfuric acid using the contact process. More than 90% of sulfuric acid is produced by this process.

The fact that sulfur dioxide gas is an intermediate in the production of H_2SO_4 has important environmental chemistry ramifications. In Section 20.9 we will find that sulfuric acid produced in an atmospheric reaction involving SO_2, O_2, and H_2O is the major acidic component of acid rain.

12.3 Nitrogen

Nitrogen is a colorless, odorless, nontoxic, diatomic gas. It is the most abundant naturally occurring gas encountered on earth, constituting 78% by volume of the atmosphere. It is the atmosphere that serves as the source for the large quantities of industrial nitrogen that are consumed.

Commercial production of pure nitrogen involves its extraction from air. Because air is a mixture and not a chemical compound, it can be separated into its components by physical processing. Chemical reactions are not needed because no chemical bonds are broken. Separation is accomplished by the use of cryogenic (low-temperature) techniques. The air is first liquefied and then its components are separated using fractional distillation.

Figure 12.4 summarizes the steps involved in obtaining nitrogen (and oxygen, at the same time) from the atmosphere. Clean air is first fed into a compressor and then cooled by refrigeration. The air is under a pressure of 200 atm by the time it reaches the expansion nozzle. During the expansion process the pressure is suddenly dropped from 200 atm to 20 atm. This results in further cooling of the gases, which is sufficient to cause them to liquefy. [Remember, from Section 8.13, that evaporation (expansion) is a cooling process.] The liquid air is filtered to removed any carbon dioxide or methane present (both are now solids) and then it is distilled. The three main components of air, nitrogen (78%), argon (1%), and oxygen (21%), have boiling points of $-196°$, $-186°$, and $-183°C$,

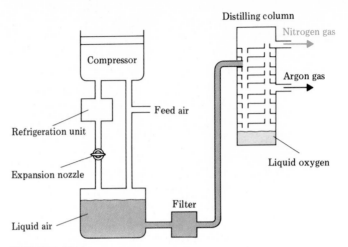

FIGURE 12.4

The steps involved in obtaining nitrogen and oxygen from the atmosphere by liquefaction and fractional distillation.

respectively. As the liquid air enters the top of the distilling column, nitrogen, which is the most volatile of the three gases (the lowest boiling point), is separated off as a gas. Gaseous argon is removed from the middle of the column and liquid oxygen, the least volatile component, collects at the bottom.

The major use for nitrogen gas obtained from the fractional distillation of air is as a chemical raw material (reactant) in the production of ammonia. (Ammonia is another of the top five chemicals). Approximately three-fifths of total N_2 production is used for this purpose. Particulars about the use of N_2 in ammonia production will be covered when we discuss ammonia (Section 12.4).

The other two-fifths of N_2 production is used in a physical rather than chemical manner; that is, the nitrogen is not a chemical reactant. The nitrogen production figure in Table 12.1 is the N_2 used in a physical manner; the nitrogen used in ammonia synthesis is included in the ammonia production figure.

Physical uses for nitrogen result mainly from two properties of this gas: (1) its lack of chemical reactivity and (2) its low boiling point. The nitrogen atoms in molecular nitrogen are joined by a triple bond (Section 5.9). This bond is the second strongest known for a diatomic molecule. [The only bond stronger is the triple bond in carbon monoxide (CO).] Because of this high bond strength, N_2 is chemically very unreactive at ordinary and low temperatures. The boiling point of N_2, $-196°C$, is the lowest of any commonly occurring gas. This means that liquid N_2 exists at temperatures much lower than that of other common liquids.

The major physical end us for industrial nitrogen is that of a "blanketing atmosphere." Chemicals that would react with oxygen or decompose in the presence of moist air are "blanketed" with nitrogen to protect them from chemical change. Chemical process industries make frequent use of "blanketing" to exclude oxygen and moisture from a chemical system. In petroleum refinery operations reactor vessels, tanks, pipelines, and other equipment are purged with nitrogen during startup, shutdown, and cleaning operations. In electronic manufacturing, semiconductors, integrated circuits, and single crystals are all produced with the use of N_2 as a blanketing agent.

The most rapidly growing use for nitrogen involves the petroleum industry and its enhanced oil recovery (EOR) projects. The pressure within oil wells, which helps bring oil to the surface, drops with the length of time the oil well has been producing. Nitrogen is being extensively used to repressurize wells (Section 19.2). Large volumes of N_2 are injected into the rock formation containing the oil.

FLASH-FREEZING OF FOOD

Use of liquid nitrogen as a cryogenic substance is also increasing. The main application here is the flash-freezing of meat, poultry, and seafood and some pastries. These substances are in high demand because of steady growth in the fast-food industry. Most foods are frozen using mechanical freezing units. Flash-freezing using liquid N_2 has been found to give a product with better appearance and taste than that obtained using the slower mechanical methods. Flash-freezing produces smaller crystals of ice than conventional freezing. Larger crystals can rupture food cells and tissue, which affects appearance and taste.

Carbon dioxide (Section 13.4) competes with liquid N_2 for use in both EOR and cryogenic freezing.

PRESSURIZATION OF BEVERAGE CONTAINERS

A new use of N_2 involves injection of a small amount of liquid N_2 into the headspace of aluminum cans used to contain noncarbonated fruit juices and other beverages. This N_2, which rapidly vaporizes, pressurizes the can in the same way that CO_2 gas does in carbonated drinks. This pressurization helps prevent dents and implosions. Previously steel cans had been used for noncarbonated beverages. Unlike CO_2, N_2 is only slightly soluble in water and therefore does not affect the beverage.

Following are some estimates of the amount of nitrogen currently being consumed in various physical end uses: in blanketing atmospheres for chemical processing, 25%; in enhanced oil recovery, 25%; in electronics manufacturing, 15%; in treating and processing of metals, 15%; in aerospace, 5%; and in freezing of meat, poultry, and seafood, 5%.

12.4 Ammonia

Ammonia (NH_3), like sulfuric acid, is a chemical whose claim to fame is closely tied to chemical fertilizers. Recall, from our discussion of sulfuric acid (Section 12.2), that the three plant nutrients in soil that are most likely to need replenishing are nitrogen, phosphorus, and potassium. Ammonia and its derivatives are the main mechanism for nitrogen replacement.

The elemental nitrogen (N_2) present in the atmosphere cannot be used directly as a nutrient by most plants. Plants cannot break the strong nitrogen–nitrogen triple bond present inthe N_2 molecule. To be useful to plants, atmospheric N_2 must be *fixed*, that is, converted into compounds in which nitrogen participates in single bonds. Nitrogen-fixing bacteria exist that can produce such nitrogen compounds in the soil from atmospheric nitrogen. These bacteria live in nodules on the roots of leguminous plants such as peas, beans, and clover. Lightning discharges produced during storms also can fix nitrogen. Here, nitrogen and oxygen in the atmosphere react, as the result of electrical discharge, to form nitric oxide (NO).

Natural methods of nitrogen fixation are not sufficient to sustain the need for fixed nitrogen demanded by modern-day agriculture. Consequently, industrial nitrogen fixation—primarily ammonia production—is used to supply plants (our food supply) with fixed nitrogen.

At room temperature and pressure, ammonia is a colorless gas that has a very sharp and irritating odor. Under increased pressure, ammonia is easily liquefied. In liquid form ammonia can be used directly as a fertilizer (see Figure 12.5). In addition, ammonia is easily converted to solid derivatives. Ammonium salts such as ammonium nitrate (NH_4NO_3), ammonium sulfate [$(NH_4)_2SO_4$], and di-ammonium phosphate [$(NH_4)_2HPO_4$] can be prepared using simple acid–base neutralization reactions (Section 10.7).

NITROGEN FIXATION

NITROGEN-CONTAINING FERTILIZERS

FIGURE 12.5
Liquid ammonia being applied to the soil as a fertilizer. [Grant Heilman.]

$$NH_3 + HNO_3 \longrightarrow NH_4NO_3$$

$$2\,NH_3 + H_2SO_4 \longrightarrow (NH_4)_2SO_4$$

$$2\,NH_3 + H_3PO_4 \longrightarrow (NH_4)_2HPO_4$$

Urea, a compound obtained from the reaction between ammonia and carbon dioxide, is also an important nitrogen fertilizer.

$$2\,NH_3 + CO_2 \longrightarrow (NH_2)_2CO + H_2O$$

At present, approximately 80% of the ammonia produced ends up as chemical fertilizer, either directly as ammonia or in ammonia derivatives.

Nonagricultural uses of ammonia account for the remaining 20% of ammonia consumption in the United States. Approximately 10% of ammonia produced is used to make intermediate products needed for the production of synthetic fibers, such as nylons and acrylonitrile (Chapter 18), and 5% is used in manufacturing explosives, primarily blasting agents used in the mining industry.

Ammonia is produced from elemental N_2 using the *Haber process,* a process in which elemental N_2 and elemental H_2 are reacted together. This process, first used in 1913, was developed by the German physical chemist Fritz Haber (1868–1934) (see Historical Profile 14).

Gaseous N_2 and H_2 do not react at room temperature and pressure. For a successful reaction to occur, high pressures (100–1000 atm), high temperatures (400–550°C), and a catalyst (usually finely divided Fe and Fe_3O_4 containing small amounts of K_2O and Al_2O_3) are required. It was Haber's studies that determined the set of conditions that maximizes the amount of ammonia produced.

$$N_2(g) + 3\,H_2(g) \xrightarrow[\text{high } T,\ \text{high } P]{\text{catalyst}} 2\,NH_3(g)$$

The raw materials for ammonia production using the Haber process are air and natural gas. Fractional distillation of liquid air (Section 12.3) supplies the needed N_2. Natural gas is the source of H_2.

You might expect that H_2O would be the source of H_2. However, the expense of obtaining H_2 from H_2O prohibits this; an enormous input of energy, usually in the form of electricity, is required to break the hydrogen–oxygen bonds in water. Currently, methane (CH_4), the main component of natural gas, is the source of almost all H_2 used in industrial processes. Hydrogen can be obtained from methane through its reaction with a limited supply of air (O_2)

$$2\,CH_4(g) + O_2(g) \xrightarrow{\text{catalyst}} 2\,CO(g) + 4\,H_2(g)$$

or through steam-reforming of methane, a process in which CH_4 reacts with water vapor at high temperatures.

$$CH_4(g) + 2\,H_2O(g) \xrightarrow[\text{high } T]{\text{catalyst}} CO_2(g) + 4\,H_2(g)$$

$$CH_4(g) + H_2O(g) \xrightarrow[\text{high } T]{\text{catalyst}} CO(g) + 3\,H_2(g)$$

[Encyclopaedia Britannica, Inc., Chicago.]

HISTORICAL PROFILE 14

Fritz Haber   1868–1934

Fritz Haber, a German chemist, was born in Breslau (now Wroclaw, Poland), the only son of a merchant dealing in pigments and dyestuffs. He found that he preferred chemistry over his father's business, after an attempt at the later. After obtaining a doctorate in organic chemistry, Haber's early years were spent in Karlsruhe; in 1912, he became director of the Kaiser Wilhelm Institute for Physical Chemistry in Berlin.

Theoretical physical chemistry rather than organic chemistry occupied his attention most of his life. He taught himself this then new subject and his major contributions to chemistry are in this field.

His most important work was carried out at Karlsruhe, where he become interested in the conditions necessary to produce ammonia from atmospheric nitrogen. Despite the skepticism of his colleagues, he was able to elucidate the conditions required to produce this gas in acceptable amounts using the process that now bears his name.

Industrialization of the Haber process came just in time to provide Germany with the raw materials for munitions needed for World War I. The British Navy cut off imports needed for munitions, and Germany would have run out of explosives by 1916 had it not been for Haber's research. When peace came, the real import of ammonia synthesis was realized—its potential for making the soil more fruitful. In 1918 Haber received the Nobel prize in chemistry for the fixing of nitrogen from the air. His process was hailed as "the means of improving agriculture and the welfare of mankind" and he became known as the one who "made bread from air."

An extremely patriotic German, Haber died a broken-hearted exile in Switzerland, in 1934. Of Jewish ancestry, early in his life he had become a Protestant. With the rise of Nazism in 1933, he was told he could remain as director of his institute but that those under him of Jewish background would have to be dismissed. This was unacceptable. Upon resigning, he stated: "For 40 years I have selected my collaborators on the basis of their intelligence and their character and not on the basis of their grandmothers, and I am not willing for the rest of my life to change this method which I have found so good." He died less than a year after leaving Germany.

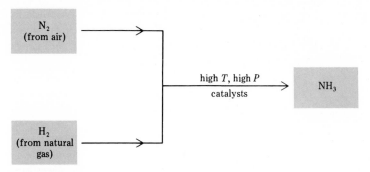

FIGURE 12.6

Schematic drawing showing how ammonia is produced from air and natural gas by the Haber process.

Figure 12.6 summarizes the basics of the production of ammonia using the Haber process.

12.5 Oxygen

Oxygen, the most abundant element on earth (Section 2.6), is the component of the atmosphere that sustains life. We can live weeks without food, days without water, but only minutes without oxygen. Oxygen is also the chief component of water and the earth's rocks and minerals (Section 13.2). It is an essential part of all fats, carbohydrates, proteins, and nucleic acids found in the human body. This colorless, odorless, tasteless, nontoxic gas is often referred to as the ''life-giving'' element.

Surprisingly, it was not until the late eighteenth century that oxygen was identified as a distinct substance. In 1774, the English chemist Joseph Priestley (1733–1804) (see Historical Profile 15) conducted experiments that showed, for the first time, that air is a mixture of gases and that one of these gases (oxygen) is involved in the process of combustion.

As evidenced by its presence in the list of top five chemicals, oxygen is important not only from a biological standpoint but also from an industrial standpoint. We have already discussed the process by which industrial oxygen is obtained. It is extracted from air in a manner similar to that for nitrogen (Section 12.3). Oxygen and nitrogen are coproducts of this process.

Most of the oxygen obtained from air separation is used as a reactant in chemical processes. The use of oxygen instead of air, which is 21% by volume O_2, in chemical processes offers a number of advantages to offset the disadvantage of increased cost, including the following.

1. The use of pure O_2 rather than air amounts to a fivefold increase in O_2 concentration, which in turn results in a large increase in production.
2. The rates of reactions are increased, and production times are decreased (Section 10.2).
3. Smaller production facilities can be used because the volume of reactant gases and exhaust gases is reduced to one-fifth of that required for air.
4. Any possibility of interfering reactions with nitrogen, the other component of air, is completely eliminated.

The oxygen molecule (O_2) itself is generally not reactive at or near room temperature. Most materials have to be provided with activation energy (Section 10.1), through heating, before they can react with oxygen at an appreciable rate. Catalysts, however, effectively lower the necessary activation energy. The actions of protein catalysts (enzymes) allow oxygen to react and release energy in living cells during metabolic processes that take place at body temperature.

The steel-making industry, discussed in detail in Section 14.2, is the dominant outlet for industrial O_2 production; approximately 50% of total production goes to this industry. Oxygen's role in steel making is to speed up, because of increased O_2 concentration, the conversion of impurities (Si, Mn, C, S, and P) in the molten iron to oxides, which then react with lime to produce slag. (Remember that the concentration of O_2 in pure oxygen is five times that in air.) The oxygen gas is blown directly into the iron melt and a complete "heat" (up to 300 tons of steel) can be produced every hour. By contrast, older furnaces using air require 10 to 12 hours per heat.

Chemical manufacturing processes that do not involve metals account for 20% of oxygen consumption. Ethylene oxide production accounts for 60% of consumption in this area.

$$2\ C_2H_4 + O_2 \longrightarrow 2\ C_2H_4O$$

$$\text{Ethylene} \qquad\qquad\qquad \text{Ethylene oxide}$$

This chemical is an intermediate in the production of ethylene glycol ($C_2H_6O_2$). Ethylene glycol is used as automobile antifreeze (Section 17.3) and as a reactant in the production of polyester films and fibers (Section 18.7).

Other important but less common applications for oxygen include metal fabrication (5%), medical and life-support uses (2%), and the space program (less than 0.1%).

HIGH-TEMPERATURE WELDING TORCHES

In metal fabrication, the cutting and welding of metals require temperatures much higher than those obtainable from combustion reactions involving air. Fuel–oxygen torches are used to obtain the needed temperatures; the fuel is acetylene or hydrogen (see Figure 12.7). In the process known as oxygen cutting, the metal is heated to its combustion temperature by the fuel-oxygen torch. The fuel gas is then shut off and a stream of pure O_2 is directed at the hot metal. This causes the metal to burn and produces a narrow cut. The torch has literally burned through the metal.

The major medical use of O_2 is in the administration of oxygen-enriched air to

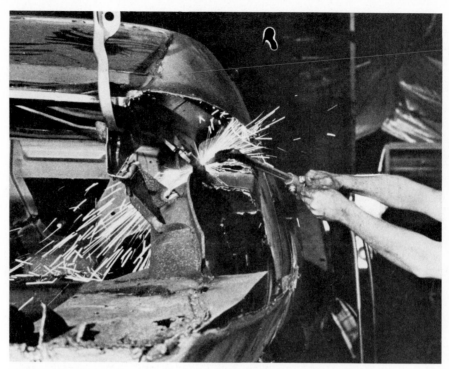

FIGURE 12.7

An acetylene–oxygen torch is used in welding operations. [© Rafael Macia/Photo Researchers, Inc.]

OXYGEN AND
RESPIRATION

persons suffering respiratory disorders. Life-support systems involving oxygen aid in maintaining or providing atmospheres necessary to sustain life in pressurized aircraft cabins, manned space vehicles, and equipment for underwater diving operations.

Humans are accustomed to breathing a mixture of oxygen and nitrogen in which the pressure of the oxygen is about one-fifth of an atmosphere (159 mm Hg). They can adapt to oxygen pressures in the range of 65 to 425 mm Hg, but cannot survive if forced to breathe oxygen outside these pressure limits. To sustain a person in environments different from normal, careful attention must be given to the oxygen content of life-support gases that are used.

The total atmospheric pressure, and consequently the oxygen pressure, drops with increasing altitude, as shown in Figure 12.8. Adjustments must therefore be made if a person is to survive at high altitudes. In high-altitude commercial aircraft the oxygen in the cabin is usually maintained at a pressure equivalent to an 8000 foot elevation (118 mm Hg). Supplemental oxygen masks are available for use in emergencies that result in decompression.

American astronauts who journey into space breathe pure oxygen at a pressure of 254 mm Hg ($\frac{1}{3}$ atm). This greater-than-minimum pressure is used to

[Burndy Library, Norwalk CT.]

HISTORICAL PROFILE 15

Joseph Priestley   1733–1804

Joseph Priestley was born in Yorkshire, England, the eldest son of a Nonconformist minister. In youth he studied languages, logic, and philosophy but never formally studied science. He developed radical religious beliefs of his own and became a Unitarian minister.

In 1766 he met Benjamin Franklin, who was in London attempting to settle the taxation dispute between the British government and the American colonists. This meeting influenced Priestley to take up a scientific career in addition to his ministerial one. At first his research involved electricity and then turned to chemistry and the study of gases.

Shortly after his meeting with Franklin, he took over a pastorate in Leeds. Next door to his quarters was a brewery. Priestley took to studying the gas (carbon dioxide) that fermenting grain produces. These studies led to the idea of dissolving the gas under pressure in water. His resulting "soda water" became famous all over Europe.

When Priestley began his gas studies, only three gases were known: air, carbon dioxide, and hydrogen (which had just recently been discovered). Numerous new gases were discovered by him, including ammonia, hydrogen chloride, sulfur dioxide, carbon monoxide, and hydrogen sulfide. In 1774, his most important discovery, that of oxygen, resulted from his heating of an oxide of the element mercury.

During all of his scientific studies he remained an outspoken man on the subject of religion and also openly supported the American colonists in their revolt against the British king. In 1791 his house and laboratory were burned down by an angry mob upset because of his sympathetic attitude toward the American and French revolutions. Priestley managed to escape, went into hiding for a time, and eventually emigrated to the United States. The last 10 years of his life were spent in Northumberland, Pennsylvania. In 1874, the hundredth anniversary of the discovery of oxygen, the American Chemical Society was organized at the residence he had occupied in Northumberland.

Priestley was the author of more than 150 books, mostly on theological subjects. He always considered theology more important than science. Concerning his scientific accomplishments, a contemporary wrote that "no single person ever discovered so many new and curious substances."

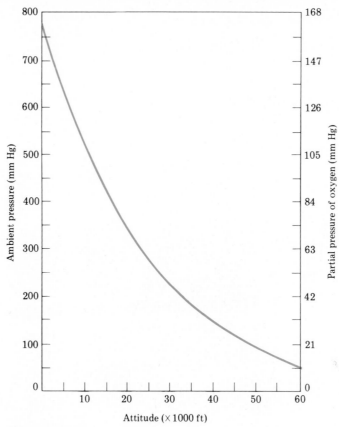

FIGURE 12.8

The change in atmospheric and oxygen pressure with altitude.

allow time for emergency action in the event of a decompression. Pure oxygen is used to avoid the added mass of nitrogen gas and the equipment needed to maintain a constant ratio of the two gases.

SCUBA DIVING

Life-support systems that operate at pressures greater than 1 atm, as in scuba diving (see Figure 12.9), require even more control. A scuba diver at a depth of 100 feet experiences approximately 3 atm of pressure and at 300 feet the pressure is nearly 10 atm. At 300 feet the pressure of oxygen is about 2 atm (21% of 10 atm) for a diver inhaling compressed air. This is too high for the human body, and consequently the oxygen must be diluted with another gas. Helium is the diluting agent used; it is less soluble in body fluids than is nitrogen.

SPACE VEHICLE
PROPELLANT SYSTEMS

Propellant systems used to launch space vehicles involve liquid O_2 use. Such propellant systems consist of an oxidant (oxidizing agent), the liquid O_2, and a fuel such as hydrazine (N_2H_4) or liquid hydrogen (H_2). The oxidant and fuel are

FIGURE 12.9
Scuba divers breathe air that has been diluted with helium to reduce oxygen pressure.
[Courtesy of Robert J. Ouellette, The Ohio State University.]

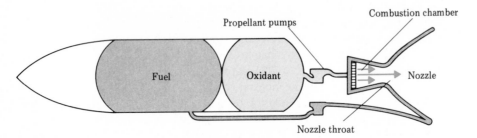

FIGURE 12.10
Components of a typical liquid propellant rocket. Fuel and oxidant are pumped from
their storage tanks to the combustion chamber. Circulation of the fuel through the
walls of the combustion chamber and nozzle cools the surfaces of these structures.

kept in separate tanks until propulsion is required. At that time streams of the two liquids are brought together and the mixture is ignited; thrust is produced by the high-speed expulsion of large amounts of gaseous combustion products. The pipes, pumps, and valves in the system must be designed to allow the entire combustion process to occur within a few seconds after the process is initiated. Figure 12.10 shows the engine components of a typical liquid propellant rocket.

12.6   Lime

The term *lime* can denote either of two substances. It is used for calcium oxide (CaO), which is specifically known as *quicklime* (or unslaked lime), and also for calcium hydroxide [$Ca(OH)_2$], which is specifically known as *hydrated lime* (or slaked lime). The ultimate source for both of these substances is limestone.

Limestone consists primarily of the mineral calcite, which is calcium carbonate ($CaCO_3$). When limestone is strongly heated, to temperatures above 900°C, it decomposes to give carbon dioxide and a residue of calcium oxide (quicklime).

$$CaCO_3(s) \xrightarrow{900°C} CaO(s) + CO_2(g)$$

$$\text{Limestone} \qquad \text{Quicklime}$$

The addition of water to quicklime, a process called slaking, produces calcium hydroxide (hydrated lime).

$$CaO(s) + H_2O(l) \longrightarrow Ca(OH)_2(s)$$

$$\text{Quicklime} \qquad\qquad \text{Hydrated lime}$$

Steel making, the dominant outlet for oxygen production (Section 12.5), is also the major outlet for lime production; it consumes approximately 40% of annual production. In the steel-making process, lime is added to a molten charge of iron that contains impurities (mainly silicon dioxide and silicates). The lime reacts with the impurities to form *slag*, a glassy waste material that floats to the top of the molten iron and is removed. The reaction between lime and the impurities (represented by the formula SiO_2) that results in slag formation is

$$CaO(s) + SiO_2(l) \longrightarrow CaSiO_3(l)$$

$$\text{Quicklime} \quad \begin{array}{c}\text{Impurities in}\\ \text{molten iron}\end{array} \quad \begin{array}{c}\text{Calcium silicate}\\ \text{(slag)}\end{array}$$

The fastest-growing uses for lime are in the environmental area. About 10% of total lime production goes to treat water in preparation for potable uses (Section 11.7), and another 15% of the total is used for various kinds of pollution control. Pollution controls include sewage treatment (Section 21.10), acid mine water drainage neutralization (Section 21.7), and the removal of sulfur oxides (Section 20.8) from exhaust gases from the combustion of fossil fuels. An

TABLE 12.2
Optimum Soil pH Values for the Growth of Selected Plants

Strongly acidic, pH 4–5	Blueberries, cranberries, ferns, holly trees, orchids, laurel, and rhododendron
Acidic, pH 5–6	Most berries and shrubs, dogwood, hemlock, and spruce trees
Near neutral, pH 6–8	Most vegetables, flowers, trees, and grass
Basic, pH 8–9	Hay and soybeans

equation for the reaction that occurs in controlling sulfur oxide pollution is

$$CaO(s) + SO_2(g) \longrightarrow CaSO_3(s)$$

Lime   Air pollutant   Calcium sulfite

The finely divided solid $CaSO_3$ is removed from the exhaust stack by electrostatic precipitators.

BRICK MORTAR   A very small amount of lime, less than 1% of total production, is consumed in building construction. Mortar used in bricklaying is made from hydrated lime, sand, and water. Initially, the mortar sets up because of the formation of solid $Ca(OH)_2$. Later, the $Ca(OH)_2$ is converted to $CaCO_3$ through reaction with CO_2 in the air.

$$Ca(OH)_2(s) + CO_2(g) \longrightarrow CaCO_3(s) + H_2O(g)$$

Hydrated lime   From air   Calcium carbonate

SOIL ACIDITY   Another small, but important, use of lime involves control of acidity in soil. Plant health is directly affected by soil pH (section 10.9), and different species of plants require different soil pH conditions for optimum growth. Optimum soil pH values for the growth of selected plants are given in Table 12.2.

If soil becomes too acidic, because of rotting plant material or from other natural sources, addition of lime is an often-used method of reducing the acidity (increasing the pH). The pH-increasing reaction that occurs may be represented as

$$CaO + 2 H^+ \longrightarrow Ca^{2+} + H_2O$$

Lime   Acidity   Neutral

Practice Questions and Problems

HIGH-VOLUME INDUSTRIAL CHEMICALS

12-1 When the '"top five" industrial chemicals are considered collectively, how many different elements are involved? List them.

12-2 Why are two elements, N_2 and O_2, listed among the top five industrial chemicals?

SULFURIC ACID

12-3 What is the relationship between sulfuric acid and the fertilizer industry?

12-4 Describe the Frasch process for mining underground sulfur deposits.

12-5 What is the relationship between air pollution control procedures, elemental sulfur, and sulfuric acid production.

12-6 Why is the industrial process for producing sulfuric acid called the *contact process*?

12-7 Write a chemical equation for each of the three steps involved in producing sulfuric acid from elemental sulfur.

12-8 Write a chemical equation for each of the following processes.
   a. The dissolving of phosphate rock by sulfuric acid.
   b. The dissolving of sulfur trioxide in water.
   c. The combustion of sulfur in air.
   d. The oxidation of sulfur dioxide to form sulfur trioxide.

NITROGEN

12-9 Summarize the steps involved in extracting nitrogen gas and oxygen gas from the atmosphere.

12-10 Uses for industrial N_2 may be divided into two broad categories. What are these two categories?

12-11 Major physical end uses for N_2 are based on what two properties of this gas?

12-12 Discuss the following uses of industrial N_2.
   a. As a blanketing agent
   b. In enhanced oil recovery projects

12-13 Compare the processes of freezing meat, poultry and seafood using liquid N_2 and freezing the same materials by mechanical methods.

12-14 A small amount of nitrogen is injected into the headspace of aluminum cans containing noncarbonated beverages. Why?

AMMONIA

12-15 Plants cannot directly use N_2 gas from the atmosphere as a nutrient. Why?

12-16 Discuss two naturally occurring nitrogen fixation processes.

12-17 Contrast the physical properties of gaseous ammonia with those of gaseous nitrogen.

12-18 Write equations for the production of the following nitrogen-containing fertilizers from ammonia.
   a. Urea
   b. Ammonium sulfate
   c. Ammonium nitrate
   d. Diammonium phosphate

12-19 List two nonagricultural uses for ammonia.

12-20 What are the sources for the two chemical reactants needed to make ammonia using the Haber process?

12-21 Describe the physical conditions necessary to maximize the production of NH_3 from elemental hydrogen and nitrogen.

12-22 Write chemical equations for the following reactions.
   a. Production of NH_3 by the Haber process.
   b. Reaction of CH_4 with a limited supply of air to produce H_2 gas.
   c. Steam-reforming of methane to produce H_2 gas.

12-23 Explain why fertilizer prices increase with natural gas prices.

OXYGEN

12-24 What are the advantages of using pure O_2 over air in industrial chemical processes?

12-25 When pure oxygen rather than air is used in the steel-making process, "heat" time is decreased. Why?

12-26 What function does O_2 (either pure or from air) serve in the steel-making process?

12-27 Describe the process of oxygen-cutting of a metal object.

12-28 What is normal oxygen pressure at the following altitudes? How does this pressure compare with the minimum required by humans?
   a. 5000 feet (Denver, Colorado)
   b. 18,000 feet (low altitude for commercial airliner)
   c. 29,000 feet (top of Mount Everest)

12-29 Why must commercial airliner cabins be pressurized?

12-30 Why must the air scuba divers breathe be diluted with helium gas?

12-31 Liquid O_2 is used in space launches. Describe its role in the process of launching a rocket.

LIME

12-32 Write a chemical formula for each of the following substances.
   a. Quicklime
   b. Slaked lime
   c. Unslaked lime
   d. Hydrated lime
   e. Limestone

12-33 Write a chemical equation for the following.
a. Production of quicklime from limestone.
b. Production of hydrated lime from quicklime.
c. Slaking of quicklime.
d. Production of slag from quicklime during steel production.

12-34 What function does lime serve in the steel-making process?

12-35 Discuss how lime is used in controlling sulfur dioxide air pollution.

12-36 The "hardening" of brick mortar involves the reaction of hydrated lime with carbon dioxide from the atmosphere. Write an equation for this process.

12-37 High soil pH can be reduced by application of quicklime. Write an equation to represent this process.

CHAPTER HIGHLIGHTS

Practice Exercises and Problems

13.1 Nonmetallic Elements

Elements can be classified as metals or nonmetals on the basis of selected physical properties (Section 4.10) The majority of elements are metals. Only 22 of the 109 elements have the properties associated with nonmetallic character.

1. Nonmetals are poor conductors of heat and electricity.
2. Nonmetals do not have a high reflectivity or a shiny metallic appearance.
3. Nonmetals may be gases, liquids, or solids at room temperature.
4. In the solid state nonmetals are generally brittle and fracture easily under stress, rather than being malleable.
5. Nonmetals generally form covalent bonds or negative ions.

All nonmetals are found in the upper right portion of the periodic table (see Figure 4.18).

The fact that the vast majority of elements are metals in no way indicates that metals are more important than nonmetals. Indeed, the opposite could possibly be true. The two most abundant elements in the earth's crust are nonmetals (see Table 2.2). These two elements—oxygen and silicon—account for 80.2% of all atoms in the "chemical world" of humans (the earth's crust). The four most abundant elements in the human body (Sections 2.6 and 4.11), which constitute more than 99% of all atoms in the body, are nonmetals—hydrogen, oxygen, carbon, and nitrogen.

In this chapter we consider further the characteristics of nonmetallic elements, with particular emphasis on the nonmetals phosphorus (Section 13.3), carbon (Section 13.4), and silicon (Section 13.5).

13.2 Abundances and Properties of Nonmetallic Elements

Nine of the 25 most abundant elements in the earth's crust are nonmetals. Besides oxygen and silicon (the two most abundant of all elements) the other seven are hydrogen, phosphorus, fluorine, carbon, sulfur, chlorine, and nitrogen. The abundances of these seven nonmetals are large when compared to abundances for most elements (they rank among the top 25), but they are

TABLE 13.1

Nonmetal Abundances in the Earth's Crust

Nonmetal	Parts per Million (Number of atoms of nonmetal per 1 million total atoms)	Atom Percent	Factor by Which Oxygen Abundance exceeds listed Nonmetal Abundance	Abundance Rank Considering All Elements
Oxygen	601,000	60.1	—	1
Silicon	201,000	20.1	3.0	2
Hydrogen	29,000	2.9	21	4
Phosphorus	680	0.068	880	11
Fluorine	500	0.050	1200	12
Carbon	350	0.035	1700	14
Sulfur	190	0.019	3200	15
Chlorine	110	0.011	5500	16
Nitrogen	29	0.0029	21000	23

small when compared to those of oxygen and silicon. A perspective of these relationships can be obtained from Table 13.1. Let us consider the forms in which the nonmetals listed in Table 13.1 occur in nature and the properties of these nonmetals in the free state.

Oxygen

Much information concerning the element oxygen has already been considered, in the context of its importance as an industrial chemical (Section 12.5). For industrial purposes it is obtained from the atmosphere through fractional distillation of liquid air.

In addition to its occurrence in the atmosphere as the free element (O_2 molecules), substantial amounts of free oxygen are also dissolved in the oceans and surface waters of the world. However, this free atmospheric and dissolved oxygen accounts for only a small fraction of total oxygen atoms. Most oxygen occurs in combined forms as both water and a constituent of most rocks, minerals, and soils. The estimated abundance of oxygen in crustal rocks is 450,000 parts per million by mass, that is, 45.5% of the mass of such rocks is oxygen. Small amounts of combined oxygen are also found in the atmosphere—in CO_2, CO, SO_2, NO, and NO_2. (The chemistry of these atmospheric oxygen compounds, all of which cause air pollution, is considered in Chapter 21.)

Silicon

Most people (except those who have had a chemistry class) are surprised when they are told that silicon is the second most abundant element. It is an element most people have never encountered in pure form in everyday life. Where is all of the silicon on the earth? Approximately 85% of the solid portion of the earth's crust is silicon-based material. Sand and sandstone are impure forms of silicon dioxide (SiO_2) and quartz is a pure crystalline form of the same substance. Silicates are additional, more complex silicon–oxygen combinations. Just as almost all biologically significant molecules contain carbon atoms, most rocks, sands, and soils of the earth's crust contain silicon. Silicon is to geology as carbon is to biochemistry. The estimated abundance of silicon in crustal rocks is 272,000 parts per million by mass.

Pure silicon, which does not occur uncombined in nature, is a shiny, silvery blue gray, brittle solid that has a definite metallic appearence but it is not a metal. This element is a semiconductor of electricity, that is, it conducts electricity to a slight extent; metals have large electrical conductivities. Pure silicon is primarily used as a semiconductor in electronic devices such as transistors and micro-computer chips (Section 13.5).

Hydrogen

The most prevalent hydrogen-containing compound is H_2O, which permeates everything and covers four-fifths of the earth's surface. It is the hydrogen atoms present in water that make hydrogen as abundant as it is. Fossil fuels (coal, petroleum, and natural gas) are also hydrogen-containing substances. The abundance of hydrogen in crustal rocks is 1520 parts per million by mass.

Pure hydrogen gas, like pure oxygen, is diatomic and both colorless and odorless. The earth's atmosphere does not contain elemental hydrogen (H_2), even though hydrogen is the most abundant element in outer space (see Table 2.2). The reason for this is the high velocities at which gaseous H_2 molecules travel. Because of their small mass (smaller than that of any other molecule), H_2 molecules are able to acquire velocities greater than that of other molecules. (There is a direct relationship between molecular velocity and formula weight—the smaller the mass the greater the velocity, at any given temperature.) In the case of gaseous H_2, the molecular speed obtained is sufficient to enable H_2 molecules to overcome the gravitational attraction of the earth and escape to outer space.

From an industrial standpoint, the most important use of H_2 gas is in the production of ammonia (Section 12.3)

Phosphorus

The predominant natural source for phosphorus, which does not occur in the free state, is phosphate rock, a material discussed previously (Section 12.2).

Phosphate rock, solubilized using sulfuric acid, is the raw material for phosphate fertilizers.

Pure phosphorus has two common solid-state forms — white phosphorus and red phosphorus. White phosphorus is a waxlike white solid that is toxic and also flammable in air. Red phosphorus, on the other hand, is a solid that is neither toxic nor flammable in air. The relationship between these two forms of phosphorus, and the process by which phosphorus is obtained from phosphate rock, are considered in Section 13.3.

Fluorine

Fluorine, the most electronegative of all elements (Section 5.11), is never found free in nature. It is the most reactive of all the elements. The three most important fluorine-containing minerals are fluorite (CaF_2), cryolite (Na_3AlF_6), and fluorapatite [$Ca_5(PO_4)_3F$]. Cryolite is an important substance in the production of aluminum (Section 14.6), and fluorapatite is a major type of phosphate rock. Elemental fluorine, a diatomic pale yellow gas with a characteristic odor, is obtained from its compounds through electrolysis (Section 10.12).

Carbon

Carbon occurs both as a free element and in combined form, mostly the latter. About half of all carbon atoms are tied up in carbonate rocks, mainly as the carbonates of magnesium and calcium. Limestone, marble, and dolomite are all carbonate rocks. Carbon is also widely distributed in the form of coal, petroleum, and natural gas. It also occurs as CO_2, a minor but crucial constituent of the atmosphere. Living systems, particularly the plant kingdom, account for the remainder of carbon atoms.

Carbon, like phosphorus, occurs as a free element in two forms — graphite and diamond. Graphite, the more common of the two, is a black, flaky powder, and diamond exists as hard colorless crystals. Further details concerning these two forms of carbon are found in Section 13.4.

Sulfur

The most stable form of pure sulfur is a yellow, soft, crystalline solid material that is easily pulverized to a powder. The Frasch process for mining underground elemental sulfur deposits was discussed in Section 12.2.

Although sulfur does occur in free form in underground deposits, the majority of sulfur atoms are found in combined form. Metal sulfides (S^{2-} ion) and metal sulfates (SO_4^{2-}) are important combined forms. Gaseous combined sulfur, in the form of H_2S, is also an important source of sulfur.

As discussed in Section 12.2, most industrial elemental sulfur is used to produce sulfuric acid, the highest-volume industrial chemical.

Chlorine

Because of its reactivity, chlorine, a greenish yellow gas with a choking odor, does not occur in the free elemental state but is widespread and abundant in the form of the Cl^- ion. Seawater and underground brines contain appreciable amounts of Cl^- ion. In addition, large mineral deposits, particularly of NaCl, exist. The principal industrial source of chlorine is the electrolysis (Section 10.12) of aqueous sodium chloride (brine) solutions.

Nitrogen

Nitrogen, a diatomic, colorless, odorless gas, is the most abundant uncombined element accessible to humans. Section 12.3 dealt with the fractional distillation of liquid air as a process for extracting N_2 from the atmosphere and also the importance of nitrogen as an industrial chemical.

Despite its ready availability in the atmosphere, nitrogen exists in relatively small quantities in the crystaline rocks and soils of the earth. The only major nitrogen-containing minerals are KNO_3 and $NaNO_3$, which occur only in small amounts. The lack of nitrogen in the solid portions of the earth's crust explains why oxygen ranks first in overall abundance and nitrogen ranks 23rd, although N_2 is four times more abundant than O_2 in the atmosphere.

Table 13.2 summarizes important characteristics of the nine nonmetals we have just considered.

TABLE 13.2

Important Characteristics of the Nine Most Abundant Nonmetals

Nonmetal	Characteristics
Oxygen	Colorless, odorless, diatomic gas
Silicon	Shiny, silvery blue-gray, brittle solid that conducts electricity to a slight extent
Hydrogen	Colorless, odorless, diatomic gas
Phosphorus	Waxy white solid flammable in air, or a red solid not flammable in air
Fluorine	Pale yellow, diatomic gas with a characteristic odor
Carbon	Black, flaky powder (graphite) or a hard colorless solid (diamond)
Sulfur	Soft, yellow, crystalline solid
Chlorine	Greenish yellow, diatomic gas with a choking odor
Nitrogen	Colorless, odorless, diatomic gas

13.3 Phosphorus and Selected Phosphorus Compounds

Phosphorus is probably unique among the elements in having been first isolated from human excreta (urine), in 1669, then from plants (1688), and finally, more than a century later (1779), found in a mineral (lead phosphate). Most elements have been initially identified in mineral sources.

The alchemist Hennig Brandt was the first to isolate phosphorus. His process involved allowing pails of urine to putrify, then boiling the resulting material until it was reduced to a thick syrup, and finally heating the syrup mixed with sand. This produced vapors of phosphorus, which were condensed under water.

Phosphorus was immediately interesting to its discoverer because of two of its properties. It glowed in the dark when exposed to moist air, and it had a low ignition temperature, so that in a warm room it smoked and finally burst into flame (unless stored under water). The first of these two properties is the basis for the name phosphorus; it comes from the Greek *phos*, "light," and *phoros*, "bringing."

In Section 13.2 we noted that there are two forms of elemental solid phosphorus—white phosphorus and red phosphorus. These two forms represent an example of the phenomenon of allotropy. **Allotropes** *are two or more different forms of an element that have a different number or arrangement of atoms in molecules.* Many nonmetallic elements besides phosphorus show allotropy, including carbon (Section 13.4) and oxygen (Section 20.6).

White phosphorus is a molecular solid in which the molecular unit is P_4. The four phosphorus atoms in such a unit are arranged at the corners of a regular tetrahedron, so that each atom is single-bonded to the other three (see Figure 13.1a). This form of phosphorus has a high chemical reactivity, igniting

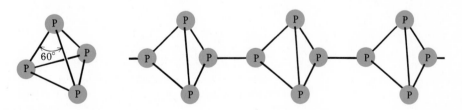

(a) White phosphorus   (b) Red phosphorus

FIGURE 13.1

White phosphorus has a structure involving individual P_4 molecules, and red phosphorus has a structure involving chains of P_4 molecules.

spontaneously in air at about 25°C. White phosphorus should not be allowed to come in contact with human skin. Human body temperature (37°C) is sufficient to ignite it spontaneously. Such ignition produces very painful burns that are slow to heal. In a chemical laboratory white phosphorus is always stored immersed in water and handled with forceps.

The heating of white phosphorus to 250°C in the absence of air (it cannot burn when O_2 is absent) produces red phosphorus. This allotrope of phosphorus is much less reactive than white phosphorus. It undergoes the same reactions, but requires higher temperatures. For example, a temperature of 260°C is required for red phosphorus to burn in air. The structure of red phosphorus involves long chains of P_4 units (see Figure 13.1*b*). If red phosphorus is heated to 600°C, its chain structure breaks up, giving P_4 molecular units, which can be condensed to produce white phosphorus.

For a century after its discovery the only source of phosphorus was urine. Today, white phosphorus is produced using a process, developed in 1867, which involves heating phosphate rock with sand and coke (a form of carbon, Section 13.4) in an electric furnace. The overall reaction that occurs may be represented as

$$3\ Ca_3(PO_4)_2(s) + 6\ SiO_2(s) + 10\ C(s) \longrightarrow 6\ CaSiO_3(s) + 10\ CO(g) + P_4(g)$$

Phosphate rock   Sand   Coke   Slag

The molten slag is tapped off from the bottom of the furnace. The gaseous products are passed through water, causing the phosphorus vapor to solidify as a white solid. The gaseous CO is only slightly soluble in water.

Most elemental phosphorus production is used to make phosphoric acid or other phosphorus-containing compounds. Phosphoric acid produced from elemental phosphorus is much purer than that obtained by dissolving phosphate rock using sulfuric acid (Section 12.2). As representative of the compounds produced from elemental phosphorus, we will consider here two derivatives: tetraphosphorus trisulfide (P_4S_3) and sodium tripolyphosphate ($Na_5P_3O_{11}$).

Tetraphosphorus Trisulfide

STRIKE-ANYWHERE
MATCHES

Heating required amounts of sulfur and phosphorus produces the phosphorus sulfide P_4S_3. This compound is important in the manufacture of matches.

The heads of so-called strike-anywhere matches contain a tip composed of P_4S_3 on top of a red portion that contains lead dioxide (PbO_2) and antimony sulfide (Sb_2S_3) (see Figure 13.2*a*). The P_4S_3 tip—the part that actually ignites— is activated by rubbing on a rough surface. Heat generated by friction causes the P_4S_3 to ignite, and the additional heat from this ignition initiates a reaction between Sb_2S_3 and PbO_2.

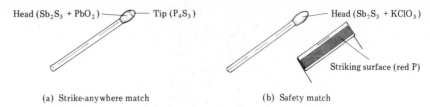

(a) Strike-anywhere match   (b) Safety match

FIGURE 13.2

Phosphorus, in the form of P_4S_3, is involved in the ignition of strike-anywhere matches, and elemental red phosphorus is present in the striking surface for safety matches.

SAFETY MATCHES

 Safety matches have a different formulation. The head is a mixture of $KClO_3$ and Sb_2S_3; no phosphorus is present. The striking surface on the side of the matchbox is a rough surface composed of a mixture of red phosphorus, glue, and abrasive (*see* Figure 13.2*b*). The friction heat from striking the safety match against the striking surface ignites the red phosphorus, which in turn ignites the reaction mixture in the matchhead. The $KClO_3$ decomposes from the heat, producing additional pure oxygen, which causes the match flame to flare momentarily.

$$2\ KClO_3\ (s) \longrightarrow 3\ KCl\ (s) + 3\ O_2(g)$$

Sodium Tripolyphosphate

Phosphoric acid can be produced from elemental phosphorus by burning the phosphorus in excess air, producing the oxide P_4O_{10}, and then immediately reacting the oxide with water.

$$P_4(s) + 5\ O_2(g) \longrightarrow P_4O_{10}(s)$$

$$P_4O_{10}(s) + 6\ H_2O(l) \longrightarrow 4\ H_3PO_4(aq)$$

When phosphoric acid is heated gently, a condensation reaction occurs. Diphosphoric acid molecules are formed from the elimination of a water molecule from a pair of phosphoric acid molecules.

$$
\underset{\text{Phosphoric acid}}{HO-\overset{\displaystyle O}{\underset{\displaystyle OH}{\|}}P-OH} + H\underset{\text{Phosphoric acid}}{O-\overset{\displaystyle O}{\underset{\displaystyle OH}{\|}}P-OH} \longrightarrow \underset{\text{Diphosphoric acid}}{HO-\overset{\displaystyle O}{\underset{\displaystyle OH}{\|}}P-O-\overset{\displaystyle O}{\underset{\displaystyle OH}{\|}}P-OH}
$$

Further condensation reactions can occur, producing longer phosphoric acid molecules. Of particular interest to us is the triphosphoric acid molecule made from condensing three phosphoric acid molecules together.

$$\begin{array}{ccc}
O & O & O \\
\parallel & \parallel & \parallel \\
HO-P-O-P-O-P-OH \\
\mid & \mid & \mid \\
OH & OH & OH
\end{array}$$

Triphosphoric acid

This molecule, $H_5P_3O_{10}$, contains five acidic hydrogen atoms (Section 10.6). From it, through neutralization the salt $Na_5P_3O_{10}$ is obtained. This compound, commonly called sodium tripolyphosphate (STPP), is a key ingredient in detergents and cleaners of various types (household, industrial, and institutional). STPP functions as a water softener, tying up hard water ions such as Mg^{2+} and Ca^{2+} (Section 11.8). Its presence also makes the pH of the cleaning solution basic, which is desired.

In recent years the level of phosphate in detergent formulations has been reduced markedly because of water pollution concerns. Phosphates discharged into wastewaters function not as toxic materials but rather as plant nutrients (fertilizers) (Section 21.4). The result is excess plant growth of species such as algae, particularly in stagnant waters. Many phosphate-free detergent formulations are now marketed in response to this problem. However, many cleansing agents still contain phosphate, at reduced levels. STPP remains an important water-softening agent.

13.4 Carbon and Selected Carbon Compounds

Two important allotropic (Section 13.3) crystalline forms of the element carbon exist, both well-known—graphite and diamond (see Figure 13.3). These two forms of carbon differ only in the way carbon atoms are bound to each other to form the crystal.

Graphite, a soft, black, slippery solid, has a structure that involves parallel sheets (or layers) of carbon atoms, with each sheet containing hexagonal arrays of carbon atoms. In these hexagonal arrays, each carbon atom is bonded to three others, forming one double bond and two single bonds (see Figure 13.4a). (Since carbon atoms have four valence electrons, they must form four bonds to gain an octet of electrons; see Sections 5.3 and 5.9.) The stacked layers of carbon atoms are held together by relatively weak intermolecular forces (Figure 13.4b). The strong bonding within graphite layers but weak interactions between layers account for graphite's softness and slipperiness; the layers can slide past one another quite readily.

The slipperiness of graphite accounts for its use as a dry lubricant. It is often used as a lubricant in locks, where oil is undesirable because it collects dirt. The

FIGURE 13.3
Elemental carbon exists in two allotropic forms—graphite, a soft black solid, and diamond, a very hard crystalline substance. [Grant Heilman.]

(a) Arrangement of atoms and bonding within a single layer of graphite

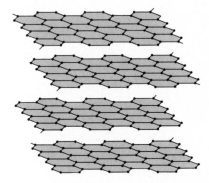

(b) Stacked layers of carbon atom sheets in graphite

FIGURE 13.4
Structure of graphite.

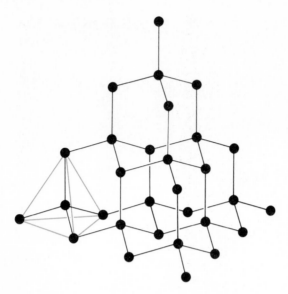

FIGURE 13.5

The crystalline structure of diamond. Each carbon atom is covalently bonded to four other carbon atoms in a gaint molecule.

PENCIL LEAD

softness of graphite accounts for its use in lead pencils. The "lead" of lead pencils is actually graphite mixed with a binder, usually clay, to make it harder. Layers of the graphite rub off from the pencil onto paper. The name graphite comes from the Greek word meaning "to write."

Diamond, in contrast to graphite, is one of the hardest substances known. Because of their hardness, nongem quality diamonds are used in industry for cutting, grinding, and polishing. Diamond hardness is related to the way in which carbon atoms are bonded to each other in this substance. Each carbon atom is attached by strong covalent bonds to four other carbon atoms, each of which in turn is bonded to four more carbon atoms, and so on. The result is a "gigantic" covalent molecule (see Figure 13.5). For a diamond to be cleaved, many bonds within the interlocking network of covalent bonds must be broken.

The fact that graphite and diamond are both forms of elemental carbon has led to numerous attempts to synthesize diamonds from graphite. This was accomplished for the first time in 1954. Today, such syntheses are carried out routinely to produce industrial diamonds. The industrial diamond market is much larger than the market for diamonds as gemstones.

SYNTHETIC DIAMONDS

The process for producing diamond from graphite involves high temperature and extraordinarily high pressures; under these conditions diamond is the more stable form of carbon. Actual production conditions involve temperatures in the 2000 to 2500°C range and pressures of 100,000 to 150,000 atm. (Special presses had to be invented to achieve such high pressures.) Conversion of graphite to diamond occurs within minutes under these conditions. The temperature and pressure are then rapidly reduced to "normal"; this traps the

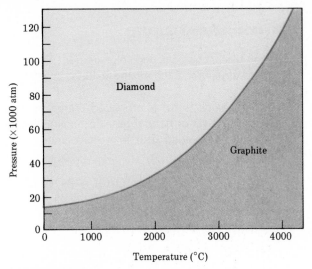

FIGURE 13.6

The graphite–diamond equilibrium curve. Diamond is more stable above the line; graphite is more stable below it.

carbon atoms in the diamond form. Diamonds so produced have exactly the same chemical composition, structure, and physical properties as diamonds formed by natural forces. Such diamonds, however, are not usually of gem quality; the transparency and clearness are not as good. This is not an important factor in industrial uses of diamond; here, hardness is the key property.

Figure 13.6 shows an equilibrium curve for graphite and diamond. Above the curve, diamond is the stable form and below the curve graphite is more stable. At room temperature and pressure graphite is the more stable form; thus, diamond should convert to graphite. The rate of this process is so slow that it is neglible. If the temperature is increased to 1500°C, the conversion of diamond to graphite occurs in minutes.

Diamond mass is specified in carats; one carat equals 200 milligrams. The largest natural diamond, the Cullinan diamond found in 1905, is 3106 carats (621.2 g). Other famous diamonds have masses in the 100 to 800-carat range. Large synthetic diamonds have masses of approximately 1 carat. The cost of producing small, synthetic, industrial-quality diamonds is now competitive with the cost of mining them, but gem-quality diamonds are still cheaper to mine than to make.

Besides graphite and diamond (crystalline forms of carbon) a number of noncrystalline carbon forms exist. These noncrystalline forms—carbon black, charcoal, and coke—plus the carbon compounds carbon monoxide and carbon dioxide occupy our attention in the remainder of this section.

Carbon Black, Charcoal, and Coke

Carbon black, charcoal, and coke are noncrystalline forms of carbon, all with the graphite structure. Each of these three substances is an important industrial chemical.

Carbon black (soot) is produced during the *incomplete* combustion of petroleum products or natural gas. For example, the burning of CH_4 (the major component of natural gas) in a very limited supply of air produces carbon black.

$$CH_4(g) + O_2(g) \longrightarrow C(s) + 2\,H_2O(g)$$

By contrast, the equation for the *complete* combustion of CH_4 to produce carbon dioxide requires twice as much oxygen.

$$CH_4(g) + 2\,O_2(g) \longrightarrow CO_2(g) + 2\,H_2O(g)$$

Deposits of carbon black on the bottoms of pans and kettles are often observed after such containers have been heated in an open flame.

The principal use of carbon black (almost two-thirds of production) is in automobile tires to strengthen, reinforce, and color the rubber. Because of its intense black color, this form of carbon is also used in small quantities as a pigment for paint, paper, and printer's ink.

Charcoal is made by heating wood to a high temperature in the absence of air. An important characteristic of this form of carbon is its very low density. (Compare the mass of a piece of charcoal with that of a similarly sized piece of wood; the charcoal is much "lighter.") This low density is the result of charcoal's extreme porosity; its structure is similar to that of a sponge, but with holes so small you cannot see them with the naked eye. This porosity results in a large surface area per unit mass, a property characteristic of good absorption agents. *Activated charcoal*, also called *activated carbon*, a pulverized form of charcoal that has been thoroughly cleaned by heating with steam, is widely used to absorb molecules, particularly "unwanted" molecules. Activated charcoal can be used to (1) remove unwanted odors from air, for example, in a refrigerator, (2) remove impurities from water in drinking water treatment (Section 11.7), and (3) remove air pollutants from exhaust gases in industrial processes.

Coke is an impure form of carbon made by heating coal in the absence of air. Heating coal under this condition produces three fractions: coal gas, coal tar (a mixture of liquids), and a solid residue (the coke). This solid residue contains 90 to 98% carbon, together with small amounts of many other elements that were present in the coal. Enormous quantities of coke are consumed in the steel-making industry (Section 14.2).

Carbon Monoxide and Carbon Dioxide

Carbon forms two extremely stable oxides, carbon monoxide (CO) and carbon dioxide (CO_2). Both of these oxides are extremely important. Both are usually produced simultaneously anytime carbon or a carbon-containing compound is heated to a high temperature in air. Carbon dioxide is always the dominant product. The amount of carbon monoxide produced depends on the air/fuel ratio. When excess air is present, very little carbon monoxide is produced. With a limited supply of air, carbon monoxide production can be significant, particularly from a health standpoint (Section 20.3). Equations for the production of these two oxides are

$$2\ C(s) + O_2(g) \longrightarrow 2\ CO(g)$$
$$C(s) + O_2(g) \longrightarrow CO_2(g)$$

Carbon monoxide is a colorless, odorless, tasteless gas that is extremely toxic to humans. Its toxicity results from its ability to combine with hemoglobin in the blood, binding itself to this biochemical substance much more strongly than oxygen does. This prevents hemoglobin from carrying out its function as an oxygen carrier. Thus, carbon monoxide poisoning produces oxygen deficiency. Further details concerning the interaction of carbon monoxide with hemoglobin are given in Section 20.3. The fact that carbon monoxide is colorless, odorless, and tasteless increases the risk of carbon monoxide poisoning; its presence is not readily detected.

Carbon monoxide is the major air pollutant produced in automobile exhaust. Section 20.3 deals with carbon monoxide in this context. Tobacco smoke at the tip of a cigarette is about 2 to 5% carbon monoxide. This topic is also considered in more detail in Section 20.3.

Carbon dioxide is colorless and odorless as well, but it has a faintly acidic (sour) taste (because of its solubility in H_2O and the resultant formation of carbonic acid).

$$CO_2(g) + H_2O(l) \longrightarrow H_2CO_3(aq)$$

Unlike carbon monoxide, carbon dioxide is an important constituent of unpolluted air. It is not normally considered an air pollutant, although there is an environmental problem related to the presence of carbon dioxide—the greenhouse effect (considered in Section 20.12).

Carbon dioxide is an integral part of many of the everyday processes we encounter. Some of these processes are biological and others are encountered in household or industrial contexts. Here we consider five areas of encounter: (1) natural environmental cycling of CO_2, and carbon dioxide's use as (2) a leavening agent, (3) a fire-extinguishing agent, (4) a carbonating agent, and (5) a refrigerant.

Natural processes involving CO_2. The presence of CO_2 in the atmosphere is an absolute necessity for sustaining life as we now know it. The growth of plant life, the source (either direct or indirect) of human food, depends on the presence of atmospheric CO_2. In *photosynthesis*, the process by which carbon is incorporated into plants, CO_2 is the carbon source. The photosynthetic reaction involves the conversion of carbon dioxide (from the atmosphere) and water to carbohydrate and oxygen (which is released to the atmosphere).

$$\text{Energy} + 6\ CO_2(g) + 6\ H_2O(l) \longrightarrow C_6H_{12}O_6(s) + 6\ O_2(g)$$
$$\text{Carbohydrate}$$

The energy for photosynthesis is supplied by sunlight and the process is catalyzed by chlorophyll, the green color matter of plants.

Natural mechanisms also exist for replenishing the atmospheric CO_2 removed by photosynthesis. Metabolism of food, the energy-producing process in humans and animals, produces CO_2 as a by-product. This process may be thought of as the reverse of photosynthesis. Carbohydrate materials react with O_2 (obtained by breathing) to produce carbon dioxide, water, and energy.

$$C_6H_{12}O_6(aq) + 6\ O_2(g) \longrightarrow 6\ CO_2(g) + 6\ H_2O(l) + \text{Energy}$$
$$\text{Carbohydrate}$$

Each time a person exhales, carbon dioxide is put back into the atmosphere.

CO_2 enters the atmosphere by means of other natural mechanisms. After plants and animals die, decomposition processes convert the carbon in their tissues back to CO_2. Also natural phenomena such as forest fires and volcanic eruptions produce carbon dioxide. Additional CO_2 is produced from human activities such as coal combustion (industrial plants and electricity production), gasoline combustion (transportation), and natural gas and wood combustion (space heating). Figure 13.7 summarizes the cycling of CO_2 in the environment.

CO_2 as a leavening agent. To leaven means to "make light and porous." Our context for considering leavening is flour mixtures for the production of baked goods. The formation of CO_2 is the principal means of leavening flour mixtures. The CO_2 source is either baking soda or baking powder.

Baking soda is sodium bicarbonate ($NaHCO_3$). This compound breaks down into sodium carbonate, water, and carbon dioxide when heated.

$$2\ NaHCO_3(s) \xrightarrow{\text{heat}} Na_2CO_3(s) + H_2O(g) + CO_2(g)$$

A problem with the use of baking soda is the soapy taste in the flour mixture produced by the residual Na_2CO_3 (the decomposition product). The presence of an acidic ingredient in the flour mixture combats this problem; a salt other than Na_2CO_3 is formed from the reaction of the acid with the $NaHCO_3$. Most acid-produced salts have not been found objectionable in flavor. Acidic

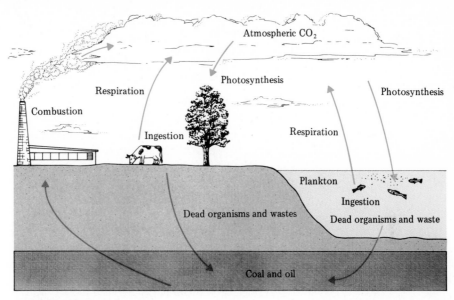

FIGURE 13.7
Many processes are involved in the cycling of carbon dioxide (CO_2) in the environment.

ingredients often used in flour mixtures include buttermilk or sour milk (lactic acid), molasses (a mixture of acids), vinegar (acetic acid), and citrus fruit juices (citric acid). The reaction between lactic acid ($HC_3H_5O_3$) and baking soda is

$$HC_3H_5O_3(s) + NaHCO_3(s) \longrightarrow NaC_3H_5O_3(s) + CO_2(g) + H_2O(g)$$

| Lactic acid | Sodium bicarbonate | Sodium lactate |
| (sour milk) | (baking soda) | (agreeable flavor) |

BAKING POWDERS   Most recipes call for baking powder instead of baking soda. Baking powders are more convenient to use. They contain one or more dry acids or acid salts as part of their formulation, in addition to baking soda and starch (added to help stablize the mixture and prevent the components—(dry acid and baking soda)—from reacting prematurely). The type of baking powder generally available for home use is a combination powder called SAS–phosphate baking powder. This is a double-acting baking powder; that is, it reacts to release carbon dioxide gas at room temperature when the dry ingredients are moistened and again when heat is applied in the process of baking. SAS–phosphate baking powder contains two acid substances, each of which reacts with soda to release carbon dioxide gas at different times in the baking process. One of the acids is a phosphate, usually calcium dihydrogen phosphate. This acid reacts with soda at room temperature as soon as liquid is added to the dry ingredients.

$$\underbrace{Ca(H_2PO_4)_2(s) + 2\,NaHCO_3(s)}_{\text{In baking powder}} \xrightarrow{\text{moisture}}$$

$$2\,CO_2(g) + 2\,H_2O(g) + CaHPO_4(s) + Na_2HPO_4(s)$$

This causes the batter to become somewhat light and porous as it is mixed. The other acid substance is sodium aluminium sulfate (SAS). It requires heat as well as moisture to complete its reaction with soda. Therefore, additional carbon dioxide gas is produced during baking. Sodium aluminum sulfate reacts in two stages. The first reaction, with water, produces sulfuric acid as heat is applied; in the second reaction, sulfuric acid reacts with soda to produce carbon dioxide gas.

$$\underbrace{Na_2SO_4 \cdot Al_2(SO_4)_3}_{\substack{\text{Sodium aluminum}\\\text{sulfate (SAS)}}} + 6\,H_2O \xrightarrow{\text{heat}} Na_2SO_4 + 2\,Al(OH)_3 + \underset{\text{Sulfuric acid}}{H_2SO_4}$$

$$\underset{\text{Sulfuric acid}}{3\,H_2SO_4} + \underset{\text{Baking soda}}{6\,NaHCO_3} \longrightarrow 6\,CO_2 + 6\,H_2O + 3\,Na_2SO_4$$

All baking powders are composed of soda plus one or more acid ingredients. Federal law requires that baking powders must contain at least 12% available CO_2 gas. Those powders manufactured for home use generally contain 14%, and some powders for commercial use have 17% available CO_2.

Carbon dioxide produced biologically, rather than chemically, is also commonly used as a leavening agent. This is what happens when yeast is used in a flour mixture. Yeast is a microorganism that contains an enzyme capable of converting simple sugars, such as glucose (Section 24.4), into CO_2 and ethyl alcohol. This process is called *fermentation*

$$\underset{\text{Glucose}}{C_6H_{12}O_6} \xrightarrow[\text{enzymes}]{\text{yeast}} 2\,CO_2 + \underset{\text{Ethyl alcohol}}{2\,C_2H_5OH}$$

The alcohol is volatilized by the heat of baking. Sugar is usually added to yeast–flour mixtures to speed up fermentation and the production of CO_2 gas. If no sugar is used, yeast can form CO_2 gas slowly from the small amount of sugar that is naturally present in flour.

CO_2 as a fire-extinguishing agent. Carbon dioxide is directly involved in the operation of two types of fire extinguishers. In the acid–soda extinguisher, it is a passive ingredient and in the liquid CO_2 extinguisher, it is active.

The oldest type of fire extinguisher in use is the acid–soda extinguisher. Although it is rapidly being supplanted with newer, better types, it is still commonly found in older buildings, particularly public ones. An acid–soda fire extinguisher contains a baking soda solution ($NaHCO_3$) and a small separate

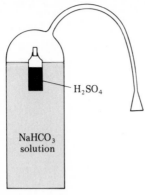

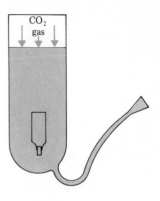

(a) Before acid and soda
   solution are mixed

(b) After mixing of acid and
   soda by inverting the
   container

FIGURE 13.8

The acid–soda fire extinguisher.

container of an acid, usually sulfuric acid (H_2SO_4). Inverting the extinguisher causes the acid and soda to mix, generating CO_2 gas from the reaction.

$$NaHCO_3 + H_2SO_4 \longrightarrow H_2O + CO_2 + NaHSO_4$$

The generated CO_2 gas acts as a propellant, which forces the water solution out of the container (see Figure 13.8). The primary extinguishing agent is water rather than CO_2. The water cools the burning fuel below its kindling temperature.

Note that water should not be used on electrical fires or fires involving burning greases or flammable liquids. Water should be avoided in electrical fires because if conducts electricity. Water often spreads grease and flammable liquid fires because the burning liquid floats on top of the water.

The liquid CO_2 extinguisher involves a steel container partially filled with liquid CO_2 under pressure of 800 to 1000 psi. When the valve on the container is opened, the sudden drop in pressure causes a mixture of gaseous CO_2 and solid CO_2 "snow" to emerge (see Figure 13.9). The "snow" is produced by the cooling effect associated with the rapid evaporation of the liquid CO_2. The solid and gaseous CO_2 affect the fire in two ways. First, the CO_2 snow produces a cooling effect as it vaporizes. More important is the blanketing effect of the CO_2 gas. Carbon dioxide is a nonflammable, nonconducting gas that is 1.5 times denser than air. Because of its higher density, CO_2 settles close to the ground and its presence dilutes the oxygen supply to the fire. If the normal 21% O_2 content of air is reduced to 17% O_2, by diluting with CO_2, most materials cease to burn. One advantage of CO_2 fire extinguishers is that no cleanup of fire extinguishing materials is needed. Also CO_2 may be used on fires for which H_2O is not desirable — burning liquids and electrical fires.

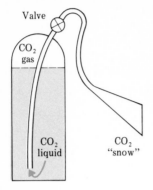

FIGURE 13.9

Liquid CO_2 fire extinguisher.

Liquid CO_2 fire extinguishers are now being replaced with more advanced, more economical types. Today, dry chemical fire extinguishers dominant the market, particularly for home use. Such extinguishers are effective on all types of small home fires.

DRY CHEMICAL FIRE EXTINGUISHERS

In a dry chemical fire extinguisher a propellant, which can be N_2, CO_2, or another nonflammable, nonconducting gas, is used to discharge a finely divided solid (a powder) onto the fire. The powder extinguishes the fire by permanently coating the surface of the burning material. This interrupts reactions within the flames and also decreases access to O_2. The most used dry chemical in such fire extinguishers is ammonium dihydrogen phosphate—$(NH_4)H_2PO_4$. This solid is ground into very small particles and treated with additives to prevent caking and the uptake of moisture. The powder is loaded into a container, and the container is pressurized with a propellant. There is a "clean-up problem" after the use of a dry-powder extinguisher.

CO_2 as a refrigerant. Carbon dioxide, in the liquid and solid states, finds extensive use as a refrigerant, particularly in the food-processing industry. Its uses here include not only the rapid cooling of loaded trucks and railcars and the freezing of food, but also chilling and processing functions. For example, in the large-scale grinding of hamburger, CO_2 snow or pellets are added to the meat to absorb processing heat and inhibit bacterial growth and discoloration. Refrigeration is the leading use for industrial CO_2, accounting for 40% of total production.

Carbon dioxide, when cooled under atmospheric pressure, changes directly to the solid state at $-78°C$, forming dry ice. When warmed, dry ice sublimes back to the gaseous state. Liquid CO_2 forms from dry ice only if the solid is warmed under pressure; the transition temperature is $-56°C$ at 5.2 atm pressure. Steel cylinders containing liquid CO_2 have internal pressures of 57 atm (at 20°C).

DRY ICE

Industrially, dry ice is obtained by expanding liquid CO_2 from cylinders. The evaporative cooling effect produces snow, which is then mechanically compressed into blocks of convenient size. Dry ice is about 1.7 times as dense as water ice and its net refrigerating effect, on a mass basis, is twice that of water ice. Until the 1960s the bulk of CO_2 refrigerant was in the form of solid CO_2, but now liquid CO_2 dominates production because of lower production costs and ease of transporting and metering the materials.

CO_2 as a carbonating agent. Beverage carbonation is the second largest end use (20%) for CO_2 production. The "bite" and "sparkle" or bubbliness of soft drinks results from the dissolution of CO_2 in water. Most of the "carbonation" in a soft drink is caused by CO_2 itself, although a very small amount of CO_2 is converted to H_2CO_3 (carbonic acid).

$$CO_2(aq) + H_2O(l) \longrightarrow H_2CO_3(aq)$$

Carbon dioxide becomes less soluble in water as the temperature increases. Thus, an open bottle of soda water loses its carbonation more rapidly as it warms up. The solubility of carbon dioxide in water at atmospheric pressure is 4.4 times greater at 0°C than at 50°C. Pressure is also an important factor in the solubility of CO_2 in water. As pressure increases, solubility also increases. Carbonated beverages are saturated carbon dioxide in water solutions under a pressure slightly greater than 1 atm. When a carbonated beverage is uncapped, the pressure is released and the carbon dioxide rushes out of solution (see Figure 13.10).

Carbonation in soft drinks is measured in terms of volume. At atmospheric pressure and 15°C, a given volume of water absorbs an equal volume of CO_2

CARBONATED BEVERAGES AND PRESSURE

FIGURE 13.10

This bottle of carbonated beverage was shaken slightly before being opened to dramatize the escape of CO_2.

TABLE 13.3
Extent of Carbonation of Various Types of Soft
Drinks

Type of Soft Drink	Volumes of Carbonation
Cola	3.5
Ginger ale	4–4.5
Root beer	4
Orange	1.5–2.5
Grape	1–2.5
Lemon-lime	3.5

and is said to contain one volume of carbonation. Table 13.3 shows the extent of carbonation of various types of soft drinks in terms of volumes of carbonation. Increased pressure is the major factor facilitating the increased volumes of carbonation.

13.5   Silicon and Selected Silicon Compounds

Silicon, the second most abundant element in the earth's crust (see Table 2.2), is not found naturally in the free form. It has a high affinity for oxygen and is always found in nature combined with oxygen. The most common simple silicon-oxygen compound is silicon dioxide (silica; SiO_2). Sand and sandstone are impure SiO_2; quartz is a pure crystalline form of SiO_2. More complex silicon–oxygen compounds, called silicates, make up the bulk of most rocks. The only rocks frequently encountered that do not contain silicon are limestone ($CaCO_3$) and dolomite ($CaCO_3 \cdot MgCO_3$).

In this section we first consider elemental silicon—its properties, production, and use as a semiconductor. Next, selected silicates are considered. Finally, two important complex silicate-based materials, glass and cement, occupy our attention.

Elemental Silicon

Elemental silicon is a gray metallic-looking solid that melts at a high temperature (1410°C). Industrially, it is prepared by heating silica (SiO_2) and coke (C) to about 3000°C in an electric arc furnace

$$SiO_2(s) + 2\ C(s) \xrightarrow{\ 3000°C\ } Si(l) + 2\ CO(g)$$

Figure 13.11 contrasts the appearance of silica with that of elemental silicon.

(a)

(b)

FIGURE 13.11

(a) Silica (SiO$_2$) is the starting material for obtaining elemental silicon (b).

[(a) PAR/NYC; (b) Barry L. Runk from Grant Heilman.]

Pure silicon has a structure identical to that of diamond (Section 13.4). Each silicon atom is attached by strong covalent bonds to four other silicon atoms, each of which, in turn, is bonded to four more silicon atoms, and so on. No silicon analog of graphite (Section 13.4) is known.

In today's technological world the word silicon turns most people's thoughts to the electronics industry with its microcomputers, electronic calculators, and so on. Indeed, two areas in the United States where large numbers of electronics companies have located are commonly known as Silicon Valley (the South Bay area of San Francisco) and Silicon Prairie (the area around Austin, Texas).

The link between silicon and the electronics industry involves the concept of semiconductors. A **semiconductor** *is a substance that conducts electricity weakly.* Semiconductor materials have electrical properties intermediate between those of metals and nonmetals. Elemental silicon has been found to be an excellent material for use in semiconductor devices—devices on which the electronics industry is based.

The silicon used in electronic devices must be ultrapure. The purity of silicon obtained from an electric arc furnace—about 98%—is not sufficient. Additional purification is required. The first additional purification step involves converting the Si to liquid $SiCl_4$ (silicon tetrachloride).

$$Si(s) + 2\ Cl_2(g) \longrightarrow SiCl_4(l)$$

Repeated distillation of the tetrachloride leaves impurities behind. The $SiCl_4$ is then converted back to elemental Si by reaction with magnesium.

$$SiCl_4(g) + 2\ Mg(s) \xrightarrow{\text{high } T} 2\ MgCl_2(s) + Si(l)$$

This Si is then purified still further using a special method of recrystallization called zone refining. Silicon with a purity of 99.9999% is possible using this technique. This ultrapure silicon is shaped into the form of cylinders, which are then sliced into thin wafers for semiconductor manufacture (see Figure 13.12).

Ultrapure silicon is a nonconductor of electricity at room temperature, as would be expected for a nonmetal. It can be converted to a semiconductor by adding selected impurity atoms. This process is called *doping*. Two types of semiconductors result, depending on the identity of the impurity used in the doping process.

DOPED SILICON SEMICONDUCTORS

A *n*-type semiconductor (*n* for negative) is produced when trace amounts of atoms with five valence electrons, such as arsenic or antimony (group VA elements) are added to the silicon. (Silicon, a group IVA element, has four valence electrons.) The arsenic or antimony atoms susbstitute for some of the silicon atoms, and their extra fifth valence electron becomes the basis for semiconductor properties. These extra electrons can move through the crystal under the influence of an electron field; they are the electric current carrier. (Electric current is the flow of electrons through a material.) The upper portion of Figure 13.13 contrasts a crystal of silicon with an *n*-type semiconductor derived from it.

A *p*-type semiconductor (*p* for positive) results when small amounts of atoms

FIGURE 13.12

Ultrapure silicon is produced in the form of a cylinder and is then sliced into thin wafers for semiconductor manufacture. [© Ellis Herwig/The Picture Cube.]

with three valence electrons, such as boron (a group IIIA element) are added to the silicon. An electron vacancy or "hole" is created at the impurity site (see Figure 13.13). Such holes serve as a means by which electrons, under the influence of an electric field, can "hop" throughout the crystal. As a neighboring electron fills a hole, it creates a new hole (its original position), which can then be occupied by another electron, creating another hole, and so on. The net effect is movement of electrical charge, which is a prerequisite for electrical conduction in a material.

Most uses for silicon-based semiconductors involve connection of *p*- and *n*-type semiconductors to give *p–n junctions*. A *p–n* junction, when attached to an external power source, conducts electrical current in only one direction, and thus can convert external alternating current (electron flow in both directions on an alternating basis) to direct current (electron flow in only one direction). Before *p–n* junctions were developed, vacuum tubes were used for this purpose. Older television sets and radios contained many such vacuum tubes, which would periodically burn out and need to be replaced.

Let us consider selected details about the operation of a *p–n* junction. Suppose the *p–n* junction is connected to a battery in such a way that the *p*-type region is in contact with the battery's positive terminal and the *n*-type region is in contact with its negative terminal (see Figure 13.14*a*). Electrons flow into the *n*-type region from the external circuit and move toward the interface with the

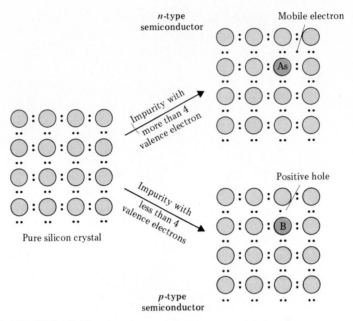

FIGURE 13.13

Two types of semiconductors, *n* and *p*, can be derived from silicon. In *n*-type semiconductors the impurity atoms furnish mobile electrons to the crystal. In *p*-type semiconductors a deficiency of electrons creates "positive holes" through which electrons can "hop" throughout the crystal.

p-type region. Simultaneously, electrons are removed from the *p*-type region and enter the external circuit. This causes the holes in the *p*-type region to move toward the interface with the *n*-type region. (The holes always move in the opposite direction to the electrons; as electrons move, holes are left where the electrons were.) At the interface between the two semiconductor types the electrons from the *n*-side enter the holes present on the *p*-side, the cycle is repeated, and current flows continuously.

If the battery is attached to the *p*–*n* junction in the opposite manner to that just discussed (see Figure 13.14*b*) current does not flow. Electrons flow out of the *n*-type region and enter from the external circuit into the *p*-type region. The entering electrons fill the holes whose direction of movement is away from the junction. Thus, near the junction there are no longer holes in the *p*-type region nor any free electrons in the *n*-type region, and flow of charge ceases. Flow of electrical charge in a *p*–*n* junction therefore has directional character; it operates in only one direction.

INTEGRATED CIRCUITS   The most important development in recent years in the electronics industry has been the production of *integrated circuits*. Such circuits are the "brains" of hand-held calculators and personal computers. Let us consider some of the chemistry involved in their production.

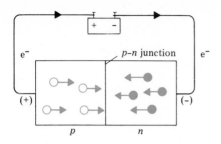

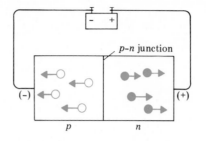

(a) The *p*-type region is connected to the positive terminal of an external power source and the *n*-type region to the negative terminal. In this situation current readily flows because both types of charge carriers are moving toward the *p–n* junction.

(b) The *p*-type region is connected to the negative terminal of an external power source and the *n*-type region to the positive terminal. In this situation the charge carriers move away from the *p–n* junction and current does not flow.

FIGURE 13.14

A *p–n* junction involves the contact of a *p*-type semiconductor and an *n*-type semiconductor. In the *n*-type region the charge carriers are electrons (●); in the *p*-type region the charge carriers are holes (○).

A very thin "doped" silicon wafer of the *n*- or *p*-type, called a *silicon chip*, is baked in oxygen. This process produces a protective coating of silicon dioxide on the surface of the chip. Next, the oxide coating is covered with a light-sensitive wax.

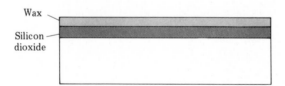

With the silicon chip in this state, it is ready for the addition of a circuit (diagram of instructions). For this, a template that lets light through in selected areas only is placed on top of the wax and the chip is exposed to light.

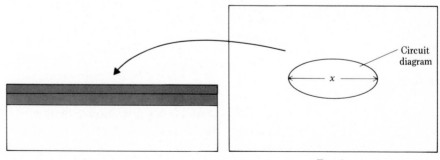

Template

Light alters the properties of the wax it contacts. In particular, the solubility characteristics change. The exposed wax is then dissolved using an appropriate solvent.

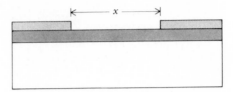

The exposed areas of SiO_2 (which is in the shape of a circuit) is then etched away using acid. The remaining wax is dissolved, leaving a silicon chip with its oxide coating intact except where etching with acid of the circuit pattern has occurred.

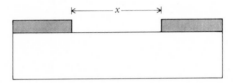

Semiconductor material of the opposite type to that of the chip is then diffused into the etched channels creating p–n junction points.

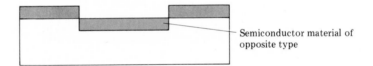

Semiconductor material of opposite type

The cycle then begins again. The chip is again coated completely with SiO_2 and wax and another template is applied, light exposure occurs, and etching of another circuit pattern is done. New p–n junctions are created. In this way thousands of pieces of information, or circuits, can be built into a single chip.

Finally, after all the circuit information has been placed on the chip, a conducting layer of aluminium is applied and etched to produce a complete circuit diagram. A single piece of silicon has been coded with thousands of "instructions."

Silicates

About 90% of the earth's crust consists of silicates, which are minerals containing negative polyatomic ions involving silicon and oxygen. The basic structure of all silicates involves the silicate ion (SiO_4^{4-}). This ion is tetrahedral in structure, with

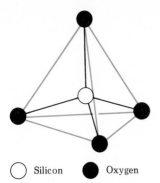

○ Silicon ● Oxygen

FIGURE 13.15

The building block for all silicates is the tetrahedral SiO_4^{4-} ion.

a central silicon atom bonded to four oxygen atoms arranged at the corners of a tetrahedron (see Figure 13.15). For charge neutrality, positive ions such as Al^{3+}, Ca^{2+}, and Na^+ must also be present.

Most silicates do not contain discrete, independent SiO_4^{4-} units. Instead, more complex units formed by the joining together of two or more SiO_4 tetrahedra by bridging oxygen atoms (that is, oxygen atoms shared by two tetrahedra) are present. The simplest of these oxygen-bridging situations, that involving two SiO_4 tetrahedra, produces the $Si_2O_7^{6-}$ ion. The structure of this ion is

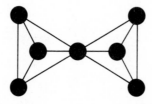

Note the notation used in depicting this structure. Only the oxygen atoms are shown in the figure; a silicon atom is in the center of each tetrahedron, but it is not shown. In Figure 13.16 this notation is used to represent selected commonly occurring higher silicate structures.

The mineral beryl ($Be_3Al_2Si_6O_{18}$) contains Be^{2+} ion, Al^{3+} ion, and the cyclic $Si_6O_{18}^{12-}$ ion. This cyclic silicate structure, shown in Figure 13.16, has two bridging oxygen atoms in each tetrahedron. High-grade beryl crystals, with trace amounts of transition metal impurities, are valued as gemstones. If chromium is the transition metal impurity, we have an emerald and if iron is present, we have aquamarine.

EMERALDS AND AQUAMARINES

The $(Si_4O_{11}^{6-})_n$ silicate unit, which involves two connected chains of silicon–oxygen tetrahedra (see Figure 13.16), is found in the mineral asbestos. The fibrous character of asbestos is a direct consequence of this chain-like structure.

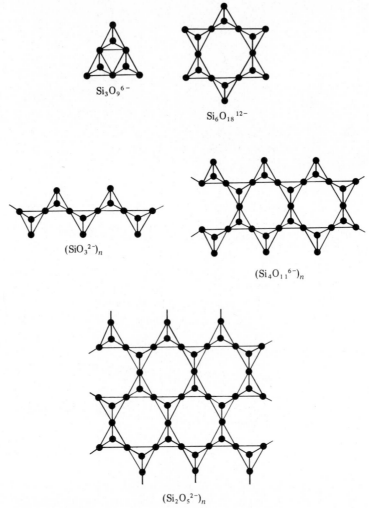

FIGURE 13.16
Schematic representation of the structures of selected higher silicate ions.

ASBESTOS

Until the last decade, asbestos was extensively used as a thermal insulator in buildings. Its high chemical and thermal stability are excellent properties for such use. It is now established that prolonged exposure to air borne asbestos fiber (asbestos dust) can damage lung tissues; the fibers become embedded in them. The use of asbestos in buildings has stopped and existent asbestos insulation is being replaced in some situations.

If three of the four oxygen atoms in each tetrahedra are involved in bridging, as is the case in the $(Si_2O_5^{2-})_n$ ion, a two-dimensional sheet-like structure results

FIGURE 13.17
A piece of mica showing its cleavage into thin sheets. [Runk/Schoenberger from Grant Heilman.]

TALCUM POWDER

(see Figure 13.16). Common minerals containing this structure are mica, talc, and kaolinite. One type of mica has the formulation $KLi_2Al(Si_2O_5)_2(OH)_2$—it contains K^+, Li^+, Al^{3+}, $(Si_2O_5^{2-})_n$, and OH^- ions. The ease with which mica can be separated into thin sheets is related to this sheet-like structure (see Figure 13.17). Talc has the composition $Mg_3(Si_2O_5)_2(OH)_2$. This very soft mineral is used to make talcum powder. Soapstone is talc in a compact form that is easily carved. Kaolinite, $Al_2Si_2O_5(OH)_4$ is a type of clay that is used extensively in pottery making. The layered structure of clay soils accounts for the "slippage" that sometimes occurs when such soils become very wet. Water penetrates between, and lubricates, the layers so that they slip over one another.

Previously we observed that when SiO_4 tetrahedra share two corners (oxygen atoms), chain-like and cyclic structures result. When SiO_4 tetrahedra share three corners, the result is a two-dimensional sheet-like structure. Is it possible for SiO_4 tetrahedra to share all four corners? The answer is yes, and the result is a three-dimensional structure. This is the diamond-like structure of crystalline SiO_2 itself.

Glass

Quartz (pure SiO_2), when heated to a high temperature (1600°C), melts to give a "sticky" liquid material. Many silicon–oxygen bonds must be broken in this transition from crystalline solid to "sticky" liquid. Rapid cooling of this liquid produces a noncrystalline glassy material known as quartz glass (or silica glass). The noncrystalline structure is the direct result of the *rapid* cooling; there was not sufficient time for the silicon–oxygen bonds to re-form in the normal pattern associated with crystalline SiO_2.

Other types of glass can be derived from SiO_2 melts by adding other substances to the melt. Common window and bottle glass contains CaO and Na_2O in addition to SiO_2. The sources for these two oxides are their naturally occurring carbonates—limestone ($CaCO_3$) and soda ash (Na_2CO_3). These inexpensive carbonates decompose at elevated temperatures to produce the oxides.

$$CaCO_3 \longrightarrow CaO + CO_2$$

$$Na_2CO_3 \longrightarrow Na_2O + CO_2$$

The oxides then combine with SiO_2 (sand) to form silicates.

$$CaO + SiO_2 \longrightarrow CaSiO_3$$

$$Na_2O + SiO_2 \longrightarrow Na_2SiO_3$$

In practice limestone, soda ash, and sand are heated together, and the resulting glassy mixture of silicate material is known as soda-lime glass.

Soda-lime glass is clear and colorless only if the purity of the raw materials needed for its production is carefully controlled. Small amounts of iron oxides, a common impurity, impart a light green color to such glass. This is the origin of the greenish tint often present in window panes in century-old homes. Other metal oxides produce other colors, and they are often purposefully added to glass formulations. Addition of cobalt (CoO) gives a blue glass, and chromium (Cr_2O_3) produces a deep green glass. The red glass in red signal lights contains zinc (ZnO) and elemental selenium (present in a suspended form).

Soda-lime glass does not have good heating characteristics. It expands when heated and contracts unevenly as it cools. The unevenness of the contraction

FIGURE 13.18

The ability of lead crystal glass to capture and refract light has led to its use by glass artisans for over 300 years. [Stenben Glass.]

BOROSILICATE GLASS

creates stress points in the cooled glass so that it shatters easily. This problem is solved by replacing some of the CaO and Na_2O in the glass formulation with boron oxide (B_2O_3) to give a borosilicate glass. Borosilicate glasses expand very little when heated. Cooking utensils and laboratory glassware are usually made of borosilicate formulations. Pyrex is a trade name for a borosilicate glass.

Glass in which lead(II) oxide (PbO) is a component has a high refractive index (it "bends" light rays well) and a high density. Such glass, which sparkles when cut or polished, is used to make decorative cut-glass articles (see Figure 13.18).

The addition of potassium oxide (K_2O) to glass formulations increases the hardness of the glass and makes it easier to grind to precise shapes. Optical glass contains significant amounts of K_2O. A recent development in optical glass manufacture involves photochromic lenses, which darken in the sun and return to their clear state in less intense light. Such glass contains small amounts of silver chloride (AgCl) or silver bromide (AgBr) dispersed throughout it. Both of these compounds are photosensitive, and sunlight causes them to decompose, giving opaque clusters of Ag atoms (which are black) and Cl (or Br) atoms.

PHOTOCHROMIC LENSES

$$\underset{\text{Clear}}{AgCl} \; \overset{\text{sunlight}}{\underset{\text{less intense light}}{\rightleftharpoons}} \; \underset{\text{Opaque}}{Ag + Cl}$$

The Ag and Cl (or Br) atoms, which are trapped in the crystal lattice, recombine into AgCl (or AgBr) in the absence of intense light.

TABLE 13.4

Oxide Composition of Selected Common Glasses

Name	Oxide (mass %)						
	SiO_2	CaO	Na_2O	K_2O	B_2O_3	Al_2O_3	PbO
Quartz glass	100	0	0	0	0	0	0
Window glass (soda–lime)	72	11	13	3.8	0	0.3	0
Heat-resistant glass (Pyrex)	76	3	5	0.5	13	2	0
Lead Glass (decorative crystal)	67	1	10	7	0	0	15
Optical glass	69	12	6	13	0.3	0	0

Table 13.4 gives the composition of various types of glasses. Glass composition is normally expressed in terms of the oxides of the elements it contains, even though these oxides have been converted to complex silicates.

SAFETY GLASS

Glass used in the automobile industry is *safety glass*. This glass is shatterproof, that is, it does not break into small *loose* fragments (shards) when impacted. There are two general ways of making safety glass. The first method involves subjecting the glass to a special heat treatment. The glass sheet is heated to a temperature just below its softening point, and the surfaces are subjected to a cold blast of air. This creates compression strains in the glass regions near the surface. These compression strains become vitally important when the glass receives a hard impact. It instantaneously develops cracks over its entire surface; the glass "crumbles" into finely divided pieces that do not fly about and therefore are relatively harmless.

Safety glass can also be made by a lamination process. A thin layer of plastic is sealed between two pieces of thin plate glass. When this laminated glass breaks, adhesion of the glass fragments to the flexible plastic reduces the danger from flying glass and jagged edges. Laminated glass, upon impact, often develops cracks that radiate from the point of impact. For bulletproof glass, several alternating layers of glass and plastic are laminated together. The thickness of such glass is usually 1.5 to 2 inches.

Glass presents a disposal problem. It is one of the most "permanent" materials known. Natural processes do not degrade discarded glass. Glass can, however, be easily recycled. Different kinds of glass must first be separated. Then, they are melted and formed into new objects. The energy savings associated with recycled glass are considerable.

POTTERY GLAZE

The *glaze* of pottery that makes it impervious to water is a coating of glass. It is formed by covering the pottery with a thin paste of the appropriate oxides (CaO, Na_2O, SiO_2, and so on) and then heating the coated pottery to a high temperature.

Cement

A cement is a material used to bind other materials together. The first use of cements is credited to the Romans. Their cements were natural mixtures of lime (CaO) and volcanic ash which, after burning, would set under water. Modern cement, also known as portland cement, was first used in the early 1800s in England. (The name portland cement, coined in 1824 in England, arose from the fact that when this powdered substance was mixed with water and sand the resulting mixture hardened into a block that resembled the natural limestone quarried at the Isle of Portland, England.) The approximate composition of portland cement, in mass percents, is 60 to 67% CaO, 17 to 25% SiO_2, 3 to 8% Al_2O_3, 1 to 5% Fe_2O_3, and small amounts of MgO, Na_2O, K_2O, and $MgSO_4$. This material is obtained by heating a mixture of limestone, clay (an aluminum-containing silicate), and iron oxide in huge kilns at temperatures of 1400 to 1500°C. In the heating process the materials lose water and carbon dioxide, leaving behind a marble-sized solid material called "clinker." This clinker, a complex mixture of aluminum, calcium, and iron silicates, is ground to a very fine powder, mixed with a small amount of calcium sulfate (gypsum), and bagged for shipment as portland cement. The added gypsum regulates the setting time of the cement.

PORTLAND CEMENT

Portland cement, when mixed with water, sand, and gravel or crushed rock, forms a semisolid mass thaat gradually sets and hardens to produce *concrete*. The reactions responsible for the setting of concrete, although not completely understood, involve a series of hydration reactions between the water and the silicates present.

Besides portland cement, a large variety of special-purpose cements, with specially designed adhesive and binding properties, are commercially available.

Practice Questions and Problems

NONMETALLIC ELEMENTS

13-1 List the general properties of nonmetallic elements.

13-2 Three of the four most abundant elements in the earth's crust are nonmetals. Identify these nonmetals and compare their abundances.

13-3 Oxygen is a colorless, odorless, nontoxic gas. Give a similar one-sentence characterization of the physical state and selected properties for each of the following nonmetals.

a. Hydrogen   b. Fluorine
c. Phosphorus   d. Sulfur
e. Silicon   f. Carbon
g. Nitrogen   h. Chlorine

13-4 Elemental hydrogen gas (H_2) is not found in the atmosphere. Explain why.

PHOSPHORUS AND ITS COMPOUNDS

13-5 Contrast the chemical properties of white phosphorus and red phosphorus.

13-6 Contrast the structures of white phosphorus and red phosphorus.

13-7 It is dangerous to allow white phosphorus to come in contact with human skin. Explain why.

13-8 Describe the change that occurs in each process.
a. White phosphorus is heated to 250°C in the absence of air.
b. Red phosphorus is heated to 600°C in the absence of air.

13-9 White phosphorus is usually stored beneath the surface of water. Explain why.

13-10 Describe, using chemical equations, each of the following processes.
a. Industrial production of white phosphorus from phosphate rock.
b. Industrial production of phosphoric acid from elemental phosphorus.
c. Production of triphosphoric acid from phosphoric acid.

13-11 Describe the role of the element phosphorus in the "chemistry" of the following items.
a. Safety matches
b. Strike-anywhere matches
c. Household detergents

13-12 Draw the chemical structure of the following.
a. Triphosphoric acid　　b. STPP

13-13 Describe the water pollution problem associated with the use of phosphates in detergent formulations.

CARBON AND ITS COMPOUNDS

13-14 Contrast the properties of diamond and graphite.

13-15 Contrast the structures of diamond and graphite.

13-16 What is the chemical composition of pencil lead, and what is the active ingredient in such lead?

13-17 What are the required physical conditions for the industrial transformation of graphite into diamond?

13-18 Describe the production method for each of the following noncrystalline forms of carbon.
a. Carbon black　　b. Charcoal
c. Activated carbon　　d. Coke

13-19 List industrial uses for each of the following forms of carbon.
a. Diamond　　b. Graphite
c. Carbon Black　　d. Activated charcoal
e. Coke

13-20 What is the basis for the toxicity of carbon monoxide to human beings?

13-21 Indicate whether carbon dioxide is a reactant or a product in each of the following processes. For parts a and b also write a chemical equation for the process.
a. Photosynthesis
b. Metabolism in human beings
c. Decomposition of animal tissue

13-22 What is the difference between baking soda and baking powder?

13-23 The most commonly used household baking powder is SAS–phosphate baking powder. What is its chemical composition?

13-24 SAS–phosphate baking powder is a "double-acting leavening agent"? Explain what this means.

13-25 Write chemical equations to describe the action of SAS-phosphate baking powder.

13-26 Identify the commonly used leavening agent whose use results in the production of the following substances.
a. Sodium carbonate
b. Ethyl alcohol
c. Sulfuric acid
d. Sodium sulfate
e. Calcium hydrogen phosphate

13-27 Describe the operation of each of the following types of fire extinguishers.
a. Acid–soda　　b. Liquid CO_2

13-28 What is the chemical formula of the compound most often used in dry chemical fire extinguishers? How does this chemical put out fires?

13-29 Compare the properties of dry ice and water ice.

13-30 What are the two largest industrial uses for carbon dioxide?

13.31 Describe the process used for industrial production of dry ice.

SILICON AND ITS COMPOUNDS

13-32 Write a chemical equation for the industrial production of the element silicon.

13-33 Contrast the structures of elemental silicon and diamond.

13-34 What is meant by the phrase "doping a silicon crystal"?

13-35 What is a semiconductor?

13-36 Describe the structures of the following.
a. A *p*-type semiconductor derived from silicon.
b. An *n*-type semiconductor derived from silicon.

13-37 What is a *p–n* junction, and what major use is made of it?

13-38 List the major chemical steps in the production of an integrated circuit.

13-39 Describe the chemical unit on which all silicate structures are based.

13-40 Describe the generalized structure or structures that result when numerous silicate tetrahedra share the following.
a. Two oxygen atoms
b. Three oxygen atoms
c. Four oxygen atoms

13-41 Name a silicate mineral that contains each of the following ions.
a. $Si_6O_{18}^{12-}$ b. $(Si_4O_{11}^{6-})_n$ c. $(Si_2O_5^{2-})_n$

13-42 Mica is a mineral that flakes off in thin layers. Relate this property of mica to its chemical structure.

13-43 Asbestos is a fibrous mineral. Relate this property of asbestos to its chemical structure.

13-44 Use of asbestos as a thermal insulator has largely been discontinued. Explain why.

13-45 What is the chemical composition of the gemstone known as emerald?

13-46 Describe the production of common window glass (soda-lime glass) using chemical equations.

13-47 What is the "added ingredient" that produces the desired properties in each of the following types of glass?
a. Heat-resistant glass
b. Decorative crystal glass
c. Optical glass

13-48 What is meant by the term photochromic glass? Explain how such glass works.

13-49 Discuss the construction and characteristics of the two major types of automobile safety glass.

13-50 Discuss the chemical composition of and manufacturing process for portland cement.

13-51 What is the purpose of the gypsum present in portland cement?

14

Some Interesting and Useful Metals

CHAPTER HIGHLIGHTS

14.1   Metallic Elements

We have seen that elements can be classified as metals or nonmetals (Section 4.10). Chapter 13 focused on selected familiar nonmetallic elements. In this chapter we take a closer look at metallic elements. After considering some general information about metals, we will discuss in further detail the important and useful metals iron, copper, gold, silver, and aluminum.

Eighty-seven of the 109 elements are metals (Section 4.10). Thus, four-fifths of all elements are metals. However, no metal is anywhere near as abundant as oxygen and silicon. These two nonmetals, with abundances of 60.1 and 20.1 atom percent, respectively, account for more than 80% of all atoms associated with the earth's surface (the earth's crust, hydrosphere, and atmosphere). The most abundant metal is aluminum, with an atom percent abundance of 6.1. As shown in Figure 14.1, only five metals have abundances greater than 1.0 atom

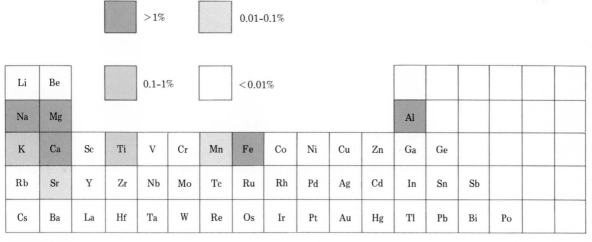

FIGURE 14.1

Periodic table color coded to show abundances (atom percent) of the metallic elements. The inner transition elements (not shown) have abundances of less then 0.01%.

TABLE 14.1
Selected Numerical Parameters for the Most Abundant Metals

Metallic Element	Abundance (atom percent)	Abundance Rank Considering Metals Only	Abundance Rank Considering All Elements
Aluminum	6.1	1	3
Calcium	2.6	2	5
Magnesium	2.4	3	6
Iron	2.2	4	7
Sodium	2.1	5	8
Potassium	0.89	6	9
Titanium	0.37	7	10
Manganese	0.037	8	13
Strontium	0.011	9	17

percent. The figure also shows that the vast majority of metals have atom percent abundances of less than 0.01 atom percent.

Numerical abundance values for the nine metals with abundances greater than 0.01% are listed in Table 14.1. To derive abundance values such as those in this table, assumptions must be made concerning the distribution of rock types in the earth's crust and scientists are not in complete agreement on the subject. This accounts for differences found in abundance tabulations from various sources. Table 14.1 reflects the most recent estimates on rock type within the earth's crust. The last column of Table 14.1 shows that 7 of the 10 most abundant elements are metals. The other three elements in the top ten are the nonmetals oxygen (1), silicon (2), and hydrogen (4) (see Section 13.2). Aluminum, the most abundant metal, is more than twice as abundant as any other metal.

The ranking of metals according to abundance (Table 14.1) does not reflect their importance in today's world. Abundance and importance (use) are two entirely different things; some metals with low abundances in the earth's crust are more widely known and used than some of the most abundant metals. Why is this so? Obviously, chemical properties of the various metals help determine their use. Another major factor is the availability of a metal's deposits in the earth's crust. Some metals have been concentrated by natural processes into localized areas. Less abundant metals that are found in "concentrated form" are usually more used than more abundant metals that are not concentrated. For example, the elements copper, silver, and gold (less abundant metals) have more uses than titanium or strontium (abundant metals). The difficulty and cost of isolating a metal from its natural sources also affect the extent of its use.

Most metals occur in nature in a chemically combined state in *ores*. An **ore** *is a naturally occurring material from which one or more metals (or other chemical substances) can be profitably extracted.* Sulfide, oxide, carbonate,

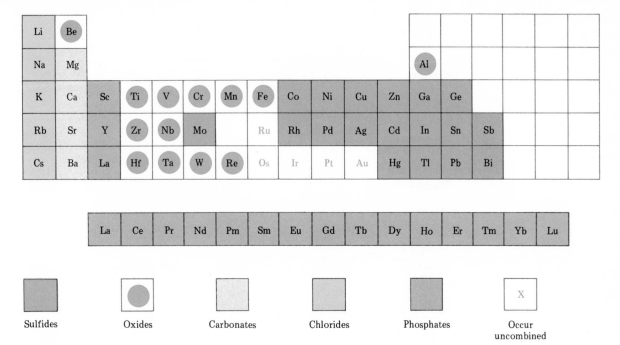

FIGURE 14.2

Periodic table color coded to show the ore form that is the source for each of the naturally occurring metallic elements. Some metals have more than one type of important ore, although only one is given here.

chloride, and phosphate ores are common. Figure 14.2 shows the type of ore source in which the various metallic elements are found in nature.

Most metals occur in nature as sulfides or oxides. The greatest tonnages of metals are obtained from oxides, both directly and indirectly. As Figure 14.2 shows, many metals occur in oxide form. In addition, numerous other metal oxides are produced by the roasting of carbonate or sulfide ores. The first step in copper metallurgy is the conversion of its sulfide (the naturally occurring form) to an oxide (Section 14.3).

Groups IA and IIA metals usually occur in nature as chlorides or carbonates. In general, such ores have greater water solubility than sulfides, oxides, and phosphates. This accounts for the high concentrations of Na^+, Mg^{2+}, K^+, and Ca^{2+} ions in seawater (see Table 11.4).

A few less reactive metals, including gold and platinum, occur in uncombined form. Such uncombined metal is found in veins of quartz (SiO_2) or other silicate material.

Although silicate minerals are very abundant in nature (Section 13.5), they are seldom used as ores to obtain metallic elements. Extraction of metals from

TABLE 14.2

World and United States Production of the Metals Used in the Greatest Amount (1980)

	Annual Production *(metric tons)*		*Factor by which Iron Production Exceeds Production of Listed Metal*	
	Worldwide	*United States*	*Worldwide*	*United States*
Iron	5.0×10^8	6.7×10^7		
Aluminum	1.6×10^7	4.5×10^6	31	15
Manganese	1.2×10^7	—	42	—
Copper	7.8×10^6	1.2×10^6	64	56
Zinc	6.1×10^6	3.7×10^5	82	181
Lead	5.4×10^6	1.1×10^6	93	61
Chromium	4.5×10^6	—	111	—
Nickel	7.6×10^5	4.0×10^4	658	1675

silicates is usually difficult and costly. Even when the chemistry of extraction is known, the cost of the process is almost always prohibitive.

Aluminum, the most abundant metal, is not the most used metal. As shown in Table 14.2, it ranks second, both worldwide and in the United States, to the metal iron in amount produced. Iron dominates today's technological world. Its use, on a worldwide basis, is 31 times greater than that of aluminum, the second-place metal. Iron production is nine times greater than that of all other metals combined; that is, 90% of all metal consumed is iron. Nearly all iron production goes into the manufacture of steel. There are many kinds of steel; all are alloys in which iron is the predominant metal. The chemistry of steel production is considered in Section 14.2.

Strategic Metals

Metallic ores are not distributed equally throughout the world. Ores of some metals are found only in particular regions of the earth's crust. Even the United States, with its wealth of natural resources, lacks domestic reserves and is a substantial importer, of certain metallic ores. For example, Table 14.2 shows that the United States has negligible domestic production of the metals chromium and manganese.

DEPENDENCE ON
IMPORTED METALS

To some observers, our nation's reliance on these imports constitutes a dangerous dependency, threatening a "materials crisis" more devastating than the "energy crisis" of the early 1970s that resulted from a petroleum embargo by countries in the Middle East. Others see the United States position as quite manageable, though not without dangers and difficulties. All observers agree that some imported metals are particularly important to the well-being of the United States and that if supplies of these metals were cut off many readjustments would have be made. These "strategic metals" are

1. Chromium
2. Cobalt
3. Manganese
4. Platinum group metals (platinum, palladium, rhodium, iridium, osmium, and ruthenium)

Three nations—South Africa, Zaire, and the USSR—hold most of the world's known reserves of these metals. Production from United States mines is negligible, but small amounts of these metals are now available domestically from recycled materials. Table 14.3 gives net import dependence and leading suppliers of these strategic metals.

Chromium, a lustrous, hard, steel-gray element, is used in a variety of applications, the most essential of which are in superalloys and stainless steel and as an alloying element in tool, spring, and bearing steels. As an alloying element, chromium increases the hardness, strength and oxidation resistance at elevated temperatures, and wear resistance of steel. These properties make chromium alloy steel essential in springs, bearings, and tools, as well as in components of automobile engines. In stainless steel, the formation of a tenacious chromium oxide film on the surface of the material provides a barrier to corrosion and oxidation that is essential in chemical processing plants, oil and gas production, power generation, and automobile exhaust systems, principally in the catalytic converter.

Cobalt is a hard, brittle metal that resembles nickel and iron in appearance. The largest and most critical end use of cobalt is in superalloys, used mainly in gas turbines that require high strength at very high temperatures—above the useful range of steel. Another application is in electrical equipment, where cobalt's strong magnetic properties are valued in small, powerful, and long-lasting magnets. Cobalt is also contained in catalysts for certain essential steps in the refining of petroleum and the manufacturing of chemicals.

TABLE 14.3

Net Import Dependence and Leading Suppliers of Strategic Metals to the United States

Metal	Net Imports 1978–1982 (percent of consumption)	Leading Suppliers
Chromium	85–91	South Africa, USSR
Cobalt	92–95	Zaire, Zambia
Manganese	97–99	South Africa, Gabon
Platinum group metals	80–90	South Africa, USSR

Li	Be											Al					
Na	Mg											Al					
K	Ca	Sc	Ti	V	Cr	Mn	Fe	Co	Ni	Cu	Zn	Ga	Ge				
Rb	Sr	Y	Zr	Nb	Mo	Tc	Ru	Rh	Pd	Ag	Cd	In	Sn	Sb			
Cs	Ba	La	Hf	Ta	W	Re	Os	Ir	Pt	Au	Hg	Tl	Pb	Bi			

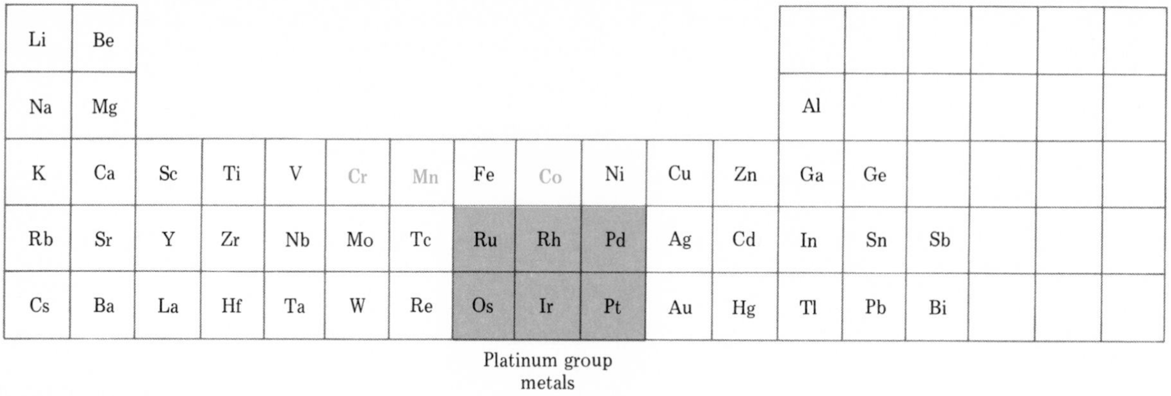

Platinum group
metals

FIGURE 14.3
The strategic metals are chromium, manganese, cobalt, and the platinum group metals.
All of these metals must be imported into the United States because this country
lacks domestic reserves.

The dominant critical use for manganese, a gray-white or silver metal, is in
steel making. The addition of manganese prevents steel from becoming brittle.
At the high temperature necessary for steel production sulfur impurities in steel
react with the iron to produce compounds that weaken the steel. When manganese is added, the sulfur reacts with it instead of with iron. The resulting manganese-sulfur compound does not make the steel weak. Manganese also has
uses as an alloying element to impart strength and hardness to all grades of steel.
For example, manganese alloy steel is used for armored vehicles that withstand
impacts.

The platinum group metals are all located together in the periodic table (see
Figure 14.3) and have similar properties. They have the ability to catalyze many
chemical reactions and withstand chemical attack, even at high temperatures.
Their leading use in the United States is in automobile catalytic converters,
which use small amounts of platinum, palladium, and rhodium. At present, there
is no satisfactory substitute for these metals in this use. Another catalytic
application is in petroleum refining to produce high-octane gasoline. The metals
are also used as contacts in telecommunication switching systems because of
their great strength and resistance to corrosion and oxidation.

14.2 Iron and Steel

In Section 14.1 we noted that iron is the most used of all metals. Indeed, 90% of
all metal consumed today is iron, but primarily as steel, not pure iron. All steels
are iron alloys. An **alloy** *is a solid material with metallic properties that has been*

produced by melting together two or more elements, at least one of which is a metal.

The production of steel from iron ore involves two separate operations.

1. The conversion of iron ore to an impure form of iron called pig iron.
2. The conversion of pig iron into various types of steel.

The Production of Pig Iron

Iron ore is an impure mixture of Fe_2O_3, Fe_3O_4, various silicates, and usually with small amounts of manganese and phosphorus. The chemical reactions whereby iron is obtained from such ore take place in a huge tower called a *blast furnace*. Blast furnaces are large, typically 30 m high, and are designed for continuous operation. A change of iron ore and other materials is added at the top of the furnace and molten impure iron is drawn off from the bottom. Once in operation, a blast furnace is run continuously for two years or longer before it must be shut down for repairs (to be rebuilt).

Figure 14.4 shows the essential features of a typical blast furnace. The *charge*—the raw materials that are introduced at the top of the furnace—is a mixture of iron ore, limestone, and coke. The limestone's function is to react with high-melting impurities to produce "slag." The coke facilitates the reduction of the iron oxides to elemental iron.

We consider first what happens near the bottom of the furnace and then work our way up to the higher regions of the furnace. To get the overall reduction process started, a blast of hot (900°C) compressed air is blown into the furnace through nozzles located at the bottom of the furnace. The result is a very exothermic (heat producing; see Section 8.12) reaction between carbon (from the coke) and oxygen (from the air).

$$C(s) + O_2(g) \longrightarrow CO_2(g)$$

The heat released from this reaction produces temperatures as high 2000°C in the lower regions of the furnace, which become the hottest regions.

The hot CO_2 gas rises and reacts with additional C to form carbon monoxide.

$$CO_2(g) + C(s) \longrightarrow 2\,CO(g)$$

Since this reaction is endothermic (Section 8.12) rather than exothermic, the temperature in this higher region of the furnace is only about 1300°C.

The carbon monoxide thus produced reacts with the iron oxides present in the charge. The reduction of the iron oxides to iron metal occurs in steps, each occurring at a different temperature (see Figure 14.4).

$$3\,Fe_2O_3(s) + CO(g) \xrightarrow{250°C} 2\,Fe_3O_4(s) + CO_2(g)$$

$$Fe_3O_4(s) + CO(g) \xrightarrow{600°C} 3\,FeO(s) + CO_2(g)$$

$$FeO(s) + CO(g) \xrightarrow{1000°C} Fe(s) + CO_2(g)$$

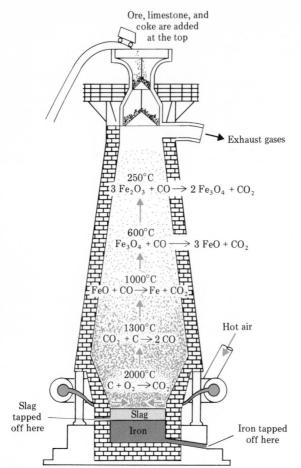

Ore, limestone, and coke are added at the top

Exhaust gases

250°C
$3 Fe_2O_3 + CO \longrightarrow 2 Fe_3O_4 + CO_2$

600°C
$Fe_3O_4 + CO \longrightarrow 3 FeO + CO_2$

1000°C
$FeO + CO \longrightarrow Fe + CO_2$

1300°C
$CO_2 + C \longrightarrow 2 CO$

Hot air

2000°C
$C + O_2 \longrightarrow CO_2$

Slag tapped off here

Slag

Iron

Iron tapped off here

FIGURE 14.4

A typical blast furnace for the production of pig iron and slag from iron ore, limestone, and coke.

As the charge reaches the bottom of the furnace, molten iron collects in a well at the base (see Figure 14.4).

The limestone present in the charge decomposes at the high temperatures of the furnace, producing calcium oxide which reacts with impurities (mostly SiO_2, but also small amounts of other oxides) to form molten *slag*.

$$CaCO_3(s) \longrightarrow CaO(s) + CO_2(g)$$

$$CaO(s) + SiO_2(s) \longrightarrow CaSiO_3(l)$$

The molten slag, which is less dense than molten iron, collects as a liquid layer on top of the molten iron and protects the molten iron from reoxidation (by air) to iron oxides.

Periodically, four to five times a day, the furnace is tapped; the iron and slag are drawn off. The iron, which still contains some impurities is called *pig iron*. (The name pig iron comes from the early practice of running the molten iron from the blast furnace through a central channel that fed into numerous sand molds. The arrangement looked a little like a litter of pigs feeding from their mother.)

The slag, in solidified form, is a valuable by-product. It is used as one of the raw materials in making portland cement (Section 13.5) and also as a base material in road construction.

The hot exhaust gases reaching the top of the furnace, which still contain some CO, are mixed with air and burned. The resulting heat is used to preheat the air that is blown in at the bottom of the blast furnace.

Cast Iron and Steel Production

The approximate composition of pig iron is 90% iron, 5% carbon, 2% manganese, 1% silicon, 0.3% phosphorus, and 0.04% sulfur (from impurities in the coke). These impurities make pig iron very brittle.

A small amount of pig iron is converted to *cast iron*. Cast iron is made by melting together scrap iron and pig iron. The melt is cast into molds and allowed to solidify into the final product. The resulting cast iron is still quite brittle and is used in items not normally subjected to sudden mechanical or thermal shock. Typical uses include iron drain pipes, steam radiators, and various stove parts.

The vast majority of pig iron production is converted into steel. In the steel-making process, two things must occur.

1. The carbon content of the pig iron must be reduced below 1.5%.
2. Silicon, manganese, and phosphorus impurities must be removed through slag formation.

SCRAP IRON

The *basic oxygen process* is the most common method for accomplishing these two tasks. This process makes use of a large pear-shaped reaction vessel mounted on pivots (see Figure 14.5). The vessel is filled with a mixture of about 70% molten pig iron and 30% scrap iron or scrap steel. Most of the scrap comes form the steel plant itself, but some (an increasing amount each year) is obtained from the recycling of steel products such as automobiles. The scrap melts in the molten pig iron. To this molten charge is added powdered limestone and pure oxygen gas. Through a tube, called an oxygen lance, pure oxygen is blown directly into the melt at a depth of 4 to 8 feet below the surface. The oxygen reacts with excess carbon, producing carbon dioxide. It also reacts with impurities, converting them to oxides that react further with the limestone to

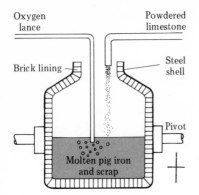

FIGURE 14.5
Basic oxygen furnace used for the production of steel.

form slag. For example, the silicon reacts as follows.

$$\text{Si} \xrightarrow{\text{O}_2} \text{SiO}_2 \xrightarrow{\text{CaO}} \underset{\text{Slag}}{\text{CaSiO}_3}$$

For many years air instead of pure oxygen was used in converting pig iron to steel. Eight to 10 hours was required to produce a batch of steel using air. A batch of steel (300 tons) can be made in less than an hour with pure oxygen. By controlling the amount of O_2 used, it is possible to adjust the carbon content of the steel within very narrow limits. In modern steel making, such control is computerized.

Approximately two-thirds of steel production in the United States is by the basic oxygen process. The other one-third of production involves the use of electric furnaces. A disadvantage of their use is the increased cost of electrical power. Advantages of the electric furnace are (1) the ability to achieve more accurate temperature control, (2) easier attainment of the high temperatures needed to make alloy steels (to be discussed shortly), and (3) the ability of furnace to accept more scrap charge; the basic oxygen furnace is limited to 35% scrap because excess scrap has too great a cooling effect on the charge.

Types of Steel

There are two major types of steel: (1) carbon steels and (2) alloy steels. Both types are metal alloys containing iron and carbon. Alloy steels contain one or more other metals in addition to iron. Approximately 85% of all steel produced in the United States is carbon steel.

The properties of steel depend to a large extent on the percentage of carbon present. Two properties very dependent on carbon content are *tensile strength*, measured by the force required to pull a rod into two pieces, and *ductility*, related to the ease with which the material may be formed into wire. Ductility decreases as carbon content increases, and tensile strength increases with increasing carbon content. Carbon is present in steel in three forms: as the chemical compound cementite (Fe_3C), as large crystals of graphite (see Section 13.4), and as very small crystals of graphite.

Low-carbon (or mild) steel (less than 0.25% C) is very ductile and is used in such products as steel cans, wire, nails, and sheet metal for automobile bodies. *Medium-carbon steel* (0.25 to 0.7% C) is less ductile and much stronger, it is used in manufacturing railroad rails, machine axles, and structural materials such as beams and girders. *High-carbon steel* (0.7 to 1.5% C) is very hard; springs, razor blades, surgical instruments, and tools contain such steel.

Alloy steels contain small amounts of other metals in addition to iron. Particular metals improve specific characteristics of the steel. Chromium improves hardness and resistance to corrosion. Molybdenum and tungsten increase heat resistance. Nickel increases toughness. Vanadium adds springiness. Manganese improves resistance to wear.

STAINLESS STEEL

The best-known alloy steel is stainless steel. This corrosion-resistant steel contains 14 to 18% chromium and 7 to 9% nickel. Table 14.4 lists the composition and properties of selected alloy steels.

TABLE 14.4
Composition and Properties of Selected Alloy Steels

Name	Composition	Characteristic Properties	Uses
Manganese steel	10–18% Mn	hard, tough, resistant to wear	railroad rails, safes, armor plate, rock-crushing machinery
Silicon steel	1–5% Si	hard, strong, highly magnetic	magnets
Duriron	12–15% Si	resistant to corrosion, acids	pipes, kettles, condensers, etc.
Invar	36% Ni	low coefficient of expansion	meter scales, measuring tapes, pendulum rods
Chrome–vanadium steel	1–10% Cr, 0.15% V	strong, adds springiness	axles
Stainless steel	14–18% Cr, 7–9% Ni	resistant to corrosion	cutlery, instruments
Permalloy	78% Ni	high magnetic susceptibility	ocean cables
High-speed steels	14–20% W or 6–12% Mo	retain temper at high temperatures	high-speed cutting tools
Nickel steel	2–4% Ni	hard and elastic, resistant to corrosion	drive, shafts, gears, cables

Corrosion of Iron and Steel

One of the biggest problems related to the use of iron and steel is *corrosion*. **Corrosion** *is the deterioration of a metal as a result of its reaction with the environment.* In corrosion processes, metals are converted to compounds of the metal. The most significant corrosion problem of iron and steel products is *rusting,* a process in which iron is changed into a hydrated oxide of iron. Rusting is a very serious problem economically. It has been estimated that as much as one-seventh of the annual production of iron simply replaces that lost by rusting.

IRON RUST

The rusting process requires the presence of both moisture and oxygen. Iron does not rust in dry air or in water from which all dissolved oxygen has been removed. Rust forms in a stepwise fashion. The iron metal is first converted to $Fe(OH)_2$, then to $Fe(OH)_3$, and finally to rust, the hydrated oxide $Fe_2O_3 \cdot H_2O$. This process can be represented by the following set of equations.

$$2\ Fe(s) + O_2(g) + 2\ H_2O(l) \longrightarrow 2\ Fe(OH)_2(s)$$

$$4\ Fe(OH)_2(s) + O_2(g) + 2\ H_2O(l) \longrightarrow 4\ Fe(OH)_3(s)$$

$$2\ Fe(OH)_3(s) \longrightarrow Fe_2O_3 \cdot H_2O + 2\ H_2O(l)$$

The overall reaction, which is the sum of these three reactions, is

$$4\ Fe(s) + 3\ O_2(g) + 2\ H_2O(l) \longrightarrow 2\ Fe_2O_3 \cdot H_2O(s)$$

The rust so produced, a reddish brown solid, does not adhere to the surface of the metal and protect it from further reaction, but peels off. This exposes a fresh surface of iron, and the rusting process occurs again.

A number of methods are used to protect iron and steel from rusting. The most obvious approach is to coat the metal surface with paint. Steel used in bridges and buildings is often painted with red lead paint, which contains Pb_3O_4. However, rust forms under the paint layer if the paint is scratched, pitted, or dented to expose even small areas of bare metal.

Another approach to rust-prevention involves exposing the iron or steel to superheated steam. This produces an adherent oxide coating of Fe_3O_4.

$$3\ Fe(s) + 4\ H_2O(g) \longrightarrow Fe_3O_4(s) + 4\ H_2(g)$$

GALVANIZED IRON AND STEEL

A third method of protecting iron and steel is to coat it with a layer of another metal. The metal most commonly used is zinc; the process is called *galvanizing.* Approximately 45% of the annual production of zinc metal is used to galvanize iron and steel products. Even if a break occurs in the zinc and exposes the iron underneath, no rusting takes place as long as the zinc remains in contact with the iron. The zinc is more easily oxidized than the iron and corrodes while the iron remains unchanged.

The use of alloying metals in the production of steel also greatly reduces the tendency of iron to rust. Chromium is particularly effective. In stainless steel, which contains both chromium and nickel, the layer of chromium oxide that forms on the surface of the steel protects it from corrosion.

14.3 Copper

Copper is one of the few elements that is used mainly in pure form rather than as an alloy or compound. It is also the only metallic element besides gold to have a natural color other than "gray." A reddish yellow, ductile, malleable metal, it is the third most used metal in the United States, exceeded in production amounts only by iron and aluminum (see Table 14.2). Copper's primary use is as a medium for the conduction of electricity. Silver is the only metal that is a better electrical conductor than copper, but the price of silver precludes its widespread use. Because it is soft and very ductile, copper is easily drawn out into the thin wires necessary for electrical wiring.

Production of Copper

Presently, most copper (88%) is obtained by open-pit mining of low-grade ore containing only 1% or less copper, which is present as copper sulfides (Cu_2S and $CuFeS_2$). The process of obtaining copper from its ore is more complicated than that for iron (see Section 14.2). Two major complicating factors are (1) the low grade of the ore and (2) the presence of sizable amounts of iron in the ore; in the compound $CuFeS_2$, copper and iron are present in a one-to-one ratio.

The metallurgy of copper consists of five steps: (1) concentration, (2) roasting, (3) smelting, (4) converting, and (5) refining.

Copper ore, in a finely crushed form, is first concentrated by *froth flotation*. The pulverized ore is mixed with water, which contains specially selected oils (frothing agents). A froth, or foam, is produced by blowing air through the mixture while it is vigorously stirred. The sulfide ore particles become coated with oil and are carried to the top of the container in the froth, where they are recovered. The sand, rock, and clay particles are wet by the water and sink to the bottom of the tank. Copper sulfide particles are not wet by water—hence, the separation. The fraction of copper in the resulting concentrate is 20 to 40%. Figure 14.6 is a diagrammatic illustration of the operation of a froth flotation unit.

After concentration, the ore is next *roasted* at a temperature below its melting point. This operation drives off moisture and also converts some iron sulfide to iron oxide, but leaves the copper as the sulfide Cu_2S.

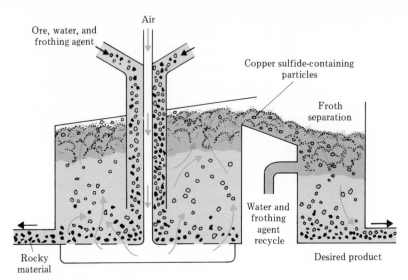

Ore, water, and
frothing agent

Air

Copper sulfide-containing
particles

Froth
separation

Water and
frothing
agent
recycle

Rocky
material

Desired product

FIGURE 14.6

Low-grade sulfide ores can be concentrated by froth flotation. The sulfide particles
have an affinity for the frothing agents, whereas the sand, rock, and clay particles
have an affinity for water.

$$2\ CuFeS_2(s) + 4\ O_2(g) \longrightarrow Cu_2S(s) + 2\ FeO(s) + 3\ SO_2(g)$$

Sulfides of iron are more easily converted to oxides than are the sulfides of
copper.

The roasted ore (FeO and Cu_2S) is transferred to a *smelting* furnace, where
temperatures are greater than the melting point of the ore. Limestone and silica
are added as fluxing agents. As the charge melts, two layers of molten materials
are produced. The bottom layer is *copper matte*, which is primarily molten
copper sulfide (Cu_2S). The top layer is silicate slag formed from the reaction of
CaO, SiO_2, and FeO.

$$CaO(s) + SiO_2(s) \longrightarrow CaSiO_3(l)$$

$$FeO(s) + SiO_2(s) \longrightarrow FeSiO_3(l)$$

The slag is drawn off.

The copper matte is transferred to another furnace, the *converter*. Additional
silicate materials are added. When air is blown through the molten material in
the converter, two reaction sequences (one involving iron and the other
involving copper) occur. Any remaining FeS is converted to oxide and then to

slag, which is removed. The Cu_2S present begins reacting with the O_2 from the air to produce Cu_2O.

$$2\,Cu_2S(l) + 3\,O_2(g) \longrightarrow 2\,Cu_2O(l) + 2\,SO_2(g)$$

As the Cu_2O is produced, it immediately reacts with remaining Cu_2S to produce metallic copper.

$$2\,Cu_2O(l) + Cu_2S(l) \longrightarrow 6\,Cu(l) + SO_2(g)$$

The copper produced is crude or "blister" copper, which is 98 to 99% copper. The term *blister copper* draws attention to the characteristic appearance of the solidified copper, which contains trapped bubbles of air and SO_2.

Blister copper can be used directly when high purity is not required, such as in copper plumbing parts. For electrical applications purification by electrolysis is necessary. Copper for this purpose must be very pure (99.95%) because even small amounts of impurities considerably reduce the electrical conductivity of copper.

The purification of copper through electrolysis is a process we discussed previously (see Section 10.12). As was shown in Figure 10.14, both anodes and cathodes in the electrolysis cell are made of copper. The impure blister copper, cast into large slabs, serves as anodes and small pieces of pure copper serve as cathodes. The net effect of the electrolysis is to transfer copper metal from the impure blister electrodes to the pure copper electrodes.

As the copper is transferred from the one electrode to the other through the electrolytic solution, metallic impurities in the blister copper that are less reactive than copper (which include trace amounts of gold and silver) do not dissolve but rather fall to the bottom of the electrolysis cell, where they collect as piles of sludge (see Figure 10.14). Because of the large amount of copper refined, these sludge piles are not "incidentals" but rather are very valuable by-products of the refining process. In 1985, 3% of United States gold production and 25% of United States silver production came from copper refining. It is the value of these by-products that offsets the high cost of the electrical power needed for electrolysis.

We should also note that each year an increasing amount of copper production comes from secondary sources (scrap copper) rather than primary sources (copper ore). The recycling of copper-containing materials is an important industry. At present, approximately one-third of copper production is recycled copper.

Gold and silver are welcome by-products of copper processing operations. An unwanted by-product is also produced—the gas sulfur dioxide (SO_2), which is a noxious air pollutant. Note the presence of SO_2 in most of the equations in this section. The release of SO_2 to the atmosphere must be controlled. Billions of dollars have been spent by the copper industry attempting to deal with this problem. Elaborate air pollution control procedures are now in effect. SO_2 as an air pollutant is discussed further in Section 20.8.

Uses of Copper

Electrical uses account for approximately 65% of copper demand. Another 18% is used in plumbing materials and other aspects of the construction industry. The third important use of copper is in the manufacture of alloys.

BRASS AND BRONZE   Two of the best-known alloys are copper alloys: brass and bronze. *Brass* is an alloy of copper containing 10 to 33% zinc. Because of its luster and attractive appearance, brass is often used as an ornamental metal in doorknobs, plumbing, hardware, and lighting. The corrosion resistance of brass is quite poor. *Bronze* is an alloy of copper containing 1 to 18% tin. It is much harder than pure copper and is employed in machine parts that require a combination of corrosion resistance and ability to withstand mechanical wear. It melts at a lower temperature than copper.

COINAGE   Copper and copper alloys have been, and still are, used as coinage metal in
FORMULATIONS   the United States. In recent years many changes in such formulations have occurred as the result of changes in metal prices. The five-cent ("nickel") coin contains 75% copper and 25% nickel. A "copper" penny, on the other hand (since 1982), is 97.5% zinc and 2.5% copper (copper plated to a zinc core). Current 10-, 25-, and 50-cent pieces are three-layer composites. The outer claddings are 75% Cu and 25% Ni, and the core is 100% Cu. Older United States "silver" coins, which have now virtually disappeared from circulation, contained 90% silver and 10% copper; the copper gave hardness and durability to the resulting silver–copper alloy.

Corrosion of Copper

Copper resists corrosion because of the formation of a protective coating. In moist air, copper first turns a dull brown, because of the formation of a thin adherent film of oxide (CuO) or sulfide (CuS). Prolonged weathering results in the formation of a green film composed of basic copper carbonate (patina). This is the familiar green coating found on many statues and buildings. The reaction is

$$2\ Cu(s) + O_2(g) + CO_2(g) + H_2O(l) \longrightarrow Cu(OH)_2 \cdot CuCO_3(s)$$

Green coating

ANTIQUING OF COPPER   The deliberate production of coatings such as those just mentioned, *called antiqueing*, is a technique used in interior decorating. The motive may be that it is difficult to prevent copper surfaces from undergoing such reactions. Often a copper surface is thinly coated with plastic to protect it. A favorite place for copper articles, however, is near fireplaces, where the plastic can be damaged by heat—or it might be removed by a zealous housekeeper with vigorous cleaning.

The surface then tends to react and antique itself. If an interior decorator or architect can convince a homeowner that artificial antiqueing is really attractive and modern (plus charge a little extra for it), many problems are solved and everyone is made happy.

14.4   Gold

Gold, a soft yellow metal, is the most malleable and ductile of all metals. It occurs naturally in elemental (native) form and was one of the first metals to be used by humans. Today, it is valued as a medium of exchange and used as a basis for monetary systems. In its earliest applications, it was valued more for its beauty—color and luster—and its complete resistance to atmospheric corrosion.

Traditionally, gold has been recovered from river sands using methods such as "panning," which depend on the high density of gold (19.3 g/cm^3) compared with that of sand (2.5 g/cm^3), but such sources have now largely been worked out. Gold production has also traditionally been a major by-product of copper production (Section 14.3). The high price of gold as well as a depressed copper market has, in recent years, decreased this dependence on copper sources markedly. (In 1979, 42% of United States gold production was derived as a by-product from copper; in 1985 this percentage had dropped to 3%.) Today, mining of low-grade gold ores (25 ppm by mass of gold) is the main source (91% in 1985) of gold production. In these ores gold is present as small particles in quartz or other silicates. The current high price of gold makes extraction from such low-grade ore economically feasible.

A hydrometallurgy process called *cyanidation* is used to recover gold from rocks containing it. The rock is crushed to a fine powder to liberate the gold particles. These are then leached (separated) from the powder with an aerated dilute solution of sodium cyanide. The action of the O_2 and CN^- ion present in the solution solubilizes the gold in the form of a complex ion, $Au(CN)_2^-$.

$$4 \, Au(s) + 8 \, CN^-(aq) + O_2(g) + 2 \, H_2O(l) \longrightarrow 4 \, Au(CN)_2^-(aq) + 4 \, OH^-$$

The gold is then recovered from the solution through a reaction involving zinc metal.

$$2 \, Au(CN)_2^-(aq) + Zn(s) \longrightarrow 2 \, Au(s) + Zn(CN)_4^{2-}(aq)$$

The production of refined gold from old scrap each year exceeds that obtained from ores by a 2 to 1 margin. Again, high gold prices encourage the recycling of gold-containing objects.

TABLE 14.5

Composition and Colors of Various 14-Karat Gold Alloys Used in Jewelry
(*in mass percent*)

Alloy Ingredients	Alloy Color			
	White	Yellow	Red	Green
Gold	58.33	58.33	58.33	58.33
Copper	17.00	25.58	27.96	6.50
Silver		13.33	6.54	35.00
Zinc	7.67	2.76	7.17	0.17
Nickel	17.00			

Excluding monetary bullion, approximately half (53%) of the current demand for gold is in the area of jewelry and arts. One-third of demand (34%) is for industrial uses including electronics (corrosion-free contacts), aerospace applications (heat reflectors), and special alloys. Dental applications account for most of the remaining 13% of production. Only a small amount (less than 0.3%) of gold production is used in items for investment (fabricated bars, medallions, and coins).

14-KARAT GOLD
JEWELRY

Pure gold is too soft for normal use in jewelry, so it is alloyed with copper, silver, and other metals with a resulting increase in hardness. The amount of gold in such alloys is specified in terms of *karats*. (This is a different unit from the carat used to specify the mass of diamonds; see Section 13.4.) Pure gold is 24-karat gold and the percentages of gold in alloys are compared with this value. Thus, 18-karat gold is an alloy with 18 out of 24 parts (by mass) gold or 75% gold. A 12-karat gold alloy would be 50% gold (12 parts out of 24). In the United States, most gold jewelry is 14-karat gold. The metals used to form gold alloys vary, depending on the color and other properties that are desired. Table 14.5 lists the composition of a few typical 14-karat gold alloys. Coinage gold is 22-karat gold.

GOLD LEAF

Because of gold's high malleability it can be beaten into extraordinarily thin sheets called *gold leaf*. Gold leaf is used in decorating buildings, edging books, and making decorative lettering. Leaf can be so thin that 300,000 thicknesses make a pile only 1 inch deep.

14.5 Silver

Silver, a white metal, has the highest electrical and thermal conductivity of any metal. This element is more malleable and ductile than any other metal except gold. Silver has a unique reflectivity. When polished, it has a brilliant luster and reflects back 95% of the light falling on it.

Silver is not attacked by oxygen in the air at ordinary temperatures, but it does tarnish quickly in the presence of sulfur-containing air pollutants such as hydrogen sulfide (H_2S) and sulfur-containing foods such as eggs and mustard. The tarnish is a thin layer of black silver sulfide (Ag_2S). The equation for its formation from H_2S, in the presence of air, is

$$4\,Ag(s) + 2\,H_2S(g) + O_2(g) \longrightarrow 2\,Ag_2S(s) + 2\,H_2O(g)$$

Many people remove silver tarnish with a scouring powder or abrasive cream and a lot of physical effort (elbow grease). Such an approach permanently removes the silver in the tarnish. In time, the silver on a silver-plated object can be worn away, exposing the underlying base metal. A much better procedure is to change the tarnish back to silver using a chemical reaction; this way, no silver is lost. To remove silverware tarnish chemically, the tarnished silver or silver-plated object is immersed in a solution of a teaspoonful of salt and a teaspoonful of baking soda in a liter (quart) of water, which is contained in an aluminum container (pan). It is important that the container is made of aluminum, a metal that is more active than silver. The sulfide tarnish slowly dissolves in the solution, producing Ag^+ ions and S^{2-} ions. Aluminum metal (from the container) replaces the silver ions in solution, which then plate back out on the silver object as neutral silver atoms.

$$3\,Ag^+(aq) + Al(s) \longrightarrow 3\,Ag(s) + Al^{3+}(aq)$$

Silver is frequently alloyed with other metals, often copper, to increase its hardness and wear resistance. *Sterling silver* is an alloy containing 92.5% silver and 7.5% copper. This silver is said to be 925 fine (925 parts Ag per 1000 parts alloy, or 92.5%). American coinage silver, used in United States coins before 1965, is 900 fine, containing 90% Ag and 10% Cu.

Production of Silver

Silver is found in both the native and combined states, although the native state is now quite rare. It is a minor component, as Ag_2S, of most sulfide ores, in mixture with larger amounts of the sulfides of lead, copper, and zinc. For many years significant amounts of silver were obtained as a by-product of the mining of these other metals. The price increase for silver that accompanied gold's dramatic price increase in the late 1970s again made the refining of low-grade silver ores economically feasible. Silver ores are easily smelted in the presence of carbon to produce silver. A typical reaction is

$$Ag_2S(s) + C(s) + 2\,O_2(g) \longrightarrow 2\,Ag(s) + CO_2(g) + SO_2(g)$$

In 1985, 60% of "new silver" came from silver ore; next was 25% from copper ore containing small amounts of silver. As with gold, recycled silver contributes heavily (35%) to annual consumption.

When silver is obtained as a by-product from another ore, the silver follows the dominant metal (lead, copper, or zinc) through the concentration and smelting processes and then is separated out by procedures determined by the identity of the dominant metal present. The silver in copper ores is recovered from the anode sludge that forms as the copper is purified electrolytically (Section 14.3). When silver is part of a lead ore, it is extracted with zinc. Silver is 3000 times more soluble in zinc than lead is. The lead (with small amounts of silver in it) is melted and mixed with a small amount of zinc. Lead and zinc do not dissolve in each other. Most of the silver leaves the lead and dissolves in the zinc, and the more volatile zinc is separated from the silver by distillation.

Uses of Silver

Photography accounts for almost half (49%) of annual silver production. Silver is important in photographic applications because certain silver salts undergo change when exposed to light. We will discuss this chemistry later in this section. Electrical and electronic products account for one-fourth (23%) of silver production. Electrical applications reflect silver's great ability—the best of any metal—to conduct electricity. Other uses for silver include brazing alloys and solders (5%), jewelry (5%), electroplated ware (3%), sterling ware (3%), and mirrors (1%). Silver mirrors involve a thin layer of silver deposited on glass. This is accomplished by reacting a solution of silver nitrate in ammonia [silver is present as $Ag(NH_3)_2{}^+$] with a reducing agent (see Section 10.10) such as glucose or formaldehyde. The equation of the reaction is

$$Ag(NH_3)_2{}^+ + e^- \text{ (from reducing agent)} \longrightarrow \underset{\substack{\text{Silver} \\ \text{mirror}}}{Ag} + 2\,NH_3$$

Silver and Photography

Photographic films and papers contain silver in the form of silver halides (AgCl, AgBr, and AgI). Silver halides slowly decompose to produce finely divided metallic silver on exposure to light; they are photosensitive compounds. This light-induced reaction, carefully controlled, is the basis for most photographic processes. (We have already encountered this reaction, in Section 13.5; it is the basis for photochromic lenses used in some eyeglass ware.)

PHOTOGRAPHIC FILM   A black-and-white photographic film consists of a thin sheet of plastic coated with a gelatin suspension of small silver halide crystals (see Figure 14.7b). The size of the silver halide particles and the relative amounts of bromide and iodide used determine the sensitivity or "speed" of the film and its "grain" or ability to show detail when the picture is enlarged.

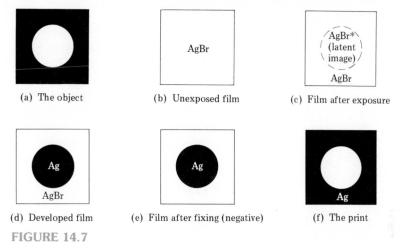

(a) The object   (b) Unexposed film   (c) Film after exposure

(d) Developed film   (e) Film after fixing (negative)   (f) The print

FIGURE 14.7

The black and white photographic process.

Upon exposure, the light reflected from an object is focused on the film surface. Light-colored objects reflect more light than darker objects. The chemical processes involved are not well understood, but apparently when light strikes a tiny silver halide crystal, a few atoms of metallic silver are produced. These atoms activate the rest of the silver halide in the crystal and make it easily reducible to free silver during the developing process. The activation reaction is represented as

$$AgX + light \longrightarrow AgX^*$$

where X represents the halogen atom in the silver halide and the asterisk refers to an activated state (see Figure 14.7c).

The film is developed by being washed in a solution (the developer), which reduces the AgX^* to metallic silver.

$$AgX^* + developing\ solution \longrightarrow Ag + X^-$$

The metallic silver produced appears black on the developed film (Figure 14.7d).

The unactivated AgX is then removed; it gradually changes to black Ag on exposure to light if allowed to remain. This process is called "fixing." The film is washed in a solution containing a chemical that dissolves only the unactivated AgX. Sodium thiosulfate ($Na_2S_2O_3$), also called "hypo," is widely used for this purpose; the thiosulfate ion, $S_2O_3^{2-}$, is the active species. The reaction is

$$AgX + 2\ S_2O_3^{2-} \longrightarrow Ag(S_2O_3)_2^{3-} + X^-$$

Both species on the right side of this equation are ions that are soluble in the water of the washing solution and are removed from the film. The developed and fixed film, called the negative, contains black metallic silver wherever light from the photographed object struck it (Figure 14.7e).

The negative is placed over a piece of paper on which a gelatin–AgX coating has been placed similar to that on the original film. Light is passed through the negative onto the paper. The dark areas of the negative, corresponding to light areas in the photographed object, block the light and prevent activation of the AgX on the paper in those regions. The exposed paper is developed and fixed in the same way as the film. The resulting dark and light areas are opposite to those on the negative and match the original photographed object (see Figure 14.7a and e).

Photographic film contains 5–15 g/m^2 of silver. A large proportion of this silver on black-and-white film and all of it on color film are removed during the film processing. Practically all of this silver is reclaimed and reenters the United States silver supply system.

14.6 Aluminum

Aluminum is the most abundant metal in the earth's crust (see Table 14.1), and it is the second most used metal (see Table 14.2). Chemically too active to occur as the free element in nature, it is usually found combined with oxygen either as an oxide or in silicates.

Pure aluminum is a silvery white metal with many desirable properties: it is light, and nontoxic, and it has excellent corrosion resistance and high thermal and electrical conductivities. Freshly cut aluminum appears silvery. Its surface quickly changes to a dull white; a thin film of aluminum oxide (Al_2O_3) forms as a result of atmospheric oxidation. This tenacious oxide coating prevents further corrosion. When household aluminum objects are "cleaned" with scouring pads or abrasive chemicals, the oxide coating is usually removed, giving the aluminum a shinier appearance. The cleaning, however, is in vain; a new oxide coating quickly forms.

CLEANING OF
ALUMINUM OBJECTS

The electrical conductivity of aluminum is 63.5% that of an equal volume of copper. However, when aluminum's lower density is considered, its conductivity is 2.1 times that of copper on a pound-for-pound basis. Ninety percent of all overhead electrical transmission lines in the United States are aluminum alloys.

Occurrence of Aluminum

A large number of aluminum compounds are known; those in greatest abundance are complex aluminum silicates. At present, no method is available for the economical extraction of aluminum from silicate materials. The source of

TABLE 14.6

Characteristics of Al_2O_3-Based Gemstones

Name of Gemstone	Transition Metal Impurity
Ruby	Chromium
White sapphire	None
Blue sapphire	Iron, titanium
Green sapphire	Cobalt
Yellow sapphire	Nickel, magnesium

almost all aluminum is the mineral *bauxite*, a hydrated aluminum oxide. Bauxite is a mixture of oxides of aluminum (Al_2O_3), iron (Fe_3O_4), and silicon (SiO_2), with 35 to 60% Al_2O_3 generally present.

Aluminum oxide is also found in nature in pure form as the mineral *corundum*. It is used as an abrasive in grinding and polishing. Another related abrasive is *emery*, a granular form of corundum contaminated with iron oxide (Fe_3O_4) and silica.

Several familiar gemstones are aluminum oxide crystals naturally colored with transition-metal impurities. The most familiar of these are the gemstones we call sapphires and rubies. Table 14.6 gives the color and transition metal impurities associated with gemstones based on Al_2O_3.

RUBIES AND SAPPHIRES

Artificial rubies and sapphires are now industrially made on a large scale by melting Al_2O_3 with small amounts of transition metal oxides (to produce the desired color) and then cooling the melt in such a way as to produce large crystals. These synthetic gemstones are indistinguishable from natural ones, except under microscopic examination. They are used as jewelry and also as bearings ("jewels") in good watches and instruments.

Another aluminum-containing gemstone is the emerald, which was discussed in Section 13.5.

Metallurgy of Aluminum

The process used to obtain aluminum from bauxite is the *Hall–Heroult process*, developed simultaneously, but independently, in 1886 by the American chemist Charles Martin Hall (1863–1914) (see Historical Profile 16) and the French chemist Paul Heroult (1863–1914). (Coincidentally, these two chemists were born in the same year and died in the same year.)

The aluminum ore (bauxite) must first be purified by removing most of the silicon dioxide and iron oxide normally present. This is accomplished by heating dried, pulverized ore with sodium hydroxide (NaOH) solution under steam pressure. The Al_2O_3 dissolves in the NaOH solution to give aluminate ion $[Al(OH)_4{}^-]$; the iron oxides and silica are insoluble and can be filtered from the solution.

Charles Martin Hall   1863–1914

[The Bettmann Archive.]

Charles Martin Hall, the son of a Protestant clergyman, was born in Ohio and raised in Oberlin, Ohio, from the age of 10. As a youngster he was intensely interested in chemistry.

Work on the process he invented, which we use today to produce aluminum metal, began while he was still a student at Oberlin College. His interest in the project was stimulated by a chance remark by his chemistry professor, "Anyone discovering a cheap way of making aluminum will grow rich and famous." (At the time aluminum cost $90 a pound and was more expensive than silver or gold. Napoleon III of France flaunted his wealth by dining with aluminum cutlery and his baby's rattle was made of aluminum.)

The then 21-year-old Hall was heard to say, "I'm going after that metal." His discovery that bauxite ore would dissolve in molten cryolite from which aluminum could be recovered by electrolysis was made in his laboratory, a woodshed near his home. His equipment was either homemade or borrowed. The container for the molten cryolite was an iron frypan; his heat source was a blacksmith's forge; and his electrical source consisted of homemade batteries made with fruit jars obtained from his mother.

Six months after his graduation in 1886, at age 22, he visited his former chemistry professor, and holding out a dozen silvery globules in the hollow of his hand, said "Professor, I've got it."

Two years later he founded a small company to produce aluminum. This later became the Aluminum Company of America. Hall founded the aluminum industry in the United States.

$$Al_2O_3(s) + 2\ OH^-(aq) + 3\ H_2O(l) \longrightarrow 2\ Al(OH)_4^-(aq)$$

The solution is then cooled and diluted with water or slightly acidified, which causes the aluminum in solution to precipitate as aluminum hydroxide [$Al(OH)_3$].

$$Al(OH)_4^-(aq) + H_3O^+(aq) \longrightarrow Al(OH)_3(s) + 2\ H_2O(l)$$

Heating the hydroxide drives off water, producing purified Al_2O_3.

$$2\ Al(OH)_3(s) \longrightarrow Al_2O_3(s) + 3\ H_2O(l)$$

The aluminum oxide is then dissolved in molten cryolite (1000°C). Cryolite is a rare naturally occurring aluminum mineral with the formula Na_3AlF_6. (The discovery that cryolite is a good solvent for Al_2O_3 was the "key" step in Hall's

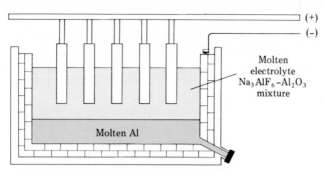

FIGURE 14.8

A diagram of the electroysis cell used to produce aluminum by the Hall–Heroult process.

and Heroult's development of their extraction processes.) Today, the demand for cryolite cannot be met from natural sources; it must be produced synthetically.

With the aluminum in solution, electrolysis is used to release it from its compound state. A molten or liquid state is a prerequisite for electrolysis; such states are necessary for the conduction of electricity. Using molten aluminum oxide itself is not practical; its melting point is too high (2050°C). A Hall–Heroult electrolysis cell is shown in Figure 14.8. The electrodes are made of graphite. The overall cell reaction may be written as

$$2\ Al_2O_3(l) \longrightarrow 4\ Al(l) + 3\ O_2(g)$$

with aluminum being produced at the anodes. Molten aluminum is denser than molten cryolite; thus, the molten aluminum produced sinks to the bottom of the cell, from which it can be tapped from time to time and run into molds.

The production of aluminum is a very energy-intensive process. Ten times more energy is required to produce a ton of aluminum than a ton of steel. The high energy requirements explain the current demand for recycled aluminum products such as aluminum beverage cans. Recycling aluminum materials requires 95% less energy than is needed to produce aluminum metal from ore. At present almost one-third of the annual aluminum production comes from recycled materials, and the amount is increasing. In 1972, 15% of aluminum beverage cans (1.3 billion cans) were recycled, and 52% (28.7 billion cans) in 1983.

Uses of Aluminum

The container and packaging industry is the largest consumer of aluminum, accounting for 26% of total end-use shipments. Eighty percent of this end use involves the manufacture of aluminum beverage cans. The transportation sector and the building and construction industry each account for 19% of aluminum consumption, ranking close seconds.

It is difficult to envision how commercial aircraft could have been developed without a plentiful supply of cheap aluminum. Magnalium, a lightweight alloy of aluminum (70 to 90%) and magnesium (10 to 30%), is used extensively in aircraft bodies; it is nearly as strong as steel. Half of the aluminum used in transportation is used in passenger cars. The aluminum content of domestically produced cars increased from 84 lb in 1975 to 137 lb in 1984.

In building and construction aluminum is found in many everyday objects including doors, windows, siding, awnings, and heating and ventilating components.

Electrical uses account for 10% of aluminum production and consumer durables for 7%. Consumer durables include appliances, cooking utensils, hardware, and lawn and patio furniture. Aluminum's light weight, its corrosion resistance, and its ability to conduct heat make it ideal for use in cooking utensils.

Practice Questions and Problems

METALLIC ELEMENTS

14-1 Only five metals have abundances greater than 1.0 atom percent. Identify these five metals.

14-2 In Chapter 13 the abundances of nonmetals were given as the number of nonmetal atoms per 1 million total atoms. Prepare a similar listing for the most abundant metals (greater than 0.01 atom percent).

14-3 List the dominant ore form for each of the metals listed in Table 14.1.

14-4 Table 14.1 lists the most abundant metals and Table 14.2 lists the most used metals. Which metals are found in both lists?

14-5 Certain metals, considered to be of strategic importance, must be imported into the United States because the United States lacks domestic ores of these metals. List four such metals and why each is considered to be strategically important.

IRON AND STEEL

14-6 What is an ore?

14-7 What is an alloy?

14-8 What is steel?

14-9 What are the two major operations in the production of steel from iron ore?

14-10 What is the *charge* used in a blast furnace?

14-11 Describe the function of each of the following in the manufacture of pig iron.
a. Coke   b. CO   c. Limestone   d. CaO

14-12 What is slag? Why is it purposely formed in a blast furnace?

14-13 What is the active reducing agent in a blast furnace and how is it chemically generated within the furnace?

14-14 Write equations for the chemical reactions in a blast furnace that involve the following.
a. Coke   b. Carbon dioxide
c. Carbon monoxide   d. Limestone
e. Calcium oxide   f. Iron oxides

14-15 What is the difference between pig iron and cast iron?

14-16 Describe how pig iron is converted to steel using the basic oxygen process.

14-17 What are the two major types of steel?

14-18 Describe the physical properties and chemical composition of the following types of steel.

a. Mild steel   b. Medium-carbon steel
c. High-carbon steel   d. Stainless steel
e. Manganese steel   f. High-speed steel

14-19 Define corrosion.

14-20 What is the composition of iron rust, and what are the steps in its formation?

14-21 Define galvanizing and state its purpose.

COPPER

14-22 What are the two "complicating factors" in copper metallurgy?

14-23 Describe how copper ore is concentrated using froth flotation.

14-24 What is accomplished when concentrated copper ore is roasted?

14-25 Describe the chemistry involved in the smelting phase of copper production.

14-26 What is copper matte?

14-27 What is blister copper?

14-28 Describe the purification of blister copper using electrolysis.

14-29 Why must copper be free of impurities when used as an electrical conductor?

14-30 Describe how gold and silver are obtained as by-products of the production of copper.

14-31 What is the difference between brass and bronze?

14-32 Explain the chemistry involved in the formation of the familiar "green coating" found on many copper-covered statues and buildings?

14-33 What are the three major uses for copper?

GOLD

14-34 Describe, with chemical equations, how gold is recovered from gold-containing rocks by the cyanidation process.

14-35 What are the three major uses for gold, excluding monetary bullion.

14-36 What is the difference in gold content of 24-karat, 18-karat, and 14-karat gold material?

14-37 What is the difference between 14-karat white gold and 14-karat yellow gold?

14-38 What is gold leaf?

14-39 What is the difference between a diamond carat and a gold-alloy karat?

SILVER

14-40 What silver compound is responsible for the dark color of tarnished silver?

14-41 What is the silver content of each of the following materials?
a. Sterling silver   b. American coinage silver
c. 850 fine silver

14-42 Describe the Parkes process for extracting silver from a lead ore.

14-43 What are the two major uses for silver?

14-44 Briefly describe the steps involved in the production of a black and white photograph, from the viewpoint of the chemistry of silver and its compounds.

ALUMINUM

14-45 What is the relationship between aluminum oxide (Al_2O_3) and the following materials?
a. Bauxite   b. Corundum   c. Emery
d. Ruby   e. Green sapphire

14-46 What is the difference between a blue sapphire and a yellow sapphire?

14-47 What is the chemical composition of the "jewels" found in expensive watches and instruments?

14-48 Explain why aluminum, the most abundant of all metals, was not commonly used until the beginning of this century.

14-49 What is the purpose or function of each of the following substances in the Hall–Heroult process for obtaining aluminum?
a. Sodium hydroxide solution
b. Molten cryolite

14-50 In the electrolysis of a molten Al_2O_3–cryolite mixture, how is the molten Al produced separated from the cryolite, which is also molten?

14-51 Explain the chemistry involved in the corrosion resistance of aluminum.

14-52 What is the major "technological reason" for the recycling of aluminum beverage cans?

14-53 What amount of annual aluminum production comes from recycled materials?

14-54 What are the three major end uses for aluminum production?

14-55 What is the chemical composition of magnalium, an aluminum alloy used in the construction of aircraft bodies?

15

Nuclear Chemistry

CHAPTER HIGHLIGHTS

15.1   Nuclear Reactions

It was once thought that chemical reactions—the rearrangement of atoms to form new compounds—were the only type of reaction possible for atoms. We now know that this is not the case. Nuclear reactions are also possible; in these reactions changes occur in the tiny nuclear centers of atoms.

In this chapter we examine the fundamentals of nuclear reactions and look at their numerous important applications. Radioactivity, radiation exposure, nuclear weapons, nuclear power plants, and nuclear medicine all fall under the umbrella of nuclear reactions.

Historians describe our time as a "nuclear age." What they mean by this phrase is that this is an age of nuclear reactions (nuclear change). Nuclear change has been a somewhat controversial subject in recent years. Problems with nuclear power plants and concerns about nuclear warfare have led to considerable negative press. However, there is also a positive side to the nuclear age—nuclear medicine. Today nuclear change is routinely used in the diagnosis and treatment of numerous diseases. Diseases once regarded as incurable can now be treated effectively using nuclear medicine. In considering the pros and cons of nuclear reactions, it has been said that "it is far more likely you will be saved by nuclear medicine than killed by nuclear weapons."

A **nuclear reaction** *is a reaction in which change occurs in the nucleus of an atom.* Nuclear reactions are not considered to be ordinary chemical reactions. All nuclei remain unchanged in an ordinary chemical reaction. The governing principles for ordinary chemical reactions deal with the rearrangement of electrons; this rearrangement occurs as the result of electron transfer or electron sharing (Section 5.1). In nuclear reactions, it is nuclei rather than electrons that are the point of focus.

A brief review of what we already know about atomic nuclei, as well as some new material concerning them, will serve as the starting point for our discussion of nuclear reactions. Atomic nuclei are the very dense, positively charged centers of atoms about which electrons move. All nuclei of the atoms of a given element contain the same number of protons. It is this characteristic number of protons that determines the identity of the element. The *atomic number* of an atom is the number of protons in its nucleus. The number of neutrons associated with the nuclei of a given element can vary within a small range. Atoms of a given element that differ in their numbers of neutrons are called *isotopes*. The *mass number* of an atom is equal to the total number of protons and neutrons present in the nucleus. Isotopes of an element have different mass numbers, but the same atomic number. Refer back to Sections 3.5 and 3.6 for further details about atomic numbers, mass numbers, and isotopes.

The term *nuclide* is used extensively in discussions about nuclear reactions. A **nuclide** *is an atom of an element that has a specific number of protons and neutrons in its nucleus.* All atoms of a given nuclide must have the same number of protons and the same number of neutrons. The term *isotope* refers to different forms of the same element. Nuclide describes atomic forms of different elements. The species $^{12}_{6}C$ and $^{13}_{6}C$ are isotopes of the element carbon. The species $^{12}_{6}C$ and $^{16}_{8}O$ are nuclides of different elements.

To uniquely identify a nucleus (or atom), both the atomic number and mass number must be specified. Two notation systems do this. Consider a nuclide of nitrogen that has seven protons and eight neutrons. This nuclide can be denoted as $^{15}_{7}N$ or as nitrogen-15. In the first notation, the superscript is the mass number and the subscript is the atomic number. In the second notation, the mass number is placed immediately after the name of the element. An advantage of the first notation is that the atomic number is shown; a disadvantage is the need for super- and subscripts. Both types of notation are used in this chapter. Note that both notations give the mass number.

Some naturally occurring nuclides, as well as all synthetically produced nuclides (Section 15.5), possess nuclei that are *unstable*. To achieve stability, these unstable nuclei spontaneously emit energy (radiation). A **radioactive nuclide** *is an atom (nuclide) that possesses an unstable nucleus and spontaneously emits energy (radiation).* This term is often shortened to *radionuclide*.

15.2 The Discovery of Radioactivity

The definition for a radioactive nuclide given in the last paragraph of Section 15.1 contains the two key concepts necessary for understanding the phenomenon of radioactivity.

1. Certain nuclides possess unstable nuclei.
2. Nuclides with unstable nuclei spontaneously emit energy (radiation).

Of the approximately 340 naturally occurring nuclides about 70 possess unstable nuclei—that is, are radioactive. In addition, numerous unstable nuclides that are not found in nature have been produced in the laboratory (Section 15.5).

The fact that certain naturally occurring nuclides are radioactive was accidentally discovered by the French physicist and engineer Antoine Henri Becquerel (1852–1908) (see Historical Profile 17) while he was studying certain minerals called phosphors, which glow in the dark (phosphoresce) after exposure to radiation such as sunlight or ultraviolet light. The rays emitted by these phosphorescing minerals, like visible light, darken a photographic plate. One day in 1896, while working with a uranium ore sample that phosphoresced in the normal manner, Becquerel was called away. As he left, he inadvertently placed the uranium ore sample on top of an unexposed photographic plate packaged to protect it from light. Later it was determined that this photographic plate had been exposed by the uranium ore despite its protective wrapping. Becquerel correctly concluded that this plate exposure was due to radiation emitted by the uranium ore without external stimulus. As a result of this incident, Becquerel is credited with the discovery of the phenomenon we now call radioactivity. Further studies by Becquerel showed that this "radioactivity" was not unique to the one uranium ore he had used, but was a characteristic of all uranium-containing substances even if they did not phosphoresce. (It seems strange now that this phenomenon was not detected earlier, since the element uranium had been isolated more than 100 years before Becquerel's discovery.)

Becquerel chose not to continue to work in this new field of study. Instead, he suggested to two of his colleagues, Pierre and Marie Curie (who had just recently been married), that they continue the project and find out what other substances possessed radioactive properties.

Pierre Curie (1859–1906) a physicist, and Marie Sklodowska (1859–1934), a chemist (see Historical Profile 18) conducted a systematic search of the then known elements to see how widespread this phenomenon was. The only other radioactive element they found was thorium. In further investigations, the Curies discovered a uranium ore sample that exhibited four times the radioactivity of a similar quantity of pure uranium or thorium, thus indicating the presence of a new substance more radioactive than either of these elements. From this ore the Curies were able to isolate, in 1898, two new radioactive elements; polonium, 400 times more radioactive than uranium, and radium, over a million times more radioactive than uranium. It was the Curies who coined the word radioactive to describe elements that spontaneously emit radiation.

The Curies spent the rest of their lifetimes studying radioactive phenomena. Notwithstanding the accidental death of her husband in 1906 (killed by a heavy horse-drawn wagon), Marie worked on tirelessly. For her monumental efforts, Madame Curie, has become one of the most celebrated scientists of all time.

HISTORICAL PROFILE 17

Antoine Henri Becquerel   1852–1908

Antoine Henri Becquerel was born, in Paris, into a distinguished French scientific family. Both his father and grandfather were members of the French Academy of Sciences and professors of physics. It was natural for Henri to follow in their footsteps, which he did in the most literal sense. In 1894, he was appointed to a physics chair in Paris held previously by his father and before that by his grandfather.

Henri's early researches dealt with optical phenomena. Once established in his profession, however, he chose not to remain active as a researcher. By 1896, at age 43, he held chairs in physics at three locations in Paris, but had not been active in research for a number of years.

Becquerel's inactive status as a researcher changed shortly thereafter as the result of a discovery by another physicist. In 1896, the German physicist Wilhelm Konrad Röntgen (1845–1923) announced the discovery of X-rays. This news caught Becquerel's fancy, mostly because his father had worked with phosphorescence shown by various materials. He wondered about any possible relationships between X-rays and phosphorescence. His curiosity stimulated, he again became active in research.

Within a year Becquerel discovered the phenomenon now known as radioactivity, his most important discovery. The uranium-containing material he was using at the time was one that had particularly interested his father.

In 1903, Becquerel and two of his colleagues (Pierre and Marie Curie) were awarded the Nobel prize in physics for work concerning radioactivity. Becquerel had gone from an inactive researcher to a Nobel prize winner in the period of 7 years.

15.3   The Nature of Emissions from Radioactive Materials

The first information concerning the nature of the radiation emanating from naturally radioactive materials was obtained by Ernest Rutherford (see Historical Profile 5 in Chapter 3) in the years 1898 through 1899 Using an apparatus similar to that shown in Figure 15.1, he found that if a beam of radiation is passed between electrically charged plates, it is split into three components; this

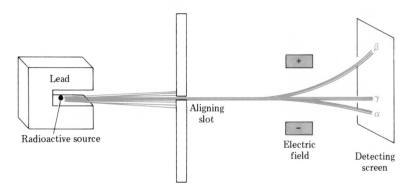

FIGURE 15.1
Effect of an electric field on radiation emanating from a naturally radioactive substance.

indicates the presence of three different types of emissions from naturally radioactive materials. A closer analysis of Rutherford's experiment reveals that one radiation component is positively charged (because it is attracted to the negative plate); a second component is negatively charged (because it is attracted to the positive plate); and the third component carries no charge (because it is unaffected by either charged plate). Rutherford chose to call the positive component alpha rays (α rays), the negative component beta rays (β rays), and the uncharged component gamma rays (γ rays). Alpha, beta, and gamma are the first three letters of the Greek alphabet. We still use these names today.

Additional research has substantiated Rutherford's conclusion that three distinct types of radiation are present in the emissions from naturally radioactive substances. This research has also supplied the necessary information for the complete characterization of each type of radiation; these complete identifications required many years of research. Early researchers in the field were hampered by the fact that many of the details concerning atomic structure were not yet known; for example, the neutron was not identified until 1932. Further information about the three types of radiation follows.

Alpha rays *consist of a stream of positively charged particles (alpha particles), each of which is made up of two protons and two neutrons.* The notation used to represent an alpha particle is $^{4}_{2}\alpha$. The numerical subscript indicates that the charge on the particle is +2 (from the two protons). The numerical superscript indicates a mass of 4 amu. On the atomic mass scale (see Section 3.7) protons and neutrons both have masses that are almost exactly equal to 1.0 amu. Thus, the total mass of an alpha particle (two protons and two neutrons) is 4.0 amu. Alpha particles are identical with the nuclei of helium-4 ($^{4}_{2}\text{He}$) atoms.

[The Bettmann Archive.]

Marie Sklodowska Curie   1859–1934

Marie Sklodowska, born in Warsaw, Poland, in 1859, was the daughter of parents formally involved with education. Her father was a physics teacher; her mother a principal of a girls' school. Times were hard for the Sklodowska family throughout Marie's younger years, mainly because of the Russian domination of Poland during this period.

Despite poverty, by 1891 Marie was able to go to France, where she remained the rest of her life. Living with the greatest frugality in Paris, she obtained additional education, and graduated at the top of her class. (Once she fainted in the classroom from hunger.)

She became involved with the field of radioactivity because she was a student of Henri Becquerel at the "right time." Her doctoral thesis dealt with radiations from uranium.

In 1894 Marie met Pierre Curie, who had just been appointed a professor of physics in Paris. They were married the next year. After their marriage the two scientists continued their research—at first, independently of each other. Marie's research proved the more productive. In 1898 Pierre abandoned his own research and started working on his wife's project as her assistant. Most of their work was carried out under miserable conditions—in a wooden shed with a leaky roof, no floor, and inadequate heat. In 1902, after 4 years of work, the couple reported the isolation of 0.1 g of radium, a new highly radioactive element they had discovered. The 0.1 g of new element was the result of processing many tons of uranium ore.

Recognition came to the Curies in 1903 with their selection as corecipients (along with Henri Becquerel) of the Nobel prize in physics. After her husband's accidental death in 1906, Marie was named to succeed him as a professor of physics, the first time a woman had occupied a professorial chair, at the Sorbonne in Paris. She continued to work, often giving up recreational and social contacts for devotion to research.

In 1911 her further work was recognized with the Nobel prize in chemistry. She is the only person to receive two Nobel prizes for scientific research. In later years Irene, one of her two daughters, worked with her as an assistant. In 1934, Marie, now respectfully called Madame Curie, died of leukemia caused by overexposure to radiation.

In 1935, one year after her mother's death, Irene Joliot-Curie (1897–1956) and her husband were awarded the Nobel prize in chemistry, the third such prize for the Curie family. The prize was in recognition of further studies concerning radioactivity, this time in the field of artificial radioactivity.

Beta rays *consist of a stream of negatively charge particles (beta particles) whose charge and mass are identical to that of an electron.* However, beta particles are not extranuclear electrons; they are particles that have been produced inside the nucleus and then ejected. We discuss how this process occurs in Section 15.4. The symbol used to represent a beta particle is $_{-1}^{0}\beta$. The numerical subscript indicates that the charge on the beta particle is -1; it is the same as that of an electron. The use of the superscript zero for the mass of a beta particle should not be interpreted as meaning that a beta particle has no mass, but rather that the mass is very close to zero amu. The actual mass of a beta particle on the atomic mass scale is 0.00055 amu.

Gamma rays *are not considered to be particles, but rather are pure energy without charge or mass.* They are very high energy radiation, somewhat like X-rays. The symbol for gamma rays is $_{0}^{0}\gamma$.

15.4 Equations for Nuclear Reactions

Alpha, beta, and gamma emissions come from the nucleus of an atom. These spontaneous emissions alter nuclei; obviously, if a nucleus loses an alpha particle (two protons and two neutrons), it will not be the same as it was before the departure of the particle. In the case of alpha and beta emissions, the nuclear alteration causes the identity of the atom to change; that is, a new element is formed. Thus, nuclear reactions differ dramatically from ordinary chemical reactions. In chemical reactions the identity of the elements is always maintained. This is not the case for nuclear reactions.

The term *decay* (or disintegration) is used to describe a nuclear process in which an element disappears; that is, an element changes into another element as a result of radiation emission. **Radioactive decay** *is the process whereby a radionuclide is transformed into an nuclide of another element as the result of the emission of radiation.*

Radioactive decay occurs only in certain ways, which are called *modes of decay*. For naturally occurring radioactive substances, the modes of decay are alpha-particle decay and beta-particle decay. Separate consideration of each mode of decay provides further insights into the nature of nuclear reactions and illustrates how equations for nuclear reactions are written.

Alpha-Particle Decay

Alpha-particle decay, which is the emission of an alpha particle from a nucleus, always results in the formation of a nuclide of a different element. The product nucleus of this type of decay has an atomic number that is 2 less than that of the original nucleus and a mass number that is 4 less than the original nucleus. We can represent alpha-particle decay in general terms by the equation

$$_Z^A X \longrightarrow {}_2^4\alpha + {}_{Z-2}^{A-4}Y$$

where X is the symbol for the nucleus of the original element undergoing decay and Y is the symbol of the element formed as a result of the decay.

To introduce actual nuclear equations, let us write equations for two alpha-particle-decay processes. Both $_{83}^{211}Bi$ and $_{92}^{238}U$ are alpha emitters; that is, they are radionuclides that undergo alpha-particle decay. The nuclear equations for these two decay processes are

$$_{83}^{211}Bi \longrightarrow {}_2^4\alpha + {}_{81}^{207}Tl$$

$$_{92}^{238}U \longrightarrow {}_2^4\alpha + {}_{90}^{234}Th$$

Let us contrast these two equations with those for ordinary chemical reactions. First of all, nuclear equations convey a different type of information from what is found in ordinary chemical equations. The symbols in nuclear equations stand for nuclei rather than atoms. (We do not worry about electrons when writing nuclear equations.) Second, mass numbers and atomic numbers (nuclear charge) are always used in conjunction with elemental symbols in nuclear equations. Third, the elemental symbols on both sides of the equation need not be (and usually are not) the same in nuclear equations.

The procedures for balancing nuclear equations are different from those used for ordinary chemical equations. In a **balanced nuclear equation** *the sums of the subscripts (atomic numbers or particle charge) on both sides of the equation are equal and the sums of the superscripts (mass numbers) on both sides of the equation are equal.* Both of our example equations are balanced. In the alpha decay of $_{83}^{211}Bi$, the subscripts on both sides total 83 and the superscripts total 211. For the alpha decay of $_{92}^{238}U$, the subscripts total 92 on both sides and the superscripts 238.

The term *parent nuclide* and *daughter nuclide* are often used in describing radioactive decay processes. The **parent nuclide** *is the nuclide that undergoes decay in a radioactive decay process.* The **daughter nuclide** *is the nuclide produced as a result of a radioactive decay process.* In our two previous equations, thallium-207 and thorium-234 are the daughter nuclides.

Beta-Particle Decay

Beta-particle decay always results in the formation of a nuclide of a different element. The mass number of the new nuclide is the same as that of the original atom. However, the atomic number has increased by one unit. The general equation for beta-particle decay is

$$_Z^A X \longrightarrow {}_{-1}^{0}\beta + {}_{Z+1}^{A}Y$$

Specific examples of beta-particle decay are

$$^{10}_{4}\text{Be} \longrightarrow {}^{0}_{-1}\beta + {}^{10}_{5}\text{B}$$

$$^{234}_{90}\text{Th} \longrightarrow {}^{0}_{-1}\beta + {}^{234}_{91}\text{Pa}$$

Both of these nuclear equations are balanced; superscripts and subscripts add up to the same sums on each side of the equation.

At this point in the discussion, we answer the question of how a nucleus, which is composed only of neutrons and protons, ejects a negative particle (beta particle) when no such particle is present in the nucleus. The accepted explanation is that a neutron in the nucleus is transformed into a proton and a beta particle through a complex series of steps; that is,

$$^{1}_{0}\text{n} \longrightarrow {}^{1}_{1}\text{p} + {}^{0}_{-1}\beta$$

Once it is formed within the nucleus, the beta particle is ejected with a high velocity. The net result of beta-particle formation is an increase by 1 in the number of protons present in the nucleus and a decrease by 1 in the number of neutrons present in the nucleus. Note in our two examples of beta emission that the daughter nuclide has one more proton than the parent; that is evidenced by the fact that the atomic number of the daughter is one unit greater than that of the parent. In each case, subtraction of the atomic number of the daughter nuclide from its mass number to get the number of neutrons reveals that the daughter nuclide contains one less neutron than the parent. Comments on why neutrons should change into protons are part of the subject matter of Section 15.6

Gamma-Ray Emission

For naturally occurring radionuclides, gamma-ray emission always takes place in conjunction with alpha- or beta-decay processes; it is never independent of these processes. These gamma rays are usually not included in the nuclear equation for the decay because they do not affect the balancing of the equation or the identity of the decay product. This can be seen from the following two nuclear equations.

Balanced nuclear equation
with gamma radiation
included:

$$^{226}_{88}\text{Ra} \longrightarrow {}^{222}_{86}\text{Rn} + {}^{4}_{2}\alpha + {}^{0}_{0}\gamma$$

Balanced nuclear equation
with gamma radiation
omitted:

$$^{226}_{88}\text{Ra} \longrightarrow {}^{222}_{86}\text{Rn} + {}^{4}_{2}\alpha$$

The fact that gamma rays are usually left out of nuclear equations does not imply they are unimportant. On the contrary, gamma rays are more important than alpha and beta particles when the effects of external radiation exposure on living organisms are considered (Section 15.14).

Note also that among synthetically produced radionuclides (Section 15.5) there are some pure "gamma emitters," radionuclides that give off gamma rays, but no alpha or beta radiations. These radionuclides are very important in diagnositic nuclear medicine (Section 15.16)

EXAMPLE 15.1

Write a balanced nuclear equation for the decay of each of the following radioactive nuclides. The mode of decay is indicated in parentheses.

(a) $^{70}_{31}Ga$ (beta emission)

(b) $^{144}_{60}Nd$ (alpha emission)

(c) $^{248}_{100}Fm$ (alpha emission)

(d) $^{113}_{47}Ag$ (beta emission)

SOLUTION

In each case, the atomic and mass numbers of the daughter nucleus are obtained by writing the symbols of the parent nucleus and the particle emitted by the nucleus (alpha or beta). Then the equation is balanced.

(a) Let X represent the daughter nuclide of the radioactive decay. Then

$$^{70}_{31}Ga \longrightarrow ^{0}_{-1}\beta + X$$

The sums of the superscripts on both sides of the equation must be equal, so the superscript for X must be 70. For the sums of the subscripts on both sides of the equation to be equal, the subscript for X must be 32. As soon as we determine the subscript of X, we can obtain the identity of X by looking at a periodic table. The element with an atomic number of 32 is Ge (germanium). Therefore,

$$^{70}_{31}Ga \longrightarrow ^{0}_{-1}\beta + ^{70}_{32}Ge$$

(b) Letting X represent the daughter nuclide of the radioactive decay, we have for the alpha decay of $^{144}_{60}Nd$

$$^{144}_{60}Nd \longrightarrow ^{4}_{2}\alpha + X$$

We balance the equation by making the superscripts on each side of the equation total 144 and the subscripts total 60. We get

$$^{144}_{60}Nd \longrightarrow ^{4}_{2}\alpha + ^{140}_{58}Ce$$

(c) Similarly, we write

$$^{248}_{100}Fm \longrightarrow ^{4}_{2}\alpha + X$$

Balancing superscripts and subscripts, we get

$$^{248}_{100}Fm \longrightarrow ^{4}_{2}\alpha + ^{244}_{98}Cf$$

In alpha emission, the atomic number of the daughter nuclide always decreases by 2 and the mass number of the daughter nuclide always decreases by 4.

(d) Finally, we write

$$^{113}_{47}\text{Ag} \longrightarrow {}^{0}_{-1}\beta + X$$

In beta emission, the atomic number of the daughter nuclide always increases by 1 and the mass number does not change from that of the parent. The balancing procedure gives us the result

$$^{113}_{47}\text{Ag} \longrightarrow {}^{0}_{-1}\beta + {}^{113}_{48}\text{Cd}$$

15.5 Synthetic Radionuclides

A **transmutation reaction** *is a nuclear reaction in which a nuclide of one element is changed into a nuclide of another element.* Radioactive decay, which was discussed in the last section, is an example of a natural transmutation process. Scientists have also developed an artifical process that causes transmutation. This process involves the use of bombardment reactions A **bombardment reaction** *is a nuclear reaction in which small particles traveling at very high speeds collide with stable nuclei, causing them to undergo nuclear change.*

The first successful bombardment reaction was carried out in 1919, 25 years after the discovery of radioactive decay. Further research carried out by many investigators has shown that numerous nuclei experience change under the stress of small particle bombardment. In most cases, the new nuclide produced as the result of the transmutation is radioactive (unstable) rather than stable.

More than 1600 synthetically produced radionuclides that do not occur naturally are now known. Included in this total is at least one radionuclide of every naturally occurring element. In addition, nuclides of 21 elements that do not occur in nature have been produced in small quantities by bombardment reactions. These synthetic elements are discussed in section 15.10. The number of synthetically produced nuclides is more than five times greater than the number of naturally occurring nuclides. Figure 15.2 shows the contrast between the number of synthetic and naturally occurring nuclides for each of the known elements.

Many of the early bombardment reactions were carried out with alpha particles obtained from naturally radioactive materials. Today, many other types of particles, generated in the laboratory by particle accelerators, are available to bombard nuclei. They include protons (hydrogen-1 nuclei), neutrons, deuterons

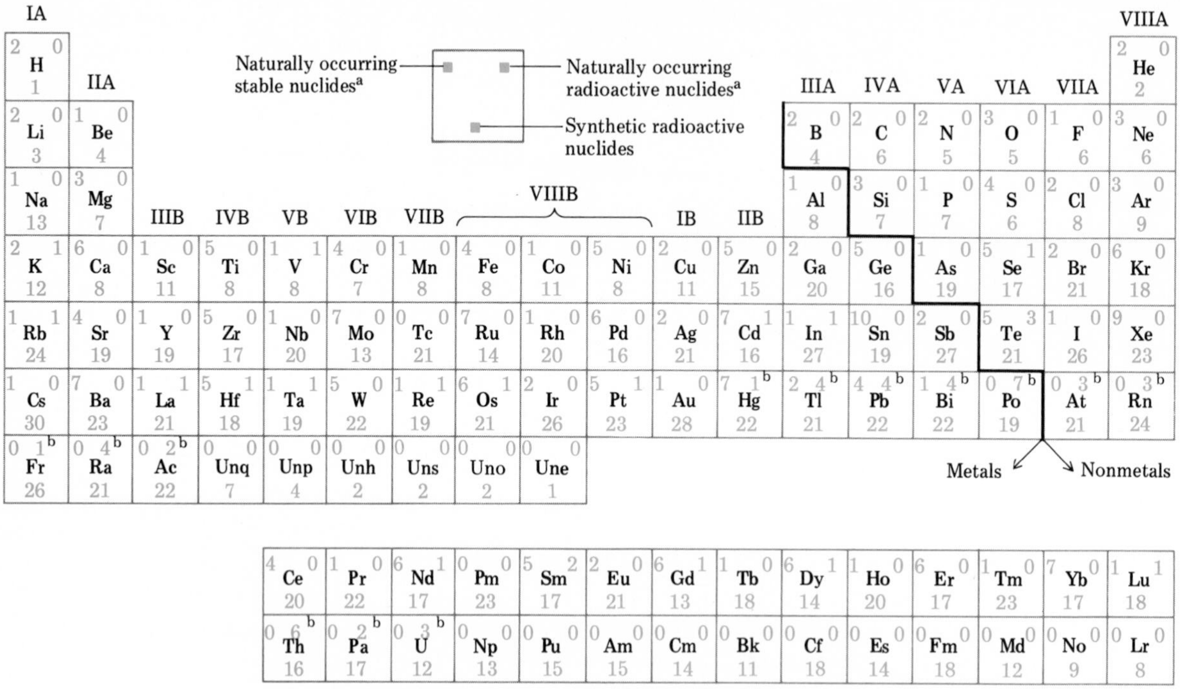

Legend:

Naturally occurring stable nuclides[a] · Naturally occurring radioactive nuclides[a] · Synthetic radioactive nuclides

Periodic table of nuclide counts (each cell: stable / naturally occurring radioactive / synthetic):

- IA: H (2, 0; 1)
- VIIIA: He (2, 0; 2)
- Li (2, 0; 3), Be (1, 0; 4), B (2, 0; 4), C (2, 0; 6), N (2, 0; 5), O (3, 0; 5), F (1, 0; 6), Ne (3, 0; 6)
- Na (1, 0; 13), Mg (3, 0; 7), Al (1, 0; 8), Si (3, 0; 7), P (1, 0; 7), S (4, 0; 6), Cl (2, 0; 8), Ar (3, 0; 9)
- K (2, 1; 12), Ca (6, 0; 8), Sc (1, 0; 11), Ti (5, 0; 8), V (1, 1; 8), Cr (4, 0; 7), Mn (1, 0; 8), Fe (4, 0; 8), Co (1, 0; 11), Ni (5, 0; 8), Cu (2, 0; 11), Zn (5, 0; 15), Ga (2, 0; 20), Ge (5, 0; 16), As (1, 0; 19), Se (5, 1; 17), Br (2, 0; 21), Kr (6, 0; 18)
- Rb (1, 1; 24), Sr (4, 0; 19), Y (1, 0; 19), Zr (5, 0; 17), Nb (1, 0; 20), Mo (7, 0; 13), Tc (0, 0; 21), Ru (7, 0; 14), Rh (1, 0; 20), Pd (6, 0; 16), Ag (2, 0; 21), Cd (7, 1; 16), In (1, 1; 27), Sn (10, 0; 19), Sb (2, 0; 27), Te (5, 3; 21), I (1, 0; 26), Xe (9, 0; 23)
- Cs (1, 0; 30), Ba (7, 0; 23), La (1, 1; 21), Hf (5, 1; 18), Ta (1, 1; 19), W (5, 0; 22), Re (1, 1; 19), Os (6, 1; 21), Ir (2, 0; 26), Pt (5, 1; 23), Au (1, 0; 28), Hg (7, 0; 22), Tl (2, 4[b]; 21), Pb (4, 4[b]; 22), Bi (1, 4[b]; 22), Po (0, 7[b]; 19), At (0, 3[b]; 21), Rn (0, 3[b]; 24)
- Fr (0, 1[b]; 26), Ra (0, 4[b]; 21), Ac (0, 2[b]; 22), Unq (0, 0; 7), Unp (0, 0; 4), Unh (0, 0; 2), Uns (0, 0; 2), Uno (0, 0; 2), Une (0, 0; 1)

Group labels: IA, IIA, IIIB, IVB, VB, VIB, VIIB, VIIIB, IB, IIB, IIIA, IVA, VA, VIA, VIIA, VIIIA

Metals ← | → Nonmetals

Lanthanides: Ce (4, 0; 20), Pr (1, 0; 22), Nd (6, 1; 17), Pm (0, 0; 23), Sm (5, 2; 17), Eu (2, 0; 21), Gd (6, 1; 13), Tb (1, 0; 18), Dy (6, 1; 14), Ho (1, 0; 20), Er (6, 0; 17), Tm (1, 0; 23), Yb (7, 0; 17), Lu (1, 1; 18)

Actinides: Th (0, 6[b]; 16), Pa (0, 2[b]; 17), U (0, 3[b]; 12), Np (0, 0; 13), Pu (0, 0; 15), Am (0, 0; 15), Cm (0, 0; 14), Bk (0, 0; 11), Cf (0, 0; 18), Es (0, 0; 14), Fm (0, 0; 18), Md (0, 0; 12), No (0, 0; 9), Lr (0, 0; 8)

[a] Nuclides that occur in nature in only trace amounts (less than 0.01% isotopic abundance) are not listed as naturally occurring.

[b] Members of the uranium-238, thorium-232, or uranium-235 decay series (see Section 15.7).

FIGURE 15.2

The number of synthetic nuclides exceeds that of naturally occurring nuclides for all elements except hydrogen and helium.

(hydrogen-2 nuclei), and gamma rays. Examples of the bombardment reactions now carried out in laboratories include

$$^{44}_{20}Ca + {}^{1}_{1}H \longrightarrow {}^{44}_{21}Sc + {}^{1}_{0}n$$

$$^{21}_{11}Na + {}^{2}_{1}H \longrightarrow {}^{21}_{10}Ne + {}^{4}_{2}\alpha$$

Gold can be produced from platinum by the bombardment technique. However, the process is astronomically expensive compared to the worth of the gold it produces. Platinum-196 is bombarded with deuterons ($^{2}_{1}H$) to produce platinum-197.

$$^{196}_{78}Pt + {}^{2}_{1}H \longrightarrow {}^{197}_{78}Pt + {}^{1}_{1}H$$

The platinum-197 decays through beta-particle emission to produce gold-197.

$$^{197}_{78}Pt \longrightarrow {}^{0}_{-1}\beta + {}^{197}_{79}Au$$

Bombardment reaction equations are balanced in the same way as radioactive decay equations. The sums of the subscripts of both sides of the equation must be equal as well as the sums of the superscripts on both sides of the equation. Bombardment reaction equations are easily differentiated from equations for decay processes; the former have two reactants and the latter only one.

There are significant uses for many synthetic radionuclides. They are not all idle laboratory curiosities. Most radionuclides used in the field of medicine are synthetic. For example, the synthetic radionuclides cobalt-60, yttrium-90, iodine-131, and gold-198 are used in radiotherapy treatments for cancer. Section 15.17 provides more information about the medical uses for radionuclides.

Synthetically produced radionuclides undergo radioactive decay just as naturally occurring radionuclides. In many cases the mode of decay involves the previously discussed (Section 15.4) alpha- and beta-particle emission processes. Frequently, however, two modes of decay not found among the naturally occurring radionuclides are encountered: positron emission and electron capture. A **positron**, *designated by the symbol* $_{1}^{0}\beta$, *is identical to an electron or beta particle except that it has a positive charge.* Its production in the nucleus is due to the conversion within the nucleus of a proton to a neutron.

$$_{1}^{1}\text{p} \longrightarrow {}_{0}^{1}\text{n} + {}_{1}^{0}\beta$$

This process is just the opposite of that occurring during beta-particle emission (Section 15.4). The net effect of positron emission is to decrease the atomic number (number of protons), while the mass number remains constant. An example of a radioactive decay process involving positron emission is

$$_{15}^{30}\text{P} \longrightarrow {}_{1}^{0}\beta + {}_{14}^{30}\text{Si}$$

In **electron capture** *an electron in a low energy orbital, such as the 1s orbital, is pulled into the nucleus, causing a proton to be converted to a neutron.*

$$_{-1}^{0}\text{e} + {}_{1}^{1}\text{p} \longrightarrow {}_{0}^{1}\text{n}$$

An example of such a process is the reaction

$$_{37}^{87}\text{Rb} + {}_{-1}^{0}\text{e} \longrightarrow {}_{36}^{87}\text{Kr}$$

More information concerning electron capture and also positron emission is given in Section 15.7.

EXAMPLE 15.2

Write a balanced nuclear equation for the decay of each of the following radioactive nuclides. The mode of decay is indicated in parentheses.

(a) $_{12}^{23}\text{Mg}$ (positron emission)   (b) $_{33}^{73}\text{As}$ (electron capture)

SOLUTION

In each case the atomic number and mass number of the daughter nucleus are obtained by first writing the symbols of the parent nucleus and the particle emitted (positron) or absorbed (electron), and then balancing the equation.

(a) Let X represent the product of the radioactive decay, that is, the daughter nuclide. Then

$$^{23}_{12}Mg \longrightarrow {}^{0}_{1}\beta + X$$

Note that the Greek letter beta is used to denote not only a beta particle but also a positron. The difference between the two particles is that the former is negatively charged ($^{0}_{-1}\beta$) and the latter is positively charged ($^{0}_{1}\beta$).

Since the sum of the superscripts on each side of the equation must be equal, the superscript for X must be 23. For the sums of the subscripts on both sides of the equation to be equal, at 12, the subscript for X must be 11. As soon as the subscript of X is determined, the identity of X is known. Looking at a periodic table, we determine that the element with an atomic number of 11 is sodium (Na). Therefore,

$$^{23}_{12}Mg \longrightarrow {}^{0}_{1}\beta + {}^{23}_{11}Na$$

(b) Similarly, letting X represent the daughter nuclide of this radioactive decay, we have for $^{73}_{33}As$ decaying by the electron capture mechanism

$$^{73}_{33}As + {}^{0}_{-1}e \longrightarrow X$$

Note that in electron capture the electron appears on the reactant side of the equation. This makes equations for electron capture different from those for alpha, beta, and positron emissions; in these latter cases the small particle involved is placed on the product side of the equation.

Balancing this equation, making the superscripts on each side of the equation total 73 and the subscripts total 32, we get

$$^{73}_{33}As + {}^{0}_{-1}e \longrightarrow {}^{73}_{32}Ge$$

15.6 Factors Affecting Nuclear Stability

Some nuclei are stable; others are not. This is true even for isotopes of the same element. For example, sodium-23 is stable, but sodium-24 is radioactive. What determines whether a given nuclide is stable or radioactive? Two key factors in determining stability are (1) the total number of nucleons in a nucleus and (2) the ratio of neutrons to protons in the nucleus. We will first look at experimental observations regarding these two factors and then consider why they should affect nuclear stability.

Every known nuclide of every element with an atomic number greater than 83 is unstable (radioactive). Thus, the number of nucleons that can be packed into a stable nucleus appears to be limited. The limit is 209, which is reached with $^{209}_{83}\text{Bi}$, the largest known stable nucleus.

It has been experimentally observed that, for stable nuclides of low atomic number, the neutron-to-proton ratios are very close to 1. For heavier elements, stable nuclides have higher neutron-to-proton ratios, with the ratio reaching approximately $3:2$ for the heaviest stable nuclides.

Figure 15.3 shows graphically the relationship between stability and neutron-to-proton ratio. In this figure, the number of neutrons is plotted against the number of protons for *stable* nuclides. The straight line (in color) represents a ratio of equal numbers of protons and neutrons and is shown for comparison. Note that all of the dots (stable nuclides) are close to the $1:1$ line for the first few elements, but this pattern holds only up to about atomic number 20.

The fact that stable nuclides fall within a rather narrow zone of stability, defined by the neutron-to-proton ratio, strongly suggests that neutrons are at least partly responsible for the stability of the nucleus. Remember that like charges repel each other and most nuclei contain many protons (like positive charges) packed together in a small volume. As the number of protons increases, the forces of repulsion between protons increases sharply. Therefore, a greater number of neutrons is required to counteract the increased repulsive forces. Finally, at element 84, the repulsive forces become so great that the nuclides are not stable, regardless of the number of neutrons present.

Radioactive decay (Section 15.4) changes the neutron-to-proton ratio. Beta decay increases this ratio and alpha decay, positron decay, and electron capture have the opposite effect. Thus, radioactive decay is nature's mechanism for changing an undesirable neutron-to-proton ratio into one that is more desirable.

15.7 Neutron-to-Proton Ratio and Mode of Decay

The mode by which a radionuclide decays depends on how the neutron-to-proton ratio for the radionuclide compares with that of stable nuclei containing approximately the same number of nucleons. The driving force for radioactive decay is the tendency of unstable nuclei to adjust neutron-to-proton ratios in such a way that stability is achieved.

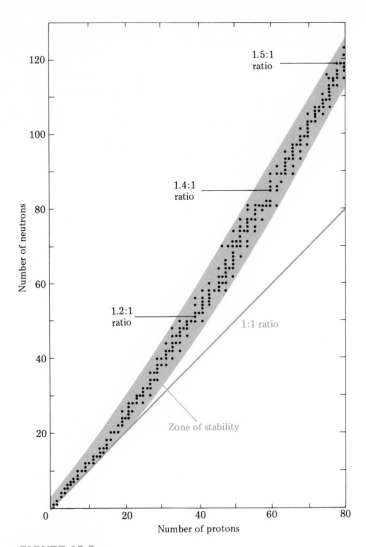

FIGURE 15.3
A graph of the number of neutrons versus the number of protons present in the nuclei of stable nuclides.

Unstable nuclei may be divided into three categories on the basis of their position on the graph in Figure 15.3 relative to the zone of stability. The *zone of stability* is the area in the figure where the stable nuclei are located. The three categories are as follows.

1. Unstable nuclei in which the neutron-to-proton ratio is too high.
2. Unstable nuclei in which the neutron-to-proton ratio is too low.

3. Unstable nuclei in which the total number of nucleons exceeds the limit for a stable nucleus, 209.

Nuclei in each of these categories have a particular mode of decay that predominates.

Unstable nuclei in which the neutron-to-proton ratio is too high lie to the left of the zone of stability in Figure 15.3. Beta emission is the predominant decay mode for such nuclides. As discussed previously (Section 15.3), beta emission involves the transformation of a neutron into a proton. This increases the number of protons, decreases the number of neutrons, and thus decreases the neutron-to-proton ratio.

Unstable nuclei in which the neutron-to-proton ratio is too low lie to the right of the zone of stability in Figure 15.3. Such nuclei decay by converting a proton into a neutron—just the opposite of the process that occurs in beta-particle emission. The conversion of a proton into a neutron may be accomplished in two ways—by positron emission or by electron capture. Both of these modes of decay were described previously in Section 15.5. The process of electron capture seems to be preferred over positron emission for nuclides of high atomic number. For lighter nuclides, numerous examples of both types of decay processes are known.

Unstable nuclei in which the total number of nucleons exceeds the limit for a stable nucleus lie beyond the zone of stability in Figure 15.3. More than one decay step is required to reach stability for most nuclides of this type. This results in the formation of a decay series. A **decay series** *is a series of elements produced from the successive emission of alpha and beta particles.* Three naturally occurring decay series are known. A fourth series was discovered after the synthesis of certain elements not found in nature (Section 15.10). Because the parent of this fourth series, plutonium-241, does not occur in nature to a measurable extent, the series is not classified as a naturally occurring series. General characteristics of the four decay series are given in Table 15.1. Figure 15.4 shows all of the members of the uranium-238 decay series. It is representative of the other three series. Note in Figure 15.4 that the order in which alpha and beta particles are emitted has no fixed pattern.

Intermediate in the $^{238}_{92}$U decay series (Figure 15.4) is the radionuclide $^{222}_{86}$Rn. This radionuclide has ''been in the news'' constantly in recent years. It is a gas at

TABLE 15.1

The Four Known Radioactive Decay Series

Parent Number	Number of Decay Steps	Final Product of Series
Uranium-238	14	lead-206
Thorium-232	10	lead-208
Uranium-235	11	lead-209
Plutonium-241	13	bismuth-209

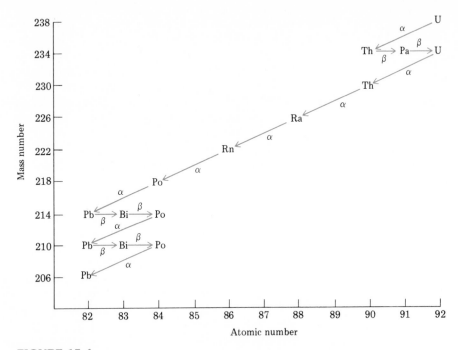

FIGURE 15.4

The $^{238}_{92}U$ decay series. Each nuclide in this series except $^{206}_{82}Pb$ is unstable, and the successive transformations continue until the stable product is obtained.

normal temperatures, and is therefore a very mobile species, and it has been detected in both aqueous and atmospheric environments. Particular concern exists about radon-226 levels in areas where spent uranium ores (uranium tailings) have been used as fill material. It is now realized that such use of tailing may produce radioactive contamination of both air and water.

15.8   Rate of Radioactive Decay

All radioactive nuclides do not decay at the same rate. Some decay very rapidly; others undergo disintegration at extremely slow rates. This indicates that all radionuclides are not equally unstable. The faster the decay rate is, the lower the stability will be.

The concept of *half-life* is used to express nuclear stability quantitatively. The **half-life** *is the time required for one-half of any given quantity of a radioactive substance to undergo decay.* For example, if a radionuclide's half-life is 12 days and you have a 4.00-g sample of it, then after 12 days (one half-life) only 2.00 g of the sample (one-half the original amount) remains undecayed and the other half will have decayed into some other substance.

Half-lives of as long as billions of years and as short as a fraction of a second have been determined. Table 15.2 contains examples of the wide range of half-life values.

Most naturally occurring radionuclides have long half-lives. However, some radionuclides with *short* half-lives are also found in nature. Because these short-lived species decay rapidly, they must be produced continually to be found in nature. Processes that result in their production are (1) the decay of naturally occurring long-lived nuclides, producing short-lived daughter nuclides, (2) the decay of short-lived nuclides produced in the previous manner, and (3) bombardment reactions involving cosmic rays that take place naturally in the upper atmosphere.

The decay rate (half-life) of a radionuclide is a constant. It is independent of outward conditions such as temperature, pressure, and state of chemical combination. For example, radioactive sodium-24 decays at the same rate whether it is incorporated into $NaCl$, $NaBr$, Na_2SO_4, or $NaC_2H_3O_2$. Once something is radioactive, nothing will stop it from decaying or increase or decrease its decay rate.

TABLE 15.2

Range of Half-lives Found for Naturally Occurring and Synthetic Radionuclides

Element	Half-Life
Naturally Occurring Radionuclides	
Vanadium-50	6×10^{15} yr
Platinum-190	6.9×10^{11} yr
Uranium-238	4.5×10^9 yr
Uranium-235	7.1×10^8 yr
Thorium-230*	7.5×10^4 yr
Lead-210*	22 yr
Bismuth-214*	19.7 min
Polonium-212*	3.0×10^{-3} sec
Synthetic Radionuclides	
Iodine-129	1.7×10^7 yr
Nickel-63	92 yr
Gold-195	200 days
Lead-200	21 hr
Silver-106	24 min
Oxygen-19	29.4 sec
Fermium-246	1.2 sec
Beryllium-8	3×10^{-16} sec

* A product in a naturally occurring decay series (Sec. 15.7)

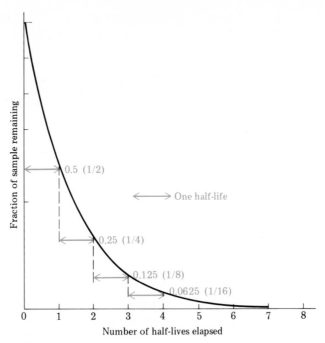

FIGURE 15.5
A general half-life decay curve for a radionuclide.

Figure 15.5 graphically illustrates the meaning of the term half-life. After one half-life has passed, half of the original atoms have decayed and half remain. During the next half-life, half of the remaining half decays, leaving one-fourth of the original atoms undecayed. After three half-lives, $\frac{1}{2} \times \frac{1}{2} \times \frac{1}{2} = \frac{1}{8}$ of the original atoms remain undecayed, and so on. Note from Figure 15.5 that a very small amount of original material (less than 1%) remains after seven half-lives have elapsed.

Calculation of the amount of radioactive material decayed, amount remaining undecayed, and time elapsed can be carried out with the following equation.

$$\begin{pmatrix} \text{Amount of radionuclide} \\ \text{undecayed after } n \text{ half-lives} \end{pmatrix} = \begin{pmatrix} \text{Original amount} \\ \text{of radionuclide} \end{pmatrix} \times \left(\frac{1}{2^n} \right)$$

EXAMPLE 15.3

Iodine-131 is a radionuclide that is frequently used in nuclear medicine. One use for it is to detect fluid buildup in the brain. The half-life of iodine-131 is 8.0 days. How much of a 0.50-g sample of iodine-131 will remain undecayed after a period of 32 days?

SOLUTION

First, we must determine the number of half-lives that have elapsed.

$$32 \text{ days} \times \left(\frac{1 \text{ half-life}}{8.0 \text{ days}} \right) = 4 \text{ half-lives}$$

Knowing the number of elapsed half-lives and the original amount of radioactive iodine present, we can calculate the amount that remains using the equation

$$\left(\begin{matrix} \text{Amount of radionuclide} \\ \text{undecayed after } n \text{ half-lives} \end{matrix} \right) = \left(\begin{matrix} \text{Original amount} \\ \text{of radionuclide} \end{matrix} \right) \times \left(\frac{1}{2^n} \right)$$

$$= 0.50 \text{ g} \times \frac{1}{2^4} \longleftarrow \text{four half-lives}$$

$$= 0.50 \text{ g} \times \frac{1}{16}$$

$$= 0.031 \text{ g}$$

EXAMPLE 15.4

Strontium-90 is a nuclide that is found in radioactive fallout from nuclear weapon explosions. Its half-life is 28.0 years. How long will it take for 94% (15/16) of the strontium-90 atoms present in a sample of material to undergo decay?

SOLUTION

If 15/16 of the sample has decayed, then 1/16 of the sample remains undecayed. In terms of $1/2^n$, 1/16 is equal to $1/2^4$, that is,

$$\tfrac{1}{2} \times \tfrac{1}{2} \times \tfrac{1}{2} \times \tfrac{1}{2} = \tfrac{1}{2^4} = \tfrac{1}{16}$$

Thus, four half-lives have elapsed in reducing the amount of strontium-90 to 1/16 of its original amount.

The half-life of strontium-90 is 28.0 years, so the total time elapsed will be

$$4 \text{ half-lives} \times \left(\frac{28.0 \text{ years}}{1 \text{ half-live}} \right) = 112 \text{ years}$$

In both Examples 15.3 and 15.4, the time elapsed was equivalent to a whole number of half-lives. To work problems involving a fractional number of half-lives, you must use more complicated equations with logarithms. Such equations are not presented in this text; you will be expected to be able to work only problems that involve a whole number of half-lives.

The areas of application for Examples 15.3 and 15.4 are respectively, nuclear medicine and nuclear fallout. These are two major areas in which half-life is of prime importance. In nuclear medicine, almost all radionuclides in use today have relatively short half-lives. Thus, their radioactive effects within the body are limited to short periods of time. The effect that nuclear fallout has on the environment is directly related to the half-lives of the species present in the fallout. The longer the half-lives, the greater the time period that the environment is affected.

15.9   Radiocarbon Dating

DATING OF
ARCHAEOLOGICAL
ARTIFACTS

The half-life concept can be used to determine the age of objects that contain a radionuclide. The most well known of radiochemical dating techniques is radiocarbon dating. Radiocarbon dating involves measuring the amount of carbon-14 present in an object. Many archaeological artifacts can be dated by carbon-14 techniques because they were made from or contain once-living carbon-containing materials (see Figure 15.6).

The isotope carbon-14, the only naturally occurring radionuclide of carbon, has a half-life of 5730 years. It is continually produced in the upper atmosphere as a result of cosmic ray bombardment.

$$^{14}_{7}\text{N} + ^{1}_{0}\text{n} \longrightarrow ^{14}_{6}\text{C} + ^{1}_{1}\text{H}$$

FIGURE 15.6
Ancient manuscripts and other cultural artifacts made from plant materials are often dated by means of carbon-14 analysis. [*Field Museum of Natural History, Chicago.*]

The steady-state (equilibrium) concentration of carbon-14, which reflects both its rate of formation and its rate of decay, is 1 carbon-14 atom per 10^{12} nonradioactive carbon atoms. This trace amount of carbon-14 in the atmosphere reacts with oxygen to give carbon dioxide in the same manner as nonradioactive carbon. Thus, approximately 1 out of every 10^{12} carbon dioxide molecules is radioactive. This radioactive carbon is incorporated into the structure of plants through photosynthesis, and into animals and human beings through the food chain. A steady-state concentration of carbon-14, equal to that found in the atmosphere, is thus found in all living organisms. Upon the death of an organism, the intake of carbon-14 ceases and the natural level of radioactive carbon present within the structure begins to decrease as the result of carbon-14 decay.

$$\ce{^{14}_{6}C} \longrightarrow \ce{_{-1}^{0}\beta} + \ce{^{14}_{7}N}$$

In carbon-14 dating the ratio of carbon-14 to total carbon in an object that contains once-living material (parchment, cloth, charcoal, etc.) is compared to that in living matter. A wooden object with a ratio of carbon-14 to total carbon one-fourth that of a living tree would be approximately 11,400 years (two half-lives) old. An important assumption in the carbon-14 dating method is that the flow of carbon-14 into the biosphere is constant over time. There is some evidence, such as the carbon-14 content of the growth rings in older trees, to indicate that this is approximately true.

15.10 Synthetic Elements

One of the most interesting facets of bombardment reaction research (Section 15.5) is the production of elements that do not occur in nature; such elements are called synthetic elements. Four synthetic elements that were produced between 1937 and 1941 filled gaps in the periodic table for which no naturally occurring element had been found. These four are technetium (Tc, element 43), an element with numerous uses in nuclear medicine; promethium (Pm, element 61); astatine (At, element 85); and francium (Fr, element 87). The reactions for their production are

$$\ce{^{96}_{42}Mo} + \ce{^{2}_{1}H} \longrightarrow \ce{^{97}_{43}Tc} + \ce{^{1}_{0}n} \qquad \text{(half-life} = 2.6 \times 10^6 \text{ yr)}$$

$$\ce{^{142}_{60}Nd} + \ce{^{1}_{0}n} \longrightarrow \ce{^{143}_{61}Pm} + \ce{^{0}_{-1}\beta} \qquad \text{(half-life} = 265 \text{ days)}$$

$$\ce{^{209}_{83}Bi} + \ce{^{4}_{2}\alpha} \longrightarrow \ce{^{210}_{85}At} + 3\,\ce{^{1}_{0}n} \qquad \text{(half-life} = 8.3 \text{ hr)}$$

$$\ce{^{230}_{90}Th} + \ce{^{1}_{1}p} \longrightarrow \ce{^{223}_{87}Fr} + 2\,\ce{^{4}_{2}\alpha} \qquad \text{(half-life} = 22 \text{ min)}$$

Bombardment reactions have also been the source for elements 93 through 109. These elements are called the *transuranium elements*, because they occur immediately following uranium in the periodic table. (Uranium is the highest atomic numbered, naturally occurring element.) All isotopes of all of the transuranium elements are radioactive.

One method used to prepare transuranium elements is to collide the nucleus of a light atom with a very large nucleus. For example, element 104 can be prepared by the reaction

$$^{249}_{98}\text{Cf} + ^{12}_{6}\text{C} \longrightarrow ^{257}_{104}\text{Unq} + 4\,^{1}_{0}\text{n} \qquad \text{(half-life = 4.5 sec)}$$

Table 15.3 provides information about the stability of the transuranium elements. Note that in many cases all the isotopes of a given element are extremely short-lived, with half-lives of only seconds. Only small amounts of these short-lived elements have been produced—in some cases only a few atoms. Because these elements quickly disappear (decay) once they are produced, they cannot be detected and identified on the basis of chemical properties. Identification of these elements is made with instruments that analyze their characteristic radiation. The more recently produced transuranium elements are just "laboratory curiosities."

TABLE 15.3
Information About the Synthetically Produced Transuranium Elements

Name	Symbol	Atomic Number	Mass Number of Most Stable Isotope	Half-life of Most Stable Isotope	Date of Discovery
Neptunium	Np	93	237	2.14×10^6 yr	1940
Plutonium	Pu	94	244	7.6×10^7 yr	1940
Americium	Am	95	243	8.0×10^3 yr	1944
Curium	Cm	96	247	1.6×10^7 yr	1944
Berkelium	Bk	97	247	1400 yr	1950
Californium	Cf	98	251	900 yr	1950
Einsteinium	Es	99	252	472 days	1952
Fermium	Fm	100	257	100 days	1953
Mendelevium	Md	101	258	56 days	1955
Nobelium	No	102	259	1 hr	1958
Lawrencium	Lr	103	260	3 min	1961
Unnilquadium	Unq	104	261	70 sec	1969
Unnilpentium	Unp	105	262	40 sec	1970
Unnilhexium	Unh	106	263	0.9 sec	1974
Unnilseptium	Uns	107	262	0.005 sec	1980
Unniloctium	Uno	108	265	0.002 sec	1984
Unnilennium	Une	109	266	0.005 sec	1982

15.11 Nuclear Fission

Nuclear fission *is the process in which a heavy element nucleus splits into two or more medium-sized nuclei as the result of bombardment.* The occurrence of such a process was not easily established or easily accepted once established. The efforts of a number of scientists, during the period 1934–1939, were needed to establish a firm base for this "revolutionary" new concept that went against all the current theories of that time.

A number of heavy element nuclei are now known to undergo fission when struck with a variety of particles, including neutrons, alpha particles, protons, and gamma rays. However, the fission reaction is considered to be of practical importance only in the case of three nuclides: naturally occurring uranium-235, synthetically produced plutonium-239, and synthetically produced uranium-233. All three of these nuclides undergo fission when bombarded by low-energy neutrons, and all have long half-lives. Many of the other fissionable nuclides require higher energy bombarding particles, are not available in large quantities, or have half-lives that are too short.

The first fissionable nucleus to be discovered, which remains the most important one, was uranium-235. Bombardment of this nucleus with neutrons causes it to split into two fragments. Characteristics of the uranium-235 fission reaction include the following.

1. There is no unique way in which the uranium-235 nucleus splits. Thus, the uranium-235 fission reaction produces many different lighter elements. The following are some examples of this fission process.

$$^{235}_{92}U + {}^{1}_{0}n \nearrow \begin{array}{l} {}^{135}_{53}I + {}^{97}_{39}Y + 4\,{}^{1}_{0}n + \text{energy} \\ {}^{139}_{56}Ba + {}^{94}_{36}Kr + 3\,{}^{1}_{0}n + \text{energy} \\ {}^{131}_{50}Sn + {}^{103}_{42}Mo + 2\,{}^{1}_{0}n + \text{energy} \\ {}^{139}_{54}Xe + {}^{95}_{38}Sr + 2\,{}^{1}_{0}n + \text{energy} \end{array}$$

2. Very large amounts of energy, which are many times greater than that from ordinary radioactive decay, are emitted during the fission process. It is this large release of energy that makes nuclear fission of uranium-235 the important process it is. **Nuclear energy** *is the energy released during a nuclear fission process.* An older term for nuclear energy is *atomic energy.*

3. Neutrons, which are reactants in the fission process, are also products. The number of neutrons produced per fission depends on how the nucleus splits, and range from two to four (as can be seen from the fission

equations given previously). On the average, 2.4 neutrons are produced per fission. The significance of the produced neutrons is that they can cause the fission process to continue by colliding with more uranium-235 nuclei. Figure 15.7 shows diagramatically the "chain reaction" that can occur once the fission process is started.

The process of nuclear fission—or "splitting the atom" as it is popularly known—can be carried out in both an uncontrolled and a controlled manner.

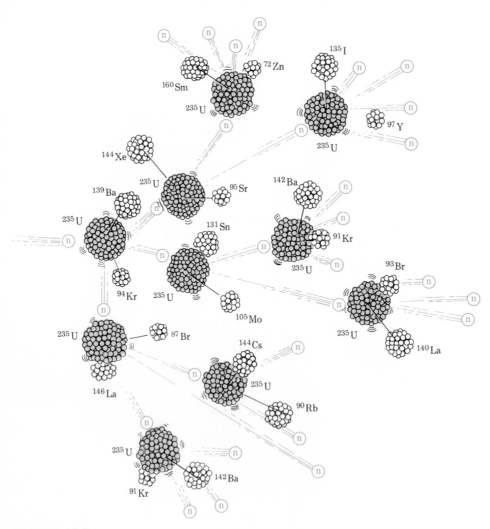

FIGURE 15.7

A fission chain reaction caused by further reaction of the neutrons produced during fission.

The key question involves what happens to the neutrons produced during fission. When small quantities of uranium-235 undergo fission, the newly produced neutrons escape from the sample into the surroundings. When a sufficiently large mass of uranium-235 is present, the neutrons collide with other uranium-235 atoms, causing more fissions. If only one of the neutrons produced each time reacts with another uranium-235 atom, the process will continue at a constant rate. When this condition prevails, the process is said to be self-propagating or *critical*. If more than one of the neutrons generated per fission produces another fission, the reaction becomes an expanding or branching chain reaction. Such a situation—known as a *supercritical* condition—leads to an explosion.

THE ATOMIC BOMB

Uncontrolled nuclear fission is the process that occurs in an atomic bomb. Intensive study of the uranium-235 fission reaction during the period 1940–1944 led to the development of the A bombs that were used to end World War II (see Figure 15.8). In an atomic bomb, small pieces of uranium, each incapable of sustaining a chain reaction, are brought together (at the appropriate time) to form one piece capable of sustaining a chain reaction.

Nuclear Power Plants

Even before the first atomic bomb was exploded, scientists began to speculate about the use of nuclear fission in a controlled manner as a source of energy for peaceful purposes. This is now a reality. Each year an increasing amount of

FIGURE 15.8

An atomic bomb of the ''Fat Man'' type, the kind detonated over Nagasaki, Japan. The bomb is 60 inches in diameter and 128 inches long; it weighs about 5 tons. [*Pictorial Parade.*]

electricity used in the United States is generated by nuclear power plants whose operation involves controlled nuclear fission.

The basic concepts in the production of electricity from a nuclear power plant are the same as those for a traditional coal-fired generating plant. In both cases heat is used to generate steam, which in turn operates a steam turbine generator. The difference between the two types of plants is that nuclear fuel replaces fossil fuel as a source of heat. The essential components of a fission-powered generating plant are shown in Figure 15.9. The uranium present in the reactor core is in the form of pellets of the oxide U_3O_8 enclosed in long steel tubes. The uranium is a mixture of all isotopes of uranium, not just uranium-235, because of the cost of isotope separation. The fission reaction is controlled with rods that absorb excess neutrons (to prevent unwanted fissions) and with moderating substances that decrease the speed of the neutrons. The energy from fission appears as heat, which is drawn out of the core by circulating primary coolant. A heat exchanger is then used to turn water to steam, which turns the electric generator. The spent steam is condensed back to water as it passes through a condenser. It then recirculates to the heat exchanger.

The use of nuclear power plants has created a lot of controversy recently. The major focus is on environmental damage and nuclear waste disposal. Large amounts of waste heat are a necessary by-product of nuclear reactor operations. How this waste heat is released into the environment is a matter of concern. This problem, often referred to as thermal pollution, is solvable through the use of cooling towers (see Figure 15.10). In addition to heat, radioactive wastes are also produced during reactor operation. They must be removed periodically, and they create disposal problems. They cannot be released into the environment, but must be stored indefinitely. Where and how to store these wastes safely is still a matter of debate.

NUCLEAR WASTE
STORAGE

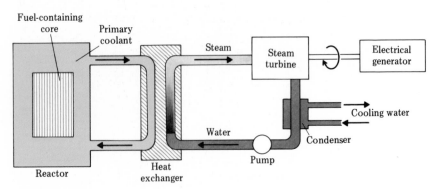

FIGURE 15.9
Components of a nuclear power plant.

FIGURE 15.10
A nuclear power plant and its associated cooling towers. It is the cooling towers, used to dissipate waste heat, that are the dominating feature of the whole area. [*AP Newsfeatures Photo.*]

Breeder Reactors

Uranium-235, the only known naturally occurring fissionable nuclide, represents only about 0.71% of naturally occurring uranium. Nonfissionable uranium-238 is the abundant isotope of uranium (99.28%). At the current and projected rate of use, known sources of uranium-235 will be exhausted shortly after the end of this century. Breeder reactors represent a solution to this problem. Such reactors produce or "breed" fissionable isotopes by nuclear transformations. Most research in breeder technology has been directed toward the use of uranium-238 and thorium-232 as sources for the fissionable fuel. Both are naturally occurring abundant isotopes that have long half-lives. Thorium is three times more abundant than uranium. The breeder reactor consumes uranium-235 as fuel, but produces fissionable plutonium-239 or uranium-233 during the operation of the reactor, as shown in the following equations.

$$^{238}_{92}\text{U} + {^1_0}\text{n} \longrightarrow {^{239}_{92}\text{U}} \xrightarrow{\beta} {^{239}_{93}\text{Np}} \xrightarrow{\beta} {^{239}_{94}\text{Pu}}$$

(from fission)   (fissionable)

$$^{232}_{90}\text{Th} + {^1_0}\text{n} \longrightarrow {^{233}_{90}\text{Th}} \xrightarrow{\beta} {^{233}_{91}\text{Pa}} \xrightarrow{\beta} {^{233}_{92}\text{U}}$$

(from fission)   (fissionable)

Experimental breeder systems are now under study in many parts of the world, although none is yet commercially producing energy. The technical problems associated with building breeder reactors have been considerable. Much research and testing still needs to be carried out. There is no question about the ability to produce these two fissionable isotopes; it is being done. The problem is in the design of an economical process that generates more fuel than it consumes.

15.12   Nuclear Fusion

Nuclear fusion *is the process in which small atoms are put together to make larger ones.* It is, thus, the opposite of nuclear fission. Nuclear fusion is a higher-energy-yielding reaction than nuclear fission.

ENERGY AND THE SUN

For fusion to occur, a very high temperature, on the order of several hundred million degrees, is required. One place that is hot enough for fusion to occur is the interior of the sun. It is postulated that the energy of the sun is derived from the conversion of hydrogen nuclei into helium nuclei by nuclear fusion. The overall reaction is thought to occur in steps.

1. Two hydrogen-1 nuclei fuse to produce a hydrogen-2 nucleus and a positron.

$$^1_1\text{H} + {^1_1}\text{H} \longrightarrow {^2_1}\text{H} + {^0_1}\beta$$

2. The hydrogen-2 nucleus fuses with another hydrogen-1 nucleus to give helium-3.

$$^2_1\text{H} + {^1_1}\text{H} \longrightarrow {^3_2}\text{He}$$

3. Pairs of helium-3 nuclei fuse to form helium-4 plus two hydrogen-1 nuclei.

$$^3_2\text{He} + {^3_2}\text{He} \longrightarrow {^4_2}\text{He} + 2\,({^1_1}\text{H})$$

The net overall reaction for the production of energy on the sun (the sum of reactions in steps 1, 2, and 3) is represented by the equation

$$4 \, (^1_1\text{H}) \longrightarrow {}^4_2\text{He} + 2 \, (^0_1\beta)$$

THE HYDROGEN BOMB

The use of nuclear fusion on earth might seem impossible because of the high temperatures required. It has, however, been accomplished in a hydrogen bomb (see Figure 15.11). In such a bomb, a *fission* device (an atomic bomb) is used to achieve the high temperatures needed to start the following fusion reaction.

$$^3_1\text{H} + {}^2_1\text{H} \longrightarrow {}^4_2\text{He} + {}^1_0n$$

FIGURE 15.11

The process of nuclear fusion is responsible for the large amount of energy generated during the explosion of a hydrogen bomb. [*Los Alamos Scientific Laboratory.*]

The hydrogen-3 needed in the reaction does not occur naturally and is produced by bombarding lithium-6 with neutrons.

$$\ce{^6_3Li} + \ce{^1_0}n \longrightarrow \ce{^3_1H} + \ce{^4_2He}$$

The necessary hydrogen-2 may be obtained from seawater. Hydrogen-1 and hydrogen-2 are the only naturally occurring isotopes of hydrogen. One out of approximately 5000 hydrogen atoms is hydrogen-2.

At the high temperature of fusion reactions, electrons completely separate from nuclei. Neutral atoms cannot exist. This high-temperature gas-like mixture of nuclei and electrons is called a *plasma* and is considered by some scientists to represent a fourth state of matter (in addition to solids, liquids, and gases).

In principle, fusion reactions can be used as controlled energy sources just as fission reactions are. Fusion reactors would have a major advantage over fission reactors. No significant amount of radioactive wastes would be produced during operation. Unfortunately, fusion reactors have not yet been proven feasible.

To start a fusion reaction, a temperature of about 100 million degrees must be achieved and the resulting plasma must be contained. Several approaches to solving the heating and containment problems are under active investigation. The use of laser beams to achieve the high temperature needed to initiate a fusion reaction is under study.

Approaches to the containment problem deal with a nonmaterial containment system called a "magnetic bottle." Charged particles (the plasma) have difficulty crossing magnetic lines of force, so a magnetic bottle can be created by surrounding the plasma with a suitable magnetic field. Magnetic fields of various shapes can be generated by passing an electric current through an appropriately arranged pattern of wires or other conductors, as shown in Figure 15.12. The use of a nonmaterial container is sometimes erroneously thought to be necessary because the hot plasma would destroy any material it contacted. However, the plasma density is so low that the total energy content is insufficient to damage container walls. On the contrary, a nonmaterial container is necessary to keep the plasma from the walls to avoid conductive heat losses that would result in plasma temperatures too low to support fusion. In addition,

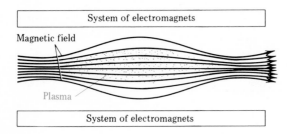

FIGURE 15.12
Magnetic containment of plasma.

contact with container walls introduces heavy contaminants into the plasma, which cause additional problems.

Although major technological advances might be made suddenly, it is now estimated that fusion power will not be available until at least the year 2000, if it proves to be feasible at all.

15.13   Interaction of Radiation with Matter

The alpha, beta, and gamma radiations produced from radioactive decay travel outward from their nuclear sources into the matter surrounding the radioactive substance. There these highly energetic radiations lose their energy through numerous interactions (collisions) with the atoms and molecules they encounter along their path. Let us consider in detail these interactions between radiation and atoms and molecules.

Because the nucleus of an atom occupies such a small portion of the total volume of an atom (Section 3.3), it is not surprising that in the great majority of radiation-atom interactions is sufficient to knock away electrons from atoms; that directly involved than the nucleus is. In many cases, energy transfer during radiation-atom interactions is sufficient to knock away electrons from atoms; that is, ionization occurs and ion pairs are formed. An **ion pair** *is the electron and positive ion that are produced during an ionization collision between an atom and radiation.* This ionization process is not the voluntary transfer of electrons that occurs during ionic compound formation, but rather a nonchemical, involuntary removal of electrons from atoms to form ions. Figure 15.13 diagrammatically shows ion-pair formation. Many ion pairs are produced by a single "radiation" because each radiation must undergo many collisions before its energy is reduced to the level of the surrounding material. The electrons ejected from an atom frequently have enough energy to bombard neighboring molecules and cause additional ionization.

Free radical formation is another effect caused by alpha, beta, and gamma radiation. A **free radical** *is a highly reactive uncharged molecular fragment, that is, an uncharged piece of a molecule.* Free radicals are very reactive entities because they always contain an unpaired electron. (Recall, from Section 5.8, that electrons occur in pairs in normal bonding situations.)

One mechanism for the formation of free radicals involves the decomposition of polyatomic positive ions present in ion pairs. After an electron has been knocked off a molecule (during ion pair formation), the resulting positive ion often does not have enough bonding electrons to hold the atoms together; hence decomposition occurs.

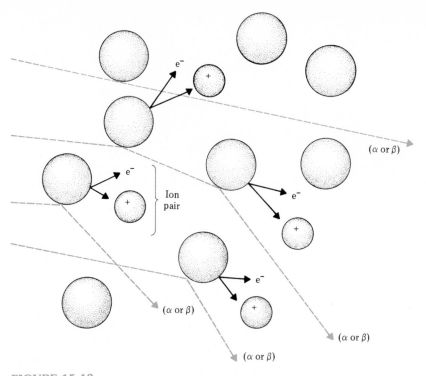

FIGURE 15.13
The interaction of radiation with atoms to form ion pairs.

$$A—B \longrightarrow (A—B)^+ + e^- \quad \text{(ion-pair formation)}$$

$$A—B^+ \longrightarrow A\cdot + B^+$$
$$A—B^+ \longrightarrow A^+ + B\cdot \quad \text{(free-radical formation)}$$

Note how free radicals are designated in the preceding equations. The unpaired electron is explicitly shown as a dot.

Free radicals can also be formed directly from molecule–radiation interactions; radiation possesses energy far in excess of that needed to break a chemical bond.

$$A—B \longrightarrow A\cdot + B\cdot$$

Here, the molecule splits apart into two free radicals.

The ionizing effect of radiation (ion-pair formation and free-radical formation) is the factor that makes radiation harmful to living matter and inert materials. In living matter, the formation of ions and free radicals disrupts cellular function.

Water molecules are the most abundant molecules in a cell; cells are approximately 80% water. Interactions of the water of a cell with radiation can result in both ion-pair and free-radical formation. Equations for these two processes for water can be written as follows.

$$H_2O \xrightarrow{\text{radiation}} H_2O^+ + e^- \quad \text{(ion-pair formation)}$$

$$H_2O^+ \longrightarrow OH\cdot + H^+ \quad \text{(free-radical formation)}$$

The highly reactive $OH\cdot$ species (the free radical) can react with a host of other molecules in a cell to produce new free radicals, which in turn react with other substances. Thus, the formation of a single free radical can instigate a large number of chemical reactions (mostly undesirable ones) that ultimately disrupt the normal operation of a cell. Note that not all $OH\cdot$ free radicals cause undesirable reactions. Some can recombine with H^+ to reproduce H_2O. Note also that an $OH\cdot$ free radical and a hydroxide ion (OH^-) are very different species. A hydroxide ion is a negatively charged species; an OH free radical has no charge.

15.14 Biological Effects of Various Types of Radiation

The extent of the biological effects of a particular source of radiation depends on three major factors.

1. *The penetrating ability of the radiation.*   Alpha particles are the most massive and also the slowest particles generated in natural radioactive decay processes; they are emitted from nuclei at a velocity of about one-tenth the speed of light and have *low* penetrating power. Beta particles are emitted from nuclei at speeds up to nine-tenths the speed of light and can penetrate matter much deeper than alpha particles; they have *moderate* penetrating power. Gamma radiations are released at a velocity equal to the speed of light; they have *high* penetrating power, entering deeply into matter. Figure 15.14 contrasts the abilities of alpha, beta, and gamma rays to penetrate paper, aluminium foil, and a thin layer of a lead-concrete mixture.

2. *The ionizing ability of the radiation.*   The ionizing power of nuclear radiation is inversely related to its penetration power; that is, the greater the penetrating power, the weaker the ionizing ability. Gamma radiation is the most penetrating but produces the fewest ion pairs (Section 15.13) per unit volume. Beta radiation, moderate in penetrating power, has an

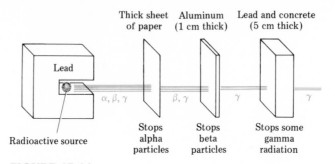

FIGURE 15.14

Penetrating abilities of alpha, beta, and gamma radiation.

intermediate ionizing ability, and the least penetrating, alpha radiation, produces the greatest ionizing effect. An alpha particle is approximately 8000 times heavier than a beta particle. It is estimated that a typical alpha particle travels 6 cm in air and produces 40,000 ion pairs, and a typical beta particle travels 1000 cm in air and produces about 2000 ion pairs. Gamma radiation causes only occasional ionization.

3. *The chemical properties of the radiation source.* Chemical properties determine how long a radiation source remains within the human body, which is an important factor in determining extent of biological damage. For example, the effects of krypton-85 and strontium-90 are not equal even though both produce beta particles. Krypton-85 is chemically unreactive (it is a rare gas; see Section 5.3) and consequently is rapidly eliminated from the body. Strontium-90, which is chemically similar to calcium (both are group IIA elements), collects in the bones where its long residence time can lead to diseases such as leukemia and bone cancer.

External human exposure to alpha particles, except in very large doses, is relatively harmless; the low-penetrating alpha radiation cannot penetrate the body's outer layers of skin. The major danger from alpha radiation occurs when alpha-emitting radionuclides are ingested, for example, in contaminated food. There are no protective layers of skin within the body.

With their greater penetrating ability, beta particles can penetrate the outer layers of human skin to damage living skin cells below the outer dead layers of skin. They can cause severe skin burns (somewhat like sunburn) if their source remains in contact with the skin for an appreciable time. As with alpha particles, if a substance that emits beta particles is ingested or inhaled, it can cause serious damage within the body.

Both external and internal exposure to gamma radiation is very dangerous. Gamma radiation penetrates human tissue, including skin, very effectively.

Contrasting *internal* exposure to the three types of radiation, alpha radiation can cause up to 10 times the damage beta or gamma radiation causes because of its greater ionizing ability.

15.15 Radiation Exposure Levels

How much radiation is harmful? The minimum radiation dosage that causes human injury is unknown. However, the effects of larger doses are known. Table 15.4 shows that very serious damage or death can result from large doses of ionizing radiation. The dosages causing the various effects listed in Table 15.4 are given in terms of the radiation unit called a *rem*, which is defined in the footnote to Table 15.4. The leakage of radiation associated with the Chernobyl nuclear power plant accident in the USSR produced many of the listed effects in the workers assigned to contain the leakage at the power plant site.

The probability of humans coming in contact with the radiation dosage necessary to cause the effects listed in Table 15.4 is extremely small. However, *low-level* exposure to ionizing radiation is something we encounter daily. In fact, there is no way we can totally avoid low-level radiation exposure because much of it derives from naturally occurring environmental sources.

Low-level radiation exposure sources fall into two categories: (1) natural or background radiation sources and (2) radiation sources associated with human activities. Before we consider major contributors to each of these source types,

TABLE 15.4

The Probable Effects on Humans of Short-Term Whole-Body Radiation Exposure

Dose (rems)	Effects
0–50	No consistent symptoms
50–200	Decreased white blood cells, nausea, vomiting; about 10% die within months at 200 rems
200–400	Loss of white blood cells, fever, hemorrhage, hair loss, nausea, vomiting, diarrhea, fatigue, skin blotches; about 20% die within months
400–500	Same symptoms as 200–400 rems but more severe, increased infections due to lack of white blood cells; 50% death rate within months, at 450 rems
500–1000	Severe gastrointestinal damage, cardiovascular collapse, central nervous system damage; doses above 700 rems fatal within a few weeks
10,000	Death in hours
100,000	Death in minutes

* A rem is the quantity of ionizing radiation that must be absorbed by a human to produce the same biological effect as 1 roentgen of high penetration X rays. A roentgen is the quantity of high penetration X rays that produces approximately 2×10^9 ion pairs per cm^3 of dry air 0°C and 1 atmosphere.

it is important to note that low-level radiation dosages are measured in milli-rems (mrem) rather than the rems used in Table 15.4. A millirem is equal to 0.001 rem. Thus, all of the low-level exposure dosages to be discussed are many times smaller than those needed to produce the effects listed in that table.

There are three major natural or background radiation sources with which humans must contend and over which they have very little control.

1. *Cosmic radiation from the upper atmosphere.*   The amount of cosmic radiation exposure a person receives depends on the altitude above sea level. The average annual dose of radiation at sea level is 40 mrem. To this exposure must be added 1 mrem for each 30 m (100 ft) a person lives above sea level.

2. *Rocks, soil, and building materials.*   The naturally occurring radionuclides potassium-40, thorium-232, and uranium-238 are present in almost all soil and rocks. They are also found in bricks, stone, and concrete used to construct houses, buildings, and roads. External radiation exposure from such sources ranges from 30 to 200 mrem, depending on location, with an average value of 55 mrem.

3. *Air, food, and water supply.*   The radionuclides most often found in air, food, and water are potassium-40, thorium-232, uranium-238, radon-222 (a gaseous decay product of uranium-238), and carbon-14. Such radio-nuclides, when in air, food, or water, are internal exposure sources. Potassium-40 is the predominant radionuclide in both food and water and accounts for over 90% of internal exposure. This naturally occur-ring potassium isotope is present in all potassium-containing materials. Annual radiation exposure from air, food, and water ranges from 20 to 400 mrem, depending on location and water supply, with 76 mrem being the average exposure value.

Exposure to low-level radiation dosages as the result of human activities occurs in many different ways, including the following.

1. *Medical and dental X-rays and tests.*   This is the largest human activity exposure source. Average exposure in the United States is 76 mrem per year. An individual X-ray film results in 22-mrem exposure, and a whole-mouth dental X-ray film involves 910 mrem exposure.

CIGARETTE SMOKING
AND RADIATION

2. *Cigarette smoking.*   Smoking a pack of cigarettes daily results in 40 mrem of annual radiation exposure. The ultimate source is small amounts of radioactive radium in the fertilizer used by tobacco growers. The radium decays into other radionuclides, which are taken up by the tobacco plant. These radionuclides vaporize when tobacco is smoked, and many lodge in the small passages of the lungs. Numerous scientists now believe that a smoker's higher risk of lung cancer is partly due to this radiation exposure.

3. *Air travel.* Air travelers receive additional cosmic ray exposure because of altitude. A transcontinental flight is equivalent to an annual 2.5-mrem dose of radiation, or 5 mrem per round trip.

4. *Fallout from nuclear weapons testing.* Open-air atmospheric testing of nuclear weapons from 1954 to 1962 resulted in increased amounts of air-borne radionuclides. Radiation exposure as the result of these tests was about 5 mrem per year in the early 1960s. With the cessation of such tests, exposure has decreased with time. It is now estimated to be about 1.5 mrem on an annual basis.

5. *Nuclear power plants.* Living within 8 km (5 miles) of a normally operating nuclear power plant adds, on the average, 0.6 mrem of radiation exposure. Living next door to a normally operating nuclear power plant would result in 4 mrem of additional annual exposure.

6. *TV and computer screens.* Two hours of viewing per day, on an annual basis, adds 4 mrem of radiation exposure to what a person otherwise receives.

7. *Brick and stone structures.* A person living in a stone or brick structure receives 40 mrem of annual radiation exposure. Having such a structure as an occupational workplace results in an additional 40 mrem of exposure.

Taking all of the preceding radiation exposure variables into consideration, the U.S. National Academy of Sciences estimates the average annual exposure per person in the United States to be 230 mrem, 130 mrem from background radiation and 100 mrem from human activities.

With low-level radiation exposure, cell damage rather than cell death usually occurs. If the damaged cells repair themselves properly, which is usually the case, no permanent damage occurs. If the damaged cells repair themselves improperly, which can occur, abnormal cells are produced when the cells replicate (divide). This leads to the concern that low-level radiation exposure may cause damage by altering genetic codes (Section 25.20). Much still needs to be learned about the long-term effects of cell damage caused by low-level radiation exposure.

Cells that reproduce at a rapid rate, such as those in bone marrow, lymph nodes, and embryonic tissue, are the most sensitive to low-level radiation damage. The sensitivity of embryonic tissue to radiation damage is the reason pregnant women need to avoid exposure to radiation from human activities. One of the first signs of overexposure to radiation is a drop in white blood cell count. This directly relates to the sensitivity of bone marrow, the site of white cell formation, to radiation.

Cancer cells, which are not subject to all of the normal controls of healthy cells, are more easily killed by radiation than normal, healthy cells. This is what makes radiation effective in treating certain types of cancers.

15.16   Detection of Radiation

You cannot hear, feel, taste, see, or smell low levels of radiation. However, there are numerous methods for detecting its presence. Becquerel initially discovered radioactivity (Section 15.2) as a result of its effect on photographic plates. Radiation affects photographic film in the same way as ordinary light; the film is exposed. Technicians and others who work around radiation usually wear film badges (see Figure 15.15) to record the extent of their exposure to radiation. The degree of darkening of the film negative indicates the extent of radiation exposure. By using different filters, various parts of the film register exposures to the types of radiation (alpha, beta, gamma, and X-rays).

Radiation can also be detected by making use of the fact that it ionizes atoms and molecules (Section 15.13). The Geiger counter operates on this principle.

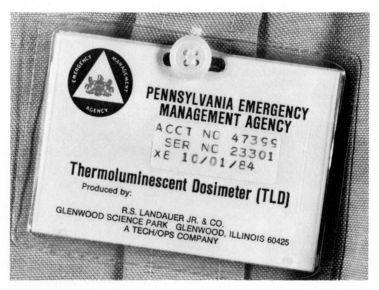

FIGURE 15.15

Film badges worn by technicians and others whose work requires them to be around radiation. The extent of exposure of all personnel is carefully logged to help prevent overexposure. [*Grant Heilman.*]

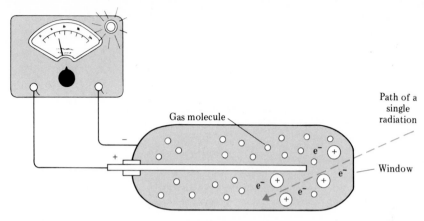

FIGURE 15.16

When radiation enters a Geiger counter through the window, it ionizes one or more gas atoms, producing ion pairs. The electrons from the ion pairs are attracted to the central wire, and the positive ions are drawn to the metal tube. This constitutes a pulse. of electric current, which is amplified and displayed on the meter or other readout.

The basic components of a Geiger counter are shown in Figure 15.16. The detector part of such a counter is a metal tube filled with a gas (usually argon). The tube has a thin-walled window made of a material that can be penetrated by alpha, beta, or gamma rays. In the center of the tube is a wire attached to the positive terminal of an electrical power source. The metal is attached to the negative terminal of the same source. Radiation entering the tube ionizes the gas, which allows a pulse of electricity to flow. This pulse is amplified and displayed on a meter or some other type of readout display.

SMOKE ALARMS Smoke detectors or smoke alarms are an important part of modern-day fire protection in a home. Many such devices involve radioactivity, and the principles of operation are similar to (the reverse of) those in a Geiger counter. Americium-241 is the radionuclide involved in smoke detectors; it decays by alpha-particle emission. Like all alpha particles, those emitted by americium-241 have a very low penetrating power, they do not escape from the detector, and thus do not constitute a health hazard.

Figure 15.17 shows a diagram of a home smoke detector. The alpha particles ionize air in the ionization chambers. The result is a flow of electric current that can be detected in an external circuit. When smoke enters the chamber, the smoke particles reduce the flow of current by impeding the movement of ions. This change is detected electronically and the alarm sounds.

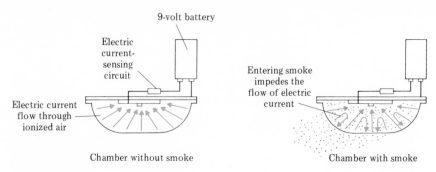

FIGURE 15.17
Americium-241 is used in many home smoke detectors; it is a source of alpha particles, which ionize air and cause a current to flow. Smoke decreases current flow, and an alarm sounds.

15.17 Nuclear Medicine

Radionuclides are used both diagnostically and therapeutically in medicine. In diagnostic applications, technicians use small amounts of radionuclides, whose progress through the body or localization in specific organs can be followed using radiation detection equipment. Larger quantities of radionuclides are used in therapeutic applications.

The fundamental chemical principle behind the use of radionuclides in diagnostic medical work is the fact that a radioactive nuclide has the same chemical properties as a nonradioactive isotope of the element. (Radioactive and nonradioactive isotopes of an element differ in nuclear but not in chemical properties.) Thus, body chemistry is not upset by the presence of a small amount of a radioactive substance that is already present in the body in a nonradioactive form.

The criteria used in selecting radionuclides for diagnostic procedures include the following.

1. At low concentrations (to minimize radiation damage) the radionuclide must be detectable by instrumentation placed outside the body. Almost all diagnostic radionuclides are gamma emitters because alpha and beta particles have a penetrating power that is too low (Section 15.14).
2. It must have a short half-life (Section 15.8), so that the intensity of the radiation is sufficient to be detected. A short half-life also limits the time of radiation exposure.

3. It must have a known mechanism for elimination from the body so that the material does not remain in the body indefinitely.

4. Its chemical properties must make it compatible with normal body chemistry and capable of being selectively transmitted to the body part or system under study.

BLOOD CLOTS AND SODIUM-24

The circulation of blood in the body can be followed with radioactive sodium-24. A small amount of this isotope is injected into the bloodstream in the form of a sodium chloride solution. The movement of this radionuclide through the circulation system can be followed easily with radiation-detection equipment. If it takes longer than normal for the nuclide to show up at a particular part of the body, this indicates the circulation is impaired in that region. Sodium-24 can also be used to locate blood clots because the amount of radiation will be high on one side and low on the opposite side of the clot.

Radiologists evaluate the functioning of the thyroid gland by administering iodine-131 to a patient, usually in the form of a sodium iodide (NaI) solution. The radioactive iodine behaves in the same manner as ordinary iodine and is absorbed by the thyroid at a rate related to the activity of the gland. In a hypothyroid condition, the amount accumulated is less than normal, and with a hyperthyroid condition a greater than average amount accumulates.

THYROID AND BRAIN SCANS

The size and shape of organs, as well as the presence of tumors, can be determined in some situations by scanning organs in which radionuclides have concentrated. Iodine-131 and technetium-99 are used to generate thyroid and brain scans. In the brain, technetium-99, in the form of a polyatomic ion (TcO_4^-), concentrates in a tumor more than in normal brain tissue; this helps radiologists determine the presence, size, and location of brain tumors. Figure 15.18 shows a brain scan obtained using technetium-99. In this figure, the bright spot at the left indicates a tumor that has absorbed more radioactive material than the normal brain tissue.

When radionuclides are used for therapeutic purposes, the objectives are entirely different from those for diagnostic procedures. The main objective for radionuclides in therapeutic use is *selective* destruction of abnormal (usually cancerous) cells. The radionuclide is often, but not always, placed within the body. There is no need to monitor the radiation produced with an external detector. Therapeutic radionuclides implanted in the body are usually alpha or beta emitters because an intense dose of radiation in a small localized area is needed.

A commonly used implantation radionuclide that is effective in the localized treatment of tumors is yttrium-90, a beta emitter with a half-life of 64 hours.

$$^{90}_{39}Y \longrightarrow {}^{90}_{40}Zr + {}^{0}_{-1}\beta$$

Yttrium-90 salts are implanted by inserting small hollow needles into the tumor.

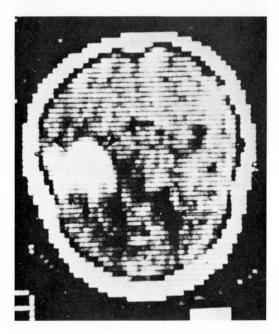

FIGURE 15.18
Brain scan obtained with radioactive technetium-99.
The bright spot on the left is a brain tumor [*Martin M. Rotker/Science Photo-Library–Taurus Photos.*]

CANCER AND
RADIATION

External, high energy beams of gamma radiation are also used extensively in treating certain cancers. Cobalt-60 is frequently used for this purpose; a beam of radiation is focused on the small area of the body where the tumor is located. This therapy usually causes some radiation sickness because normal cells are also affected, though to a lesser extent. The operating principle here is that abnormal cells are more susceptible to radiation damage than normal cells are. Radiation sickness is the price paid for destruction of the abnormal cells.

Practice Questions and Problems

NOTATION FOR NUCLIDES

15-1 Write an alternative nuclear symbol for each of the following nuclides.
  a. Nitrogen-14   b. $^{121}_{50}Sn$
  c. Chlorine-35   d. $^{24}_{11}Na$
  e. Boron-10   f. $^{197}_{79}Au$

15-2 Use two different notations to denote each of the following radioactive nuclides.
  a. Contains 4 protons, 4 electrons, and 6 neutrons

  b. Contains 20 protons, 20 electrons, and 18 neutrons
  c. Contains 41 protons, 41 electrons, and 55 neutrons
  d. Contains 99 protons, 99 electrons, and 157 neutrons

TYPES OF RADIATION

15-3 Three distinctive types of radiation are present in the emissions given of by naturally occurring radioactive substances.
  a. What are the names of these types of radiation?
  b. What notation is used to represent each?

c. What is the makeup of each of these radiations in terms of protons, neutrons, and electrons?

d. What charge is associated with each type of radiation?

EQUATIONS FOR RADIOACTIVE DECAY PROCESSES

15-4 Write balanced nuclear equations for the alpha decay of the following.

a. $^{200}_{84}Po$ b. Thorium-229

c. Fermium-252   d. $^{244}_{96}Cm$

15-5 Write balanced nuclear equations for the beta decay of the following.

a. $^{25}_{11}Na$ b. Germanium-77

c. Uranium-237   d. $^{104}_{45}Rh$

15-6 Write balanced nuclear equations for the positron decay of the following.

a. $^{8}_{5}B$ b. Rubidium-79

c. Barium-127   d. $^{103}_{47}Ag$

15-7 Write balanced nuclear equations for the electron capture decay of the following.

a. $^{59}_{28}Ni$ b. Palladium-100

c. Gold-189   d. $^{161}_{67}Ho$

15-8 What is the effect (change) on the mass number and atomic number of the parent nuclide when the following nuclear transmutations occur?

a. An alpha particle is emitted.

b. A beta particle is emitted.

c. An electron is captured.

d. A positron is emitted.

15-9 Supply the missing symbol in each of the following radioactive decay equations.

a. $^{10}_{4}Be \longrightarrow ^{10}_{5}B + ?$

b. $^{210}_{83}Bi \longrightarrow ^{4}_{2}\alpha + ?$

c. $^{41}_{20}Ca + ? \longrightarrow ^{41}_{19}K$

d. $^{15}_{8}O \longrightarrow ^{15}_{7}N + ?$

e. $^{44}_{22}Ti + ^{0}_{-1}e \longrightarrow ?$

f. $? \longrightarrow ^{4}_{2}\alpha + ^{222}_{86}Rn$

15-10 Write the nuclear equation for each of the following radioactive decay processes.

a. Tantalum-181 is formed by electron capture.

b. Cesium-139 is formed by beta emission.

c. Bromine-84 undergoes beta emission.

d. Copper-59 undergoes positron emission.

e. Mercury-200 is formed by alpha emission.

f. Sulfur-33 is formed by positron emission.

BOMBARDMENT REACTIONS

15-11 Identify the missing symbol in each of the following bombardment reactions.

a. $^{24}_{12}Mg + ? \longrightarrow ^{27}_{14}Si + ^{1}_{0}n$

b. $^{27}_{13}Al + ^{2}_{1}H \longrightarrow ? + ^{4}_{2}\alpha$

c. $? + ^{1}_{1}H \longrightarrow 2 (^{3}_{2}He)$

d. $^{14}_{7}N + ^{4}_{2}\alpha \longrightarrow ? + ^{1}_{1}H$

e. $? + ^{11}_{5}B \longrightarrow ^{257}_{103}Lr + 4 (^{1}_{0}n)$

f. $^{9}_{4}Be + ? \longrightarrow ^{12}_{6}C + ^{1}_{0}n$

15-12 Write equations for the following nuclear bombardment processes.

a. Beryllium-9 captures an alpha particle and emits a neutron.

b. Nickel-58 is bombarded with a proton and an alpha particle is emitted.

c. Bombardment of a radionuclide with an alpha particle results in the production of curium-242 and one neutron.

d. Bombardment of curium-246 with a small particle results in the production of $^{254}_{102}No$ and four neutrons.

15-13 How does the number of synthetically produced radionuclides compare with the number of naturally occurring nuclides?

15-14 Using Figure 15.2 as your source of information, determine the following.

a. Elements for which no stable isotopes exist.

b. Elements for which only one stable isotope exists.

c. Elements for which more stable than unstable isotopes exist.

d. Element for which the greatest number of stable isotopes exists.

e. Element for which the greatest number of unstable isotopes exist.

f. Element for which the greatest total number of isotopes exist (both stable and unstable).

NUCLEAR STABILITY

15-15 List two key factors that determine nuclear stability.

15-16 Calculate the neutron-to-proton ratio before and after each of the following processes takes place. Then indicate whether the ratio has increased or decreased as a result of the decay process.

a. Beta decay of neon-24

b. Alpha decay of $^{218}_{85}At$

c. Position decay of copper-59

15-17 What is the predominant mode of decay for a radionuclide in each of the following situations?

a. The neutron-to-proton ratio is too high for stability.

b. The neutron-to-proton ratio is too low for stability.

RATE OF RADIOACTIVE DECAY

15-18 What is meant by the term half-life when it is applied to a radionuclide?

15-19 Explain the fallacy in the conclusion that the whole-life of a radionuclide is equal to twice the half-life.

15-20 Technetium-99 has a half-life of 6.0 hours. What fraction of the technetium-99 atoms in a sample will remain after the following times?

a. 12 hours   b. 36 hours

c. 3 days   d. 3 half-lives

e. 6 half-lives   f. 10 half-lives

15-21 Sodium-24, which can be used to locate blood clots in the human circulatory system, has a half-life of 15.0 hours. If you start with 10.0 g of sodium-24, how much will be left in 60.0 hours?

15-22 Some objections to nuclear power plants are based on the need to store the radioactive wastes that result. One radionuclide found in reactor core wastes is strontium-90, which has a half-life of 28 years. How long would this nuclide have to be stored to decrease its amount to about 1/1000 of what was originally in the waste?

NUCLEAR FISSION AND NUCLEAR FUSION

15-23 Identify which of the following characteristics apply to the fission process, the fusion process, or both processes.

a. A high temperature is required to start the process.

b. An example of the process occurs on the sun.

c. Transmutation of elements occurs.

d. Radiation is emitted as a result of the process.

e. Neutrons are involved.

f. The process is now used to generate some electrical power in the United States.

15-24 Identify the following as fission or fusion reactions or as neither.

a. $^{3}_{2}He + ^{3}_{2}He \longrightarrow ^{4}_{2}He + 2\,(^{1}_{1}H)$

b. $^{239}_{92}U \longrightarrow ^{239}_{93}Np + ^{0}_{-1}\beta$

c. $^{235}_{92}U + ^{1}_{0}n \longrightarrow ^{144}_{55}Cs + ^{90}_{37}Rb + 2\,(^{1}_{0}n)$

d. $^{230}_{90}Th + ^{1}_{1}H \longrightarrow ^{223}_{87}Fr + 2\,(^{4}_{2}\alpha)$

e. $^{3}_{1}H + ^{2}_{1}H \longrightarrow ^{4}_{2}He + ^{1}_{0}n$

15-25 Why is uranium-235, which is much less abundant than uranium-238, considered more important than uranium-238 in nuclear fission processes?

15-26 What role do each of the following nuclides play in breeder reactor technology?

a. Uranium-233   b. Uranium-235

c. Uranium-238   d. Thorium-232

e. Plutonium-239

15-27 What role do each of the following nuclides play in fusion technology and from where are they obtained?

a. Hydrogen-2   b. Hydrogen-3

15-28 A "fourth state of matter" called a plasma is encountered in studying fusion reactions. What are the characteristics of matter in this state?

IONIZING EFFECTS OF RADIATION

15-29 What are ion pairs and how are they produced?

15-30 What is a free radical and how is it produced?

15-31 Write equations that show ion-pair formation and free-radical formation involving water.

BIOLOGICAL EFFECTS OF RADIATION

15-32 Contrast the penetrating powers of alpha, beta, and gamma rays into matter.

15-33 If only a heavy cardboard partition separated you from a radioactive source, would you be endangered more by alpha rays or gamma rays? Explain.

15-34 Suggest an explanation for the fact that alpha particles produce more ion pairs per unit volume than beta particles do.

15-35 Why is internal exposure to alpha or beta rays more dangerous to human health than external exposure to the same rays?

15-36 Contrast the speed at which alpha, beta, and gamma rays are emitted by a nucleus.

15-37 What are the three major factors that determine the extent of biological damage caused by radiation?

RADIATION EXPOSURE

15-38 What are the three major natural radiation sources of low-level human radiation exposure?

15-39 List five radionuclides that contribute to radiation exposure through air, food, and water.

15-40 In each of the following pairs of low-level radiation sources indicate which one would most likely be the greater radiation source per annum.

  a. Two-pack-a-day cigarette smoking or 8 hours/day of computer screen viewing

  b. Three dental X-rays or living near a nuclear power plant

  c. Nuclear weapon testing fallout or living in a brick home

  d. Working in a brick building or flying once a month from New York to Los Angeles and back.

NUCLEAR MEDICINE

15-41 The radionuclides used for diagnostic procedures are almost always gamma emitters. Why is this so?

15-42 The radionuclides used in diagnostic procedures almost always have short half-lives. Why is this so?

15-43 How do the radionuclides used for therapeutic purposes differ from the radionuclides used for diagnostic purposes?

15-44 Contrast the different manners in which cobalt-60 and yttrium-90 are used in radiation therapy.

Introduction to Organic Chemistry— Hydrocarbons

CHAPTER HIGHLIGHTS

16.1 Organic Chemistry— A Historical Perspective

In 1675, Nicholas Lémery (1645–1715), a French scientist and philosopher and an early writer of chemistry textbooks, published a book entitled *Cours de Chymie*. This book's significance, in the context of this chapter, is its inclusion, for the first time, of a classification scheme that distinguished between substances derived from plant or animal sources and those obtained from mineral constituents.

During the 1700s, this classification scheme became almost universally accepted by scientists. The term *organic* was introduced to refer to substances obtained from living matter (*organisms*) and the term *inorganic* to denote materials originating from nonliving (inanimate) matter such as minerals. At the same time, the belief developed that living organisms contained some mysterious "vital force" that was necessary to produce organic substances. This belief arose because no scientists of the time could synthesize any known organic material from inorganic materials. With each additional failure at such a synthesis, the "vital force theory" became more firmly entrenched. By 1800, it was universally accepted.

The vital force theory is now known to be incorrect. Its demise began in 1828 as a result of an experiment performed by the German chemist Friedrich Wöhler (1800–1882) (see Historical Profile 19). While heating two inorganic salts in an attempt to produce a new inorganic salt, Wöhler found that, instead, he had produced urea, a very well known organic compound found in urine. Wöhler's successful synthesis was the stimulus for renewed efforts by many scientists to synthesize organic substances from inorganic materials. This time, after a century of negative results, better techniques produced numerous successful reactions. By 1850 the "vital force theory" was laid to rest.

489

[Burndy Library, Norwalk CT.]

HISTORICAL PROFILE 19

Friedrich Wöhler   1800–1882

German chemist Friedrich Wöhler, the son of a schoolmaster in Escherheim (near Frankfurt-am-Main), obtained his formal training in the field of medicine at the Universities of Marburg and Heidelberg. After obtaining a medical degree, specializing in gynecology, he chose not to practice medicine. Instead, he became a student in the chemistry laboratory of Jöns Jacob Berzelius (Historical Profile 1 in Chapter 4) in Sweden. The rest of Wöhler's life was spent as a chemist.

At an early age Wöhler showed pleasure in experimenting and collecting. He did not get good grades in his early school years because he would rather conduct chemical experiments or collect minerals than study. He gradually transformed his room at home into a laboratory full of glasses, retorts, mortars, and pestles. He used the coal ovens in the kitchen in conducting experiments. His university landlord was very displeased upon finding that Wöhler had turned his rented quarters into a laboratory.

Wöhler is best known for his discovery, in 1828, of the transformation of the inorganic compound ammonium cyanate (NH_4OCN) to urea ($(NH_2)_2CO$). The impact of his urea synthesis was not dramatic; it did not cause the immediate downfall of the vital force theory. The discovery's significance is that it inspired other chemists to tackle again the problem of synthesizing organic compounds from inorganic starting materials. The cumulative effects from many inorganic to organic syntheses caused the ultimate downfall of the theory. The synthesis of urea was, however, a very exciting moment for Wöhler. In correspondence written right after the experiment, he wrote: "I must tell you that I can make urea without the use of kidneys, either man or dog."

Although Wöhler worked at times in the field of organic chemistry, his primary interest was inorganic chemistry. He was the first to isolate the elements aluminum and beryllium. He discovered the substance calcium carbide. He did extensive work with silicon, titanium, and boron compounds, discovering silicon hydrides.

In addition to research, he loved teaching. He spent considerable time translating and publishing Berzelius's works in German, and he wrote his own textbooks.

The significance of this portion of chemical history, in terms of modern chemistry, is that the terms *organic* and *inorganic*, whose origins lie in the vital force theory, are still used. The definitions, however, have changed.

Today, **organic chemistry** *is defined as the study of hydrocarbons (binary compounds of hydrogen and carbon) and their derivatives.* Interestingly, almost all compounds found in living organisms still fall within the category of organic chemistry when this modern definition is applied. In addition, many compounds synthesized in the laboratory, which have never been found in nature or in living organisms, are considered to be organic compounds.

In a less rigorous sense, organic chemistry is often defined as the study of carbon-containing compounds. It is true that almost all carbon-containing compounds qualify as organic compounds. There are, however, some exceptions. The oxides of carbon, carbonates, cyanides, and metallic carbides are all considered to be inorganic rather than organic compounds. The field of *inorganic chemistry*, from a modern viewpoint, encompasses the study of all noncarbon-containing compounds (the other 108 elements) plus the few carbon-containing compounds just mentioned.

In essence, organic chemistry is the study of one element (carbon), and inorganic chemistry the study of 108 elements. Why are the elements so unequally divided between these fields of study? The answer is simple. The chemistry of carbon is so much more extensive than that of the other elements. Approximately 6 million organic compounds are known; fewer than 500,000 inorganic compounds exist—approximately a 12-to-1 ratio.

Why does carbon form approximately 12 times as many compounds as all of the other elements combined? The reason is that carbon possesses the unique ability to bond to itself in long chains, rings, and complex combinations of both. Chains and rings of all lengths are possible. All such chains and rings may contain carbon-atom side chains as well.

Literally, the number of possible arrangements for carbon atoms bonded to each other is limitless. It has been calculated that 20 carbon atoms can be arranged in 366,319 different ways, based on a chain of atoms and allowing for side chains.

Figure 16.1 illustrates some of the possible ways of arranging carbon atoms to form organic molecules. Each of the carbon atoms in these structures is also involved in additional bonds to those shown—most often, to hydrogen atoms.

Organic compounds are the chemical basis for life itself, as well as the basis for our current high standard of living. Not only are proteins, carbohydrates, enzymes, and hormones organic molecules, but also natural gas, petroleum, coal, gasoline, and many synthetic materials, such as plastics, dyes, and fibers such as rayon, nylon, and dacron.

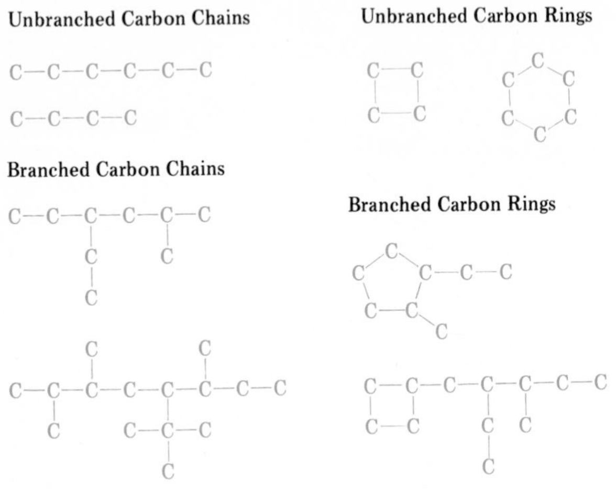

Unbranched Carbon Chains

C—C—C—C—C—C

C—C—C—C

Branched Carbon Chains

Unbranched Carbon Rings

Branched Carbon Rings

FIGURE 16.1

Simple and complex chains and rings of carbon atoms.

16.2   Hydrocarbons

The formal definition of organic chemistry, given in Section 16.1, suggests a logical way of organizing our study of the many known organic compounds. We will first consider hydrocarbons (this chapter) and then derivatives of hydrocarbons (Chapter 17)

As the name implies, **hydrocarbons** *are compounds that contain only the two elements hydrogen and carbon.* With such a limiting restriction on composition (only two elements), you might suppose only a few such compounds would exist. However, this is not the case; literally thousands of hydrocarbons exist (as we will see shortly).

There are three distinctly different families (or classes) of hydrocarbons: (1) saturated, (2) unsaturated, and (3) aromatic hydrocarbons. We discuss saturated hydrocarbons first (Sections 16.3 through 16.7), then unsaturated hydrocarbons (Sections 16.8 through 16.10), and finally aromatic hydrocarbons (Section 16.11).

16.3   Saturated Hydrocarbons

Saturated hydrocarbons *are compounds in which all of the carbon—carbon bonds are single bonds.* Saturated hydrocarbons may be divided into two groups—the alkanes and the cycloalkanes—based on the arrangement of carbon atoms. **Alkanes** *are saturated hydrocarbons in which the carbon atom arrangement is that of a unbranched or branched chain.* Cyclic arrangements of carbon atoms cannot be present in an alkane. **Cycloalkanes** *are saturated hydrocarbons in which at least one cyclic arrangement of carbon atoms is present.*

Before considering the structures of specific alkanes and cycloalkanes, let us note one general bonding characteristic of all carbon atoms in any saturated hydrocarbon, or for that matter in any hydrocarbon or hydrocarbon derivative: *carbon atoms that are present in hydrocarbons or hydrocarbon derivatives must always have four bonds.* We can understand this generalization when we consider the number of valence electrons that carbon possesses. Carbon is the first member of group IVA in the periodic table and thus possesses four valence electrons. To obtain a stable octet of electrons (recall the octet rule from Section 5.8) through covalent bonding, a carbon atom needs to share four electrons with other atoms; that is, carbon must form four covalent bonds.

Carbon can meet this four-bond requirement in three ways.

1. Bonding to four other atoms. This situation requires the presence of four single bonds

$$-\overset{|}{\underset{|}{C}}-$$

Four single bonds

2. Bonding to three other atoms. This situation requires the presence of two single bonds and one double bond.

$$-\overset{|}{C}=$$

Two single bonds and
one double bond

3. Bonding to two other atoms. This situation requires the presence of either two double bonds or a triple bond and a single bond.

        Two double bonds   One triple bond and
                                         one single bond

It is the first of these three possibilities, four single bonds, that is relevant to our discussion of saturated hydrocarbons. [The other two possibilities become important when we discuss unsaturated hydrocarbons (Section 16.8).] The word *saturated* indicates that the carbon atoms in saturated hydrocarbons are bonded to the maximum number of atoms possible; that is, the carbon atoms are "saturated" with respect to the number of other atoms to which they can be bonded.

16.4 Alkanes

The simplest alkane hydrocarbon is *methane*, which has the formula CH_4. The methane molecule has a tetrahedral structure. The carbon atom is found at the center of the tetrahedron, and the four hydrogen atoms bonded to the carbon are at the corners of the tetrahedron. Different representations of this tetrahedral structure for methane are shown in Figure 16.2.

The next member of the alkane hydrocarbon series is *ethane*, which has the molecular formula C_2H_6. This molecule may be thought of as a methane molecule with one hydrogen atom removed and a —CH_3 group put in its place. Perspective drawings of the ethane molecule, shown in Figure 16.3, illustrate that the bonds about the carbon atoms still have a tetrahedral orientation.

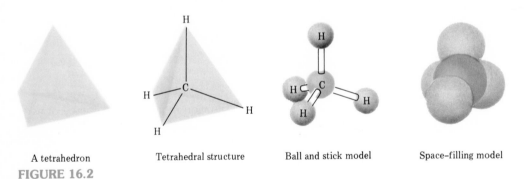

| A tetrahedron | Tetrahedral structure | Ball and stick model | Space–filling model |

FIGURE 16.2

Different ways of showing the tetrahedral arrangement of atoms in the methane (CH_4) molecule.

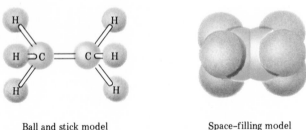

| Ball and stick model | Space–filling model |

FIGURE 16.3

Ball and stick and space-filling models of the ethane (C_2H_6) molecule.

Ball and stick model   Space–filling model

FIGURE 16.4
Ball and stick and space-filling models of the propane (C_3H_8) molecule.

Propane, the third member of the alkane series, has the molecular formula C_3H_8. Once again, we can produce this formula by removing a hydrogen atom from the preceding compound (ethane) and substituting a —CH_3 group. Because all six hydrogen atoms of ethane are equivalent, it makes no difference which one we replace. Both ball and stick and space-filling models of the propane molecule are shown in Figure 16.4.

All alkanes have molecular formulas that fit the generalized formula C_nH_{2n+2}, where *n* is the number of carbon atoms present. This general formula indicates that the number of hydrogen atoms present in an alkane is always double the number of carbon atoms plus two more. This mathematical formula can be applied to any alkane, regardless of its size. For example, an alkane with 20 carbon atoms would have a molecular formula of $C_{20}H_{42}$. The number of hydrogen atoms present is always two more than twice the number of carbon atoms. The formulas for methane, ethane, and propane are all in agreement with the generalized formula for alkanes.

16.5   Structural Isomerism

It should be apparent that the procedures outlined in the last section in establishing the structures of the ethane and propane molecules (replacement of a hydrogen with a —CH_3 group) can be used to generate other members of the alkane series. A complication arises, however, when four or more carbon atoms are present—different structures may be obtained, depending on which hydrogen is replaced.

Two different structural arrangements of four carbons can be produced by removing a hydrogen atom from propane and replacing it with a —CH_3 group. This is the result of all the hydrogen in propane not being geometrically equivalent. The two hydrogens attached to the central carbon atom in propane

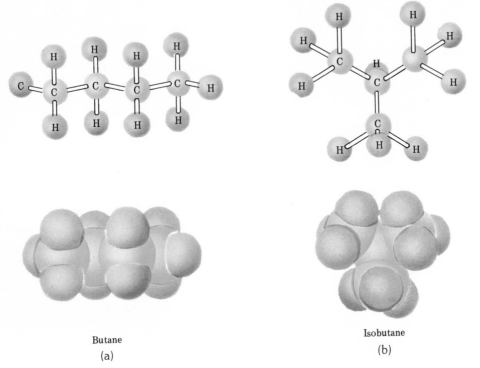

Butane

(a)

Isobutane

(b)

FIGURE 16.5

Ball and stick and space-filling models of (a) butane and (b) isobutane.

(Figure 16.4) are equivalent to each other but distinct from the six associated with the end carbons, which in turn are all equivalent to each other. Replacing a hydrogen on an end carbon gives *butane*, the compound shown in Figure 16.5a. The compound in Figure 16.5b, *isobutane*, is the result of a —CH_3 group replacing a hydrogen on the central carbon atom.

Butane and isobutane, although they both have the same molecular formula of C_4H_{10}, are two different compounds with different properties. The melting point of butane is $-138.3°C$ and that of isobutane is $-160°C$. The boiling points of the two compounds are, respectively, $-0.5°C$ and $-12°C$. Their densities at $20°C$ also differ: 0.579 g/mL for butane and 0.557 g/mL for isobutane.

Compounds such as butane and isobutane are called structural isomers. **Structural isomers** *are compounds that have the same molecular formula but different structural formulas, that is, different arrangements of atoms within the molecule.* Structural isomers, as we will see, are not rare in organic chemistry; in fact, they are the rule rather than the exception. The phenomenon of structural isomerism is the major reason why there are so many organic compounds.

In Figure 16.5, three-dimensional representations were used for the structures of butane and isobutane. Such representations, which give both the arrangement and spatial orientation of the atoms in a molecule, are often difficult to draw, especially for those with no artistic talent, and they are time-consuming to draw. Because of these drawbacks, an easier system to indicate structure has been developed. This alternative system involves structural formulas. **Structural formulas** *are two-dimensional (planar) representations of the arrangement of the atoms in molecules.* They include complete information about the arrangement of the atoms in a molecule, but not their spatial orientation. The structural formulas for butane and isobutane are

$$CH_3\text{—}CH_2\text{—}CH_2\text{—}CH_3 \qquad CH_3\text{—}\underset{\underset{\displaystyle CH_3}{|}}{CH}\text{—}CH_3$$

Butane   Isobutane

Note that in structural formulas each carbon atom is followed by the attached hydrogens. The number of attached hydrogens is determined by the number of other carbon atoms to which a given carbon atom is bonded. Because each carbon atom must have four bonds (Section 16.4), carbon atoms attached to only one other carbon need three hydrogen atoms (CH_3). Carbons bonded to two other carbons need two hydrogens (CH_2), those bonded to three other carbons need only one hydrogen (CH), and those bonded to four other carbons need no hydrogens.

In alkanes with five carbon atoms, we find three structural isomers.

$$CH_3\text{—}CH_2\text{—}CH_2\text{—}CH_2\text{—}CH_3 \qquad CH_3\text{—}\underset{\underset{\displaystyle CH_3}{|}}{CH}\text{—}CH_2\text{—}CH_3 \qquad CH_3\text{—}\overset{\overset{\displaystyle CH_3}{|}}{\underset{\underset{\displaystyle CH_3}{|}}{C}}\text{—}CH_3$$

Pentane   Isopentane   Neopentane

These three C_5H_{12} isomers, like the two C_4H_{10} isomers, are distinctly different compounds with different properties.

The number of possible structural isomers increases rapidly with the number of carbon atoms in an alkane. There are five alkane isomers with six carbons (all with the formula C_6H_{14}) and nine isomers with seven carbons (all C_7H_{16}). A listing of the number of possible alkane isomers, as a function of the number of carbon atoms present, is given in Table 16.1. Obviously, no one has prepared all the isomers for compounds that contain a large number of carbon atoms. However, methods are available for their synthesis if the need arises.

The three isomers of C_5H_{12} are pentane, isopentane, and neopentane. What prefixes do we use to name the five isomers of C_6H_{14}? What about names for the 9 C_7H_{16} isomers and the 75 $C_{10}H_{22}$ isomers? Obviously, using prefixes to distinguish between various isomers is not workable when the number of isomers becomes large; we rapidly run out of prefixes to use. Even if we had a sufficient number of prefixes, how could we remember the exact meaning of

TABLE 16.1

Possible Number of Isomers for Selected Alkanes

Molecular Formula	Possible Number of Isomers
CH_4	1
C_2H_6	1
C_3H_8	1
C_4H_{10}	2
C_5H_{12}	3
C_6H_{14}	5
C_7H_{16}	9
C_8H_{18}	18
C_9H_{20}	35
$C_{10}H_{22}$	75
$C_{15}H_{32}$	4,347
$C_{20}H_{42}$	336,319
$C_{30}H_{62}$	4,111,846,763

each prefix? It would be an impossible task. There must be a better way, and there is. A systematic nomenclature system called the IUPAC system (for the International Union of Pure and Applied Chemistry) assigns to organic compounds names that relate directly to their structures. This system keeps use of prefixes to a minimum. The logic of the system also minimizes rote memorization. The IUPAC rules for naming alkanes (both branched and unbranched) are discussed in Section 16.6.

16.6   IUPAC Nomenclature for Alkanes

When only a relatively few organic compounds were known, chemists used what are known today as their common names, which were selected arbitrarily. Isopentane and neopentane (Section 16.5) are examples. As more and more compounds became known, it became obvious that a systematic method for naming organic compounds was needed. The IUPAC nomenclature system meets that need.

For nomenclatural purposes, we classify alkanes in two categories: normal chain and branched chain. In **normal chain** *hydrocarbons, all carbon atoms are connected in a continuous nonbranching chain.* In **branched-chain** *hydrocarbons, one or more side chains of carbon atoms are attached at some point to a continuous chain of carbon atoms.* The two isomeric four-carbon alkanes (Section 16.5) illustrate this classification system.

$$CH_3-CH_2-CH_2-CH_3 \qquad CH_3-\overset{\displaystyle |}{\underset{\displaystyle \underset{CH_3}{|}}{CH}}-CH_3$$

Normal chain   Branched chain

In the branched-chain compound the longest continuous chain of carbon atoms is three, to which is attached (on the middle carbon of the three) a —CH_3 group (the branch).

Let us first consider IUPAC names for normal chain alkanes because these compounds are structurally the simplest of all alkanes. IUPAC names for such alkanes with 1 through 10 carbons are given in Table 16.2. Note that all of the names end in -*ane*, the characteristic ending for all alkanes. The first four names do not indicate the number of carbon atoms in the molecule, but beginning with five carbon atoms the Greek numerical prefixes give the number of carbon atoms present. It is important to memorize the names for these normal chain alkanes because they are the basis for the entire IUPAC nomenclature system. (The IUPAC system also includes names for normal chain alkanes with more than 10 carbon atoms, but we will not consider them. The names given in Table 16.2 are sufficient for our purposes.)

Branched-chain alkanes always contain alkyl groups. The alkyl groups are the branches. An **alkyl group** *is an alkane from which one hydrogen atom has been removed.* It is the removal of the one hydrogen atom that allows the branches (alkyl groups) to be attached to a main carbon chain. The general formula for an alkyl group is C_nH_{2n+1}. Alkyl groups do not lead a stable, independent existence, but rather are always attached to another group of atoms.

The two most commonly encountered alkyl groups are the two simplest ones—the methyl group and the ethyl group.

$$-CH_3 \qquad\qquad -CH_2-CH_3$$

Methyl group   Ethyl group

(The bond on the left in the alkyl group structures denotes the point of attachment to the carbon chain.) Note how alkyl groups are named: the stem of the name of the parent alkane plus the ending -*yl*.

Two three-carbon alkyl groups exist, differing from each other in their points of attachment to the main carbon chain.

$$-CH_2-CH_2-CH_3 \qquad\qquad -\overset{\displaystyle |}{\underset{\displaystyle \underset{CH_3}{|}}{CH}}-CH_3$$

Propyl group   Isopropyl group

A propyl group is attached to a carbon chain through an end carbon atom, whereas an isopropyl group is attached by the middle carbon of the three propyl group carbons. The isopropyl group is an example of a branched chain alkyl

TABLE 16.2

IUPAC Names for Normal Chain Alkanes with 1 Through 10 Carbon Atoms

CH_4	methane
C_2H_6	ethane
C_3H_8	propane
C_4H_{10}	butane
C_5H_{12}	pentane
C_6H_{14}	hexane
C_7H_{16}	heptane
C_8H_{18}	octane
C_9H_{20}	nonane
$C_{10}H_{22}$	decane

group, that is, a branched branch. Other branched alkyl groups, containing more carbon atoms, also exist. Rules for naming these more complex groups are not considered in this text.

To name branched-chain alkanes—that is, alkanes in which the carbon atoms are not arranged in a continuous chain—the following rules are used.

RULE 1   *Select the longest continuous carbon atom chain in the molecule as the base for the name.*

This longest carbon atom chain is named as in Table 16.2.

Example: $CH_3-CH_2-CH-CH_2-CH_2-CH_2-CH_3$

$$| \atop CH_3$$

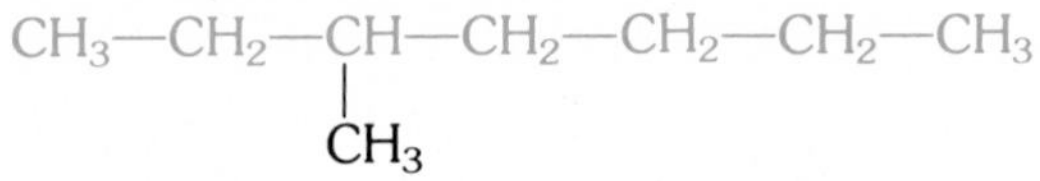

In this example the longest continuous carbon atom chain (shown in color) consists of seven carbon atoms. Therefore, the base name (but not the complete name) for this compound is *heptane*.

Example: $CH_3-CH_2-CH-CH_2-CH-CH_3$

$$CH_3 \qquad CH_2$$
$$CH_2$$
$$CH_3$$

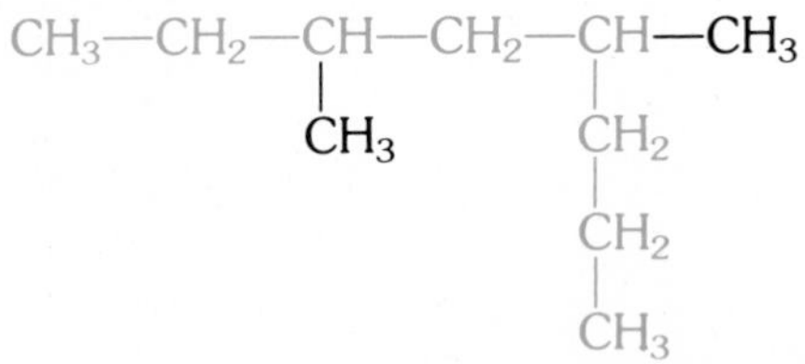

In this example the longest continuous carbon chain (shown in color) possesses eight carbon atoms. Note that the carbon atoms in the longest continuous chain do not necessarily lie in a straight line. The base name (but not the complete name) for this alkane is *octane*.

RULE 2   *The carbon atoms in the longest continuous chain of carbon atoms are numbered consecutively from the end nearest a branch (alkyl group).*

Since a chain always has two ends, there are always two ways to number the chain: either from left to right or from right to left. The effect of this rule is to give the first-encountered alkyl group the lowest number possible.

$$CH_3 \longleftarrow \text{Alkyl group}$$

Example: $CH_3-CH_2-CH_2-CH-CH_2-CH_3$

| 1 | 2 | 3 | 4 | 5 | 6 | (left to right) |
| 6 | 5 | 4 | 3 | 2 | 1 | (right to left) |

Carbon atom to which the alkyl group is attached

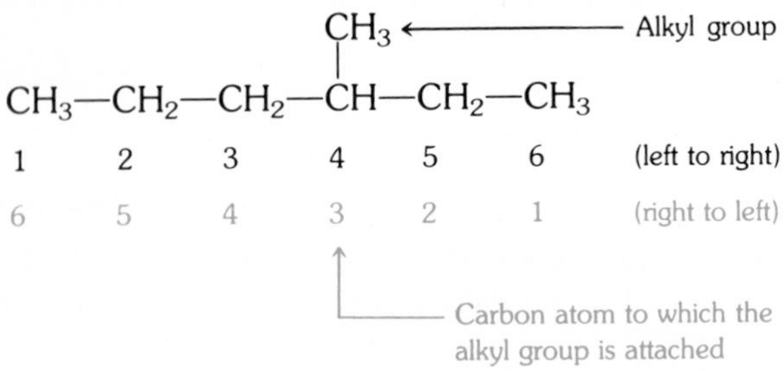

In this example, the right-to-left numbering system is used because the alkyl group is nearer the right end of the chain than the left end.

If two or more alkyl groups are present on a carbon chain and both numbering systems (left–right and right–left) give the same number for the first-encountered alkyl group, then the second-encountered alkyl group is used to distinguish between numbering systems.

$$\text{CH}_3$$
$$|$$
Example: $\text{CH}_3-\text{CH}_2-\overset{}{\underset{\underset{\text{CH}_3}{|}}{\text{C}}}-\text{CH}-\text{CH}-\text{CH}_2-\text{CH}_3$

with $\text{CH}_3 \quad \text{CH}_3$ below the CH positions

| 1 | 2 | 3 | 4 | 5 | 6 | 7 | (left to right) |
| 7 | 6 | 5 | 4 | 3 | 2 | 1 | (right to left) |

In this example there are four alkyl groups (side chains) attached to the main chain. Both numbering systems give carbon 3 as the position of the first alkyl group. Considering the first two encountered alkyl groups the left–right numbering systems gives 3 and 3 for the alkyl group locations and the right-left numbering system gives 3 and 4. (Note that when a carbon atom carries two alkyl groups that carbon's number must be counted twice.) Thus, the left–right numbering system (3,3) is used instead of the right–left numbering system (3,4).

RULE 3   *The complete name for an alkane contains the location by number and the name of each alkyl group and the name of the longest carbon chain.*

The alkyl group names, with their locations, always precede the name of the base chain of carbon atoms.

Example: $\overset{1}{\text{CH}_3}-\overset{2}{\underset{\underset{\text{CH}_3}{|}}{\text{CH}}}-\overset{3}{\text{CH}_2}-\overset{4}{\text{CH}_2}-\overset{5}{\text{CH}_3}$ is 2-methylpentane

Example: $\overset{1}{\text{CH}_3}-\overset{2}{\text{CH}_2}-\overset{3}{\underset{\underset{\text{CH}_3}{|}}{\text{CH}}}-\overset{4}{\text{CH}_2}-\overset{5}{\text{CH}_3}$ is 3-methylpentane

Note that each name is written as one word, with a hyphen between the number (location) and the name of the alkyl group.

RULE 4   *If two or more of the same kind of alkyl group are present in a molecule, the number is indicated with a Greek numerical prefix. In addition, a number specifying the location of each identical group must be included. These position numbers, separated by commas, precede the numerical prefix. Numbers are separated from words by hyphens.*

Example: $\overset{1}{CH_3}-\overset{2}{CH}-\overset{3}{CH_2}-\overset{4}{CH}-\overset{5}{CH_3}$ is   2,4-dimethylpentane

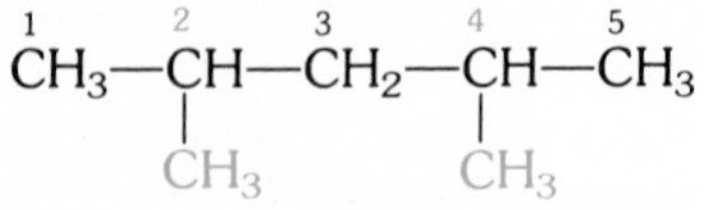

                         CH_3 CH_3

Example: $\overset{1}{CH_3}-\overset{2}{CH_2}-\overset{3}{\underset{CH_3}{\overset{CH_3}{C}}}-\overset{4}{CH_2}-\overset{5}{CH_3}$ is   3,3-dimethylpentane

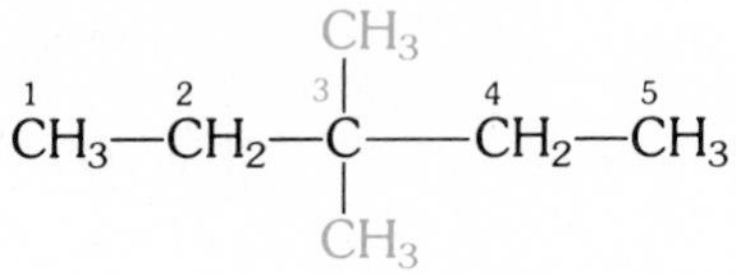

Note that the prefix "di" must always be accompanied by two numbers, "tri" by three, and so on, even if the same number is written twice, as in 3,3-dimethylpentane.

RULE 5   *When two kinds of alkyl groups are present on the same carbon chain, each group is numbered separately and the names of the alkyl groups are listed in alphabetical order.*

Example: $\overset{5}{CH_3}-\overset{4}{CH_2}-\overset{3}{CH}-\overset{2}{CH}-\overset{1}{CH_3}$ is   3-ethyl-2-methylpentane

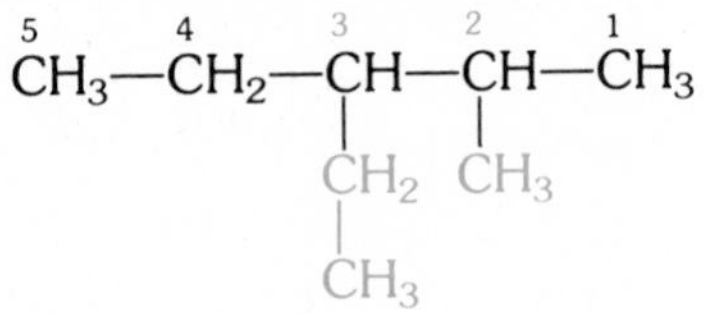

                           CH_2 CH_3

                           CH_3

Note that ethyl is named first in accordance with the alphabetical rule determining the order in which alkyl groups are listed.

Example: $\overset{1}{CH_3}-\overset{2}{CH_2}-\overset{3}{CH}-\overset{4}{CH}-\overset{5}{CH}-\overset{6}{CH_2}-\overset{7}{CH_2}-\overset{8}{CH_3}$

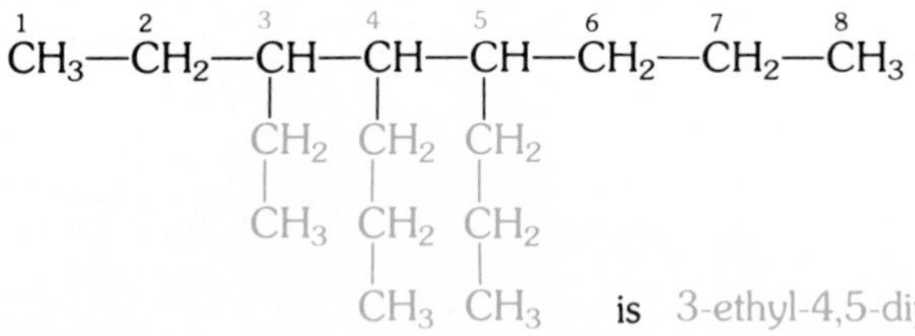

                         CH_2 CH_2 CH_2

                         CH_3 CH_2 CH_2

                         CH_3 CH_3 is   3-ethyl-4,5-dipropyloctane

Note that the prefix di- does not affect the alphabetical order for the alkyl groups; ethyl precedes propyl.

EXAMPLE 16.1

In Section 16.5 common names were used in naming the four-carbon alkane isomers (butane and isobutane) and the five-carbon alkane isomers (pentane, isopentane, and neopentane). What is the IUPAC name for each of these compounds.

SOLUTION

(a) Butane, CH_3—CH_2—CH_2—CH_3

The IUPAC name is butane. Thus, the common and IUPAC names correspond.

(b) Isobutane, CH_3—$\underset{\underset{\displaystyle CH_3}{|}}{CH}$—$CH_3$

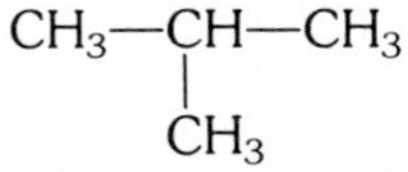

The longest chain contains three carbon atoms, and the attached methyl group is at carbon 2, so the name is 2-methylpropane.

(c) Pentane, CH_3—CH_2—CH_2—CH_2—CH_3

The IUPAC name is pentane. Thus, the common and IUPAC names correspond.

(d) Isopentane, CH_3—$\underset{\underset{\displaystyle CH_3}{|}}{CH}$—$CH_2$—$CH_3$

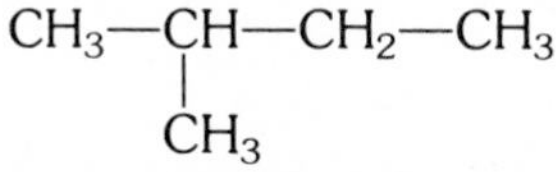

The longest chain contains four carbon atoms, and a methyl group is attached to the chain at carbon 2. Thus, the IUPAC name is 2-methylbutane.

(e) Neopentane, CH_3—$\overset{\overset{\displaystyle CH_3}{|}}{\underset{\underset{\displaystyle CH_3}{|}}{C}}$—$CH_3$

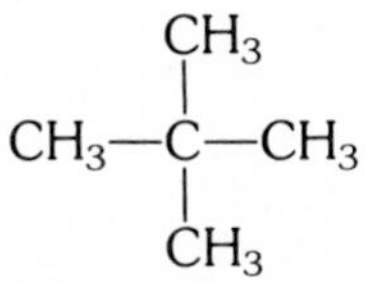

The longest chain contains only three carbons and two alkyl groups are attached to it. Both alkyl groups are methyl groups and both are atttached to carbon 2. Thus, the IUPAC name is 2,2-dimethylpropane

Structural formulas are easily obtained from correct IUPAC names since the name contains all the information necessary to draw a structure. Example 16.2 illustrates the process of converting an IUPAC name to a structural formula.

EXAMPLE 16.2

Draw the structural formula of the alkane whose IUPAC name is 4,4-diethyl-2,5-dimethyloctane.

SOLUTION

STEP 1   The IUPAC name indicates that the base chain contains eight carbon atoms (octane). Draw an octane skeleton (no hydrogens) and number it.

$$\overset{1}{C}-\overset{2}{C}-\overset{3}{C}-\overset{4}{C}-\overset{5}{C}-\overset{6}{C}-\overset{7}{C}-\overset{8}{C}$$

STEP 2   Place two ethyl groups on carbon number 4.

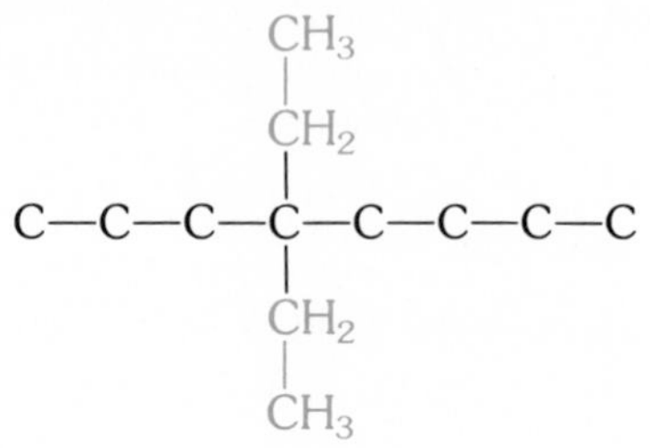

$$CH_3$$
$$|$$
$$CH_2$$
$$|$$
$$C—C—C—C—C—C—C—C$$
$$|$$
$$CH_2$$
$$|$$
$$CH_3$$

STEP 3   Place methyl groups on carbons 2 and 5.

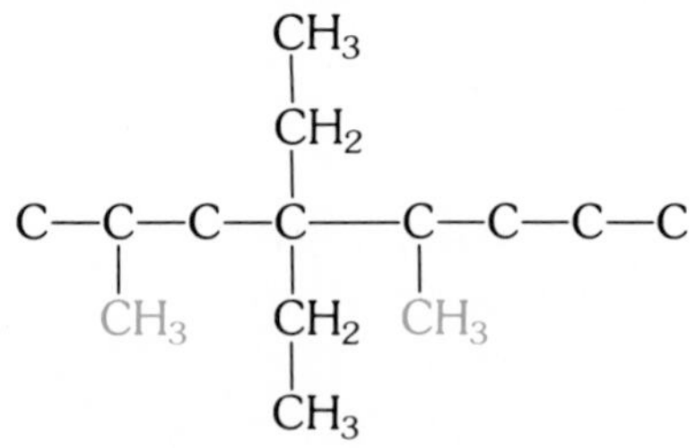

$$CH_3$$
$$|$$
$$CH_2$$
$$|$$
$$C—C—C—C———C—C—C—C$$
$$\quad\quad|\quad\quad\ |\quad |$$
$$\quad\quad CH_3\quad CH_2\ CH_3$$
$$\quad\quad\quad\quad |$$
$$\quad\quad\quad\quad CH_3$$

STEP 4   Add necessary hydrogen atoms to the carbon base chain so that each carbon atom has four bonds.

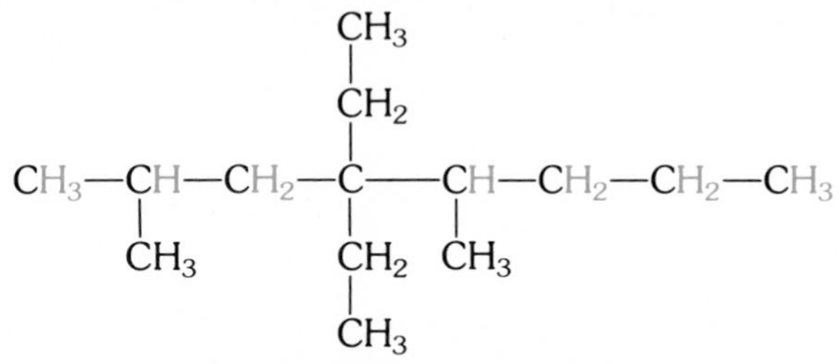

$$CH_3$$
$$|$$
$$CH_2$$
$$|$$
$$CH_3—CH—CH_2—C———CH—CH_2—CH_2—CH_3$$
$$\quad\quad\quad\ |\quad\quad\ |\quad\ |$$
$$\quad\quad\quad\ CH_3\quad CH_2\ CH_3$$
$$\quad\quad\quad\quad\quad\ |$$
$$\quad\quad\quad\quad\quad\ CH_3$$

This structure has 14 carbon atoms. It should therefore have 30 hydrogen atoms (C_nH_{2n+2}). Counting the number of hydrogen atoms enables you to check the correctness of the structure.

16.7   Cycloalkanes

In Section 16.3 we noted that saturated hydrocarbons can be subclassified into two families: alkanes and cycloalkanes. Recall, from Section 16.3, that cycloalkanes contain cyclic arrangements of carbon atoms.

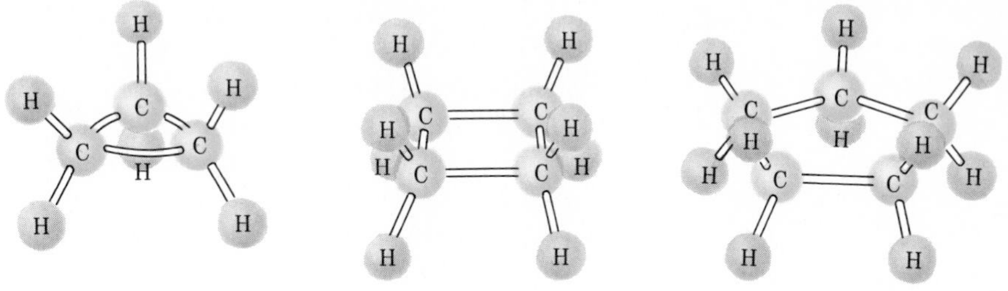

FIGURE 16.6

Ball and stick models of simple cycloalkanes.

The simplest cycloalkane is cyclopropane, which is a cyclic arrangement of three carbon atoms; it takes a minimum of three carbon atoms to form a cyclic system. Figure 16.6 shows ball and stick models of cyclopropane, cyclobutane (a four-membered carbon ring), and cyclopentane (a five-membered carbon ring).

The IUPAC nomenclature for cycloalkanes is very similar to that for alkanes (Section 16.6). There are only two modifications to the alkane nomenclature rules we have already discussed.

1. The prefix *cyclo-* is placed before the name that corresponds to the non-cyclic chain with the same number of carbon atoms as the ring.
2. When alkyl groups are present, they are located by numbering the carbons in the ring according to a system that yields the lowest numbers for the carbons at which alkyl groups are attached.

EXAMPLE 16.3

Assign IUPAC names to each of the following cycloalkanes.

(a) 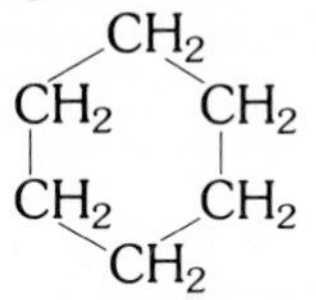
$$\begin{array}{ccc} & CH_2 & \\ CH_2 & & CH_2 \\ CH_2 & & CH_2 \\ & CH_2 & \end{array}$$

(b) 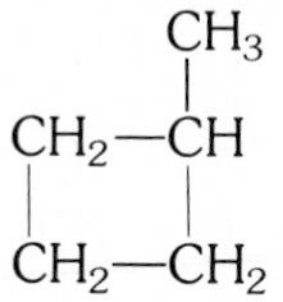
$$\begin{array}{c} CH_3 \\ | \\ CH_2-CH \\ | \quad\quad | \\ CH_2-CH_2 \end{array}$$

(c)
$$\begin{array}{cc} CH_2 & CH_3 \\ CH_2 & CH \\ CH_2 & CH_2 \\ & CH \\ & | \\ & CH_3 \end{array}$$

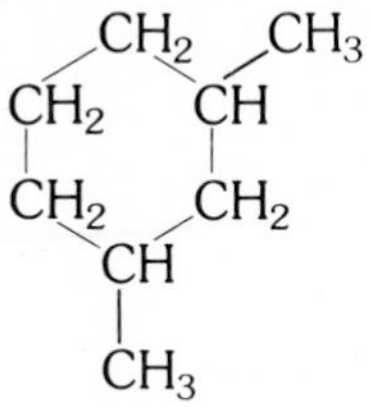

SOLUTION

(a) The six-carbon alkane is called *hexane*. Six carbon atoms in a cyclic arrangement is called cyclohexane.
(b) The molecule is a cyclobutane (a four-membered carbon ring) to which a methyl group is attached. Its name is simply methylcyclobutane The

position of a single alkyl group on a cycloalkane ring does not need to be located with a number because all positions on the ring are equivalent to each other and, thus, there is only one way to attach the methyl group.

(c) This compound is a dimethylcyclohexane. The positions of the two methyl groups, relative to each other, must be specified with numbers. The ring is numbered so that the methyl groups are on carbons 1 and 3. We want to keep the numbers as low as possible.

This compound's full name is 1,3-dimethylcyclohexane Note that in numbering a ring to which alkyl groups are attached, the number 1 is always assigned to one of the carbon atoms that bears an alkyl group.

For brevity and simplicity in drawing cycloalkane structures, a geometrical figure is often used to represent the cyclic part of a cycloalkane structure: a triangle for a three-membered ring, a square for a four-carbon ring, and so on. When these geometrical figures are used, each corner is assumed to represent a carbon atom, along with the number of hydrogens needed to give the carbon four bonds. In the following example, this geometrical figure notation is used in drawing structural formulas for cyclobutane, 1,2-dimethylcyclopropane, and 1-ethyl-4-methylcyclohexane.

Cyclobutane   1,2-Dimethylcyclopropane   1-Ethyl-4-methylcyclohexane

The general formula for cycloalkanes is C_nH_{2n}. Thus, any given cycloalkane contains two fewer hydrogen atoms than an alkane with the same number of carbon atoms. (Recall, from Section 16.4, that the general formula for an alkane is C_nH_{2n+2}.) Thus, butane (C_4H_{10}) and cyclobutane (C_4H_8) are not isomers; isomers must have the same molecular formula (Section 16.5).

It is easy to visualize why cycloalkanes contain two fewer hydrogen atoms than the corresponding alkanes. Consider what would have to occur for butane to be converted to cyclobutane. First, the terminal hydrogen atom from both ends of the butane chain would have to be removed. Then, the two end carbons could be connected together, giving the cyclic structure. Cyclization would not be possible without the removal of the hydrogen atoms, because the two end carbon atoms would already have four bonds.

As with alkanes, structural isomers are possible for cycloalkanes. For example, there are four cycloalkane structural isomers that have the formula C_5H_{10}: one based on a five-membered ring, one based on a four-membered ring, and two based on a three-membered ring. These isomers are

Note that the last isomer has both methyl groups attached to the same carbon atom. The name of this isomer is 1,1-dimethylcyclopropane.

16.8 Unsaturated Hydrocarbons

In saturated hydrocarbons all carbon–hydrogen and carbon–carbon bonds are single bonds. A single bond is the only type of bond possible between carbon and hydrogen. No such restriction exists for bonds between two carbon atoms. With carbon atoms, single, double, and triple bond formation is possible. An **unsaturated hydrocarbon** *is a hydrocarbon that contains one or more carbon–carbon double or triple bonds.*

Unsaturated hydrocarbons are divided into two groups: those that contain double bonds and those that contain triple bonds. We will discuss the hydrocarbons containing carbon–carbon double bonds first (Section 16.9). The principles we learn about these compounds are applicable, with only slight modification, to hydrocarbons containing carbon–carbon triple bonds (Section 16.10).

16.9 Alkenes

Alkenes *are noncyclic hydrocarbons that contain one or more carbon–carbon double bonds.* The simplest alkene is ethene (common name, ethylene), which has the formula C_2H_4. Obviously, a one-carbon alkene cannot exist because

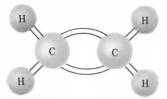

Ball and stick model

Space–filling model

FIGURE 16.7

Ball and stick and space-filling models of ethene, the simplest alkene.

two carbon atoms are required to have a carbon–carbon double bond. Ball and stick and space-filling models of ethene are shown in Figure 16.7. Note from the three-dimensional ball and stick model in Figure 16.7 that the arrangement of bonds about the carbon atoms is not tetrahedral as it is in alkanes; it is planar. The two carbon atoms and four hydrogen atoms all lie in a plane with an angle of 120 degrees between bonds. Any carbon involved in multiple bonding always has a nontetrahedral arrangement of bonds about it. (The term *multiple bond* is commonly used collectively to refer to both double and triple bonds.)

The general formula for an alkene containing only one double bond is C_nH_{2n}. Thus alkenes contain two fewer hydrogen atoms than the maximum number possible in hydrocarbons (alkanes). The reduced number of hydrogen atoms is caused by the presence of the double bond; two hydrogen atoms must be lost for a double bond to form.

In Section 16.7, we noted that the general formula for a cycloalkane is C_nH_{2n}. Thus, alkenes containing one double bond and cycloalkanes have the same general formula. This means that compounds from these two families with the same number of carbon atoms are isomeric.

Cycloalkenes *are hydrocarbons that have a ring structure and also one or more double bonds within the ring structure.* Cycloalkenes containing one double bond have the general formula C_nH_{2n-2}. This general formula reflects the loss of four hydrogen atoms from maximum (C_nH_{2n+2}); two hydrogen atoms are lost because of the double bond and two more as a result of cyclization.

The simplest cycloalkene is the compound cyclopropene, C_3H_4, a three-membered carbon ring containing one double bond.

The rules for naming alkanes and cycloalkanes (Sections 16.6 and 16.7) can be used, with slight modifications, to name alkenes and cycloalkenes. The modifications needed to name alkenes and cycloalkenes containing only one double bond are as follows.

1. The *-ane* ending characteristic of alkenes is changed to *-ene* for alkenes. The *-ene* ending means that a double bond is present.
2. The root name for the alkene is derived from the longest continuous chain (or ring) of carbon atoms *containing the double bond.*
3. For noncyclic molecules containing more than three carbon atoms, the position of the double bond must be specified because it may occupy more than one position. Carbon atoms in the longest chain containing the double bond are numbered consecutively from the end nearest the double bond. The double-bond position is given by a single number (the number of the lower-numbered carbon involved in the double bond), which is placed immediately in front of the base chain name.

Examples: $\overset{1}{C}H_2\!=\!\overset{2}{C}H\!-\!\overset{3}{C}H_2\!-\!\overset{4}{C}H_3$ is 1-butene

and

$\overset{1}{C}H_3\!-\!\overset{2}{C}H\!=\!\overset{3}{C}H\!-\!\overset{4}{C}H_3$ is 2-butene

The compounds 1-butene and 2-butene are isomers. Note that in the case of 1-butene the double bond involves carbons 1 and 2, but we only write the lower of the two numbers in the name. For 2-butene, the double bond involves carbons 2 and 3 and we only write the number 2 in the name.

4. For noncyclic molecules where alkyl groups are present in addition to the double bond, the numbering system chosen is the one that assigns the lowest possible number to the carbon atom at which the double bond originates. Numbering the alkyl groups is of lower priority than numbering the double bond.

Example: $\overset{4}{C}H_3\!-\!\overset{3}{C}H\!-\!\overset{2}{C}H\!=\!\overset{1}{C}H_2$ is 3-methyl-1-butene
                      |
                    CH_3

The chain is always numbered in such a way as to give the double bond the lowest number possible, *even if this means that alkyl groups must get higher numbers.*

5. For cycloalkenes containing one double bond, the ring is numbered to give the double-bonded carbons the numbers 1 and 2. The direction of numbering is chosen so that any alkyl groups present receive the lowest numbers possible. The position of the double bond is not given because we know it is between the number 1 and 2 carbon atoms.

Example: 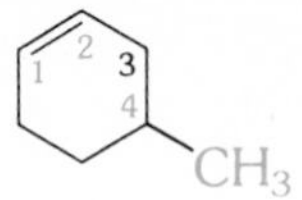   is   4-methylcyclohexene

---

EXAMPLE 16.4

Assign IUPAC names to each of the following alkenes or cycloalkenes.

(a) CH_2=CH—CH_2—CH_2—CH_2—CH_3

(b) CH_3—$\overset{\displaystyle |}{\underset{\displaystyle CH_3}{CH}}$—$CH$=$CH$—$CH_3$

(c) CH_3—CH_2—$\overset{\displaystyle |}{\underset{\displaystyle \overset{\displaystyle CH_2}{\underset{\displaystyle |}{CH_3}}}{C}}$=$CH_2$

(d)

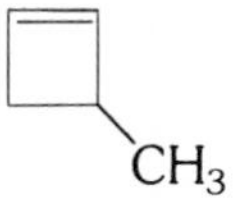

SOLUTION

(a) The longest continuous chain containing the double bond is six carbons long. If this were an alkane, the base name would be *hexane*. In this case, the base name is *hexene* because the *-ane* ending must be changed to *-ene* to indicate the presence of the double bond. The double bond is located between carbons 1 and 2. Therefore, the IUPAC name is 1-hexene.

(b) The longest carbon chain containing the double bond has five carbons. This gives the molecule the base name *pentene*. Numbering the chain from right to left so that the double bond receives the lowest number possible results in the name 4-methyl-2-pentene.

$$\overset{5}{CH_3}—\overset{4}{\underset{\displaystyle \underset{\displaystyle CH_3}{|}}{CH}}—\overset{3}{CH}=\overset{2}{CH}—\overset{1}{CH_3}$$

If we had numbered the chain in the opposite direction, from left to right, we would have obtained the name 2-methyl-4-pentene. This name is incorrect even though the numbers are the same as in the correct name. The correct numbering system is the one that assigns the lowest number possible to the double bond.

(c) You may be tempted to identify the longest carbon chain, which is five carbons long, as the base for the name of this compound. That is incorrect. The base name for an alkene is derived from the longest carbon chain *that contains the double bond*. The longest chain containing the double bond has four carbon atoms. Therefore, the proper base name is *butene*. The chain is numbered from right to left— from the end closest to the double bond. An ethyl group is attached on carbon 2. The complete name for this alkene is 2-ethyl-1-butene

$$\overset{4}{C}H_3-\overset{3}{C}H_2-\overset{2}{C}=\overset{1}{C}H_2$$
$$|$$
$$CH_2$$
$$|$$
$$CH_3$$

(d) The base name for a cyclic alkene is determined by the number of carbon atoms in the ring; in this case, it is *cyclobutene*. The IUPAC rules state that we must number the ring so that the first carbon of the double bond is number 1. We then proceed to number the ring in the direction of the double bond, so that the other carbon of the double bond is number 2. This numbering system gives the number 3 to the carbon to which the methyl group is attached.

The complete IUPAC name for this cycloalkene is 3-methylcyclobutene The location of the double bond is not explicitly given because it is understood that it will always involve carbons 1 and 2.

Numerous unsaturated hydrocarbons contain more than one double bond. Compounds containing two double bonds are properly referred to as *alkadienes* and those containing three double bonds as *alkatrienes*.

The general family names just mentioned give us the key to naming compounds containing more than one double bond. A prefix is added to the base chain ending to indicate the number of double bonds—*diene* for two double bonds, *triene* for three double bonds, and so on. In addition, a separate number locates *each* double bond. If there are three double bonds, three numbers are needed.

EXAMPLE 16.5

Name the following compounds, each of which contains more than one multiple bond.

(a) CH_2=CH—CH=CH_2 (b) 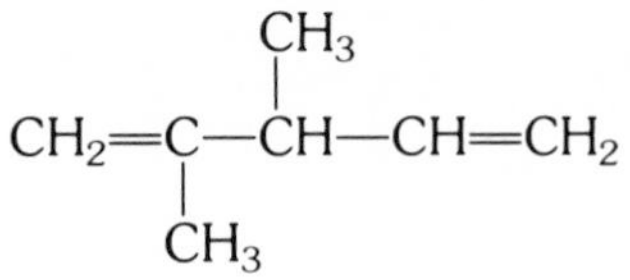
$$CH_2=C-\overset{\overset{\displaystyle CH_3}{|}}{C}H-CH=CH_2$$
with CH_3 below the C.

SOLUTION

(a) The base name for this hydrocarbon is *butadiene*, since two double bonds are present in a chain of four carbons. It does not matter which way we number the chain because of the symmetrical nature of the molecule. Either way, the double bonds involve carbons 1 and 3, and the compound is therefore named 1,3-butadiene

(b) 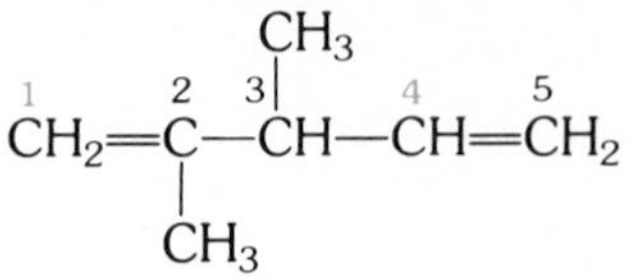
$$\overset{1}{CH_2}=\overset{2}{C}-\overset{\overset{\displaystyle CH_3}{\overset{|}{3}}}{C}H-\overset{4}{C}H=\overset{5}{C}H_2$$
with CH_3 below carbon 2.

This molecule is a *pentadiene* with two methyl groups on it. It does not matter which end we number from, in relation to the double bonds, because of their symmetrical positioning within the molecule. It does, however, matter where we start in relation to the alkyl groups. The first-encountered alkyl group is on carbon 2 when numbering begins at the left and on carbon 3 when numbering begins at the right. According to the left–right numbering system, the IUPAC name for this compound is 2,3-dimethyl-1,4-pentadiene

---

16.10 Alkynes

Alkynes *are noncyclic hydrocarbons that contain one or more carbon–carbon triple bonds.* Alkynes possessing one triple bond have two fewer hydrogen atoms than the corresponding alkene and four fewer hydrogen atoms than the corresponding alkane. Structural formulas for the two-carbon entities in each of these families are as follows.

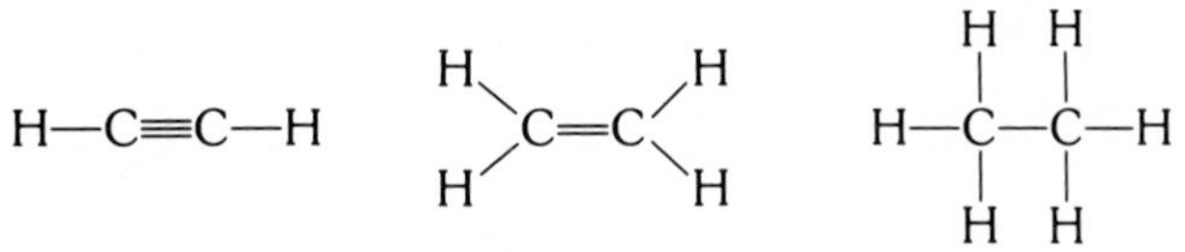

$$H-C\equiv C-H \qquad \underset{H}{\overset{H}{\diagdown}}C=C\underset{H}{\overset{H}{\diagup}} \qquad H-\underset{\underset{H}{|}}{\overset{\overset{H}{|}}{C}}-\underset{\underset{H}{|}}{\overset{\overset{H}{|}}{C}}-H$$

Ethyne (C_2H_2),   Ethene (C_2H_4),   Ethane (C_2H_6),
the simplest alkyne   the simplest alkene   the two-carbon alkane

The general formula for alkynes containing a single triple bond is C_nH_{2n-2}. Thus, the simplest member of this series has the formula C_2H_2 and the next member, $n = 3$, the formula C_3H_4. The presence of a carbon–carbon triple bond in a molecule always results in a linear arrangement for the two atoms attached to the carbons of the triple bond; that is, all four atoms will always lie in a straight line. Figure 16.8 shows ball and stick and space-filling models that illustrate the linear geometry for C_2H_2, the simplest alkyne. The IUPAC name for this compound is ethyne and its common name is acetylene. (Unfortunately, the common name ends in -ene, which incorrectly suggests that a double bond is present.)

The rules for naming alkynes are identical to those used for naming alkenes (Section 16.9) except the ending -*yne* is used to denote the presence of a triple bond. The -*yne* ending replaces the -*ene* ending used in alkene nomenclature. Cycloalkynes, molecules containing a triple bond as part of a ring structure, are known, but they are not common. The same statement applies to alkynes containing more than one triple bond.

Ball and stick model

Space–filling model

FIGURE 16.8
Ball and stick and space-filling models of ethyne, the simplest alkyne.

16.11   Aromatic Hydrocarbons

Cyclic alkenes can contain more than one double bond. For example, introducing a second double bond into the cyclohexene molecule produces either 1,3-cyclohexadiene or 1,4-cyclohexadiene

1,3-Cyclohexadiene   1,4-Cyclohexadiene

depending on where the additional double bond is positioned relative to the first one. These two compounds are isomers and differ only in the positions of the double bonds (1,3- versus 1,4-).

The introduction of a third double bond into 1,3-cyclohexadiene produces the compound

This compound contains a cyclic system of alternating double and single bonds. Using the IUPAC nomenclature rules for alkenes presented in Section 16.9, we would expect to call this compound *1,3,5-cyclohexatriene*. However, instead, this compound is named *benzene*. The reason for this rule violation is that benzene does not possess the chemical properties normally associated with cyclic hydrocarbons containing double bonds. Its properties are different enough from those of these other compounds that it is considered to be the first member of a new series of hydrocarbons called aromatic hydrocarbons. **Aromatic hydrocarbons** *are hydrocarbons that contain the characteristic benzene ring as part of their structure.*

Benzene, which has the molecular formula C_6H_6, is considered the parent molecule for all aromatic hydrocarbons. An understanding of the chemical bonding that occurs within the benzene molecule is the key to understanding aromatic hydrocarbon chemistry. We have noted that the structure given as

meaning

represents a compound that is not named as a cyclohexatriene. What is the reason for this?

The cyclohexatriene interpretation of the benzene structure implies that alternating single and double carbon–carbon bonds are present about the carbon ring. With this interpretation, we would predict that not all carbon–carbon bonds in the ring are of the same length because carbon–carbon double bonds are known to be shorter than carbon–carbon single bonds. This prediction is contrary to the experimental information available on bond lengths in the benzene molecule. Experimental studies indicate that all carbon–carbon bonds in the benzene molecule are of equal length. Thus, the bonding present in the carbon ring must be something different from alternating single and double bonds.

If all of the carbon–carbon bonds in benzene were of the same kind—that is, all single bonds or all double bonds—we would have a molecule whose carbon–carbon bond lengths were consistent with experimental bond length data. However, neither of these structures is acceptable; in both structures the carbon atoms violate the octet rule (Section 5.3). The all-single-bond suggestion leaves each carbon atom with only three bonds instead of the needed four bonds. In an all-double-bond molecule, each carbon atom would have too many bonds—five per carbon atom.

To obtain a structure for benzene consistent with both the octet rule and experimental bond length information, a new type of bonding concept must be invoked—that of delocalized bonding. In **delocalized bonding** *three or more atoms share the same valence electrons.* (Up to now all bonding discussions in the text have involved localized bonds, that is, bonds in which electrons are shared between *two* atoms.) Let us now consider how the delocalized bond concept helps explain the characteristics of benzene and other aromatic compounds.

Carbon atoms have four valence electrons for use in bonding. Three of the four electrons on each carbon atom in benzene will be used in forming the "structural framework" of a benzene molecule, as shown in the following diagram.

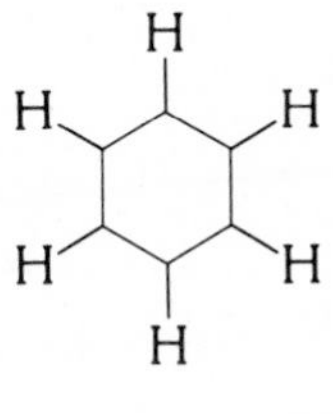

In this structure each carbon atom has formed three bonds. Thus, each carbon atom still has one more electron available for bonding, which is shown as a dot in the following structure.

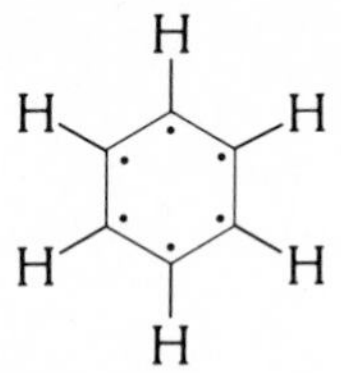

This additional electron, in each case, is in a *p* orbital (Section 4.6) whose orientation is such that it may interact simultaneously with two other *p* orbitals—the *p* orbitals on each side of it in the carbon ring. This results in a continuous interaction around the ring and a delocalized bond that "runs" completely around the ring, as shown in Figure 16.9. In this delocalized bond, the six electrons (one from each carbon atom) are considered to be shared equally by all six carbon atoms.

A circle drawn inside the hexagon is used to denote the delocalized bond present in benzene.

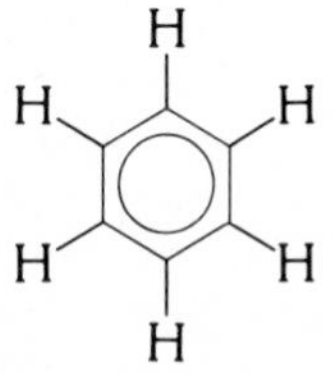

Note that in this structure each carbon atom has four bonds: one bond to an H atom, two localized bonds to other C atoms, and one bond from involvement in the delocalized bond.

Delocalized bonding is possible only for specific orientations of atomic orbitals. Aromatic hydrocarbons are the only hydrocarbon class for which proper orbital orientation is present.

The presence of a delocalized bond gives a molecule extra stability. Because of this extra stability, benzene and the other aromatic hydrocarbons have chemical properties that are different from those of other hydrocarbons.

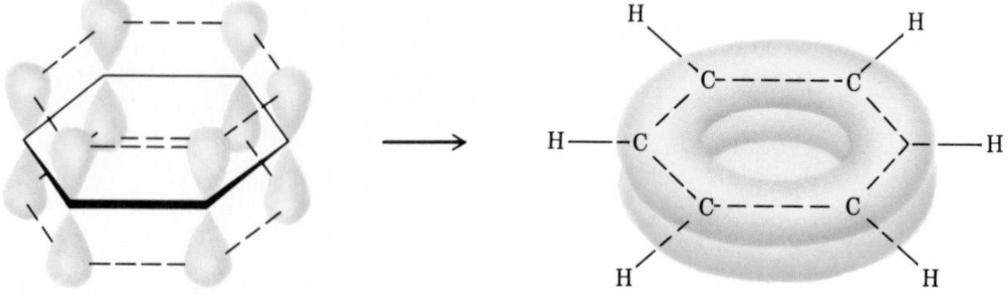

FIGURE 16.9

The six *p* orbitals of benzene interact to form a delocalized bond "running" completely around the ring.

The hydrogen atoms on benzene may be replaced by alkyl groups without destroying the delocalized bonding present in the carbon ring. Replacements of this type give rise to numerous other members of the aromatic hydrocarbon family. The simplest aromatic hydrocarbon (next to benzene) is methylbenzene (common name, toluene), a compound in which one hydrogen on the ring has been replaced by a methyl group.

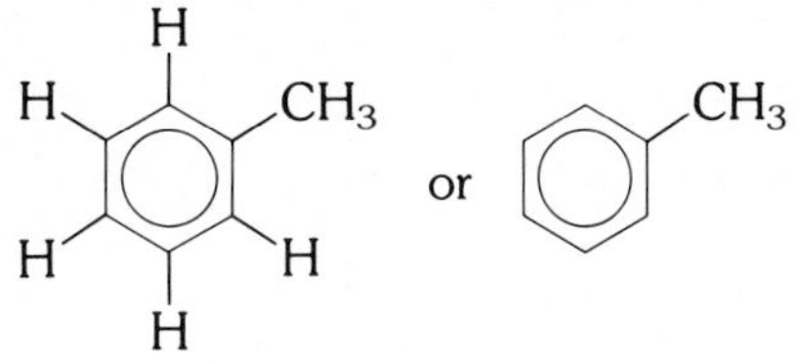

Methylbenzene
or
Toluene

Three isomeric compounds are possible when two methyl groups are placed on a benzene ring. IUPAC names for these compounds are 1,2-,1,3-, and 1,4-dimethylbenzene. Their structures are, respectively,

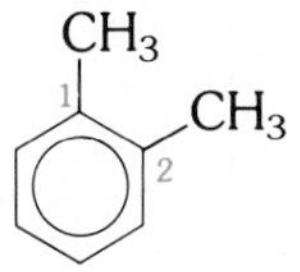

1,2-Dimethylbenzene 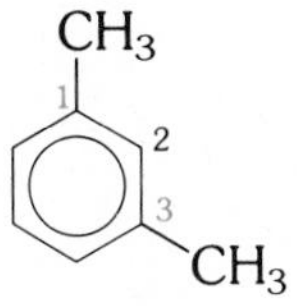   1,3-Dimethylbenzene 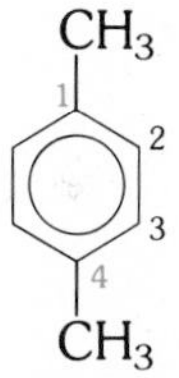   1,4-Dimethylbenzene

A common nomenclature system for indicating positioning on the carbon ring in disubstituted benzenes uses the prefixes *ortho-* (substituents that are adjacent to each other—that is, 1,2-), *meta-* (substituents that are one carbon removed from each other—that is, 1,3), and *para-* (substituents that are two carbons removed from each other—that is, 1,4-). These prefixes are often abbreviated as *o-*, *m-*, and *p-*. Thus, for the three dimethylbenzenes (common name xylene) we have the following additional acceptable names.

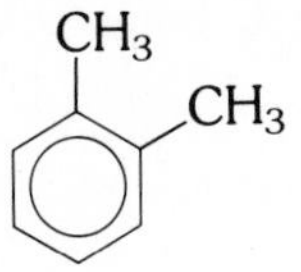

ortho-Dimethylbenzene   meta-Dimethylbenzene   para-Dimethylbenzene
ortho-Xylene   meta-Xylene   para-Xylene
o-Xylene   m-Xylene   p-Xylene

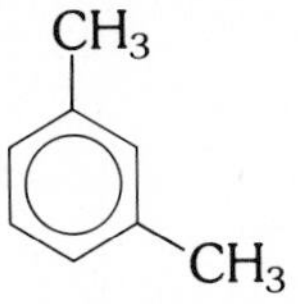

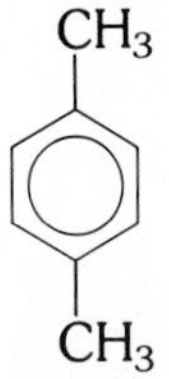

When three or more groups are attached to a benzene ring, the carbon atoms in the ring are given numbers, which indicate the position of the groups. The numbering system that gives the lowest possible sum is the system selected.

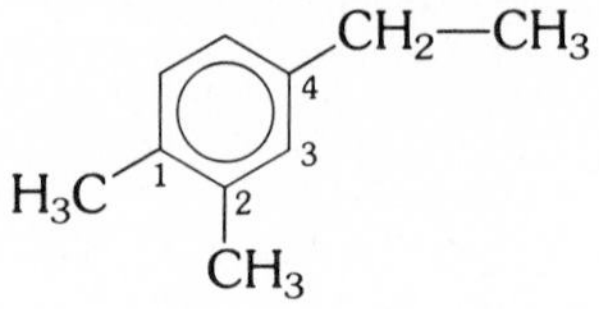

4-Ethyl-1,2-dimethylbenzene

Note that attachments to a benzene ring need not be identical, as the previous example illustrates.

Sometimes it is necessary to consider the benzene ring as a substituent group bonded to a chain or ring. When such is the case, the benzene side chain is called a *phenyl group*. For example, the following compound is 3-phenyl-heptane.

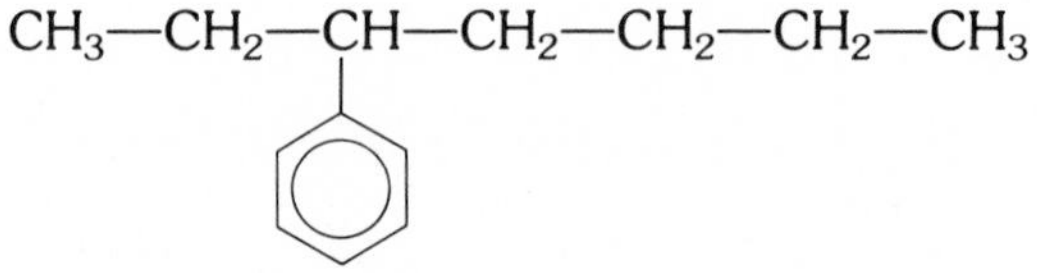

Another type of aromatic hydrocarbon contains *fused* benzene rings. The simplest example of such a compound is naphthalene, commonly sold as moth repellent.

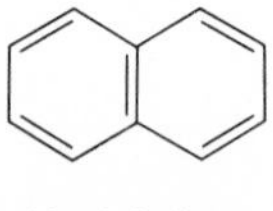

Naphthalene

Another fused-ring aromatic hydrocarbon is benzpyrene, which has been identified as one of the cancer-causing agents in tobacco smoke.

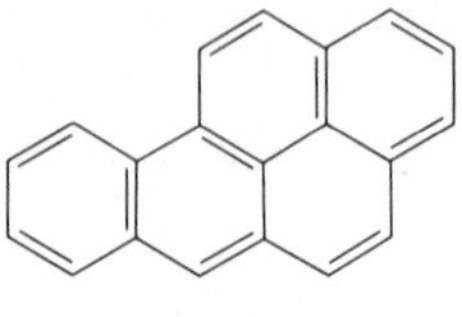

Benzpyrene

16.12 Sources and Uses of Hydrocarbons

There are three major natural sources of hydrocarbons: natural gas, crude oil (petroleum), and coal. Collectively, these substances are known as *fossil fuels*.

Unprocessed natural gas contains methane (50–90%) and ethane (1–10%), with smaller amounts of propane and butanes. Processed natural gas, the "natural gas" used in homes for cooking and as a heating fuel, is mainly methane.

Crude petroleum is an extremely complex mixture of hydrocarbons. Some crudes consist chiefly of noncyclic alkanes; others contain as much as 40% of other types of hydrocarbons—cycloalkanes and aromatics. In its natural state, crude petroleum has very few uses, but this complex hydrocarbon mixture can be separated into various useful fractions through refining (Section 19.4). The resulting fractions are still hydrocarbon mixtures, but each one is simpler (fewer compounds are present). Useful hydrocarbon fractions obtained from crude petroleum include petroleum ether, gasoline, kerosene, fuel oil, lubricating oils, greases, and asphalt.

Although petroleum is by far the leading natural source of hydrocarbons, coal is another significant source. When coal is heated in the absence of air, the coal breaks down into three components: (1) coal gas, (2) coke, and (3) coal tar (a liquid). Coal gas, mainly methane and hydrogen, is used as a fuel. Coke, (Section 13.4), the major product, is essentially pure carbon. Coal tar is a mixture of numerous hydrocarbons, many of which are aromatic. Coal tar is a significant source for aromatic hydrocarbons.

Fossil fuels (hydrocarbons) are an absolute "must" in the world in which we now live. Our high standard of living is directly tied to the availability of these substances. Because of their importance, an entire chapter of the text (Chapter 19) is devoted to them.

16.13 Properties of Selected Hydrocarbons

In this section we consider briefly the properties and uses of selected individual hydrocarbons: methane (CH_4), the simplest alkane; ethene (C_2H_4), the simplest alkene; ethyne (C_2H_2), the simplest alkyne; and benzene (C_6H_6), the simplest aromatic hydrocarbon.

Methane is a colorless, odorless gas under normal conditions. Beside being the chief constituent of natural gas, it is also formed naturally by the anaerobic (oxygen-free) decomposition of organic materials by microorganisms. Such oxygen-free conditions are found in swamps and marshlands, where decaying plant matter is covered with water. In this context, methane is called "swamp gas." Anaerobic decomposition also forms methane gas in sewage treatment plants. We previously mentioned (Section 12.4) that methane is the source for the hydrogen used to make ammonia.

Ethene (ethylene) is also a gas under normal conditions. The reactivity of this compound, caused by the presence of the double bond, results in negligible amounts of ethene being present in fossil fuels. Large quantities of ethene, produced from reactions involving alkanes, are used to manufacture motor fuels, polymers (Chapter 18), and other hydrocarbons. Ethene also has a biological source. It is produced as fruit ripens and causes the change in the color

FIGURE 16.10

Once unripe picked fruit reaches its destination, treatment with ethene (C_2H_4) hastens the ripening process. [*Courtesy of Robert J. Ouellette, The Ohio State University.*]

of the skin of ripened fruit. Fruit growers and marketers must pick fruit green and firm and ship it to its destination in that state in order to have it arrive in acceptable condition. Such shipped fruit is treated with ethene at its destination, which helps give the fruit the "correct" appearance in its ripened state (see Figure 16.10).

Ethyne (acetylene), also a gas at room temperature and pressure, is used mainly as fuel for oxyacetylene cutting and welding torches and as an intermediate in the manufacture of other substances. Years ago, miners used helmet lanterns in which acetylene, produced by mixing small amounts of calcium carbide and water, was burned. These carbide lamps were used extensively until the development of today's safer and more efficient battery-powered lamps.

Benzene, unlike methane, ethene, and ethyne, is a colorless liquid at room temperature. It is a very good solvent for numerous substances, and many of its uses depend on this property. In addition, benzene is the starting material for the production of hundreds of other valuable aromatic compounds. Use of benzene in chemical laboratories as a solvent was discontinued in the late 1970s based on findings that continued inhalation of benzene vapors can reduce white blood cell counts in humans and that it has caused leukemia in laboratory test animals.

16.14 Hydrocarbons: A Summary

This chapter has focused on hydrocarbons, the simplest organic compounds. Because of the unique properties of carbon, almost limitless variations in hydrocarbon structure are possible. Beyond structural isomers for saturated hydrocarbons, the ability of carbon to form multiple bonds further extends the range of possible compounds.

Classification of the many hydrocarbons into families (alkanes, alkenes, cycloalkenes, and so on) eases the task of dealing with their diversity. It is appropriate, as we end our consideration of hydrocarbons, to summarize the relationships among hydrocarbon families.

Hydrocarbons fall into two broad categories: saturated and unsaturated. Both saturated and unsaturated hydrocarbons are of two types: open chain (noncyclic) and cyclic. Cyclic unsaturated hydrocarbons can be further classified as aromatic or nonaromatic. Aromatic hydrocarbons are of two types: those based on a single ring of carbon atoms and those containing fused carbon rings. Figure 16.11 shows diagrammatically the interrelationships among the hydrocarbon categories.

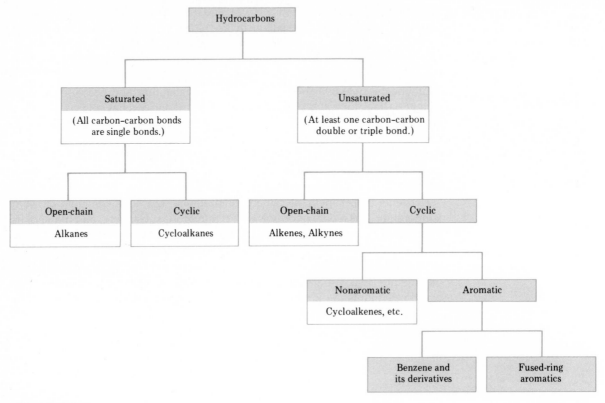

FIGURE 16.11
Relationships among various families of hydrocarbon compounds.

Practice Questions and Problems

ORGANIC CHEMISTRY—GENERAL CONSIDERATIONS

16-1 Contrast the modern-day and historical definitions of *organic chemistry*.

16-2 List the main reason for the large number of organic compounds.

16-3 How many covalent bonds do we expect a carbon atom in an organic compound to form? Explain your answer.

16-4 What is a hydrocarbon and what are the three major classes of hydrocarbons?

ALKANES

16-5 Draw structures for the three simplest alkanes: methane, ethane, and propane.

16-6 Using the general formula for an alkane, calculate the number of hydrogen atoms that would be present in an alkane molecule containing the following.
a. 8 carbon atoms   b. 14 carbon atoms
c. 22 carbon atoms   d. 35 carbon atoms

16-7 The following structural formulas for alkanes are incomplete in that hydrogen atoms attached to each carbon are not shown. Complete each of these formulas by writing in the correct number of hydrogen atoms attached to each carbon atom.

a. 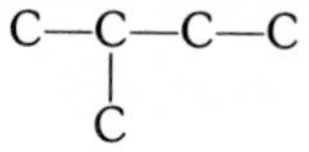

b. 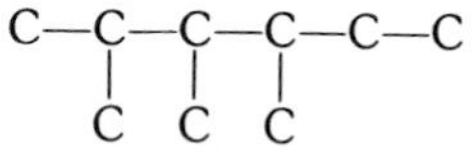

c. C—C—C—C—C—C

d. 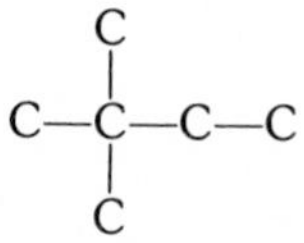

16-8 The first step in naming an alkane is to identify the longest continuous carbon atom chain. Give the number of carbon atoms in the longest continuous chain in each of the following carbon atom arrangements.

a. 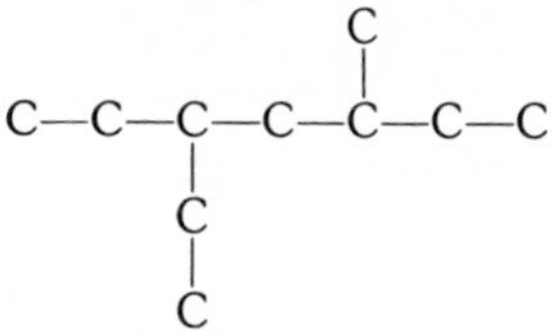

b. 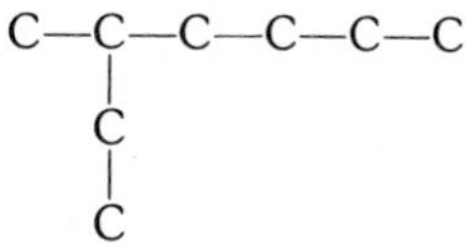

c. 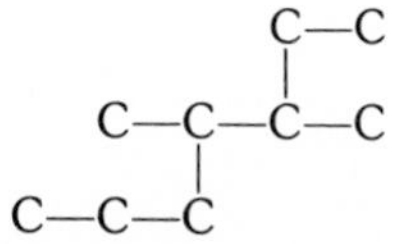

d. 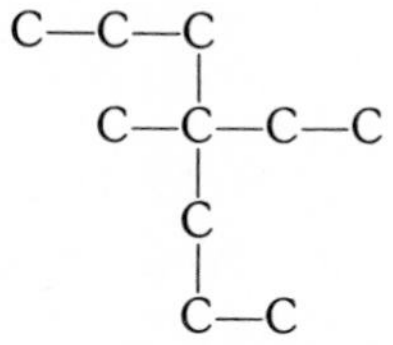

e. 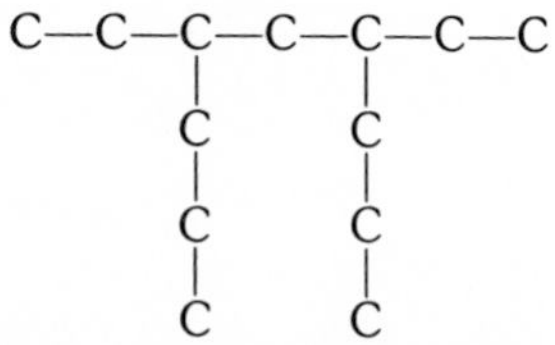

16-9 Using the IUPAC system, name the following alkanes.

a. 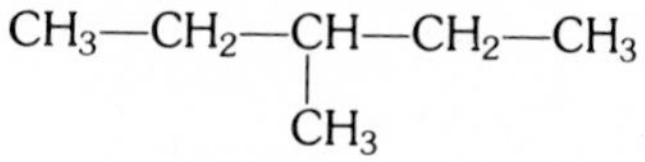

b. 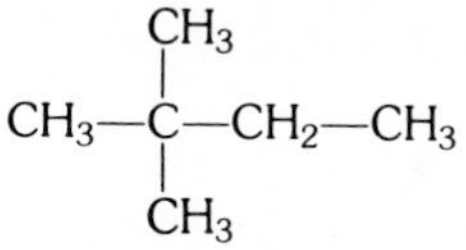

c. 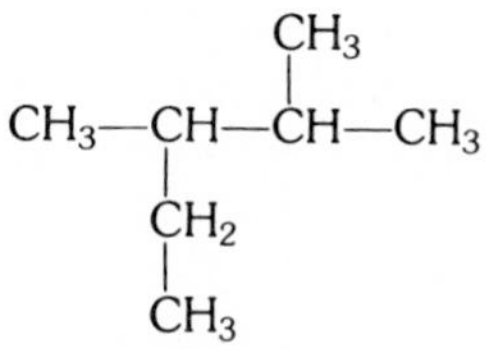

d. 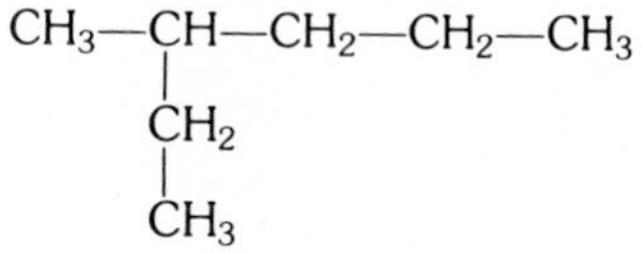

e. 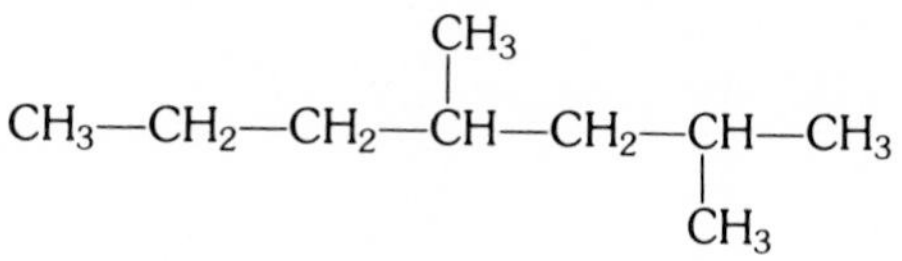

f. 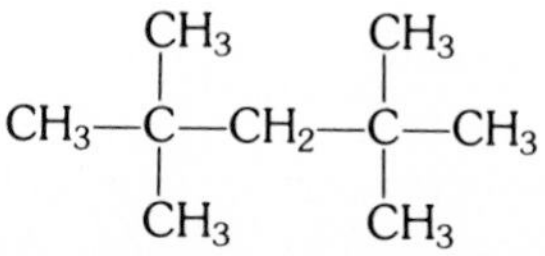

16-10 Indicate the total number of alkyl groups present in each of the molecules in Problem 16-9.

16-11 Draw the structural formula for each of the following compounds.

a. 3-Methylpentane
b. 2,2-Dimethylpropane
c. 3-Ethyl-4,4-dimethyl-6-propyldecane
d. 2,4-Dimethyloctane
e. 3-Isopropylheptane

16-12 What are structural isomers?

16-13 Write structural formulas (showing only carbon atoms) for all isomers for each of the following molecular formulas. Also assign an IUPAC name to each isomer drawn.

a. C_5H_{12} (three isomers are possible)
b. C_6H_{14} (five isomers are possible)
c. C_7H_{16} (nine isomers are possible)

16-14 For each of the following pairs of molecules, indicate whether the members of each pair are structural isomers.

a. 2-Methylpentane and 3-methylpentane
b. 2-Methylpentane and 2-methylhexane
c. 2,3-Dimethylbutane and 2,2-dimethylbutane
d. 2,3-Dimethylbutane and 2-methylpentane
e. Hexane and 2-methylhexane
f. Hexane and 2,2-dimethylbutane

CYCLOALKANES

16-15 All of the bonds in both butane (C_4H_{10}) and cyclobutane (C_4H_8) are single bonds, yet the latter compound contains two fewer hydrogens than the former. Why?

16-16 How many hydrogen atoms are present in each of the following molecules?

a. (triangle) CH₃
b. (square) CH₂—CH₃ / CH₂—CH₃
c. (hexagon)
d. (pentagon) CH₃ / CH₃

16-17 Assign IUPAC names to each of the compounds in Problem 16-16.

16-18 Write structural formulas for each of the following cycloalkanes using geometrical figures to denote rings of carbon atoms.
a. 1,2,4-Trimethylcyclohexane
b. 3-Ethyl-1,1-dimethylcyclopentane
c. Propylcyclobutane
d. Isopropylcyclobutane

16-19 Draw and name the three possible dimethylcyclobutane isomers.

ALKENES AND ALKYNES

16-20 Classify each of the following hydrocarbons as unsaturated or saturated. In addition, classify each unsaturated compound as an alkene, alkyne, cycloalkene, or cycloalkyne.

a. $CH_3—CH=CH_2$
b. (hexagon) CH₃
c. $CH_2—CH—CH_2—C≡CH$ / CH₃
d. (hexagon)

e. $CH=CH$ / $CH_2—CH_2$
f. $CH_3—CH_2—CH_2—CH_2—CH_3$

16-21 Using the IUPAC system, name each of the following unsaturated hydrocarbons.

a. $CH_3—CH_2—CH=CH—CH_3$
b. $CH_3—C≡C—CH—CH_3$ / CH_3
c. $CH_3—CH—CH=CH_2$ / CH_3
d. $CH_3—CH—C≡C—CH_3$ / CH_2 / CH_3
e. (hexagon) CH₃
f. (square) CH₃
g. $CH_3—CH=CH—CH=CH_2$
h. (hexagon) CH₃

16-22 Draw structural formulas (show carbon atoms only) for the following unsaturated compounds.
a. 2-Methyl-3-hexyne
b. 3-Ethyl-1,4-pentadiene
c. 3-Methylcyclopentene
d. 4,4,5-Trimethyl-2-heptyne
e. 1,3-Butadiene
f. 3,3-Dimethyl-1,4-pentadiene

16-23 How does the general formula for an alkene containing one double bond differ from that for an alkane?

16-24 Cycloalkanes and alkenes with one double bond containing the same number of carbon atoms are isomeric. Draw all isomers for both types of compounds that have the formula C_4H_8. (There is a combined total of five isomers.)

AROMATIC HYDROCARBONS

16-25 Classify each of the following cyclic hydrocarbons into the categories aromatic, nonaromatic unsaturated, or saturated.

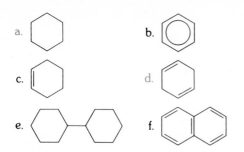

a.

b.

c.

d.

e.

f.

16-26 How many hydrogen atoms are present in each of the molecules in Problem 16-25?

16-27 Give the IUPAC name for each of the following aromatic hydrocarbons using the word benzene in the name.

a.

b. CH₃ / CH₃

c. CH₃ / CH₃

d. CH₃ / CH₃ / CH₃

e. CH₃ / CH₃ / CH₃

f. CH₂—CH₃ / CH₃

16-28 When and how are the prefixes ortho-, meta-, and para- used in naming aromatic compounds?

16-29 Write structural formulas for the following aromatic compounds.
a. 1,3-Diethylbenzene
b. *o*-Xylene
c. Toluene
d. *p*-Ethylmethylbenzene
e. 1,4-Diethyl-2,5-dimethylbenzene
f. meta-Ethylpropylbenzene

16-30 Draw structural formulas for all of the isomeric tetramethylbenzenes (three isomers are possible).

16-31 Describe how the bonding in benzene differs from that in cyclohexane. Both compounds have a ring of six carbon atoms.

Hydrocarbon Derivatives

CHAPTER HIGHLIGHTS

17.1   Functional Groups

Hydrocarbons, as numerous as they are, make up only a small fraction of known organic compounds. The majority of organic compounds are *hydrocarbon derivatives*. A **hydrocarbon derivative** *is a hydrocarbon molecule in which one or more of the hydrogen atoms has been replaced with a new atom or group of atoms.* In this way oxygen, nitrogen, sulfur, and other elements are introduced into hydrocarbon molecules. This new nonhydrocarbon part of the hydrocarbon derivative, called the *functional group*, serves as a basis for characterizing the derivative. A **functional group** *is an atom or group of atoms that serves as a basis for characterizing the chemical behavior of a type of hydrocarbon derivative.* The term functional group is appropriate because it is at the functional group site that most chemical reactions occur. The functional group literally determines how the hydrocarbon derivative functions in chemical reactions.

The ability to recognize and understand the behavior of functional groups is a powerful tool in organic chemistry. Although millions of hydrocarbon derivatives exist, they can be grouped into a relatively small number of categories, based on the functional groups they contain. Once we understand how a particular functional group behaves, we can immediately understand how almost all molecules containing the same functional group behave. Table 17.1 lists the classes of hydrocarbon derivatives most commonly encountered in organic chemistry as well as the functional group that characterizes each class.

Note that the element oxygen is a particularly common component of functional groups. The element nitrogen is the next most frequently encountered functional group constituent. The symbol R that is used in the general formulas for hydrocarbon derivatives (third column of Table 17.1) is a designation for the hydrocarbon part of any hydrocarbon derivative. For example, the

TABLE 17.1

Classes of Hydrocarbon Derivatives

Class Name	Functional Group	General Formula for Class*	Example
Halide	—X (X = F, Cl, Br, or I)	R—X	CH_3—Br
Alcohol	—OH	R—OH	CH_3—OH
Ether	—O—	R—O—R′	CH_3—O—CH_3
Thioalcohol	—SH	R—SH	CH_3—SH
Thioether	—S—	R—S—R′	CH_3—S—CH_3
Aldehyde	$\overset{\text{O}}{\overset{\|}{-C-H}}$	$\overset{\text{O}}{\overset{\|}{R-C-H}}$	$\overset{\text{O}}{\overset{\|}{CH_3-C-H}}$
Ketone	$\overset{\text{O}}{\overset{\|}{-C-}}$	$\overset{\text{O}}{\overset{\|}{R-C-R'}}$	$\overset{\text{O}}{\overset{\|}{CH_3-C-CH_3}}$
Carboxylic acid	$\overset{\text{O}}{\overset{\|}{-C-OH}}$	$\overset{\text{O}}{\overset{\|}{R-C-OH}}$	$\overset{\text{O}}{\overset{\|}{CH_3-C-OH}}$
Ester	$\overset{\text{O}}{\overset{\|}{-C-O-}}$	$\overset{\text{O}}{\overset{\|}{R-C-O-R'}}$	$\overset{\text{O}}{\overset{\|}{CH_3-C-O-CH_3}}$
Amine	$-NH_2$	$R-NH_2$	CH_3-NH_2
Amide	$\overset{\text{O}}{\overset{\|}{-C-NH_2}}$	$\overset{\text{O}}{\overset{\|}{R-C-NH_2}}$	$\overset{\text{O}}{\overset{\|}{CH_3-C-NH_2}}$

* The symbol R′ represents a hydrocarbon group that may or may not be the same as R.

general formula for a carboxylic acid

$$R-C\overset{\displaystyle O}{\underset{\displaystyle OH}{}}$$

collectively designates the following compounds,

$$CH_3-C\overset{O}{\underset{OH}{}} \qquad CH_3-\overset{\underset{\displaystyle CH_3}{|}}{CH}-C\overset{O}{\underset{OH}{}} \qquad CH_3-CH_2-C\overset{O}{\underset{OH}{}} \qquad \bigcirc\!\!\!\!\!\!\!\text{—}C\overset{O}{\underset{OH}{}}$$

$$R = CH_3- \qquad R = CH_3-\overset{\underset{\displaystyle CH_3}{|}}{CH}- \qquad R = CH_3-CH_2- \qquad R = \bigcirc-$$

as well as hundreds of other similar compounds that differ from each other only in the identity of the hydrocarbon part of the molecule.

Brief considerations of the various hydrocarbon derivative classes listed in Table 17.1 constitute the remainder of the material in this chapter. The various classes are considered in the order that they are listed in Table 17.1

17.2 Halogenated Hydrocarbons (Organic Halides)

Halogenated hydrocarbons (*or organic halides*) *are hydrocarbon derivatives in which one or more halogen atoms have replaced hydrogen atoms in the parent hydrocarbon.* (The group name for the elements in Group VIIA of the periodic table—fluorine, chlorine, bromine, and iodine—is *halogen.*)

In the IUPAC system for naming organic halides, the prefixes fluoro-, chloro-, bromo-, and iodo- are used to designate the various halogen atoms. These prefixes (and location numbers for the halogen atoms, if isomers are possible) are attached to the name of the longest continuous hydrocarbon chain. Thus, the compounds

$$CH_3\text{—}\underset{\underset{\displaystyle Cl}{|}}{CH}\text{—}CH_2\text{—}CH_3 \quad \text{and} \quad CH_3\text{—}\underset{\underset{\displaystyle F}{|}}{CH}\text{—}\underset{\underset{\displaystyle F}{|}}{CH_2}$$

would be named, respectively, 2-chlorobutane and 1,2-difluoropropane.

Many halogenated hydrocarbons are commercially important. The structures for some of these compounds are shown in Table 17.2.

Numerous Freons exist. (Freon is the Du Pont trade name for fluorinated methanes.) Such compounds are used as refrigerants (for air conditioning, refrigeration, and so on) and propellants in aerosol cans. Considerable controversy has developed recently over this latter use for Freons. There is a possibility that these compounds are negatively affecting the ozone layer in the earth's upper atmosphere. (The function of the ozone layer is to filter out most of the ultraviolet radiation given off from the sun.) The use of Freons, which are nontoxic, odorless, nonflammable compounds, to pressurize aerosol cans has decreased markedly because of the concern relative to the ozone layer. More details concerning possible interaction of Freons with the ozone layer are given in Section 20.7.

DRYCLEANING SOLVENTS

Chlorinated ethenes (trichloroethylene and tetrachloroethylene) are important solvents in the drycleaning industry. Carbon tetrachloride is used as a stain remover because of its ability to dissolve greases.

Chlorinated aromatic hydrocarbons have been used for more than 40 years as pesticides. One of the simplest of such compounds is paradichlorobenzene, a

TABLE 17.2

Some Commercially Important Halogenated Hydrocarbons

Structural Formula	Common Name	IUPAC Name	Uses
F—C—Cl with F and Cl	Freon-12	dichlorodifluoromethane	refrigerant, aerosol can propellant
C=C with Cl Cl / Cl H	trichloroethylene	trichloroethene	drycleaning solvent
Cl—C—Cl with Cl, Cl	carbon tetrachloride	tetrachloromethane	solvent
benzene ring with Cl at 1,4	p-dichlorobenzene	1,4-dichlorobenzene or p-dichlorobenzene	moth repellent
DDT structure	DDT (dichlorodiphenyl trichloroethane)	1,1,1-trichloro-2,2-bis-(p-chlorophenyl)ethane	insecticide
PCB structure	PCBs (polychlorinated biphenyls—various numbers of chlorine atoms in various positions)	—	insulation in electrical equipment; fire retardant

compound we call mothballs. The most "famous" of chlorinated hydrocarbons is DDT; its use exceeds that of any other insecticide. Discovered shortly before World War II, DDT was hailed as a miracle chemical. Its widespread and successful use during the 1940s and 1950s helped dramatically in controlling insect-borne malaria, as well as most agricultural crop pests. Because it was chemically inert, cheap to make, and posed little threat to humans, DDT seemed to be the ultimate answer to insect control.

We now know there are also problems associated with DDT use—its effects on nontarget organisms such as birds and fish. For example, DDT interferes with the calcium metabolism cycle in birds such as eagles, hawks, and falcons. The net result is eggs whose shells are thinner than normal—so thin that they cannot withstand the rigors of incubation, and reproductive failure results (see Figure 17.1). Because of these as well as other environmental considerations, DDT is no longer used as an insecticide. Other insecticides that have lower toxicities for mammals and birds, but are still extremely toxic for certain species of insects, have replaced DDT.

Polychlorinated biphenyls (PCBs) are another group of chlorinated hydrocarbons that have made "headlines" in the last few years. These compounds are

FIGURE 17.1
This crushed egg in the nest of a peregrine falcon had such a thin shell that the mass of the nesting parent's body destroyed it. High levels of DDT were associated with this egg. [*R. T. Smith/Arden London Ltd.*]

derived from the chlorination of biphenyl (two benzene rings bonded together through a carbon–carbon single bond).

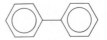

Biphenyl

Such chlorination results in a mixture of many different chlorinated biphenyls. The number of chlorine atoms on the product can vary from 1 to 10. Chemical inertness, thermal stability, desirable electrical properties, and low vapor pressure make PCBs useful in a variety of applications, including fire retardants, plasticizers, electrical equipment insulation, carbonless copy paper component, and some epoxy paints. It was not known until years after their introduction that PCBs act as a nerve poison and accumulate in the fatty tissues of organisms with repeated exposure (see Figure 17.2). Their use has been restricted since 1977. However, many materials and objects manufactured before 1977 still contain PCBs.

The plastics polyvinyl chloride (PVC) and Teflon are made from the halogenated hydrocarbon starting materials chloroethene and tetrafluoroethene, respectively.

FIGURE 17.2
This photo shows the extreme precautions now taken with PCBs. Contaminated soil is covered and labeled. [*Cathy Cheney/EKM-Nepenthe.*]

$$
\begin{array}{cc}
\underset{\mid}{\overset{\mid}{C}}=\underset{\mid}{\overset{\mid}{C}} & \underset{\mid}{\overset{\mid}{C}}=\underset{\mid}{\overset{\mid}{C}}
\end{array}
$$

Chloroethene   Tetrafluoroethene
(vinyl chloride)

The structures of the plastics derived from these compounds are considered in Section 18.4.

17.3 Alcohols

Alcohols *are hydrocarbon derivatives in which one or more hydroxyl groups (—OH groups) have replaced hydrogen atoms in the parent hydrocarbon.* Besides being hydrocarbon derivatives, alcohols may also be thought of as derivatives of water. If one of the hydrogen atoms in water is replaced with an R group, an alcohol results.

$$H—O—H \qquad R—O—H$$

Water   An alcohol

The IUPAC name of an alcohol containing only one hydroxyl group is obtained from the name of the parent alkane by replacing the -*e* ending with -*ol*. When necessary, the position of the —OH functional group is specified by a number. The simpler alcohols are also often known by common names, in which the alkyl group name is followed by the word alcohol.

$$
CH_3—OH \qquad \underset{\underset{OH}{\mid}}{CH_2}—CH_2—CH_2—CH_3
$$

Methanol   1-Butanol
or   or
Methyl alcohol   Butyl alcohol

Alcohols that contain more than one hydroxyl group are called *polyhydric alcohols*. Dihydric alcohols (two —OH groups) are also called glycols or diols. Trihydric alcohols (three —OH groups) are called triols.

Five of the most common alcohols are methanol, ethanol, 2-propanol, 1,2-ethanediol, and 1,2,3-propanetriol. Structural formulas and common names (which are used more often than the IUPAC names) for these alcohols are listed in Table 17.3.

Methanol is the simplest alcohol; it has one C atom and one OH group. It is a colorless liquid, with a characteristic odor similar to "duplicating fluid." Its

TABLE 17.3
Some Important Common Alcohols

IUPAC Name	Structural Formula	Common Names
Methanol	CH_3—OH	methyl alcohol wood alcohol
Ethanol	CH_3—CH_2—OH	ethyl alcohol grain alcohol drinking alcohol

2-Propanol	CH_3—CH—CH_3
	        |
	       OH
		isopropyl alcohol
		rubbing alcohol

1,2-Ethanediol	CH_2—CH_2 | | OH OH	ethylene glycol
1,2,3-Propanetriol	CH_2—CH—CH_2 | | | OH OH OH	glycerol glycerin

WOOD ALCOHOL

common name, wood alcohol, comes from the fact that its principal source for many years was wood. It was obtained by heating wood to a high temperature in the absence of oxygen. Methanol is one of the compounds produced as the wood decomposes. Today, methanol is synthetically produced by reacting carbon monoxide with hydrogen gas.

$$CO + 2 H_2 \xrightarrow[\text{catalysts}]{\text{heat, pressure}} CH_3OH$$

Methanol is an excellent solvent, and is often used in shellacs and varnishes as well as in the production of some plastics. Approximately half of industrial methanol produced is used in polymer manufacture (Chapter 18). Such polymers include adhesives, fibers, and plastics. Methanol is also used as an antifreeze in automobile windshield washer fluid.

Methanol is extremely toxic. Once inside the body, methanol is quickly oxidized by a liver enzyme to produce formaldehyde. Formaldehyde is even more toxic than methanol. Within the body, formaldehyde disrupts protein function. Temporary blindness, permanent blindness, or death can result from the ingestion of only small amounts of methanol.

The two-carbon alcohol ethanol is the compound known to the layperson as "alcohol." In addition to being the active ingredient in alcoholic beverages, it is also used in the pharmaceutical industry as a solvent (tinctures are ethanol solutions), as a medicinal ingredient (cough syrups often contain as much as 20% by volume ethanol), and as an industrial solvent. A 70% by volume ethanol solution is an excellent antiseptic and is used extensively for skin disinfection.

Like methanol, ethanol is toxic when taken internally. If rapidly ingested, as little as one pint of pure ethanol will kill most people. Nevertheless, many people still consume large quantities of alcoholic beverages in spite of the long-term adverse effects. Excessive use is known to cause such undersirable effects as cirrhosis of the liver and loss of memory and can lead to strong physiological addiction. In recent years, a link has been established between certain birth defects and the ingestion of ethanol by mothers during pregnancy.

All alcohol destined for human consumption (alcoholic beverages) is produced by carefully controlled fermentation of sugars obtained from various plant sources. *Fermentation* is a process in which sugars, in the absence of oxygen, are converted to ethanol by a series of reactions involving enzymes. The synthesis of ethanol from the fermentation of grains such as corn, rice, or barley is the basis of its common name, *grain alcohol*. The maximum concentration of alcohol obtainable from fermentation processes is 12 to 14%, because enzymes cannot function above this level. Beverages with a higher concentration of alcohol than this are prepared by either distillation or fortification with alcohol obtained by the distillation of another fermentation product. Table 17.4 lists the alcohol content of common alcoholic beverages.

GRAIN ALCOHOL

Wine is made by fermenting the natural sugars present in grape juice and other fruit juices. Malt sugar, obtained from barley and wheat is used in making beer. Whiskey is made from various grains including barley, rye, and corn. Vodka is obtained from sugar sources in potatos, and rice is the basic sugar source for several strong Asian drinks. The distinctive taste of a given type of alcoholic beverage is related to characteristic compounds, other than ethanol, that are not removed by fermentation or other treatment steps.

The alcohol content of alcoholic beverages is sometimes stated using a system based on *proof*. The proof of a beverage is twice its percent of alcohol. An 86-proof brandy, for example, contains 43% alcohol. Pure ethanol (100% ethanol) is 200 proof. The origin of the proof system for concentration dates back to the seventeenth century, when dealers of alcoholic beverages would pour some of the beverage over a small amount of gunpowder to prove to skeptical customers that their products had not been watered down. The mixture had to ignite as "proof" that the beverage was strong. When the concentration

TABLE 17.4

Ethanol Content of Common Alcoholic Beverages

Name	Percent of Ethanol
Beer	3.2–9
Wine	12 maximum, unless fortified
Brandy	40–45
Whiskey	45–55
Rum	45

of alcohol is 50% (m/v), gunpowder ignites; this is why 100 proof is actually 50% ethanol.

Ethanol destined for human consumption in alcoholic beverages is heavily taxed by the United States government. However, ethanol that is used as a solvent in industry is exempt from taxes. Industrial ethanol is usually rendered unfit for human consumption by the addition of small amounts of toxic substances. Ethanol treated in this way is called *denatured alcohol*. Some denaturants that do not affect the solvent properties of the ethanol are methanol, benzene, and formaldehyde. Alcohol used in industry is produced by the direct hydration of ethene rather than by fermentation.

DENATURED ALCOHOL

$$CH_2{=}CH_2 + H_2O \xrightarrow[\text{catalyst}]{\text{heat, pressure}} CH_3{-}CH_2{-}OH$$

Absolute alcohol is ethanol from which all traces of water have been removed. Industrially produced ethanol, as well as that obtained from fermentation (after distillation), contains about 5% by volume water. The presence of this small amount of water does not affect most of the uses for ethanol. In certain chemical laboratory settings, however, no water can be tolerated because of possible side reactions. In these cases, chemists use absolute alcohol.

Gasohol, a mixture of 10% ethanol and gasoline, is an automobile fuel developed in response to the energy crisis. The ethanol is prepared by fermenting cereal grain residues.

Isopropyl alcohol (2-propanol) is commonly known as *rubbing alcohol*. Drugstore rubbing alcohol is a 70% by volume solution of isopropyl alcohol. Rubbing alcohol's rapid evaporation rate creates a dramatic cooling effect when it is applied to the skin; hence, alcohol rubs are often used to combat high body temperature. A 70% by volume isopropyl alcohol solution is as *effective* as ethanol in disinfection procedures (see Figure 17-3).

Although more toxic than ethanol, isopropyl alcohol seldom causes serious problems if ingested. It is restricted to external use because it induces vomiting, which eliminates the alcohol from the system. Other (longer-carbon-chain) alcohols also induce vomiting when taken internally.

Ethylene glycol (1,2-ethanediol) is the simplest alcohol containing two —OH groups. This compound is the main ingredient in most brands of permanent antifreeze for automobile radiators. It is nonvolatile (boiling point = 197°C), completely miscible with water, noncorrosive, and relatively inexpensive to produce. All of these characteristics are ideal properties for an antifreeze.

Ethylene glycol is also an important chemical raw material. Both Dacron and Mylar film are made from it (see Section 18.7). Dacron is an important polyester fiber used in clothing and Mylar film is used in tapes for recorders and computers.

Glycerol (glycerin or 1,2,3-propanetriol) is a clear, thick liquid with the consistency of honey. It is a sweet-tasting liquid and is not toxic. As we will see in Section 24.10, glycerol plays an important role in the structure of fat molecules.

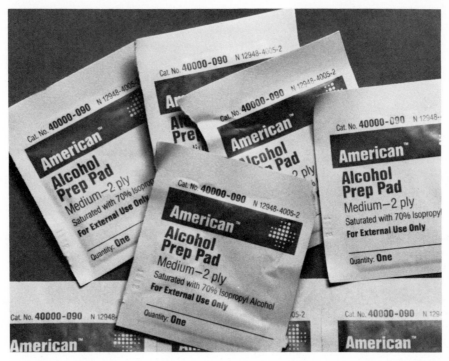

FIGURE 17.3
A 70% by volume aqueous solution of isopropyl alcohol is used as a disinfectant for the skin. [*PAR/NYC.*]

Glycerol has a great affinity for water vapor (moisture). Because of this, it is often added to pharmaceutical preparations such as skin lotions and soaps. Florists sometimes use glycerol on cut flowers to help them retain water and maintain freshness. Its lubricative properties also makes it useful in shaving creams and applications such as glycerin suppositories for rectal administration of medicines.

A class of compounds that is closely related to the alcohols is the phenols. A **phenol** *is a compound in which an —OH group is bonded to a carbon atom of a benzene ring.* Phenols are, thus, aromatic alcohols. The simplest of all phenols is the compound whose name is phenol.

OH

Phenol

All other phenols differ from this compound with respect to the groups that are bonded to the aromatic ring, and are named as derivatives of phenol.

p-Chlorophenol   3,5-Dimethylphenol

We will encounter the term phenol in a number of discussions in later chapters.

17.4 Ethers

Ethers *are compounds in which both hydrogens of water are replaced by carbon chains or rings.* Alternatively, ethers may be treated as derivatives of alcohols in which the hydrogen of the hydroxyl group has been replaced by an R group.

Water   An alcohol   An ether

Thus, the general formula for an ether is R—O—R′, where the R groups may or may not be the same. When both R groups are the same, the compound is called a simple ether. When R and R′ are different, a mixed ether results.

$$CH_3\text{—O—}CH_3 \qquad CH_3\text{—}CH_2\text{—O—}CH_3$$

A simple ether   A mixed ether

Common names for ethers consist of the names of the two hydrocarbon groups attached to the oxygen followed by the word *ether*. If the two hydrocarbon groups are identical, the prefix *di-* is often omitted.

$$CH_3\text{—O—}CH_3 \qquad CH_3\text{—}CH_2\text{—O—}CH_3$$

Dimethyl ether   Ethyl methyl ether
(methyl ether)

In the IUPAC system the —O—R group is called an alkoxy group; for example, —O—CH_3 is the methoxy group. The alkoxy group is then treated as

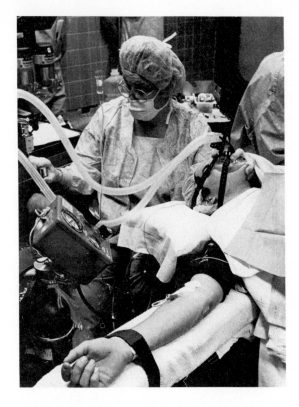

FIGURE 17.4
The use of ether general anesthetics during surgery revolutionized medicine. [*MacDonald Photography/ The Picture Cube.*]

a substituent on a parent hydrocarbon molecule. In mixed ethers, the smaller alkyl group becomes the alkoxy group.

$$CH_3—O—CH_3 \qquad CH_3—CH_2—CH—CH_3$$
$$\underset{\displaystyle O—CH_3}{\big|}$$

Methoxymethane   2-Methoxybutane

GENERAL ANESTHETICS   Ethers are used extensively as general anesthetics. A *general anesthetic* is a compound or mixture of compounds that acts on the brain to produce both unconsciousness and insensitivity to pain. The availability of general anesthetics has revolutionized surgery and greatly reduced the risks from shock during and after surgery (see Figure 17.4).

The most familiar, and also the first (in 1846) of the general anesthetic ethers to be used, is diethyl ether; often simply called *ether*. Diethyl ether is relatively safe because the gap between the effective level for anesthesia and the lethal dose is wide. The disadvantages to its use are its high flammability and its side

TABLE 17.5

Some Common Inhalation Anesthetics That Are Ethers

Structure	Chemical Name	Trade Name
$CH_3\!-\!O\!-\!CH_2\!-\!CH_2\!-\!CH_3$	methyl propyl ether	Neothyl
$CH_2\!\!=\!\!CH\!-\!O\!-\!CH\!\!=\!\!CH_2$	divinyl ether	Vinethene
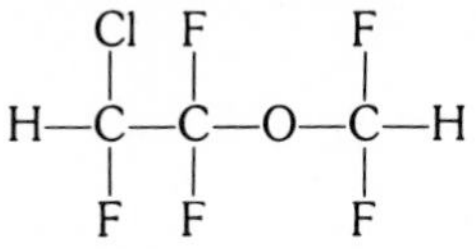 (structure)	2-chloro-1,1,2-trifluoroethyl difluoromethyl ether	Enflurane or Ethrane
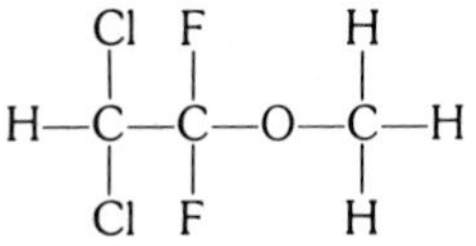 (structure)	2,2-dichloro-1,1-difluoroethyl methyl ether	Methoxyflurane or Penthrane
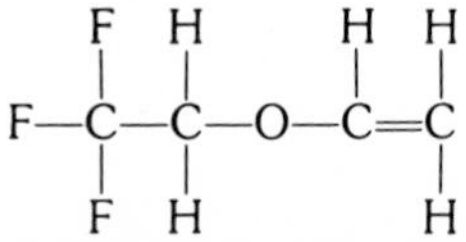 (structure)	2,2,2-trifluoroethyl vinyl ether	Fluoxene or Fluoromar

effect of postoperative nausea. Other anesthetics without these disadvantages have taken its place. Many of these newer anesthetics are halogenated ethers. In many cases the introduction of halogen atoms into an ether cuts down on the flammability of the compound. Table 17.5 lists the structures and names of some of the ethers now used as inhalation anesthetics.

The mechanism by which a general anesthetic exerts its effects is related to its solubility in fat (a relatively nonpolar substance). The anesthetic molecules dissolve in the fatlike membranes of nerve cells. This affects the permeability of the membranes to other substances and depresses the nerve cells' conductivity.

The gradual phaseout of lead additives in gasoline (Section 19.5) has created a market for new gasoline additives that raise octane rating. The most prominent new octane booster in use is the ether methyl tertiary-butyl ether (MTBE).

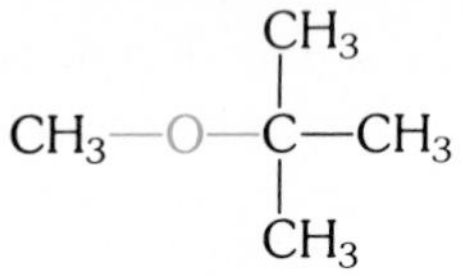

MTBE

$$CH_3\!-\!O\!-\!\underset{\underset{CH_3}{|}}{\overset{\overset{CH_3}{|}}{C}}\!-\!CH_3$$

MTBE

A particularly important market for this small mixed ether is premium unleaded gasoline, a gasoline that is becoming increasingly popular.

17.5 Thioalcohols and Thioethers

Just as alcohols and ethers are organic derivatives of water (H_2O), thioalcohols and thioethers are organic derivatives of hydrogen sulfide (H_2S). The general structural formulas for these two classes of compounds are

R—S—H   and   R—S—R′

Thioalcohol   Thioether

These general structural formulas are identical to those for alcohols and ethers, except that sulfur has replaced oxygen. The prefix thio- means that an oxygen atom has been replaced by a sulfur atom. The —SH functional group is called a *sulfhydryl* group.

The replacement of an oxygen atom with a sulfur atom should not be surprising to you. Oxygen and sulfur are in the same group in the periodic table; sulfur is the element directly below oxygen in Group VIA of the periodic table. Hence, the two elements have similar electronic properties and their bonding behavior should be similar (Section 4.8).

To write the name of a thioalcohol add the suffix -thiol to the name of the parent hydrocarbon

CH_3—SH CH_3—CH—CH_3
                              |
                             SH

Methanethiol   2-Propanethiol

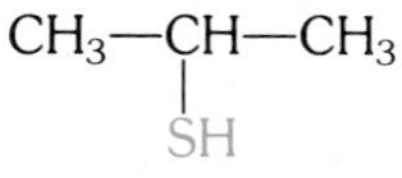

We can derive the common name of a thioether by specifying the alkyl groups attached to the sulfur atom and adding the word *sulfide*. Two examples are

CH_3—CH_2—S—CH_3 and CH_3—CH_2—S—CH_2—CH_3

Ethyl methyl sulfide   Diethyl sulfide

Most thioalcohols and thioethers have strong (usually unpleasant) odors. The strong, distinctive odor of skunks originates from a mixture of compounds, two of which are

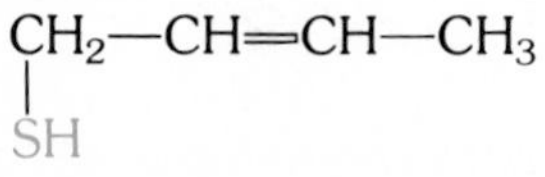

SKUNK ODOR

CH_2—CH=CH—CH_3 and CH_2—CH_2—CH—CH_3
 | | |
SH   SH CH_3

2-Butene-l-thiol   3-Methyl-l-butanethiol

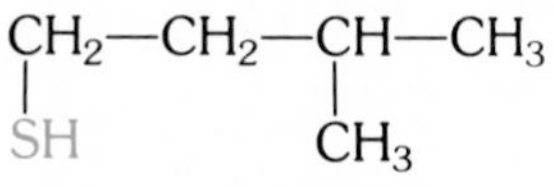

The familiar odor of natural gas—for example, from a gas range—comes from the addition of small amounts of ethanethiol (CH_3—CH_2—SH) to the gas. Natural gas is odorless, and the trace amount of thioalcohol odorant is necessary to facilitate detection of leaks. The odor of ethanethiol is detectable at a concentration of 1 part ethanethiol per 50 billion parts air. Its odor is one of the most potent known.

ONIONS AND GARLIC

Many of the strong odors associated with foods are caused by the presence of thio compounds. Major contributors to the odor and flavor of garlic are the compounds

$$CH_2{=}CH{-}\underset{\underset{\displaystyle SH}{|}}{CH_2} \quad \text{and} \quad CH_2{=}CH{-}S{-}CH{=}CH_2$$

2-Propene-1-thiol   Divinyl sulfide

1-Propanethiol is the eye irritant that is released when fresh onions are chopped up.

$$CH_3{-}CH_2{-}CH_2{-}SH$$

1-Propanethiol

This compound also contributes to the onion's characteristic smell.

DMSO

Oxidation of the thioether dimethyl sulfide produces the compound dimethyl sulfoxide (DMSO). An oxygen atom is added to the sulfur atom.

$$CH_3{-}S{-}CH_3 \xrightarrow{\text{oxidation}} CH_3{-}\overset{\overset{\displaystyle O}{\|}}{S}{-}CH_3$$

Dimethyl sulfide   Dimethyl sulfoxide

DMSO has an unusually great ability to penetrate skin. This has led to its use in treating numerous conditions including arthritis, gout, inflammation, burns, and athletic injuries such as sprains with positive results. These medical applications for DMSO have not, however, received FDA approval because of concerns about possible side effects. The only approved medical use for DMSO is treatment for a particular type of bladder problem.

17.6 Aldehydes and Ketones

Aldehydes and ketones are similar in that both classes of compounds contain a *carbonyl group*. A **carbonyl group** *consists of a carbon atom bonded to an oxygen atom through a double bond.* Because any carbon atom in an organic molecule must form four bonds, the carbon atom of a carbonyl (pronounced "carbon-EEL") group must also be bonded to two other atoms

Carbonyl group

These two other bonds determine whether the compound is an aldehyde or ketone.

Aldehydes *are compounds containing a carbonyl group in which the carbonyl carbon atom is bonded to at least one hydrogen atom.* The other group attached to the carbonyl carbon atom can be a hydrogen atom or a carbon chain or ring.

$$H-C\!\!\overset{\displaystyle O}{\underset{\displaystyle H}{\Big\langle}} \quad \text{or} \quad R-C\!\!\overset{\displaystyle O}{\underset{\displaystyle H}{\Big\langle}}$$

The aldehyde functional group, which is the structural feature common to both of the preceding structures, is

$$-C\!\!\overset{\displaystyle O}{\underset{\displaystyle H}{\Big\langle}}$$

Ketones *are compounds containing a carbonyl group in which the carbonyl carbon atom is bonded to two other carbon atoms.* The bonded groups may be any combination of alkyl or aromatic groups. The ketone functional group, which is the structural feature common to all ketones is

$$\underset{\displaystyle C}{\overset{\displaystyle O}{\parallel}}$$

From an alternative viewpoint, we can distinguish between aldehydes and ketones by noting the location of the carbonyl group within a structural formula. If the carbonyl group is located at the end of a hydrocarbon chain, we have an aldehyde. Ketones possess carbonyl groups that are not at the ends of the carbon chain.

$$C-C-C-C-C-C\!\!\overset{\displaystyle O}{\underset{\displaystyle H}{\Big\langle}} \qquad C-C-C-\overset{\displaystyle O}{\overset{\displaystyle \parallel}{C}}-C-C$$

An aldehyde   A ketone

The simpler aldehydes and ketones usually go by their common names rather than IUPAC names. The simplest aldehyde possible is formaldehyde, the aldehyde in which two H atoms are attached to the carbonyl group. The common name of the two-carbon aldehyde is acetaldehyde. One- and two-carbon ketones cannot exist. A minimum of three carbon atoms is required in a ketone—one C for the carbonyl group and one C for each R group. The simplest ketone, in which both R groups are methyl, is acetone.

$$CH_3-\overset{\overset{\displaystyle O}{\|}}{C}-CH_3$$

Acetone, the simplest ketone

When formaldehyde, a colorless gas, is bubbled through water, highly concentrated aqueous solutions, commonly called formalin, are produced. A formalin solution is the biological preservative whose pungent odor is familiar to anyone with experience in a biology laboratory.

Acetone is an excellent solvent because it is miscible with both water and nonpolar solvents. Acetone is the main ingredient in gasoline treatments designed to solubilize water in the gas tank, allowing it to pass through the engine in miscible form. Acetone can also be used to remove water from glassware in the laboratory or to strip old paint and varnish from furniture.

Many naturally occurring aldehydes and ketones have been isolated from plants. Many flowers release small quantities of ketones, which are responsible for their odor. Many plant flavors are aldehydes or ketones. Benzaldehyde, a colorless liquid, is responsible for the flavor we call "almond." Cinnamaldehyde is responsible for the characteristic odor we call "oil of cinnamon." The active ingredient in vanilla flavoring is the aldehyde vanillin. At one time the only source for vanillin was the pod-like capsules of certain climbing orchids. Today, most vanillin is produced synthetically.

ALMONDS, CINNAMON, AND VANILLIN

Benzaldehyde   Cinnamaldehyde   Vanillin

A closer examination of the structure of vanillin shows that along with aldehyde group are two other functional groups—the methoxy group (ether) and a hydroxyl group (alcohol). Many naturally occurring molecules contain more than one functional group. Glucose, also known as blood sugar (Section 24.4), a most important compound in biochemical discussions, is a polyhydroxy aldehyde.

$$CH_2-CH-CH-CH-CH-C\overset{O}{\underset{H}{}}$$
$$\underset{OH}{|}\quad\underset{OH}{|}\quad\underset{OH}{|}\quad\underset{OH}{|}\quad\underset{OH}{|}$$

Glucose

17.7 Carboxylic Acids

Carboxylic acids *are compounds whose characteristic functional group is the carboxyl group.* Such compounds were among the first organic compounds studied in detail because of their wide distribution and abundance in natural products. A **carboxyl group** *is a group containing a carbon atom that is double bonded to an oxygen atom and is also bonded to a hydroxyl group, that is,*

$$-C\overset{O}{\underset{OH}{}}$$

Although a carboxyl group structurally appears to be a combination of a carbonyl group and a hydroxyl group, from which comes the name carboxyl, this group should not be thought of in terms of the properties of its component parts. Instead, we must think of it as a whole—a group containing two oxygen atoms directly bonded to the same carbon atom. The chemical properties of a carboxyl group are distinctly different from those of either an alcohol or a carbonyl group.

The carbon atom in a carboxyl group must form one more bond to have the required four bonds for a carbon atom. This fourth bond may be to a hydrogen atom, an alkyl group, or an aromatic group.

In the case of the carboxyl group being bonded to a hydrogen atom, we have the simplest of all carboxylic acids—the one-carbon carboxylic acid, commonly called formic acid. Formic acid is a naturally occurring substance; it is the irritant in the "sting" of red ants. (*Formica* is the Latin word for "ant.")

The simplest situation in which the fourth bond on the carboxyl carbon involves an alkyl group is that in which the alkyl group is a methyl group. The result is the two-carbon carboxylic acid, acetic acid.

$$CH_3-C\overset{O}{\underset{OH}{}}$$

Acetic acid

Acetic acid is the most commonly encountered carboxylic acid. Vinegar is a dilute solution [approximately 5% (v/v)] of acetic acid; the Latin word for vinegar is *acetum*). A salt and vinegar solution is used as a preservative for pickled vegetables because of the ability of acetic acid to inhibit microbial growth at moderate concentrations.

The simplest of the aromatic carboxylic acids is benzoic acid, a carboxylic acid in which the fourth bond of the carboxyl carbon atom is to a phenyl group.

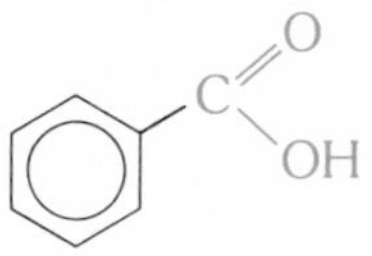

Benzoic acid

Carboxylic acids are known that contain two or more carboxyl groups. Molecules with two carboxyl groups are called dicarboxylic acids and those with three are tricarboxylic acids. The smallest dicarboxylic acid is oxalic acid.

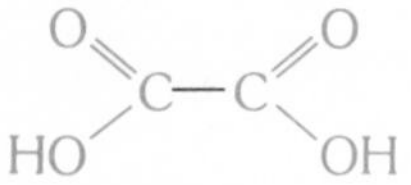

Oxalic acid

Oxalic acid is found in various plants, including spinach, cabbage, and rhubarb. Commercially, oxalic acid is used in rust-removing preparations, for bleaching straw and leather, and to remove ink stains. In high concentrations, it is toxic to living organisms. The amount of oxalic acid present in the previously mentioned vegetables, except for rhubarb *leaves*, is usually not harmful. Rhubarb leaves should not be eaten because their high oxalic acid levels.

Many carboxylic acids contain one or more hydroxyl groups in addition to the acid groups. Representative of such compounds are lactic acid (one carboxyl group and one hydroxyl group) and citric acid (three carboxyl groups and one hydroxyl group).

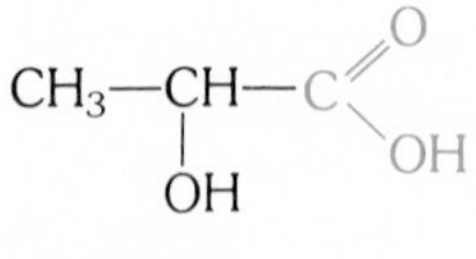

Lactic acid 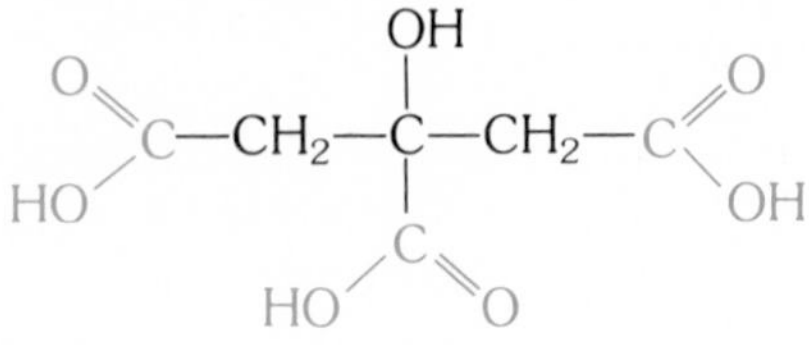   Citric acid

Lactic acid is formed in muscles during exercise and is also found in sour milk. Citric acid, as the name implies, is present in all citrus fruits; the sour (tart) taste of such fruits is related to the presence of this acid.

The newest nonprescription pain reliever on the market, approved by the FDA in 1984 for nonprescription distribution, is ibuprofen. It is the active ingredient in Advil and Nuprin. It is a monocarboxylic acid with a complex carbon chain–ring R group.

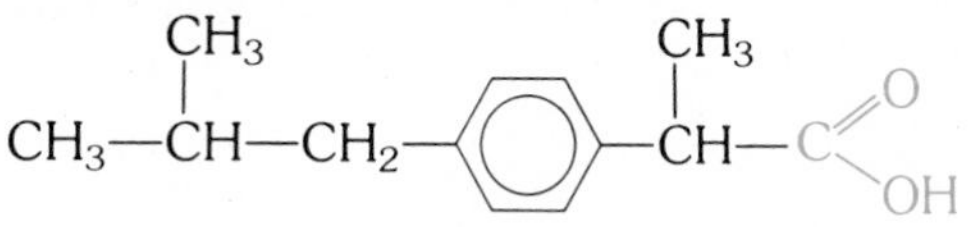

$$CH_3-CH-CH_2- \underset{}{\bigcirc} -CH-C \underset{OH}{\overset{O}{\diagup}}$$

Ibuprofen

IUPAC names for carboxylic acids are derived by replacing the *-e* of the name of the longest carbon chain containing the functional group with *-oic* and then adding the word acid. The functional group carbon atom is counted as part of the longest chain and is assigned the number 1 when a numbering system is needed. Examples of IUPAC nomenclature and common names for selected carboxylic acids of common interest are given in Table 17.6.

Carboxylic acids, although the most acidic of all organic compounds, are weak compared to inorganic acids such as HCl and HNO_3. Most carboxylic acids are less than 2% ionized in water. HCl and HNO_3 are virtually 100% ionized in water (Section 10.5). The acidic hydrogen atom in a carboxylic acid is the hydrogen atom attached to the oxygen in the carboxyl group; it is the hydrogen that leaves the molecule during ionization.

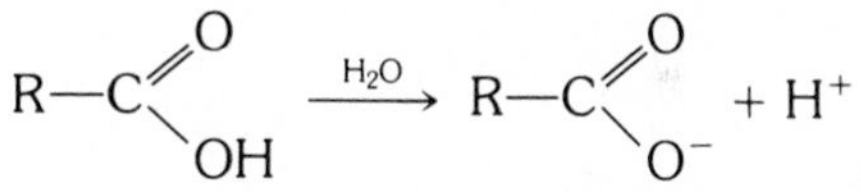

$$R-C\overset{O}{\underset{OH}{\diagup}} \xrightarrow{H_2O} R-C\overset{O}{\underset{O^-}{\diagup}} + H^+$$

Carboxylate
ion

The negative organic ion produced from this dissociation is called a carboxylate ion.

17.8 Esters

Esters *are derivatives of carboxylic acids in which the —OH of the carboxyl group has been replaced by an —OR group.* The ester functional group thus contains a carbonyl group and an ether link involving the carbonyl carbon atom.

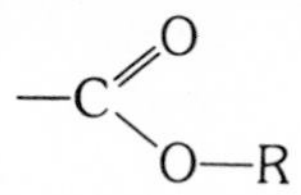

$$-C\overset{O}{\underset{O-R}{\diagup}}$$

TABLE 17.6

Structure, Nomenclature, and Uses of Selected Carboxylic Acids

Common Name (IUPAC Name)	Structural Formula	Characteristics and Typical Uses
Formic acid (methanoic acid)		stinging agent of red ants, bees, and nettles
Acetic acid (ethanoic acid)		active ingredient in vinegar
Propionic acid (propanoic acid)		salts of this acid used as mold inhibitor in breads and cereals
Butyric acid (butanoic acid)		odor-causing agent in rancid butter; present in human perspiration
Caproic acid (hexanoic acid)		characteristic odor of limburger cheese
Lactic acid (2-hydroxypropanoic acid)		found in sour milk and sauerkraut; formed in muscles during exercise
Ibuprofen (2-(p-isobutylphenyl)- propanoic acid)		pain reliever in Advil, Nuprin (nonprescription strength) and Motrin, Rufen (prescription strength)
Oxalic acid (ethanedioic acid)		poisonous material in leaves of some plants such as rhubarb: used as cleaning agent for rust stains on fabric and porcelain
Citric acid (3-hydroxy-3-carboxy- pentanedioic acid)		present in citrus fruits, used as a flavoring agent in foods

The carbon–oxygen single bond that involves the carbonyl carbon atom is called an *ester linkage*. This particular bond, highlighted in the following diagram, becomes very important in later discussions concerning esters (Section 18.7).

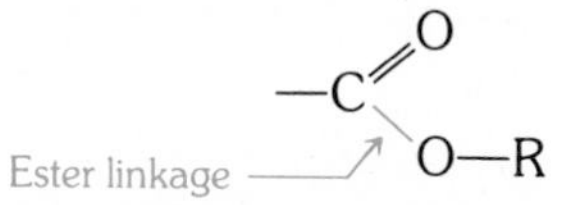

Ester linkage

Names of esters are directly related to the names of the acids from which they may be considered to be derived. The name of the ester is formed by changing the *-ic* or *-oic* ending of the acid name (either common or IUPAC name) to *-ate* and preceding this name with that of the R group attached to the oxygen atom, as a separate word. This method of naming is illustrated with some derivatives of acetic acid.

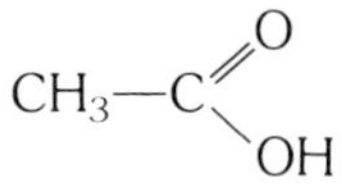

| Acetic acid | Methyl acetate | Ethyl acetate | Phenyl acetate |

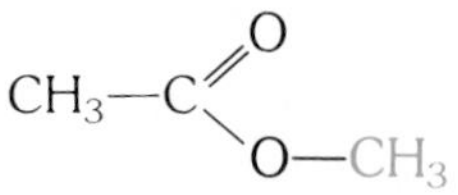

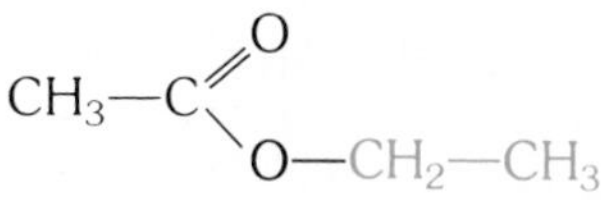

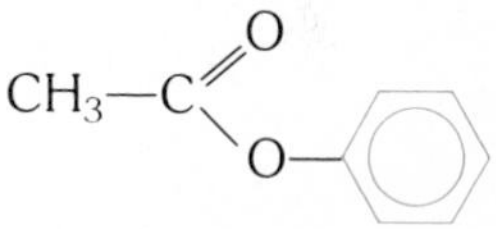

One of the preceding compounds, ethyl acetate, is the solvent in many nail polish removers and is responsible for their characteristic odor.

Esters are found widely distributed in nature. Many of the characteristic tastes of fruits and fragrances of flowers are imparted by esters. Many of these esters can be synthesized in the laboratory and are used to flavor foods and drinks artificially. Table 17.7 lists common ester flavoring agents and the odors or tastes we associate with them. Note, from the table, how a relatively small change in the size of the R group of the ester functional group significantly alters our perception of flavor. A five-carbon R group (pentyl acetate) is perceived as banana flavor, whereas an eight-carbon R group (octyl acetate) registers as orange flavor. The difference between apple and pineapple flavors is that of a methyl group versus ethyl group (see Table 17.7).

A number of esters are used in the field of medicine. The most used of all medicinal esters is acetylsalicyclic acid, more commonly known as aspirin. The structure of aspirin is

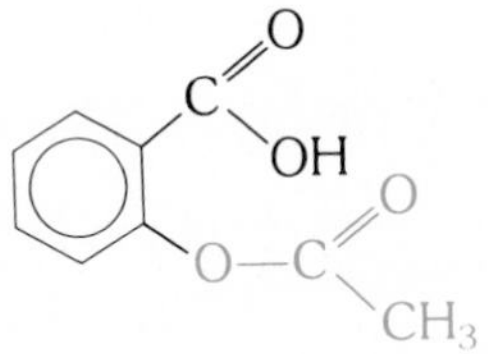

Note that an acid functional group, in addition to the ester functional group, is present in this molecule. We consider how aspirin actually functions in the human body in Section 22.2

FLAVORING AGENTS

TABLE 17.7

Fruit Odor or Flavor Associated with Selected Ester Flavoring Agents

IUPAC Name (Common Name)	Structural Formula	Characteristic Flavor and Odor
Isobutyl methanoate (isobutyl formate)	$\underset{\displaystyle H}{\overset{\displaystyle O}{\|\|}}$ H—C—O—CH$_2$—CH—CH$_3$ with CH$_3$ branch	raspberry
Pentyl ethanoate (pentyl acetate)	CH$_3$—C(=O)—O—(CH$_2$)$_4$—CH$_3$	banana
Octyl ethanoate (octyl acetate)	CH$_3$—C(=O)—O—(CH$_2$)$_7$—CH$_3$	orange
Pentyl propanoate (pentyl propionate)	CH$_3$—CH$_2$—C(=O)—O—(CH$_2$)$_4$—CH$_3$	apricot
Methyl butanoate (methyl butyrate)	CH$_3$—(CH$_2$)$_2$—C(=O)—O—CH$_3$	apple
Ethyl butanoate (ethyl butyrate)	CH$_3$—(CH$_2$)$_2$—C(=O)—O—CH$_2$—CH$_3$	pineapple
Methyl 2-aminobenzoate (methyl anthranilate)	benzene ring with C(=O)—O—CH$_3$ and NH$_2$ substituents	grape

17.9 Amines and Amides

The four most abundant elements in living organisms are hydrogen, oxygen, carbon, and nitrogen (Section 2.6). None of the hydrocarbon derivatives we have discussed to this point contain nitrogen. We now turn our attention to this element. Two important classes of nitrogen-containing hydrocarbon derivatives are the *amines* and the *amides*. Amines are carbon-hydrogen-nitrogen compounds, and amides contain oxygen in addition to these three elements.

Amines *are organic derivatives of ammonia (NH_3) in which one or more of the hydrogen atoms on the nitrogen has been replaced by an alkyl or aromatic group.* Depending on the degree of substitution—that is, the number of alkyl or aromatic groups—amines are classified as primary, secondary, or tertiary amines. The general formulas for these three types of amines are

$$R-NH_2 \quad \text{or} \quad R-N\begin{matrix}H\\\\H\end{matrix} \qquad R-N\begin{matrix}H\\\\R'\end{matrix} \qquad R-N\begin{matrix}R''\\\\R'\end{matrix}$$

Primary amine   Secondary amine   Tertiary amine

Amines have the same relationship to ammonia as alcohols and ethers to water, and thioalcohols and thioethers to hydrogen sulfide.

The three simplest amines are methylamine (a primary amine), dimethylamine (a secondary amine), and trimethylamine (a tertiary amine). Their structures are as follows.

$$CH_3-NH_2 \qquad CH_3-N\begin{matrix}H\\\\CH_3\end{matrix} \qquad CH_3-N\begin{matrix}CH_3\\\\CH_3\end{matrix}$$

Methylamine   Dimethylamine   Trimethylamine

(a primary amine)   (a secondary amine)   (a tertiary amine)

As just illustrated, assigning common names to amines is very simple: list the alkyl or aromatic group or groups attached to the nitrogen atom, in alphabetical order when different groups are present, followed by the suffix *-amine*, all as one word. Prefixes such as di- and tri- are added when identical groups are bonded to the nitrogen atom.

Amines may also be named as hydrocarbon derivatives. The location of the —NH_2 functional group, called an *amino* group, is specified in the same way as for other hydrocarbon derivatives. The IUPAC name for methylamine is aminomethane.

Methylamine, dimethylamine, and trimethylamine are all fishy-smelling gases at room temperature. One of the most notable properties of small amines is their "foul" smell. Most decaying matter, especially that high in protein, produces amines. Part of the odor of rendering plants, meat packing houses, and sewage treatment plants is due to amines. Two naturally occurring amines partly responsible for the smell of decaying animals are putrescine and cadaverine (whose names match their odors).

$$H_2N-CH_2-CH_2-CH_2-CH_2-NH_2 \qquad H_2N-CH_2-CH_2-CH_2-CH_2-CH_2-NH_2$$

Putrescine   Cadaverine

(1,4-diaminobutane)   (1,5-diaminopentane)

These compounds are formed by bacterial decomposition of proteins through enzymatic reactions. Note that both compounds contain two amino groups, one at each end of the carbon chain.

Two simple amines that also contain hydroxyl groups are very important in the functioning of the human body—the adrenal gland secretions epinephrine (adrenaline) and norepinephrine.

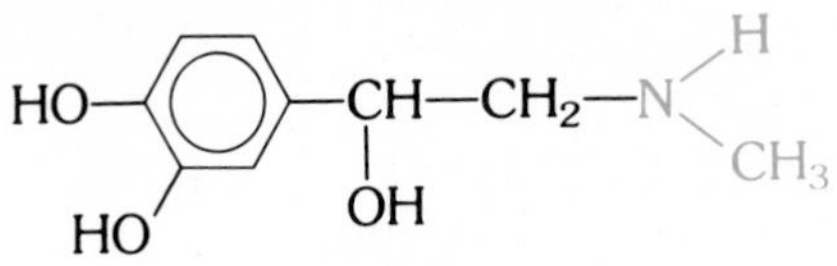

Epinephrine
(adrenaline)

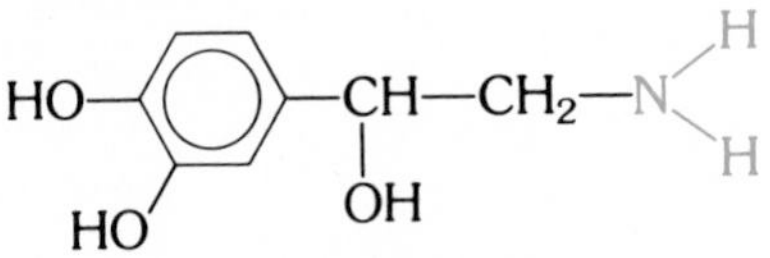

Norepinephrine

When these two substances are secreted into the blood they increase cardiac output, raise the blood pressure, stimulate the central nervous system, increase heat production, and elevate the blood sugar concentration. Excitement or fear triggers the release of these substances into the blood.

Amides *are derivatives of carboxylic acids in which the —OH of the carboxyl group has been replaced by an amino group or substituted amino group.* Amides may be simple, monosubstituted, or disubstituted. The general formulas for these three types of amides are

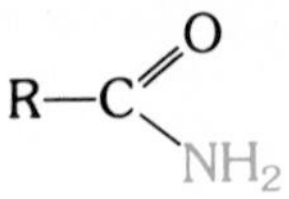

Simple amide 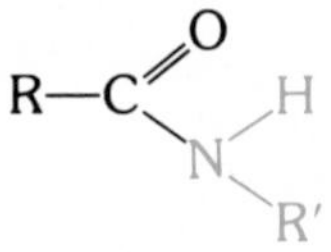　　　Monosubstituted amide 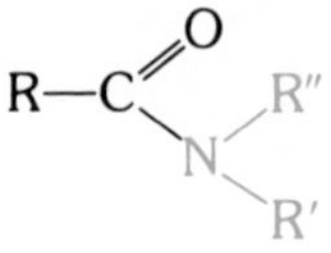　　　Disubstituted amide

The carbon–nitrogen bond in the amide functional group is often called an *amide linkage.*

Amide linkage

Such terminology parallels that used for esters, with which we have an ester linkage. An amide linkage is also sometimes called a *peptide bond.* This term is used in Section 25.5, when proteins are discussed; amide linkages are present in proteins.

Amides are named by dropping the *-ic acid* or *-oic acid* ending from the name of the acid and replacing it with *-amide*

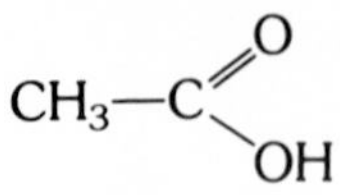

Acetic acid　　　　and　　　　Acetamide

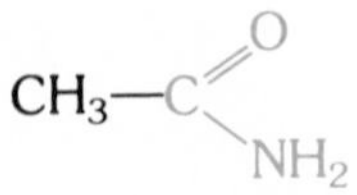

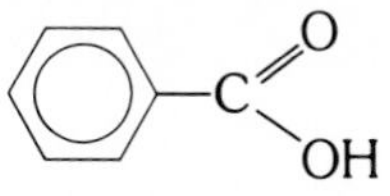
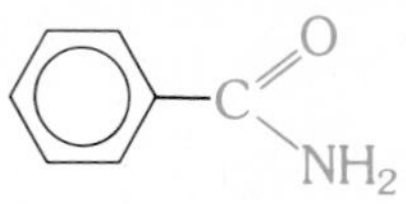

Benzoic acid and Benzamide

UREA

One of the simplest amides is also one of the most important—the compound urea, whose synthesis led to abandonment of the vital-force theory (Section 16.1). Urea is the normal end-product of human metabolism of nitrogen-containing foods (proteins). It is also an industrially important chemical, with its most important industrial use being fertilizer manufacture (Section 12.4). When added to soil, urea reacts with water to produce CO_2 and NH_3. The structure of the urea molecule is

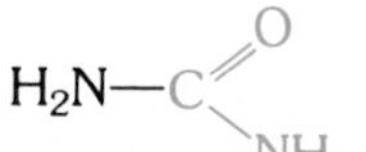

A number of synthetic amides exhibit physiological activity and are used as drugs in the human body. One of the best known of such compounds is acetaminophen, a pain reliever that is the active ingredient in nonprescription Tylenol formulations. It is frequently recommended for people who are allergic to aspirin.

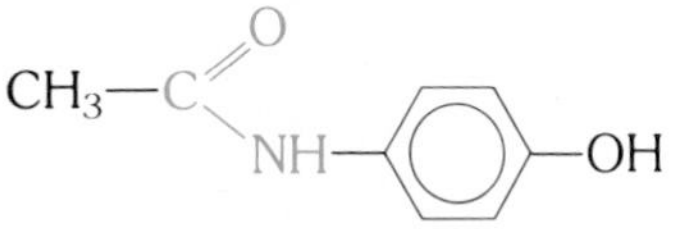

Acetaminophen

Practice Questions and Problems

FUNCTIONAL GROUPS

17-1 What is a functional group?

17-2 Identify all the functional groups listed in Table 17.1 with the following characteristics.

a. Contains the element oxygen
b. Contains the element nitrogen
c. Contains the element sulfur
d. Contains a carbon–oxygen double bond
e. Contains two oxygen atoms

17-3 In each of the following hydrocarbon derivatives identify by name the functional group present.

a. $CH_3-\underset{\displaystyle Br}{\underset{|}{CH}}-CH_2-CH_3$

b. $CH_3-CH_2-C\overset{\displaystyle O}{\underset{\displaystyle OH}{\lessgtr}}$

c. $CH_3-CH_2-O-CH_2-CH_3$

d. $CH_3-CH_2-\underset{\displaystyle OH}{\underset{|}{CH_2}}$

e. $CH_3-CH_2-\overset{\displaystyle O}{\overset{\|}{C}}-O-CH_2-CH_3$

f. $CH_3-CH_2-\overset{\displaystyle O}{\overset{\|}{C}}-CH_2-CH_3$

g. $CH_3-CH_2-C\overset{\displaystyle O}{\underset{\displaystyle NH_2}{}}$

h. $CH_3-CH_2-CH_2-C\overset{\displaystyle O}{\underset{\displaystyle H}{}}$

HALOGENATED HYDROCARBONS

17-4 What is a halogen atom?

17-5 Write all possible products resulting from the chlorination of methane.

17-6 Chlorofluorohydrocarbons of low molecular weight are marketed under the trade name Freons. Draw the structures of the following Freons.
a. Trichlorofluoromethane
b. Dichlorodifluoromethane
c. 1,2-Dichloro-1,1,2,2-tetrafluoroethane

17-7 Name the following halogenated hydrocarbons according to IUPAC rules.

a. $H-\overset{\displaystyle H}{\underset{\displaystyle H}{C}}-\overset{\displaystyle H}{\underset{\displaystyle Br}{C}}-Br$ b. $H-\overset{\displaystyle H}{\underset{\displaystyle Br}{C}}-\overset{\displaystyle H}{\underset{\displaystyle H}{C}}-Br$

c. $CH_3-\underset{\displaystyle CH_3}{CH}-\underset{\displaystyle I}{CH}-CH_3$ d. CH_3-I

e. $H-\overset{\displaystyle H}{\underset{\displaystyle H}{C}}-\overset{\displaystyle Cl}{\underset{\displaystyle Br}{C}}-\overset{\displaystyle Cl}{\underset{\displaystyle H}{C}}-H$ f. $H-\overset{\displaystyle H}{\underset{\displaystyle H}{C}}-\overset{\displaystyle H}{\underset{\displaystyle H}{C}}-F$

17-8 How many products (isomers) can result from the substitution of a chlorine atom for a hydrogen atom in each of the following alkanes?
a. Propane   b. Butane
c. 2-Methylbutane   d. 2,2-Dimethylbutane

17-9 Each of the following halogenated hydrocarbons has made headlines in recent years. In each case, discuss briefly what the concern was or is
a. DDT   b. PCBs   c. Freons

ALCOHOLS

17-10 Explain the relationship between the structure of a simple alcohol and the structure of water.

17-11 Name the following alcohols according to IUPAC rules.

a. $CH_3-CH_2-\underset{\displaystyle CH_3}{CH}-OH$

b. $CH_3-CH_2-CH_2-\overset{\displaystyle CH_3}{\underset{\displaystyle OH}{C}}-CH_3$

c. $\underset{\displaystyle CH_3}{CH_2}-\underset{\displaystyle OH}{CH}-\underset{\displaystyle CH_3}{CH_2}$

d. $CH_3-\underset{\displaystyle CH_3}{CH}-CH_2-\overset{\displaystyle CH_3}{\underset{\displaystyle CH_3}{C}}-OH$

17-12 Write structural formulas for the following alcohols.
a. 4-Methyl-2-pentanol
b. 2,4,4-Trimethyl-2-heptanol
c. Ethyl alcohol
d. Isopropyl alcohol

17-13 Assign IUPAC names to the following polyhydric alcohols.

a. $\underset{\displaystyle OH}{CH_2}-CH_2-\underset{\displaystyle OH}{CH_2}$

b. $\underset{\displaystyle OH}{CH_2}-CH_2-\underset{\displaystyle OH}{CH}-\underset{\displaystyle OH}{CH_2}$

c. $CH_3-\underset{\displaystyle OH}{CH}-\underset{\displaystyle CH_3}{CH}-OH$

17-14 Define the following terms.
a. Absolute alcohol   b. Grain alcohol
c. Wood alcohol   d. Denatured alcohol
e. Rubbing alcohol   f. Drinking alcohol
g. 40-proof alcohol

17-15 List a specific alcohol used in each of the following ways.
a. Primary component of "duplicating fluid"
b. Moistening agent in many cosmetics
c. "Skin coolant" for the human body
d. Antifreeze in windshield washer fluids
e. Antifreeze in car radiators
f. Ingredient in cough syrups and tinctures

17-16 What is the purpose of denaturing ethanol and what chemical substances are used as denaturants?

17-17 The maximum concentration of ethanol obtainable from fermentation is about 12 to 14%. Why is this so?

17-18 What are phenols?

17-19 Name each of the following compounds using the word phenol as part of the name.

a.

OH

CH_2—CH_2—CH_3

b.

OH

c.

OH

H_3C

d.

Cl

Cl

OH

ETHERS

17-20 What is the relationship between the structure of an alcohol and an ether?

17-21 Assign a common name and an IUPAC name to each of the following ethers.
a. CH_3—O—CH_2—CH_2—CH_3
b. CH_3—CH_2—CH_2—O—$\overset{\displaystyle CH_3}{\underset{|}{CH}}$—$CH_3$
c. CH_3—CH_2—O—CH_2—CH_3

17-22 Classify each of the ethers in problem 17-21 as a simple ether or a mixed ether.

17-23 Draw the structure of each of the following ethers.
a. Isopropyl propyl ether
b. Ethyl phenyl ether
c. 2-Ethoxypentane
d. 1-Methoxy-2,2-dimethylpropane

17-24 Most "modern" ethers used as inhalation anesthetics are halogenated ethers. Why is this so?

17-25 What is the structure of the compound MTBE and what is the significance of this ether?

17-26 Draw structural formulas for the six possible isomeric ethers that fit the molecular formula $C_5H_{12}O$.

THIOALCOHOLS AND THIOETHERS

17-27 Describe the structural similarities and differences between the following.

a. An ether and a thioether
b. An alcohol and a thioalcohol

17-28 According to law, thioalcohols must be added to natural gas before the natural gas is distributed to the public. Why?

17-29 What does the prefix thio- mean when it is used in a chemical name?

17-30 Write the structure of a sulfur-containing hydrocarbon derivative associated with each of the following.
a. Odor of a frightened skunk
b. Eye irritant released from chopped onions
c. Odor of garlic

17-31 What is the structure of the compound DMSO, and what are its uses?

ALDEHYDES AND KETONES

17-32 Indicate whether each compound contains a carbonyl group.

a. CH_3—CH_2—$\overset{\displaystyle O}{\underset{\displaystyle H}{C}}$

b. CH_3—$\overset{\displaystyle O}{\overset{\|}{C}}$—O—$CH_3$

c. CH_3—CH_2—O—CH_3

d. CH_3—CH_2—CH_2—OH

e. CH_3—CH_2—$\overset{\displaystyle O}{\underset{\displaystyle OH}{C}}$

f. CH_3—CH_2—$\overset{\displaystyle O}{\underset{\displaystyle CH_3}{C}}$

g. CH_3—CH_2—$\overset{\displaystyle O}{\underset{\displaystyle NH_2}{C}}$

h. CH_3—CH_2—$\overset{\displaystyle CH_3}{\underset{\displaystyle H}{C}}$

17-33 What structural feature distinguishes aldehydes from ketones?

17-34 Identify which compounds in problem 17-32 are aldehydes and which are ketones.

17-35 How many carbon atoms are present in the following?
a. The simplest aldehyde
b. The simplest ketone
c. Acetaldehyde

17-36 What is formalin and what are its uses?

17-37 Draw the structure of an aldehyde responsible, at least in part, for each of the following flavorings.
a. Almond   b. Cinnamon   c. Vanilla

17-38 Draw the structure of acetone and indicate some of its uses.

CARBOXYLIC ACIDS

17-39 What is the difference between a carbonyl group and a carboxyl group?

17-40 Contrast the general formulas for an aldehyde, a ketone, and a carboxylic acid.

17-41 Using IUPAC rules, name the following carboxylic acids.

a. $CH_3-CH_2-CH_2-CH_2-C\overset{\displaystyle O}{\underset{\displaystyle OH}{}}$

b. $CH_3-CH_2-\underset{\underset{\displaystyle CH_3}{|}}{CH}-CH_2-C\overset{\displaystyle O}{\underset{\displaystyle OH}{}}$

c. $CH_3-\underset{\underset{\displaystyle CH_3}{|}}{CH}-CH_2-\underset{\underset{\underset{\displaystyle CH_3}{|}}{\underset{\displaystyle CH_2}{|}}}{CH}-CH_2-C\overset{\displaystyle O}{\underset{\displaystyle OH}{}}$

d. $CH_3-C\overset{\displaystyle O}{\underset{\displaystyle OH}{}}$

17-42 Draw structural formulas to represent the following carboxylic acids.
a. 2,3-Dimethylbutanoic acid
b. Acetic acid (common name)
c. Ethanoic acid
d. Lactic acid (common name)
e. Oxalic acid (common name)
f. Citric acid (common name)

17-43 Draw the structure and state the significance of the compound Ibuprofen.

17-44 Draw the structures and give names of carboxylate ions formed when each of the following carboxylic acids is dissolved in water.

a. $CH_3-C\overset{\displaystyle O}{\underset{\displaystyle OH}{}}$ b. $\underset{}{\bigcirc}-C\overset{\displaystyle O}{\underset{\displaystyle OH}{}}$

c. $H-C\overset{\displaystyle O}{\underset{\displaystyle OH}{}}$

ESTERS

17-45 What is the difference between a carboxylic acid functional group and an ester functional group?

17-46 Draw structural formulas for each of the following esters.
a. Ethyl propanoate   b. Propyl acetate
c. Ethyl methanoate   d. Methyl ethanoate

17-47 Using IUPAC rules, name the following esters.

a. $CH_3-C\overset{\displaystyle O}{\underset{\displaystyle O-CH_3}{}}$

b. $CH_3-CH_2-C\overset{\displaystyle O}{\underset{\displaystyle O-CH_2-CH_2-CH_3}{}}$

c. $CH_3-CH_2-CH_2-C\overset{\displaystyle O}{\underset{\displaystyle O-CH_2-CH_2-CH_3}{}}$

17-48 Draw the structure of aspirin and identify the two functional groups present in this structure.

17-49 Using the structural information given in Table 17-7, describe the differences in structure between molecules responsible for the following.
a. Apple and pineapple flavoring
b. Banana and apricot flavoring

AMINES AND AMIDES

17-50 Explain the relationship between the compound ammonia and the three types of amines.

17-51 Classify each of the following amines as primary, secondary, or tertiary amine.
a. $CH_3-NH-CH_2-CH_3$
b. $CH_3-CH_2-\underset{\underset{\displaystyle CH_3}{|}}{N}-CH_2-CH_3$
c. $CH_3-CH_2-NH_2$
d. $CH_3-\underset{\underset{\displaystyle CH_3}{|}}{\overset{\overset{\displaystyle CH_3}{|}}{C}}-NH_2$
e. $CH_3-N\overset{\displaystyle H}{\underset{\displaystyle CH_3}{}}$
f. $CH_3-N\overset{\displaystyle CH_2-CH_3}{\underset{\displaystyle CH_3}{}}$

17-52 Name each of the amines in problem 17-51.

Each end of a polyethylene molecule has a terminating group, derived from the initiator. The presence of these two end groups has no appreciable effect on the composition and properties of the polyethylene and they are not even shown in the polymer formula; they are a trivial fraction of the atoms present in the molecule. Thus, an overall equation for the polymerization of ethylene can be written schematically as

$$n\ CH_2{=}CH_2 \longrightarrow {+\!}CH_2{-}CH_2{\,\overline{)}}_n$$

Monomer　　　　　　Polymer

The net effect of the polymerization reaction is a gigantic chain of singly bonded carbon atoms; the double bonds of the monomers have reverted to single bonds.

$$\cdots\underset{\underset{\textstyle H}{|}}{\overset{\overset{\textstyle H}{|}}{C}}{-}\underset{\underset{\textstyle H}{|}}{\overset{\overset{\textstyle H}{|}}{C}}{-}\underset{\underset{\textstyle H}{|}}{\overset{\overset{\textstyle H}{|}}{C}}{-}\underset{\underset{\textstyle H}{|}}{\overset{\overset{\textstyle H}{|}}{C}}\cdots$$

Polyethylenes formed under various temperatures, pressures, and catalytic conditions have different properties because of different carbon chain arrangements. Two general categories of polyethylenes are recognized: (1) high-density polyethylenes (HDPE) and (2) low-density polyethylenes (LDPE).

HDPE

High-density polyethylenes ($0.94\ g/cm^3$ or above) have carbon chains that are essentially linear; there are few branches off the carbon chains. Such molecules can assume a fairly ordered solid-state arrangement; the net result is high density. HDPE production accounts for approximately one-third of total polyethylene production. Most of this type of polyethylene is used in molded items, primarily bottles and containers. Almost all 1-gallon milk containers are made of HDPE. The containers for many other consumer goods, including bleaches, detergents, and motor oils, are also made of HDPE.

LDPE

Low-density polyethylenes (less than $0.94\ g/cm^3$) account for the other two-thirds of polyethylene production. They have many carbon branches off the main carbon chain and the carbon chains are often cross-linked.

A **cross-link** *is an atom or group of atoms, which are simultaneously bonded to two polymer chains.* The large number of side-chains present in LDPE prevent the molecules from assuming a "compact" arrangement; the result is a lower density. Low-density polyethylenes are waxy, semirigid, transparent materials. Their major use, two-thirds of LDPE production, is film and sheet materials used mainly for packaging. A major new market for LDPE involves the shift from paper bags to polyethylene in the grocery and other retail areas.

18.4 Addition Polymers Involving Substituted Ethylenes

Derivatives of ethylene in which one or more of the hydrogen atoms have been replaced by other atoms or groups of atoms can also be polymerized in a manner similar to that for polyethylene—that is, addition polymerization occurs. The double bond is the key structural feature involved in addition polymerization reactions, and substituted ethylenes still contain the original double bond. Polymers of substituted ethylenes have significantly different characteristics (properties) than polyethylene, and thus have different uses.

The three most produced substituted polyethylenes are polyvinyl chloride (also known simply as PVC), polypropylene, and polystyrene. Annual production figures for these substances, in billions of pounds, are 7, 5.5, and 4, respectively; the annual production figure for polyethylene is 16 billion pounds.

The monomers for these polymers are all simple monosubstituted ethylenes.

Monomer for polyethylene   Monomer for polyvinylchloride   Monomer for polypropylene   Monomer for polystyrene

PVC

Uses for polyvinyl chloride, the monochloro derivative of polyethylene, are concentrated in building construction. Along with familiar PVC pipe and conduit, floor coverings is an important end-use. Plastic money (credit cards) is of polyvinyl chloride formulation. Other lesser uses include garden hoses and phonograph records.

Polypropylene, the methyl derivative of polyethylene, is used in transportation applications (mainly battery casings and automotive trim parts), and in fibers for carpets, artificial turf, rope, and fishing nets.

Polystyrene, which contains a six-membered ring of carbon atoms (a phenyl group; see Section 16.11) in place of a hydrogen atom, is used in shipping containers and packing (where polystyrene's light mass is ideal), egg cartons, thermal insulation, toys, and appliance parts. Styrofoam, a low-density form of polystyrene, is obtained by adding a foaming agent (which produces gas bubbles) to molten polystyrene and then letting it solidify.

STYROFOAM

Another polyethylene derivative, which is well known though produced in smaller volume, is Teflon, whose name comes from the name of its monomer, *tetrafluoroethylene*. It has a tetrasubstituted monomer; all four hydrogen atoms of ethylene have been replaced with fluorine atoms.

TEFLON

$$n\,CF_2{=}CF_2 \longrightarrow +CF_2{-}CF_2 \rightarrow_n$$

Tetrafluoroethylene   Teflon

The best-known use for Teflon is as a nonstick coating for cooking utensils. Other uses include gasket and valve materials and as an insulation.

The final substituted polyethylene we consider in this section, Saran, has a more complicated structure than the preceding examples. It is a *copolymer*. A **copolymer** *is a polymer containing two different kinds of monomers.* Saran is a copolymer of vinyl chloride (the monomer for PVC) and 1,1-dichloroethene.

SARAN

```
  H   H              H   Cl
  |   |              |   |
  C = C              C = C
  |   |              |   |
  H   Cl             H   Cl

Vinyl chloride     1,1-Dichloroethene
```

The repeating unit for this polymer contains one unit of each of the monomers.

```
        1st       2nd
      monomer   monomer
     ⌒⌒⌒⌒    ⌒⌒⌒⌒
    / H   H   H   Cl \
    | |   |   |   |  |
  +-C---C---C---C-+
    | |   |   |   |  |
    \ H   Cl  H   Cl /n
```

Saran is extensively used in the food packaging industry.

Figure 18.2 illustrates some of the end-product uses for polyethylene and polyethylene derivatives.

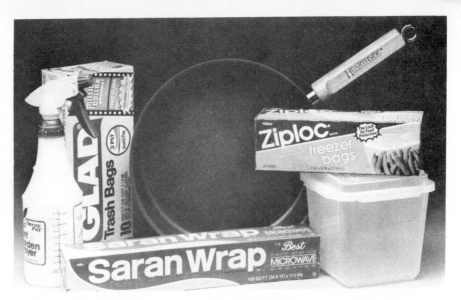

(a) Food storage bags (polyethylene), Saran Wrap (a chlorinated polyethylene), Teflon-coated frying pan and spatula, and polyethylene storage container and sauce dispenser.

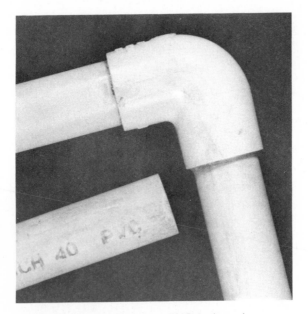

(b) Drainage pipe made from PVC (polyvinyl chloride). [John Schultz, PAR/NYC.]

FIGURE 18.2
Polymers based on ethylene and its derivatives find extensive use in many areas.

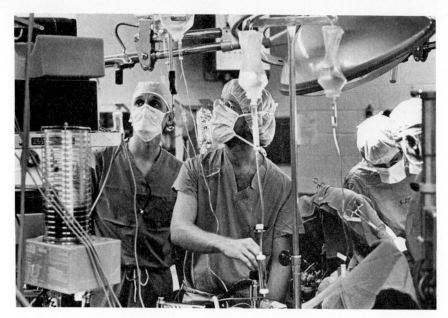

(c) Polyethylene tubing used for intravenous therapy and for in-dwelling catheters. [Will McIntyre/Photo Researchers, Inc.]

(d) Foamed polystyrene (Styrofoam) used to insulate buildings. [Arthur Grace/Stock, Boston.]

FIGURE 18.2 (*Continued*)

18.5   Butadiene Addition Polymers

Addition polymers based on butadiene or substituted butadiene monomers constitute an important class of polymeric materials. These polymers are elastomers. An **elastomer** *is a material that can be stretched without undergoing a permanent change and, immediately upon release of the stress, can return with force to its original shape and size.* Natural rubber and many types of synthetic rubber are butadiene elastomers.

Recall, from Section 16.9, that a butadiene is a four-carbon alkene containing two double bonds. The structure of 1,3-butadiene, the simplest monomer for this type of polymer, is

$$\overset{1}{CH_2}{=}\overset{2}{CH}{-}\overset{3}{CH}{=}\overset{4}{CH_2}$$

The polymerization of 1,3-butadiene to give a butadiene polymer chain can be represented by the equation

$$n\ CH_2{=}CH{-}CH{=}CH_2 \longrightarrow +CH_2{-}CH{=}CH{-}CH_2\rangle_n$$

1,3-Butadiene   Polybutadiene

Unlike polymers derived from ethylene (Sections 18.3 and 18.4), butadiene polymers are unsaturated, that is, they have double bonds within the polymeric structure. Each repeating unit in the polymer contains one double bond. Note that the double bond in the polymer repeating unit is in a different location from either of the double bonds in the monomer molecules. The two double bonds in a monomer are in positions 1 and 3; the double bond in the polymer repeating unit is in position 2. This shift in double bond formation can be considered to result from the following movement of electrons caused by free radical initiation (Section 18.3) of the polymerization reaction.

$$CH_2{::}\overset{\overset{\displaystyle H}{|}}{C}{-}\overset{\overset{\displaystyle H}{|}}{C}{::}CH_2 \longrightarrow \cdot CH_2{-}\overset{\overset{\displaystyle H}{|}}{C}{=}\overset{\overset{\displaystyle H}{|}}{C}{-}CH_2\cdot$$

Butadiene monomer   Free radical with
                                                      unpaired electrons

Natural rubber is a polymer involving the monomer isoprene, which is the 2-methyl derivative of 1,3-butadiene.

$$CH_2{=}CH{-}\overset{\overset{\displaystyle CH_3}{|}}{C}{=}CH_2$$

Isoprene
(2-methyl-1,3-butadiene)

The addition polymerization reaction equation is

$$n\ CH_2=CH-\overset{\overset{\textstyle CH_3}{\textstyle |}}{C}=CH_2 \longrightarrow +CH_2-CH=\overset{\overset{\textstyle CH_3}{\textstyle |}}{C}-CH_2+_n$$

Isoprene   Natural rubber

This polymer occurs naturally as a suspension of particles in the sap (latex) of the rubber tree. It oozes from rubber trees when they are cut. The suspended rubber "particles" coagulate into a gummy mass when the latex is acidified.

Natural rubber has some undesirable properties; it becomes sticky in hot weather and stiff in cold weather. This problem is solved by vulcanizing the **VULCANIZED RUBBER** rubber, a process in which the rubber is heated with sulfur. *Vulcanization* involves the formation of sulfur atom cross-links (Section 18.3) between polymer chains. The number of sulfur atoms in a cross-link varies usually between 1 and 4. Figure 18.3 shows a segment of vulcanized rubber.

The process of vulcanization was discovered accidentally in 1839, by the American chemist Charles Goodyear (1800–1860). Goodyear accidentally spilled a sulfur–rubber mixture on a hot stove; the resulting rubber was stronger, more elastic, and more resistant to heat and cold. His vulcanization process made rubber usuable in countless applications and opened the door for its use in many industries.

Natural rubber accounts for approximately one-fourth of the rubber consumed annually in the United States. Almost all of it is imported; three countries—Indonesia, Malaysia, and Liberia—supply most of these imports. The remaining three-fourths of the demand for rubber in the United States is met by a variety of synthetic rubbers. Synthetic rubbers are a product of research carried out during World War II. During the war the United States found itself cut off from imported supplies of natural rubber necessitating the search for synthetic substitutes.

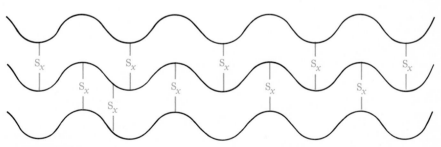

FIGURE 18.3

In vulcanized rubber, sulfur atoms cross-link the polymer chain. The S_x notation indicates that the number of sulfur atoms in a cross-link varies.

Most synthetic rubbers have structures similiar to that for natural rubber. Neoprene, the first commercially successful general-purpose synthetic rubber, is produced from the monomer chloroprene (2-chloro-1,3-butadiene).

$$n \; CH_2{=}CH{-}\underset{\underset{Cl}{|}}{C}{=}CH_2 \longrightarrow +CH_2{-}CH{=}\underset{\underset{Cl}{|}}{C}{-}CH_2{+}_n$$

Chloroprene   Neoprene
                                              (polychloroprene)

Notice that the difference between chloroprene and isoprene monomers is that a chloro group has been substituted for a methyl group. Neoprene has excellent resistance to deterioration from gasoline and oils as well as to abrasion. Its major use is in mechanical parts that undergo severe service in industry. It is not a tire rubber.

Polybutadiene, made by polymerizing the monomer 1,3-butadiene, is another important synthetic rubber. The equation for this polymerization reaction was given earlier in this section. The relationship between isoprene and 1,3-butadiene monomers is simple; a hydrogen atom has replaced the methyl group. The major use for this rubber is in tire treads, for which is provides long wear and less resistance to rolling than most lower-cost synthetic rubbers.

The dominant synthetic rubber in use today, accounting for almost half of total production, is styrene–butadiene rubber (SBR), which is a copolymer (Section 18.4) made by mixing butadiene and styrene (Section 18.4) in a 3 : 1 ratio. A typical segment of an SBR molecule contains three butadiene units and one styrene unit.

SBR

$$75\% \; CH_2{=}CH{-}CH{=}CH_2 + 25\% \; CH_2{=}CH$$

1,3-Butadiene   Styrene

$$+CH_2{-}CH{=}CH{-}CH_2{+}CH_2{-}CH{+}CH_2{-}CH{=}CH{-}CH_2{+}CH_2{-}CH{=}CH{-}CH_2{+}$$

| Butadiene | Styrene | Butadiene | Butadiene |
| unit | unit | unit | unit |

SBR polymer is more resistant to oxidation and abrasion than natural rubber, but it has less satisfactory mechanical properties. Its greatest advantage over natural rubber is in economics of production. As with natural rubber, vulcanization is used to improve the properties of this synthetic rubber (as well as other synthetic rubbers). Tires and tire components such as tread rubber consume more than 70% of SBR produced in the United States. Tire treads involve SBR blended with polybutadiene.

Synthetic polyisoprene—a substance identical to natural rubber, except that is comes from petroleum refineries rather than from rubber trees—can now be produced on an industrial scale. Use of synthetic polyisoprene is limited in the United States for economic reasons; raw material costs make it much more expensive than natural rubber imports.

Tires and tire products account for more than half of the world's rubber consumption. On a worldwide basis, two-thirds of all natural rubber and slightly over half (54%) of all synthetic rubber is used for this purpose.

18.6   Condensation Polymerization Reactions

A **condensation reaction** *is a reaction in which two molecules react to give a larger molecule with the elimination of a small molecule such as water.* Many reactions between hydrocarbon derivatives are of this type. Two examples of condensation reactions, both of which produce an ester (Section 17.8), are

$$CH_3-C\!\!\begin{array}{c}O\\\\OH\end{array} + HO-CH_3 \longrightarrow CH_3-C\!\!\begin{array}{c}O\\\\O-CH_3\end{array} + H_2O$$

A carboxylic acid   An alcohol   An ester

$$CH_3-C\!\!\begin{array}{c}O\\\\Cl\end{array} + HO-CH_2-CH_3 \longrightarrow CH_3-C\!\!\begin{array}{c}O\\\\O-CH_2-CH_3\end{array} + HCl$$

An acid chloride   An alcohol   An ester

In the first reaction, the small molecule produced is water, and in the second reaction it is hydrogen chloride. Note that the small molecule always contains atoms from both reacting functional groups.

A condensation reaction does not depend on the presence of a double bond, as was the case for addition reactions that occur during addition polymerization (Section 18.3). The necessary requirement for a condensation reaction is the presence of two different functional groups (Section 17.1), which react with each other, on two different molecules.

Condensation reactions can be used to produce polymers called condensation polymers. A **condensation polymer** *is a polymer produced by reacting bifunctional monomers to give large molecules and some small molecules such as water.* *Bifunctional* monomers possess two functional groups. Dicarboxylic acids, dialocohols, and diamines are three of many types of such molecules that exist.

A four-carbon dicarboxylic acid   A two-carbon dialcohol   A two-carbon diamine

Let us consider the first few steps in the formation of a polymer from a three-carbon diacid and a two-carbon dialcohol. Reaction between one diacid molecule and one dialcohol molecules produces a dimer (two units).

A diacid   A dialcohol

From diacid   From dialcohol
A dimer

Note that the dimer still possesses an acid functional group (left end of the molecule) and an alcohol functional group (right end of the molecule), which are available for reaction with an additional monomer of the appropriate type; that is, the dimer is bifunctional. Letting the dimer functional groups react further produces a tetramer (4 units).

The tetramer still contains two functional groups, one on each end; hence, the reaction process can continue until a large polymer molecule is produced. The acid group on the one end of the polymer chain always reacts with another alcohol monomer and the alcohol group on the other end of the chain always reacts with another acid monomer.

A condensation polymer differs from an addition polymer in that the latter contains *all* of the atoms present in the reacting monomers; the polymer is the only product in addition polymerization.

The general formula for the condensation polymer, whose dimer and tetramer were previously shown, is

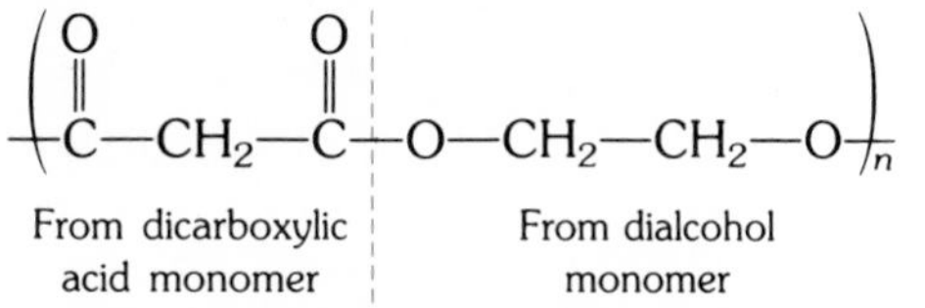

$$\left(\begin{matrix} O \\ \| \\ C-CH_2-C \end{matrix} \begin{matrix} O \\ \| \\ \end{matrix} O-CH_2-CH_2-O \right)_n$$

From dicarboxylic   From dialcohol
acid monomer   monomer

Many different types of condensation polymers have been synthesized. Important classes include the polyamides (Section 18.7), polyesters (Section 18.7), and formaldehyde-based network polymers (Section 18.8).

18.7 Polyamide and Polyester Fibers

A **polyamide** *is a condensation polymer that contains amide linkages between monomer units.* Recall from Section 17.9 that the general formula for an amide is

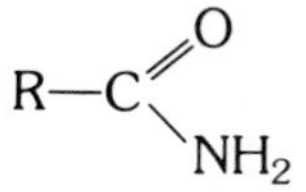

$$R-C \overset{\displaystyle O}{\underset{\displaystyle NH_2}{<}}$$

and that an amide linkage is the carbon–nitrogen bond in the structure

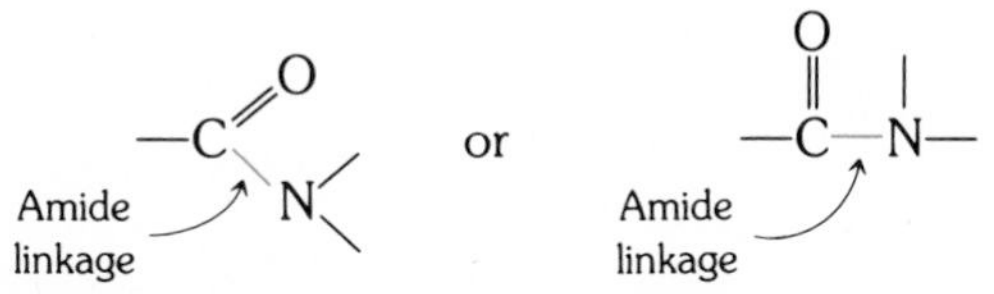

$$-C \overset{\displaystyle O}{<} N \overset{\displaystyle }{<} \qquad or \qquad -\overset{\displaystyle O}{\overset{\displaystyle \|}{C}}-\overset{\displaystyle |}{N}-$$

Amide   Amide
linkage   linkage

In commerce, polyamides are called *nylons*. Many kinds of nylon have been prepared and tried on the consumer market, but the two most successful are nylon 66 and nylon 6.

The monomers used in the production of Nylon 66 are 1,6-diaminohexane (a diamine) and adipic acid (a dicarboxylic acid).

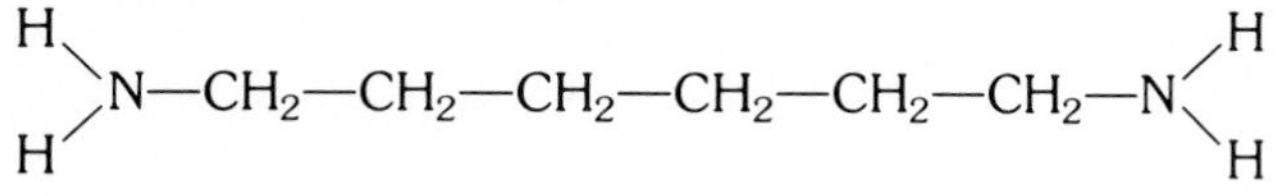

$$\overset{\displaystyle H}{\underset{\displaystyle H}{>}}N-CH_2-CH_2-CH_2-CH_2-CH_2-CH_2-N\overset{\displaystyle H}{\underset{\displaystyle H}{<}}$$

1,6-Diaminohexane

$$\overset{\displaystyle O}{\underset{\displaystyle HO}{>}}C-CH_2-CH_2-CH_2-CH_2-C\overset{\displaystyle O}{\underset{\displaystyle OH}{<}}$$

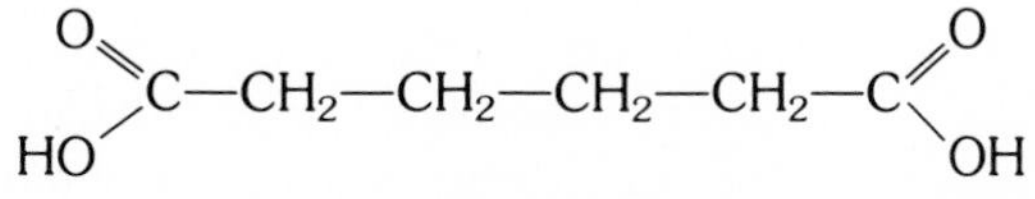

Adipic acid

(The name Nylon 66 comes from the fact that each monomer has 6 carbon atoms.)

Formation of Nylon 66 (or any other polyamide) follows the general principles outlined in Section 18.6 for condensation polymer formation. The reaction of one acid group of the diacid with one amine group of the diamine initially produces a dimer molecule with an acid group left over on one end and an amine group on the other end.

- Leftover acid group, which can react further
- Amide linkage
- Leftover amine group, which can react further

This species then reacts further, with the process continuing until a long polymeric molecule, nylon, has been produced.

A portion of the polyamide Nylon 66

A water molecule is formed every time an amide linkage is formed. The term polyamide draws attention to the amide linkages that connect monomer units. Note that the polymer "backbone" is not just a chain of carbon atoms; it is a chain containing both carbon and nitrogen atoms.

Nylon 6 is prepared from the monomer caprolactum, which is a cyclic form of aminocaproic acid.

$$H_2N-CH_2-CH_2-CH_2-CH_2-CH_2-C\!\!\begin{array}{c}{}^{\displaystyle O}\\{}_{\displaystyle OH}\end{array}$$

Aminocaproic acid

Only this one monomer is needed because it contains both an amino group and a carboxylic acid group. The amino end of one monomer reacts with the carboxylic acid end of the next monomer, and so on. A portion of the structure of Nylon 6 is as follows.

$$\cdots -\underset{\underset{H}{|}}{N}-(CH_2)_5-\underset{\underset{\|}{O}}{C}-\underset{\underset{H}{|}}{N}-(CH_2)_5-\underset{\underset{\|}{O}}{C}-\cdots$$

A portion of Nylon 6

Nylons are used in clothing and hosiery as well as in carpets, tire cord, rope, and parachutes. Other uses include paint brushes, electrical parts, valves, and fasteners. Nylon fibers are tough, strong, nontoxic, nonflammable materials that are resistant to most chemicals. Surgical suture (Figure 18.4) is made of nylon because it is such a strong fiber.

Home furnishings, mostly carpet and rug fiber, account for 60% of nylon consumption, apparel for 20% of production, and industrial uses, mainly tire cord, for the remaining 20%.

NYLONS, SILK, AND WOOL

Silk and wool are naturally occurring polyamide fibers. Both of these substances are proteins, and proteins are polyamides. A significant portion of Chapter 25 is devoted to the subject of proteins. Figure 18.5 shows the chemical kinship between the two natural fibers, silk and wool, and nylon, which could be called "artificial silk."

A **polyester** *is a condensation polymer that contains ester linkages between monomer units.* Recall from Section 17.8 that the general formula for an

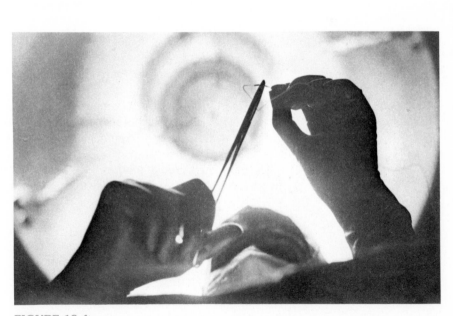

FIGURE 18.4
Nylon fibers are used in sewing up surgical incisions. [*Stan Levy/Photo Researchers, Inc.*]

Silk
(fibroin)

$$-\overset{\overset{\textstyle O}{\textstyle \|}}{C}-NH-CH_2-\overset{\overset{\textstyle O}{\textstyle \|}}{C}-NH-CH_2-\overset{\overset{\textstyle O}{\textstyle \|}}{C}-$$

Nylon 6
(polycaproamide)

$$-\overset{\overset{\textstyle O}{\textstyle \|}}{C}-NH-(CH_2)_5-\overset{\overset{\textstyle O}{\textstyle \|}}{C}-NH-(CH_2)_5-\overset{\overset{\textstyle O}{\textstyle \|}}{C}-$$

Wool
(beta-keratin)

$$-\overset{\overset{\textstyle O}{\textstyle \|}}{C}-NH-\underset{\underset{\textstyle R}{\textstyle |}}{CH}-\overset{\overset{\textstyle O}{\textstyle \|}}{C}-NH-\underset{\underset{\textstyle R}{\textstyle |}}{CH}-\overset{\overset{\textstyle O}{\textstyle \|}}{C}-$$

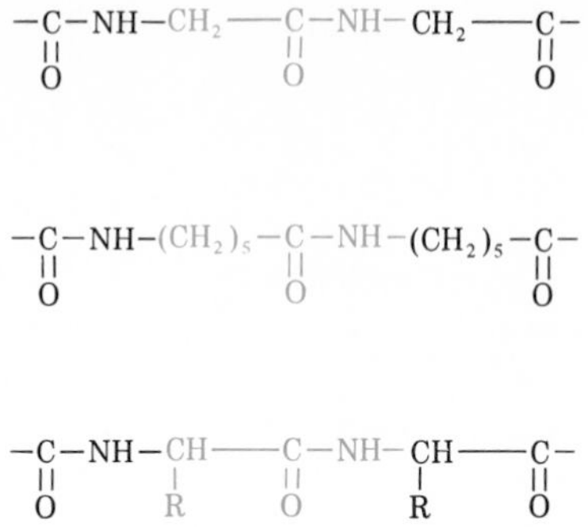

FIGURE 18.5

The natural fibers silk and wool and the synthetic nylon fibers have chemically similar structures.

ester is

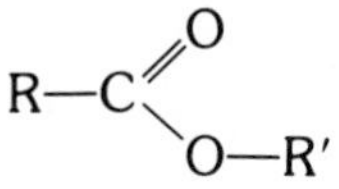

$$R-C\overset{\displaystyle\nearrow O}{\underset{\displaystyle O-R'}{}}$$

and that an ester linkage is the carbon–oxygen single bond that involves the carbonyl carbon atom.

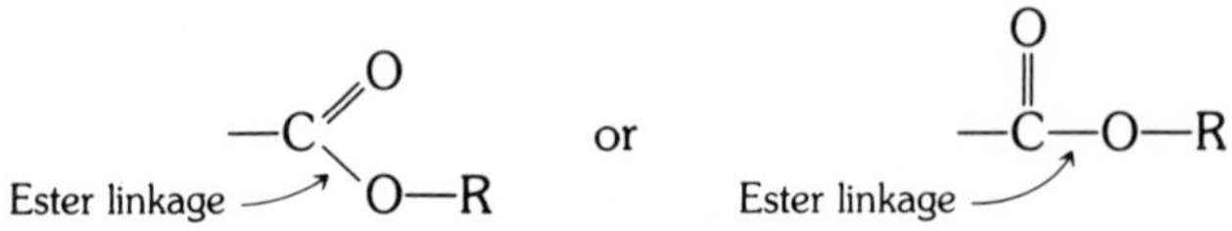

$$-C\overset{\displaystyle\nearrow O}{\underset{\displaystyle O-R}{}} \qquad or \qquad -\overset{\overset{\textstyle O}{\textstyle \|}}{C}-O-R$$

Ester linkage ⟶   Ester linkage ⟶

The monomers needed to form a polyester are a dicarboxylic acid and a dialcohol.

The most used and best known of all polyesters is poly(ethylene glycol terephthalate) (PET), which is marketed as the textile fiber Dacron. Incidentally, the trade name *Dacron* is commonly mispronounced; the correct pronunciation is Day-kron.

DACRON

The monomers used to produce PET, which explains its long chemical name, are ethylene glycol (a dialcohol) and terephthalic acid (a carboxylic acid).

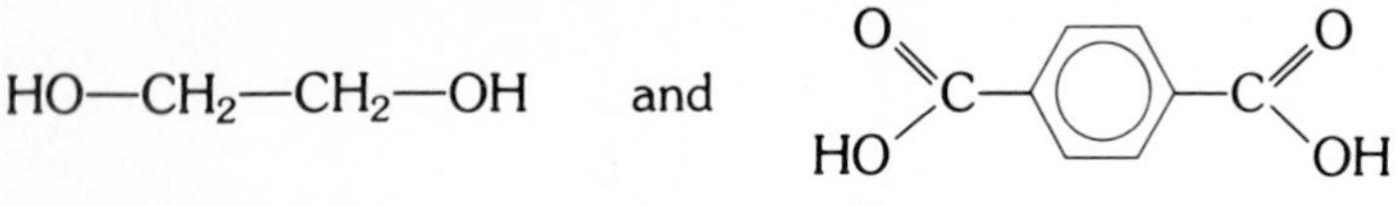

$$HO-CH_2-CH_2-OH \qquad and \qquad \underset{HO}{\overset{O}{\diagdown}}C-\underset{}{\bigcirc}-C\underset{OH}{\overset{O}{\diagup}}$$

Ethylene glycol   Terephthalic acid

The reaction of one acid group of the diacid with one alcohol group of the dialcohol initially produces an ester molecule with an acid group left over on one end and an alcohol group on the other end.

$$\text{(structure: terephthalic acid)} + HO-CH_2-CH_2-OH \longrightarrow$$

$$\text{(ester product structure)} + H_2O$$

Leftover acid group, which can react further

Ester linkage

Leftover alcohol group, which can react further

This species can react further. The remaining acid group can react with another alcohol group (from another monomer), and the alcohol group can react with another acid group (from another monomer). This process continues until an extremely large molecule, the polyester, is produced.

$$\cdots C - \bigcirc - C - O - (CH_2)_2 - O - C - \bigcirc - C - O - (CH_2)_2 - O - C - \bigcirc - C \cdots$$

Ester linkages

Ester linkages

A portion of the polyester Dacron

Note that the polymer backbone contains oxygen atoms in addition to carbon atoms. This contrasts with the C—N backbone in polyamides.

Dacron fibers are used extensively in various types of wash-and-wear clothing. They are usually blended in a roughly 2-to-1 ratio with cotton because the resulting blended fibers are softer and more permeable to moisture than pure Dacron. Bed linens, at one time a 50–50 polyester–cotton blend, now contain more polyester (a 65–35 blend).

PERMANENT PRESS CLOTHING

The term *permanent press* is often used to describe textile materials, such as Dacron-cotton blends, that require no ironing after laundering. Permanent press fabrics are fabrics in which the polymeric fibers present have been treated with chemicals that cross-link adjacent molecular chains together and pull them back into position after the fibers are bent. The concept of cross-linking has already been encountered twice in this chapter—with low-density polyethylene (Section 18.3) and with vulcanized rubber (Section 18.5).

Approximately 50% of polyester fiber production is used in wearing apparel. Another major use is automotive upholstery. Cloth interiors of mostly polyester for new automobiles now account for about 90% of auto interiors, in contrast with about 90% ethylene-based addition polymers a few years ago.

PET can also be produced in the form of a film rather than fiber. In film form, PET is marketed as Mylar. When magnetically coated, Mylar tape is used in audio and video recordings.

18.8 Formaldehyde-Based Network Polymers

In discussions on condensation polymers to this point (Sections 18.6 and 18.7), all monomers have been bifunctional molecules. Polymer chain extension occurs by adding monomers to the two ends of the existent chain. We now consider a variation of this process, which involves monomers with more than two functional groups—that is, monomers with more than two reaction sites. With such monomers polymer chains can extend in many directions at the same time. The result is a condensation network polymer. A **condensation network polymer** *is a high-molecular-weight molecule in which monomers are connected in a three-dimensional cross-linked network.*

The cross-linking in network polymers is different from that previously encountered for vulcanized rubber (Section 18.5) or permanent press polyesters (Section 18.7). In the latter two cases, cross-linking results from the purposeful addition of a cross-linking agent to already formed polymer mix. With network polymers, cross-linking is part of the polymerization process, and such cross-links are identical with other monomer linkages in the polymer.

Two important classes of network condensation polymers are the *phenolics* and the *melamines*. These two types of macromolecules share a common feature. The simple molecule formaldehyde, which has the structure

$$\underset{\text{H}-\overset{\overset{\displaystyle \text{O}}{\|}}{\text{C}}-\text{H}}{}$$

is one of the monomers used in their production. Phenolics and melamines are distinguished from each other by the identity of the second monomer; it is phenol in phenolics and melamine in melamines. The structures of these two monomers are

Phenol   Melamine

The arrows in these structural diagrams point to the reactive sites in the monomers. Both phenol and melamine exhibit trifunctionality in polymerization reactions, that is, a given monomer may be linked to three other monomers. In the case of phenol, remember that a hydrogen atom is present at each of the unsubstituted positions on the benzene ring (Section 16.11).

To better understand the network polymerization process, let us consider details involved in the formation of the phenol-formaldehyde polymer. The condensation process involves the splitting out of water molecules, with the hydrogen atoms coming from the benzene ring of phenol and the oxygen atom from formaldehyde.

As a first step in the process, a formaldehyde molecule reacts with two phenol molecules with the production of one water molecule.

Each of the phenol groups in this species still contains two reactive sites. Similar reactions occuring at these sites lead to the formation of a large formaldehyde-phenol aggregate. This process continues, giving the polymer, a typical section of which is as follows. Each CH_2 bridge comes from a formaldehyde molecule.

Similarly, in melamines a given melamine monomer can be bonded to three other melamine monomers through CH_2 bridges derived from formaldehyde molecules.

FIGURE 18.6
Bakelite electric switch.
[PAR/NYC.]

Because of the extensive cross-linking in phenolics and melamines, they are thermosetting (Section 18.2) polymers. They are fused during production but become permanently hard when cured. The hardness and insulating properties of such polymers make them excellent as moldings for radios and other containers for electrical circuits. Bakelite, used in electric switches, is a phenolic polymer (see Figure 18.6).

The major use for phenolic polymers is not in molded items (10% of production) but rather in adhesives (65% of production). Half of the phenolics produced are used in the production of plywood. Particle board is another use for phenolic adhesive.

MELMAC AND FORMICA

Melmac (dishware) and Formica (countertops) are two trade names associated with melamine-formaldehyde polymers. The starting material in the production of Formica-type materials and other decorative laminates is pigmented paper with a design or photograph. The paper is saturated or impregnated with various partly polymerized network polymers, and the polymers are cured (hardened). Then several layers of a melamine-formaldehyde polymer are used as a top coat for wear-resistance. Thus, when you look at a piece of decorative laminate or synthetic veneer, you are looking through a thin, clear layer of formaldehyde-based polymer at a paper (photograph or design) saturated with formaldehyde-based polymer (see Figure 18.7).

FIGURE 18.7
Melamine-formaldehyde polymers are important components of laminated countertops such as Formica.

18.9 Rearrangement Polymers: Polyurethanes

The urethane functional group has structural features of both an ester and an amide. Its general formula is

$$-\overset{\underset{\displaystyle H}{|}}{N}-\overset{\underset{\displaystyle O}{||}}{C}-O-$$

Amide linkage   Ester linkage

The simplest molecule containing such a structural feature is

$$\overset{\displaystyle H}{\underset{\displaystyle H}{>}}N-\overset{\underset{\displaystyle O}{||}}{C}-O-CH_3$$

A **polyurethane** *is a polymer that contains urethane linkages between monomer units.* A urethane linkage is

$$-N-C-O-$$
$$|\parallel$$
$$HO$$

The monomers needed to form a polyurethane molecule are a dialcohol and a diisocyanate. The isocyanate functional group has the structure

$$-N{=}C{=}O$$

The interaction between an isocyanate group and a hydroxyl group, to form a urethane linkage, occurs through a *rearrangement* process rather than a *condensation* process. In a condensation process (Section 18.6), two interacting functional groups join together with a small molecule, such as water, being produced also. In a rearrangement process two interacting functional groups join together without the production of a small molecule.

Polymers resulting from rearrangement processes are called rearrangement polymers. Polyurethanes are the most important class of such polymers. A **rearrangement polymer** *is a high-molecular-weight molecule formed from the reaction between bifunctional monomer units in which no other products (small molecules) are produced from the functional group interactions.* Thus, the difference between a condensation polymer and a rearrangement polymer is that small-molecule formation accompanies the synthesis of the former but not of the latter.

The rearrangement reaction that occurs to produce a urethane linkage can be illustrated by considering the reaction between the dialcohol ethylene glycol and the aromatic diisocyanate paraphenylene diisocyanate.

$$O{=}C{=}N-\!\!\left\langle\bigcirc\right\rangle\!\!-N{=}C{=}O \;+\; H{-}O{-}CH_2{-}CH_2{-}OH \longrightarrow$$

p-Phenylene diisocyanate   Ethylene glycol

$$O{=}C{=}N-\!\!\left\langle\bigcirc\right\rangle\!\!-\underset{\underbrace{}}{\overset{\overset{\textstyle H}{|}\;\overset{\textstyle O}{\parallel}}{N-C-O}}-CH_2{-}CH_2{-}OH$$

Urethane
linkage

In this reaction a hydrogen atom is shifted (rearranged), moving from the alcohol to the nitrogen atom of the isocyanate; the remainder of the alcohol molecule (all of it except the shifted hydrogen) becomes attached to the carbon atom of the isocyanate. The net effect of this rearrangement is the conversion of a carbon–nitrogen double bond to a carbon–nitrogen single bond, with each of these two atoms now bearing an added substituent.

The dimer formed by this rearrangement process is still bifunctional; it can therefore react with additional monomers. The ultimate result of continued reaction is a polyurethane polymer. The general structure of a segment of a polyurethane molecule, based on the two monomers we have just discussed, would be

$$\cdots O{-}(CH_2)_2{-}O{-}\overset{\overset{O}{\|}}{C}{-}\overset{\overset{H}{|}}{N}{-}\!\!\bigcirc\!\!{-}\overset{\overset{H}{|}}{N}{-}\overset{\overset{O}{\|}}{C}{-}O{-}(CH_2)_2{-}O{-}\overset{\overset{O}{\|}}{C}{-}\overset{\overset{H}{|}}{N}{-}\!\!\bigcirc\!\!{-}\overset{\overset{H}{|}}{N}{-}\overset{\overset{O}{\|}}{C}{-}O\cdots$$

Many polyurethanes are soft elastic (rubbery) materials that are used in foam rubber for furniture upholstery, packaging materials, life preservers, and many other products. A polyurethane fabric (trademark Biomer) is used as the diaphragm in the Jarvik-7 artificial heart (see Figure 18.8a), and a polyurethane material is used as a skin substitute in treating burn victims (see Figure 18.8b). Membrane skin substitutes help patients recover more rapidly by allowing only oxygen and water to pass through.

FOAM RUBBER

Foamed polyurethane is made by adding water to the reaction mixture. The water reacts with some of the isocyanate monomers to produce carbon dioxide gas that causes the polymer mixture to foam. The result is a spongelike polymer containing numerous tiny gas-filled cavities.

Petroleum is the ultimate source for polyurethane monomers as well as for the monomers used in condensation (Section 18.7) and addition (Section 18.4) polymerization reactions. Such monomers are not actually present in petroleum but rather are synthesized from petroleum hydrocarbons. Thus, the price of petroleum (Section 19.12) is an important factor in the "economics" of polymer production and the extent of polymer use.

18.10 Paints: Polymeric Coatings

The chemical basis for the use of paints as protective or decorative coverings (or both) is polymer formation. The type of polymer involved is usually an addition polymer (Section 18.3).

There are three major types of ingredients in a paint: (1) a pigment, (2) a binder, and (3) a solvent.

The *pigment* supplies the desired color. The pigment titanium dioxide (TiO_2) is usually used to produce a white base, which is then tinted various colors.

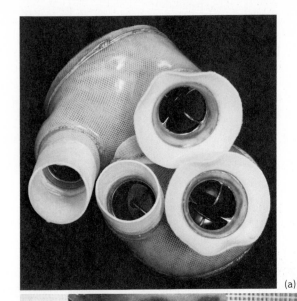

(a)

(b)

FIGURE 18.8

(a) The Jarvik-7 artificial heart. The base is aluminum, and the valves are pyrolytic graphite and titanium. The base, valves, and valve-holding rings are coated with the polyurethane biomer. Layers of this polymer serve as the diaphragm in the device. (b) Membranes made of polyurethane are being used as a skin substitute in treating severe burns. Such membranes help patients recover more rapidly by passing only oxygen and water. [(a) *Brad Nelson/University of Utah Health Sciences Center.* (b) © *Dan McCoy/Rainbow.*]

Additional pigments include carbon black (C), chrome yellow ($PbCrO_4$), and oxides of iron (brown or red).

The *binder*, which becomes polymeric as the paint dries, serves as the protective coating and also as a "trapping agent" to hold the pigments.

The *solvent* makes possible the "mixing" of the various paint ingredients. In most paints the solvent-binder-pigment mix is an emulsion (Section 9.7). Recall that an emulsion is composed of two or more immiscible liquids, one dispersed as tiny droplets in the other. An emulsifying agent (such as soap) is usually required to keep the small droplets (usually binder) dispersed in the solvent.

There are two major types of paint with classification determined by the nature of the solvent: (1) water-based latex paints and (2) oil-based paints. At present, approximately 60% of paint used for architectural purposes is water-based and 40% is oil-based. The first commercial water-based paint was introduced in 1948.

WATER-BASED PAINTS

A partially polymerized synthetic addition polymer with rubberlike properties serves as the binder in water-based paints. From these rubberlike properties comes the name latex paint. The binder also contains sufficient monomer that additional polymerization can occur as the paint dries. The binder is dispersed as tiny droplets in the water base (an emulsion).

After application of a latex paint, the water (solvent) begins to evaporate. Loss of water causes the emulsion to break down. This increases the rate of water evaporation. The binder, as a film (with pigments trapped within it), is left behind. With time, completion of the polymerization process occurs within the binder. Remember that unreacted monomer was present in the emulsion.

Substituted ethylenes are used extensively as binders in latex paints. Two such binders, used in large amounts, are vinyl acetate and acrylonitrile. The equations for their polymerization are

Vinyl acetate   Polyvinyl acetate

Acrylonitrile   Polyacrylonitrile

Paints containing polyacrylonitrile are often called acrylic paints. These paints are more washable and much more resistant to light damage than other latex paints.

Water-based latex paints have advantages over oil-based paint in reducing fire hazards and air pollution resulting from solvent evaporation.

The solvent for older, "more traditional" oil-based paints is usually turpentine or mineral spirits, both of which are mixtures of hydrocarbons. Binders in such paints include boiled linseed, soybean, castor, and coconut oils. With evaporation of the solvent, such oils (often called drying oils) react with oxygen from the air to form a polymeric material. As with water-based paints, the polymerized binder traps and holds the pigments.

Drying oils are triesters, that is, molecules containing three ester lingkages.

$$
\begin{array}{c}
\quad\quad\quad\quad O \\
\quad\quad\quad\quad \| \\
H_2C-O-C-R \\
| \quad\quad\quad O \\
| \quad\quad\quad \| \\
HC-O-C-R \\
| \quad\quad\quad O \\
| \quad\quad\quad \| \\
H_2C-O-C-R
\end{array}
$$

The R groups present are long unsaturated chains of cabon atoms. The chemical structure of the predominant triester (61%) in boiled linseed oil is

$$
\begin{array}{c}
\quad\quad\quad\quad O \\
\quad\quad\quad\quad \| \\
H_2C-O-C-(CH_2)_8-CH{=}CH-CH{=}CH-(CH_2)_4-CH_3 \\
| \quad\quad\quad O \\
| \quad\quad\quad \| \\
H-C-O-C-(CH_2)_8-CH{=}CH-CH{=}CH-(CH_2)_4-CH_3 \\
| \quad\quad\quad O \\
| \quad\quad\quad \| \\
H_2C-O-C-(CH_2)_8-CH{=}CH-CH{=}CH-(CH_2)_4-CH_3
\end{array}
$$

The polymerization reaction (drying reaction) involves carbon atoms next to C=C double bonds in the long carbon chains. Such carbon atoms in two different oil molecules can become cross-linked through reaction with O_2. In the process, these carbon atoms lose a hydrogen and become involved in an ether linkage, as shown.

Part of one molecule

$$-CH_2-CH_2-CH{=}CH-CH_2-$$

$$-CH_2-CH_2-CH{=}CH-CH_2-$$

Part of another molecule

$$+ O_2 \longrightarrow$$

$$-CH_2-CH-CH{=}CH-CH_2-$$
$$\quad\quad\quad | \quad\quad\quad\quad\quad\quad\quad\quad\quad\quad + H_2O$$
$$\quad\quad\quad O \quad \text{Ether linkage}$$
$$\quad\quad\quad |$$
$$-CH_2-CH-CH{=}CH-CH_2-$$

Once paints have hardened (dried), they can be removed only by destroying the polymeric structure; this requires the breaking of chemical bonds. Paint removers usually contain strong bases that can attack the chemical bonds.

Practice Questions and Problems

GENERAL CHARACTERISTICS OF POLYMERS

18-1 In many manufactured products, synthetic polymers are now used in place of natural materials. Give three reasons why this change has taken place.

18-2 Distinguish between the following pairs of terms.
a. Monomer and polymer
b. Thermoset and thermoplastic polymer

18-3 List three major types of polymers based on the type of reaction that takes place during polymerization.

ETHYLENE-BASED ADDITION POLYMERS

18-4 What structural feature must be present in an organic molecule for it to undergo addition polymerization?

18-5 What are the three general steps in an addition polymerization reaction?

18-6 Describe, using equations, the role of peroxide free radicals in the initiation of the addition polymerization reaction involving ethylene.

18-7 Describe, using equations, the role of free radicals in the termination of an addition polymerization reaction.

18-8 In referring to the molecular weight of polyethylene (or any other polymer), we can speak only of the "average molecular weight." Explain why this is so.

18-9 How do HDPE and LDPE differ in structure and properties?

18-10 Draw the structure of a portion of the polymer chain consisting of four repeating units for the following addition polymers.
a. Teflon   b. PVC
c. Polypropylene   d. HDPE

18-11 Give the chemical notation used to denote the following addition polymers.

a. Polyvinyl Chloride   b. Polypropylene
c. Teflon   d. Polystyrene

18-12 The fiber marketed as Orlon is an addition polymer with the following structure.

$$-CH_2-CH-CH_2-CH-CH_2-CH-CH_2-CH-$$
$$\qquad\; CN \qquad\quad CN \qquad\quad CN \qquad\quad CN$$

What is the monomer from which this polymer is made?

18-13 Define the term copolymer and give an example of an addition copolymer?

18-14 The monomer vinyl chloride is used in the production of both PVC and Saran. What is the difference between the two substances?

18-15 What are major end-uses for the following addition polymers?
a. Teflon   b. Saran
c. Polypropylene   d. Polyvinyl chloride
e. Polyethylene   f. Polystyrene

BUTADIENE ADDITION POLYMERS

18-16 What is an elastomer?

18-17 What is the chemical structure of the monomer or monomers used in the production of the following?
a. Natural rubber   b. Polybutadiene
c. Polychloroprene   d. Neoprene
e. SBR   f. Synthetic natural rubber

18-18 Describe in chemical terms the vulcanization of natural rubber

18-19 Many synthetic rubbers are copolymers. Write the basic polymer repeating unit in a one-to-one copolymer formed from ethylene and propylene.

18-20 Draw the structure of a three repeating unit portion of polymer chain for the polymer that can be formed from the addition polymerization of 2,3-dimethyl-1,3-butadiene.

18-21 What are the relative amounts of natural rubber and synthetic rubber consumed in the United States at present?

CONDENSATION POLYMERS

18-22 What type of monomers are required for a condensation polymerization reaction?

18-23 In what general way does a condensation polymer differ from an addition polymer?

18-24 Define the terms polyamide and polyester.

18-25 Give the identity, in general terms, of the monomers needed to produce the following.
a. Polyamide   b. Polyester

18-26 What feature do all condensation polymerization reactions have in common?

18-27 Draw the structure of a two-repeating-unit portion of the polyamide Nylon 66, and then indicate where the amide linkages are in the structure.

18-28 What is the structural difference between Nylon 6 and Nylon 66?

18-29 Contrast the differences in structure of Nylon 6, silk, and wool.

18-30 Why is Dacron called a polyester?

18-31 What are the two monomers used in the production of Dacron?

18-32 Draw the structure of a two-repeating-unit portion of Dacron and then indicate where the ester linkages are in the structure.

18-33 Draw a two-repeating-unit segment of the structure of a polymer made by the elimination of water using the following monomers.

$$HO-CH_2-CH_2-OH$$

and

18-34 Kevlar, a polyamide used to make bulletproof vests, is made from terephthalic acid and phenylene

diamine, whose structure is

Write the structure for a two-repeating-unit segment of a Kevlar molecule.

18-35 Could a polymer be formed by the reaction of terephthalic acid with ethyl alcohol? Explain.

18-36 What is the relationship between vulcanized rubber and permanent press fabrics?

18-37 What type of monomer is needed to produce a network condensation polymer?

18-38 Draw the structural formula for the species that results from the reaction between one formaldehyde molecule and two phenol molecules.

18-39 Describe the general structural differences between phenolics and melamines, both of which are network condensation polymers.

18-40 Both phenol and melamine are used as monomers in the production of formaldehyde-based network polymers. Indicate the locations of the "reaction sites" in each of these monomers.

18-41 Use of a *p*-methylphenol in place of phenol to form a phenol-formaldehyde polymer gives a nonnetwork polymer rather than a network polymer. Explain this. The structure of *p*-methylphenol is

REARRANGEMENT POLYMERS

18-42 What do rearrangement polymers and addition polymers have in common?

18-43 What do rearrangement polymers and condensation polymers have in common?

18-44 Give the identity, in general terms, of the monomers needed to produce a polyurethane.

18-45 Show, using structural formulas, the rearrangement that occurs when a dialcohol reacts with a diisocyanate.

18-46 Show the position of the amide linkage and the ester linkage within a urethane linkage.

18-47 How is the foam introduced into polyurethane foam or sponge rubber materials?

PAINTS: POLYMERIC COATINGS

18-48 What are the three major types of ingredients in a paint?

18-49 What are the two major types of architectural paints?

18-50 What type of chemical change takes place during the drying of the following?
a. Water-based paint   b. Oil-based paint

18-51 What are the solvent and binder in the following?
a. Water-based paint   b. Oil-based paint

18-52 In what proportions are oil-based paint and water-based paint consumed in the United States at present for architectural purposes?

18-53 A water-based paint would dry in a "vacuum," but an oil-based paint would not. Explain.

CHAPTER HIGHLIGHTS

19.1 Fossil Fuels—Origins and Usage

Fossil fuels *are naturally occurring hydrocarbon deposits, with some derivatives present, formed from the remains of once living organisms, that are used as fuels.* Petroleum, natural gas, and coal are all fossil fuels.

The process by which fossil fuels were produced in nature is not completely understood. Chemists have debated the question of primary origin for many years. Did fossil fuels originate from animals or plants? It is now generally accepted that they were derived from the decomposition of marine organisms, both plants and (to a much lesser extent) animals, in the absence of oxygen in the environment of deep, ocean-bottom sediments. With time, older layers of bottom sediments were buried deeper and deeper by an increasing overburden of silt and sediment, and the pressure on them increased significantly. This pressure, together with the heat of the earth and chemical and bacterial action, is thought to have converted the biological debris into hydrocarbon materials.

Fossil fuels are currently used to meet almost all of the energy needs within the United States, whether in the industrial, transportation, commercial, or residential sector. This has been the case for almost 100 years, as can be seen from the information graphically portrayed in Figure 19.1.

Into the early 1800s most useful energy was obtained from windmills, waterwheels, and the burning of wood. Wood was the fuel used in early steam engines, including those of the riverboats and railroad locomotives of the early 1800s. The use of coal did not begin until 1830. One hundred years ago wood was still the source of almost half of the nation's energy supply. However, by 1900 the percentage of energy obtained from wood fuel had dropped to only 20%, and coal had become the nation's main energy source.

In this century we have gone from a wood-and-coal-based energy supply to a system in which two-thirds of the energy comes from petroleum products and natural gas. Petroleum did not come on the scene until 1860, and its growth as an energy source has paralleled the increasing use of automobiles, aircraft, and other modern forms of transportation. It provided only 2% of the total energy used in 1900, 18% in 1925, peaked at 46% in 1960, and now provides slightly over 40% of total energy.

The use of natural gas started to increase about 1930 and increased especially rapidly after World War II, when the development of large-diameter pipelines allowed the gas to be moved easily throughout the country. Natural gas usage peaked at 33% in 1970 and has since decreased to under 25% of total energy demand. In the last decade, the contribution of coal to the total energy supply has began to increase again, at the expense of natural gas and petroleum.

Figure 19.2 shows the contribution of the various fossil fuels, as well as hydro and nuclear power, to the United States energy mix (consumption of energy) for

591

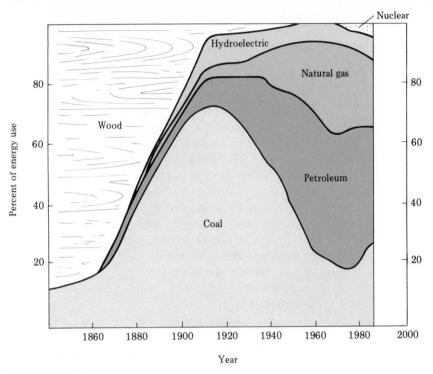

FIGURE 19.1

Contributions of fossil fuels to the total energy demands in the United States as a function of time.

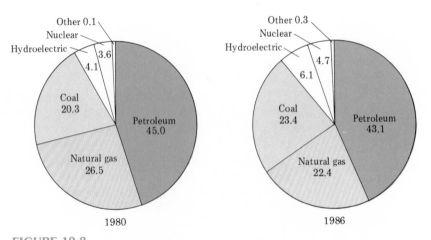

FIGURE 19.2

Consumption of energy by source in the United States for the years 1980 and 1986.

the years 1980 and 1986. In comparing the two "energy pictures," we note that consumption of both natural gas and petroleum has decreased slightly at the expense of the other energy sources. Also, over the same period, coal has displaced natural gas as the No. 2 fossil fuel.

In the remainder of this chapter we consider the chemistry of petroleum, natural gas, and coal—the three dominant sources of energy in the United States. These three fossil fuels, which are considered in the listed order, accounted for 92.1% of energy consumption in 1980 and 88.9% of energy consumption in 1986.

19.2 Occurrence of Petroleum

Products obtained from petroleum (crude oil) are the fuels used in the greatest amounts in the United States. **Petroleum** *is a complex liquid mixture of hydrocarbons, with small quantities of other materials, that is found underground, trapped under pressure in rock strata.* Most petroleum is found, as a dark thick liquid, at a depth of several thousand feet in dome-shaped rock strata. When a hole is drilled into such a rock formation, it is possible to recover some of the crude petroleum found there. The pressure released during the drilling of the oil well, supplemented with surface pumping when needed, is sufficient to bring petroleum to the surface. Natural gas is often associated with the petroleum in the rock formation.

Figure 19.3 shows a petroleum–natural gas deposit and the geology associated with it. Three geological features are required for petroleum to accumulate in sufficient amounts to make recovery economically feasible.

1. Reservoir rock—a layer of porous, permeable rock.
2. Cap rock—an overlaying layer of nonporous, nonpermeable rock.
3. Anticline trap—an upward fold in the earth's strata that forms an arch.

Reservoir rock, the actual rock in which the petroleum is found, must be porous and permeable. Porous rock contains voids or open spaces within its structure. If the pores or open spaces are interconnected, then the rock is also permeable. Crystalline rocks may have porosities of less than 1%, whereas some sandstones and limestones have values of 40% or higher. Gases and liquids can migrate readily through porous, permeable rock.

Cap rock, because of its impermeability, when located above reservoir rock, serves as a "lid" that prevents upward migration of the petroleum. Liquids or gases cannot readily move through such rock.

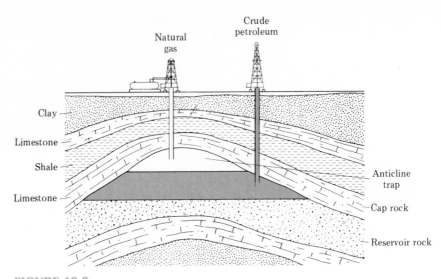

FIGURE 19.3

Geological features associated with petroleum and natural gas deposits.

The anticline trap, an upward fold in the rock strata that creates a "dome," is the structural feature that causes the petroleum (and natural gas, if it is present) to accumulate in a localized area. Without such a trap, there would be no curtailment of the lateral movements of the petroleum and natural gas.

As ever-present underground fluids (mixtures of salt water, petroleum, and gases) move through porous, permeable rock, a separation determined by density differences occurs in the vicinity of an anticline trap. Petroleum and natural gas, being less dense than water, have a greater tendency to move upward. This upward movement traps some petroleum and natural gas at the top of the dome. Over a long period of time, significant amounts of petroleum can accumulate. Some anticline traps span many miles.

A common misconception is that liquid petroleum occurs underground in liquid pools. This is not true, as we have just discussed. Liquid petroleum is found dispersed in solid rock formations. Oil wells are drilled into solid rock, and the petroleum must be recovered from this rock. Recovery efficiencies are low, averaging approximately 30%; that is, 70% of the petroleum must be left in the rock formation.

A major cause for the low efficiency is the decrease in internal pressure within the rock formation as the oil field is worked. This internal pressure is needed to help force the crude petroleum up the well bore hole. Finally, even with surface pumping, it is no longer economically feasible to bring more petroleum to the surface. On the average, only 30% of the petroleum has been recovered when this point is reached.

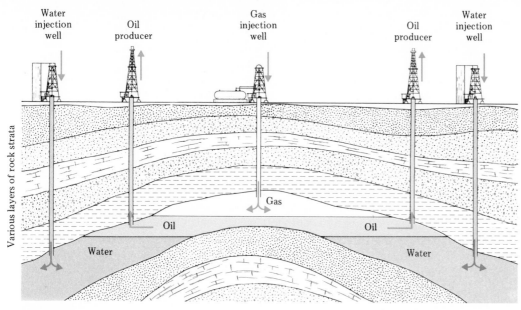

FIGURE 19.4

Repressurization of an oilfield allows an additional amount of petroleum to be recovered.

REPRESSURIZING OF OILFIELDS

Repressurizing of oilfields is a method now increasingly used to boost well efficiency and prolong the producing life of a field. Repressurizing an oilfield involves drilling auxiliary wells and injecting liquid (water) or gases (carbon dioxide, nitrogen, and steam) into the rock formation. Figure 19.4 shows the schematics involved in repressurizing an oilfield. Petroleum recovered from a repressurized well is called secondary petroleum production and in some areas it accounts for almost 20% of production.

19.3 Chemical Composition of Petroleum

Crude petroleum is a complex mixture containing hundreds of different compounds. The vast majority of these compounds are simple hydrocarbons (Chapter 16). Small amounts of other substances, mostly hydrocarbon derivatives, are also present, as reflected in elemental analysis results for crude petroleum samples, which show nitrogen (up to 0.5%), sulfur (up to 6%), and oxygen (up to 3.5%).

No crude petroleum sample has yet been completely characterized because of the complexity of the chemical mixture. Extensive analysis on one particular sample of United States crude resulted in the positive identification of 175 different hydrocarbons—70 alkanes, 48 cycloalkanes, and 57 aromatic hydrocarbons. These are only the compounds with low boiling points. The higher-boiling compounds are yet to be studied. Table 19.1 lists the most abundant of the 175 compounds already identified from this particular sample. Note that many of these most abundant compounds have simple unbranched carbon chains.

Crude petroleums from different locations contain essentially the same hydrocarbons, but the proportions in which the various molecules occur vary considerably. This is not surprising, because the local distribution of plant and animal life probably varied from region to region at the time petroleum was being formed, just as it varies today. Differences in the percentage of the various hydrocarbons influence the physical properties of petroleum. Consequently, the properties of crude petroleums are extremely diverse. For example, some petroleums are almost colorless whereas others are pitch black, amber, brown, or green.

TABLE 19.1

The Most Abundant Hydrocarbons in a Particular U.S. Crude Petroleum Sample

Formula	Name	Structural Formula	Amount (% by volume)
C_7H_{16}	heptane	$CH_3-(CH_2)_5-CH_3$	2.3
C_8H_{18}	octane	$CH_3-(CH_2)_6-CH_3$	1.9
C_6H_{14}	hexane	$CH_3-(CH_2)_4-CH_3$	1.8
C_9H_{20}	nonane	$CH_3-(CH_2)_7-CH_3$	1.8
$C_{10}H_{22}$	decane	$CH_3-(CH_2)_8-CH_3$	1.8
C_7H_{14}	methylcyclohexane		1.6
$C_{11}H_{24}$	undecane	$CH_3-(CH_2)_9-CH_3$	1.6
$C_{12}H_{26}$	dodecane	$CH_3-(CH_2)_{10}-CH_3$	1.4
$C_{13}H_{28}$	tridecane	$CH_3-(CH_2)_{11}-CH_3$	1.2
$C_{10}H_{14}$	1-methyl-3-isopropylbenzene		1.1
$C_{14}H_{30}$	tetradecane	$CH_3-(CH_2)_{12}-CH_3$	1.0

19.4 Petroleum Refining

In its natural state, crude, petroleum has few uses; it is just too complex a mixture of hydrocarbons. **Petroleum refining** *is the process in which crude petroleum is separated into various useful components, called fractions.* The various fractions resulting from petroleum refining are still hydrocarbon mixtures, but each one is simpler (contains fewer compounds) than the original crude petroleum.

There are three major steps to the petroleum refining process.

1. Physical separation of compounds using fractional distillation to produce various fractions.
2. Chemical alterations and conversions to adjust the amounts of and properties of the various fractions.
3. Removal of sulfur-containing compounds.

Let us consider these three refining steps in detail.

Fractional distillation *is a process used to separate liquid components of a solution that is based on boiling point differences.* The different boiling points of the various hydrocarbons present in crude petroleum are the basis for the separation of the crude into fractions, using fractional distillation. Figure 19.5 shows the structural components of a fractional distillation unit. Heating of the crude petroleum to a temperature of 350 to 400°C vaporizes the majority of the hydrocarbons present. The resulting mixture of hot vapors rise in the fractionating column, condense at various collecting points, and are drawn off. Because the temperature within the fractionating column decreases with height, hydrocarbons with high boiling points condense out near the bottom of the column (very little cooling of vapor is required) and those with low boiling points rise to higher levels before condensing (extensive cooling of vapor is required). Each condensed fraction contains a variety of hydrocarbons with the common characteristic of similar boiling points. The boiling point values for the various fractions correlate with the temperatures of the distillation tower at the particular points where condensation of the fractions occurred.

Hydrocarbons that did not vaporize in the atmospheric distillation tower (left side of Figure 19.5) are passed on to the vacuum distillation tower (right side of Figure 19.5). The vacuum distillation second stage allows high-boiling hydrocarbons to be vaporized at reduced temperatures (which saves on energy costs) because of reduced pressure within the unit. Recall, from Section 8.15, that reducing external pressure lowers the boiling point.

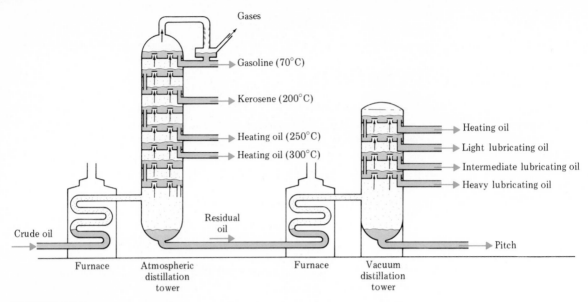

FIGURE 19.5

The complex mixture of hydrocarbons present in crude petroleum is separated
into simpler mixtures (fractions) by the process of fractional distillation.

The properties and uses of the various fractions obtained from a fractional
distillation unit are listed in Table 19.2. Note the many important and familiar
products obtained from crude petroleum fractionation.

The relative amount of each fraction obtained from fractional distillation
depends on the composition of the original crude petroleum, and usually does

TABLE 19.2

Products Obtained from the Fractional Distillation of Crude Petroleum

Fraction	Molecular-Size Range	Boiling-Point Range (°C)	Typical Uses
Gas	C_1–C_4	−164–30	gaseous fuel
Petroleum ether	C_5–C_7	30–90	solvent, dry cleaning
Straight-run gasoline	C_5–C_{12}	30–200	motor fuel
Kerosene	C_{12}–C_{16}	175–275	fuel for stoves, diesel and jet engines
Heating oil	C_{15}–C_{18}	up to 375	furnace oil
Lubricating oils	C_{16}–C_{20}	350 and up	lubrication, mineral oil
Greases	C_{18} and up	semisolid	lubrication, petroleum jelly
Paraffin (wax)	C_{20} and up	melts at 52–57	candles
Pitch and tar	high	residue in boiler	roofing, paving

not correspond to the amounts needed to satisfy consumer demands. Chemical alterations and conversions enable refiners to change molecules found in the fractions in little demand (greases and waxes) into molecules corresponding to fractions in high demand (gasoline and fuel oils).

Four basic alteration-conversion processes are used to increase the yield of desired fractions at the expense of low-demand fractions: (1) cracking, (2) polymerization, (3) reforming, and (4) isomerization.

Cracking *is a petroleum-refining process in which long-chain hydrocarbon molecules are broken into short-chain hydrocarbon molecules as the result of breaking carbon–carbon bonds.* When alkane vapors at 500°C are passed over a heated catalyst of alumina (Al_2O_3) and silica (SiO_2), they decompose into a mixture of smaller alkanes and alkenes.

$$CH_3-CH_2-CH_2-CH_3 \xrightarrow[\Delta]{Al_2O_3 + SiO_2} CH_4 + CH_2{=}CH-CH_3$$

$$\text{Butane} \qquad\qquad\qquad\qquad \text{Methane} \qquad \text{Propene}$$

$$\xrightarrow[\Delta]{Al_2O_3 + SiO_2} CH_3-CH_3 + CH_2{=}CH_2$$

$$\text{Ethane} \qquad \text{Ethene}$$

The alkenes produced are converted to alkanes by adding hydrogen or by allowing the alkenes to undergo polymerization reactions.

Polymerization *is a petroleum-refining process in which small hydrocarbon molecules are combined to produce larger hydrocarbon molecules.* It is the reverse of cracking. Usually, two, three, or four alkenes combine to produce a larger hydrocarbon molecule.

$$3\ CH_3-CH{=}CH_2 \xrightarrow{catalyst} CH_3-\underset{\underset{CH_3}{|}}{CH}-CH_2-\underset{\underset{CH_3}{|}}{CH}-CH{=}CH-CH_3$$

$$\text{Propene} \qquad\qquad\qquad \text{4,6-Dimethyl-2-heptene}$$

Polymerization is usually not a separate refinery process, but takes place in connection with cracking and reforming.

Reforming *is a petroleum refining process in which an alkane or cycloalkane is converted to an aromatic compound.* This process is important for gasoline because it improves octane rating (Section 19.5). Reforming is accomplished by heating hydrocarbon vapor in the presence of catalysts containing platinum and one or more other metals.

$$CH_3-CH_2-CH_2-CH_2-CH_2-CH_2-CH_3 \xrightarrow[\Delta]{Pt} \underset{\text{Methylbenzene}}{\overset{\overset{\displaystyle CH_3}{\big|}}{\bigcirc}} + 4\ H_2$$

$$\text{Heptane}$$

$$\text{Cyclohexane} \xrightarrow[\Delta]{\text{Pt}} \text{Benzene} + 3H_2$$

Cyclohexane   Benzene

Isomerization *is a petroleum-refining process in which straight-chain hydrocarbons are converted to branched-chain hydrocarbons.* This is accomplished by heating hydrocarbon vapor, with $AlCl_3$ as a catalyst.

$$CH_3{-}CH_2{-}CH_2{-}CH_2{-}CH_3 \xrightarrow[\Delta]{AlCl_3} CH_3{-}\underset{\underset{CH_3}{|}}{CH}{-}CH_2{-}CH_3$$

Pentane   2-Methylbutane

$$CH_3{-}CH_2{-}CH_2{-}CH_2{-}CH_2{-}CH_2{-}CH_3 \xrightarrow{AlCl_3} CH_3{-}\underset{\underset{CH_3}{|}}{CH}{-}\overset{\overset{CH_3}{|}}{CH}{-}CH_2{-}CH_3$$

Heptane   2,3-Dimethylpentane

This process, like reforming, is very important in improving the octane rating of gasoline (Section 19.5).

Sulfur-containing compounds present in petroleum must be removed for several reasons. Some of these compounds impart unpleasant odors to petroleum products; others are corrosive to storage facilities and automobile engines; still others "poison" or deactivate gasoline additives. During the combustion of petroleum products, these compounds produce sulfur oxides, a serious type of air pollutant (Section 20.8).

Hydrogen gas (under pressure) is used to desulfurize petroleum and petroleum products. It reacts with the sulfur to produce gaseous hydrogen sulfide (H_2S), which is easily removed from the reaction mixture. An equation representing such desulfurization is

$$C_4H_4S + 4\,H_2 \xrightarrow[\text{catalysts}]{\text{high pressure}} C_4H_{10} + H_2S$$

19.5   Gasoline and Octane Rating

The gasoline fraction obtained from crude petroleum contains those hydrocarbons with from 6 to 12 carbon atoms per molecule that have a boiling point ranging from about 60 to 180°C (Table 19.2). The supply of gasoline fraction obtainable from crude petroleum is insufficient to meet the demand of the millions of automobiles on United States highways. Thus, other petroleum fractions less in demand must be converted to gasoline. The cracking and polymerization reactions discussed in Section 19.4 are used mainly for this purpose.

The burning (combustion) of gasoline in an automobile engine involves oxidation of the hydrocarbons present. This oxidation reaction is initiated by a spark from a spark plug. The chemical equation for this process, where gasoline is represented by one of its heptane components, is

$$C_7H_{16} + 11\ O_2 \longrightarrow 7\ CO_2 + 8\ H_2O + heat$$

This reaction is idealized in that it assumes sufficient oxygen is present for complete combustion. This is not generally true, as small amounts of CO are also produced. This is the origin of the carbon monoxide pollution associated with automobile operation (Section 20.3).

For a gasoline to function properly in an automobile engine, it should not begin to burn before it is ignited by the spark plug. Premature ignition, before pistons are in proper position, leads to loss of engine power and overheating; the engine is said to "knock." Knocking is a problem in modern high-compression automobile engines if adjustments are not made in the composition of the gasoline fraction obtained from fractional distillation. An understanding of the concept of octane number helps us understand the adjustments that are needed in gasoline composition.

The tendency of a gasoline to cause knocking during engine operation is rated using octane numbers. An **octane number** *is a measure of the burning efficiency of a hydrocarbon or mixture of hydrocarbons such as gasoline.* High values for octane numbers indicate low knocking tendencies for a fuel.

The octane number scale is an arbitrary scale that was established in 1927. An octane number of 100 is assigned to 2,2,4-trimethylpentane (isooctane), an excellent fuel that burns without knocking. Heptane, an unbranched-chain hydrocarbon that is a poor fuel causing much knocking, is assigned an octane number of zero. An octane number for a gasoline is determined by comparing its knocking tendencies in a standard engine with those of mixtures of the two reference compounds (isooctane and heptane). The octane number of a gasoline is set as the percentage of isooctane in a reference mixture with the same knocking tendencies. Thus, a gasoline would be assigned an octane number of 91 if its knocking pattern was the same as a mixture containing 91% isooctane and 9% heptane.

Table 19.3 gives the octane numbers for a number of individual hydrocarbons, all of which are present in gasoline. The following two generalizations can be made based on the information in this table.

1. Octane numbers decrease with increasing molecular weight. Comparison of the octane numbers of butane, pentane, hexane, and heptane illustrates this generalization.

TABLE 19.3

Octane Numbers of Selected Hydrocarbons

Formula	Compound	Octane Number
C_4H_{10}	butane	94
C_5H_{12}	pentane	62
	2-methylbutane	94
C_6H_{14}	hexane	25
	2-methylpentane	73
	2,2-dimethyl-butane	92
C_7H_{16}	heptane	0
	2-methylhexane	42
	2,3-dimethylpentane	90
C_8H_{18}	2-methylheptane	22
	2,3-dimethylhexane	71
	2,2,4-trimethylpentane	100

2. Octane numbers increase for isomeric compounds (same molecular weight) as the degree of branching off the carbon chain increases. Comparison of the octane numbers of the isomeric six-carbon alkanes hexane, 2-methylpentane, and 2,2-dimethylbutane illustrates this generalization.

Straight-run gasoline, the gasoline fraction as it comes from a fractionation column, has an octane number of 50 to 55. Gasoline with such a low octane rating cannot be used in present-day automobile engines; they require 87 to 90 octane gasoline. A number of methods exist for increasing the octane rating of straight-run gasoline.

1. Conversion of normal chain (unbranched) hydrocarbons in the straight-run gasoline to branched chain hydrocarbons. This is the isomerization process discussed in the last section. Branched-chain hydrocarbons have higher octane numbers than the corresponding normal-chain hydrocarbon (see Table 19.3).
2. Conversion of nonaromatic hydrocarbons in the straight-run gasoline to aromatic hydrocarbons. This is the reforming process discussed in the last section. Aromatic hydrocarbons have very high octane numbers. The octane number of benzene is 106 and that of toluene is 110. (How octane numbers with values greater than 100 are obtained is considered later in this section.)
3. Addition of hydrocarbon molecules with high octane numbers obtained from sources other than the straight-run gasoline.

4. Addition of nonhydrocarbon antiknock agents and blending agents. A number of substances, including many lead-containing compounds and oxygenated hydrocarbons (alcohols and ethers), have been used for this purpose.

TETRAETHYLLEAD

The most widely used of all antiknock agents has been the compound tetraethyllead (TEL), whose structure is

$$
\begin{array}{c}
CH_3 \\
| \\
CH_2 \\
| \\
CH_3-CH_2-Pb-CH_2-CH_3 \\
| \\
CH_2 \\
| \\
CH_3
\end{array}
$$

No antiknock agent is as effective at as low a cost as TEL. As little as 0.01% TEL in gasoline can increase the octane rating.

The use of TEL, as well as all other lead-containing gasoline additives, to increase the gasoline octane rating, has dropped dramatically during the last 10 years, as is shown in Figure 19.6. There are two major reasons for this rapid turnaround in lead-additive use.

1. Mandatory installation of catalytical converters on all new automobiles to control air pollution.
2. Increased concerns about the health effects of airborne lead compounds.

To control air pollution from automobiles, cars manufactured from 1975 on have been equipped with catalytic converters. Lead compounds in gasoline deactivate the catalysts in such converters. This problem has resulted in production of nonleaded gasoline. To make up for the drop in octane rating caused by the removal of TEL, greater amounts of branched and aromatic hydrocarbons are added to nonleaded gasoline. The expense of preparing the additional branched and aromatic hydrocarbons exceeds the cost, by a wide margin, of TEL additive; hence, nonleaded gasoline is more expensive than leaded gasoline.

Recognition of the health hazards associated with airborne lead compounds, whose major source is automobile exhaust, has led the U.S. Environmental Protection Agency to restrict the use of TEL in leaded gasoline. The allowable limit of lead in leaded gasoline was reduced to 1.10 g of lead per gallon of leaded gasoline in 1982 and further restricted to 0.10 g in 1986.

Blending agents are now also being used to help make up for the octane number loss caused by reducing the amount of lead in gasoline. Blending

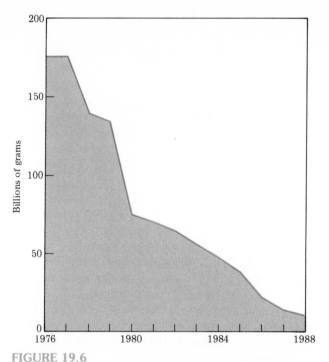

FIGURE 19.6

The use of lead in gasoline has decreased dramatically in a decade.

agents, which are themselves fuels, are distinguished from additives by the amounts used. Blending agents are used in large amounts (several percent by volume) and additives are used in small amounts (tenths of a percent).

The blending agent used in the greatest quantity is methyl *tert*-butyl ether (MTBE), whose structure is

$$CH_3—O—\underset{\underset{CH_3}{|}}{\overset{\overset{CH_3}{|}}{C}}—CH_3$$

This compound is of such superior octane number (115) that blending it with gasoline in substantial amounts results in a greater antiknock quality than the base gasoline. (To rate the octane level of fuels with octane numbers higher than 100, the fuels are compared with reference fuels containing isooctane, whose octane number is 100, plus known amounts of TEL, and a conversion chart is used to obtain octane numbers.)

Next to MTBE, the most used blending agent is ethyl alcohol. Both MTBE and ethyl alcohol are oxygenated hydrocarbons.

19.6 Oil Shale

Oil shale *is a fine-grained sedimentary rock that contains a complex hydrocarbon-based solid material called kerogen.* The chemical difference between petroleum and kerogen is primarily one of cross-linking. Petroleum molecules are typically composed of linear chains with some attached rings and branches (Section 19.3). Very little cross-linking occurs between chains. In kerogen, on the other hand, the chains are cross-linked to a significant extent.

When kerogen is heated to 480°C, the links between chains break and the solid is chemically transformed into a liquid (representing about 60% of the kerogen mass), a fuel gas (9%) and a coke-like solid (25%). The resulting liquid materials are heavy and stiff and have a high sulfur and nitrogen content. However, after further treatment, they become as good a refinery feedstock as most crude petroleum.

The United States has vast oil shale deposits, which have the potential of being a significant energy resource. Problems preventing the development of the oil shale industry at present are both economical and environmental.

The kerogen trapped in oil shale rock is not a fluid and thus cannot be pumped. The shale rock must be mined, crushed, and then heated to release the kerogen. The economics of such processing is not favorable with crude petroleum prices at their present level. In addition, processing yields large quantities of waste (the stripped shale) that must be disposed of. (The shale actually expands upon heating and the stripped shale has a volume at least 12% greater than the original mined shale; thus, not all of the stripped shale can be returned to the original mine or excavation.)

One process under study for obtaining petroleum refinery feedstock from oil shale involves a rotary kiln in which heat is transferred to crushed shale by circulating heated ceramic balls (see Figure 19.7). The solid coke-like residue left on the stripped shale is used as the fuel for heating the ceramic balls.

19.7 Natural Gas

Natural gas *is a mixture of gaseous hydrocarbons, with small amounts of other gases, that is found underground, trapped under pressure in rock strata.* The geological conditions under which natural gas is found are similar to those for petroleum deposits (Section 19.2). Because of the similarities in reservoir

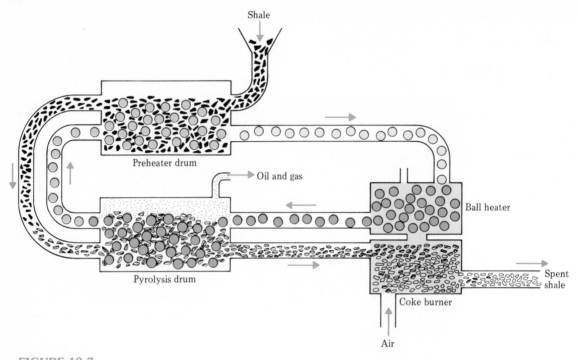

FIGURE 19.7

One process under study for obtaining petroleum feedstock from oil shale involves
circulating hot ceramic balls to heat the crushed oil shale.

conditions, natural gas and petroleum are sometimes, though not always, found
together in the same reservoir. When they are found together, the natural gas is
referred to as associated natural gas, and natural gas found alone is called
nonassociated natural gas. At present approximately three-fourths of United
States natural gas production comes from nonassociated natural gas deposits.

Methane (CH_4), the simplest of all hydrocarbons, is the predominant component present in *commercial* natural gas. It constitutes 85 to 95% by volume of
the natural gas, as can be seen from the numbers in Table 19.4. This table also
shows that ethane is always the second most abundant component of commercial natural gas and that smaller amounts of propane, butanes (two isomers),
and pentanes (three isomers) are also present. Small amounts of two nonhydrocarbon gases, carbon dioxide and nitrogen, are also almost always present.
Table 19.4 also shows that natural gas composition varies from one producing
area to another.

Natural gas as obtained from underground deposits generally has a composition significantly different from that of the familiar commercial fuel. The crude
gas usually contains some undesirable impurities as well as some heavy,

TABLE 19.4

Commercial Natural Gas Composition and Heating Value

| | Composition (% by volume) | | | | | | | Heating |
| | | | | | | | | Value |
Source	Methane (CH$_4$)	Ethane (C$_2$H$_6$)	Propane (C$_3$H$_8$)	Butanes (C$_4$H$_{10}$)	Pentanes (C$_5$H$_{12}$)	CO$_2$	N$_2$	(Btu/ft^3)
Birmingham, Alabama	93.14	2.50	0.67	0.32	0.12	1.60	2.14	1024
Columbus, Ohio	93.54	3.58	0.66	0.22	0.06	0.85	1.11	1028
Dallas, Texas	86.30	7.25	2.78	0.48	0.07	0.63	2.47	1093
New Orleans, Louisiana	93.75	3.16	1.36	0.65	0.66	0.42	0.00	1072
San Francisco, California	88.69	7.01	1.93	0.28	0.03	0.62	1.43	1086

condensable hydrocarbons. Appropriate processing eliminates or reduces the amount of the undesirable impurities and allows the condensable hydrocarbons to be collected as a separate fraction, which has marketable value.

Hydrocarbons of higher molecular weight than ethane are removed by bubbling the gas through a high-boiling hydrocarbon oil. The extracted hydrocarbons are then distilled from the oil in a separate operation. The recovered condensable hydrocarbons are called *natural gas liquids* (NGL). The propane and butanes obtained from these liquids are marketed as *liquefied petroleum gas* (LPG or LP gas) and can be used for heating homes. Pentanes and heavier hydrocarbons obtained from natural gas liquids are referred to as *natural gasoline*. This material evaporates readily and is usually blended with gasoline obtained from petroleum refineries to produce a more volatile gasoline, which ignites readily in cold weather. The removal of natural gas liquids during processing is never 100% complete; hence the small amounts of these substances present in commercial natural gas (see Table 19.4).

The undesirable impurities found in crude natural gas include carbon dioxide, nitrogen, water vapor, hydrogen sulfides, and thioalcohols or other organic sulfur compounds. Carbon dioxide and nitrogen reduce the heating value of the gas. Water vapor is troublesome, if present in high enough concentrations, because it can condense to the liquid state during pipeline transportation of the natural gas (decreasing the efficiency of the pipeline) as a result of sudden drops in temperature and pressure. Sulfur compounds, mainly H$_2$S and thioalcohols, are objectionable impurities because they create both corrosion and odor problems—they are acidic substances with pungent odors. Helium, found in some natural gases, is considered a valuable by-product of natural gas processing. All of the commercial helium, whose prime use is in lighter-than-air vessels, comes from natural gas.

Water is removed from natural gas by adsorption onto the surface of activated solid drying agents or by absorption into hygroscopic (moisture-retaining) liquids. Polyhydroxy alcohols (Section 17.3) are often used for this purpose. Carbon dioxide and sulfur compounds are removed by passing the crude through solutions of hydroxyamines (Section 17.9). No economical process has been found to remove nitrogen from most natural gases. In those few cases in which helium is removed by cryogenic (very low temperature) means, nitrogen can be removed as well.

Natural gas is often called the "ideal fuel," for the following reasons.

1. *Natural gas is a clean-burning fuel.*   As noted previously, impurities that would cause air-pollution problems (sulfur compounds) are removed by processing. (The fuel's gaseous state greatly simplifies this removal process.) Also, no solid residues result from natural gas combustion.
2. *Natural gas is easily transported, with a minimum energy expenditure.* Very little energy is needed to get natural gas from processing plants to consumer because of the vast pipeline distribution system in existence. The gas flows freely to where it is needed, with only occasional use of pumping stations. Figure 19.8 shows the major natural gas pipelines in the coterminous United States. Note the dominant role played by the Texas–Louisiana–Oklahoma–New Mexico region in the distribution of natural gas. These are the leading producing states.
3. *Natural gas burns with a high heat output.*   Natural gas has a higher heat output than most other fuels, as can be seen from the heating values given in Table 19.5.

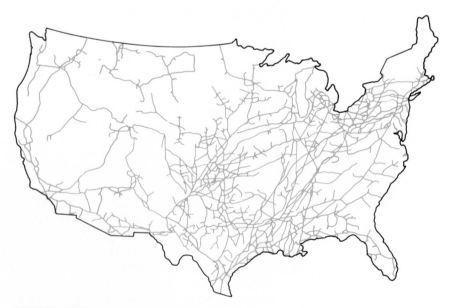

FIGURE 19.8
Major natural gas pipelines in the conterminous United States.

TABLE 19.5

Fuel Values of Some Common Fuels

Fuel	Fuel Value (kcal/g)
Wood (pine)	4.1
Anthracite coal (Pennsylvannia)	7.4
Bituminous coal (Pennsylvania)	7.6
Charcoal	8.1
Crude oil (Texas)	10.8
Gasoline	11.5
Natural gas	11.7

19.8 Origin and Grades of Coal

Coal *is a solid black or brownish black hydrocarbon-like material found in underground deposits at varying depths from the earth's surface.* It is thought to have been produced by the action of various chemical and physical processes on buried, *partially* decomposed vegetation. The plant origin of coal is verified by fossil remains and stratification found in coal beds.

Partially decomposed vegetation suitable for coal formation is obtained only under specific conditions. Accumulated plant remains are usually in contact with the atmosphere, or are found in highly aerated water. Under these conditions, decomposition is continuous and complete. Only under anaerobic (oxygen-free) conditions does partial decomposition takes place. Such conditions often prevail in stagnant swamp water, where fallen vegetation is soon covered by other debris.

Heat and pressure, acting for long periods of time, change the chemical composition of partially decomposed plant material into that characteristic of coal. The heat and pressure result primarily from the steady accumulation of overburden, which is composed of more recent sediments and debris.

Coal is graded (or ranked), based on the degree of exposure to the high temperature and pressure conditions associated with its formation. Four broad grades of coal are recognized, each representing an increase in the carbonization (percent carbon) of the original plant material. Figure 19.9 shows these grades of coal with some of their characteristics. It is generally agreed that the pressures and temperatures resulting from increasing overburden are sufficient to cause carbonization only to the rank of bituminous coal. Further increase in rank to anthracite requires the higher temperature and pressure reached only during geologic, mountain-building processes.

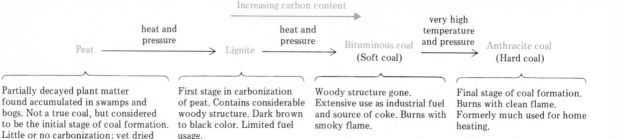

Increasing carbon content

| Peat | heat and pressure → | Lignite | heat and pressure → | Bituminous coal (Soft coal) | very high temperature and pressure → | Anthracite coal (Hard coal) |

Partially decayed plant matter found accumulated in swamps and bogs. Not a true coal, but considered to be the initial stage of coal formation. Little or no carbonization; yet dried and used as fuel in some places.

First stage in carbonization of peat. Contains considerable woody structure. Dark brown to black color. Limited fuel usage.

Woody structure gone. Extensive use as industrial fuel and source of coke. Burns with smoky flame.

Final stage of coal formation. Burns with clean flame. Formerly much used for home heating.

FIGURE 19.9

Stages in the formation of coal.

The natural coal-forming process has been partially duplicated in laboratory experiments. Materials of grades lignite and bituminous coal have been made in laboratories. However, such processes have no immediate practical significance because more energy is required to produce the coal than can be recovered by burning it.

19.9 Chemical Composition of Coal

The chemical composition of coal is affected by the origins of the incorporated plant materials, the degree of carbonization that has occurred, and the impurities absorbed from the surroundings during the formation process. The approximate percentages of the three major elements present in coal—carbon, hydrogen, and oxygen—are shown in Table 19.6. Note that the compositions are given on a dry, mineral-free basis. Also included are the normal moisture content and the heat yielded upon combustion (calorific values). The two major changes in composition that occur in the carbonization process are readily apparent—the

TABLE 19.6

The Chemical Composition of Coal*

Type of Coal	Carbon (%)	Hydrogen (%)	Oxygen (%)	Moisture as Found (%)	Calorific Value (Btu/lb)
Peat	45–60	3.5–6.8	20–45	70–90	7,500–9,600
Brown coal and lignites	60–75	4.5–5.5	17–35	30–50	12,000–13,000
Bituminous	75–92	4.0–5.6	3.0–20	1.0–20	12,600–16,000
Anthracite	92–95	2.9–4.0	2.0–3.0	1.5–3.5	15,400–16,000

* Dry, mineral-free basis.

percentage of carbon increases, and the percentage of oxygen decreases. Small amounts of nitrogen and sulfur are present in coal. The presence of the element sulfur in coal is a "problem" that is considered in the next section.

Coal is chemically a complex mixture of organic compounds that behaves in a manner *unlike* that of any other class of organic materials. In spite of many investigations, the molecular structure of coal is not completely understood. In general, coals consist of loosely defined regions of more or less similar structure. The regions are held together by a type of bonding that is different from, and weaker than, that characteristic of the better-defined regions. Thus, coal structure involves a macrostructure superimposed on a more well-defined microstructure.

Figure 19.10 shows a possible model for the macrostructure for bituminous coal. The five-membered rings in the structure *do not represent chemical structure*, but rather are notation for clusters of molecules and atoms. The shaded circular areas are branch clusters that are attached to more than one

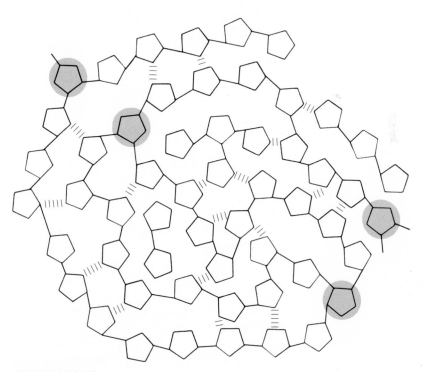

FIGURE 19.10

A possible model for the "macrostructure" of bituminous coal. The five-sided polygons in this model are not chemical structures but a notation for "clusters" of molecules and atoms.

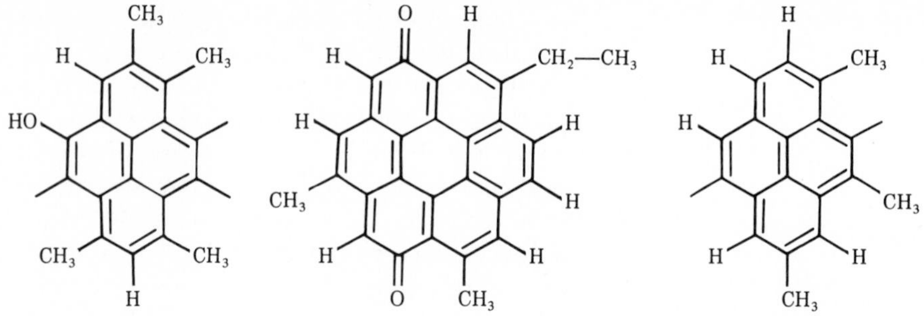

FIGURE 19.11
Typical cluster units associated with the microstructure of coal. Such units always involve a number of fused benzene rings.

branch of the network. The bonding between clusters (cross-links) involves hydrogen-bonding types of interactions (Section 8.16).

Within a cluster the basic chemical unit is the benzene ring, with a number of benzene rings usually found associated (fused) together. Figure 19.11 depicts three typical cluster units; note the presence of oxygen atoms. The number of fused rings in a cluster unit ranges from about 4 in low-grade coals, and 5 to 10 in bituminous coal, up to 30 in anthracite.

19.10   Problems with Coal Usage

For many years coal was the dominant fuel in the United States. Coal usage, as a percent of total United States energy consumption, peaked around 1920, dropped below 50% around 1940, and continued to decrease until the early 1970s (see Figure 19.1). Since that time, the percentage of coal usage has increased slightly. What is the basis for this marked decrease in the use of coal?

The mining and use of coal as a fuel present three problems.

1. Burning of coal causes air pollution, mostly from the sulfur compounds present in the coal.
2. Mining of coal can severely disrupt the environment.
3. Coal's solid form makes it inconvenient to handle, transport, and use.

The chief objection to the use of coal, from an air-pollution standpoint, is its sulfur content. All coals contain sulfur as a normal impurity. This is not surprising in view of the origin of coal. Plant life and natural waters are both necessary for coal formation, and both contain sulfur. Some of this sulfur survives the carbonization process and remains in the resulting coal. The sulfur content of coal ranges from less than 1% to about 7% by mass. During the combustion of coal, any sulfur present is converted to gaseous sulfur oxides—mainly sulfur doixide. We consider the serious effects of sulfur oxide air pollution in Section 20.8.

We should note that sulfur is found in all fossil fuels. However, during processing (natural gas) and refining (petroleum) the sulfur present in other fossil fuels is removed. Because of coal's solid state it is not economical to remove sulfur from coal.

Two major methods of mining coal exist: deep mining (underground) and strip mining (surface). At the present time, approximately 60% of United States coal production is obtained by strip mining, which can greatly disrupt the environment. Usually landscapes and soil must be moved to access the near-surface coal seams. In the 1960s, the past history of such disruptions made strip mining a national issue. Strip-mining problems are being solved through reclamation. Laws now mandate reclamation as a part of any strip-mining project. Good reclamation reconstructs the land, after the mining operation, to restore and possibly improve on its premining productivity.

STRIP MINING

Coal's solid state increases transportation and handling costs. It is much cheaper to put natural gas into a pipeline than to load coal into railroad cars or barges. In addition, as a solid, coal is unsuitable for much energy-consuming equipment—automobile and aircraft engines and most home furnaces. To be used in such markets coal must be converted to a gas or liquid. Considerable research is being devoted to these tasks. The next section discusses the topic of coal gasification.

19.11　Coal Gasification

Coal, as will be discussed in Section 19.12, is the most abundant fossil fuel in the United States. A promising possibility for greater utilization of coal reserves is to convert the coal to gaseous fuel. Coal gasification would not only provide the United States with a major "new" energy source but would also diminish use of the less abundant fossil fuels (petroleum and natural gas).

To convert coal from a solid to a gas requires reducing the size of the molecules. This involves chemical reactions in which many carbon–carbon bonds are broken.

A number of coal-gasification processes are in various stages of development. They all use the same basic set of reactions, although not always in the same sequence. Typically, the coal, in the form of a ground powder, is heated in a reactor (gasifier). Reactor conditions involve pressures from 20 to 70 atm and temperatures up to 1500°C.

Initial heating of the coal drives off volatile constituents, producing small amounts of the gases methane (CH_4) and hydrogen (H_2), and leaves a solid residue called *char*, which has a high carbon content. Reaction of the char with superheated steam produces a gaseous product that is a mixture of carbon monoxide (CO), hydrogen, and methane. All three of these gases can be used as fuel because each reacts with O_2, in a combustion reaction, to product heat. The gaseous mixture so obtained is further treated, catalytically, to maximize methane production; CO and H_2, under proper conditions, react with each other to produce additional methane. Impurities present (sulfur compounds), as well as any unreacted CO_2 and H_2O, are removed. The gaseous products, a mixture containing predominantly methane but also some H_2 and CO, is called *synthesis gas* (or syngas). A simplified diagram showing some of the reactions that occur during syngas production is presented in Figure 19.12.

At the present time, synthesis gas cannot compete economically with natural gas or petroleum. Any increases in the cost of these two substances would make

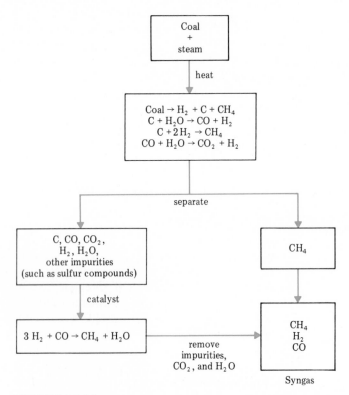

FIGURE 19.12

The basic chemical reactions in coal gasification.

synthesis gas more competitive. The technology is available for synthesis gas production, but its use must wait for the right economic climate.

Syngas can be used directly as a fuel. It can also be directly converted to gasoline-type molecules, or used to produce other chemicals such as methyl alcohol.

19.12 Fossil Fuels and an "Energy Crisis"

The current per-capita energy demands in the United States are the highest in the world. At the present time, each person in the United States uses, either directly or indirectly, an amount of energy equivalent to approximately 100 times the average human caloric intake. This can be visualized as the equivalent of 100 slaves working for each United States citizen.

In the early 1970s many people became concerned about the ability of the United States to maintain such an energy-use pattern. This concern was touched off by short-term supply and demand imbalances, the impact of energy production and use on the environment, projected long-term problems related to the depletion and ultimate exhaustion of fuel sources, and, in 1973, an oil embargo declared by the Organization of Petroleum Exporting Countries (OPEC) against all nations that supported Israel, which included the United States. As a result, in the early 1970s energy supply and demand became a matter of public concern, and from this concern came the term *energy crisis*.

In this section we consider the concept of an energy crisis as it relates to fossil fuels. First, we look at quantitative parameters (with only limited commentary) concerning the following two major factors contributing to United States fuel concerns.

1. The United States dependence on imported fossil fuels to help meet energy demands.
2. The dwindling size of proven reserves of fossil fuels in the United States.

Then we look at the chemical ramifications of the situation.

Fossil Fuel Imports

The extent of United States dependence on fossil fuel imports to meet its energy needs varies with the type of fuel. For many years, the United States has been an exporter rather than importer of coal. In 1986, 4.4% of natural gas consumed was imported and all of these imports came from Canada. Imports of crude petroleum and petroleum products for the year 1986 amounted to 33% of consumption. Thus, import discussions center almost completely on crude petroleum and petroleum products. Because petroleum is currently the dominant fuel in the United States energy mix (see Figure 19.2), import dependence on this fuel is of considerable importance.

A graphical view of petroleum import dependence for the period 1960–1986, including information on area of import origin, is given in Figure 19.13. Dependence on imports, after a steady increase, began decreasing in 1980. To put import dependence in better perspective, the percentage contribution of imports to total petroleum consumption is given for selected years, along the uppermost graph line in the figure.

Fossil Fuel Proven Reserves

A common statement heard in discussions of the energy crisis is that the United States is "running out of fossil fuels". Let us look at numbers pertaining to this area of concern.

We begin with coal, the most abundant of the fossil fuels in the United States. Proven reserves of coal, in 1986, were 488.3 billion short tons. *Proven reserves* comprise that portion of coal deposits that geological and engineering data

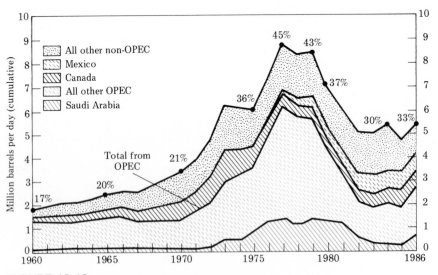

FIGURE 19.13

Imports of crude oil and petroleum products into the United States by country of origin, 1960–1986. The percentages are the contributions of imports to total consumption.

demonstrate can be recovered with reasonable certainty under existing economic and operating conditions. Proven reserves fluctuate from year to year. They increase when exploration results in additions to the proven reserve base and decrease as a result of production from the reserve base during the year.

An estimate of the significance of proven reserve size is obtained by determining the ratio of total proven preserves at year-end to production during the year. This ratio, called the *life index*, gives the estimated lifetime of the proven reserves *assuming* that no additional reserves are found and that the production level remains constant.

Consumption of domestic coal in 1986 was 806 million short tons. Division of the proven reserve amount by this production amount, taking into account that the units are different (billions of short tons and millions of short tons) gives the life index for coal.

FOSSIL FUEL LIFE INDEXES

$$\text{Life index for coal} = \frac{488{,}300 \text{ million short tons}}{806 \text{ million short tons/year}}$$

$$= 606 \text{ years}$$

The United States has plentiful coal supplies, enough to last for many years.

Let us now consider natural gas, using similar procedures. Proven reserves of natural gas at the end of 1986 were 185 trillion cubic feet. Domestic natural gas production (and consumption) was 16.0 trillion cubic feet. Thus, the life index for natural gas is

$$\text{Life index for natural gas} = \frac{185 \text{ trillion cubic feet}}{16.0 \text{ trillion cubic feet/year}}$$

$$= 11.6 \text{ years}$$

This life index is startlingly small compared with that for coal. Again, what does this number mean? If natural gas exploration ceased, resulting in no addition to the reserves, the existent reserves would be depleted in 11.6 years. But natural gas exploration has not ceased, and new reserves are added each year, approximately compensating for the amount consumed. This is confirmed by the fact that the life index for natural gas in the year 1975 was 11.9 years.

Estimated proven domestic reserves of petroleum and natural gas liquids at the end of 1986 were 36.3 billion barrels. Consumption of petroleum and natural gas liquids in 1986 was 3937 million barrels. Using these two figures, we obtain a life index for petroleum, again taking into account the difference in units (billion and million), of 9.2 years.

$$\text{Life index for petroleum} = \frac{36{,}300 \text{ million barrels}}{3937 \text{ million barrels/year}}$$

$$= 9.2 \text{ years}$$

The life index for petroleum is about the same as that for natural gas. Both are very small compared with coal's life index. The life index of petroleum in 1975 was 10.7 years. Again, as with natural gas, this indicates significant additions to the proven reserve base over the 11 year period.

It should not be forgotten, however, that the United States is dependent on petroleum imports, which account for approximately one-third of consumption. Use of this imported petroleum decreases the rate of depletion of United States proven reserves. Without imports, consumption would have to be cut or domestic reserves would rapidly dwindle.

Chemical Ramifications of Imports and Life Indexes

Of the many problems associated with reliance on fossil fuel imports, the following are major.

1. Imports are subject to interruption for many reasons including wars, changing of governments in exporting nations, and numerous other political factors. The United States has already experienced one such interruption in 1973–1974.
2. The cost of imports is subject to sudden change, determined solely by exporting nations. Large price increases can dramatically alter the economy of the importing nation. The United States has experienced numerous large price increases for imported petroleum over the last two decades, as is shown in Figure 19.14.

With heavy reliance on fossil fuel imports, the interruption factors and/or cost increases affect almost all areas of our life. Let us consider, from a chemical viewpoint, selected examples of these effects.

1. *Fossil fuels and food production are related.* In Section 12.4 we discussed the production of NH_3 (ammonia), a "top-five" chemical, whose predominant end use is nitrogen fertilizer. Hydrocarbons (fossil fuels) are the source of the H_2 gas needed for NH_3 production. Thus, food production is affected by fossil fuel shortfalls and price increases.
2. *Fossil fuels and wearing apparel are related.* In Chapter 18 we discussed various addition, condensation, and rearrangement polymers that have become part of our high standard of living. Many of these synthetic polymers are marketed in the form of fibers, such as polyesters and polyamides, which are used in clothing manufacture. The monomers used in producing these polymers are hydrocarbon derivatives whose ultimate source is fossil fuels. Thus, clothing production is affected by fossil fuel shortfalls and price increases.

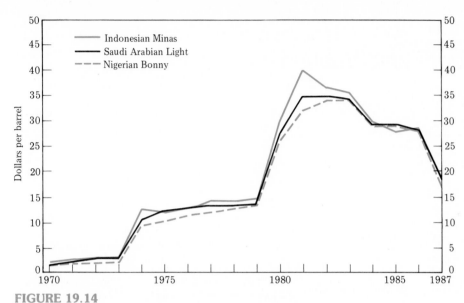

FIGURE 19.14

Official prices of selected foreign oil crudes, 1970–1987.

3. *Fossil fuels and vehicles for transportation are related.* Almost all transportation vehicles have rubber tires. Synthetic rubber (Section 18.5) is necessary for production of rubber tires, and such rubber requires hydrocarbon-based monomers for its production. Thus, vehicle production is affected by fossil fuel shortfalls and price increases.

4. *Fossil fuels and fuels for transportation are related.* Hydrocarbons are used directly as fuels for operation of automobiles, aircraft, and other forms of transportation. Thus, transportation is one of the first areas to be affected by fossil fuel shortfalls and price increases.

The smallness of the life indexes for petroleum and natural gas and the largeness of coal's life index are also major factors in United States energy considerations. Petroleum's and natural gas's share of the United States energy mix has decreased from approximately 70% in 1972 to 65% in 1986 (see Figure 19.1). Domestic natural gas production peaked in 1972, and consumption is now only three-fourths that of the peak value. Coal's share of the United States energy mix is again increasing (see Figure 19.1) after many years of decline. The life indexes for the three fossil fuels have a direct bearing on these changes.

The extensive interest in using coal in the form of a gaseous fuel (Section 19.11) is directly tied to its large life index and the smallness of the other two life indexes. Interest in alternative fuels such as oil shale (Section 19.6) would not exist if the life indexes for petroleum and natural gas were large.

Practice Questions and Problems

FOSSIL FUELS—GENERAL CONSIDERATIONS

19-1 What is a fossil fuel?

19-2 What was the approximate percentage contribution of fossil fuels to the total energy demand in the United States in each of the following years?
a. 1860   b. 1880   c. 1900   d. 1920
e. 1940   f. 1960   g. 1980   h. 1986

19-3 What was the dominant fuel in use in the United States in each of the following years?
a. 1920   b. 1940   c. 1960   d. 1980

OCCURRENCE OF PETROLEUM

19-4 What are the three geological features required for the accumulation of petroleum in sufficient amounts to make recovery economically feasible?

19-5 Explain what is wrong with the statement that "petroleum occurs underground in liquid pools."

19-6 What is the average recovery efficiency for petroleum from underground deposits?

19-7 What is the major cause for low recovery efficiencies for petroleum and what is one method for overcoming this problem?

19-8 Explain what is meant by secondary petroleum production.

CHEMICAL COMPOSITION OF PETROLEUM

19-9 What are the predominant types of hydrocarbons found in petroleum?

19-10 Table 19.1 lists the most abundant hydrocarbons in a specific petroleum sample. How many of these compounds are the following?
a. Alkanes
b. Cycloalkanes
c. Aromatic hydrocarbons

PETROLEUM REFINING

19-11 What is the difference between crude petroleum and a petroleum fraction?

19-12 What are the three major parts to the petroleum refining process?

19-13 Describe the process of fractional distillation.

19-14 What are two major differences between the gasoline fraction and the kerosene fraction obtained from fractional distillation of petroleum?

19-15 Describe how hydrocarbon molecules are affected by each of the following petroleum refining operations.
a. Cracking   b. Polymerization
c. Reforming   d. Isomerization

GASOLINE AND OCTANE RATING

19-16 What is meant by the term *knocking*?

19-17 A gasoline produces the same knocking tendency in a standard engine as a test mixture containing 40% heptane and 60% isooctane. What is the octane number of the gasoline?

19-18 For each of the following pairs of hydrocarbons, indicate which one has the higher octane number.
a. Butane and hexane
b. 2-Methylpentane and 2,2-dimethylbutane
c. 2,2-Dimethylbutane and 2,2-dimethylhexane

19-19 List four methods by which the octane number of a gasoline could be increased.

19-20 Why is the use of TEL as an antiknock agent in gasoline being phased out?

19-21 What is the difference between an antiknock additive and a blending agent?

19-22 What is the chemical structure of the most used blending agent, and what is its octane number?

19-23 How are octane numbers greater than 100 for gasoline determined?

OIL SHALE

19-24 What is oil shale?

19-25 What is the difference between kerogen and petroleum?

19-26 What steps are involved in converting kerogen to petroleum refinery feedstock?

NATURAL GAS

19-27 What is the difference between associated and nonassociated natural gas?

19-28 What is the difference in chemical composition between commercial natural gas and crude natural gas?

19-29 Discuss what is meant by each of the following terms.
a. Natural gas liquids
b. Liquefied petroleum gas
c. Natural gasoline

19-30 Why is natural gas often called the "ideal fuel"?

ORIGIN AND GRADES OF COAL

19-31 What are the physical conditions thought to be necessary for coal formation?

19-32 Discuss the physical characteristics of each of the following grades of coal.
a. Peat   b. Lignite
c. Bituminous coal   d. Anthracite

19-33 What are the principal differences in chemical composition between anthracite and bituminous coal?

19-34 Describe each of the following aspects of coal structure.
a. The macrostructure
b. The microstructure
c. The major elements present in coal

PROBLEMS WITH COAL USAGE

19-35 What are the two major methods for mining coal, and which one is used to the greatest extent at the present time?

19-36 Why is sulfur oxide air pollution a problem in burning coal but not in burning petroleum and natural gas? All fossil fuels contain sulfur.

COAL GASIFICATION

19-37 Why are attempts being made to discover economical processes for converting coal into gaseous fuel?

19-38 Describe the general steps involved in coal gasification processes.

19-39 What are the differences in composition between commercial natural gas and synthesis gas?

19-40 What is preventing synthesis gas from being used as an energy source at the present time?

FOSSIL FUELS AND AN ENERGY CRISIS

19-41 Contrast import dependence for coal, petroleum, and natural gas.

19-42 Contrast petroleum import dependence for the years 1965, 1975, and 1985 in terms of the following.
a. Percentage of total petroleum consumption
b. Country of origin

19-43 What are the two major problems associated with petroleum imports?

19-44 What is meant by the phrase "proven reserves of a fossil fuel"?

19-45 What is meant by the phrase "life index of a fossil fuel"?

19-46 Discuss the fallacy in the statement "Petroleum supplies will be depleted in 9.2 years because the life index of petroleum is 9.2 years."

19-47 How is a life index for a fossil fuel calculated?

19-48 By what factor did the price of imported crude oil increase over the period 1970–1985?

19-49 Why did the United States shift from coal to petroleum and natural gas when we have much larger proven reserves of coal than of the other two fossil fuels?

19-50 How many times greater is the life index of coal than the collective life indexes of petroleum and natural gas?

19-51 What is the relationship between fossil fuel supplies and the following?
a. Food production
b. Clothing manufacture
c. Vehicles for transportation
d. Operation of vehicles for transportation

Air Pollution

CHAPTER HIGHLIGHTS

20.1   The Atmosphere

The **atmosphere** *is a mixture of gases held to earth by gravity.* A consideration of the general characteristics of this gaseous body, the lower portions of which we call *air,* is a logical starting point for the topic of air pollution.

The two most important regions of the atmosphere, from a human activity viewpoint, are the regions closest to earth—the troposphere and the stratosphere. The *troposphere* extends to an average height of about 7 miles (10 miles at the equator) and the *stratosphere* extends from this point to about 30 miles altitude. Naturally, the troposphere, being closer to the earth, is affected to a greater degree by human activities than the stratosphere. Almost all weather-related phenomena occur within the troposphere. Most air pollutants remain within the lower regions of the troposphere. Most commercial airline flights traverse the upper parts of the troposphere. Very little human activity touches the stratosphere. Supersonic aircraft fly in the lower regions of the stratosphere. Some very unreactive (inert) air pollutants occasionally reach the stratosphere, as do materials from large volcanic eruptions and atmospheric nuclear explosions. At the present time, there is concern about the extent of the interaction of air pollutants reaching the stratosphere with ozone naturally present in the stratosphere. We consider this subject in Section 20.7.

Measurements show that about 50% of the mass of the atmosphere is within 4 miles of the earth's surface, 90% within 10 miles, and 99% within 20 miles. This is because there is a rapid decrease in atmospheric density with increasing altitude. Together, the troposphere and stratosphere account for 99.9% of the mass of the atmosphere, with 75% of the mass found in the troposphere. Near the top of the stratosphere, the atmosphere has a density of only about 1% that at sea level.

The temperature of the troposphere decreases steadily with increasing altitude to the top of this region, where the temperature reaches about −70°F. Within the stratosphere, the temperature increases with altitude, warming to about 30°F in the upper stratosphere. It is the change in temperature gradient, from decrease to increase, that defines the beginning of the stratosphere.

Circulation (or lack of circulation) is a major area of difference between the troposphere and stratosphere. Much turbulence exists within the troposphere, and there is strong vertical mixing of the various gases present in this region. The stratosphere, on the other hand, is relatively calm, with little vertical mixing. This

623

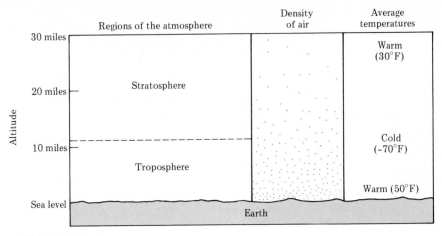

FIGURE 20.1

Various characteristics of the troposphere and stratosphere regions of the atmosphere.

distinction between the two regions becomes important in later considerations in this chapter. Figure 20.1 summarizes some of the differences between the troposphere and the stratosphere.

Life, as we now know it on this planet, is very dependent on the chemical species found within the troposphere and stratosphere. The oxygen content of the lower atmosphere is an absolute necessity for normal respiration in human beings and animals. Carbon dioxide in the air is the basis for the process of photosynthesis that occurs within plants. Atmospheric water, which is deposited on earth in the form of dew, rain, and snow, is the water source that sustains human, animal, and plant life. Atmospheric nitrogen, which is the predominant chemical species in lower atmospheric regions, is the ultimate source of the nitrogen fertilizers that assist in producing the food necessary to sustain the world's inhabitants (Section 12.4). Ozone in the stratosphere screens us from the effects of harmful ultraviolet radiation from the sun.

The actual composition of the atmosphere is remarkably uniform from place to place on the earth's surface and at different altitudes in the troposphere and stratosphere. Table 20.1 gives the composition of clean, dry air. As indicated in this table, nitrogen and oxygen are the predominant gases in the atmosphere, and together constitute 99% of the mixture by volume. Nearly all of the remainder of the atmosphere is made up of argon and carbon dioxide. The total volume percent of these four components in clean, dry air is 99.99%. Also included in Table 20.1 are other components of clean, dry air whose volume percentage exceeds 0.0001.

Two different concentration units—percent by volume and parts per million specify concentration in Table 20.1. Percent by volume is the volume of the

TABLE 20.1

Composition of Clean, Dry Air

Gaseous Component	Formula	Percent by Volume	Parts per million (by volume)
Major Components			
Nitrogen	N_2	78.08	780,800
Oxygen	O_2	20.95	209,500
Minor Components			
Argon	Ar	0.934	9,340
Carbon dioxide	CO_2	0.0314	314
Neon	Ne	0.00182	18
Helium	He	0.000524	5
Methane	CH_4	0.0002	2
Krypton	Kr	0.000114	1

specific component contained in 100 volumes of air, which is equivalent to parts per hundred. Parts per million (ppm) represents the volume of a specific component in one million volumes of air. Parts per million is especially useful in expressing the very low concentrations common to air pollutants. Note that concentrations in ppm are 10,000 times the volume percentages.

Table 20.1 lists the components of *dry* air and thus neglects any water present. Water is not included in Table 20.1 because, unlike the other components, water vapor is found in variable amounts in clean air. Depending on the temperature and the evaporation rate from available water sources, the amount of water in the atmosphere varies from 6% down to a few tenths of 1%. Typically, the value is between 1 and 3%. Water vapor is usually the third most abundant component of air.

20.2   Types and Sources of Air Pollutants

Very small amounts of many different substances enter the atmosphere daily. Some of these substances result from ongoing natural processes and others are generated from human activities, that is, from *anthropogenic* sources. Are all of these substances air pollutants? The answer is no. A more limiting definition of the term *air pollutant* is usually used. An **air pollutant** *is a substance added to the atmosphere in amounts sufficient to cause measurable undesirable effects on humans, animals, vegetation, or materials.*

Three frequently asked questions concerning air pollutants and air pollution are the following.

1. What air pollutants are discharged into the air in the greatest amounts?
2. What anthropogenic sources put the largest amounts of air pollutants into the atmosphere?
3. How rapidly are the concentrations of air pollutants decreasing as the result of air pollution control procedures, or are concentrations still increasing?

We consider some general answers to these questions in this section before discussing the chemistry associated with important individual pollutants in later sections.

Five types of substances account for more than 90% of the nationwide air pollution problem.

1. Carbon monoxide
2. Nitrogen oxides
3. Volatile organic compounds
4. Sulfur oxides
5. Particulates

Table 20.2 lists the major anthropogenic sources, and amounts they emit, for each of these five types of air pollutants. It should be noted that the pollutant emission values used in this table are based on nationwide data. Such information is useful as a general indication of pollutant levels, but it has definite limitations. For example, national totals, averages, and trends are not always useful for estimating pollution problems in a specific locality.

Table 20.2 indicates that transportation is the main source of anthropogenic air pollution; its 65.9 million metric tons is almost equal to the amounts from

TABLE 20.2

National Air Pollutant Emissions Estimates, by Pollutant and Anthropogenic Source, 1983 (*million metric tons per year*)

Source	Carbon Monoxide	Nitrogen Oxides	Volatile Organic Compounds	Sulfur Oxides	Particulates	Total Mass of Pollutants Produced from Each Source
Transportation	47.7	8.8	7.2	0.9	1.3	65.9
Stationary source fuel combustion	7.0	9.7	2.1	16.8	2.0	37.6
Industrial processes	4.6	0.6	7.5	3.1	2.3	18.1
Solid waste	2.0	0.1	0.6	0.0	0.4	3.1
Miscellaneous	6.3	0.2	2.5	0.0	0.9	9.9
Totals	67.6	19.4	19.9	20.8	6.9	134.6

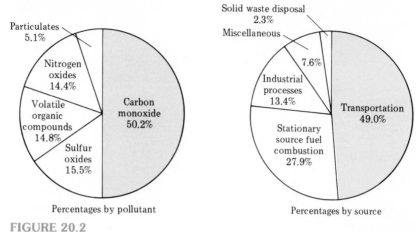

Percentages by pollutant

Percentages by source

FIGURE 20.2

National air pollutant emission estimates, 1983, by pollutant percentages and source percentages.

all other sources combined. Carbon monoxide is the major individual pollutant, its tonnage being three times greater than that of any other pollutant. Now we know why carbon monoxide and transportation are often the first items mentioned when air pollution is discussed.

Figure 20.2 is a presentation of the information in Table 20.2 in terms of percentages. Again, it is obvious that transportation is the dominant anthropogenic source and carbon monoxide emissions greatly exceed those of any other pollutant.

The tonnage of carbon monoxide emitted nationwide is slightly greater than the total of all other pollutants combined. Does this mean that carbon monoxide pollution is the greatest air pollution problem nationwide? Not necessarily. There is one serious drawback to looking at pollutants only in terms of tonnage (see Table 20.2 and Figure 20.2). The possibility that one pollutant might be more harmful or dangerous than another is overlooked. Small amounts of a very toxic pollutant are more worrisome than large amounts of a less toxic pollutant. Pollutant toxicities differ widely; carbon monoxide is actually one of the least toxic pollutants. Toxicity levels are considered in the discussions of individual pollutants.

Public and governmental concern over air pollution during the late 1960s led to the passage of the federal Clean Air Act of 1970 as well as to additional state regulations. These laws placed restrictions on the amount of emissions from automobiles, industries, and stationary fuel combustion sources (mainly electrical-generating plants). Many of these controls required several years to be phased in. The results have been positive. The nation's air quality is improving, as is shown in the statistics of Table 20.3, which cover the period 1940–1983.

TABLE 20.3

National Air Pollutant Emissions Estimates, by Pollutant, 1940–1983 (*millions of metric tons per year*)

	Carbon Monoxide	Nitrogen Oxides	Volatile Organic Compounds	Sulfur Oxides	Particulates	Total Mass of Pollutants Produced
1940	79.4	6.7	17.7	18.0	22.4	144.2
1950	84.8	9.3	20.3	20.3	24.2	158.9
1960	87.5	12.8	23.3	20.0	20.9	164.5
1970	98.3	18.1	27.0	28.2	18.0	189.6
1980	75.0	20.3	22.3	23.2	8.3	149.1
1983	67.6	19.4	19.9	20.8	6.9	134.6
Percent change, 1970–1983	−31	+7	−26	−26	−62	−29

20.3 Carbon Monoxide

Carbon monoxide (CO), which is a colorless, odorless, tasteless gas that is not appreciably soluble in water, is produced as a by-product of incomplete combustion of any material containing the element carbon.

The main chemical process resulting in formation of carbon monoxide is the combustion of carbon-containing fuel, especially gasoline in motor vehicles. In simplified terms, the combustion of carbon in gasoline (or any other carbon-containing fuel) is a two-step process. The C in the fuel is first converted to CO, which then reacts with more O_2 to produce CO_2, the final product of the combustion.

$$2\,C + O_2 \longrightarrow 2\,CO$$

$$2\,CO + O_2 \longrightarrow 2\,CO_2$$

The first step (CO production) occurs about 10 times faster than the second step (CO_2 production). Thus, CO is an intermediate in the overall combustion reaction and can show up as an end product if insufficient O_2 is present to complete the second reaction. CO may be an end product even with sufficient O_2 present in the combustion mixture if the fuel and air are poorly mixed. Poor mixing leads to localized areas of oxygen deficiency in the air–fuel mixture. Oxygen deficiency and poor mixing are more likely to happen when combustion occurs in a closed system (interior of an automobile engine) than in a system open to the atmosphere (furnance or fireplace).

The transportation sector is the dominant source of anthropogenic CO emissions (Table 20.2). More than two-thirds of anthropogenic CO entering the atmosphere comes from motor vehicle exhaust. Stationary fuel combustion sources, such as power plants, rank a distant second as an anthropogenic CO source, accounting for slightly over 10% of emissions (Table 20.2).

Total anthropogenic CO emissions peaked in the year 1970—at 98.3 million metric tons—and have now decreased to less than 70 million metric tons (Table 20.3). Motor vehicle emission control systems (catalytic converters) are the major cause of this reduction. Catalytic converters became mandatory equipment on all new automobiles in 1975.

In principle, once carbon monoxide is released into the atmosphere it should be rapidly converted to CO_2 because plenty of oxygen is present in the atmosphere.

$$2\,CO + O_2 \longrightarrow 2\,CO_2$$

CATALYTIC
CONVERTERS

However, this reaction does not appreciably decrease the CO concentrations because the rate at which it occurs is extremely slow at normal atmospheric temperatures. Within a catalytic converter (Figure 20.3) this reaction is accelerated. The high temperature of the CO-containing exhaust gases coming from the engine, coupled with the catalyst, effectively speeds up the reaction. [The catalyst in catalytic converters usually consists of small amounts of the metal platinum (1 to 3 g) mixed with small amounts of other heavy metals such as rhodium or palladium.] Thus, catalytic converters convert CO to CO_2, releasing CO_2 to the atmosphere. Use of catalytic converters has resulted in approximately a 20% decrease in highway vehicle CO emissions in the last decade, despite a 20% increase in total vehicle miles traveled.

The current *average* atmospheric CO concentration at monitoring sites in the United States is 8 ppm. The typical average for the more clean locations is 5 ppm, and for the more polluted locations 14 ppm. The *national ambient air quality standard* (NAAQS) for CO, as specified by the Clean Air Act, is 9 ppm. (To meet this NAAQS, the average CO concentration over an 8-hour period at a given location should not exceed 9 ppm.) Air quality data for the period

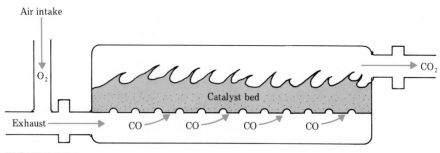

FIGURE 20.3
An automobile catalytic converter reduces CO emissions by converting CO to CO_2.

1984–1986 show that a number of major metropolitan areas are still not in compliance with the standard. The 10 least-complying areas for this period are

1. Denver, Colorado
2. Los Angeles–Long Beach, California
3. Las Vegas, Nevada
4. Phoenix, Arizona
5. Fresno, California
6. New York, New York
7. Fairbanks, Alaska
8. Provo, Utah
9. Sacramento, California
10. Raleigh–Durham, North Carolina

CARBON MONOXIDE
AND HUMAN BLOOD

Although carbon monoxide does not pose a significant threat to vegetation and materials, it does affect humans. It can impair human health by reducing the oxygen-carrying capacity of the blood. Carbon monoxide does this by interacting with the hemoglobin of red blood cells. Normally, hemoglobin picks up oxygen in the lungs and distributes it to oxygen-deficient cells throughout the body. The O_2 is carried to the cells in the form of an oxygen–hemoglobin complex called oxyhemoglobin.

$$\text{hemoglobin} + O_2 \longrightarrow \text{oxyhemoglobin}$$

Carbon monoxide also combines with hemoglobin, but it does so much more strongly; its affinity for hemoglobin is about 200 times that of O_2.

$$\text{hemoglobin} + CO \longrightarrow \text{carboxyhemoglobin}$$

Once carboxyhemoglobin forms in a red blood cell, the cell loses its ability to carry oxygen. A relatively small quantity of CO can inactivate a substantial fraction of the hemoglobin in the blood. This reduces the amount of oxygen delivered to all tissues of the body.

The extent of the adverse effects of CO on the oxygen-transport system of the body is determined by the amount of air breathed, the concentration of the CO, and the length of exposure. Symptoms of CO poisoning, such as headache, fatigue, and dizziness are present when 10% of a person's hemoglobin is tied up as carboxyhemoglobin. When the fraction rises to 20%, death can result unless the victim is removed from the poisonous atmosphere. The tying up of hemoglobin by CO is a reversible process. However, this reversal requires a considerable amount of time (5–8 hours) in an unpolluted atmosphere.

It should be noted that cigarette smoke also contains carbon monoxide. Therefore, cigarette smokers have a portion of their hemoglobin inactivated by this source as well as by external air pollution. Numerous studies have shown that the concentration of carboxyhemoglobin in the blood of people who smoke is two to five times higher than that in nonsmokers. Table 20.4 shows the

TABLE 20.4

Carboxyhemoglobin as Percent of Total Hemoglobin in the
Blood of Humans under Various Conditions

	Percent Carboxyhemoglobin
Continuous exposure, 10 ppm CO	2.1
Continuous exposure, 50 ppm CO	8.5
Nonsmokers, Detroit	1.6
Smokers, Detroit	5.6
Nonsmokers, Salt Lake City	1.2
Smokers, Salt Lake City	5.1

percentages of carboxyhemoglobin in blood typical of various situations.

The background concentration of CO in relatively clean, unpolluted air rarely exceeds 1.0 ppm. By comparison, in city traffic, concentrations of 50 ppm are often reached, and CO concentration may go as high as 140 ppm in a traffic jam. The inhaled smoke from cigarettes contains about 400 ppm CO.

Calculations indicate that enough CO is discharged annually into the atmosphere of the United States to cause a doubling of the ambient CO level within 5 to 7 years. This has not occurred. Ambient levels of CO have changed very little over the past two decades. This suggests the existence of natural processes that remove CO from the atmosphere. Recent research has shown that the major mechanism for CO removal from the atmosphere involves soil microorganisms. Fourteen species of fungi have been identified as the active agents. The capacity for CO uptake by the soils of the United States far exceeds the amount of CO present in the air. Nevertheless, CO is still found at significant concentrations in the air. Why is this so? The primary reason is that CO and soil are not distributed uniformly. The largest CO-producing areas are large cities, areas with much concrete and asphalt and little soil.

Natural sources also contribute to atmospheric levels of CO. How do natural and anthropogenic sources compare in size. Surprisingly, approximately nine times as much CO enters the atmosphere from natural sources as from all human activities. A major natural source of CO, as can be seen in Figure 20.4, is the atmospheric oxidation of methane (CH_4) gas. Almost 80% of all atmospheric CO comes from this process. The decomposition of submerged organic materials in swamps, rice paddies, and tropical regions of the world generate large quantities of methane gas. The oceans constitute the second largest source of atmospheric CO. Algae and other biological sources contribute substantial amounts of CO to surface waters. This CO is subsequently released into the atmosphere.

The impact of anthropogenic CO sources, despite their comparatively small contribution to worldwide CO emissions (9.4%), must not be minimized. The emission data of Figure 20.4 fail to account for a very important fact—natural

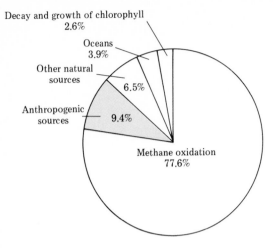

FIGURE 20.4
Worldwide natural and anthropogenic CO sources

CO sources are distributed throughout the world, whereas anthropogenic sources are concentrated in very small areas. For example, 95 to 98% of the atmospheric CO in a metropolitan area comes from human activities. The resulting CO levels are commonly 50 to 100 times higher than the characteristic worldwide values of 0.1 to 0.5 ppm.

In addition to the local nature of their emissions, anthropogenic CO sources have another undesirable characteristic—they emit CO at a high rate. This rate overwhelms the capacity of existing natural mechanisms, such as soil fungi that remove CO from the atmosphere.

20.4 **Nitrogen Oxides**

Eight different nitrogen oxides are known, but only three are normally detected in the atmosphere. The three are nitrous oxide (N_2O), nitric oxide (NO), and nitrogen dioxide (NO_2). Important considerations about each of these three gases are as follows.

1. *Nitrous oxide (N_2O).* N_2O is a colorless, nonflammable, nontoxic gas with a slightly sweet taste. Until recently its presence in the atmosphere was not considered an air pollution problem because its sources are almost exclusively natural processes. Recently, evidence has accumulated that N_2O functions as a "greenhouse gas" (Section 21.12) in the atmosphere.

2. *Nitric oxide (NO)*. NO is a colorless, odorless, nonflammable gas that is toxic. This is the nitrogen oxide that enters the atmosphere in the greatest amounts as a result of human activities.

3. *Nitrogen dioxide (NO$_2$)*. NO$_2$, in pure form, is a nonflammable reddish brown gas that is toxic and characterized by a strong choking odor. However, at the typical low concentrations present in the atmosphere it is colorless and odorless. In terms of air pollution NO$_2$ is the major nitrogen oxide species present in the atmosphere. Its chemistry is closely tied to that of NO since it is formed from this species within the atmosphere. The notation NO$_x$ is often used to represent collectively the NO and NO$_2$ present in the atmosphere because of this connection between them.

The two major components of air are N$_2$ and O$_2$ (Section 20.1). At normal atmospheric temperatures these two gases have little tendency to react with each other because their reaction is very endothermic. However, at high temperatures the necessary heat energy required is available and N$_2$ and O$_2$ become slightly reactive toward each other. It is this slight reaction between the two components of air, at high temperatures, that is the origin of NO$_x$ pollution. The amount of nitrogen oxides produced is small but significant in terms of atmospheric pollution concentrations.

The high temperatures associated with fuel combustion are sufficient to make N$_2$ and O$_2$ slightly reactive toward each other. Thus, all types of fuel combustion, both stationary source and mobile source (transportation) generate nitrogen oxides. As Table 20.2 indicates, transportation and stationary source fuel combustion contribute heavily to emissions, stationary fuel combustion being the slightly greater source.

Nitrogen oxide production from N$_2$ and O$_2$ is a two-step process.

$$N_2 + O_2 \longrightarrow 2\,NO$$

$$2\,NO + O_2 \longrightarrow 2\,NO_2$$

Almost all nitrogen oxide pollution entering the atmosphere is in the form of NO rather than NO$_2$. This is the opposite of the situation with carbon oxides, in which the dioxide is dominant (see Section 20.3). With nitrogen NO is the dominant oxide because NO$_2$ is unstable at high temperatures (flame temperatures) and any that is formed decomposes back to NO.

$$2\,NO_2 \xrightarrow{\text{high temperature}} 2\,NO + O_2$$

Once in the atmosphere the NO reacts rapidly with atmospheric O$_2$ to form the dioxide.

$$2\,NO + O_2 \longrightarrow 2\,NO_2$$

At atmospheric temperatures NO$_2$ is very stable.

The NAAQS for NO_2 is 0.053 ppm, measured as an annual arithmetic mean. The current *average* annual mean at monitoring sites in the United States is about 0.025 ppm. The typical average for the more clean locations is 0.011 ppm, and for the more polluted locations 0.047 ppm. Thus, even the most polluted sites are in compliance with the NO_2 standard. The NAAQS for NO_2 (0.053 ppm) differs from that for CO (9 ppm) by a factor of 170.

Health problems associated with NO_2 can include lung irritation, lower resistance to respiratory infections such as influenza and increased susceptibility to bronchitis and pneumonia. Very seldom, however, do NO_2 concentrations reach the level where such effects are a major concern.

The major impact of NO_2 pollution is an indirect health effect. Nitrogen dioxide plays a major role in the atmospheric reactions that produce photochemical smog and is itself primarily responsible for smog's yellow-brown color. The formation and health effects of photochemical smog are the subject of Section 20.6.

Nitrogen oxide pollution has proven to be the most difficult of the major air pollution types to control, as evidenced by an actual increase in emissions during the period 1970–1983 (Table 20.3). Formation of these pollutants requires only air and a high temperature. Combustion processes always involve air and high temperatures.

Beginning with 1981 model-year cars, catalytic converters have been modified to reduce NO emissions. With rhodium (an expensive transition metal) as an added catalyst, such converters catalyze the reaction

$$2\,CO + 2\,NO \longrightarrow 2\,CO_2 + N_2$$

and also reactions between hydrocarbons (unburned gasoline) and NO that produce CO_2, H_2O, and N_2. NO_2 emissions are now beginning to decrease slightly.

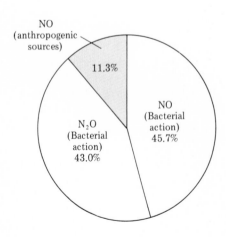

FIGURE 20.5
Worldwide natural and anthropogenic nitrogen oxide emissions, in volume percent.

The major process responsible for the removal of NO_2 from the atmosphere involves its conversion to nitric acid and nitrate salts, which are then carried out of the atmosphere in rainfall or dust. Nitric acid in rainfall contributes to the problem called acid rain, which is discussed in Section 20.9.

As with carbon monoxide (Section 20.3), natural and anthropogenic sources both contribute to current atmospheric levels of nitrogen oxides. Also, as with CO, natural sources contribute significantly more than human activities. Bacterial action in the soil, resulting from the decomposition of nitrogen-containing compounds, is the major source of nitrogen oxides (Figure 20.5). Combustion is the major source of anthropogenic nitrogen oxide emissions. Again, we must not be misled by these data and conclude that anthropogenic emissions are not important. Natural sources are distributed throughout the world, while anthropogenic sources are concentrated in small areas.

20.5 Volatile Organic Compounds

Most volatile organic compounds (VOC) enter the atmosphere from natural sources, as was true for CO and NO_x pollution. Almost all of these sources involve biological processes, although a few include geothermal activity and processes taking place in coal, petroleum, and natural gas fields.

Methane (CH_4), the simplest organic compound, is emitted into the atmosphere in quantities larger than any other volatile organic compound. Bacterial decomposition reactions represent its primary source, with much of this decomposition occurring in swamps, marshes, and other water bodies. Plants, particularly trees, are also natural sources of volatile organic compounds. Most of these compounds are hydrocarbons known as terpenes.

Worldwide anthropogenic VOC emissions are estimated at 15.5% of total VOC emissions (Figure 20.6). A large part of this total comes from activities involving petroleum, which is itself a complex mixture of organic compounds (Section 19.3). Typical activities in this category are refining and transporting of petroleum and the combustion of refined products. Exhaust gases from the incomplete combustion of refined products always contain small amounts of uncombusted fuel.

Gasoline, the most used petroleum fraction (Section 19.5), is a very volatile substance that evaporates readily. Gasoline vapors enter the atmosphere during the loading of tank trucks, the filling of service station storage tanks, and the filling of automobile fuel tanks. Such evaporation, plus the emission of unburned fuel in automobile exhaust, makes gasoline the major anthropogenic source of VOC emissions.

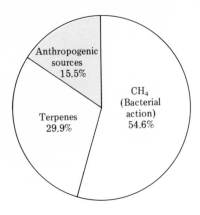

FIGURE 20.6
Worldwide natural and anthropogenic volatile organic compound emissions, in volume percent.

Another major anthropogenic VOC source, though a much smaller one than gasoline, is the evaporation of organic solvents. These solvents are important ingredients in paints, varnishes, lacquers, undercoatings, and similar products. These products consist of 40 to 80% solvent, which usually evaporates during or after application. The processes involved in the manufacture of these products also constitute VOC sources.

The Clean Air Act, when formulated in 1970, contained a national ambient air quality standard for hydrocarbons. This standard was eliminated in 1983 because a bettter measure of VOC pollution was found—the amount of ozone present in the atmosphere near the earth's surface. Ozone formation in the atmosphere, as is discussed in the next section, requires the presence of volatile organic compounds. An NAAQS for ozone has been added to the Clean Air Act.

20.6 Ozone and Other Photochemical Oxidants

Ozone is a colorless, highly reactive gas with the molecular formula O_3. It is a form of the element oxygen in which there are three atoms per molecule rather than the usual two atoms. Unlike the other air pollutants discussed in this chapter, ozone is not emitted directly into the atmosphere by specific sources. Instead, it is formed in the atmosphere by chemical reactions involving pollutants already there (nitrogen oxides and volatile organic substances). Ozone is a secondary pollutant rather than a primary pollutant. A **primary air pollutant** *is a pollutant directly emitted into the air from natural or anthropogenic sources. A* **secondary air pollutant** *is a pollutant produced within the atmosphere from the interaction of primary pollutants.*

Ozone is the main "irritant" in photochemical smog, a type of smog that first gained public attention in the 1940s in the Los Angeles, California, area. It has long been a major air pollution problem in that city, as well as in most major metropolitan areas of the United States. Such smog is called *photochemical* smog because sunlight is necessary for its formation.

PHOTOCHEMICAL SMOG

Photochemical smog production results from an "unbalancing" of a naturally occurring sequence of atmospheric reactions involving NO and NO_2, called the nitrogen dioxide photolytic cycle. The unbalancing influence is volatile organic compounds. To understand the "upsetting" of the NO_2 photolytic cycle, we must first understand the cycle itself.

There are three reactions in the NO_2 photolytic cycle.

REACTION 1 NO_2 molecules absorb energy in the form of ultraviolet light from the sun. The absorbed energy causes the NO_2 molecules to break apart into NO molecules and oxygen atoms (O), with the latter being an extremely reactive species.

$$NO_2 + \text{sunlight} \longrightarrow NO + O$$

REACTION 2   The atomic oxygen (O) reacts with atmospheric oxygen (O_2) to produce ozone (O_3).

$$O + O_2 \longrightarrow O_3$$

REACTION 3   The O_3 reacts with NO to give NO_2 and O_2.

$$O_3 + NO \longrightarrow NO_2 + O_2$$

These three reactions constitute a naturally occurring cycle in which NO_2 is continually broken apart (by sunlight) and then reformed (reaction 3). The cyclic nature of this reaction system is shown in Figure 20.7.

The presence of volatile organic compounds in the atmosphere "unbalances" the NO_2 photolytic cycle. They react with some of the atomic oxygen generated in the first step in the cycle to produce hydrocarbon free radicals. These hydrocarbon free radicals then react with a number of substances including NO, NO_2, and O_2. The result is the production of a number of very reactive species, collectively called *photochemical oxidants*. A **photochemical oxidant** *is a substance produced by a photochemical process that oxidizes materials not readily oxidized by atmospheric oxygen (O_2).* Figure 20.8 shows the disruption by volatile organic compounds of the NO_2 photolytic cycle.

Photochemical oxidants are the "active ingredients" in smog. Ozone itself is the most abundant of these oxidants. A family of compounds called peroxyacyl nitrates (PANs) are also present in relatively large amounts. The general

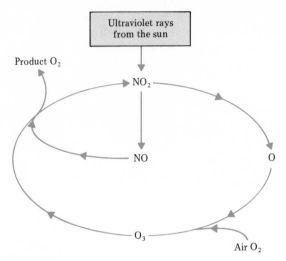

FIGURE 20.7
The nitrogen dioxide photolytic cycle.

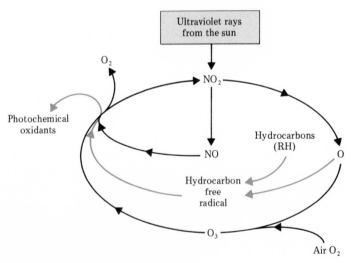

FIGURE 20.8
Volatile organic compound disruption of the NO_2 photolytic cycle.

structure of PAN is

$$R—\overset{\overset{\displaystyle O}{\|}}{\underset{\underbrace{\qquad}_{acyl}}{C}}—\underbrace{O—O—N}_{nitrate}\overset{\displaystyle O}{\underset{\displaystyle O}{\diagdown}}$$ (R = hydrocarbon chain or ring)

Ozone, PANs, and the other photochemical oxidants cause the eye irritation that is characteristic of exposure to photochemical smog.

Ozone, by itself, severely irritates the mucous membranes of the nose and throat. It impairs normal functioning of the lungs and reduces the ability to perform physical exercise. Its effects are more severe in individuals with chronic lung disease. The length of exposure, frequency of exposure, and ozone concentration are significant factors in determining the effects. Individuals with asthma or diseases of the heart and circulatory system experience symptoms at lower ozone concentrations. It also appears that ozone in combination with sulfur dioxide (Section 20.8) has a greater effect on respiratory function than either pollutant alone.

Ozone also damages leafy plants such as lettuce, cabbage, tobacco, and spinach and certain materials such as rubber, plastics, dyes, and nylon. Ozone damage to vegetation and materials in the United States is a multibillion-dollar problem.

The NAAQS for ozone is 0.12 ppm as a daily 1-hour average. This standard and the CO standard (Section 20.3) are the two NAAQSs metropolitan areas are having the most trouble meeting. The ozone NAAQS is the least-complied-with standard at present. The current *average* reading for monitoring stations is 0.13 ppm, which is greater than the NAAQS. The typical average for the more clean locations is 0.08 ppm, and for the more polluted locations 0.19 ppm. Based on air quality data for the period 1984–1986, the ten least-complying metropolitan areas are

1. Los Angeles, California
2. San Diego, California
3. Houston, Texas
4. New York, New York
5. Greater Connecticut
6. Providence, Rhode Island
7. Sacramento, California
8. Atlantic City, New Jersey
9. Chicago, Illinois
10. Philadelphia, Pennsylvania

Atmospheric ozone levels are significantly influenced by meteorology. Concentration levels are usually higher during the summer months, when there is more sunlight, than during winter months. High ozone levels can, however, occur during winter months as a result of air stagnation (temperature inversion; see Section 20.11). The weather affects air quality levels for all pollutants, but ozone levels are the most sensitive to meteorological conditions.

20.7   The Ozone Layer and Stratospheric Pollution

In the upper atmosphere (stratosphere) ozone is a naturally occuring species. The photodissociation of molecular oxygen represents the principal mechanism for its formation in the upper atmosphere. Energy for the process comes from ultraviolet solar radiation.

$$O_2 \xrightarrow{\text{UV radiation}} O + O$$

$$O + O_2 \longrightarrow O_3$$

Ozone formed in this way is continually decomposed back to ordinary oxygen. In the stratosphere, the rates of ozone formation and decomposition are such that a net amount of ozone is always present.

The equilibrium concentration of ozone in the atmosphere varies with altitude, as shown in Figure 20.9, and with the seasons in a way that is strongly related to the activity of the sun. The upper regions of the stratosphere, where ozone concentrations maximize, is often called the "ozone layer." Note that the maximum ozone concentrations in the ozone layer are around 200 parts per billion. A concentration of 200 ppb is equal to 0.2 ppm.

The stratospheric ozone layer screens out 95 to 99% of ultraviolet solar radiation. High-intensity radiation of this type is harmful to nearly all forms of life. Ultraviolet radiation can disrupt carbon–hydrogen bonds of organic molecules as well as dissociate water molecules. All living organisms contain many molecules with carbon–hydrogen bonds; physiological change can result when such bonds are disrupted. Thus, the shielding ozone layer is very important to life on the earth.

Ozone thus presents a paradox. In the troposphere it is considered an unwanted, dangerous pollutant (Section 20.6). In the stratosphere its presence is vital to life on earth.

In the early 1970s concern began to be expressed about the possibility of very unreactive chemical species of anthropogenic origin traversing the troposphere, entering the stratosphere, and negatively affecting the ozone layer. Of particular concern was the possibility that such species would catalyze the ozone decomposition reaction.

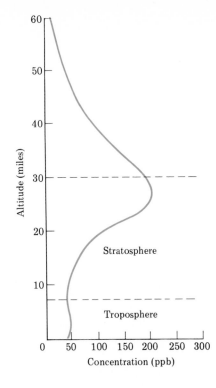

FIGURE 20.9
Variation of atmospheric ozone concentration with altitude (at 35° North latitude).

$$O_3 \xrightarrow{\text{catalyst}} O + O_2$$

One particular family of compounds, chlorofluorocarbons (CFCs), has been singled out as having potential to affect the ozone layer negatively. The two best-known as well as most used CFCs are Freon-11 and Freon-12, whose structures are

Freon-11   Freon-12

These compounds, simple halogenated methanes, have been used extensively for many years as aerosol propellants. They have excellent properties for such use: they are nonflammable, nonreactive (even with photochemical oxidants), and nontoxic. In the atmosphere, they do not cause tropospheric pollution problems because of their nonreactivity.

This lack of chemical reactivity also means that no reactions remove these compounds from the troposphere. Thus, CFCs have long lifetimes in the

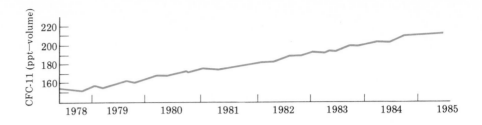

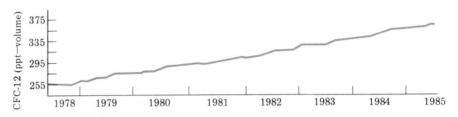

FIGURE 20.10
Ground level concentrations of CFC-11 and CFC-12 measured at an atmospheric gases monitoring station in Barbados.

atmosphere, and over time their concentrations should increase. Experimental measurements indicate that this is the case. Figure 20.10 shows CFC-11 and CFC-12 concentrations measured since mid-1978 at an atmospheric gas monitoring station in Barbados. The measurements are done at ground level. Note that the concentrations are in parts per trillion rather than the normal parts per million. A concentration of 200 ppt is equal to 0.0002 ppm.

CFCs AND THE OZONE
LAYER

In 1974 balloon instrument packages detected trace amounts of CFCs in the lower stratosphere. This evoked immediate concern by many scientists. CFCs, because of their nonreactivity, were beginning to enter the stratosphere. What effect would they have on the ozone layer as they continued to rise?

It has been theorized, but not definitely proven, that CFCs could decrease the amount of ozone through catalysis of the ozone decomposition reaction. Laboratory experiments indicate that ultraviolet radiation "activates" CFC molecules. In a manner similar to what happens to NO_2 in the NO_2 photolytic cycle (Section 20.6), CFCs break apart under ultraviolet radiation absorption. One of the species produced is atomic chlorine. For example, with Freon-11 the following reaction could occur.

$$CCl_3F \longrightarrow CCl_2F + Cl$$

Atomic chlorine is a known catalyst (from laboratory experiments) for accelerating the decomposition of ozone. Would it also function as a catalyst under stratospheric conditions? As CFC molecules rise in the stratosphere, the possibility of their being activated by ultraviolet radiation increases because more ultraviolet radiation is encountered.

Concern about possible ozone depletion by CFCs has led to decreased use of these substances. In 1978 the United States banned the use of CFCs as aerosol propellants in nearly all consumer products. The ban did not affect other applications for CFCs, which include use as a refrigerant and in air conditioning (for both homes and automobiles). (Not all nations followed the United States lead in banning aerosol uses for CFCs.) Figure 20.11 shows worldwide production levels for aerosol and nonaerosol CFCs since 1960. Note that, despite decreases in aerosol propellant use, the total production of CFCs is rising again.

Are the CFCs now present in the stratosphere already affecting ozone concentrations? The answer is not a definite yes or a definite no.

Determination of the effects of CFCs on ozone concentration is a difficult experimental problem. The ozone layer is not a directly accessible location to humans. Data must be obtained by satellite observation, unmanned instrument balloons and from spectrophotometers at ground-level locations.

To uncover any changes in ozone that may be caused by human influences, raw experimental data must first be analyzed to filter out the huge natural variations in ozone concentration. Ozone concentrations naturally change from season to season and as output from the sun varies. Natural events such as

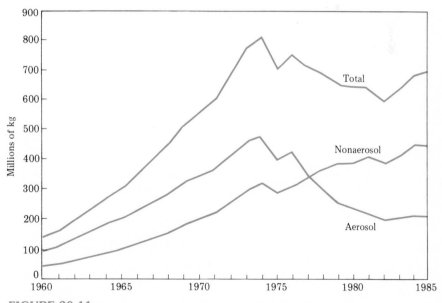

FIGURE 20.11

Total reported production of Freon-11 and Freon-12 peaked in 1974 and declined fairly steadily until 1983 when it began rising.

FIGURE 20.12

Natural fluctuations in global average total ozone.

volcanic eruptions also have their effect. Figure 20.12 shows the fluctuations observed (from ground-based data) in average global total ozone since 1956.

Computer models of the atmosphere used to predict CFC effects are very complex. Such models contain about 160 chemical reactions and more than 40 reactive species. (Ozone is not the only chemical substance present in the ozone layer.) Each time a reaction rate is remeasured, a new reaction or new species discovered, or an additional product found for a new reaction, the whole interlocking system is affected. With such complexities, modelers are forced to rely on simplifications and assumptions. Whether these assumptions are appropriate is a question constantly being asked.

Data from NASA's Nimbus 7 satellite show total global ozone declined about 3% between the end of 1978 and 1984. Ground station measurements, at about 40 locations, using spectrophotometers, show no statistically significant change from 1970 to 1985. The ground observations involve the atmosphere only directly above the ground station. Is the reported 3% loss real? If so, is the ozone decrease due to natural phenomena or human activity?

Another question involves "self-healing." If ozone is destroyed in the upper stratosphere, more ultraviolet radiation can pass through to dissociate oxygen molecules and produce ozone at lower altitudes. This increase in ozone in the lower atmosphere is called the self-healing affect. Would this affect actually occur?

At present, much is yet to be learned concerning the earth's ozone layer and the effects that trace gases such as CFCs have on ozone levels.

20.8 Sulfur Oxides

The air pollution chemistry of the element sulfur centers on two compounds—sulfur dioxide (SO_2) and sulfur trioxide (SO_3). Collectively, these two oxides are designated as SO_x.

Sulfur dioxide is a colorless, nonflammable gas that is odorless in typical atmospheric concentrations (less than 0.1 ppm). It has a pungent and choking odor at concentrations above about 3 ppm.

Sulfur trioxide is an exceedingly reactive colorless gas that is easily condensed to a liquid. Under normal conditions, very little SO_3 is found in the atmosphere because it reacts rapidly with moisture (water vapor) to form sulfuric acid (H_2SO_4).

The main chemical process resulting in sulfur oxide pollution is the combustion of coal. Sulfur is an impurity in all coals (Section 19.10). (Almost all of the sulfur impurities are removed from petroleum and natural gas during processing; see Sections 19.4 and 19.7.)

In simplified terms, the combustion of sulfur in fuel is a two-step process. The sulfur in the fuel is first converted to SO_2, which then reacts with additional O_2 to produce SO_3.

$$S \text{ (fuel)} + O_2 \longrightarrow SO_2$$

$$2\,SO_2 + O_2 \longrightarrow 2\,SO_3$$

However, what enters the atmosphere as the result of coal combustion is primarily SO_2 rather than SO_3. Very little SO_3 forms during the combustion process because of the instability of SO_3 at high temperatures (combustion temperatures). A situation similar to this was encountered with NO_x pollution (Section 20.4); the lower rather than the higher oxide is dominant.

Total emissions of SO_x pollution to the atmosphere resulting from human activities are decreasing, as is shown in Table 20.3. Note from this table the dominance of stationary source fuel combustion as a source of SO_x. Approximately 80% of sulfur-containing emissions come from coal combustion at power plant locations. Metallurgical processes that involve the metals copper, zinc, mercury, and lead rank a distant second as an overall source of SO_x pollution. (Recall, from Section 14.3, that most copper ores are copper sulfides.)

Sulfur dioxide concentrations are decreasing because of (1) a reduction in the average sulfur content of fuels consumed and (2) the installation of flue gas control equipment at coal-fired electric-generating stations. Sulfur dioxide control is done mainly "after the fact"; the gas is removed from the flue gases after its formation. In power plant operations, powdered limestone is often added to a furnace where coal is burning. The high temperature decomposes the limestone.

$$CaCO_3(s) \longrightarrow CaO(s) + CO_2(g)$$

The CaO so produced reacts with SO_2 to produce calcium sulfite, $CaSO_3$, which is easily removed from the combustion gases.

$$CaO(s) + SO_2(g) \longrightarrow CaSO_3(s)$$

Figure 20.13 shows the details of an actual SO_2 control system. $CaSO_3$ and any unreacted SO_2 leave the furnace and enter a purification chamber where a

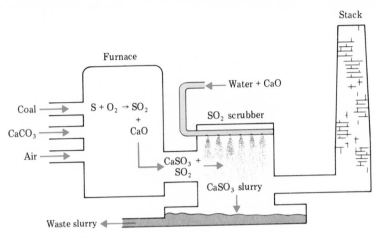

FIGURE 20.13

A system designed to remove SO_2 from flue gases generated in coal combustion.

shower of CaO and water precipitates the $CaSO_3$ into a watery residue (a slurry) and converts the remaining SO_2 to $CaSO_3$.

The NAAQS for SO_2 is 0.03 ppm SO_2, measured as an annual arithmetic mean. This standard appears to have been achieved at almost all sites. The *current* average reading for monitoring stations is 0.01 ppm. The typical average for the more clean locations is 0.003 ppm, one-tenth the NAAQS, and for the more polluted locations 0.02 ppm, two-thirds the NAAQS.

Sulfur-containing compounds are naturally present in the atmosphere, even in unpolluted air. Natural sources include volcanic eruptions and the bacterial decay of natural organisms. The major sulfur-containing compound that enters the air from such natural sources is not an oxide of sulfur but rather hydrogen sulfide (H_2S). Within the atmosphere H_2S is oxidized to SO_2. Figure 20.14 shows the percentage of emissions from natural and anthropogenic SO_x sources;

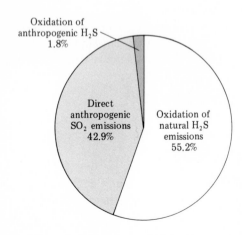

FIGURE 20.14

Worldwide natural and anthropogenic sulfur oxide sources, in volume percent.

note that the two are about equal. This means that sulfur oxides are different from previously discussed pollutants where natural sources dominate over anthropogenic emissions by a wide margin. The contribution of anthropogenic sources is greater for SO_x pollution than for any other type of air pollution.

Within the atmosphere, a large part of the SO_2 present is oxidized to SO_3, which then reacts with water vapor to form sulfuric acid (H_2SO_4). The sulfuric acid reacts with other available substances to form sulfates. For example, ammonium sulfate [$(NH_4)_2SO_4$] results when the acid reacts with ammonia. The sulfate salts settle out of the atmosphere or are washed out by rain. Figure 20.15 summarizes this sequence of events and also shows the major sources and means of removal for other sulfur compounds. The wide arrows indicate the most significant ways in which sulfur is transferred between the earth and the atmosphere.

The presence of sulfuric acid makes rainwater more acidic than normal. The ramifications of this increased acidity are discussed in Section 20.9, where the subject of acid rain is considered.

Both plants and human beings are susceptible to SO_2, with plants being affected at lower levels. The harmful effects of elevated sulfur dioxide concentrations on sensitive vegetation have been extensively documented. Crop yield reductions have been experienced in areas close to large emission sources.

Most of the effects of SO_2 on human health are related to the irritation of the respiratory system and eyes. Of particular concern are the effects of SO_2 exposure on individuals who suffer from chronic respiratory diseases such as bronchitis and asthma. These individuals exhibit diminished lung functions at much lower SO_2 levels than healthy individuals.

Either SO_2 or SO_3 can be an active participant in the formation of *industrial smog*. Industrial smog is a mixture of SO_2, soot, fly ash, smoke, and volatile

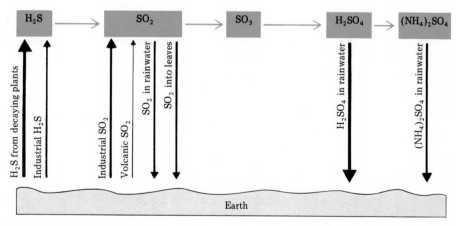

FIGURE 20.15
The atmospheric sulfur cycle.

organic compounds. It is a type of smog associated with areas where large amounts of coal are consumed. Industrial smog was much more of a problem in the early 1900s than now because coal was the dominant fuel for heating at that time (Section 19.1). Photochemical smog (Section 20.6) differs from industrial smog in that it is essentially free of SO_2 and contains fewer suspended particles (soot, fly ash, etc.). Serious episodes of either type of smog can result from a temperature inversion (Section 20.11).

20.9 Acid Rain

Rainfall or precipitation in a clean atmosphere is intrinsically slightly acidic. This acidity results from the presence of carbon dioxide in the atmosphere, which reacts with water to produce carbonic acid (a weak acid).

$$CO_2(g) + H_2O \longrightarrow H_2CO_3(aq)$$

This reaction produces rainwater with a pH of 5.6. (The pH scale for specifying acidity was discussed in Section 10.9.)

Acid rain, rainfall with a pH less than 5.6, has been observed with increasing frequency in recent years in many areas of the world. Rainfall with pH values between 4 and 5 is common, and occasionally rainfall has a pH as low as 2.

Within the United States this acid rain phenomenon has most often been observed in the northeastern states (see Figure 20.16). The maritime provinces of Canada have also been greatly affected.

The causes of acid rain are the air pollutants SO_x (Section 20.8) and NO_x (Section 20.4). After being discharged into the atmosphere, these pollutants can be chemically converted into sulfuric acid (H_2SO_4) and nitric acid (HNO_3) through oxidation processes. There are several complicated pathways or mechanisms by which oxidation can occur. Which pathway is actually taken depends on numerous factors, such as the concentration of heavy metal pollutants, the intensity of sunlight, and the amount of ammonia present (see Figure 20.17).

Environmental damage caused by acid rain has been most clearly associated with certain materials and structures and some aquatic ecosystems. The effects of acid rain on forest growth, crops, and human health have been studied less.

A pH of 5.5 has been shown to be stressful to certain sensitive cold-water game fish, and in a laboratory environment these fish cannot survive in water that has a pH of 5.0. Fortunately, low-pH rainfall does not automatically mean that the water in lakes and streams will have low pH values. A large dilution factor affects rain falling directly into a body of water.

An important factor in determining the impact of acid rain on the environment is the ability of the natural ecosystem to neutralize or buffer incoming acidity. For

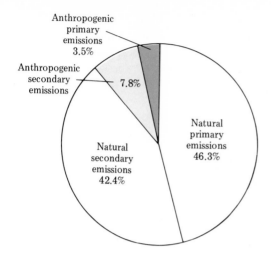

FIGURE 20.20

Worldwide natural and anthropogenic particulate sources, in volume percent.

they obey the same optical laws as macroscopic objects and intercept or scatter light. This property is important in determining the effects of atmospheric particulates on solar radiation (climate) and visibility.

Table 20.3 indicates that industrial processes are the main source of primary anthropogenic particulates, and also that both industrial particulate emissions and total particulate emissions are decreasing. We noted previously (Section 20.2) that during the period 1970–1983 a 62% reduction in particulate emissions was accomplished, the greatest percentage reduction for any pollutant (see Table 20.3).

Techniques for controlling particulate emissions are based mainly on the idea of capturing the particulates before they enter the atmosphere. Depending on the size of particles involved, several processes and devices are used—gravity settling chambers, wet scrubbers, electrostatic precipitators, and cyclone collectors, among others.

The NAAQS for particulates is 75 μg/m^3, as an annual geometric mean. Particulate levels at most monitoring sites are now in compliance with this standard. The current *average* reading for monitoring stations is approximately 50 μg/m^3. The typical average for the more clean locations is 35 μg/m^3, and for the more polluted locations 80 μg/m^3.

Particulate pollutants enter the human body almost exclusively by way of the respiratory system. The health effects they exert depend on their size and composition. The larger particles are usually filtered out in the nose and throat and rapidly cleared from the body. Smaller particles may be carried deeper into the lungs. Particles reaching sensitive deep lung areas are considered relatively more important for health purposes. Particle composition is also important because some compounds are relatively harmless whereas others—such as asbestos and beryllium—can result in serious health problems.

Particles that enter and remain in the lungs can exert toxic effects in three ways.

1. The particles themselves may be intrinsically toxic because of chemical and physical characteristics. Such particles include trace metals, asbestos fibrils, and sulfuric acid aerosols.
2. The particles may be inert but, once in the respiratory tract, may interfere with the removal of other more harmful materials.
3. The particles may carry adsorbed or absorbed irritating gas molecules and thus enable these molecules to reach and remain in sensitive areas of the lungs.

Nonhealth effects of particulates include soiling clothes and surfaces and, in combination with some gases such as sulfur dioxide, corroding materials.

Particulates in the atmosphere also exert a definite influence on the amount and type of sunlight that reaches the earth. A principal effect is decreased visibility, which may be dangerous (interfering with aircraft or automobile operation) or merely annoying (interfering with sightseeing).

PARTICULATES AND CLIMATE

The climate is affected by particulate pollution in two ways. Particulates, by acting as condensation sites, influence the formation of clouds, rain, and snow. The reduction in the amount of sunlight reaching the earth's surface is another effect of particulates. Less sunlight could upset the earth's heat balance and cause the atmosphere to cool. Large volcanic eruptions are known to have a definite influence on climate in the months following the eruption.

Ultimately, all atmospheric particulates are deposited on the earth's surface. The most important atmospheric removal process is wet deposition (precipitation), which is classified into two categories: rainout and washout. *Rainout* involves processes in clouds in which particulates serve as sites (nuclei) for water condensation or ice formation. During *washout*, falling rain or snow collects particulates from the atmosphere and carries them to the earth's surface. Washout is most effective in removing particles larger than 1 μm. Smaller particles are carried out of the way of the falling droplets with the parting air. Rainout, however, accomplishes the removal of significant amounts of the smaller particles.

20.11 Temperature Inversions

Temperature inversions can cause serious air pollution problems, not because they represent a source of pollution, but because they cause air pollutants to accumulate in the lower troposphere instead of dispersing. Many of the most serious air-pollution episodes (occurrences of extremely adverse health effects) have occurred in this country during temperature inversions.

Air in the troposphere can move vertically or horizontally. Horizontal movement is governed mainly by prevailing winds. If these winds are active and of sufficient force, pollutants have little chance to build up before they are dispersed. Surrounding mountains, hills, and even buildings in a large city slow down and break up winds and lessen the horizontal movement of air. With limited horizontal movement, pollutant dispersion depends on the vertical movement of air.

The temperature of air in the troposphere decreases with altitude. Air closest to the earth's surface is warmed by the earth, expands, and becomes less dense than the cooler air above it. The warm, less dense air then rises through the cooler air, which flows in to replace it. This new lower air is warmed, expands, and in turn rises. Air currents are created this way, and pollutants are dispersed.

Meteorological conditions can cause a reversal or an *inversion* in the normal pattern of temperature change within the troposphere. The result is the formation of an inversion layer. When this occurs, the air temperature will be found to decrease from the earth's surface to some altitude, for example, 3000 feet. This normal behavior then gives way to an abnormal one in which the air temperature increases with altitude from 3000 feet to 5000 feet. Beyond this layer, normal behavior again occurs, as the air temperature decreases with altitude. The warm layer starting at 3000 feet and extending to 5000 feet is the inversion layer, which effectively creates a lid for the air under it (see Figure 20.21).

The presence of an inversion layer prevents vertical atmospheric circulation, because the cooler air cannot rise through the warm inversion layer. Pollutants in the air are then trapped in the lower, noncirculating air. Situations of this type may continue for days until weather conditions change and the inversion layer breaks up. Inversion layers cause the added pollution problem of increased photochemical activity. The inversion layer is usually warm, dry, and cloudless, and so transmits a maximum amount of sunlight, which interacts photochemically with the trapped pollutants to form great amounts of photochemical smog. Thus, high levels of photochemical smog are usually associated with air pollution involving temperature inversions.

20.12   The Greenhouse Effect

In the early 1970s, and again in the mid-1980s, an environmental problem known as the *greenhouse effect* or *global warming* was a focus of much attention. This problem involves the concept that increased concentrations of

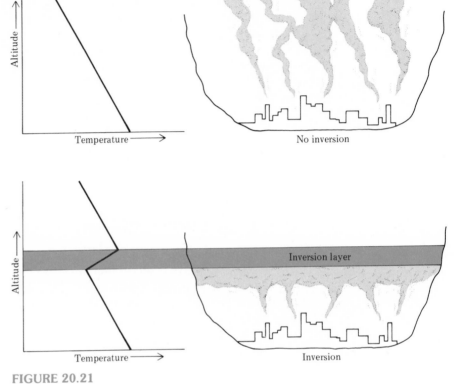

FIGURE 20.21

In a temperature inversion a layer of warm air over a layer of cooler air holds air pollution close to the ground.

certain gases in the atmosphere, caused by anthropogenic activities, can cause climatic changes by affecting the surface temperature of the earth.

A number of gases are potential contributors to this greenhouse effect, and many are not ordinarily considered to be air pollutants. These gases include carbon dioxide, methane, nitrous oxide, ozone, and CFCs.

The most studied of the "greenhouse gases" is carbon dioxide. Let us use carbon dioxide as a representative greenhouse gas to examine this global warming phenomenon in detail. Later we will look at the effects of other greenhouse gases.

Carbon dioxide is not normally considered an air pollutant, because it is a natural component of the air. This gas is continually being cycled into and out of the atmosphere by natural processes. Volcanic eruptions, decay of organic matter, and human and animal respiration all result in CO_2 entering the atmosphere. Plant photosynthesis and dissolution of CO_2 into the oceans both remove CO_2 from the atmosphere. This CO_2 cycle in nature (described here in a simplified form) results in an essentially constant atmospheric CO_2 level, if it is not upset by human activities.

Two major human activities have disrupted this cycle. Clearing of land decreases the available plants needed to remove CO_2 from the atmosphere. The burning of fossil fuels increases the amount of CO_2 entering the atmosphere. The net effect of these activities is an increasing atmospheric CO_2 level, as is shown in Figure 20.22. The average global concentration of CO_2 increased from about 315 ppm in 1958 to about 345 ppm in 1985.

The greenhouse effect results from an interaction between the increasing amount of atmospheric CO_2 and solar radiation. Although sunlight consists of many wavelengths, much of the radiation reaching the earth's surface is in the range of visible light. Ozone, in the stratosphere, filters out most ultraviolet light (wavelengths shorter than visible; see Section 20.6) and atmospheric water vapor and CO_2 absorb much of the incoming infrared light (wavelengths longer than visible), which we detect on our skin as heat.

Approximately one-third of the light reaching the earth's surface is reflected back into space. Most of the remaining two-thirds is absorbed by such inanimate

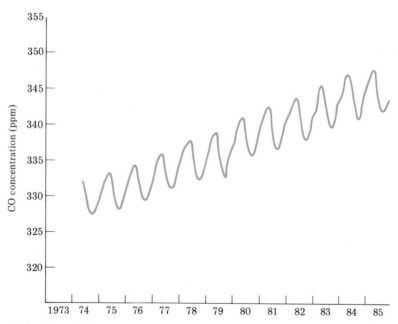

FIGURE 20.22
Mean monthly atmospheric carbon dioxide concentrations, in ppm, at the Mauna Loa, Hawaii, observatory.

matter as rocks and concrete. This warms the earth's surface. When the earth cools, it emits heat (infrared) radiation. This infrared radiation is absorbed by atmospheric CO_2 (and water vapor). Hence, instead of passing through the atmosphere, the radiated energy is absorbed by the atmosphere, and the temperature of the atmosphere is increased. This phenomenon is called the *greenhouse effect* because it is similar to the trapping of the sun's heat in a greenhouse.

Thus, carbon dioxide effectively behaves as a one-way filter, allowing visible light to pass through in one direction, but preventing light of a longer wavelength (infrared) from passing in the opposite direction. This behavior is represented in Figure 20.23. (Although water acts as a filter in the same way as CO_2, its concentration is not appreciably increased by human activities, and so its contribution to the atmospheric temperature remains constant.)

The one-way filtering action of CO_2 has led some scientists to predict that the temperature of the atmosphere and earth will increase as CO_2 concentration increases. Is a global warming effect actually occurring? The answer is maybe. The average global temperature appears to have increased about 0.5 degree Celsius over the past century. A change of that small magnitude, however, is within the natural variability of the climate, so the increase cannot be conclusively assigned to the greenhouse effect. Figure 20.24 gives temperature data for the earth's surface since the advent of the Industrial Revolution and the accompanying use of fossil fuels. The temperatures are shown as deviations from the 1951–1970 reference period, represented as zero.

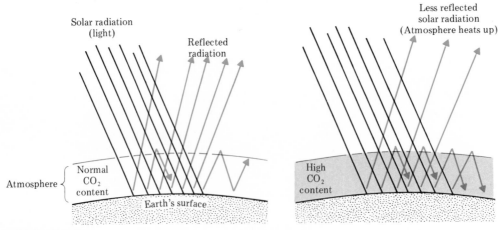

FIGURE 20.23

Carbon dioxide permits the passage of visible radiation from the sun to the earth but stops some of the infrared radiation leaving the earth, resulting in an increased atmospheric temperature.

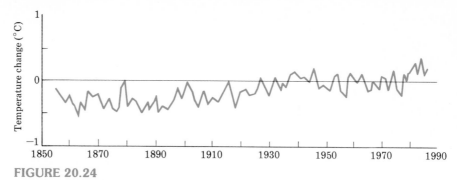

FIGURE 20.24

Average global air temperature has risen 0.5°C in the past century. Temperature changes are annual averages, relative to 1951–1970 reference period.

Various computer models for the greenhouse effect indicate that a doubling of CO_2 concentration would produce a 2 to 5°C warming of the earth's climate. The wide range in projections is due to uncertainty about what factors can dampen or accelerate the warming.

Why the concern about increases of a few degrees in the earth's average surface temperature? It should certainly make winters more pleasant. The concern is that average temperature increases of 2 to 5°C could lead to increased melting of polar ice caps and glaciers. Computer models project that release of the water in the polar ice caps could increase the average depth of all oceans by 200 to 250 feet. This would put many coastal areas, including cities, under water, as is shown in Figure 20.25.

In the early 1980s a new concern about the greenhouse effect surfaced. Studies showed that other gases besides carbon dioxide, whose concentrations were increasing in the atmosphere, could also exert a greenhouse effect. These new greenhouse gases included Freon-11 and Freon-12, other chlorofluoro-carbons, nitrous oxide (N_2O), and methane (CH_4). Concentrations of all of these gases are increasing in the atmosphere. We have previously considered increased Freon-11 and Freon-12 concentrations (Section 20.7) in another context, that of ozone depletion in the stratosphere.

Nitrous oxide and methane both have mostly natural sources. Nitrous oxide was briefly discussed in Section 20.4. Bacteria produce most nitrous oxide in oxidation–reduction processes that occur in soils. Humans may be increasing N_2O emissions from soils by using more and more nitrogen fertilizers. Burning coal, which contains organic nitrogen compounds, is another anthropogenic source of N_2O. Nitrous oxide levels have increased from 300 to 310 ppb in the period 1978–1985.

Bacteria also produce methane in decomposition processes in wet locations where oxygen is scarce: swamps, paddies, and natural wetlands. Oil and natural

Seattle

Bangor

Boston

New York

Philadelphia

Baltimore

Washington, D.C.

San Francisco

Los Angeles

New Orleans

Houston

—— Present United States coastline
- - - - Coastline if oceans rose 250 feet

FIGURE 20.25

The impact on the coastlines of the United States if the oceans were to rise 250 feet as a result of global warming (2–5°C rise)—the greenhouse effect.

gas exploitation may also be a significant source. Methane concentrations are now about 1650 ppb.

Computer models show that these other greenhouse gases collectively have a warming effect equal to that of CO_2, as is shown in Figure 20.26. The miscellaneous contribution in this figure includes ozone, other CFCs than Freon-11 and Freon-12, and stratospheric water.

Obviously, a number of variables are involved in all of the predictions concerning greenhouse gases, and the net effects of these gases on the climate and weather remain to be seen. They could prove to be very profound, or they might even cancel one another out. One calculation shows that a 25% increase in particulates in the air would counteract a 100% increase in CO_2. Another study shows that a 1% increase in cloud density will counteract an 0.5°C rise in temperature caused by increased atmospheric greenhouse gases.

No direct evidence exists that definitely ties increased greenhouse gas levels in the atmosphere to climatic changes. On the other hand, it would be foolish to discount completely the possibility that human activities could bring about atmospheric changes that might cause temperature changes. Much remains to be learned in this area.

A final note is that the greenhouse effect is not a modern-day phenomenon. Carbon dioxide has always exerted a warming effect on the earth's climate. If the atmosphere were only molecular nitrogen and oxygen, which hardly absorb any infrared radiation, the radiation leaving the earth's surface during cooling would go into space without any attenuation. The result would be an average temperature of about −20°C for the earth surface.

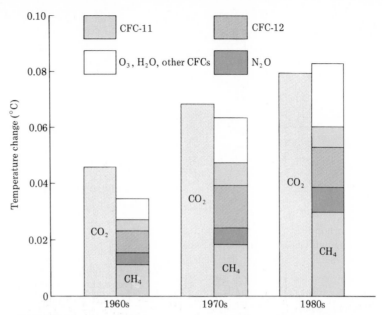

FIGURE 20.26

Other trace gases are as important as CO_2 in greenhouse warming. This graph compares the relative effects on temperature, neglecting feedback effects, of increases in various greenhouse gases.

The earth's actual average surface temperature is about 15°C. Thus, the greenhouse effect is responsible for an average increase of about 35°C in the temperature of the earth's surface. Life would not be possible on the earth without the greenhouse effect. What we have considered in this section is an unwanted enhancement of the greenhouse effect.

Practice Questions and Problems

THE ATMOSPHERE

20-1 Explain why the stratosphere, which is over 20 miles thick, contains a smaller total mass than the troposphere, which is under 10 miles thick.

20-2 Contrast the temperature patterns in the troposphere and the stratosphere.

20-3 How do the troposphere and stratosphere differ in regard to air circulation patterns?

20-4 What is the concentration, in parts per million, of each of the following substances in clean, dry air?
a. Nitrogen
b. Oxygen
c. Carbon dioxide
d. Argon

20-5 What is the range for the concentration of water vapor in air, in parts per million?

TYPES AND SOURCES OF AIR POLLUTANTS

20-6 What are the five major types of air pollutants that account for more than 90% of air pollution in the United States?

20-7 What is the dominant type of air pollution associated with each of the following source categories?

a. Transportation
b. Stationary source fuel combustion
c. Industrial processes
d. Solid waste
e. Miscellaneous

20-8 What is the dominant source category for each of the following types of air pollution?

a. Carbon monoxide
b. Nitrogen oxides
c. Volatile organic compounds
d. Sulfur oxides
e. Particulates

20-9 Nationwide, carbon monoxide is emitted into the air in the greatest tonnage. Why is it not necessarily the most serious air pollution problem in this country?

20-10 Air pollution control procedures are reducing total air pollutant emissions significantly. During the period 1970–83 what were the percentage changes in emissions for the following?

a. Carbon monoxide
b. Nitrogen oxides
c. Volatile organic compounds
d. Sulfur oxides
e. Particulates

CARBON MONOXIDE

20-11 What is the relationship between CO formation and CO_2 formation in the combustion of a carbon-containing substance?

20-12 What is the source of the largest part of total anthropogenic CO emissions?

20-13 What chemical reaction that occurs in an automobile catalytic converter reduces CO emissions?

20-14 What is the NAAQS for CO and what are approximate average levels of CO in the air at the present time?

20-15 What is the major effect carbon monoxide has on the human body, and how does this effect occur?

20-16 What natural process exists for removing CO from the atmosphere?

20-17 Discuss the major natural sources for CO emissions.

20-18 Natural sources of CO are many times larger than anthropogenic sources of CO. Why, then, is there so much concern about anthropogenic CO?

NITROGEN OXIDES

20-19 What does the notation NO_x stand for?

20-20 What is the relationship between NO formation and NO_2 formation in the combustion of fuels?

20-21 Why is NO rather than NO_2 the predominant nitrogen oxide species released into the air by anthropogenic processes?

20-22 What are the two chemical reactants required for nitrogen dioxide production during combustion?

20-23 Both stationary source and mobile source combustions are important sources of NO_x pollution, but only mobile source combustion contributes to CO pollution. Explain.

20-24 What is the NAAQS for NO_2 and what are approximate average levels of NO_2 in the air at the present time?

20-25 What is the chemistry associated with the reduction of NO levels in automobile exhaust?

20-26 Discuss the major natural sources of nitrogen oxide emissions.

20-27 What natural process exists for removing NO_x pollution from the atmosphere?

VOLATILE ORGANIC COMPOUNDS

20-28 What are the major sources of anthropogenic VOC emissions?

20-29 An NAAQS for volatile organic compounds was eliminated in 1983. Why?

OZONE AND PHOTOCHEMICAL OXIDANTS

20-30 What is the difference between a primary air pollutant and a secondary air pollutant?

20-31 What are the three reactions that occur in the NO_2 photolytic cycle?

20-32 Draw a diagram that shows the cyclic nature of the NO_2 photolytic cycle.

20-33 What is the relationship between photochemical smog and the NO_2 photolytic cycle?

20-34 What is a photochemical oxidant?

20-35 Describe the chemistry involved in the production of photochemical smog.

20-36 What are PANs?

20-37 What is the NAAQS for ozone and what are approximate average levels of ozone in the air at the present time?

THE OZONE LAYER AND STRATOSPHERIC POLLUTION

20-38 What is the ozone layer and what are the maximum ozone concentrations in this layer?

20-39 What is the mechanism for the formation of ozone in the ozone layer?

20-40 What are CFCs and what is the relationship between them and the ozone layer?

20-41 Is the amount of ozone in the ozone layer decreasing? Explain your answer.

SULFUR OXIDES

20-42 What does the notation SO_x stand for?

20-43 What is the relationship between SO_2 and SO_3 formation in the combustion of sulfur-containing fuels?

20-44 Explain why more SO_2 than SO_3 is discharged into the atmosphere from anthropogenic sources.

20-45 Explain why stationary source fuel combustion is the dominant source of SO_x pollution.

20-46 Discuss the chemistry involved in the control of SO_x pollution using limestone.

20-47 What is the NAAQS for SO_2 and what are approximate average levels of SO_2 in the air at the present time?

20-48 Discuss the major natural sources of SO_x pollution and how they compare in magnitude to anthropogenic sources.

20-49 What are the adverse effects of SO_2 on plants and human beings?

ACID RAIN

20-50 What is acid rain?

20-51 What two types of pollutants are the cause of acid rain?

20-52 Discuss the effects of acid rain on the environment.

20-53 Write a chemical equation for the reaction for the deterioration of limestone and marble structures by acid rain.

PARTICULATES

20-54 What are particulates?

20-55 What are the reactants for secondary particulate pollution?

20-56 Compare the magnitudes of primary particulate pollution and secondary particulate pollution.

20-57 What is the difference between the following sorption processes: adsorption, chemisorption, and absorption?

20-58 What are the three ways in which particulates that enter and remain in the lungs can exert toxic effects?

20-59 What is the NAAQS for particulates and what are approximate average levels of particulates in the air now?

20-60 What are the two ways in which particulates can affect climate?

20-61 What is the difference between rainout and washout?

TEMPERATURE INVERSIONS

20-62 What is a temperature inversion?

20-63 What are the pollution effects resulting from temperature inversions?

THE GREENHOUSE EFFECT

20-64 What is the potential problem with the continuous increase in carbon dioxide concentrations in the atmosphere?

20-65 What is the origin of the greenhouse effect?

20-66 Is the average temperature of the earth's surface increasing? Explain your answer.

20-67 What are the physical ramifications of global warming?

20-68 What gases besides carbon dioxide are greenhouse gases?

20-69 How do the possible effects of other greenhouse gases compare with those of carbon dioxide?

Water Pollution

CHAPTER HIGHLIGHTS

21.1 Types of Water Pollutants

Water is considered polluted when it contains any substance in sufficient amount to prevent it from being used for purposes regarded as normal. The water pollution problem is complex, partly because the normal uses of water are so varied. Normal uses include recreation, public water supply, habitat for fish and other aquatic life including some animals, agriculture, and industry. Water that is suited for some uses, and therefore is unpolluted, may have to be considered polluted for other uses.

Many different "substances" and "conditions" can cause water pollution. In this chapter we discuss eight of the more commonly reported categories of water pollutants and their effects.

1. Oxygen-demanding wastes
2. Disease-causing agents
3. Plant nutrients
4. Suspended solids (turbidity)
5. Dissolved solids (salinity)
6. Acids (pH)
7. Heat (thermal pollution)
8. Toxic substances

21.2 Oxygen-Demanding Wastes

Dissolved oxygen is a fundamental requirement for the plant and animal population in any given body of water. Their survival depends on the water's ability to maintain certain minimal concentrations of this vital substance. Fish require the highest levels, invertebrates lower levels, and bacteria the least. For a diversified warm-water biota, including game fish, the dissolved-oxygen (DO) concentrations should be at least 5 ppm. For a cold-water biota, DO concentrations at or near saturation values are desirable. The minimum level should be no lower than 6 ppm. The amount of DO at saturation varies with water temperature and altitude, as is shown in Figure 21.1. High mountain lakes may contain 20 to 40% less DO than similar lakes at sea level.

A body of water is classified as polluted when the DO concentration drops below the level necessary to sustain a normal biota for that water. The primary

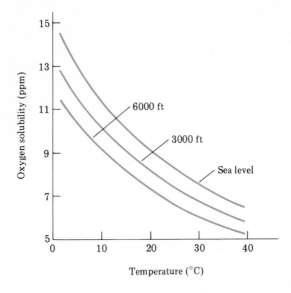

FIGURE 21.1

Solubility of oxygen in water at various temperatures and altitudes.

cause of water deoxygenation is the presence of substances collectively called oxygen-demanding wastes. **Oxygen-demanding wastes** *are substances that are easily broken down or decayed by bacterial activity in the presence of oxygen.* The presence of such substances in water quickly depletes dissolved oxygen.

Most oxygen-demanding wastes are organic compounds that come from plants, animals, and humans. Typical sources include (1) sewage, both domestic and animal, (2) industrial wastes from food-processing plants, papermills, and tanneries, and (3) effluent from slaughterhouses and meat-packing plants. The effects of these wastes are a function of the amount of water available for dilution as they enter a body of water. For this reason, it is not surprising to find that low-DO problems are especially common in late summer and early fall, when water levels are normally low.

The carbon atoms in oxygen-demanding wastes are oxidized, with bacterial help, by the dissolved oxygen to carbon dioxide.

$$C + O_2 \longrightarrow CO_2$$

In this reaction, 32 g of oxygen (the molecular mass of O_2 is 32 amu) is required to oxidize 12 g of carbon (the atomic mass of C is 12 amu). The carbon can thus be thought of as demanding nearly three times its mass in oxygen. On this basis, 9 ppm of oxygen is needed to react with approximately 3 ppm of dissolved carbon. This amounts to a reaction between the dissolved oxygen from a gallon of water and a small drop of oil. It becomes easy to see how waters can quickly be depleted of dissolved oxygen.

Because oxygen-demanding wastes rapidly deplete the DO, it is important to

TABLE 21.1

BOD Values for Various Types of Oxygen-Demanding Wastes

Source	BOD Range (ppm)
Untreated municipal sewage	100–400
Runoff from barnyards and feed lots	100–10,000
Food-processing wastes	100–10,000

be able to estimate the amount of oxygen-demanding wastes in a given body of water—the *biological oxygen demand* (BOD). The **biological oxygen demand** *is the amount of oxygen required to decompose all of the biodegradable organic wastes in a given amount of water.* It is measured by incubating a sample of water for 5 days at 20°C.

A BOD of 1 ppm is characteristic of nearly pure water. Water is considered to be of doubtful purity when the BOD value reaches 5 ppm. Public health authorities object to runoffs entering streams if the BOD of the runoff exceeds 20 ppm. Table 21.1 gives the BOD levels of various types of oxygen-demanding wastes. Obviously, these wastes must be highly diluted when they enter the water or the DO is rapidly and completely depleted.

The disappearance of plant and animal life, whether killed or forced to migrate, is an obvious result of the oxygen depletion of water. A less obvious but important result is a shift in water conditions from those favoring aerobic activity (oxygen required) to those that support anaerobic activity (oxygen not needed). This occurs when the oxygen levels become so low that the aerobic microorganisms are destroyed or driven away and replaced by anaerobic ones. The products of decomposition following these two pathways are quite different, as is shown in Table 21.2

FOUL-SMELLING WATERS

Methane (CH_4) is odorless and flammable; ammonia (NH_3) has a very irritating odor, and amines have a fishy smell; hydrogen sulfide (H_2S) is bad smelling and toxic; and some phosphorus compounds have unpleasant odors. The shift from aerobic to anaerobic conditions of decomposition resulting from oxygen depletion is not a favorable one for humans who like to breathe fresh air or for other species of life. The unpleasant odors associated with such a body of water leave no doubt as to what has occurred.

TABLE 21.2

Comparison of Decomposition End-Products Under Different Conditions

Aerobic Conditions	Anaerobic Conditions
C $\longrightarrow$ CO_2	C $\longrightarrow$ CH_4
N $\longrightarrow$ $NH_3 + HNO_3$	N $\longrightarrow$ NH_3 + amines
S $\longrightarrow$ H_2SO_4	S $\longrightarrow$ H_2S
P $\longrightarrow$ H_3PO_4	P $\longrightarrow$ PH_3 and phosphorus compounds

21.3   Disease-Causing Agents

Water is a potential carrier of pathogenic microorganisms that can endanger health and life. The pathogens most frequently transmitted through water are those responsible for infections of the intestinal tract (typhoid and paratyphoid fevers, dysentery, and cholera) and those responsible for polio and infectious hepatitis. Historically, the prevention of waterborne diseases was the primary purpose of water treatment. Modern disinfection techniques have greatly reduced this danger in the United States. This is not true for a large part of the world, where, for example, cholera epidemics still occur. The fact that such disease-causing agents are under control must not result in a sense of false security. The occurrence of a polluted water supply leading to an outbreak of disease is always a possibility. The responsible organisms are present in the feces or urine of infected people, and these organisms are ultimately discharged into a water supply.

Although it might seem desirable, a direct check for these organisms, is not routinely performed on water supplies. In developed countries indirect methods are used instead, for the following reasons. Pathogens are likely to gain entrance into water only sporadically, and once in the water they do not survive for long periods of time. Consequently, their presence could easily be missed by routine sampling. Laboratory procedures are likely to fail to detect pathogens that are present in very small numbers. Also, it takes 24 hours or longer to obtain results from a laboratory examination. If pathogens were found in a water sample, it is likely that many people would already have used the water and be subject to infection.

An indicator organism, a coliform bacterium, is the basis of the indirect method commonly used. Such benign organisms live in the large intestine of humans. They cause no diseases and are always present in feces. Their presence in water is an indication of fecal discharge into the water. These organisms are present in large numbers, making their detection quite easy. Billions of coliform bacteria are excreted daily by an average person.

These natural inhabitants of the human bowel do not find environmental conditions in natural waters suitable for multiplication and they begin to die rapidly. Their presence in water samples therefore permits a rough diagnosis of the time elapsed since fecal contamination took place. If fecal contamination is recent, it can be assumed that pathogenic organisms may be present along with the harmless coliform bacteria. The absence of coliform bacteria implies that recent intestinal discharges are not present in the water, and presumably the water is free of pathogens. Coliform bacteria count is widely used as a measure of the "swimmability" of a body of water.

"SWIMMABILITY" OF
NATURAL WATERS

21.4   Plant Nutrients

The principal adverse effect of plant nutrients on water quality is excessive growth of aquatic plants. The ramifications of such unwanted plant growth are widespread. Murky water, floating algae, dense mats of rooted and floating aquatic plants, and depletion of dissolved oxygen (Section 21.2) associated with decaying plant material are some of the problems. Recreational use of such water is usually hampered. Treatment costs for municipal and industrial users of the water often increase.

Excessive plant growth problems are more severe in lakes, reservoirs, and estuaries fed by rivers than in the rivers themselves, where velocities of flow reduce the adverse effects.

The term *eutrophication* is frequently used in referring to water pollution caused by excessive plant growth. In a pollution context **eutrophication** *is the enrichment of a body of water with plant nutrients such that plant growth makes the water unfit for human purposes.* The term eutrophication comes from two Greek words meaning "well nourished."

Nutrients are an important limiting factor in the growth of all plants. All other factors being equal, the rate and profuseness of plant growth are proportional to the amount of nutrient available.

Many elements are needed for plant growth, and most are available to plants in excess of their needs. But two of the "fertilizer elements"—nitrogen and phosphorus (Sections 12.2 and 12.4)—are usually present in amounts very close to the minimum required and therefore used by plants almost to the point of exhaustion. The growth of a plant ceases when the least available of these two elements has been depleted. Eutrophication pollution problems are caused by anthropogenic sources increasing the amounts of these growth-limiting elements.

The most important sources of human-generated phosphorus and nitrogen are domestic sewage (phosphate-containing detergents and nitrogen-containing body wastes), runoff from agricultural land (fertilizers containing both nitrogen and phosphorus), runoff from livestock areas (animal wastes containing nitrogen), and various industrial wastes.

Natural sources also exist for these "fertilizer" elements. The dissolution of phosphate-containing rocks (abundant in some parts of the country) is an additional source of phosphorus-containing waters. Nitrogen fixed by bacteria, algae, and lightning, as well as atmospheric nitrogen-containing compounds (acid rain), are additional sources of nitrogen that enters the water supply.

A major emphasis of eutrophication control programs involves limiting the amount of phosphorus that enters the environment. This has been done for a number of reasons.

1. Phosphorus sources are not as scattered as those of nitrogen. Domestic sewage and certain industrial processes account for a large percentage of the total phosphorus discharge. Point sources of this type are much easier to control than scattered sources such as farmland runoff.
2. Phosphorus in water is almost entirely in the form of phosphates and polyphosphates. The phosphates are easily precipitated and removed by a variety of wastewater treatment processes. Nitrogen is usually found in the form of very soluble nitrates that are extremely difficult to remove.
3. Gaseous atmospheric sources exist for nitrogen. Such sources are impossible to control. No sources of atmospheric phosphorus are known.

The phosphate content of household detergents has been a prime target in eutrophication control procedures. The chemical formulation of a detergent involves numerous compounds. The two main ingredients are a surfactant and a builder. The *surfactant* functions as a wetting agent. It increases the ability of water to dissolve nonpolar substances such as greases and oils (Section 24.10). The most common surfactants used are linear alkyl sulfonates. The structure of a representative linear alkyl sulfonate is as follows.

$$CH_3-CH_2-CH_2-CH_2-CH_2-CH_2-CH_2-CH_2-CH_2-CH_2-CH-\bigcirc-SO_3^-\ Na^+$$
$$\underset{CH_3}{|}$$

The primary role of the *builder* in a detergent formulation is to act as a sequestering agent, tying up hard-water ions in the form of large water-soluble ions. Builders also make wash water alkaline; the alkalinity aids in the effective removal of dirt. The most commonly used builders are polyphosphates, such as sodium tripolyphosphate ($Na_5P_3O_{10}$). The active sequestering agent, the triphosphate ion, has the following chemical structure.

$$\left[\begin{array}{ccc} & O & O & O \\ & \| & \| & \| \\ O-&P-O-&P-O-&P-O \\ & | & | & | \\ & O & O & O \end{array}\right]^{5-}$$

Polyphosphates are ideal builders because they are readily broken down into simple phosphates, which are nontoxic to aquatic life and pose no health hazards to humans. In addition, polyphosphates are both safe to use on colors, fibers, and fabrics and noncorrosive to metals (e.g., washing machines). Their only drawback is that they happen to be plant nutrients.

NONPHOSPHATE
DETERGENTS

Many states now restrict the amount of phosphate builders in detergents. The limit is usually set at 0.5% by mass total phosphorus in the detergent for-

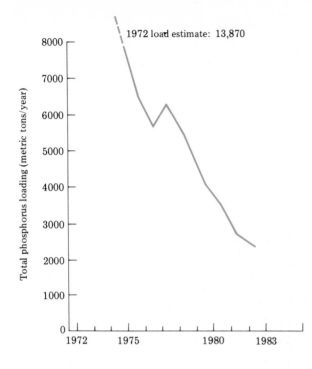

FIGURE 21.2

Municipal phosphorus loads discharged into Lake Erie, from the United States side, on an annual basis for the period 1972–1983.

mulation. Nonphosphorus-containing builders, with less ideal properties than the polyphosphates, have totally replaced the polyphosphates in many detergents formulations.

Laws also now require the reduction of phosphorus in municipal sewage effluent to a concentration of 1.0 mg/L in many communities. The combination of phosphate reduction in detergents and phosphorus control in sewage effluent has markedly reduced the amount of phosphorus entering water bodies such as lakes. Figure 21.2 shows the municipal phosphorus load entering Lake Erie for the period 1972–1983. A sevenfold decrease in phosphorus load has been accomplished, and this effort is reflected in the improvement of the water quality for this lake.

21.5 Suspended Solids

Suspended solids is a general term for undissolved organic and inorganic particulate matter in water. Such suspended solids cause turbidity problems while in the suspended state and sedimentation problems when they leave the suspended state. The term *sediments* is also used to describe suspended solids.

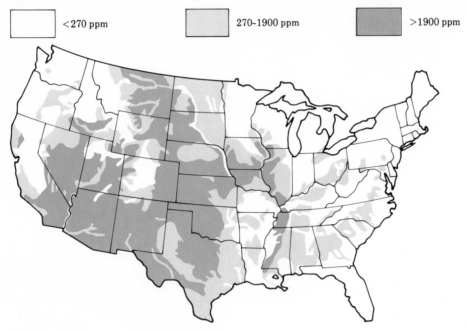

| | <270 ppm | | 270–1900 ppm | | >1900 ppm |

FIGURE 21.3
Suspended solid concentrations in streams.

Suspended solids are almost taken for granted because of the naturally occurring process of erosion. Erosion is a greater problem in some parts of the United States than in others. This variation is not all a result of human activities; also involved are the type of soils, geology, topography, precipitation, and vegetation cover. As can be seen in Figure 21.3, the Southwest and Midwest have the greatest erosion problems.

Human activities have an enormous effect on erosion rates. Erosion rates of land are increased four to nine times by agricultural development and may be increased by a factor of 100 as a result of construction activities. Coal strip-mining activities greatly influence the rate of erosion in an area. The amount of sediment washed from an area depends very much on the land's ground cover. This is indicated by the data in Figure 21.4. These data are based on 2.4 inches of rain in 1 hour.

Turbid water interferes with recreational use and aesthetic enjoyment of water. Turbid waters can be dangerous for swimming, especially if diving facilities are provided, because of the possibility of unseen submerged hazards. Turbid water also reduces light penetration into water. Reduced light penetration means a reduced rate of photosynthesis by plants, which decreases production of the oxygen that is needed for normal stream balance. Turbid water is more expensive to treat to make it fit for drinking. Sediment causes serious abrasion and wear as it passes through industrial pumps and turbines.

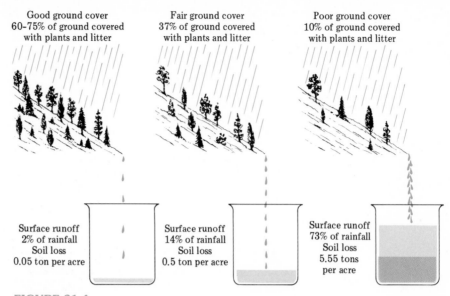

FIGURE 21.4

Effects of ground cover on erosion.

Sedimentation causes channels to overflow more easily, alters flow rates and depths of channels, and reduces the useful life of reservoirs. Expensive dredging is required to counteract these effects. Fish and shellfish populations are affected by sedimentation. Sediment settling on the bottom disrupts fish habitats, reduces the abundance of food available to the fish, and interferes with the development of fish eggs. Toxicants (pesticides, radionuclides, and toxic metals) adsorbed onto particulate matter and deposited in bottom sediments, may have adverse effects. Because most potentially toxic trace materials adhere strongly to sediment particles, concentrations of contaminants and trace materials in bottom sediments may range from 10 to 1000 times higher than concentrations of the same substances in the overlying water.

21.6 Dissolved Solids

Dissolved solids generally consist of inorganic salts and small amounts of organic matter. The major inorganic components of dissolved solids found in rivers are sodium, potassium, calcium, magnesium, carbonate, hydrogen carbonate, chloride, and sulfate ions. Water containing large amounts of these ions is called saline water. Thus, increased water salinity is the major problem associated with dissolved solids in water. Increased salinity can restrict water use for many purposes.

TABLE 21.3

Suggested Guidelines for Salinity in Irrigation Water

	Total Dissolved Inorganic Solids (mg/L)
Water for which no detrimental effects are usually noticed	<500
Water that can have detrimental effects on sensitive crops	500–1000
Water that may have adverse effects on many crops; requires careful management practices	1000–2000
Water that can be used only for salt-tolerant plants on permeable soils with careful management practices	2000–5000

The main sources of increased salinity in fresh waters are the dissolution of rock and soil, atmospheric deposition, and human activities. Human activities contribute dissolved solids through the discharge of wastewater from such point sources as municipal and industrial waste treatment plants and through runoff and drainage from such nonpoint sources as agricultural and urban areas. One of the most important sources of dissolved solids is irrigation return flow to streams by direct surface runoff or by subsurface drainage.

High water salinity levels cause problems in addition to rendering water unfit for drinking. Dissolved inorganic and mineral substances exert adverse effects on aquatic animal and plant life and cause many irrigation problems in the agricultural industry. Damage to aquatic life is primarily related to the osmosis process (Section 9.6), assuming the dissolved substances are nontoxic. Generally, the concentration of dissolved materials in body fluids is the maximum that an aquatic organism can tolerate. When these organisms are in contact with water containing higher concentrations, there is a tendency for water to move out of the cells (osmosis) of the organism into the surrounding waters. The resulting increase in concentration within the cells of the organism can lead to death. Many freshwater species disappear when waters become brackish.

REUSE OF IRRIGATION WATER

One of the more serious increased salinity problems involves the use and reuse of water in irrigation. It is estimated that about 25% of the irrigated land of the United States is now affected to some degree by water salinity. Irrigation water brought onto a field always contains some dissolved salts. Plants extract water from the irrigated field, but most of the dissolved salts are excluded by the roots. Water that evaporates from the soild surface, a major process during hot weather, also leaves dissolved salts behind. These two processes, water uptake by plants and evaporation, cause residual salts to accumulate in the soil.

To preserve the salt balance of the soil and avoid damage to crops, the excess salt accumulation must be leached from the soil with excess irrigaton water. Thus, drainage water from the soil contains increased concentrations of salts, which are carried back to the general water supply. Irrigation does not actually produce a pollutant in the form of dissolved salts but merely returns the salts to the general water supply in a more concentrated form.

Average total dissolved solid (TDS) concentrations in the Colorado River range from about 50 mg/L in the headwater areas of the Colorado Rockies to about 800 mg/L at Imperial Dam in California, near the Mexican border. The contributions of salinity to the river are estimated as 47% from natural sources, 37% from irrigation, and 16% from other sources including municipal and industrial consumption.

Table 21.3 gives general crop responses to various concentrations of total dissolved inorganic solids in irrigation water. To put these figures into perspective, it is worth noting that the recommended TDS level for drinking water is less than 200 mg/L and the permissible maximum is 500 mg/L.

Plants differ markedly in their tolerance to water salinity. Figure 21.5 shows the salt tolerances of various vegetable crops. Note that the most sensitive crops show decreases in growth and yield at 1000 mg/L (1 g/L) TDS.

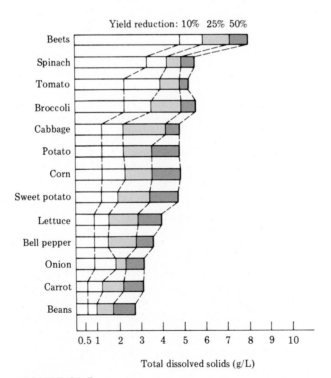

FIGURE 21.5

Response of various vegetable crops to increasing water salinity. Crops are ranked in order of decreasing salt tolerance. Crosslines are placed at 10, 25, and 50% yield reductions.

In geographical areas with long and severe winters, tons of deicing salt (NaCl and $CaCl_2$) are used on roads and highways. In cities the resulting salt solutions flow into sewers and are then discharged into rivers or other receiving bodies of water. Outside the cities, most of the salt is absorbed into the soil adjacent to the road. The effects of deicing salt on public water supplies has not been studied in detail. The limited information available indicates that the effects are minimal in large cities, as a result of dilution from the large quantities of water used and discharged into sewer systems. Some problems have been found in rural areas that obtain culinary water from wells located near roads. Salt deposited in roadside soil percolates through the soil and enters the well water. The resulting increases in salt concentration do not render the water unpalatable, but can make it unsuitable for use by individuals who must limit their intake of salt for health reasons.

21.7   Acids

The most important sources of increased acidity in natural waters are acid mine drainage and acidic rainfall. The topic of acidic rainfall (acid rain) was considered in Section 20.9. Acid mine drainage is a phenomenon associated with sulfur-bearing deposits. It can originate from the mining of sulfur-bearing lead, zinc, and copper ores, but it is associated primarily with coal deposits that contain varying amounts of iron sulfide (pyrite).

The actual pollutants present in mine drainage are sulfuric acid (H_2SO_4) and iron compounds. These substances are formed as the result of a reaction between air, water, and pyrite (FeS_2) present in coal seams. Certain types of bacteria are also involved in the reaction. This reaction can take place in both underground and surface mines. The equation for the overall reaction is

$$2\,FeS_2 + 7\,O_2 + 2\,H_2O \longrightarrow 2\,FeSO_4 + 2\,H_2SO_4$$

During coal-mining operations in underground mines, the strata between the coal seam and the surface are invariably disturbed. Fissures appear through which water drains into the mine from many surface areas. This water, after reaction with pyrite, is eventually discharged into surface streams, either naturally or through anthropogenic processes. Acid mine drainage is formed in surface coal mines (strip-mining operations) if the surface runoff water comes in contact with pyrite-containing coal.

The drainage water becomes still more acidic when small amounts of Fe^{2+} are oxidized to Fe^{3+} and additional sulfuric acid is produced.

$$4\,FeSO_4 + 10\,H_2O + O_2 \longrightarrow 4\,Fe(OH)_3 + 4\,H_2SO_4$$

As acid mine drainage flows over mineral deposits and rocks, the sulfuric acid within it dissolves calcium and magnesium compounds that may be present. A typical reaction is

$$CaCO_3 + H_2SO_4 \longrightarrow CaSO_4 + H_2O + CO_2$$

Insoluble   Soluble

Although such reactions neutralize some of the acid, they increase the hardness of natural waters by increasing the concentration of hard water ions (Ca^{2+} and Mg^{2+}). The H_2SO_4 can also dissolve other metals, with the result that a large variety of toxic metals may enter the water.

At the present time, much effort is directed toward either preventing the formation of acid mine drainage or removing the acid by chemical treatment before releasing the discharge into natural waters. Three methods are commonly used.

1. *Sealing of abandoned mines.* Sealing abandoned mines to keep out air or water helps to prevent the fundamental reactions by eliminating at least one of the required reactants. The inability to attain tight seals has been a problem with this method.
2. *Drainage control.* Attempts have been made to minimize the contact time between water and pyrite by quickly removing water from mines. It has been difficult to gather the water from many sources available to an underground mine.
3. *Chemical treatment.* This method is being used in many active mines. The mine discharge is sent to a nearby treatment plant where hydrated lime is added. This is followed by aeration of the water. The water is then placed in large lagoons, where the sludge, created by the process, settles to the bottom and a clear overflow is discharged into the natural waters.

Because of the presence in solution of carbonate ($CO_3{}^{2-}$) and bicarbonate ($HCO_3{}^-$) ions, the pH of most productive, fresh natural waters is between 6.5 and 8.5. These two ions also form a system that can neutralize (buffer) small amounts of acid (or base) that enter natural waters, thus preventing pH change. Large influxes of acid, however, can overwhelm this buffering ability and cause drastic drops in pH values. The effects of these changes depend on the magnitude of the pH drop involved. Some of these effects are as follows.

1. *Destruction of aquatic life.* At pH levels below 4.0, all vertebrates, most invertebrates, and many microorganisms are destroyed. Most higher plants are eliminated, leaving only a few algae and bacteria. Acid mine drainage is one of the primary causes of fish kills in the United States. Excessive precipitation increases the mine drainage output and compounds the problem.

2. *Corrosion.*   Water with a pH lower than 6.0 can cause excessive corrosion of plumbing systems, boats, piers, and related structures.
3. *Agricultural crop damage.*   Acidity and alkalinity of irrigation water are
   usually of little consequence over a pH range of 4.5 to 9.0 because the
   soil is a system that can accommodate such varability in pH. Problems
   may occur if the pH drops below 4.5. Such acidic water increases the
   solubility of such substances as Fe, Al, and Mg salts. These ions, at the
   resulting high concentrations, are sometimes toxic to plants.

One mechanism by which fish and other aquatic animals are destroyed by
acid pollution involves the equilibrium between CO_2, HCO_3^-, and CO_3^{2-}. At
low pH, HCO_3^- and CO_3^{2-} are both converted to CO_2, which dissolves in
water.

$$HCO_3^- + H^+ \longrightarrow H_2O + CO_2$$

$$CO_3^{2-} + 2\,H^+ \longrightarrow H_2O + CO_2$$

This excessive "free" CO_2 in the water interferes with the processes by which
CO_2 is eliminated from aquatic organisms. As a result of metabolic activity, CO_2
is produced in the cells and moves by way of the blood to organs (gills), where it
leaves the blood and diffuses into the water. An increase in external CO_2 slows
the diffusion rate from the blood. The result is an accumulation of CO_2 in the
blood of the organism, with the result that less oxygen can be carried by the
blood and the pH of the blood goes down. These conditions can eventually lead
to the death of the organism.

21.8 Heat

Heat is not ordinarily thought of as a pollutant, at least not in the same sense
as a toxic chemical. However, the addition of excess heat energy to a body of
water—called *thermal pollution*—causes numerous adverse effects. Thermal
pollution originates primarily from the use of water as a coolant in industrial
processes. Most water used for this purpose is returned, with heat added, to the
original sources. Approximately 70% of the water diverted to industrial use
serves as a cooling medium.

The most obvious result of thermal pollution is to make the water less efficient
for further cooling applications. Far more important, however, are the biological
and biochemical implications.

Used coolant water frequently has a temperature 10 to 20°C higher than the
river or stream to which it is returned. This added heat raises the temperature of
the natural waters. The possible effects of this are (1) a decrease in the amount

of dissolved oxygen in the water, (2) an increase in the rates of chemical reactions, (3) false temperature cues to aquatic life, and (4) lethal temperature levels.

The solubility of oxygen in water decreases with increasing temperature, as shown in Figure 21.1. Some of the effects of low DO levels on bodies of water were discussed in Section 21.1 when oxygen-demanding wastes were considered.

The addition of heated water to a cooler body of water, unless it is fast-moving, may accelerate the lowering of DO levels because of density differences between the two. The less dense warm water tends to form a layer on top of the cooler, denser water. This takes place particularly when the body of cool receiving water is deep. The resulting blanket of "hot" water cannot dissolve as much atmospheric oxygen as the underlying cold water, which is denied contact with the atmosphere. Normal biological reduction of the DO level of the atmospherically unreplenished lower layer may lead to anaerobic conditions.

Another effect of this stratification may show up downstream from a dam. When the oxygen-deficient lower level is discharged through the lower gates of a dam, there may be serious effects on downstream fish life. Also, the ability of the stream below the dam to assimilate oxygen-demanding wastes will be curtailed.

The effects of heat in water sometimes show up in nature without human interference. On hot summer days the temperature of shallow waters sometimes reaches a point that critically reduces the DO level; under these conditions, suffocated fish are often found on the surface.

A rough rule of thumb often used by chemists is that the rate of any chemical reaction, including those of respiration, approximately doubles with every 10°C increase in temperature (Section 10.2). In thermally polluted water, fish require more oxygen because of an increased respiration rate. However, the available oxygen in such water has been decreased. Thus, thermal pollution has a double-barreled effect on fish.

Other reactions are also influenced. Trout eggs hatch in 165 days when incubated at 3°C (a normal situation). When water temperatures are 12°C, only 32 days are required, and no hatching occurs at water temperatures in excess of 15°C. Such a result can be disastrous to fish populations. If the fish hatch early and find no natural food organisms available, they do not survive. The natural food of such hatchlings depends on a food chain originating with plants whose abundance is a function of day length as well as temperature.

The life cycle and natural processes of many aquatic organisms are closely and delicately geared to water temperature. Fish often migrate, spawn, and are otherwise distributed in response to water temperature cues. Shellfish such as oysters spawn within a few hours after their environment reaches a critical temperature. These normal life patterns of aquatic organisms can be completely disrupted by artificial changes in water temperatures.

There are several solutions to the thermal pollution problem. Various industries, including nuclear power plants (Section 15.11), have incorporated cooling

towers into their operations to remove heat from cooling water before return-
ing it to the natural water supply. The use of cooling ponds or lakes is another
alternative. These ponds could serve as ice-free wintering areas for waterfowl
in northern locations or as a means of extending the range (northward) of
certain fish.

21.9   Toxic Substances

Toxic water pollutants are generally defined as substances that, by themselves or
in combination with other chemicals, are harmful to animal life or human health.
All water pollutants in some sense are toxic—that is, harmful—in excessive
concentrations. For example, untreated sewage and animal wastes threaten
human health because they carry potentially harmful bacteria and viruses, and
sedimentation from agriculture or construction can smother, and therefore
destroy, aquatic life. Toxic water pollutants, as a category of pollutants, differ
from these substances in that they may have impact on human health or aquatic
environment at *relatively low* concentrations.

Largely through advances in analytical techniques and improvements in
detection instruments, scientists are now aware of low levels of many "toxic
chemicals" in all bodies of water. These chemicals include some of the metals,
pesticides, and other synthetic organic compounds. Concentrations of these
substances range from parts per million down to fractions of a part per trillion.

Table 21.4 shows known environmental effects for 15 of the most studied
"toxics." We should note that in most cases the known effects are caused by
toxic concentrations much greater than the low level concentrations found in
most waters. The presence of trace amounts of these substances in water,
however, is a warning of potential problems if steps are not taken to eliminate
anthropogenic sources of these substances that enter water supplies. The
long-term human health effects of most of these toxic substances, at the low
levels now present in waters, are not known. This basic lack of knowledge makes
it extremely difficult for responsible agencies to establish control requirements.

BIOACCUMULATION OF
TOXICS

A major concern today about toxics such as those listed in Table 21.4 is that
some of these substances have been shown to bioaccumulate through the
aquatic food chain. In this way concentration levels could build up to the point of
concern about human consumption of such aquatic life. It is bioaccumulation
that led to the banning of DDT and other structurally similar insecticides in the
mid-1960s. Figure 21.6 shows a typical bioaccumulation pattern for DDT.

One of the most common ways of monitoring toxic pollutants in water is to
monitor bioaccumulation in fish. Large fish can accumulate easily detectable
levels of such toxics in their tissue. The FDA, EPA, and many states have

TABLE 21.4

Known Environmental Effects of Common Toxic Substances Found in Bodies of Water

Chemical	Type of Substance	Environmental Effects
Aldrin	pesticide	toxic to aquatic organisms, reproductive failure in birds and fish, bioaccumulation in aquatic organisms
Arsenic	nonmetal	toxic to legume crops
Benzene	SOC*	toxic to some fish and aquatic invertebrates
Cadmium	metal	toxic to fish, bioaccumulates significantly in bivalve mollusks
Chromium	metal	toxic to some aquatic organisms
Copper	metal	toxic to juvenile fish and other aquatic organisms
DDT	pesticide	reproductive failure in birds and fish, bioaccumulates in aquatic organisms
Dibutyl-phthalate	SOC	eggshell thinning in birds, toxic to some fish
Dioxin	SOC	bioaccumulates, lethal to aquatic organisms, birds and mammals
Lead	metal	toxic to domestic plants and animals, biomagnifies to some degree in food chain
Methyl mercury	metal compound	reproductive failure in fish species, inhibits growth and kills fish; biomagnifies
PCBs	SOC	liver damage in mammals, kidney damage and eggshell thinning in birds, suspected reproductive failure in fish
Phenols	SOC	reproductive effects in aquatic organisms, toxic to fish
Toxaphene	pesticide	decreased productivity of phytoplankton communities, birth defects in fish and birds, toxic to fish and invertebrates

* Synthetic organic compound.

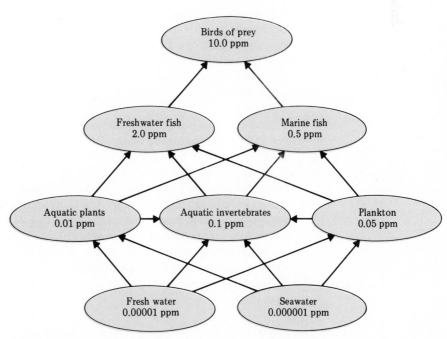

FIGURE 21.6

A food-chain bioaccumulation pattern for the insecticide DDT.

TABLE 21.5

Average Concentrations of Heavy Metals in Street
Sweepings and Shale
(*milligrams per kilogram*)

Heavy Metal	Street Sweepings	Average Composition of Shale
Cadmium (Cd)	3.4	0.3
Chromium (Cr)	211	100
Copper (Cu)	104	57
Iron (Fe)	22,000	47,000
Lead (Pb)	1,810	20
Manganese (Mn)	418	850
Nickel (Ni)	35	95
Zinc (Zn)	370	80

established "action" levels—concentrations of specific chemicals in fish tissue at which there is potential for harm to humans. One common response to the detection of toxics at "action" levels is a fish consumption advisory or ban for species obtained from the affected waterway.

Toxics can come from a variety of disparate sources, which makes their control extremely difficult. Mining is often a source of metals. Certain substances, such as arsenic and mercury, occur naturally in the strata of some geologic formations and may make their way into surface water and groundwaters. Sediments may hold toxics that were discharged in the past; long after the discharge has ceased, a toxic can be released from sediments to the water column and to aquatic organisms. A change in the pH of water activates sediments in some cases. Pesticides and herbicides used in farming often enter water bodies through runoff. In addition, municipal facilities, especially those that receive industrial waste that has not been pretreated, can be a significant source of toxic substances. Landfills that are not properly sited, lined, or controlled can contribute many varieties of toxic pollutants, as can spills from trucks, ships, and storage facilities.

Urban runoff is a common source of metals, particularly lead (from the deposition of auto emissions), entering surface waters. Concentrations of some heavy metals can be higher in street sweepings than in naturally occurring soils, rocks, and sediments. As shown in Table 21.5, higher than average concentrations of cadmium, chromium, copper, lead, and zinc can occur in street sweepings than in shale (a sedimentary rock used here to represent the composition of sediments deposited in the absence of human influences). All of the listed metals are used in common industrial processes or in domestic materials.

TABLE 21.6

Average Concentrations of Heavy Metals in Urban
Runoff Compared with Water-Quality Criteria for
Public Water Supplies
(*micrograms per liter*)

Heavy Metal	Urban Runoff	Criteria for Public Water Supplies
Cadmium (Cd)	18	10
Chromium (Cr)	33	50
Copper (Cu)	45	1000
Lead (Pb)	235	50
Nickel (Ni)	24	—
Zinc (Zn)	236	5000

Average concentrations of cadmium and lead in urban runoff exceed criteria
for public water supplies, as shown in Table 21.6. Concentrations of all heavy
metals listed in Table 21.6 far exceed criteria for protection of marine and
freshwater aquatic life.

21.10   Wastewater and Its Treatment

Wastewater *is water used by a community which is discharged into natural
bodies of water.* Such water is a major pollution source for natural waters
(oceans, lakes, rivers, etc.). Consequently, the cleaning of wastewater effluents is
an important aspect of present water pollution control efforts.

Wastewater has four important sources: (1) sewage, (2) commercial wastes,
(3) industrial wastes, and (4) storm runoff. The term *sewage* includes not only
human waste but also everything else that makes its way from various drains,
sinks, and washing machines in the home to sewer systems. It is estimated, on
average, that each individual in the United States contributes approximately
100 gal of wastewater per day to a community's sewage flow. Commercial
wastes from office buildings include human wastes and water from cleaning
and other processes. Industrial wastes usually involve large volumes of water
used in the manufacture of industrial products.

Wastewater averages 99.94% water by mass and 0.06% by mass dissolved or
suspended materials. The dissolved and suspended materials of most concern,
from a water pollution standpoint, are (1) oxygen-demanding wastes (Section
21.2), disease-causing agents (Section 21.3), and plant nutrients (Section 21.4).

Wastewater treatment is generally classified into three stages or steps: primary, secondary, and tertiary. We now consider each of these stages of wastewater treatment.

Primary Treatment Processes

Primary wastewater treatment *consists of processes involving filtering and sedimentation.* The solids are allowed to settle, and any floating scum is removed. The process normally consists of a series of steps.

1. *Screening.* Large floating objects are removed by passing the wastewater through screens. The debris caught on the upstream surface of the screen can be raked off manually or mechanically. The debris removed from the screen is usually buried in a landfill.

2. *Grit removal.* Sand, grit, cinders, and small stones are allowed to settle to the bottom of a grit chamber. A grit chamber is highly important for cities with combined storm and municipal sewage systems because of the grit or gravel that washes off streets or land during a storm and ends up at treatment plants. The grit obtained in this process is disposed of by using it for landfill.

3. *Sediment removal.* Wastewater, even after removal of grit, still contains suspended solids. These will settle out if the speed of the sewage flow is reduced. This is accomplished in a sedimentation tank. The suspended solids settle out and the solid mass, called raw sludge, is collected for disposal. Sedimentation tanks are usually 10 to 12 feet deep, and hold the wastewater for periods of 2 to 3 hours. Chemicals such as aluminum sulfate are sometimes added to speed up sedimentation. This process, called flocculation, was previously discussed in Section 11.7 at the time drinking water purification was considered.

The three major steps in primary wastewater treatment are shown diagramatically in Figure 21.7. The effluent from primary treatment then enters

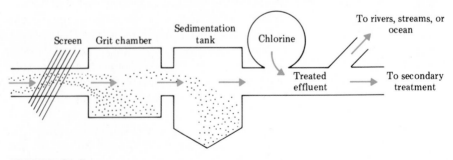

FIGURE 21.7
Steps in the primary treatment of wastewater.

secondary treatment processes or is treated with chlorine gas (to destroy disease-causing bacteria; see Section 11.7) prior to entering rivers, lakes, or streams. Primary treatment operations remove about 60% of the suspended solids and about one-third of the oxygen-demanding wastes. Very little of the dissolved nitrogen and phosphorus content of the water (plant nutrients) is removed.

Secondary Treatment Processes

Secondary wastewater treatment *consists of biological processes that use bacteria to break down soluble organic materials.* Secondary wastewater treatment is required of municipalities by federal legislation (Clean Water Act).

Two processes find extensive use in accomplishing secondary treatment objectives: (1) trickling filters and (2) an activated-sludge process. A trickling filter is simply a bed of stones and gravel 3 to 10 feet deep through which the wastewater passes slowly (see Figure 21.8). Bacteria gather and multiply on the stones and gravel until they become numerous enough to consume most of the organic matter in the wastewater. The water, after passing through the bed of solid materials, trickles out through pipes in the bottom of the filter. Filtration is not the important part of the process. It is the absorption

FIGURE 21.8

A trickle filter used in secondary wastewater treatment. The wastewater is distributed over the surface of the rocks by a rotating arm. [Grant Heilman.]

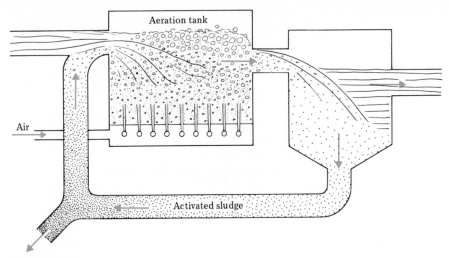

FIGURE 21.9

The activated-sludge process is a closed-system environment in which organic substances are degraded by bacteria found in sludge that is continuously recycled.

of organic matter on slimy stones and gravel followed by bacterial decomposition that cleans the water. The essential overall reaction that takes place can be represented as

$$\text{Organic matter} + \text{bacteria} + O_2 \longrightarrow CO_2 + H_2O + NH_3$$

Most impurities are removed from wastewater as solids. The resulting solid residue is often called *sludge*. In the activated-sludge process, air (or pure oxygen) and bacteria-laden sludge (saved from previous treatment cycles) are brought into contact with primary treatment effluent. The essentials of the process are illustrated in Figure 21.9. The sludge and effluent remain in contact for several hours in the aeration tank. During this time, the organic wastes are broken down by bacterial action. From the aeration unit, the effluent flows to a secondary settling tank where the biologically active sludge is collected. Part of the sludge is used to seed the next batch of wastes from the primary settling tanks. This sludge recycling is essential; without it, there would be insufficient biological activity in the aeration unit. The effluent from the secondary settling tank can be discharged into natural waters or sent on for tertiary treatment.

The relative effectiveness of primary and secondary wastewater treatment is evident from the comparative data given in Table 21.7. Note that 70% of the phosphorus (mostly as phosphates) and 50% of the nitrogen (mostly as nitrates) still remain after secondary treatment.

TABLE 21.7

Approximate Performance of Conventional Treatment Methods on Municipal Wastewater
(*based on raw waste concentrations*)

	Removal Efficiency of Treatment* (%)	
	Primary	Primary Plus Secondary
Biochemical oxygen demand	35	90
Refractory organics	20	60
Suspended solids	60	90
Total nitrogen	20	50
Total phosphorus	10	30
Dissolved minerals	—	5

* Figures may differ significantly in specific instances.

Tertiary Treatment Processes

Tertiary wastewater treatment *consists of one or more specialized chemical and/or physical processes performed on the effluent from secondary wastewater treatment.* Many tertiary treatment processes are in various stages of application and development. All such processes are more expensive to carry out than those involved in primary and secondary treatment. Indeed, cost is the major factor preventing more extensive use of tertiary treatment systems. Only a small percentage of wastewater facilities now use any tertiary treatment processes. As representative of tertiary treatment processes we now consider (1) coagulation-filtration and (2) carbon adsorption.

1. *Coagulation-filtration.* Secondary treatment effluent still contains suspended solids, primarily from sludge that was not removed in previous settling processes. These solids account for a large part of the remaining BOD of the effluent. One tertiary method now in use for their removal involves coagulation followed by filtration. Alum [$Al_2(SO_4)_3 \cdot 12H_2O$], a typical coagulant, reacts with a basic solution of bicarbonate ions as follows.

$$Al_2(SO_4)_3 + 6\,HCO_3^- \longrightarrow 2\,Al(OH)_3 + 3\,SO_4^{2-} + 6\,CO_2$$

The resulting $Al(OH)_3$ is a gelatinous material that envelopes suspended particles and settles with them to the bottom of a settling tank or pond. Alum also reacts with PO_4^{3-} ions to form a solid precipitate of aluminum phosphate, which settles out with the other solids. Approximately 90% of the phosphorus in wastewater can be removed using this process.

TABLE 21.8

Percentages of United States Population Served by Various Types of Wastewater Systems

	1978	1984
No wastewater system	30%	27%
Wastewater system with no treatment	1%	—
Wastewater system with primary treatment	21%	16%
Wastewater system with secondary treatment	25%	31%
Wastewater system with advanced secondary treatment or tertiary treatment	22%	26%

2. *Carbon adsorption.* Some organic materials that are resistant to biological breakdown persist in secondary wastewater effluent. These persistent materials are often referred to as "refractory organics," and they are responsible for the color and odor problems associated with secondary effluent. The most practical method available for removing these materials is the use of *activated carbon*. Activated carbon is a product formed by heating carbonaceous materials (wood, pulp-mill char, peat, low-grade coal, etc.) in the absence of oxygen (Section 13.4). This activated carbon removes organic contaminants from water by adsorption, which is the attraction and accumulation of one substance on the surface of another. The activated carbon used may be either in a granular form (the size of a fairly course sand) or in a powdered form. The powdered form requires less contact time but is more difficult to handle and regenerate. The carbon gradually loses its adsorptive capacity as organic materials accumulate on its surface. Regeneration of the spent carbon is essential to favorable economics because of the large quantities of carbon needed.

Extent of Wastewater Treatment

In 1984 the wastewater of 73% of the United States population was processed in treatment facilities. Twenty-two percent of this treated wastewater received only primary treatment, 42% received secondary treatment, and the remaining 37% received advanced secondary or tertiary treatment. Wastewater from the other 27% of the population, mostly those living in rural areas, was degraded in septic tanks or cesspools or went directly into rivers, streams, and lakes in untreated form. Wastewater treatment in the United States is being ungraded steadily; Table 21.8 shows the improvement from 1978 to 1984.

Practice Questions and Problems

OXYGEN-DEMANDING WASTES

21-1 Describe qualitatively how each of the following factors affects the solubility of oxygen in water.
a. An increase in water temperature.
b. A decrease in external pressure caused by an increase in altitude.

21-2 Define an oxygen-demanding waste.

21-3 What are typical sources for oxygen-demanding wastes?

21-4 What is the BOD of the following.
a. Nearly pure water
b. Untreated municipal sewage
c. Runoff from barnyards and feedlots
d. Food-processing wastes

21-5 What is the difference between aerobic decomposition conditions and anaerobic decomposition conditions?

21-6 What are the oxidation products for the elements carbon, nitrogen, and sulfur present in organic compounds when the oxidation occurs under the following conditions.
a. Aerobic conditions   b. Anaerobic conditions

DISEASE-CAUSING AGENTS

21-7 Why are tests for direct detection of the presence of disease-causing organisms in water supplies usually not carried out?

21-8 What are coliform bacteria and how are they used in determining water quality?

21-9 The "swimmability" of a body of water is often determined by a coliform bacteria count. Explain why.

PLANT NUTRIENTS

21-10 In a pollution context, define the term eutrophication.

21-11 What two elements are most likely to be limiting factors in plant growth and what are the natural and anthropogenic sources of these two elements?

21-12 A major emphasis of eutrophication control programs has been to limit the amount of phosphorus entering the environment. Why is this particular element singled out in control programs?

21-13 What single consumer product has been most affected by attempts to solve the plant nutrient problem?

21-14 How does eutrophication contribute to decreased DO levels?

21-15 How does eutrophication limit the recreational uses of affected waters?

21-16 What function do phosphorus compounds serve in detergent formulations?

SUSPENDED SOLIDS

21-17 What are the ecological consequences of turbidity caused by suspended solids?

21-18 What are the economic consequences of turbidity caused by suspended solids?

21-19 What are the ecological consequences of sedimentation resulting from suspended solids?

21-20 What are the economic consequences of sedimentation resulting from suspended solids?

DISSOLVED SOLIDS

21-21 List the major inorganic components of dissolved solids.

21-22 What is the relationship between irrigation practices and dissolved solids in waterways?

21-23 Dissolved solid effects on aquatic life are primarily related to what process?

21-24 What is the recommended TDS level for drinking water and the permissible maximum TDS level for drinking water?

ACIDS

21-25 What is the origin of acid mine drainage?

21-26 Write chemical equations for two reactions through which sulfuric acid is produced from acid mine drainage.

21-27 Discuss methods used to prevent and/or control acid mine drainage.

21-28 What is the relationship between increased water hardness and acid mine drainage?

21-29 What are the ecological effects of acid mine drainage?

21-30 Describe a mechanism involving carbon dioxide by which fish and other aquatic animals are destroyed by acid pollution.

HEAT

21-31 What are some ecological consequences of thermal pollution?

21-32 What is the relationship between thermal pollution and DO levels?

21-33 What is the relationship between rates of chemical reactions and thermal pollution?

TOXIC SUBSTANCES

21-34 All water pollutants are toxic if their specific effects occur to a great enough extent. Why, then, are certain water pollutants singled out as "toxic" pollutants?

21-35 List three toxic water pollutants of each type.
a. Metals
b. Pesticides
c. Synthetic organic compounds other than pesticides

21-36 What does the term *bioaccumulate* mean?

21-37 What are the dangers associated with bioaccumulation of toxic substances?

21-38 How is the bioaccumulation of toxic substances monitored?

21-39 List general sources for
a. Anthropogenic toxic substances
b. Naturally occurring toxic substances

WASTEWATER TREATMENT

21-40 What are the four major sources for wastewater?

21-41 What are the three major steps in primary treatment of wastewater?

21-42 Describe what occurs during each of the following secondary wastewater treatment processes.
a. Trickling filter process
b. Activated sludge process

21-43 In a wastewater treatment context, what is "sludge"?

21-44 Contrast the pollutant removal effectiveness of primary and secondary wastewater treatments.

21-45 Describe what occurs during each of the following tertiary wastewater treatment processes.
a. Coagulation using alum   b. Carbon adsorption

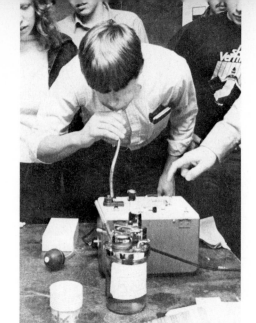

22

Chemistry and Medicine: Drugs

CHAPTER HIGHLIGHTS

22.1   Drugs: Their Effect on Our Lives

Presently, the average life expectancy in the United States is more than 70 years (almost 80 years for women). In 1850 life expectancy was about 35 years and in 1900 about 45 years. What has caused a doubling of life expectancy in the United States since the turn of the century?—principally, the use of chemicals to control and cure sickness and disease. Indeed, health care is the area in which chemistry has probably contributed the most to human well-being.

"Modern medicine" is an accepted fact of life for most people. They expect physicians to be able to prescribe medications to control pain (even mild pain), bleeding, infection, inflammation, high blood pressure, and many other conditions.

In this chapter we consider many of the drugs now commonly used by humans. A **drug** *is any substance taken into the body or used on the body to produce a desired effect.* We consider not only prescription medicinal drugs and nonprescription medicinal drugs (over-the-counter drugs) but also selected nonmedicinal drugs (alcohol, nicotine, and caffeine) and "illegal" drugs.

TABLE 22.1

The Top Ten Prescribed Medicinal Drugs for the Years 1986 and 1982

1986	1982
1. Tagamet (stomach ulcers)	1. Tagamet (stomach ulcers)
2. Zantac (stomach ulcers)	2. Inderal (heart disease)
3. Dyazide (hypertension)	3. Valium (tranquilizer)
4. Naprosyn (antiinflammatory)	4. Motrin (antiinflammatory)
5. Tenormin (hypertension)	5. Dyazide (hypertension)
6. Inderal (heart disease)	6. Keflex (antibiotic)
7. Keflex (antibiotic)	7. Naprosyn (antiinflammatory)
8. Procardia (heart disease)	8. Aldomet (hypertension)
9. Feldene (antiinflammatory)	9. Clinoril (antiinflammatory)
10. Cardizem (heart disease)	10. Orthonovum (birth control)

Any discussion of drugs, particularly prescription medicinal drugs, is outdated almost from the time it is printed. New and better drugs are continually reaching the market-place as a result of chemical research. Very few drugs that were used in hospitals 50 years ago, or even 25 years ago, are in use today. This constant change (for the better) continues in the 1980s. Table 22.1 lists the most prescribed drugs in the United States for the years 1982 and 1986. Note that only 5 of the 10 medications on the 1986 list appear on the 1982 list.

22.2 Mild Pain Relievers

Pain is any unpleasant sensation that a person experiences within his or her body. It is an entirely subjective and personal experience that involves complex interrelationships between mind and body, and it is the most common symptom for which patients seek medical attention. Pain may be considered a protection mechanism alerting an individual to some bodily malfunction.

Analgesics *are drugs used to relieve pain.* Mild analgesics, considered here, are usually nonprescription medications; we consider stronger, prescription pain relievers in Section 22.3. A **mild analgesic** *is a substance that relieves muscle, joint, and minor headache pain, usually without producing drowsiness or interfering with mental alertness.* The three most widely used mild analgesics are aspirin, acetaminophen, and ibuprofen.

Aspirin

Aspirin is probably the most widely used drug in the world. It has the ability to decrease pain (analgesic properties), to lower body temperature (antipyretic properties), and to reduce inflammation (antiinflammatory properties).

Chemically, aspirin is a derivative of salicylic acid, a hydroxy derivative of benzoic acid.

Benzoic acid   Salicylic acid

Reaction of salicylic acid with acetic acid produces the ester (Section 17.8) acetylsalicylic acid, which is commonly known as aspirin.

Salicylic acid   Acetic acid   Acetylsalicylic acid (aspirin)

Aspirin is a term coined from *a*, for acetyl, and *spir* from the spirea family of plants. Historically, salicylic acid (the starting material for aspirin) was first isolated from plants of the genus *Spirea*. For over 2000 years herbal preparations containing salicylic acid were used to relieve pain and fever, despite side effects of mouth, throat, and stomach irritation. The salicylic acid ester now called aspirin, first synthesized in the 1890s, proved to be more effective and more palatable as a pain reliever than salicylic acid.

Aspirin is most frequently taken in tablet form. A tablet usually contains 325 mg of aspirin held together with an inert starch binder. Aspirin, itself, a white crystalline solid with a bitter taste, is moderately soluble in water and readily soluble in the acidic medium (digestive juices) of the stomach. After ingestion, aspirin undergoes hydrolysis (reaction with water) to produce the two carboxylic acids from which it is formed.

Aspirin   Salicylic acid   Acetic acid

SIDE EFFECTS OF ASPIRIN

Salicylic acid is the "active form" of aspirin—the substance that exerts a pain-relieving effect. The carboxylic acids produced by hydrolysis can cause heartburn or increased stomach acidity. Excessive use of aspirin can produce internal bleeding and, ultimately, ulcers. Recent research indicates that some intestinal bleeding occurs every time aspirin is ingested. The blood loss is usually minor, but it may be substantial in some people; hypersensitivity to aspirin occurs in 0.2% of the general population.

Despite aspirin's long history of use, only in the last 20 years have scientists begun to understand how it works in the body. Research indicates that aspirin inhibits the synthesis of a class of hormones called prostaglandins. These hormones, when present in higher than normal levels in the bloodstream, cause pain, headache, and inflammation.

A number of over-the-counter combination pain relievers have aspirin as the major ingredient. Some of these combination pain relievers contain basic

TABLE 22.2
Selected Aspirin-Containing Analgesics

Product (Trade Name)	Ingredients
Bufferin	Aspirin, magnesium carbonate
Excedrin	Aspirin, caffeine, acetaminophen
Excedrin P.M.	Aspirin, acetaminophen, pyrilamine maleate (an antihistamine)
Empirin	Aspirin, caffeine
Anacin	Aspirin, caffeine

substances, such as magnesium carbonate ($MgCO_3$), to help neutralize the acidic products of aspirin hydrolysis and prevent stomach upset. Table 22.2 lists the ingredients in selected over-the-counter aspirin-containing analgesics.

Acetaminophen

A second widely used mild analgesic is acetaminophen, a substituted amide (Section 17.9) whose structure is

Acetaminophen

This compound is the active ingredient in Tylenol and Datril. Some combination pain relievers (Table 22.2) contain acetaminophen as well as aspirin.

Acetaminophen is often used as an aspirin substitute because it has no irritating effect on the intestinal tract and yet has comparable analgesic and antipyretic effects. Unlike aspirin, however, it is not effective against inflammation and is of limited use for the "aches and pains" of arthritis.

Acetaminophen is available as tablets and capsules, as well as in a stable liquid form that is used extensively for pain relief and fever reduction in small children and other patients who have difficulty taking solid tablets or capsules. Its extensive use with children has a drawback; it is the drug most often involved in childhood poisonings. Acetaminophen is toxic to the liver if an overdose is taken.

Ibuprofen

Ibuprofen is a "newcomer" to the over-the-counter pain reliever market. It was cleared by the Food and Drug Administration (FDA) in 1984 for nonprescription sales. Prior to that time (since 1974) it was available only by prescription. Advil and Nuprin are over-the-counter ibuprofen-containing formulations containing 200 mg per tablet of this pain reliever. Tablets of Motrin and Rufen, available only by prescription, contain larger amounts of ibuprofen.

Chemically, ibuprofen is a substituted form of 2-phenylpropanoic acid.

$$CH_3-CH-C\overset{O}{\underset{OH}{\Big\langle}}$$

2-Phenylpropanoic acid

$$CH_3-CH-C\overset{O}{\underset{OH}{\Big\langle}}$$
$$CH_2$$
$$CH_3-CH-CH_3$$

Ibuprofen

Numerous studies have shown that nonprescription-strength ibuprofen relieves minor pain and fever as well as aspirin or acetaminophen. Like aspirin, ibuprofen reduces inflammation. (Prescription-strength ibuprofen has extensive use as an antiinflammatory agent for the treatment of rheumatoid arthritis.) There is evidence that ibuprofen is more effective than either aspirin or acetaminophen in reducing dental pain and menstrual pain. Both aspirin and ibuprofen can cause stomach bleeding in some people, although ibuprofen seems to cause fewer problems. Ibuprofen is more expensive than either aspirin or acetaminophen.

22.3 Strong Pain Relievers

Treatment of moderate to severe pain arising from postsurgical trauma or the chronic, severe, intractable pain of terminal illness requires the use of strong pain relievers called *narcotic analgesics*. A **narcotic analgesic** *is both an analgesic (pain-relieving) and a sedative (sleep inducing).* The term *narcosis* is Greek for "sleep" or "stupor." A most important family of narcotic analgesics are the *opiates*. An **opiate** *is any natural or synthetic drug that exerts actions upon the human body similar to those induced by morphine, the major pain-relieving agent obtained from the opium poppy plant.*

Through much of history the opium poppy plant has served as a source of pain-relieving materials. The hardened, dried juice of unripe seeds from this plant is called *opium,* a crude extract containing numerous compounds, including the painkillers morphine and codeine (see Figure 22.1). Morphine is the major pain-relieving drug found in opium, constituting approximately 10%

(a)   (b)

FIGURE 22.1

(a) The opium poppy plant and (b) opium gum obtained from the poppy plant. Morphine makes up the largest fraction, about 10% by mass, of the substances present in opium. [(a) Steve Patten/Black Star; (b) Charles Marden Fitch/Taurus Photos.]

by mass of crude opium. Codeine amounts to only 0.5% by mass of the opium extract.

Structurally, morphine and codeine are closely related. Both are nitrogen-containing multiring compounds in which the nitrogen atom is present as an amine nitrogen (Section 17.9) that is part of a six-membered heterocyclic ring.

Morphine   Codeine

TABLE 22.3

Contents of Selected Narcotic Analgesic Combination Products

Trade Name	Narcotic	Other Ingredients
Empirin #2	15 mg codeine phosphate	325 mg aspirin
Empirin #3	30 mg codeine phosphate	325 mg aspirin
Empirin #4	60 mg codeine phosphate	325 mg aspirin
Tylenol #1	7.5 mg codeine phosphate	300 mg acetaminophen
Tylenol #2	15 mg codeine phosphate	300 mg acetaminophen
Tylenol #3	30 mg codeine phosphate	300 mg acetaminophen
Tylenol #4	60 mg codeine phosphate	300 mg acetaminophen
Percodan	4.5 mg oxycodone hydrochloride, 0.38 mg oxycodone terephthalate	325 mg aspirin
Percoset-5	5 mg oxycodone hydrochloride	325 mg acetaminophen
Tylox	4.5 mg oxycodone hydrochloride	500 mg acetaminophen
Darvocet-N-50	65 mg propoxyphene napsylate	325 mg acetaminophen
Darvocet-N-100	100 mg propoxyphene napsylate	650 mg acetaminophen

The two —OH groups of morphine are particularly important because many morphine derivatives can be made by modifications of either or both of these groups. Codeine is methylmorphine, with the methyl group replacing the hydrogen atom of the phenolic —OH group. Almost all codeine used in modern medicine is produced by methylating the more abundant morphine.

Named after Morpheus, the ancient Greek god of dreams, morphine is one of the most effective painkillers known and is used widely in medicine for that purpose. Its analgesic properties are about 100 times greater than those of aspirin. Morphine acts by blocking the process in the brain that interprets pain signals coming from the peripheral nervous system. The major drawback to use of morphine, or any opiate, is that it is addictive.

Morphine is usually administered to patients in the form of its hydrochloride or hydrogen sulfate salt, since these salts are more soluble in water than morphine itself. The solubility of morphine in water is 0.25 g/L and that of morphine hydrochloride is 57 g/L. (Many medications, not just morphine, are administered in salt form because of the enhanced water solubility.)

Morphine hydrochloride

Morphine hydrogen sulfate

Codeine is a less potent painkiller than morphine with an analgesic activity about one-sixth that of morphine. The most used of all narcotic analgesics, codeine produces less sedation and depression of the central nervous system. Many prescription cough syrups contain codeine; its function is to decrease throat discomfort. Numerous analgesics combining codeine with such other ingredients as aspirin, acetaminophen, muscle relaxants, and cough medications are available by prescription. As shown in Table 22.3, many brand-name formulations are marketed with various amounts of codeine.

CODEINE-CONTAINING
COMBINATION
ANALGESICS

Another narcotic analgesic similar in structure to morphine and codeine is oxycodone, the active ingredient in formulations such as Percodan, Percoset-5, and Tylox (see Table 22.3).

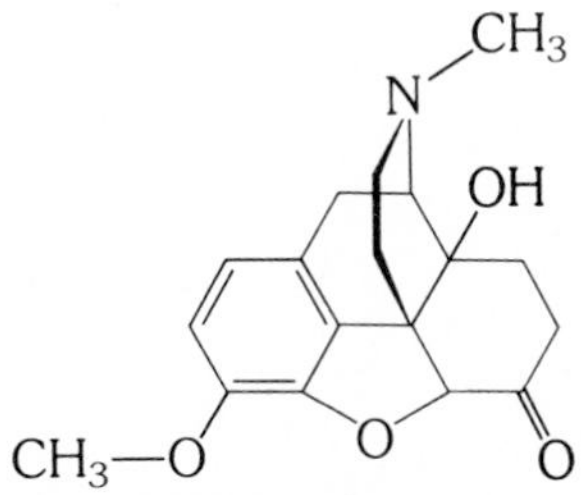

Oxycodone

Heroin is a synthetic derivative, the diacetyl ester (Section 17.8), of morphine. This chemical modification increases the potency, heroin being three times as potent an analgesic as morphine; 3 mg of heroin produces the same analgesic effect as 10 mg of morphine.

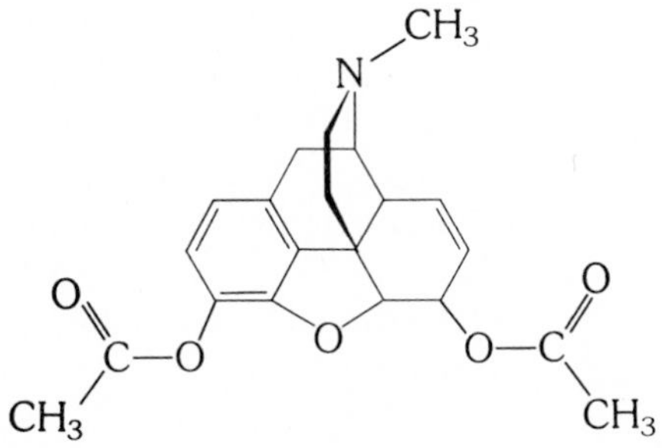

Heroin

Heroin was initially developed to help morphine addicts "kick the habit." Unfortunately, heroin is no less, and probably more, addictive than morphine. Addiction develops faster and is also harder to cure. Heroin has no medical use, and its possession is illegal in the United States.

Many nonopium-based drugs with physiological actions similar to morphine have now been synthesized. The first totally synthetic narcotic analgesic was meperidine (Demerol). It is a more potent analgesic than codeine, but a significantly weaker analgesic than morphine.

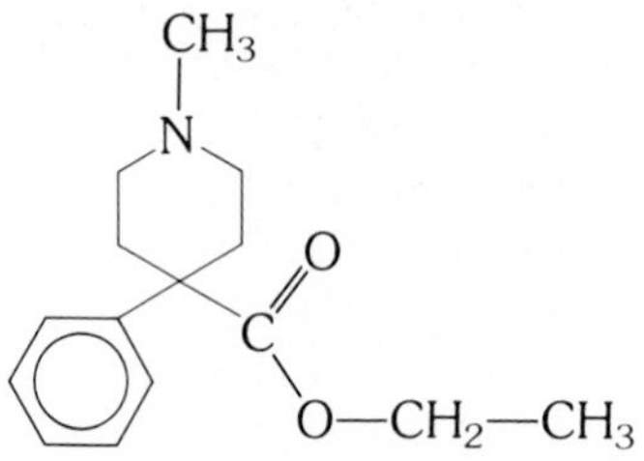

Meperidine (Demerol)

The structural similarity between meperidine and morphine can be seen when the structures are redrawn in the following manner.

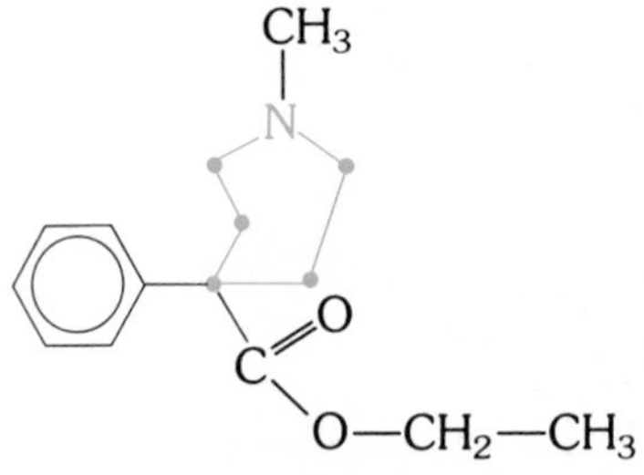

Meperidine

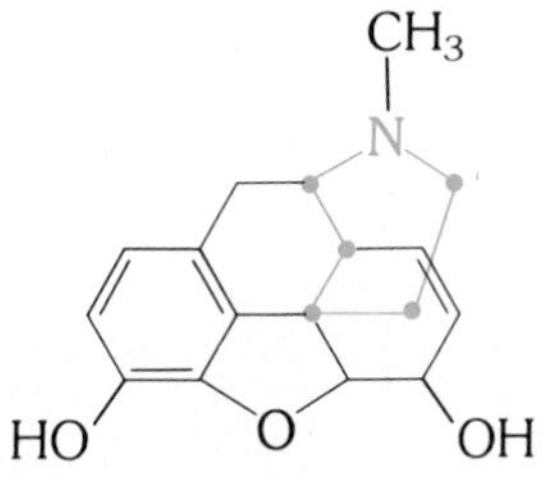

Morphine

Meperidine is probably the narcotic analgesic most frequently involved in physician-induced addiction of patients because of its extensive use for pain control in hospitalized patients.

HEROIN ADDICTION AND METHADONE

Another synthetic analgesic is methadone. One of the important uses of methadone today is in treating heroin addiction. Methadone, which is also addictive, is substituted for the heroin and the patient is later slowly withdrawn from methadone. The withdrawal is milder and less acute than that from heroin. Unlike an heroin addict, a person on methadone maintenance can often hold a productive job since methadone does not have the sleep-inducing characteristics of heroin intoxication.

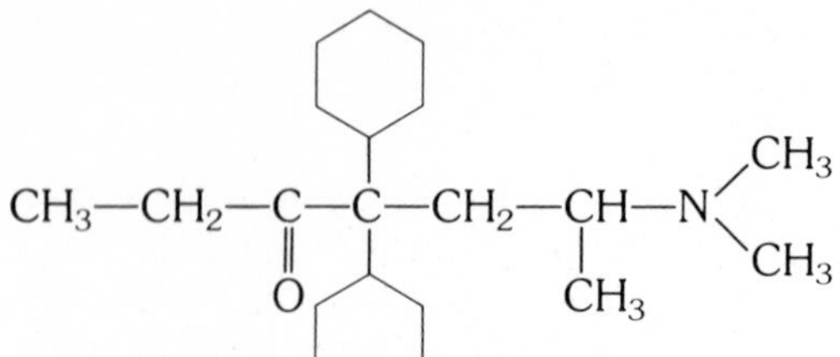

Methadone (Dolophine)

Propoxyphene (Darvon) is a narcotic analgesic with a structure very similiar to that of methadone. As an analgesic, its potency is considerably less than codeine but greater than therapeutic doses of aspirin. In large doses, opiate-like effects are seen.

Propoxyphene (Darvon)

22.4 Gastrointestinal Drugs

We consider in this section three types of drugs used in treating problems associated with the human gastrointestinal system: (1) antacids, (2) laxatives, and (3) antiulcer agents.

Antacids

Hydrochloric acid (HCl), necessary for proper digestion of food, is present in the gastric juices of the human stomach. Normal hydrochloric acid levels result in gastric juices with pH values between 1 and 3 (Section 10.9). Overeating and emotional factors can cause the stomach to produce too much hydrochloric acid, the resulting hyperacidic condition being known as "sour" stomach, upset stomach, heartburn, or indigestion. Substances called *antacids* provide symptomatic relief from such hyperacidity. An **antacid** *is an over-the-counter drug containing one or more alkaline (basic) substances capable of neutralizing hydrochloric acid present in gastric juice.*

Only a few different neutralizing agents are present in the wide variety of brand-name antacids available.

1. Slightly soluble hydroxides such as magnesium hydroxide [$Mg(OH)_2$] and aluminum hydroxide [$Al(OH)_3$].
2. Very slightly soluble calcium carbonate ($CaCO_3$).

3. Water-soluble sodium bicarbonate ($NaHCO_3$).
4. Complex hydroxide compounds such as $AlMg(OH)_5$ and $AlNaCO_3(OH)_2$.

Magnesium and aluminum hydroxide react with hydrochloric acid by the following acid–base reactions (Section 10.7).

$$2\ HCl + Mg(OH)_2 \longrightarrow MgCl_2 + 2\ H_2O$$

$$3\ HCl + Al(OH)_3 \longrightarrow AlCl_3 + 3\ H_2O$$

The reaction involving sodium bicarbonate produces a gas in addition to a salt and water,

$$HCl + NaHCO_3 \longrightarrow NaCl + CO_2 + H_2O$$

and the reaction of calcium carbonate with hydrochloric acid is similar.

$$2\ HCl + CaCO_3 \longrightarrow CaCl_2 + CO_2 + H_2O$$

In the complex hydroxide $AlNaCO_3(OH)_2$ both the hydroxide ion and carbonate ion consume acid.

An "ideal" antacid would have the following four characteristics.

1. It decreases stomach acidity rapidly (instant relief.)
2. It does not decrease stomach acidity too much—this causes *acid rebound* where the body responds by secreting more hydrochloric acid.
3. It maintains *normal* stomach acidity for 30 minutes to an hour (lasting relief).
4. It does not cause side effects such as constipation or diarrhea.

ANTACIDS AND ACID
REBOUND

No antacid formulation is "ideal". Aluminium and calcium preparations tend to cause constipation, whereas magnesium preparations act as laxatives. (This is one reason for using formulations containing more than one neutralizing ingredient.) Sodium bicarbonate reacts quickly, but has a short relief time (5 minutes) and tends to cause acid rebound. Calcium carbonate gives longer relief time (30 minutes), but also tends to cause acid rebound. Note that because of acid rebound, taking more antacid than needed can cause rather than relieve hyperacidity. Table 22.4 lists the neutralizing agents present in selected brand-name antacids.

Laxatives

Constipation involves impairment in the normal capacity of the colon to produce properly formed stools at regular intervals. Causes include emotional stress, change in diet, traveling, and the side effects of prescription medications being used to treat other conditions. Many cases of constipation cure themselves without the use of laxatives, although use of a mild laxative for a day or two is

TABLE 22.4

Neutralizing Agents Present in Selected
Antacid Formulations

Brand Name	Neutralizing Agents
Alka-Seltzer	$NaHCO_3$
BiSoDol	$NaHCO_3$
DiGel	$Mg(OH)_2$, $Al(OH)_3$
Gaviscon	$Al(OH)_3$, $NaHCO_3$
Gelusil	$Mg(OH)_2$, $Al(OH)_3$
Maalox	$Mg(OH)_2$, $Al(OH)_3$
Milk of Magnesia	$Mg(OH)_2$
Mylanta	$Mg(OH)_2$, $Al(OH)_3$
Riopan	$AlMg(OH)_5$
Rolaids	$AlNaCO_3(OH)_2$
Tums	$CaCO_3$

often recommended. A **laxative** *is a drug that changes fecal consistency, speeds passage of feces through the colon, and aids in elimination of stool from the rectum.*

A very heterogeneous group of substances find use as laxatives. They are classified into categories based on mechanism of action, three of which are

1. Bulk-forming laxatives.
2. Hyperosmolar and saline laxatives.
3. Irritant laxatives.

Bulk-forming laxatives absorb water and expand, increasing both the bulk and moisture content of the stool. The increased bulk stimulates contraction of specialized muscles in the gastrointestinal tract that force materials through the intestine and rectum. Pysillium preparations, derived from psyllium seeds, find extensive use in bulk-forming laxatives. Whole or powdered seeds of the psyllium plant function as indigestible, nonabsorbable, instantly miscible dietary fiber. Metamucil, Naturacil, and Fiberall are brand names associated with this type of laxative preparation.

Hyperosmolar and saline laxatives produce an osmotic effect (Section 9.6) in the colon, distending the bowel from water drawn into it (through osmosis) and promoting contraction of muscles that move materials through the bowel. Milk of Magnesia, (either as a suspension of $Mg(OH)_2$ in water or as tablets), and Phospho-Soda, a mixture of sodium hydrogen phosphate (Na_2HPO_4) and sodium phosphate (Na_3PO_4), are commonly used laxatives of this type.

Irritant laxatives exert their effect by local irritation. Dulcolax and Babylax are such laxatives. A colorless, tasteless, practically water-insoluble compound called bisacodyl is the active ingredient in Dulcolax. It functions as a contact laxative, directly irritating the colon lining.

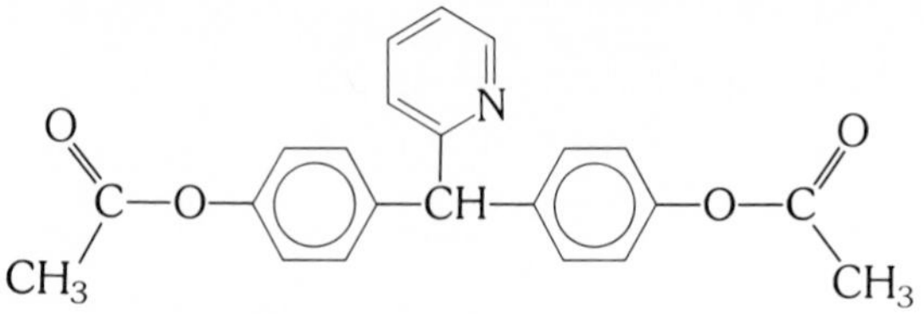

Bisacodyl

Babylax, used in young children, has glycerin (Section 17.3) as its active ingredient. Administered rectally, the glycerin dehydrates exposed tissues to produce an irritant effect.

Antiulcer Agents

The two most prescribed drugs in the United States—Tagamet and Zantac (see Table 22.1)—are both used in the treatment of stomach ulcers. Both drugs act by blocking the the action of histamine, which helps regulate gastric acid secretion, at receptor sites in the gastric-acid-secreting cells of the gastrointestinal tract lining. The net effect is decreased amounts of gastric secretion in the stomach. This lowered acidity allows for healing of ulcerated tissue.

Tagamet, whose chemical name is cimetidine, is a *planned* drug that was produced by making substitutions in the histamine molecule structure until a compound was found that blocked the action of the parent compound. The relationship between the structures of histamine and cimetidine is as follows.

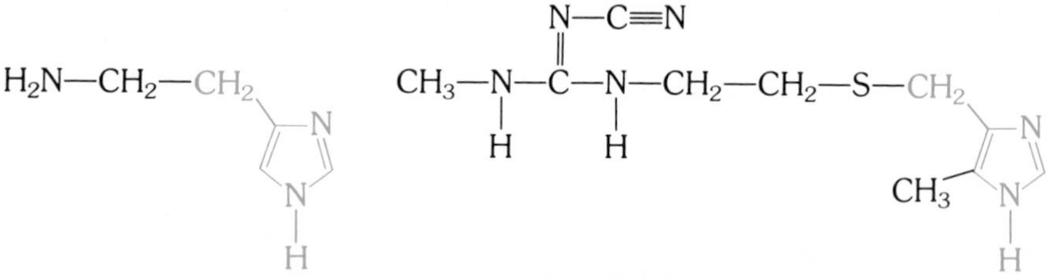

Histamine   Cimetidine (Tagamet)

Zantac, whose chemical name is ranitidine, is a newer product than Tagamet and in many cases is preferred over Tagamet because of fewer interactions with other body chemicals. The heterocyclic ring in ranitidine contains oxygen rather than nitrogen, as in histamine.

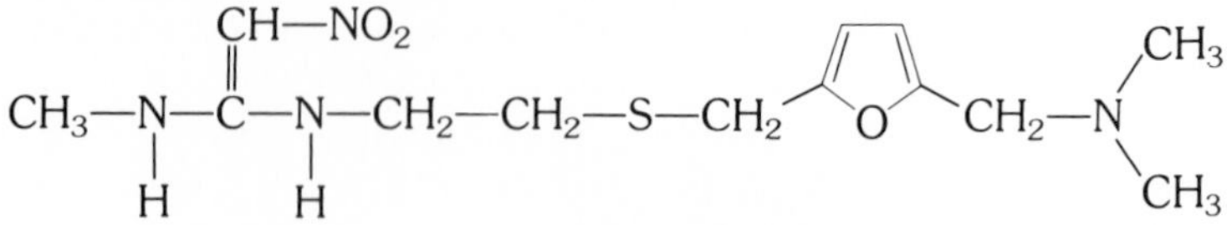

Ranitidine (Zantac)

Note that the long side chains in Zantac and Tagamet are almost identical.

22.5 Cardiovascular Drugs

The cardiovascular system, sometimes simply called the circulatory system, consists of the heart (a muscular pumping device) and a closed system of vessels called arteries, veins, and capillaries. As implied by the name circulatory system, blood is pumped by the heart around a closed circle or circuit of vessels, passing again and again through the various parts of the body.

Heart disease is the number one cause of death among Americans, accounting for almost 40% of total yearly deaths. In the last 25 years, the development of many new drugs has made heart disease much more treatable, dramatically improving the quality and length of life for those having cardiovascular problems. Table 22.1 reflects the importance of these new drugs; five of the ten most prescribed drugs relate to the cardiovascular system (heart disease and hypertension).

Two conditions commonly associated with heart disease are atherosclerosis and hypertension. **Atherosclerosis** *is the buildup of plaque (cholesterol and fatty acid deposits) on the inner walls of arteries.* This plaque buildup reduces the flow of blood to the heart (see Figure 22.2). Substantial blockage of a coronary artery by plaque results in a heart attack (sudden damage to heart muscle caused by a cutoff of coronary circulation). About 98% of all heart attack victims have atherosclerosis. **Hypertension** (*high blood pressure*) *is an intermittent or sustained blood pressure over 140/90.* This condition causes a narrowing of blood vessels, which in turn requires the heart to work harder to pump blood. A harder-working heart eventually enlarges. The enlarged heart requires more oxygen, and if it cannot meet the demands placed on it (because of lack of oxygen), angina pectoris (heart pains resulting from insufficient oxygen) or a heart attack may result.

The treatment of heart disease with drugs concentrates on improving the rate of blood flow to the heart. Cardiovascular system drugs act at many sites in the

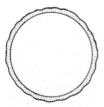   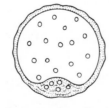   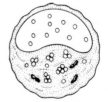

Normal blood vessel   Beginning atherosclerosis   Advanced atherosclerosis

FIGURE 22.2

A major cause of heart disease is the narrowing of the passageways of the blood vessels due to atherosclerosis.

body and through numerous mechanisms. Three important types of such drugs are

1. Diuretics
2. Beta-blocking agents
3. Calcium-channel-blocking agents

One of the five cardiovascular drugs listed in Table 22.1 is a diuretic (Dyazide), two are beta blockers (Tenormin and Inderal), and two are calcium channel blockers (Procardia and Cardizem).

Diuretics

One method for treating hypertension involves reducing extracellular fluid volume using diuretics. A **diuretic** *is a drug that acts to eliminate fluid from the body by increasing the amount of urine released.* The most used diuretic antihypertensive agent is Dyazide, a combination drug. Its components are hydrochlorothiazide and triamterene, present in a 1:2 mass ratio.

Hydrochlorothiazide   Triamterene

Hydrochlorothiazide blocks the normal flow of sodium and chloride ions into the kidneys, producing an osmotic effect that draws water out of tissue fluids. The function of triamterene is to prevent exchange of K^+ and H^+ ions for the "blocked out" Na^+ ions, which would lead to potassium deficiencies.

Beta-Blocking Agents

Heart muscle contains receptor sites called beta receptors. Stimulation of such sites by the adrenal gland secretions epinephrine and norepinephrine (Section 17.9) causes increased cardiac activity. **Beta-blocking agents** *are drugs that block the action of epinephrine and norepinephrine at beta receptor sites.* Use of beta blockers decreases the heart rate, allowing an overworked heart (from, for example, the buildup of plaque in the arteries) to relax and thereby strengthen itself.

Blood vessels also contain beta receptor sites. Use of beta-blocking agents can cause relaxation of blood vessels and a resulting decrease in blood pressure.

The heart drugs Inderal and Tenormin, with the following structures, function as beta-blocking agents.

$$\text{O—CH}_2\text{—CH—CH}_2\text{—N—CH}$$

Inderal (Propranolol)

$$\text{O—CH}_2\text{—CH—CH}_2\text{—N—CH}$$

Tenormin (Atenolol)

Note that both structures contain an identical long side chain. Inderal, also called propranolol, has been on the market the longest of all beta blockers. It is a nonselective beta blocker, exerting its effect at all beta receptor sites. Tenormin, also called atenolol, works predominantly at heart muscle beta receptor sites.

Calcium-Channel-Blocking Agents

Calcium ions in heart muscle trigger the interaction that causes heart muscle contraction. These Ca^{2+} ions move into heart muscle cells by means of holes or channels in the membrance surrounding each cell. After the contraction, these

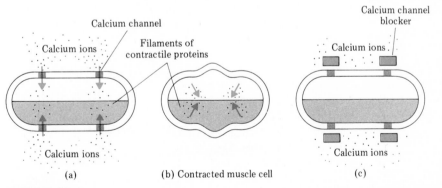

FIGURE 22.3

(a) Calcium ions flowing into a muscle cell. (b) Contraction of the muscle cell caused by the presence of the calcium ions. (c) Relaxed muscle cells with calcium channels blocked by drug molecules.

same Ca^{2+} ions move out of the cell, causing it to relax. Blocking the flow of Ca^{2+} ions into a cell causes the muscle to relax (see Figure 22.3). Relaxation of muscles in the walls of the coronary arteries results in their expansion (dilation) and an increase in the supply of blood to the heart. Some calcium blockers also decrease the force of heart muscle contraction and thus decrease the oxygen requirements of the heart.

Calcium channel blockers became available in the United States in the early 1980s. (Use in European countries had begun many years earlier.) Two of the most used drugs of this type are Procardia and Cardizem. Calcium channel blockers are not without side effects such as headaches, dizziness, and light-headedness.

Procardia (Nifedipine)   Cardizem (Diltiazem)

22.6 Antibacterial Drugs

Less than a century ago infectious diseases were the leading causes of death for Americans (Figure 22.4a). Today only one out of the 10 leading causes of death—pneumonia and influenza—is an infectious disease (Figure 22.4b). What caused the change? The major factor is the advent of antibacterial drugs, the first of which came to be called "miracle drugs." **Antibacterial drugs** *are compounds that inhibit the growth of microorganisms.* We consider three types.

1. The sulfa drugs
2. Penicillins and cephalosporins
3. Broad-spectrum antibiotics

The Sulfa Drugs

The compound prontosil, the prototype for what are now called sulfa drugs, was discovered in 1935. Used to stain (dye) bacteria, it was found to be a selective

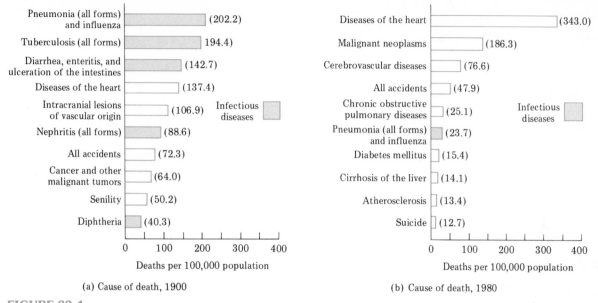

(a) Cause of death, 1900

(b) Cause of death, 1980

FIGURE 22.4

Causes of death for Americans in (a) 1900 and (b) 1980. By 1980 only one infectious disease was among the top 10 causes of death.

killer of certain types of bacteria. Further research showed that prontosil was converted to another compound, sulfanilamide, within living organisms and that this latter compound was the actual antibacterial agent.

Prontosil   Sulfanilamide

Sulfanilamide and some of its derivatives make up the group of compounds called sulfa drugs. Before sulfa drug were in use, the death rate from streptococcal meningitis was higher than 90%; after their introduction it dropped to less than 10%. During World War II soldiers often carried packets of "sulfa" for sprinkling into wounds to prevent infection; this practice saved many lives.

Sulfa drugs work by preventing bacteria from synthesizing folic acid, a chemical essential to their growth. This inhibition relates to the similiarity of sulfa drugs to para-aminobenzoic acid (PABA), a key ingredient in the bacterial synthesis of folic acid.

PABA   Sulfanilamide

Bacteria mistakenly try to convert the drug instead of PABA into folic acid. Humans also need folic acid, but we obtain it from our diets. Hence, sulfa drugs have no effect on human cells.

Only a few sulfa drugs are still in use today. For most purposes, they have been replaced by newer antibacterial drugs. They still, however, are a drug of choice in treatment of many types of urinary tract infections. The development of resistance in formerly susceptible organisms was a major factor in the shift away from sulfa drugs. Also some sulfa drugs can cause kidney damage through a crystallization process. Two sulfa drugs still in use are sulfadiazine and sulfisoxazole.

Sulfadiazine   Sulfisoxazole
(Microsulfone)   (Gantrisin)

Penicillins and Cephalosporins

Penicillins and cephalosporins are antibiotics. An **antibiotic** *is a chemical produced by a living organism—usually a mold or fungus—that kills or inhibits growth of other microorganisms.*

Penicillin was the first antibiotic to be discovered and used. Its discovery, in 1928, by the Scottish microbiologist Alexander Fleming was "accidental." While studying cultures of an infectious staphylococcus bacterium he spotted something unusual about one contaminated culture about to be discarded—a zone of inhibition of staphylococcal growth around a blue-colored mold (the contaminant). Fleming's genius was in recognizing the potential importance of this chance observation: a living organism, the mold, was producing a chemical toxic to the bacteria. But his attempts to isolate the active ingredient, which he named penillicin (from the name of the mold species), were unsuccessful. A decade later, the Oxford University scientists Howard Flory and Ernst Chain did acccomplish this feat. Further testing showed penicillin to be a powerful agent in the treatment of bacterial infections. By 1943 penicillin was available for clinical use, and by 1945 there was a large enough supply for general use throughout the world. A different strain of *Penicillium* mold (discovered on a moldy cantaloupe in a Peoria, Illinois, market) proved to be a particularly good source of the antibiotic. By the late 1940s, many lives threatened by pneumonia, bone infections, gangrene, and other infectious diseases were saved through penicillin use.

Many different penicillin-type antibiotics, most of them semisynthethic, are now known. All of them have common basic structural characteristics of the following general form.

The identity of the R group at the left of the structure determines the specific penicillin. Table 22.5 gives the structures of selected important penicillins. In each structure the R group is highlighted. The original penicillin isolated from mold is now known as penicillin G. It had to be administered by injection because it was destroyed by stomach acid when taken orally. Chemical modification of penicillin G has solved this problem as well as decreased, but not eliminated, allergic reactions some people have to it. The other penicillins in Table 22.5 are such semisynthetic compounds.

Following the clinical introduction of penicillin, the search began for other antibiotic-producing organisms. This search became particularly important when it was found that some bacteria were developing resistance to penicillins. (Such penicillin-resistant bacteria produce an enzyme called penicillinase that converts penicillin to an inactive form.) From this search came the cephalosporin antibiotics—isolated from a mold originally found in seawater near a sewage outfall off Sardinia. Many cephalosporin antibiotics are in current use. Keflex, the only antibiotic listed in the "top ten" prescription drugs (Table 22.1), is a cephalosporin.

PENICILLIN-RESISTANT BACTERIA

Structurally, cephalosporins are close relatives of penicillins. Each has the same four-membered nitrogen-heterocyclic ring fused to a sulfur-containing ring. This latter ring is five-membered in penicillins and six-membered in cephalosporins.

Penicillin ring system   Cephalosporin ring system

The structure of Keflex (which is also called cephalexin), the most used

TABLE 22.5
Structures of Five Types of Penicillin*

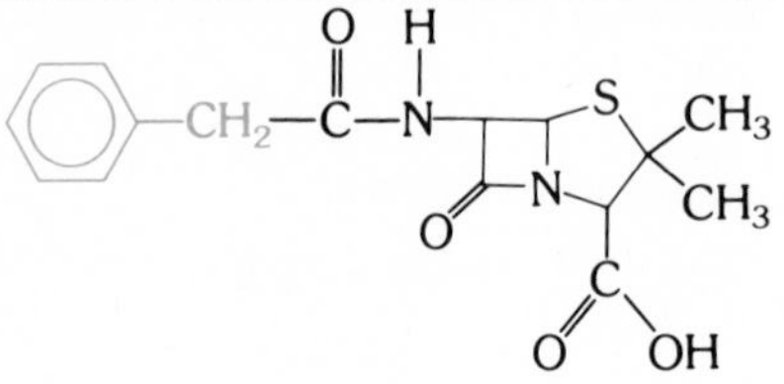

Penicillin G
(original penicillin)

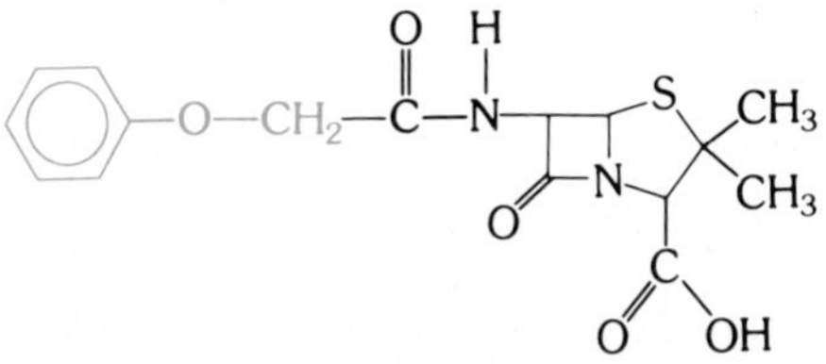

Penicillin V
(acid resistant)

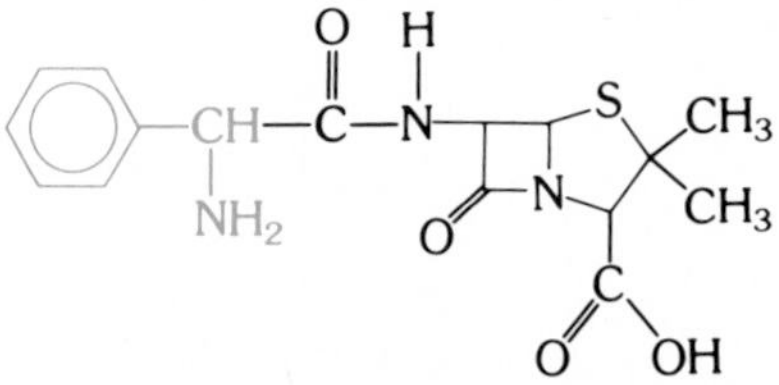

Ampicillin
(acid resistant)

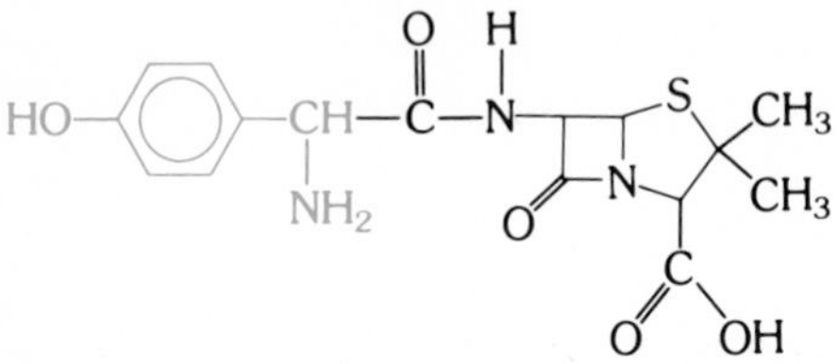

Amoxycillin
(acid resistant)

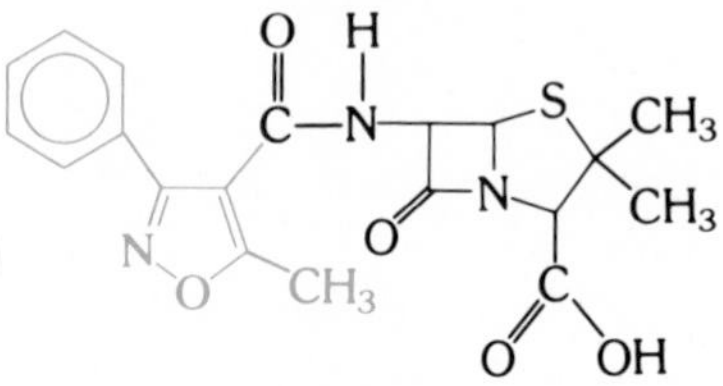

Cloxacillin
(acid and penicillinase resistant)

* Penicillin G is the original penicillin. The other penicillins are semisynthetic derivatives of it.

cephalosporin, is

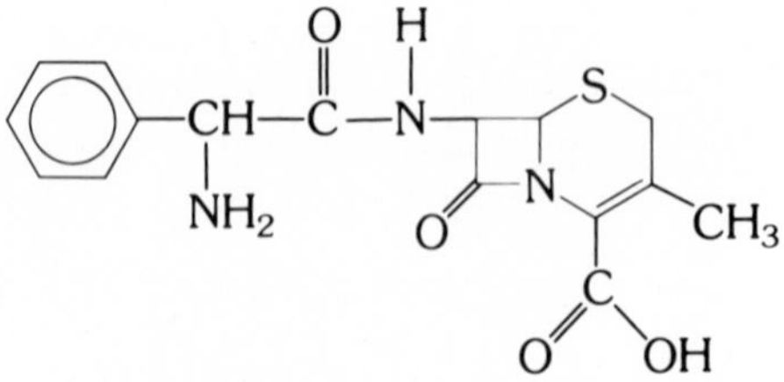

Keflex (Cephalexin)

Note that the side chains at the left in the structures of Keflex and Ampicillin (Table 22.4) are identical.

Broad-Spectrum Antibiotics

The penicillins and cephalosporins, like the sulfa drugs, are effective against only certain types of bacteria. In the 1950s a new group of antibiotic compounds, effective against a wide variety of bacteria, were isolated. These *broad-spectrum* antibiotics were the tetracyclines, so called because of a basic ring structure involving four six-membered rings fused together. Two important tetracyclines are Aureomycin and Terramycin.

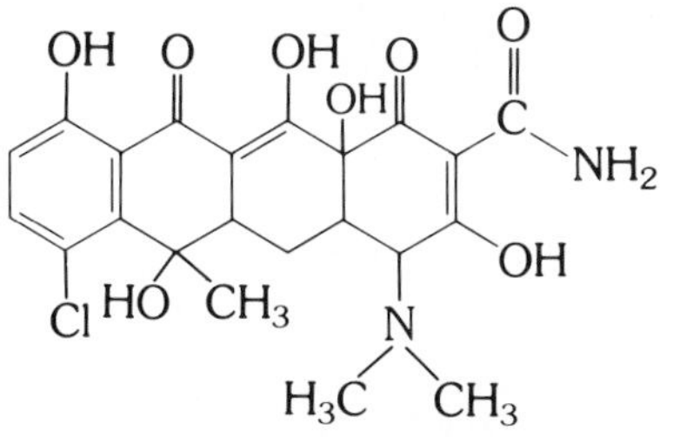

Aureomycin

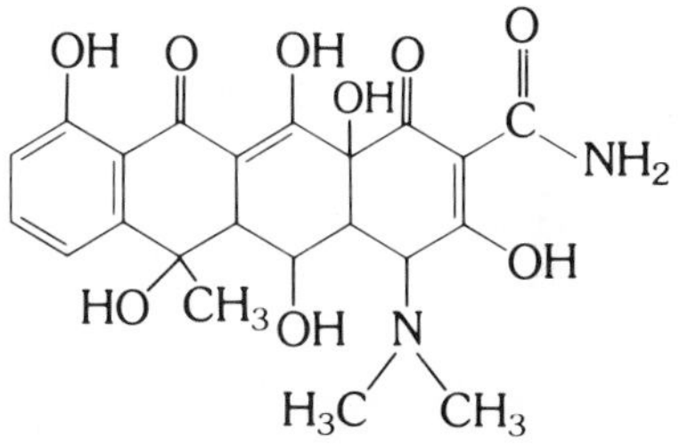

Terramycin

Aureomycin, the first tetracycline to be marketed, was isolated from a gold-colored fungus. Terramycin was discovered after testing 116,000 different soil samples. One side effect of these drugs is diarrhea, caused by the killing of the bacteria that normally reside in a person's intestinal tract.

Synthetic broad-spectrum antibiotics have now also been produced. Chloromycetin, the first such compound to be produced, is used in the treatment of eye and ear infections.

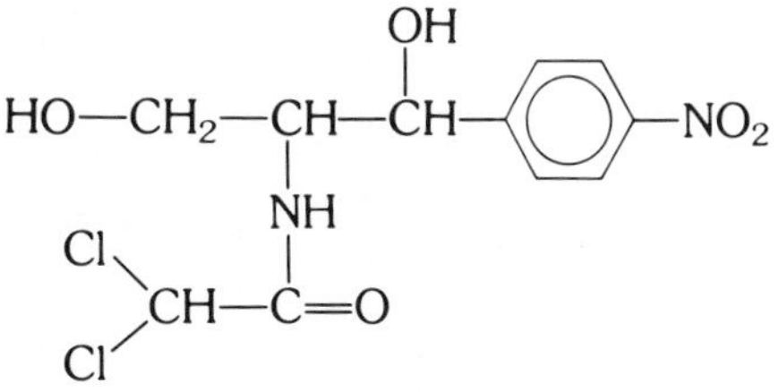

Chloromycetin

22.7 Central Nervous System Stimulants

Central nervous system (CNS) stimulants *are drugs that increase behavioral activity, thought processes, and alertness or elevate the mood of an individual.* Such stimulants differ widely in their molecular structures and in their mechanisms of action. In this section we consider the group of CNS stimulants called amphetamines plus three individual stimulant compounds—caffeine, nicotine, and cocaine.

Amphetamines

Amphetamines *are a class of central nervous system stimulants structurally similar to the neurotransmitter norepinephrine.* The structural relationship between norepinephrine, a chemical naturally produced by the body (Section 17.9), and benzedrine, the simplest amphetamine, is as follows.

Norepinephrine　　　　　　　　　　　　Amphetamine

Benzedrine, also often simply called amphetamine, the parent compound in the amphetamine family, was first synthesized in 1927 as a drug to simulate the actions of the hormone adrenaline (Section 17.9). Following benzedrine's synthesis a number of derivatives also came onto the market. Two of the best-known derivatives are the compounds methamphetamine and methoxyamphetamine.

Methamphetamine (Methedrine)　　　　　　Methoxyamphetamine

Generally, amphetamines increase both the heart and respiratory rates. They reduce fatigue and diminish hunger by raising the glucose level in the blood. Physicians use them to treat mild depression and narcolepsy (the tendency to fall asleep at any time). At one time, they were widely used as appetite suppressants in the treatment of obesity, but because of many adverse effects, their use in weight control has diminished.

A great problem with amphetamines has been the diversion of large quantities of these relatively inexpensive drugs into the illegal drug market (see Figure 22.5). In this market they are known as bennies, pep pills, reds, red devils, speed, STP, dexies, and uppers. Methamphetamine is "speed," and methoxyamphetamine is "STP." Abuse of these drugs can produce severe physiological reactions. Once the drugs wear off, the user tends to "crash" into a state of

FIGURE 22.5
Large quantities of amphetamine tablets are consumed in the illegal drug market. [Alfred Pasieka/Taurus Photos.]

physical and mental exhaustion. Withdrawl produces fatigue and profound and prolonged sleep.

Caffeine

Caffeine, whose structure involves a six-membered heterocyclic ring fused to a five-membered heterocyclic ring, is the most widely used nonprescription central nervous system stimulant.

Caffeine

In addition to stimulating the central nervous system, caffeine can produce a variety of other effects, depending upon the amount consumed. It increases heartbeat and basal metabolic rate, promotes secretion of stomach acid, and steps up production of urine. The overall effect an individual experiences is usually interpreted as a "lift," a feeling of being wide awake and able to focus on mental or manual tasks. Caffeine is probably the CNS stimulant most frequently used to assist individuals in staying awake or to overcome fatigue.

CAFFEINE
INTOXICATION

Excessive consumption of caffeine, medically called caffeine intoxication, leads to restlessness and disturbed sleep, irritation of the stomach, and diarrhea.

TABLE 22.6

Amount of Caffeine Present in Selected Caffeine-
Containing Products

Product	Caffeine (mg)
Coffee	
Drip (5 oz)	146
Percolated (5 oz)	110
Instant, regular (5 oz)	53
Decaffeinated (5 oz)	2
Tea	
One-minute brew (5 oz)	9–33
Five-minute brew (5 oz)	20–46
Canned iced tea (12 oz)	22–36
Cola drinks	
Coca Cola (12 oz)	40
Pepsi Cola (12 oz)	43
Royal Crown Cola (12 oz)	40
Cocoa and chocolate	
Cocoa beverage (water mix, 6 oz)	10
Milk chocolate (1 oz)	6
Baking chocolate (1 oz)	35
Nonprescription drugs (standard dose)	
Stimulants	
Caffedrine capsules	200
No Doz tablets	200
Pain relievers	
Anacin	64
Excedrin	130
Cold remedies	
Dristan	32
Triaminicin	30
Diuretics	
Aqua-Ban	200
Permathene H_2Off	200

What constitutes an excessive intake varies among individuals. It ranges from as low as 200 mg per day to 750 mg per day depending upon the adult.

Caffeine is mildly addicting. People who ordinarily consume substantial amounts of caffeine-containing beverages or drugs experience withdrawal symptoms if caffeine is eliminated. Such symptoms include headache and depression for a period of several days. Many people need a cup of coffee before they "feel good" each morning as a result of this dependence on caffeine.

Coffee beans and tea leaves are the best known of numerous natural sources for caffeine. While coffee is the major source of caffeine for most Americans, substantial amounts may be consumed in soft drinks, tea, and numerous nonprescription medications including combination pain relievers (Anacin, Midol, Empirin), cold remedies (Dristan, Triaminicin), and antisleep agents (No Doz, Vivarin). Caffeine is used in combination with antihistamines to combat the sedative properties of the antihistamine. Table 22.6 gives caffeine amounts derived from several common sources.

A compound structurally similar to caffeine, theobromine, occurs in cocoa drinks. It is a weaker stimulant than caffeine. The structural relationship between the two compounds is as follows.

Caffeine   Theobromine

Nicotine

Next to caffeine, nicotine is the most widely used central nervous system stimulant in our society. Since tobacco leaves naturally contain this substance, it is found in both smoking and chewing tobacco (see Figure 22.6). Nicotine's

FIGURE 22.6
Both smoking and chewing tobacco, obtained from dry tobacco leaves, have as an active ingredient the heterocyclic amine nicotine. [Daniel S. Brody/Stock, Boston.]

chemical structure, like that of caffeine, involves both a six-membered and a five-membered nitrogen-containing heterocyclic ring system. Unlike caffeine, however, the two ring systems in nicotine are not fused together. Also, the nitrogen content of the ring systems is less than that in caffeine.

Nicotine

Nicotine is readily and completely absorbed from the stomach after oral administration and from the lungs upon inhalation. Its mild effect on the central nervous system is rather transient in nature. After an initial response, depression follows. Most cigarettes contain between 0.5 and 2.0 mg of nicotine, depending upon the brand. Conservative estimates indicate that approximately 20% (between 0.1 and 0.4 mg) of this nicotine will actually be inhaled and absorbed into the bloodstream. The physiological effects of smoking a single cigarette can be closely duplicated by the intravenous injection of these amounts of nicotine.

Nicotine is quickly distributed throughout the body, rapidly penetrating the brain, all body organs, the fetus, and, in general, all body fluids. There is now strong evidence that cigarette smoking affects a developing fetus.

As many smokers know, nicotine induces both physiological and psychological dependence. It is, in fact, the most widespread example of drug dependence in the country. Withdrawal is accompanied by headache, stomach pain, irritability, and insomnia. Withdrawal appears to be quite prolonged, as evidenced by the individual who relapses after successfully completing the early stages of withdrawal. Full withdrawal can take six months or longer. The number of cigarettes one must smoke per day to be considered addicted is unclear, but those who smoke 15 or more are very likely cigarette dependent.

In large doses nicotine is a potent poison. Nicotine poisoning causes vomitting, diarrhea, nausea, and abdominal pain. Death occurs from respiratory paralysis. The lethal dose of the drug is considered to be approximately 60 mg. Nicotine was once used as an insectide.

Cocaine

Cocaine, extracted from the coca plant, which grows on the eastern slopes of the Andes mountains in South America, is a very powerful central nervous system stimulant. Mental awareness is increased, and fatigue is decreased. The stimulating effect is short-lived (lasting about an hour) and is often followed by a period of deep depression. In the last decade cocaine has become a dominant chemical in the illegal drug market. Street names for this substance are "coke" and "snow." Recent studies indicate that cocaine is one of the most addictive drugs known, much more addictive than previously thought.

As a street drug, cocaine is usually consumed in the form of its hydrochloride salt, a white, powdery, water-soluble form of the substance.

Cocaine hydrochloride Cocaine

COCAINE AND NASAL "STUFFINESS"

Because of its rapid absorption through the mucous membranes of the nose and throat, it is generally "sniffed" by users. This causes relaxation of the muscles of the bronchial airways of the lungs and constriction of the blood vessels of the nose, making breathing easier. When the drug wears off, however, the bronchial muscles contract and the blood vessels relax, causing regular users nasal stuffiness and tender, bleeding nasal membranes. As a result a cocaine user will often seek relief by sniffing more drug, delaying but intensifying the symptoms. Repeated administrations in this manner lead to a deterioration of the mucous linings, chronic runny nose, and holes in the nose cartilage.

The behavioral and toxic effects produced by cocaine are generally similar to those produced by amphetamines since the modes of action of the drugs are similar, involving simulation of the actions of norepinephrine. The effects of an inhaled dose of about 16 mg of cocaine closely resemble those of an inhaled dose of about 10 mg of amphetamine, except that the effects of amphetamine last longer. Oral doses of cocaine are well absorbed, although so slowly that the effects are often missed. Intravenous administration increases the effects several-fold, though the duration of action is very short (5 to 15 minutes) and the potential for over dosage is great. An overdose of cocaine may cause death through cardiac or respiratory arrest.

22.8 Central Nervous System Depressants

Central nervous system depressants *are drugs that decrease behavioral activity, alertness, and thought processes.* Such compounds are often called *sedative-hypnotic* compounds. Sedatives are drugs that cause relaxation.

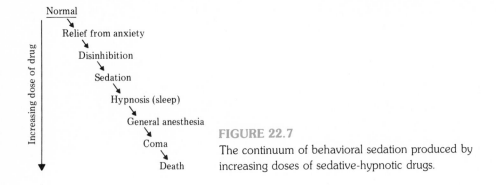

Normal
  Relief from anxiety
    Disinhibition
      Sedation
        Hypnosis (sleep)
          General anesthesia
            Coma
              Death

Increasing dose of drug

FIGURE 22.7

The continuum of behavioral sedation produced by increasing doses of sedative-hypnotic drugs.

Hypnotics produce sleep in the user. The actual effect of a sedative-hypnotic type compound is directly related to dosage. Figure 22.7 shows the continuum of behavioral sedation that can be produced by a central nervous system depressant. Sedative-hypnotics resemble general anesthetics (Section 17.4) in inducing sleep, but they differ in being solid or liquid (rather than gaseous) and in having a longer action than general anesthetics.

In this section we consider barbiturates, a well-known type of CNS depressant, and benzodiazepines, which find extensive use as tranquilizer drugs. Section 22.9 deals with ethyl alcohol, the most popular CNS depressant.

Barbiturates

Barbiturates are a heavily used group of prescription drugs that cause relaxation (tranquilizer), sleep (sedative), and death (overdose). All barbiturates are derivatives of barbituric acid, a cyclic amide first synthesized from urea and malonic acid.

Urea   Malonic acid   Barbituric acid

(The researcher who first synthesized this compound named it after his girlfriend Barbara.)

A large group of physiologically active compounds is obtained when various alkyl groups are substituted for hydrogen atoms at the hydrogen-bearing carbon atom in barbituric acid. Table 22.7 lists common barbiturates and the R groups that are bonded to the barbituric acid structure of each.

Barbiturates depress the central nervous system in a manner similar to alcohol and general anesthetics. The duration of the effects varies with the barbiturate. Phenobarbital is long-lasting, putting people to sleep for 10 to 12 hours. Pentobarbital and secobarbital are short-acting drugs (3 to 4 hours). Amobarbital is an intermediate drug (6 to 8 hours).

Barbiturates are one of the most abused classes of prescription drugs. Because they are sedatives, they are called "downers," in contrast to "uppers" (amphetamines). Barbiturates are especially dangerous when combined with alcohol use (which often occurs). The combined effect (called a synergistic effect) is usually greater than the sum of the effects from the two drugs separately.

Barbiturates induce physical dependence. Withdrawal symptoms appear when administration of the drug is stopped. Physical dependence on barbiturates is in some respects similar to the physical dependence induced by the opiate narcotics (Section 22.3), but differs in two important respects. First, the dose of barbiturate required to induce physical dependence is much higher than the dose usually required to induce sleep. Second, barbiturate withdrawal also differs from opiate withdrawal in that the former may lead to life-threatening convulsions. Withdrawal from the opiate narcotics is not usually as dangerous.

TABLE 22.7

Identity of the R Groups Present in Common Barbiturates

Barbiturate	R'	R'
Barbital	$CH_3{-}CH_2{-}$	$CH_3{-}CH_2{-}$
Phenobarbital (Luminal)	$CH_3{-}CH_2{-}$	$C_6H_5{-}$
Amobarbital (Amytal)	$CH_3{-}CH_2$	$\begin{array}{c}CH_3\\ \end{array}{>}CH{-}CH_2{-}CH_2{-}$ CH_3
Pentobarbital (Nembutal)	$CH_3{-}CH_2{-}$	$CH_3{-}CH_2{-}CH_2{-}\overset{\textstyle |}{CH}{-}CH_3$
Secobarbital (Seconal)	$CH_2{=}CH{-}CH_2{-}$	$CH_3{-}CH_2{-}CH_2{-}\overset{\textstyle |}{CH}{-}CH_3$
Aprobarbital	$CH_2{=}CH{-}CH_2{-}$	$\begin{array}{c}CH_3\\ \end{array}{>}CH{-}$ CH_3

TABLE 22.8
Structures of Selected Prescription Benzodiazepine Tranquilizer Compounds

Diazepam (Valium)
antianxiety effect

Chlordiazepoxide (Librium)
antianxiety and appetite-stimulating effects

Lorazepam (Ativan)
antianxiety effect

Flurazepam (Dalmane)
treatment of insomnia

Benzodiazepines

Benzodiazepines are structurally unrelated to the barbiturates, but, pharmacologically, closely resemble them. The central feature of the benzodiazepine structure is a seven-membered cyclic amide ring containing two nitrogen atoms. Table 22.8 gives the structures of four commonly used benzodiazepines. Valium, Librium, and Ativan are three popular tranquilizers. Dalmane is used primarily as a sleeping pill (to treat insomnia).

22.9   Alcohol: The Most Popular Drug

Ethyl alcohol, often simply called alcohol (Section 17.3), is as powerful as the more "notorious" CNS depressants. Yet it is so accepted in our society that groups of "learned" men and women meeting to discuss the abuse of drugs see no irony in consuming alcoholic beverages prior to and during their meetings.

Most authorities will agree that alcoholism represents the major drug problem in the United States and most other countries, especially when one views alcoholism in terms of productivity loss, accidents, damaged health, crime, and death.

Alcohol is a general central nervous system depressant similar to the barbiturates. The drinker's excited behavior actually stems from depression of brain control areas that ordinarily exert inhibitory control over psychomotor activity and behavior. Because of these effects, many people believe alcohol is a stimulant, which it is not.

Alcohol is rapidly and completely absorbed from the entire gastrointestinal tract. Already a liquid, alcohol does not have to dissolve in the stomach as does a drug in tablet form. The effects of alcohol on the human body that accompany alcoholic beverage consumption are summarized in Table 22.9. Note that the numbers in this table are only average values and can depend on many factors such as a person's weight, drinking history, and the amount of food in his or her stomach. Drinking history is very important. Chronic drinkers develop a tolerance for alcohol that permits them to sustain higher alcohol blood levels without achieving the desired euphoric state created at an earlier age.

The numbers in Table 22.9 can be related to the problem of drunk driving by considering what constitutes a "drunk driver" in terms of blood-alcohol level. Drivers are considered to be drunk in most states (46 of the 50) when the blood-alcohol level reaches 0.10% (m/v). Four states have more stringent guidelines, 0.08% (m/v) in three and 0.05% (m/v) in one. A blood-alcohol level of 0.10% (m/v) is approximately equivalent to 1 drop of alcohol in 1000 drops of blood.

Once in the body, alcohol is broken down by the liver. The product from alcohol oxidation is the two-carbon aldehyde acetaldehyde.

$$CH_3-CH_2-OH \xrightarrow[\text{oxidation}]{\text{enzymatic}} CH_3-C\overset{O}{\underset{H}{\diagup\diagdown}}$$

Ethanol (Alcohol)   Acetaldehyde

Acetaldehyde, in turn, is oxidized to the two-carbon carboxylic acid acetic acid, which is a normal constituent of cells.

$$CH_3-C\overset{O}{\underset{H}{\diagup\diagdown}} \xrightarrow[\text{oxidation}]{\text{enzymatic}} CH_3-C\overset{O}{\underset{OH}{\diagup\diagdown}}$$

Acetaldehyde   Acetic acid

ALCOHOLICS AND ANTABUSE

Drugs, such as disulfiram (Antabuse), that are used to treat alcoholics prevent oxidation of acetaldehyde to acetic acid (the second oxidation step). Buildup of acetaldehyde in the body causes dizziness, headaches, nausea, vomiting, and difficulty in breathing.

TABLE 22.9

Approximate Relationship Between Drinks Consumed, Blood-Alcohol Level, and Behavior for a Moderate Drinker Weighing 150 Pounds

Number of Drinks*	Blood-Alcohol Level (mass/volume %)	Behavior
2	0.04	mild sedation; tranquility
4	0.08	lack of coordination
6	0.12	obvious intoxication
10	0.20	unconsciousness
20	0.40	possible death

* Rapidly consumed 1-oz shots of 90-proof whiskey or 12 oz of beer.

Long-term, excessive use of alcohol is known to cause undesirable effects such as cirrhosis of the liver, loss of memory, and strong physiological addiction. In recent years, links have been established between certain birth defects and the ingestion of alcohol by mothers during pregnancy.

Despite a knowledge of the harmful effects of alcohol, approximately two-thirds of the adult population of the United States drink alcoholic beverages, at least occasionally.

22.10 Marijuana

The substance marijuana, commonly known as "pot" or "grass," is a preparation made from the leaves, flowers, seeds and small stems of a hemp plant called *Cannabis sativa* (see Figure 22.8). These are generally dried and smoked.

The major "active" ingredient in marijuana is a complex phenol (Section 17.3) known as tetrahydrocannabinol, called THC for short.

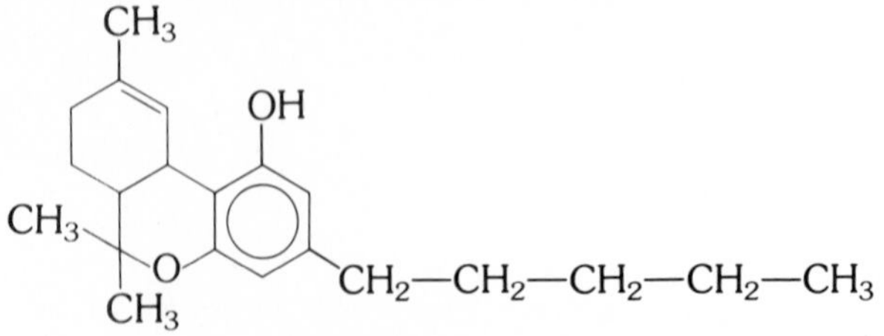

Tetrahydrocannibinol

The THC content of marijuana varies considerably depending on the genetic

(a)

(b)

FIGURE 22.8

(a) The marijuana plant, *Cannabis sativa*. (b) Marijuana pressed into a brick and broken, with pipe (left) and cigarette papers (right). [(a) Jerry Howard/Stock, Boston; (b) Richard Lawrence Stack/Black Star.]

variety of plant. Most THC sold in the illegal drug market of North America has a THC content of 1 to 2%.

THC is a difficult compound to classify. At low to moderate doses, THC is a mild central nervous system depressant, resembling alcohol and antianxiety agents such as Valium and Librium in its pharmacological effects. Unlike CNS depressants, however, higher doses of THC may produce (in addition to sedation) hallucinations and heightened sensation, effects similar to those caused by hallucinogenic drugs such as LSD. Also unlike the sedatives, high doses of THC do not produce anesthesia, coma, or death. The structure of THC does not resemble that of any known compound associated with nerve-impulse transmission in the human body. Marijuana is one of the least understood of all natural drugs, and its mechanism of action in the body remains unknown.

In the United States, THC is usually administered in the form of a hand-rolled marijuana cigarette, the average cigarette containing between 1 and 2 grams of plant material. Thus, if a marijuana cigarette contained 1.5 grams of plant materials with a THC content of 2%, the cigarette would contain approximately 30 mg of THC. In general, approximately one-half of the THC present in a marijuana cigarette, about 15 mg of THC, is actually available in the smoke. It is extremely unlikely that 100% of this available THC is actually absorbed into the bloodstream.

The onset of action of THC is usually within minutes after smoking begins, and peak concentrations in plasma occur in from 10 to 30 minutes. Unless more is smoked, the effects seldom last longer than 2 or 3 hours. Since THC is only very slightly soluble in water, once it enters the bloodstream and is distributed to the various organs of the body, it tends to be deposited in the tissues, especially those that have significant concentrations of fatty material. Because THC is soluble in fat, it readily penetrates the brain; the blood-brain barrier does not appear to hinder its passage. Similarly, THC readily crosses the placental barrier and reaches the fetus. Unlike alcohol, THC persists in the bloodstream for several days, with the products of its breakdown remaining in the blood for as long as 8 days.

The commonly observed responses to the acute use of either orally administered or smoked marijuana are seen as changes in the functioning of both the CNS and the cardiovascular system. Increases in pulse rate and slight increases in blood pressure are commonly encountered. Blood vessels of the cornea also dilate, resulting in the bloodshot eyes usually associated with indulgence in alcohol. THC users frequently report increased appetite, dry mouth, occasional dizziness, increased visual and auditory perception, and some nausea.

It was previously thought that physical dependence on THC did not develop, and this may be true in recreational users of the drug. Drug withdrawal symptoms, however, are seen in some individuals exposed repeatedly to high doses.

Practice Questions and Problems

MILD PAIN RELIEVERS

22-1 What is a mild analgesic?

22-2 What are the three most widely used mild analgesics?

22-3 What is the chemical structure of aspirin?

22-4 Write the chemical equation for the hydrolysis of aspirin within the human body, and identify the product that is the actual pain-relieving agent.

22-5 What is the major side effect of taking aspirin?

22-6 What is the mechanism by which aspirin exerts its pain-relieving effect within the body?

22-7 What is the advantage of using acetaminophen rather than aspirin as a pain-reliever? What is the disadvantage?

22-8 Name the major mild pain reliever that is a(n)
a. Substituted carboxylic acid
b. Substituted amide
c. Ester of salicylic acid

22-9 What is the identity of the mild analgesic present in each of the following brand-name medications?
a. Advil   b. Nuprin   c. Tylenol
d. Excedrin   e. Bufferin   f. Empirin

STRONG PAIN RELIEVERS

22-10 What is an opiate?

22-11 What are the two major pain-relieving drugs found naturally in opium extracts?

22-12 How does codeine differ from morphine in (a) chemical structure and (b) physiological effects?

22-13 How does heroin differ from morphine in (a) chemical structure and (b) physiological effects?

22-14 What is the mechanism by which morphine exhibits its pain-relieving effects.

22-15 Why are morphine salts, rather than morphine itself, the usual form in which morphine is administered to patients?

22-16 What is the narcotic analgesic present in each of the following?

a. Tylenol #2   b. Empirin #2
c. Percodan   d. Darvocet
e. Tylox

22-17 What is methadone maintenance and how does it work?

GASTROINTESTINAL DRUGS

22-18 What is an antacid?

22-19 List five neutralizing agents used in over-the-counter antacid formulations and identify a "brand-name" product containing each agent.

22-20 What are the "ideal" characteristics for an antacid?

22-21 List the mode of action and the "active ingredient" in each of the following laxative formulations
a. Metamucil   b. Milk of Magnesia
c. Phospho-Soda   d. Dulcolax

22-22 Describe the mode of action for the antiulcer medications Tagamet and Zantac.

22-23 What structural similarities exist between each pair?
a. Histamine and cimetidine (Tagamet)
b. Cimetidine (Tagamet) and ranitidine (Zantac)

CARDIOVASCULAR DRUGS

22-24 What is atherosclerosis, and what are the two major ingredients in atherosclerotic plaque?

22-25 What is hypertension?

22-26 Explain how a beta-blocker drug lowers blood pressure.

22-27 What is the structural similarity between the beta-blocking agents Inderal and Tenormin?

22-28 What effect does calcium ion have on the heart?

22-29 Explain how a calcium channel blocker works and how it helps in treating the effects of heart disease.

ANTIBACTERIAL AGENTS

22-30 What part of the molecular structure is common to all compounds in each group?
a. Sulfa drugs   b. Penicillins
c. Cephalosporins   d. Tetracyclines

22-31 Describe how the sulfa drug sulfanilamide works.

22-32 How was penicillin discovered?

22-33 What advantages do semisynthetic penicillins have over naturally occurring penicillin (penicillin G)?

22-34 What is penicillinase?

22-35 What is the structural similarity between penicillins and cephalosporins?

22-36 What advantages do tetracyclines have over penicillins and cephalosporins?

CNS STIMULANTS

22-37 How do amphetamines stimulate the central nervous system?

22-38 What is the basic structural feature common to various amphetamines?

22-39 Draw the complete structure of the caffeine molecule.

22-40 What are the best known natural sources of caffeine?

22-41 What are the symptoms of "caffeine intoxication"?

22-42 Indicate which substance contains more caffeine in each of the following pairs of substances.
 a. 5 oz of percolated coffee or 5 oz of tea
 b. 12 oz of cola drink or 1 cup of coffee
 c. 1 cup of coffee or 1 No Doz tablet
 d. 1 oz of milk chocolate or 1 oz of baking chocolate

22-43 Draw the complete structure of the nicotine molecule.

22-44 What is the major natural source of nicotine?

22-45 Describe the effect of nicotine on the CNS system.

22-46 What are the symptoms of nicotine withdrawal?

CNS DEPRESSANTS

22-47 What is the structural feature common to all barbiturate molecules?

22-48 How is the basic barbiturate structure modified to change the properties of individual barbiturate drugs?

22-49 What is the chemical structural feature common to the benzodiazepine tranquilizers?

ALCOHOL

22-50 What human behavior is associated with each of the following blood-alcohol levels?
 a. 0.04% (m/v)   b. 0.08% (m/v)
 c. 0.12% (m/v)   d. 0.20% (m/v)

22-51 What constitutes a legally "drunk driver" in terms of blood-alcohol level?

22-52 Write the chemical equations for the two-step process by which alcohol is broken down through oxidation within the human liver.

22-53 Describe how the drug disulfiram (Antabuse) functions within the body when used in treating alcoholics.

MARIJUANA

22-54 THC, the major active ingredient in marijuana, belongs to which class of organic compounds?

22-55 What are the similarities and differences between the actions of marijuana and of central nervous system depressants such as alcohol and barbiturates?

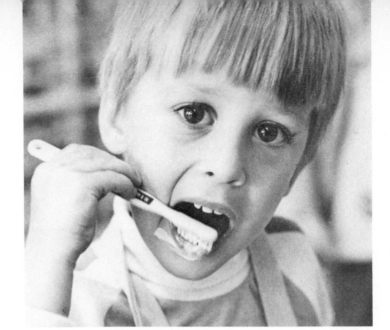

23

Personal Care Products

CHAPTER HIGHLIGHTS

23.1   Cosmetics

A **cosmetic** *is a chemical preparation (usually a mixture of substances) intended to be rubbed, poured, sprinkled, or sprayed on the human body to cleanse, beautify, promote attractiveness, or alter appearance.* Use of cosmetics goes back thousands of years. For example, evidence indicates that ancient Egyptians used powdered colored substances as eye shadow and also used perfumed hair oils. Such early cosmetic use, however, is insignificant when compared to that which occurs in our day. Almost all women use powders and creams to alter their complexion and perfumes to create "interesting" aromas. Men consume great quantities of products such as aftershave lotion, deodorants, and antiperspirants. The cosmetics industry in the United States at present is a $15 billion a year industry.

Cosmetics are distinguished from drugs in that drugs are used to alter body function. Technically, antiperspirants and antidandruff shampoos, two types of substances we normally think of as cosmetics, should be classified as drugs.

In this chapter we consider major classes of chemical preparations used as cosmetics. The active ingredients present in such preparations are of concern to us as well as the chemical effects produced by these ingredients.

23.2   Deodorants and Antiperspirants

Perspiration formation is a natural body function that helps regulate body temperature, eliminate certain waste products, and protect the skin against dryness. Normal skin secretions do not have objectionable odors. However, when skin bacteria interact with such secretions, the result may be unpleasant odors.

Up to an environmental temperature of about 88 to 90°F, the sweat glands secrete continuously but so slowly that the droplets of moisture evaporate before they can become visible. If the environmental temperature rises above this

730

critical range, there is a sharp increase in the number of sweat glands functioning and the volume of perspiration they produce. Drops of visible moisture break out over the skin surface. When heavy labor is performed under extremely hot conditions, as much as 3 gallons of perspiration may be produced in 24 hours.

The major source of perspiration is the eccrine glands, found on almost all surfaces of the body. Chemically, eccrine perspiration resembles a very dilute urine solution. It consists of about 99% water, about 0.5% sodium chloride, and the rest organic substances, mainly urea (Section 17.9), along with uric acid, glucose, lactic acid, amino acids (Section 25.3), and fatty acids (Section 24.9). The exact composition of perspiration is determined partly by heredity and to some degree by the content of the diet. Both these factors contribute to a real but often unconsidered problem of people of different races or cultures sometimes smelling offensive to one another.

Both deodorants and antiperspirants are used to control the problem of undesirable body odor resulting from the interaction of bacteria with perspiration. The difference between these two types of substances is in their mode of action. A **deodorant** *is a formulation that kills odor-causing bacteria on the skin.* Two of the most used antibacterial agents in deodorant formulations are the compounds benzethonium chloride and zinc phenolsulfonate, whose structures are as follows.

Benzethonium chloride

Zinc phenolsulfonate

Sometimes baking soda, which has very mild antibacterial activity, is also present in a formulation. Usually, deodorants also contain fragrances (Section 23.6) whose function is to mask body odors.

An **antiperspirant** *is a formulation that stops or retards the operation of sweat glands and thereby lowers the amount of perspiration produced.* Since such action is a modification of body function, antiperspirants are considered drugs by the Food and Drug Administration (FDA).

The active ingredient in almost all antiperspirants is one of the aluminum chlorohydrates, $Al_2(OH)_5Cl$ or $Al_2(OH)_4Cl_2$. These two water-soluble ionic compounds produce Al^{3+} ion in solution. The Al^{3+} ion is an *astringent*, a chemical species that acts by constricting the opening of sweat glands. By this mechanism the flow of perspiration is reduced or, for some glands, prevented altogether. Aluminum chlorohydrates also have some antibacterial properties.

23.3   Hair Care Products

The average human scalp contains about 100,000 hairs. Each hair consists of a *root*, enclosed within a hair follicle in the skin, and a *shaft*, the visible portion. Only the root portion of a hair is alive. The shaft, composed of a fibrous protein called keratin (Section 25.9), is dead; it has no blood vessels or nerves of its own.

Each hair follicle has two or more sebaceous glands opening into it that secrete an oily substance (sebum). This secretion spreads out to form a protective film over the entire shaft that helps maintain softness and pliability. Too little sebum results in dry hair, and too much sebum produces greasy hair. Sebum secretion contains fats (Section 24.10), soaps (Section 24.11), cholesterol (Section 24.13), and inorganic salts. It is the accumulation of dirt and bacteria adhering to sebum secretion that makes hair look dull instead of shiny.

A hair shaft cannot repair itself. If any hair shaft becomes dry, cracked, or loses its softness, the human body has no mechanism to correct the problem. Thus, the need for proper hair care becomes apparent. In this section we consider three cosmetic-type formulations used in hair care—shampoos, conditioners, and hair spray. In addition, the problem of dandruff is considered.

Shampoos

Shampoos *are detergent solutions to which have been added coloring agents, scents, acidifiers, foam stabilizers, and selected other ingredients.* Their primary function is to remove dirt and excess sebum from hair.

The detergent present in a shampoo is the actual cleansing agent. Its mode of action is enhancement of the solubility of greases and oils in water solution (Section 24.11). Sodium lauryl sulfate (SLS) is the most used shampoo detergent.

$$CH_3-CH_2-CH_2-CH_2-CH_2-CH_2-CH_2-CH_2-CH_2-CH_2-CH_2-CH_2-O-\overset{\overset{\displaystyle O}{\|}}{\underset{\underset{\displaystyle O}{\|}}{S}}-O^-\ Na^+$$

Sodium lauryl sulfate (SLS)

SLS is a mild cleansing agent compared to those present in laundry detergents. A strong detergent would solubilize all sebum present, a situation not desired. Shampoos for oily or dry hair differ in the relative amounts of detergent present in a given volume. The greater the amount of detergent present, the greater the amount of sebum that dissolves.

ACIDIFIERS IN SHAMPOOS

Extremes in acidity or alkalinity can cause hair protein to decompose. The ideal pH (Section 10.9) for wet hair is between 4 and 6. Since detergent solutions, even those of mild detergents, are basic, shampoo formulations include an acidifier to lower the pH. Citric acid (the same acid found in citrus fruits) and phosphoric acid are commonly used.

Citric acid   Phosphoric acid

FIGURE 23.1
Foam stabilizers are additional ingredients in all hair shampoos.

FIGURE 23.2
Hair conditioner formulations help combat hair problems such as split ends, static electricity, and snarling.

Advertising claims such as "pH-controlled," "nonalkaline," and "acid-balanced" draw attention to the presence of an acidifier.

All shampoos contain additional ingredients that have very little effect on hair cleanliness or appearance. For example, a foam stabilizer, such as lauramide diethylamine, is usually added so that a thick lather forms (see Figure 23.1).

$$CH_3\!-\!CH_2\!-\!CH_2\!-\!CH_2\!-\!CH_2\!-\!CH_2\!-\!CH_2\!-\!CH_2\!-\!CH_2\!-\!CH_2\!-\!CH_2\!-\!\underset{\underset{O}{\|}}{C}\!-\!\underset{\underset{\underset{CH_3}{|}}{CH_2}}{N}\!-\!CH_2\!-\!CH_3$$

Lauramide diethylamine

Although "foaming" shampoo does not get hair any cleaner, it does make rinsing easier.

Hair Conditioners

Cleanliness and shininess are not the only desirable properties for hair. Most people appreciate hair that is easy to comb (no snarls), has no split ends, and is not unruly (static electricity). Hair conditioners are formulations that help accomplish these objectives (see Figure 23.2). Four key ingredients in hair conditioner formulations are

1. An *oil* or *resin*, which may be as common as mineral oil or lanolin (grease from sheep's wool) or as exotic as aloe vera or jojoba oil. The function of the oil or resin is to mimic the lubricating effect of sebum, preventing moisture loss.

2. A **humectant**, *a substance that attracts water from the atmosphere to itself*. The most common humectants are glycerin, propylene glycol, and sorbitol. All three of these compounds are alcohols (Section 17.3) containing multiple hydroxyl groups.

$$CH_2\text{—}CH\text{—}CH_2 \qquad CH_2\text{—}CH_2\text{—}CH_2 \qquad CH_2\text{—}CH\text{—}CH\text{—}CH\text{—}CH\text{—}CH_2$$
$$\;\;|\qquad\;\;|\qquad\;\;| \qquad\qquad\;\;|\qquad\qquad\;\;| \qquad\qquad\;\;|\qquad\;\;|\qquad\;\;|\qquad\;\;|\qquad\;\;|\qquad\;\;|$$
$$OH \quad OH \quad OH \qquad\quad OH \qquad\qquad OH \qquad\quad OH \quad OH \quad OH \quad OH \quad OH \quad OH$$

Glycerin   Propylene glycol   Sorbitol

Hydrogen bonding (Section 8.16) between alcohol hydroxyl groups and water molecules is the basis for the water-attracting ability of these humectants (see Figure 23.3)

3. *Protein fragments*, which help in combating the problem of hair damage (split ends). Usually derived from animal hides and hoof, they function much like the spackling compound used to repair a damaged wall; they fill in the cracks and dents. Damaged areas of hair possess relatively more negative charge. (The inner core of hair has a different composition from that of its outer layer and tends to possess greater negative charge.

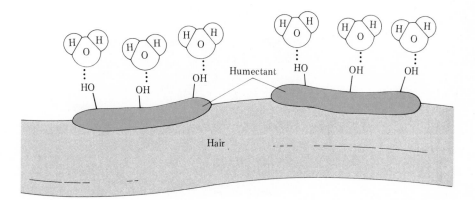

FIGURE 23.3
The hydroxyl groups present in commonly used humectants hydrogen bond to water molecules obtained from air.

Damage exposes this inner core.) The protein fragments, which are polar (Section 5.12), are attracted to the more negative (damaged) areas of hair. This causes "recombining" of split ends and smoothing out of "rough" spots.

4. *Substituted ammonium salts*, are perhaps the most significant ingredient in hair conditioners. The positive ion in such salts is a large ion produced by substituting one or more hydrocarbon groups (—R groups; see Section 17.1) for the hydrogen atoms in an ammonium ion.

$$\left[\begin{array}{c} H \\ | \\ H-N-H \\ | \\ H \end{array}\right]^{+} \qquad \left[\begin{array}{c} H \\ | \\ H-N-(CH_2)_{15}-CH_3 \\ | \\ (CH_2)_{15}-CH_3 \end{array}\right]^{+}$$

Ammonium ion   Disubstituted ammonium ion

STATIC ELECTRICITY AND HAIR CONDITIONERS

Such large positive ions counteract the static electricity (negative charge) that naturally builds up in hair, thereby reducing "fly-away" problems.

Hair Sprays

Hair sprays coat the hair with a "film" of a polymeric material (see Figure 23.4). It is this film that holds the hair in place.

Major ingredients in an aerosol hair spray are a propellant–solvent mixture, a resin (the polymeric material), a plasticizer (to make the polymeric material more pliable), and a water repellent. The resin concentration of hair sprays ranges from 4 to 6% by mass. The solvent is specially denatured alcohol (Section 17.3),

FIGURE 23.4
The "film" of polymeric material that forms when hair spray is allowed to dry on a surface and then is pulled up with a metal object.

the denaturants in which are compatible with biological skin processes. The evaporation of solvent causes the resin to "set."

One of the most used hair spray resins is the addition polymer polyvinylpyrrolidone (PVP).

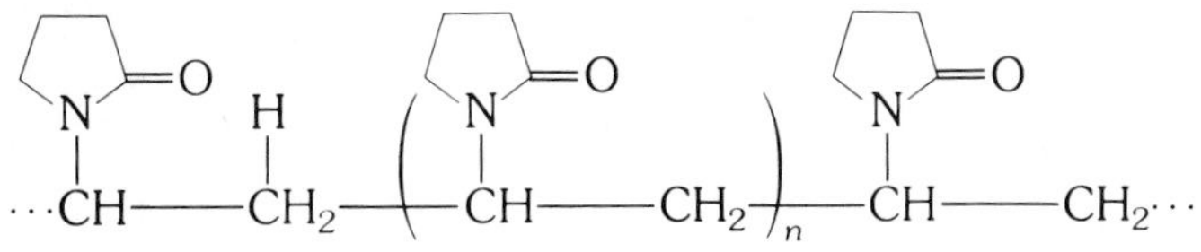

Polyvinylpyrrolidone (PVP)

An undesirable property of PVP is the tendency to pick up moisture. Better moisture control is obtained from a copolymer (Section 18.4) which is a 60:40 blend of vinylpyrrolidone and vinyl acetate (Section 18.10).

Dandruff

Dandruff is a scalp problem, not a hair problem. All over the body, skin cells continually slough off unnoticed to make way for new ones. In the scalp area, such cells are often caught by hair and combine with sebum and dirt. The result is the telltale flakes we know as dandruff.

Dandruff shampoos contain ingredients that either slow the "sloughing off" process, usually zinc pyrithione, or break up the flakes into insignificant pieces, usually salicyclic acid (Section 22.2).

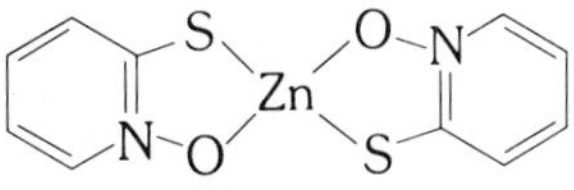

Zinc pyrithione

23.4 Skin Care Products: Creams and Lotions

Cosmetic preparations designed for the skin act only on its outermost layer, the epidermis. This very thin layer, less than 0.12 mm (1/200 inch) thick, is composed mostly of dead cells that have been converted to the moisture-repellent protein keratin (Section 25.9). The moisture content of the epidermis is approximately 10%, much less than the 70% water associated with most living cells.

DRY SKIN

Epidermal keratin has considerable resistance to damage, but it can become cracked, irritated, or diseased if it is not taken care of. Dry skin results when the moisture content of the epidermis drops below 10%. Such a change can result

TABLE 23.1

Typical Hand Cream and Moisturizing Hand Lotion Formulations

Hand Cream	Composition (% by mass)	Moisturizing Hand Lotion	Composition (% by mass)
Mineral oil (emollient)	1	Petrolatum (emollient)	10
Stearic acid (emollient)	5	Mineral oil (emollient)	18
Emulsifier	12	Emulsifier	3
Sorbitol (humectant)	10	Propylene glycol (humectant)	3
Water	72	Water	63

from too much exposure to sun and wind or too much washing (which removes sebum). Excess moisture can also cause problems. A moisture content significantly greater than 10% creates suitable conditions for the growth of undesirable microorganisms.

Cosmetic creams and lotions are moisturizing products designed to increase or maintain the water content of the epidermis. Such cosmetics contain both humectants (Section 23.3) and emollients as active ingredients. An **emollient** *is a substance that prevents water loss through formation of a waterproof coating.* Emollients are often called "skin softeners" because the presence of an emollient layer causes the skin to feel smooth and slick to the touch.

Fats and oils (Section 24.10) from many sources find use as emollients. They include lanolin (grease from wool), avocado oil, soybean oil, and safflower oil as well as numerous synthetic oils.

Cosmetic creams and lotions are emulsions (Section 9.7), that is, colloidal suspensions of one liquid in another. Most commonly the preparation is an oil-in-water emulsion in which small droplets of emollients and humectants are dispersed throughout a water solution. Such a preparation coats the skin with a water-repellent surface. Table 23.1 gives typical formulations for a hand cream and a moisturizing hand lotion.

The emollients listed in Table 23.1 all involve either hydrocarbons or long-chain hydrocarbon derivatives. Petrolatum, also called petroleum jelly, is a colorless to amber semisolid mixture of saturated and unsaturated hydrocarbons (alkanes and alkenes; see Sections 16.4 and 16.9). The molecular size of hydrocarbons in petrolatum mixtures is in the C_{16} to C_{32} range. Mineral oil, also called liquid petrolatum, is an almost tasteless and odorless mixture of hydrocarbons obtained from the distillation of high-boiling petroleum fractions (Section 19.4). Stearic acid, a major component of beef tallow and lard and a minor component of vegetable oils (Section 24.10), is simply an 18-carbon monocarboxylic acid.

$$CH_3-(CH_2)_{16}-C\overset{\displaystyle O}{\underset{\displaystyle OH}{\diagup}}$$

Stearic acid

23.5 Face Makeup Formulations

In this section we consider three types of cosmetics used as face makeup. They are (1) face powder, (2) lipstick, and (3) eye makeup.

Face Powder

The purpose of face powder use is twofold: (1) to give skin a smooth appearance by covering up the "shiny" appearance caused by skin oil (sebum) and (2) to cover up minor skin blemishes.

Numerous ingredients are needed to give face powder the desired spreading ability, sticking properties, absorbency (for sebum and other oily skin materials), and proper appearance (not too dull and not too shiny). A typical powder formulation is given in Table 23.2. Note the dominance of inorganic materials over organic materials in the formulation. Talc, the major ingredient, is a naturally occurring soft mineral with the composition $Mg_3(Si_2O_5)_2(OH)_2$ (Section 13.5). In powdered form this mineral is an excellent absorbent. Precipitated chalk, the other absorbent present, is a finely divided form of calcium carbonate $(CaCO_3)$ that is synthetically produced.

Lipstick

The most obvious component of a lipstick, its coloring agents, is the principal factor is consumer selection of the product. Lipstick contains sizable amounts of coloring agent, compared to most cosmetics (Table 23.3).

As with creams and lotions (Section 23.4), hydrocarbon or hydrocarbon-like molecules make up the base, in which the coloring agents are dissolved or

TABLE 23.2
A Typical Face Powder Formulation

	Composition (% by mass)
Talc (absorbent)	56
Precipitated chalk (absorbent)	10
Zinc oxide (astringent)	20
Zinc stearate (bonder)	6
Perfume (odor)	trace
Dye (color)	trace

TABLE 23.3

Composition of a Typical Lipstick

Ingredient	Function	Composition (% by mass)
Dye	provides color	4–8
Castor oil, Alkanes	dissolves dye	50
Lanolin	emollient	25
Waxes	causes stiffness	18
Perfume	imparts pleasant odor	trace

suspended. Castor oil, a common component of lipstick base, is obtained from the seeds of the castor bean. Waxes (Section 24.12) are also mixed with the base material. They make the lipstick harder so it retains its shape in the dispensing tube. Perfumes are needed to cover up the slightly unpleasant odor of the hydrocarbon base. The dyes used in most modern lipsticks are dibromofluorescein (yellow-red) and tetrabromofluorescein (purple), mixed in various proportions.

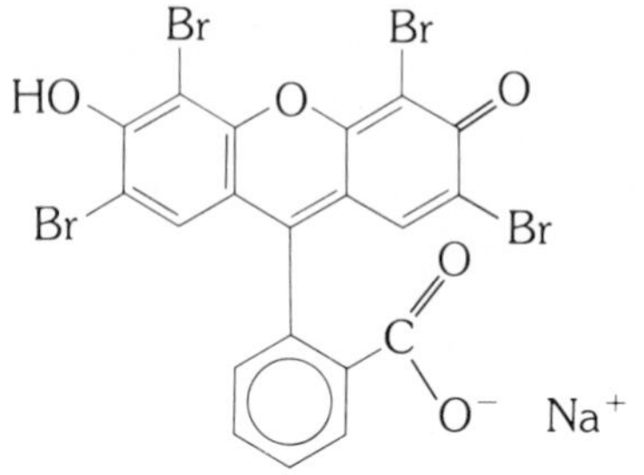

Sodium tetrabromofluorescein

The major function of lipstick is as a beauty aid. A secondary function is to protect the lips from drying out (chapping). Here it functions as an emollient. The surface area of the lips dries out easier than most other skin surfaces. The normal moisture content of the lip surfaces is maintained primarily by the mouth.

Eye Makeup

As with lipstick, consumer selection of an eye makeup, another beauty aid, is linked primarily to color. Eye makeup is marketed for application to the eyebrows (eyebrow pencils), the eyelashes (mascara), and eyelids (eye shadow).

Eyebrow pencils have formulations similar to that for lipstick. The pigments used for color, however, differ. Lampblack, a soot obtained by burning heavy oils in insufficient air, is the major black pigment. Shades of brown are obtained by using iron oxide pigments in combination with lampblack.

Soaps (Section 24.11) are major ingredients in mascara bases. A typical formulation consists of 50% soaps, 40% waxes, 5% lanolin, and 5% coloring matter. Waxes, which include beeswax and paraffin wax, control the hardness of the product. Mineral coloring agents—chromic oxide (Cr_2O_3) and ultramarine (a complex sodium aluminum silicate), respectively—give the greenish and bluish tints.

Both petrolatum (Section 23.4) and titanium dioxide, a white powder, are used as eyeshadow bases. A typical petrolatum-based formulation would be 60% petrolatum, 6% lanolin, 10% fats and waxes, and the balance zinc oxide (white) plus tinting and coloring agents.

23.6 Sunscreens

Both sunburn and suntan are caused by the interaction of ultraviolet (UV) radiation from the sun with skin. Sudden high levels of UV radiation exposure can cause the skin to burn. Steady low-level exposure to UV radiation can have a different effect, that of a suntan.

Sunscreen formulations are used to moderate the amount of UV radiation that interacts with skin, thus preventing burning and encouraging tanning. Although they are not usually considered so, sunscreens are cosmetics. They are applied to the skin, and they do not alter body functions. It is their function to filter out (by absorption) UV radiation present in sun rays.

Human skin has a built-in defense system to protect itself against ultraviolet radiation. It is a brown-colored skin pigment called melanin, which acts as a protective barrier by absorbing and scattering UV light. Structurally, melanin is a polymeric substance involving many interconnected units. A representation of a portion of its structure is

The darker that natural skin color is, the greater the number of melanin molecules already present in the upper layers of the skin. Thus, a fair-skinned person has less melanin pigment than does a dark-complexioned person. Correspondingly, dark-complexioned people have more built-in protection against sunburn (UV radiation) than do fair-skinned people.

When melanin-producing cells deep in the skin are exposed to UV radiation, melanin production increases. The presence of this extra melanin in the skin gives the skin an appearance that we call a "tan." The larger the melanin molecules so produced, the deeper the tan. People who tan readily have skin that can produce a large amount of melanin.

Sunscreen products contain substances that absorb UV radiation in a manner similar to that of melanin. Sunscreen use allows sunbathers to control UV radiation dosage and thus avoid sunburn (too much UV radiation) and yet develop a suntan (a small amount of UV radiation). Fair-skinned people must use stronger sunscreen protection since they have less natural protection (melanin).

The UV-absorbing agents in most sunscreens are compounds that are derivatives of either para-aminobenzoic acid (PABA) or benzophenone.

PABA   Benzophenone

PABA itself, initially used in sunscreen formulations, has been replaced by ester derivatives because they are less soluble in water. Decreased water solubility helps with the problem of maintaining protection while swimming. Benzophenone derivatives have various attachments on its benzene rings. Figure 23.5 shows the extent of UV radiation absorption (as a function of wavelength) by the two parent compounds. Note that absorption is maximized for wavelengths that pose the greatest danger for skin.

A sunscreen's ability to protect skin is indicated by a "sun protection factor" or SPF number—a number between 2 and 15. The higher the number, the more protection the product offers. Sunscreens with an SPF of 15 filter out so much UV radiation that they are essentially complete sun blocks. Table 23.4 gives FDA recommended sunscreen strength to use at the start of the "tanning" season. If a person tans at all, he or she can gradually decrease the strength of the sunscreen as the natural sunscreen, melanin, increases. No sunscreen product can make the skin tan faster or more deeply than the melanin-producing ability of the skin

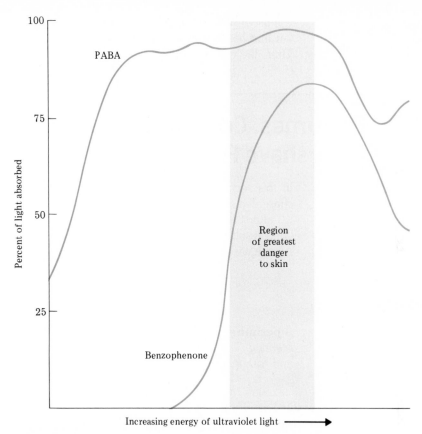

FIGURE 23.5
PABA and benzophenone, the active ingredients in sunscreen formulations, absorb most of the ultraviolet light most harmful to human skin.

TABLE 23.4
FDA-Recommended Sun Protection Factors

Complexion	Sun Sensitivity	Eventual Color	SPF Recommended
Very fair, often freckled	Always burns easily	Same, no tan	8–15
Fair	Always burns easily	Almost the same, little tan	6–7
Light to medium	Burns moderately	Tans gradually to light brown	4–5
Medium	Burns a little	Tans readily, to moderate brown	2–3
Dark brown or black	Rarely or never burns	Deeply pigmented skin may or may not darken	2–3 or none

will allow. When a person sunburns, the skin peels. When peeling occurs, any tan that has been built up (excess melanin) will slough off with the dead skin. Thus, the tanning process must begin anew.

23.7   Perfumes, Colognes, and Aftershave Preparations

In this section we consider three types of cosmetic products where the purchase decision is based primarily on fragrance: (1) perfumes, (2) colognes, and (3) aftershave preparations. For the first two of these products fragrance is the "end result" rather than a quality of a product having another major purpose. Aftershave preparations have other purposes beside fragrance.

Perfumes

A **perfume** *is a chemical formulation with a pleasant odor that is used for body scenting.* The word perfume comes from the Latin words *per*, meaning through, and *fumus*, meaning smoke, and derives from the pleasant-smelling smoke of

TABLE 23.5
Compounds with Flowery and Fruity Odors Used as Fragrances in Perfumes

Name	Structure	Odor
Citral	$CH_3-C=CH-CH_2-CH_2-C=CH-C \overset{O}{\underset{H}{\diagdown}}$ with CH_3 branches	Lemon
Ionone		Violet
Jasmone		Jasmine
Geraniol	$CH_3-C=CH-CH_2-CH_2-C=CH-CH_2-OH$ with CH_3 branches	Rose

burning incense, a feature of religious ceremonies from earliest times. Perfumes are among the most ancient and the most widely used of cosmetics. Perfumes have been found in the tombs of Egyptian pharaohs.

Most perfumes are complex mixtures of many fragrances dissolved in a specially denatured alcohol solvent. Some of the fragrances are from plant sources; most are synthetic. Animal secretions (or their synthetic analogs) are among the most interesting and valued perfume fragrances. Typically, perfumes are 10–25% fragrances and 75–90% ethyl alcohol. Table 23.5 gives the formulas of several synthetic fragrances, with fruity or flowery odors, produced in large quantities for perfume use. Citral is a carboxylic acid (Section 17.7), ionone and jasmone are ketones (Section 17.6), and geraniol is an alcohol (Section 17.3).

Many fruity or flowery fragrances are "too sweet," even in dilute solution. Musks, compounds originally obtained from animal secretions but now available in synthetic form, are often added to moderate the fragrance. Such compounds have extremely disagreeable odors in concentrated form, but often are interestingly pleasant at extreme dilution. The two most used "animal scents" are civetone and muscone, whose structures are given in Table 23.6. The civet cat, a skunk-like animal of eastern Africa, is the original source for civetone. Muscone is a secretion from the male musk deer.

AGING OF PERFUMES

When a group of perfume fragrances is first dissolved in alcohol, the resulting fragrance appears chaotic, in that many of the individual components can be detected by the nose and readily identified. However, as the oils remain in alcohol solution, this chaotic condition slowly disappears and a skillfully compounded perfume develops a well-rounded "bouquet" after a period of time, usually between 2 and 3 months.

TABLE 23.6

Unpleasant-Smelling Compounds from Animal Sources Used to "Moderate" Odors in Perfumes

Compound	Structure	Natural Source
Civetone		Civet cat
Muscone		Musk deer

Colognes

Colognes *are dilute perfumes.* Fragrances constitute only 1 to 2% by mass of cologne formulations. Dilution of perfumes to produce colognes can be made with alcohol or with alcohol–water mixtures. Most often the latter is used. The amount of water present in the alcohol is determined by the type of fragrance used and its solubility characteristics. Most fragrances have poor solubility in water and in water-alcohol mixtures of less than 65% alcohol content. Often a cologne is only 75% alcohol. Use of water instead of alcohol saves on formulation costs. With their fragrance concentration of only 1 to 2%, colognes are approximately 10 times weaker than perfumes.

Aftershave Preparations

Aftershave preparation purchase decisions are based primarily on fragrance. It is for this reason that aftershave preparations are grouped with perfumes and colognes in this section. However, the function of an aftershave preparation goes far beyond that of face scenting. An **aftershave preparation** *is a chemical formulation designed to refresh the skin, soothe minor irritations, and impart a feeling of well-being.*

The solvent in aftershave preparations is specially denatured alcohol in a concentration between 40 and 60% by volume. Concentrations of alcohol higher than 60% cause excessive sting and smarting. Concentrations below 40% are difficult to perfume because of limited solubility of perfume oils. The alcoholic solvent produces a mild astringency and a refreshing coolness.

Most aftershave formulations also contain menthol or a menthol-like compound for its cooling effects.

Menthol

COOLING EFFECT OF MENTHOL

The concentration of this cyclic nonaromatic alcohol varies, but is never more than 0.2%. Such a concentration gives ample cooling effect to allay pain from a "close" shave. Menthol gives the sensation of coolness because it is a differential anesthetic. Differential anesthetics anesthetize only certain sensations. Menthol partially anesthetizes most skin sensations except cold. Higher concentrations of menthol cannot be used because its odor would overpower the perfume of the product. [Medicinally, menthol is one of the most widely used of antipruritics (anti-itching agent) in dermatological therapy.]

After a shave with soap, the skin of the face has become alkaline and only slowly recovers its normal acidity. Small amounts of weak acids (Section 10.5), such as boric, lactic, or benzoic, in an aftershave preparation help to neutralize this alkalinity and restore the normal, slightly acid, condition of skin.

The addition of an aluminum or zinc salt helps increase astringency and styptic (stopping of bleeding) action. Aluminum chlorohydrates (section 23.2) are often used for this function. It should be noted that styptic action sufficient to immediately stanch bleeding from minor razor cuts is difficult to achieve because of the solvent action of alcohol on blood clots.

Emolliency (Section 23.4) is imparted most readily by the use of humectants. Low concentrations (up to 3%) of polyhydroxyl alcohols such as glycerin, propylene glycol, and sorbital generally are employed for this purpose. By controlling the moisture exchange between the skin and air, they assist in softening the skin through rehydration.

23.8 Nail Polish and Polish Remover

A **nail polish** is *a colored chemical formulation used as a "coating" for covering fingernails.* The characteristics of nail polish formulations must include: (1) easy to apply by brushing, (2) fast drying, (3) proper adhension to the nail, and (4) resistance to chipping and abrasion.

Five key ingredients in fingernail polishes, besides colorants, which are usually dyes, are

1. *A film-forming agent.* The most used film-forming agent is the polymer nitrocellulose (cellulose nitrate), a fluffy white fibrous material resulting from the reaction of cellulose with nitric acid. Cellulose (Section 24.7), chiefly obtained from wood pulp, is a polymer in which the monomers can be considered to be cyclic trihydroxy alcohols. A segment of a cellulose polymer, $[C_6H_7O_2(OH)_3]_n$, is

Under suitably controlled conditions, cellulose reacts with nitric acid producing nitrocellulose, a polymer in which the hydrogen atoms on the —OH groups have been replaced with —NO$_2$ (nitro) groups.

$$—O—H \xrightarrow[\text{(nitric acid)}]{\text{H—NO}_2} —O—NO_2$$

2. *A plasticizer.* This additive makes the polymeric nitrocellulose less brittle. Butyl stearate, an ester (Section 17.8), is often the plasticizer of choice.

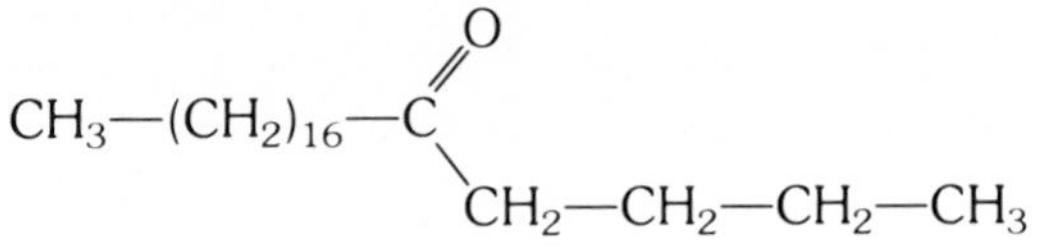

Butyl stearate

3. *A resin.* A resin (Section 18.1) is added to make the polymeric film adhere to the nail better and prevent flaking. *Ester gum*, a resin obtained from rosin, is often present in nail polish. *Rosin* is the sticky residue that remains after pine tree gum is subjected to distillation. When this residue is treated with alcohol, ester gum is produced. Ester gum is a mixture of esters—glyceryl, methyl, and ethyl—of abietic acid. Rosin is 80 to 90% abietic acid.

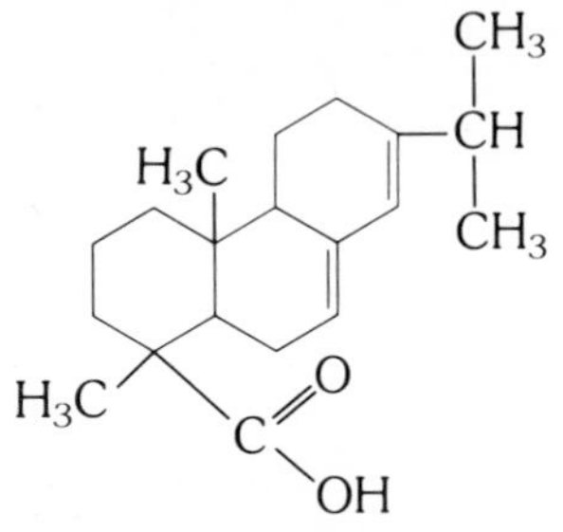

Abietic acid

TABLE 23.7
A Typical Fingernail Polish Formulation

Ingredient	Function	Composition (% by mass)
Nitrocellulose	Film-forming agent	15
Acetone	Solvent	45
Amyl acetate	Solvent	30
Butyl stearate	Plasticizer	5
Ester gum	Resin	5
Perfumes	Odorant	trace
Dyes	Colorant	trace

TABLE 23.8

Typical Fingernail Polish Remover Formulations

Formula A	Composition (% by mass)	Formula B	Composition (% by mass)
Methyl ethyl ketone	85	Ethyl acetate	20
Diethylene glycol monomethyl ether	10	Acetone	66
Butyl stearate	5	Butyl acetate	5
Perfume and colorant	trace	Water	8
		Lanolin derivative	1
		Perfume and colorant	trace

4. *A solvent.* Acetone (Section 17.6), the simplest of all ketones, is the solvent of choice. The evaporation of solvent, upon application of the polish, leaves a film of nitrocellulose, plasticizer, resin, and dye.

5. *A mixture of perfumes.* Perfumes are added to cover up the odors of the other fingernail polish components.

Table 23.7 gives the amounts of the various types of substances present in a typical fingernails polish formulation.

Nail polish removers are solvents for nitrocellulose combined with emollients designed to prevent or reduce the drying of skin and nails, owing to oil extraction by the solvents. Common antidrying additives are butyl stearate and diethylene glycol monomethyl ether.

$$CH_2—CH_2—O—CH_2—CH_2$$
$$OH \qquad\qquad\qquad O—CH_3$$

Diethylene glycol monomethyl ester

Table 23.8 gives two typical fingernail polish remover formulations. Acetone, ethyl acetate, and methyl ethyl ketone are all common solvents. Often mixtures of these compounds serve as the solvent.

Both nail polish and nail polish remover are very flammable formulations primarily because of the volatile solvents used. They should not be used in the presence of open flames or lighted cigarettes.

23.9 Mouthwashes

As noted in Section 23.1, it is sometimes difficult to distinguish between cosmetics and drugs, and this is particularly true for mouthwashes. Mouthwashes that have only a cleansing, refreshing, and/or deodorizing action are defined as cosmetics. Those that aid in dental hygiene because of the presence of plaque reducing chemicals may be classified as drugs. Many oral mouthwash formulations serve both cosmetic and drug functions.

Bad breath has numerous causes. Whether a mouthwash can help depends on the cause. Bad breath is sometimes a symptom of a mouth or throat infection. Mouthwashes cannot cure such infections.

BAD BREATH FROM
ONIONS AND GARLIC

Bad breath oftentimes originates from the lungs. The aroma of garlic and onions has such an origin. Odor-causing chemicals in these two substances (Section 17.5) are absorbed into the bloodstream in the stomach and are eventually released into the lungs. Similarly, smoker's breath originates from the lungs rather than from the mouth. Mouthwashes may mask such odors temporarily, but the effect lasts no more than 15 to 20 minutes.

Mouthwashes do affect one common cause of bad breath — bacteria. Bacteria normally present in the mouth and around teeth interact with leftover bits of food to produce smelly sulfur compounds. (This process is most pronounced at night while a person sleeps. The result is the odor and taste often called "morning mouth.") Simply brushing your teeth or rinsing out your mouth with water can eliminate or reduce the odor. A mouthwash, however, gives a longer-lasting effect against this problem. Some mouthwashes contain compounds that actually react with the sulfur-containing compounds present. Also antibacterial agents present in a mouthwash can reduce the mouth's bacterial population for as long as 3 or 4 hours. After that time, however, the bacterial population gradually returns to normal.

A mouthwash has very little effect on a sore throat. It may make the sore throat feel better temporarily, but it has no curative effect. The temporary relief is due to ingredients such as menthol (Section 23.7) that have a brief anesthetic effect.

Most mouthwashes fall into three color/flavor groups. The products within a group have similar "active" ingredients.

1. *The amber/medicinal group* has the highest alcohol content — 22 to 30%. (Alcohol at this concentration does not serve as a germicidal ingredient; see Section 25.11.) The flavoring agents in an amber/medicinal mouthwash are compounds originally obtained from plants. Among these are eucalyptol, the chief constituent of oil of eucalyptus; thymol, the flavoring constituent of thyme: menthol, from peppermint oil (Section 23.7); and methyl salicylate, from oil of wintergreen.

Eucalyptol   Thymol   Methyl salicylate

These flavoring agents also exert antibacterial action and function as the active ingredients in this type of mouthwash. Amber/medicinal mouthwashes also contain benzoic acid or sodium benzoate as additional antibacterial agents.

Listerine, which carries the name of Sir Joseph Lister, discoverer of the medical benefits of antiseptics, is the prototype for amber/medicinal mouthwashes. It was first marketed as a mouthwash in 1921 and is the oldest product of this type.

2. *The green/minty group* has an alcohol content of 14 to 22%, two-thirds that of the amber/medicinal group. Active ingredients are the dental plaque (Section 23.10) reducers cetylpyridinium chloride and domiphen bromide.

Cetylpyridinium chloride   Domiphen bromide

Plaque reductions in the range of 25%, on the average, can occur. Addition ingredients include a benzoic acid-type antibacterial agent, glycerin (which adds a sweet taste and serves as a soothing agent), and saccharin (as a sweetener).

Scope is the most popular green/minty mouthwash. The mint version of Cepacol is another important mouthwash of this type.

3. *The red/spicy group*, the cinnamon-flavored mouthwashes, has a low alcohol content, from 5 to 12%. The active ingredients are glycerin and either zinc chloride or sodium zinc citrate. The zinc compound acts as an astringent and odor neutralizer. Saccharin is also used as a sweetener in this type of mouthwash. Lavoris is the leading brand-name mouthwash of this type.

23.10 Tooth Decay and Toothpaste

Tooth enamel, the hard outer covering of a tooth, is made up of a three-dimensional network of calcium ions (Ca^{2+}), phosphate ions (PO_4^{3-}), and hydroxide ions (OH^-) arranged in a regular pattern (see Figure 23.6). The formula for this material is $Ca_{10}(PO_4)_6(OH)_2$, and its name is hydroxyapatite. Fibrous protein (Section 25.9) is dispersed in the spaces between the ions.

An equilibrium process (Section 8.14) involving the dissolving of and reformation of hydroxyapatite is continually occurring within the mouth. Tooth enamel

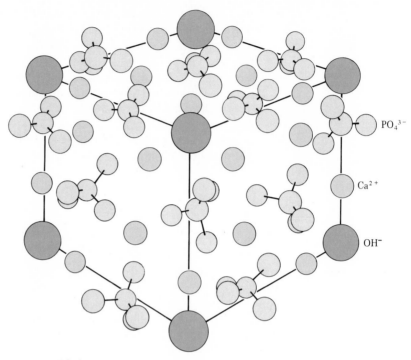

FIGURE 23.6

Hydroxyapatite, the hard outer covering of a tooth, has a structure involving a network of Ca^{2+}, PO_4^{3-}, and OH^- ions.

is continually dissolving to a slight extent to give a water solution in saliva of Ca^{2+}, PO_4^{3-}, and OH^- ions. This process is called *demineralization*. At the same time, however, the ions in the saliva solution are recombining to deposit enamel back on the teeth. This process is called *mineralization*. As long as demineralization and mineralization occur at equal rates, no net loss of tooth enamel occurs.

$$Ca_{10}(PO_4)_6(OH)_2 \underset{\text{mineralization}}{\overset{\text{demineralization}}{\rightleftharpoons}} 10\,Ca^{2+} + 6\,PO_4^{3-} + 2\,OH^-$$

Tooth decay results when chemical factors within the mouth cause the rate of dimineralization to exceed the rate of mineralization. The chemical species that most often causes the demineralization process to dominate is the acidic hydrogen ion (H^+; see Section 10.4). Hydrogen ions react with the OH^- ions produced during demineralization, upsetting the equilibrium process, and the Ca^{2+} ions and PO_4^{3-} ions diffuse away in the saliva. The continuation of this process over an extended period of time results in the formation of irregularities (pits or cavities) in tooth enamel. Eventually the enamel is breached, allowing bacteria to enter the tooth structure and cause decay.

Hydrogen ions, the species needed to start tooth decay, are readily available in the mouth. Their predominant source is colonies of acid-producing bacteria present in plaque, a substance that forms on teeth after food intake if the teeth are not throughly brushed and flossed. **Plaque** *is a soft, white or off-white, sticky film, consisting of about 70% bacteria, that forms on teeth within a few hours after eating.* The bacteria in plaque act on sucrose and other carbohydrates (food residues) also present in plaque, converting them to carboxylic acids. Lactic acid (Section 17.7) is the acid present in plaque in the greatest amount. The normal pH for saliva in the mouth is about 6.8. When plaque-produced lactic acid lowers the pH to 5.5 or less, the concentration of H^+ ions is sufficient to significantly upset the demineralization-mineralization equilibrium.

Plaque does not accumulate equally on all tooth surfaces. Areas in constant contact with the lips and tongue have very little plaque. It is the out-of-the way cracks and crevices where plaque accumulation is maximized. Such plaque can combine with Ca^{2+} and PO_4^{3-} ions in saliva and harden into *tartar* (also called calculus), a white or yellowish deposit that can only be "scaled" from teeth during professional cleaning. Tartar consists mainly of calcium phosphate $[Ca_3(PO_4)_2]$ and calcium carbonate $(CaCO_3)$. Under such hardened deposits hydrogen ions can work on enamel without being disturbed.

TARTAR

Toothpaste *is a chemical formulation whose main purpose is to remove plaque from teeth.* Secondary purposes include freshening of breath, helping to prevent tartar formation, and giving teeth added resistance to decay (if fluoride ion is present).

The two major ingredients in any toothpaste are an abrasive and a detergent. Abrasives typically make up approximately 50% by mass of a toothpaste formulation. Their hardness is such that they can "scratch off" plaque without harming the enamel. Tartar is too hard to be removed by toothpaste abrasives. Abrasive compounds used in toothpaste formulations include special forms of silicon dioxide (SiO_2) and various calcium compounds—chalk (powdered $CaCO_3$), calcium monohydrogen phosphate $(CaHPO_4)$, and calcium pyrophosphate $(Ca_2P_2O_7)$.

The purpose of the detergent in a toothpaste is to assist in suspending particles removed by abrasion in a water medium so they can be rinsed away. Sodium lauryl sulfate, a detergent often used in shampoos (Section 23.3), is a detergent of choice for toothpastes.

To make the abrasive and detergent more palatable, toothpastes always contain a number of other ingredients. They include colors, aromas, sweet-tasting compounds, and flavors, such as mint and peppermint.

Some toothpastes contain an additive to help control tartar formation. This compound, sodium pyrophosphate $(Na_4P_2O_7)$, interferes with the formation of the tartar crystalline structure.

Some toothpastes, particularly gels, contain additives for breath odor control. They are antibacterials that affect other bacteria in the mouth besides those in plaque. Sodium *N*-lauroyl sarcosinate is an additive of this type.

$$CH_3-CH_2-CH_2-CH_2-CH_2-CH_2-CH_2-CH_2-CH_2-CH_2-CH_2-\underset{\underset{O}{\|}}{C}-\underset{\underset{CH_3}{|}}{N}-CH_2-\underset{O^-}{\overset{O}{C}}\quad Na^+$$

Sodium N-lauroyl sarcosinate

This compound also has detergent properties.

The most publicized additives in toothpastes are the fluoride-containing compounds. Almost all toothpastes now contain them. It is now definitely documented that persons using fluoridated toothpastes experience fewer cavities than people using identical toothpastes without fluoride.

The three commonly used fluorine-containing toothpaste additives are stannous fluoride (SnF_2), sodium monofluorophosphate or MFP (Na_2PO_3F), and sodium fluoride (NaF). Monofluorophosphate ions release fluoride ions when they react with water in saliva.

$$PO_3F^{2-} + H_2O \longrightarrow H_2PO_4^- + F^-$$

The choice of fluoride compound depends on the abrasive and other additives present. The toothpaste formulation must be such that reactions between the fluoride-containing additive and other components do not "inactivate" the fluoride. Inactivation occurs when the soluble fluoride is converted to insoluble calcium fluoride (CaF_2).

Fluoride ion appears to have two major actions: (1) it strengthens enamel, and (2) it reduces acid formation by bacteria. Tooth enamel is strengthened when fluoride ion exchanges with hyroxide ion in the hydroxyapatite structure.

TOOTH ENAMEL AND FLUORIDES

$$Ca_{10}(PO_4)_6(OH)_2 + 2\ F^- \longrightarrow Ca_{10}(PO_4)_6F_2 + 2\ OH^-$$

Hydroxyapatite   Fluoroapatite

This replacement of hydroxide by fluoride in the apatite crystal produces an enamel that is less soluble in acidic medium.

Fluoride ion readily mixes with saliva and diffuses into the plaque on teeth. Here it affects the metabolism of bacteria, impairing their ability to produce organic acids from sucrose and other sugars.

Fluoride ion in drinking water (Section 11.10) exhibits the same effects on tooth enamel and plaque as does fluoridated toothpaste.

Practice Questions and Problems

DEODORANTS AND ANTIPERSPIRANTS

23-1 What are the purposes of perspiration formation by the human body?

23-2 Describe the odor of normal skin secretions.

23-3 Distinguish between an antiperspirant and a deodorant.

23-4 What is the function of an aluminum chlorohydrate in an antiperspirant?

23-5 What is the function of benzethonium chloride or zinc phenol sulfonate in a deodorant?

HAIR CARE PRODUCTS

23-6 What is sebum and what does it do for human hair?

23-7 What are the effects of too much sebum and too little sebum on human hair?

23-8 What is the structural formula for SLS, and what is its function in a hair shampoo?

23-9 Why is an acidifier needed in a hair shampoo?

23-10 What advertising phrases are used to draw attention to the presence of acidifiers in shampoo?

23-11 What are the four key ingredients in a hair conditioner formulation?

23-12 What is a humectant?

23-13 Name three humectants commonly found in hair care product formulations.

23-14 By what mechanism do protein fragments "bond" to damaged hair strands?

23-15 What type of chemical substance is used to combat static electricity in hair?

23-16 What is the structural formula for PVP and what is its function in a hair spray?

CREAMS AND LOTIONS

23-17 What is an emollient?

23-18 What is an emulsion and what type is most commonly associated with cosmetic creams and lotions?

23-19 Describe the chemical composition of the following emollients.
a. Petrolatum   b. Mineral oil

FACE MAKEUP FORMULATIONS

23-20 What are the two major reasons for face powder use?

23-21 What is the chemical nature of the major mineral ingredient present in face powder formulations?

23-22 What is the function of waxes present in a lipstick formulation?

23-23 What chemicals are used to obtain the following?
a. Black and brown colors of eyebrow pencils
b. Greenish and bluish tints of eye mascara

SUNSCREENS

23-24 Describe the natural chemical process by which skin tans.

23-25 Explain the chemical basis for a fair or dark skin complexion.

23-26 What are the structural formulas for the two compounds from which most sunscreen agents are derived?

23-27 Under what conditions should a person choose a sunscreen with a low SPF? a high SPF?

PERFUMES, COLOGNES AND AFTERSHAVE PREPARATIONS

23-28 What is a perfume?

23-29 What is the difference between a perfume and a cologne?

23-30 What is the purpose of "musks" in perfumes? What were the original sources for such compounds?

23-31 Perfumes must "age." Explain

23-32 Most perfumes and colognes are alcohol based rather than water based. Explain why.

23-33 Menthol is used in aftershave preparations for its cooling effect. Explain the mechanism by which this cooling effect occurs.

23-34 Why do aftershave preparations contain small amounts of weak acids?

23-35 Aftershave preparations are not very effective in stanching bleeding from minor razor cuts. Explain why.

NAIL POLISH AND POLISH REMOVER

23-36 What are the five key ingredients, besides colorants, in a fingernail polish?

23-37 What is the function of nitrocellulose in a fingernail polish formulation?

23-38 What is ester gum and what is its function in a fingernail polish formulation.

23-39 What are the names of the three solvents commonly used in fingernail polish removers?

MOUTHWASHES

23-40 Mouthwashes have very little effect on garlic and onion odors on the breath. Explain why.

23-41 What is "morning mouth"?

23-42 Name the "active ingredients" and list their functions in the following types of mouthwashes.
  a. Amber/medicinal   b. Green/minty
  c. Red/spicy

23-43 Compare the alcohol contents for the three major types of mouthwashes.

TOOTH DECAY AND TOOTHPASTE

23-44 What is the chemical formulation for hydroxyapatite, the principal structural material in tooth enamel.

23-45 Contrast the processes of demineralization and mineralization.

23-46 Describe the role of hydrogen ions in the tooth decay process.

23-47 What is plaque?

23-48 Describe how lactic acid, the major acid present in plaque, is produced.

23-49 What is tartar?

23-50 What are the two major types of ingredients present in a toothpaste?

23-51 List four compounds that are used as abrasives in toothpaste formulations.

23-52 Identify the function of each of the following compounds in a toothpaste formulation.
  a. Sodium pyrophosphate
  b. Sodium *N*-lauroyl sarcosinate
  c. MFP
  d. Calcium monohydrogenphosphate

23-53 What are the two major actions by which fluoride ions protect teeth from decay?

23-54 What are the three compounds commonly used as fluoride additives in toothpaste formulations?

23-55 Contrast the solubility characteristics of hydroxyapatite and fluoroapatite.

Biochemistry: Carbohydrates and Lipids

24.1 Types of Biochemical Substances

Biochemistry *is the study of the chemistry of living systems.* It is a constantly changing field in which new discoveries are made almost daily about how life is maintained, how diseases occur, how diseases may be prevented, and how cells manufacture the molecules needed for life.

The substances found in living organisms, biochemicals, may be divided into two general groups: bioinorganic substances and bioorganic substances. Water and inorganic salts are the major bioinorganic substances. There are four major groups of bioorganic substances: carbohydrates, lipids, proteins, and nucleic acids. Figure 24.1 schematically shows the major types of biochemical substances present in living systems. The type of living system does not matter; be it a microorganism, an elephant, or a human being, the basic chemicals of life are similar.

The major types of biochemical substances are not equally abundant. As pointed out previously (Section 11.1), water is the most abundant substance in a living organism. Approximately two-thirds of the mass of a human body is water. Another 4 to 5% of human body mass is inorganic salts. The principal ions present in body fluids, derived from inorganic salts, are shown in Figure 24.2. Within cells (intracellular fluid) K^+ is the predominant positive ion and HPO_4^{2-} is the principal negative ion. Outside of cells (interstitial fluid and blood plasma) Na^+ is the most abundant positive ion and Cl^- is the most abundant negative ion.

We tend to think of the human body as being made up of organic substances.

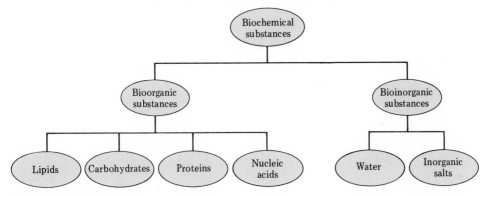

FIGURE 24.1

Biochemical substances can be divided into two major groups, bioorganic substances and bioinorganic substances. Further, there are four groups of bioorganic substances and two groups of bioinorganic substances.

Actually bioorganic substances constitute only about one-fourth of the mass of a human body. Proteins are the most abundant (approx. 15%). Next come lipids, a diverse group of compounds (approx. 8%). Carbohydrates and nucleic acids are found in relatively small amounts in most tissues (approx. 2%). Composition data for the human body, in terms of general types of substances present is summarized in Figure 24.3.

We consider the four major types of bioorganic substances associated with living organisms in two chapters. Carbohydrates and lipids are the subject of this chapter and proteins and nucleic acids are considered in Chapter 25.

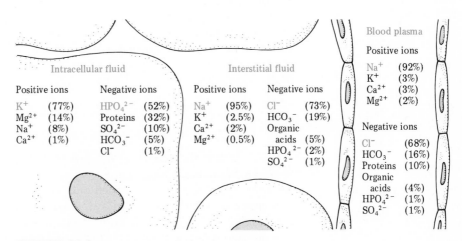

FIGURE 24.2

The principal ions present in body fluids.

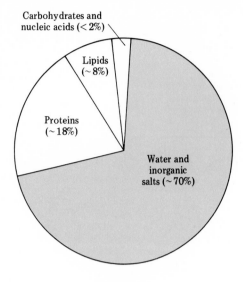

FIGURE 24.3

The percent by mass composition of the human body in terms of major types of biochemical substances present.

24.2 Occurrence of Carbohydrates

Carbohydrates, which include such familiar substances as glucose, table sugar (sucrose), starch, and cellulose, are of major importance to both plants and animals. Most of the matter in plants, not counting the water, is made up of carbohydrates.

The process of photosynthesis, carried out by green plants, that is, chlorophyll-containing plants, is the method by which almost all carbohydrates are produced. In this process, plants combine carbon dioxide from the air with water from the soil, using sunlight as the source of energy, to give simple carbohydrates (mainly glucose, $C_6H_{12}O_6$).

$$6\ CO_2 + 6\ H_2O \xrightarrow[\text{chlorophyll}]{\text{sunlight}} C_6H_{12}O_6 + 6\ O_2$$
A carbohydrate

Within plants, the carbohydrates produced from photosynthesis serve two major functions. In the form of cellulose, they are structural elements, and in the form of starch, they provide needed nutritional reserves (energy) for the plant.

Animals and humans obtain carbohydrates by eating plants. Typically, carbohydrates constitute 65% of the human diet. Carbohydrate food intake in humans serves two purposes: (1) to provide energy, obtained through carbohydrate oxidation, and (2) to supply carbon atoms for the synthesis of other biochemical substances (proteins, lipids, and nucleic acids).

24.3 Classifications of Carbohydrates

Carbohydrates *are polyhydroxy aldehydes or ketones, or substances that yield such compounds upon hydrolysis (reaction with water).* Compounds classified as carbohydrates vary in structure from those consisting of a few carbon atoms to gigantic (polymeric) molecules having molecular masses in the hundreds of thousands.

As the definition specifies, three functional groups are prevalent in carbohydrates: hydroxyl groups (—OH) and aldehyde (—$\overset{\displaystyle O}{\underset{\displaystyle H}{C}}$) or ketone (—$\overset{\displaystyle O}{C}$—) groups (see Sections 17.3 and 17.6). The carbohydrates glucose and fructose are specific examples, respectively, of a polyhydroxy aldehyde and polyhydroxy ketone. One way of denoting their structures is as follows.

Glucose
(a polyhydroxy aldehyde)

Fructose
(a polyhydroxy ketone)

The notation CH_2OH at the bottom of each structure denotes a carbon atom bonded to two hydrogen atoms and a hydroxyl group.

Carbohydrates are often referred to as *saccharides* (from the Latin *saccharum* meaning "sugar") because of the sweet taste of most of the simpler carbohydrates. Names for various classes of carbohydrates involve the use of this term. Three important classes of carbohydrates are the monosaccharides, disaccharides, and polysaccharides.

A **monosaccharide** *is a carbohydrate that cannot be broken down into simpler units by hydrolysis reactions.* Naturally occurring monosaccharides with five and six carbon atoms are particularly common. Glucose, the major product formed when carbohydrates from plants are metabolized in the human body, is a most important monosaccharide.

A **disaccharide** *is a carbohydrate composed of two monosaccharide units.* Common table sugar, sucrose, is a disaccharide. Monosaccharides and disaccharides are often called *sugars* because of the sweet taste of many of these compounds.

A **polysaccharide** *is a carbohydrate composed of many, usually hundreds, monosaccharide units.* Polysaccharides are the most abundant carbohydrate form in plants. Both cellulose and starch are polysaccharides. We encounter various forms of these two substances throughout our world. The paper on which this book is printed is mainly cellulose, as are the cotton in our clothes and the wood in our houses. Starch is a component of the flour used to make bread and of many other foodstuffs such as potatoes, rice, corn, beans, and peas.

24.4   Important Monosaccharides

In this section we consider four monosaccharides: glucose, galactose, fructose, and ribose. Glucose, galactose, and fructose all have the same molecular formula, $C_6H_{12}O_6$; they are isomers (Section 16.5). The formula for ribose, which has one less carbon atom, is $C_5H_{10}O_5$.

Glucose

Glucose, a white, crystalline water-soluble solid, is the most abundant and also the most important of the monosaccharides. Natural sources for it are grapes, figs, and dates, as well as other fruits. Ripe grapes contain 20 to 30% glucose. Glucose is also found naturally as a component of the disaccharides sucrose, maltose, and lactose (see Section 24.5) and is the monomer of the polysaccharides starch, cellulose, and glycogen (see Section 24.7). Two other names for glucose are dextrose and blood sugar.

The term blood sugar draws attention to the fact that blood contains dissolved glucose. The concentration of glucose in human blood is fairly constant, being in the range of 70 to 100 mg per 100 mL of blood. Cells use glucose as a primary energy source.

A 5% (m/v) solution of glucose in water is frequently used, in hospitals, as an intravenous (IV) source of nourishment for patients who cannot take food by mouth (see Figure 24.4). The solution concentration is 5% (m/v) so that it will be isotonic with body fluids (Section 9.6). A 5% (m/v) glucose–water solution is also the source of nourishment given newborn babies during their first few hours (sometimes days) of life.

Certain body conditions, such as those caused by the disease diabetes mellitus, result in too much glucose in the blood. People with above-normal

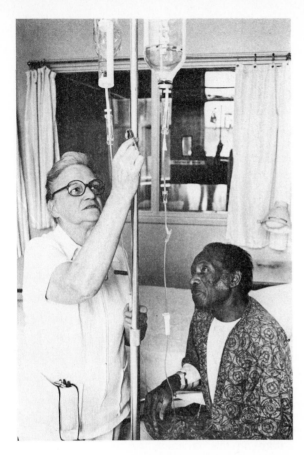

FIGURE 24.4
A 5% (m/v) glucose solution is frequently used in a hospital as an intravenous (IV) source of nourishment for patients who cannot take food by mouth. [Stock, Boston.]

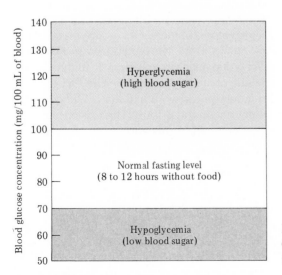

FIGURE 24.5
Conditions related to the concentration of glucose in the blood.

blood glucose levels are said to be *hyperglycemic* (see Figure 24.5). Use of the substance insulin, as a medication, is often needed to control this condition. The opposite condition—lower than normal blood glucose levels—is called *hypoglycemia* (see Figure 24.5). Symptoms of hypoglycemia include nausea and dizziness. Hypoglycemia is treated by regulating the dietary intake of glucose; more frequent meals are often required.

To adequately represent the structure of glucose (or most other monosaccharides) three structural representations are required. For glucose the three structures are

Open-chain glucose
Structure 1

α-Glucose
Structure 2

β-Glucose
Structure 3

Structure 1 (previously given in Section 24.3) represents the open-chain form of the molecule. (The carbon atoms in this structure are numbered for later reference.) The open-chain form of glucose exists in equilibrium with two isomeric cyclic forms, which actually are the predominant species (both in aqueous solution and in the solid state). The cyclic structures (2 and 3) form from the open-chain through an *intra*molecular reaction in which carbon 1 bonds to carbon 5 through the oxygen atom (of the —OH group) on carbon 5 (see Figure 24.6). The difference between the two cyclic forms, which are labeled α and β, involves the —OH group attached to carbon 1. In the α-form, the —OH group is on the opposite side of the ring from the —CH$_2$OH group at carbon 5, whereas the —CH$_2$OH group on carbon 5 and the —OH group on carbon 1 are on the same side of the ring in the β-form. The existence of two cyclic forms (α and β) of a monosaccharide influences the properties of polysaccharides (see Section 24.7).

Galactose

Galactose does not occur free in nature. It is most often obtained by hydrolysis of the disaccharide lactose (which consists of a glucose unit and a galactose unit; see Section 24.5).

The structures of galactose and glucose differ only in the configuration of the

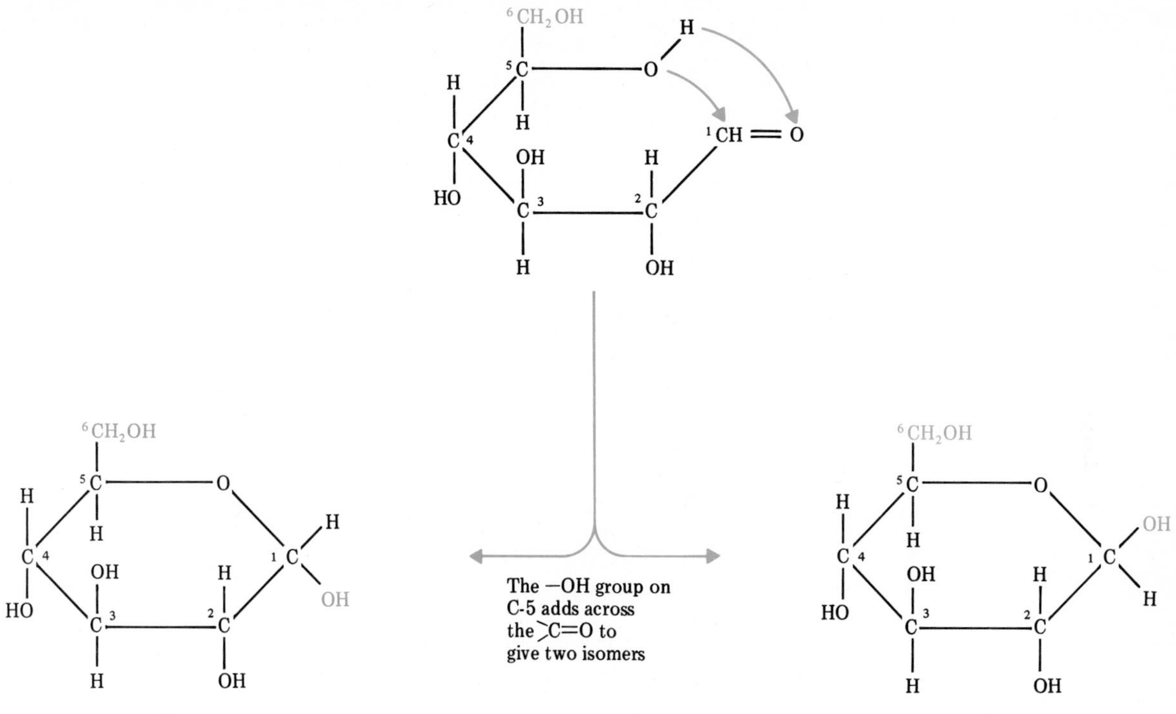

α-Glucose β-Glucose

FIGURE 24.6
The cyclization of the open-chain form of glucose.

—OH group on carbon 4. It is to the left in the galactose structure and to the right in the glucose structure.

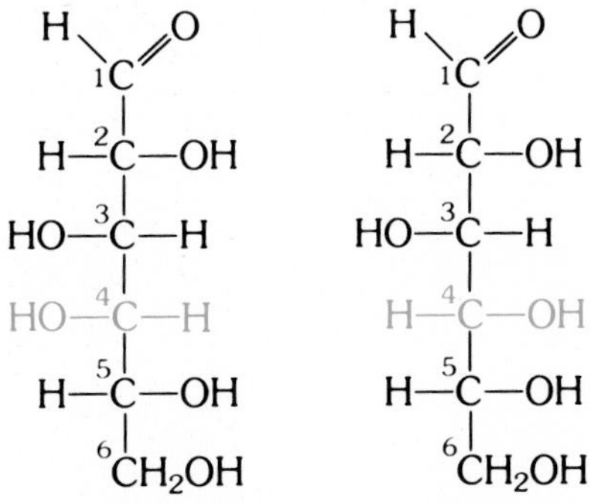

Galactose Glucose

Galactose is synthesized from glucose in the mammary glands. It is then used to make the lactose present in milk. Galactose is also found in brain and nerve tissue as part of very complex molecules.

Like glucose, galactose has three structural forms—an open-chain form and α and β cyclic forms.

Fructose

The structure of fructose is identical to that of glucose from the third to sixth carbon.

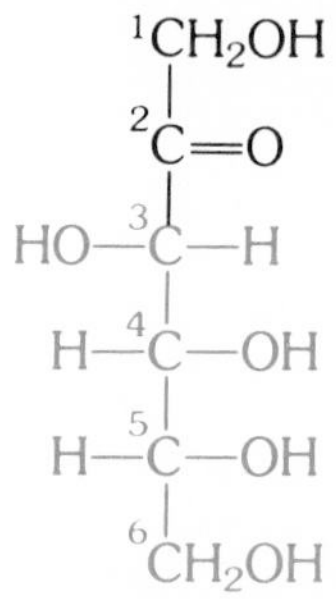

Fructose

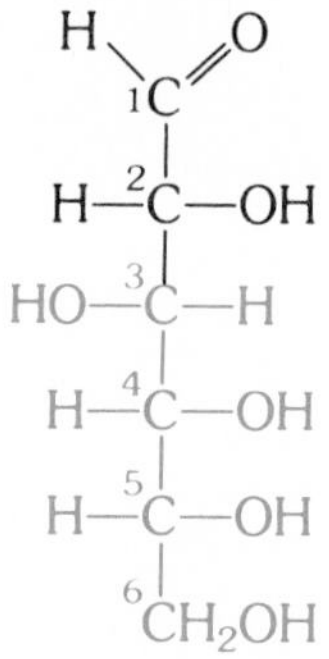

Glucose

The differences at carbons 1 and 2 result from the presence of a ketone group in fructose and an aldehyde group in glucose.

Fructose, which is also known as levulose, is the sweetest of all sugars (Section 24.6). It is found in many fruits and is present in honey in equal amounts with glucose. Fructose is sometimes used as a dietary sugar, not because it has few calories per gram than other sugars, but because less is needed for the same amount of "sweetness."

FRUCTOSE AS A
DIETARY SUGAR

As with glucose and galactose, cyclic forms of fructose exist. However, the cyclic forms of fructose involve a five-membered ring, rather than a six-membered ring, because ring closure involves carbon 2 (where the carbon–oxygen double bond is located).

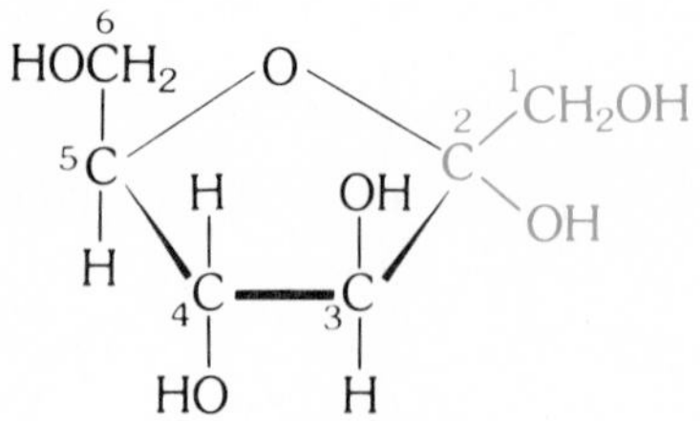

α-Fructose

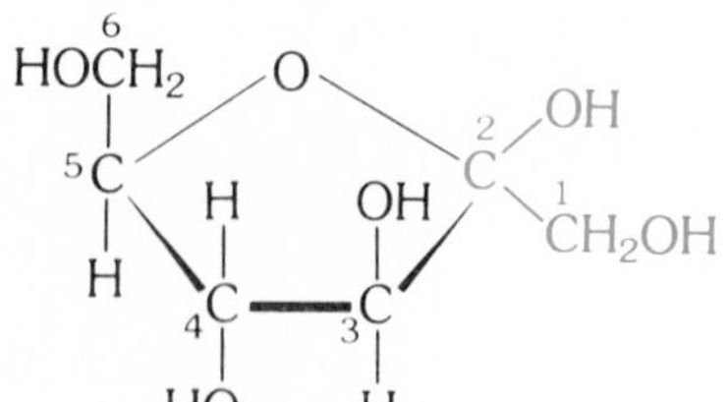

β-Fructose

Ribose

The three monosaccharides previously discussed in this section have all contained six carbon atoms. Ribose has only five carbon atoms. If carbon 3, with its H and OH, was eliminated from the structure of glucose, the remaining structure would be that of ribose.

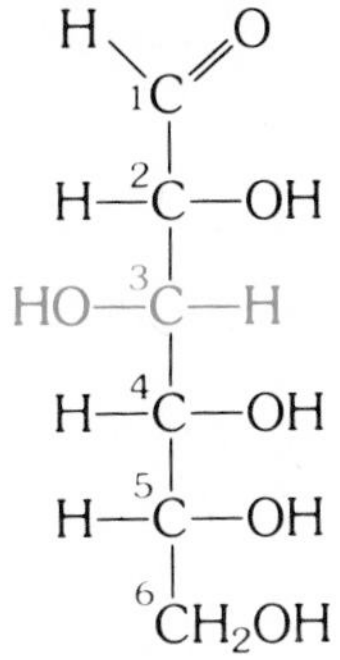

Glucose 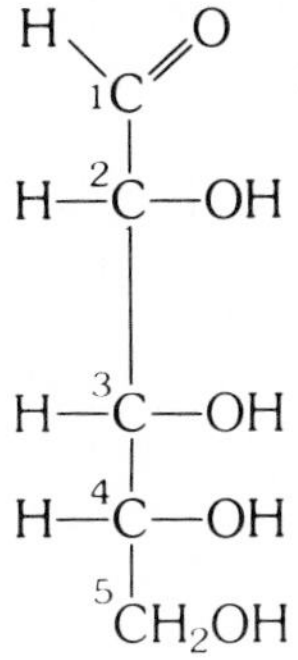   Ribose

Ribose is a component of a variety of complex molecules including ribonucleic acids (RNAs), which will be discussed in Chapter 25.

A compound closely related to ribose, 2-deoxyribose, is also an important component in nucleic acid structures. This compound, as its name implies, has one less oxygen atom than does ribose. On the second carbon atom, 2-deoxyribose has two H atoms compared to an H and OH in ribose.

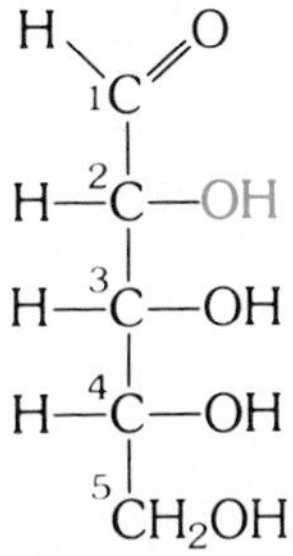

Ribose 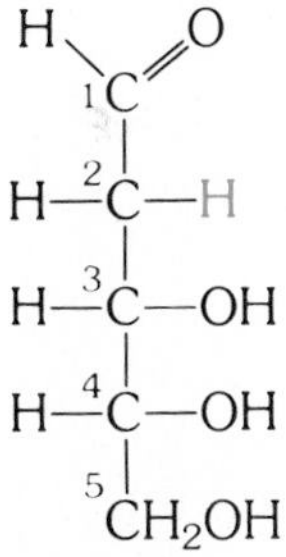   2-Deoxyribose

2-Deoxyribose is a component of the structures of deoxyribonucleic acids (DNAs). DNA molecules control protein synthesis in cells and carry hereditary information (Chapter 25).

24.5 Important Disaccharides

The three most commonly encountered disaccharides (two monosaccharide units bonded together) are

1. Sucrose (from sugarcane or sugar beets), which consists of a glucose unit and a fructose unit.

2. Maltose (from starch), which consists of two glucose units.
3. Lactose (from milk), which consists of a glucose unit and a galactose unit.

Sucrose

Sucrose is the carbohydrate most people refer to as simply "sugar"; it is our common table sugar. Although found in most fruits and vegetables, two plants dominate as sources of sucrose—sugarcane and sugar beets (see Figure 24.7). Sugarcane contains up to 20% sucrose by mass and sugar beets up to 17% by mass.

Average per capita consumption of sucrose in the United States is approximately 150 grams per day, which provides between a fourth and a third of the calories consumed daily. Much of the sucrose we consume is added to foods, as Table 24.1 shows, rather than sugar that is "naturally" present.

FIGURE 24.7
Much of the sucrose (table sugar) used in the United States is obtained from sugarcane plants that grow in semitropical climates. [Foto du Monde/The Picture Cube.]

TABLE 24.1
Amounts of Sugar (Sucrose) Added to Selected Foods

Canned or Packaged Food	Sugar (g) / Serving	Percent of Mass	Percent of Calories
Beverages			
Cola type	37 / 12 oz	10	100
Fruit juice drink	20 / 6 oz	12	85
Kool-Aid	25 / 8 oz	11	98
Desserts			
Peaches, light syrup	9 / $\frac{1}{2}$ cup	7	48
heavy syrup	16 / $\frac{1}{2}$ cup	12	61
Pudding, starch type	25 / 5 oz	18	50
Gelatin dessert	26 / 5 oz	18	97
Milk chocolate candy	12 / 1 oz	44	32
Brownies	12 / each	50	30
Coconut cream pie	24 / $\frac{1}{4}$ pie	68	66
Ready-to-eat cereals			
Corn or wheat flakes	2–3 / 1 oz	7–11	11–15
Presweetened flavored	8–14 / 1 oz	29–44	26–45
"100% natural"	6 / 1 oz	19	15

The two monosaccharide units found within a sucrose molecule are glucose and fructose. Carbon 1 of glucose is bonded to carbon 2 of fructose through an oxygen link.

Sucrose

Sucrose is readily hydrolyzed to produce a 1:1 mixture of glucose and fructose, called *invert sugar*.

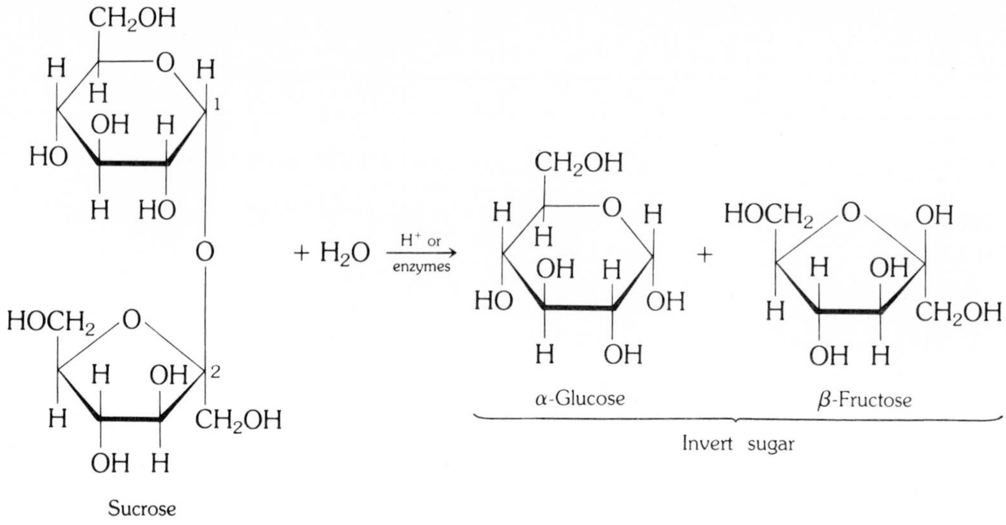

$$\text{Sucrose} + H_2O \xrightarrow[\text{enzymes}]{H^+ \text{ or}} \alpha\text{-Glucose} + \beta\text{-Fructose}$$

Sucrose

Invert sugar

When sucrose is cooked with acid-containing foods, such as fruits or berries, partial hydrolysis takes place, and some invert sugar forms. Jams and jellies prepared in this manner have a "sweetness" greater than that of pure sucrose from the presence of the invert sugar (Section 24.6).

Maltose

Maltose is composed of two molecules of glucose. Carbon 1 of one glucose molecule is bonded to carbon 4 of the other glucose molecule through an oxygen link.

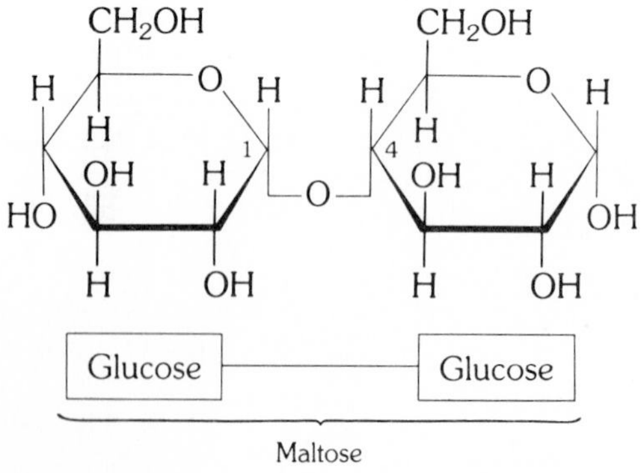

Glucose — Glucose

Maltose

Maltose, also called malt sugar, does not occur abundantly in nature, except in germinating grain. Nonetheless, maltose is an important disaccharide because it is the main product obtained from the digestion of starch (a polysaccharide; see Section 24.7). Once produced in the digestive tract (from starch hydrolysis), maltose is further hydrolyzed (with the help of the enzyme *maltase*) to give two molecules of glucose.

$$\text{Maltose} + H_2O \xrightarrow{\text{maltase}} \text{Glucose} + \text{Glucose}$$

Maltose is an ingredient in formulas for feeding infants and in corn syrup and is important in the production of beer.

Lactose

The disaccharide lactose is made up of a galactose unit and a glucose unit linked from carbon 1 (of galactose) to carbon 4 (of glucose).

Lactose is the major sugar found in milk. This accounts for its common name, milk sugar. Enzymes in mammary glands take glucose from the bloodstream and synthesize lactose in a two-step process. First, a glucose molecule is converted to a galactose molecule, and then a bond is formed between the galactose and another glucose molecule.

Lactose is an important ingredient in commercially produced infant formulas that are designed to simulate mother's milk. The souring of milk is due to the conversion of lactose to lactic acid by bacteria in the milk.

Lactose taken into the human body, from the drinking of milk, is first broken apart into its two constituent monosaccharides. An enzyme called *lactase* is required for this hydrolysis process.

$$\text{Lactose} + H_2O \xrightarrow{\text{lactase}} \text{Galactose} + \text{Glucose}$$

The free galactose so produced is then converted to glucose through another reaction pathway.

There are two methods or pathways by which free galactose in the human body (formed from lactose hydrolysis) is converted to glucose. During infancy only one method is functional. Later in life, a second pathway also begins to function. Some infants are born with a genetic disease called *galactosemia*, which is an inability to convert galactose to glucose. If this disease is not

GALACTOSEMIA

detected, milk or other foods containing milk cause vomiting and diarrhea and, in the extreme, cataract formation and even mental retardation. Excluding milk from the diet by feeding the infant formulas based on other disaccharides usually eliminates the problem. In later years, as enzymes in the second pathway are produced, the problem usually disappears and milk may again be included in the diet.

LACTOSE INTOLERANCE

Another type of ''lactose problem'' found in some people is *lactose intolerance*, a condition in which the enzyme *lactase*, which is needed to hydrolyze lactose to galactose and glucose, is lacking. Deficiency of lactase can be due to a genetic defect, physiological decline with age, or injuries to the mucosa lining of the intestines. The consequences of an inability to hydrolyze lactose in the upper small intestine are (1) inability to absorb lactose and (2) bacterial fermentation of ingested lactose further along the intestinal tract. Bacterial fermentation results in the production of gas (distension of the gut) and osmotically active solutes that draw water into the intestines, resulting in diarrhea.

The level of the enzyme lactase in humans varies with age and race. Most children have sufficient lactase during the early years of their life when milk is a much needed source of calcium in their diet. In adulthood, the enzyme level decreases and lactose intolerance results. This explains the change in milk-drinking habits of many adults. Some experts estimate that as many as one out of three adult Americans suffer a degree of lactose intolerance.

FIGURE 24.8
Human mother's milk has a lactose content almost double that of cow's milk. [Peter Vandermark/Stock, Boston.]

The amount of lactose in milk varies with species. Human mother's milk obtained by nursing infants (see Figure 24.8) contains 7 to 8% lactose, almost double the 4 to 5% lactose found in cow's milk.

24.6 Artificial Sweeteners

Because of the high caloric value of sucrose, it is often difficult to satisfy a demanding "sweet tooth" without adding a few pounds or inches to the waistline. To combat this problem, scientists have sought to find artificial sweeteners with little or no caloric value as alternatives to sucrose. Consumers for more than two decades have been in the middle of a seemingly unending controversy over three well-known artificial sweeteners: saccharin, sodium cyclamate, and aspartame. All have received wide publicity over the years. All have been subjected to long review, scrutiny, and debate by industry, scientists, and the Federal Drug Administration (FDA).

Saccharin   Sodium cyclamate   Aspartame

Saccharin, the sweetest of the three, contains no calories and has been around the longest. Discovered in 1879, it was used initially as an antiseptic and a food preservative. Soon diabetics began to use it, and from the turn of the century until the 1950s saccharin dominated the market.

When approved for commercial use in 1951, sodium cyclamate quickly became popular. Also having no caloric value, sodium cyclamate was widely used in canned fruits and many other foods and in chewing gum, toothpastes, and mouthwashes.

During the 1960s sales of both saccharin and sodium cyclamate increased as the demand for diet soft drinks appeared. Although saccharin is much sweeter, mixtures of the two were often used because sodium cyclamate cut the "aftertaste" of saccharin. In 1970 sodium cyclamate was banned when questions over its safety (cancer-causing potential) came to light. By the late 1970s the only available sweetner, saccharin, was being consumed at the rate of 6 million pounds annually (the majority in soda pop). During the late 1970s warning labels about potential health hazards of saccharin appeared in grocery

TABLE 24.2

Comparative Sweetnesses of Selected Sugars and Artificial Sweeteners

Sugar or Artificial Sweetener	Sweetness Relative to Sucrose	Type
Lactose	0.16	Disaccharide
Galactose	0.32	Monosaccharide
Maltose	0.33	Disaccharide
Glucose	0.74	Monosaccharide
Sucrose	1.00	Disaccharide
Invert sugar	1.25	Mixture of glucose and fructose
Fructose	1.73	Monosaccharide
Sodium cyclamate	30	Artificial sweetener
Aspartame	180	Artificial sweetener
Saccharin	450	Artificial sweetener

stores because certain studies suggested that saccharin might cause cancer. Although later studies provided no confirmation, the shadow of doubt still clouds the economic future of saccharin.

Aspartame (Nutra-Sweet) came into the marketplace in 1981. It was accidentally discovered in 1965 by a research team working on ulcer drugs, but 15 years elapsed before it acquired all necessary approvals for marketing. Aspartame has the same food value (4 calories per gram) as sucrose, but it is 180 times sweeter and, unlike saccharin, has no bitter aftertaste. Aspartame has quickly found its way into almost every diet food on the market today. So little aspartame is needed to sweeten foods that there is less than 1 calorie per serving.

The safety of aspartame lies with its hydrolysis products: the amino acids aspartic acid and phenylalanine. These amino acids are identical to those obtained from digestion of proteins (Section 25.10) and hence are considered safe. The only danger of aspartame is that phenylalanine can lead to mental retardation among young children suffering from PKU (phenylketonuria). Labels on all products containing aspartame warn of the danger to phenylketonurics.

Sweetness values for common sugars and artificial sweeteners are given in Table 24.2. All values are based on a value of 1.00 for sucrose.

24.7 Polysaccharides

Polysaccharides (defined in Section 24.3) are polymers in which the monomer units are monosaccharides. In some polysaccharides, the monomer units are linked together to form a linear (unbranched) chain. In other polysaccharides

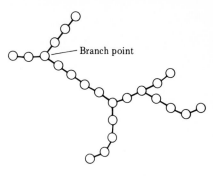

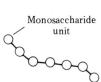

(a) Linear polysaccharide   (b) Branched polysaccharide

FIGURE 24.9

Monosaccharide units in a polysaccharide can form linear or branched-chain structures.

extensive chain branching occurs (see Figure 24.9). The number of monomer units in a polysaccharide chain varies from a few hundred up to 12,000 to 15,000, depending on the identity of the polysaccharide.

Unlike monosaccharides and most disaccharides, polysaccharides are not sweet. They have limited water solubility because of their size. However, individual —OH groups can become hydrated by water molecules. The result is usually a thick colloidal suspension (Section 9.7) of the polysaccharide in water. Polysaccharides are often used as thickening agents in sauces, gelatins, desserts, and gravies because of their ability to form thick colloidal suspensions.

Many naturally occurring polysaccharides are known. Three of the most important are cellulose, starch, and glycogen, substances that play vital roles in living systems. Both cellulose and starch are important in plants, and glycogen is a key substance in animals and humans.

Although differing in structural details, these three substances have one common characteristic. All three, when hydrolyzed (broken up into monosaccharide units), yield glucose as the *only* product. Thus these three polysaccharides—cellulose, starch, and glycogen—are polymers containing only glucose units.

Cellulose

Cellulose is the most abundant of all polysaccharides. It is the structural component of the cell walls of plants. It is estimated that approximately half of all the carbon atoms contained in the plant kingdom are found within cellulose molecules.

Structurally, cellulose is a linear (unbranched) glucose polymer in which the glucose units are all in the cyclic β-form (Section 24.4). A portion of such a polymer is as follows.

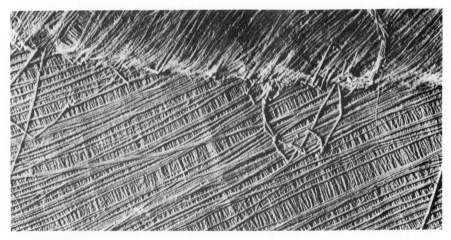

CH$_2$OH

β-Glucose

β-Glucose

β-Glucose

β-Glucose

A typical cellulose chain contains about 5000 glucose units and is a macromolecule with a molecular mass of about 400,000 amu. The long chains usually line up side by side, held together by hydrogen bonding between chains. These bundles of cellulose chains are called *fibrils*. This structural feature makes cellulose fibrous, tough, and insoluble in water. Figure 24.10 is an electron micrograph picture of cellulose fibers. In addition to interchain hydrogen bonding a noncarbohydrate glue-like substance called *lignin* also helps hold the fibers together to give even greater strength.

Cotton is almost pure cellulose (95%), and wood is about 50% cellulose. Chemical treatments of cellulose yield many important commercial products.

FIGURE 24.10
Electron micrograph of cellulose fibers. [Biophoto Associates/Photo Researchers, Inc.]

Paper and paper products derive from wood treated with various chemicals. Rayon fiber is made from treating cellulose with sodium hydroxide (NaOH) and carbon disulfide (CS_2). Treatment with nitric acid produces cellulose nitrate, which is used in "smokeless" gun powder and fingernail polish (Section 23.8). Cellulose acetate is the raw material for some film used in the motion picture industry.

Despite being a polymer of glucose, cellulose is not a nutrition source for human beings. Humans lack enzymes capable of catalyzing the hydrolysis of the bonds between β-glucose units. Even grazing animals lack these necessary enzymes. However, the intestinal tracts of such animals, including horses, cows, and sheep, contain microorganisms (bacteria) that produce *cellulase*, an enzyme that can hydrolyze these linkages and produce free glucose from cellulose. Thus, grasses and other plant materials are a source of nutrition for grazing animals. The intestinal tracts of termites contain these same microorganisms, enabling termites to use wood as their source of food. Microorganisms in the soil can also metabolize cellulose, thus allowing the biodegradation of dead plants.

DIETARY FIBER

Most food in our diet of plant origin does contain some cellulose. Even though we get no nourishment from it, it still has an important role to play. Cellulose provides the digestive tract with "bulk" or "fiber," which helps move food through the intestinal tract and facilitates the excretion of solid wastes. Cellulose absorbs much water, leading to softer stools and frequent bowel action. Since links have been found between the length of time a stool spends in the colon and possible colon cancer, many diet-conscious professionals recommend a high-fiber content in the foods we eat. Bran, celery, green vegetables, and fruits are rich in dietary fiber.

Starch

Starch, like cellulose, is a polymer containing only glucose units. It differs from cellulose, however, in that its units are all α-glucose rather than β-glucose. Starch is the storage polysaccharide in plants. If excess glucose enters a plant cell, it is converted to starch and stored for later use. When the cell cannot get sufficient glucose from outside, it hydrolyzes starch to release glucose.

Two different polyglucose polysaccharides can be isolated from most starches—amylose and amylopectin. *Amylose*, a straight-chain glucose polymer, usually accounts for 15 to 20% of the starch, and *amylopectin*, a highly branched glucose polymer, for the remaining 80 to 85%. The number of glucose units present in an amylose chain depends on the source of the starch, with 300 to 500 monomer units usually being present. Amylose polymers adopt a coiled structure called a helix. Such a structure resembles a stretched-out spring (see Figure 24.11). Amylopectin, in contrast, has a branch about once every 25 to 30 glucose units (see Figure 24.12) and therefore has a larger average molecular mass than the linear amylose. The average molecular mass of amylose is 50,000 or more, compared to 300,000 or more for amylopectin.

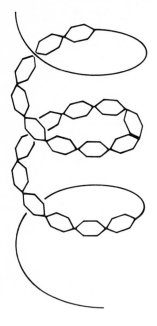

FIGURE 24.11

The helical structure of the amylose component of starch.

Enzymatic hydrolysis of starch produces first intermediate-sized polysaccharides called *dextrins*. The dextrins are then further hydrolyzed to maltose and ultimately to glucose.

Starch	⟶	Dextrins	⟶	Maltose	⟶	Glucose
(Polysaccharide)		(Polysaccharide)		(Disaccharide)		(Monosaccharide)

DEXTRINS

Dextrins have numerous commercial uses. Mucilage and some pastes contain dextrins. Such adhesives are used on postage stamps and envelopes. A mixture of dextrins and maltose is used in baby foods and infant formulas.

Heat can also convert starch to dextrins. When bread is toasted, the change in texture (and taste) is the result of some starch being converted to dextrins (see Figure 24.13).

The bonds between α-glucose units in amylose and amylopectin can be broken through hydrolysis reactions within the human digestive tract, with the help of the enzyme *amylase*. Hence, starch has nutritional value for humans. The starches present in potatoes, wheat, rice, and cereal grains account for approximately two-thirds of the world's food consumption.

Glycogen

Glycogen is the storage polysaccharide in animals and humans. Its function is much like that of starch in plants, and sometimes it is referred to as *animal starch*.

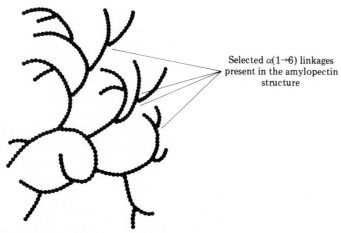

All glucose units are α-glucose

(a)

Selected α(1→6) linkages
present in the amylopectin
structure

(b)

FIGURE 24.12

The branched structure of the polysaccharide amylopectin. In (b) each circle represents a glucose unit.

FIGURE 24.13
The process of toasting bread converts some starch to dextrins. This changes the texture and taste of the bread.

Although glycogen is found in many tissues in the human body, it is mainly concentrated in the liver and in muscles.

Like amylopectin, glycogen is a branched α-glucose polymer but glycogen is even more highly branched, having a branch every 8 to 12 α-glucose units. Molecular masses of up to 3 million may be reached for glycogen polymers.

When excess glucose is present in the blood (normally from eating too much starch), the liver and muscle tissue convert the excess glucose to glycogen, which is then stored in these tissues. Whenever the glucose blood level drops (from exercise, fasting, or just normal activities), some stored glycogen is hydrolyzed back to glucose. These two opposing processes are called *glycogenesis* and *glycogenolysis*, the formation and decomposition of glycogen.

$$\text{Glucose} \xrightleftharpoons[\text{glycogenolysis}]{\text{glycogenesis}} \text{Glycogen}$$

The amount of stored glycogen in the human body is relatively small. Muscle tissue contains approximately 1% glycogen, and liver tissue 2 to 3% glycogen. This amount, however, is sufficient to take care of normal activity glucose demands for about 15 hours. During strenuous exercise glycogen supplies can be exhausted rapidly. At this point the body begins using the oxidation of fat (Section 24.10) as a source of energy.

CARBOHYDRATE
LOADING

Many marathon runners (Figure 24.14) eat large quantities of starchy foods (such as spaghetti) the day prior to a race. This practice, called *carbohydrate loading*, maximizes body glycogen reserves, which may then be drawn upon during the race.

FIGURE 24.14
Carbohydrate loading, the consumption of large amounts of starch, is practiced by many marathon runners the day prior to a race. This maximizes glycogen reserves, which may be drawn upon during the race. [Bonnie Freer/Photo Researchers, Inc.]

24.8 Classification and Occurrence of Lipids

Lipids, in contrast to carbohydrates and most other classes of biochemical compounds, cannot be defined in terms of structure. A variety of functional groups and structural features are found in molecules classified as lipids. What lipids have in common are their solubility properties. **Lipids** *are a structurally heterogeneous group of compounds that are not soluble in water but are soluble in nonpolar organic solvents such as chloroform, acetone, diethyl ether, and carbon tetrachloride.* When biological material (animal and plant tissue) is homogenized in a blender and then mixed with a nonpolar organic solvent, the substances that dissolve in the solvent are the lipids.

Lipids may be divided into subclasses based on common structural features. In the remainder of this chapter we consider three lipid subclasses: (1) fats and oils, (2) waxes, and (3) steroids.

The term *lipid* comes from the Greek *lipos*, meaning *fat* or *lard*. In humans and many animals excess carbohydrates (glucose) and other energy-yielding foods are converted to and stored in the body as fats (a type of lipid). These fat reservoirs constitute a major storage form of chemical energy and carbon atoms in the body. Fats and other lipids also surround and insulate numerous vital body organs, providing protection from mechanical shock and helping to maintain the correct body temperature. Other lipids serve as basic structural components of all cell membranes. Many "chemical messengers" in the human body, substances we call hormones, are lipids.

According to some estimates, as much as one-third of the average American diet consists of edible lipids. Since the body cannot synthesize some lipids, certain lipids in the diet are necessary for good health.

24.9 Fatty Acids

Fats and oils (Section 24.10) and waxes (Section 24.11) all contain one or more fatty acid molecules as a structural building block. Thus, a consideration of fatty acids is a natural starting point for a discussion of lipids.

Fatty acids *are monocarboxylic acids that contain long unbranched hydrocarbon chains.* It is the long carbon chains present in fatty acids that makes them and most of their derivatives water-insoluble. The fatty acids of most biological interest have carbon chain lengths in the range 12 to 26. The functional group of a carboxylic acid (see Section 17.7) is the carboxyl group,

$$-C\overset{\displaystyle O}{\underset{\displaystyle OH}{\diagup}}$$

Fatty acids are further classified as *saturated* or *unsaturated.* A **saturated fatty acid** *has a carbon chain in which all carbon–carbon bonds are single bonds.* An example is lauric acid, a 12-carbon acid.

$$CH_3-CH_2-CH_2-CH_2-CH_2-CH_2-CH_2-CH_2-CH_2-CH_2-CH_2-C\overset{\displaystyle O}{\underset{\displaystyle OH}{\diagup}}$$

Lauric acid

The structural formula of lauric acid, or any other fatty acid, can be, and usually is, written in a more condensed form. For lauric acid this condensed form is

$$CH_3-(CH_2)_{10}-C\overset{\displaystyle O}{\underset{\displaystyle OH}{\diagup}}$$

An **unsaturated fatty acid** *has a carbon chain containing one or more carbon–carbon double bonds.* One example is oleic acid, an 18-carbon acid with one double bond, whose structural formulas (expanded and condensed) are

$$CH_3\!-\!(CH_2)_7\!-\!CH\!=\!CH\!-\!(CH_2)_7\!-\!C\underset{OH}{\overset{O}{\diagup}}$$

Oleic acid

Note the angled hydrocarbon chain in oleic acid that results from the presence of the double bond. Such "bends" are always found at double bond locations in naturally occurring biologically important fatty acids.

The term *polyunsaturated fatty acid* is often applied to fatty acids containing more than one double bond. Up to four double bonds are found in biologically important unsaturated fatty acids.

The naturally occurring fatty acids most often encountered in lipid structures are shown in Table 24.3. Two features to note about the given structural formulas are that (1) all hydrocarbon chains are linear rather than branched, and (2) every chain has an even number of carbon atoms. The reason for the even number of carbon atoms is related to how these acids are synthesized. The synthesis process involves joining together residues containing two carbon atoms.

The melting points of fatty acids depend on both the length of the hydrocarbon chain and the degree of unsaturation (number of double bonds per molecule), as exemplified by the melting points of the fatty acids listed in Table 24.3. High molecular mass fatty acids have higher melting points, whereas more double bonds mean lower melting points. The effect of unsaturation is clearly shown by the 18-carbon molecules stearic acid (unsaturated), oleic acid (one double bond), and linoleic acid (two double bonds), which have, respectively, melting points of 70, 16, and −5°C.

Molecules of saturated fatty acids fit very closely together in a neat orderly

TABLE 24.3

Structures, Melting Points and Sources for Selected Saturated and Unsaturated Fatty Acids.

Name	Number of Carbon Atoms	Condensed Structural Formula	Melting Point (°C)	Common Sources
		Saturated Fatty Acids		
Lauric acid	12	$CH_3(CH_2)_{10}$—C(=O)OH	44	Coconut oil
Myristic acid	14	$CH_3(CH_2)_{12}$—C(=O)OH	54	Butterfat, coconut oil, nutmeg oil
Palmitic acid	16	$CH_3(CH_2)_{14}$—C(=O)OH	63	Lard, beef fat, butterfat, cottonseed oil
Stearic acid	18	$CH_3(CH_2)_{16}$—C(=O)OH	70	Lard, beef fat, butterfat, cottonseed oil
Arachidic acid	20	$CH_3(CH_2)_{18}$—C(=O)OH	76	Peanut oil
Cerotic acid	26	$CH_3(CH_2)_{24}$—C(=O)OH	97	Beeswax, wool fat
		Unsaturated Fatty Acids		
Oleic acid	18	$CH_3(CH_2)_7CH=CH(CH_2)_7$—C(=O)OH	16	Lard, beef fat, olive oil, peanut oil
Linoleic acid	18	$CH_3(CH_2)_4(CH=CHCH_2)_2(CH_2)_6$—C(=O)OH	−5	Cottonseed oil, soybean oil, corn oil, linseed oil
Linolenic acid	18	$CH_3CH_2(CH=CHCH_2)_3(CH_2)_6$—C(=O)OH	−11	Linseed oil, corn oil
Arachidonic acid	20	$CH_3(CH_2)_4(CH=CHCH_2)_4(CH_2)_2$—C(=O)OH	−50	Corn oil, linseed oil, animal tissues

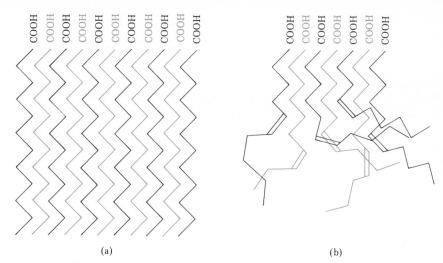

(a)   (b)

FIGURE 24.15

Diagrammatic view of molecular stacking of saturated and unsaturated fatty acids. (a) Orderly stacking of saturated fatty acids gives them high melting points. (b) Disorderly stacking of unsaturated fatty acids gives them low melting points.

fashion as the result of intermolecular attractions between carbon chains (see Figure 24.15a). Longer hydrocarbon chains allow for more attractions between molecules, and the result is a higher melting point. The presence of double bonds disrupts orderly stacking because of the bends in the chains (see Figure 24.15b). Hence, unsaturated fatty acids have lower melting points than saturated ones.

24.10   Fats and Oils

Fats and oils are the most abundant of naturally occurring lipids. Fats are generally solids at room temperature and usually are obtained from animal sources. Oils are generally liquids at room temperature and usually are obtained from plant sources. Fats and oils have identical structural building blocks. In this section we first discuss their common structural features and then consider the "fine points" that distinguish fats from oils.

Fats and oils *contain triesters formed from the condensation reaction between three fatty acid molecules and a glycerol molecule.* Glycerol (see Section 17.3) is an alcohol possessing three hydroxyl groups.

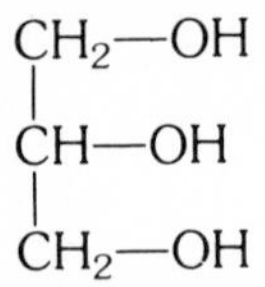

$$CH_2—OH$$
$$CH—OH$$
$$CH_2—OH$$

Each one of these hydroxyl groups can react with the carboxyl group of a fatty acid molecule; the result is the triester structure characteristic of fats and oils.

The general formula and a "block diagram" showing structural components of such triesters are as follows.

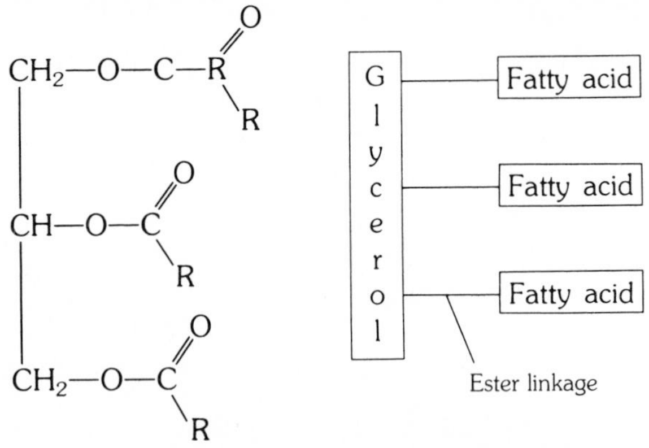

Ester linkage

Triesters of this formulation are commonly called *triglycerides*, a practice we will follow. Each of the fatty acid molecules in a triglyceride is bonded to glycerol via an ester linkage (Section 18.7), hence the designation triester.

The reaction between glycerol and three molecules of stearic acid produces a triglyceride with the following structural features.

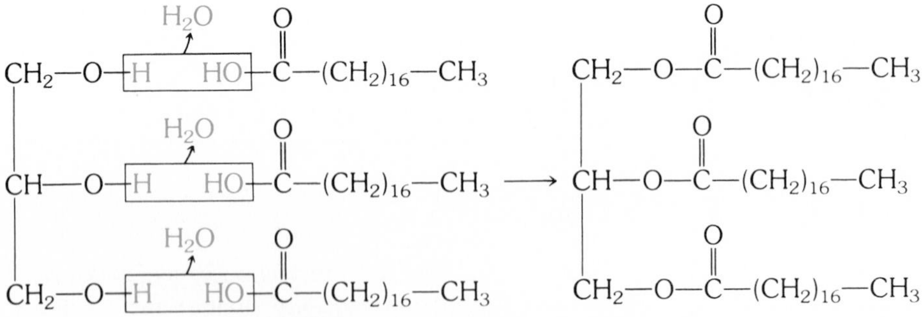

Note that as each ester linkage forms, a molecule of water is also produced. The triester produced from glycerol and stearic acid is an example of a simple triglyceride. **Simple triglycerides** *are triesters formed from the reaction of glycerol and three molecules of the same fatty acid.*

The three fatty acid molecules present as structural components of a triglyceride need not be identical. In the following reaction three different fatty acid molecules are involved in the esterification process.

$$\text{CH}_2-\text{O}-\boxed{\text{H} \quad \text{HO}}-\overset{\text{O}}{\overset{\|}{\text{C}}}-(\text{CH}_2)_{16}-\text{CH}_3 \qquad \text{CH}_2-\text{O}-\overset{\text{O}}{\overset{\|}{\text{C}}}-(\text{CH}_2)_{16}-\text{CH}_3$$

$$\text{CH}-\text{O}-\boxed{\text{H} \quad \text{HO}}-\overset{\text{O}}{\overset{\|}{\text{C}}}-(\text{CH}_2)_{14}-\text{CH}_3 \quad\longrightarrow\quad \text{CH}-\text{O}-\overset{\text{O}}{\overset{\|}{\text{C}}}-(\text{CH}_2)_{14}-\text{CH}_3$$

$$\text{CH}_2-\text{O}-\boxed{\text{H} \quad \text{HO}}-\overset{\text{O}}{\overset{\|}{\text{C}}}-(\text{CH}_2)_{10}-\text{CH}_3 \qquad \text{CH}_2-\text{O}-\overset{\text{O}}{\overset{\|}{\text{C}}}-(\text{CH}_2)_{10}-\text{CH}_3$$

This triester is a mixed triglyceride. **Mixed triglycerides** *are triesters formed from the reaction of glycerol with more than one kind of fatty acid molecule.* The block diagrams in Figure 24.16 contrast simple and mixed triglycerides. In nature, simple triglycerides are rare. Most naturally occurring triglycerides are mixed triglycerides.

Fats and oils that are of biological importance are complex mixtures of mixed triglycerides. No single triglyceride structure adequately describes a naturally occurring fat or oil. The acid components of the triglycerides vary with the source of a fat or oil, as can be seen from the composition data given in Table 24.4.

Such composition data (Table 24.4) lead to an important generalization about fats and oils. Triglyceride mixtures obtained from animal sources contain more saturated fatty acid components than do triglycerides from plant sources. This generalization leads to more specific definitions for fats and oils. **Fats** *are triglyceride mixtures having a relative high percentage of saturated fatty acid components.* **Oils** *are triglyceride mixtures having a relative high percentage of unsaturated fatty acid components.* Such definitions are consistent with the fact that fats are solids and oils are liquids. Recall, from Section 24.9, how degree of unsaturation affects packing of molecules, which in turn affects melting points for free fatty acids. This same principle applies to triglycerides. Note how unsaturation (double bond) affects the "packing" of the hydrocarbon chain portions of the triglyceride represented in Figure 24.17b.

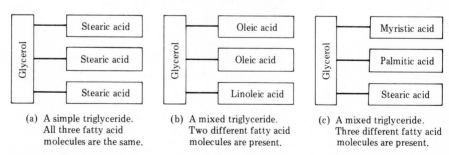

(a) A simple triglyceride. All three fatty acid molecules are the same.

(b) A mixed triglyceride. Two different fatty acid molecules are present.

(c) A mixed triglyceride. Three different fatty acid molecules are present.

FIGURE 24.16

Simple triglycerides have three identical fatty acid molecules, whereas mixed triglycerides contain two or three different kinds of fatty acid molecules.

TABLE 24.4

Fatty Acid Composition of Selected Naturally Occurring Fats and Oils
(*percent by mass*)

	Saturated				Unsaturated		
	Myristic	*Palmitic*	*Stearic*	*Arachidic*	*Oleic*	*Linoleic*	*Linolenic*
Animal fats							
Butter	11	29	9	2	27	4	—
Lard	1	28	12	—	48	6	—
Human fat	3	24	8	—	47	10	—
Beef tallow	5	28	23	—	40	3	—
Plant oils							
Corn	1	10	3	—	50	34	—
Cottonseed	1	23	1	1	23	48	—
Linseed	—	6	2	1	19	24	47
Olive	—	7	2	—	84	5	—
Peanut	—	8	3	2	56	26	—
Soybean	—	10	2	—	29	51	6

The percentages can vary widely for some fats and oils. Environmental factors affect fatty acid composition in plant materials, and diet affects animal fat composition.

The percentages total less than 100% in some cases because of significant amounts of fatty acids above C_{18} or below C_{14}.

A number of factors influence the degree of unsaturation among fatty acids in plants and animals. For example, linseed oil from flaxseed grown in warm climates often contains up to twice as many double bonds as oil obtained from seed grown in cold climates. The degree of unsaturation in lard from hog fat depends on diet; the fat of corn-fed hogs is more saturated than that of peanut-fed hogs.

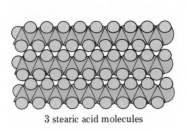

(a)
All fatty acid components
are saturated.

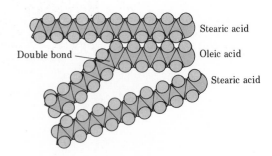

(b)
Both saturated and unsaturated
fatty acid components are present.

FIGURE 24.17

Triglycerides containing unsaturated fatty acid components have lower melting points than those with only saturated fatty acid components because it is more difficult to pack the molecules together in the solid state.

(a)

(b)

FIGURE 24.19

Carboxylate salts obtained from the saponification of animal fats (a) are the active ingredients in soap (b).

the other end, the carboxylate portion, is polar. This dual polarity for the fatty acid salt sodium stearate, which is representative of that for all soap molecules, is as follows.

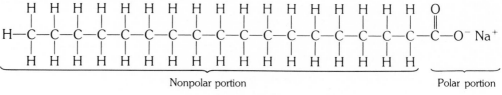

Nonpolar portion   Polar portion

Sodium stearate

Soap "solubilizes" oily and greasy materials because the nonpolar portion of the molecule dissolves in the oil or grease, while the polar portion maintains its solubility in water. Let us look at this "solubilizing" process more closely.

The penetration of the oil or grease by the nonpolar end of the soap molecules is followed by the formation of *micelles*, which are small, spherical grease–soap droplets that are soluble in water as the result of the polar groups on their surface (the heads of the soap molecules; see Figure 24.20). These soap heads, the carboxylate groups (COO⁻), and water molecules are attracted to each other, causing the "solubilizing" of the micelle. The micelles do not combine into larger drops because their surfaces are all negatively charged; like

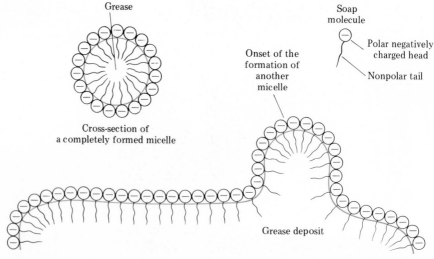

FIGURE 24.20
The cleansing action of soap is due to the ability of soap molecules to form micelles that encapsulate grease and carry it away.

charges repel each other. The water-soluble micelles are subsequently rinsed away, leaving a material devoid of oil and grease.

DETERGENTS

Detergents act by the same mechanism as soaps. Indeed, the structures of the active ingredients in detergents are patterned after those in soaps. The major difference is that salts of sulfonic acids rather than fatty acids are involved.

The most widely used sulfonic acid salts in detergents are salts of substituted benzenesulfonic acids such as

$$H-\overset{\overset{\displaystyle H}{|}}{\underset{\underset{\displaystyle H}{|}}{C}}-\overset{\overset{\displaystyle H}{|}}{\underset{\underset{\displaystyle H}{|}}{C}}-\overset{\overset{\displaystyle H}{|}}{\underset{\underset{\displaystyle H}{|}}{C}}-\overset{\overset{\displaystyle H}{|}}{\underset{\underset{\displaystyle H}{|}}{C}}-\overset{\overset{\displaystyle H}{|}}{\underset{\underset{\displaystyle H}{|}}{C}}-\overset{\overset{\displaystyle H}{|}}{\underset{\underset{\displaystyle H}{|}}{C}}-\overset{\overset{\displaystyle H}{|}}{\underset{\underset{\displaystyle H}{|}}{C}}-\overset{\overset{\displaystyle H}{|}}{\underset{\underset{\displaystyle H}{|}}{C}}-\overset{\overset{\displaystyle H}{|}}{\underset{\underset{\displaystyle H}{|}}{C}}-\overset{\overset{\displaystyle H}{|}}{\underset{\underset{\displaystyle H}{|}}{C}}-\overset{\overset{\displaystyle H}{|}}{\underset{\underset{\displaystyle H}{|}}{C}}-\overset{\overset{\displaystyle H}{|}}{\underset{\underset{\displaystyle H}{|}}{C}}-\underset{\underset{\displaystyle O}{\|}}{\overset{\overset{\displaystyle O}{\|}}{S}}-O^- \ Na^+$$

These molecules have a water-soluble end and an oil-soluble end, as do soap molecules. Sulfonic acids are synthetic materials, whereas fatty acid salts must be obtained from natural products (animal fats). Today's demand for cleansing agents could not be met just with soap; we do not have sufficient animal fat supplies.

Hydrogenation

Hydrogenation involves the addition of hydrogen across a carbon–carbon multiple bond. Through this process carbon–carbon multiple bonds are converted to carbon–carbon single bonds.

Hydrogenation of fats and oils decreases the degree of unsaturation of such substances, as some double bonds are converted to single bonds. With this change there is a corresponding change (increase) in the melting point of the fat or oil.

The hydrogenation process is used in the production of many food products. The peanut oil in many popular brands of peanut butter has been partially hydrogenated to convert the oil to a solid, which does not separate out of the mixture as the oil would. Hydrogenation is used in the production of solid cooking shortenings or margarines from liquid vegetable oils (see Figure 24.21). It is important not to complete the reaction and totally saturate all of the double bonds. If this is done, the product is hard and waxy (like beef tallow) instead of the desired smooth, creamy product.

SOFT-SPREAD MARGARINE

"Soft-spread" margarine spreads are partially hydrogenated oils. The extent of hydrogenation is carefully controlled, so as to make the margarine soft at refrigerated temperatures (4°C). The irony of some advertising is that these same spreads are often proclaimed as "high in polyunsaturates," even though many

FIGURE 24.21
Hydrogenation is used in the production of many consumer food products from vegetable oil bases.

of the double bonds have purposely been changed to single bonds through controlled hydrogenation.

$$
\begin{array}{l}
CH_3(CH_2)_{14}\!-\!\overset{\displaystyle O}{\overset{\|}{C}}\!-\!O\!-\!CH_2 \\[2mm]
CH_3(CH_2)_7CH\!=\!CH(CH_2)_7\!-\!\overset{\displaystyle O}{\overset{\|}{C}}\!-\!O\!-\!CH \ +\ 2\,H_2 \ \xrightarrow[\text{pressure}]{\text{Ni catalyst}} \\[2mm]
CH_3(CH_2)_4(CH\!=\!CHCH_2)_2(CH_2)_6\!-\!\overset{\displaystyle O}{\overset{\|}{C}}\!-\!O\!-\!CH_2
\end{array}
$$

A cottonseed oil triglyceride

$$
\begin{array}{l}
CH_3(CH_2)_{14}\!-\!\overset{\displaystyle O}{\overset{\|}{C}}\!-\!O\!-\!CH_2 \\[2mm]
CH_3(CH_2)_7CH\!=\!CH(CH_2)_7\!-\!\overset{\displaystyle O}{\overset{\|}{C}}\!-\!O\!-\!CH \\[2mm]
CH_3(CH_2)_{16}\!-\!\overset{\displaystyle O}{\overset{\|}{C}}\!-\!O\!-\!CH_2
\end{array}
$$

Solid shortening or margarine

Rancidity

Fats and oils, upon exposure to air at room temperature, often develop unpleasant odors and/or flavors. Such affected fats and oils have become *rancid*. Rancidity results from two kinds of unwanted reactions: (1) hydrolysis of triglyceride ester linkages and (2) oxidation of carbon–carbon double bonds in the fatty acid chains of triglycerides.

Hydrolytic rancidity results from the exposure of fats and oils to *moist* air. (All air contains some moisture.) Microorganisms present in air supply necessary enyzmes to catalyze the hydrolysis. Volatile fatty acid molecules, especially those of low molecular mass, freed as the result of triglyceride hydrolysis contribute to the disagreeable odors associated with rancid foods. For example, butyric acid is the source of the characteristic odor of rancid butter. Hydrolytic rancidity is prevented by storing fats and oil in closed containers (to minimize contact with microorganisms) and by refrigeration (all reactions occur more slowly at lower temperatures). The unpleasant odor of perspiration results from the hydrolysis of fats and oils on the skin.

Oxidative rancidity, in most cases, is a more important factor in the overall process of rancidity than is hydrolytic rancidity. In oxidative rancidity, double bonds in the unsaturated fatty acid components "rupture," and low molecular mass aldehydes are produced. Many of these aldehydes have objectionable odors. In addition, these aldehydes can be further oxidized to give equally or more offensive (from an odor viewpoint) low molecular mass carboxylic acids. Warmth and exposure to atmospheric oxygen induce oxidative rancidity. To avoid unwanted oxidation, the food industry adds antioxidants to foods. Two naturally occurring antioxidants are vitamin C (ascorbic acid) and vitamin E (α-tocopherol). Two synthetic oxidation inhibitors are BHA and BHT.

Butylated hydroxyanisole (BHA)

Butylated hydroxytoluene (BHT)

24.12 Waxes

Waxes *are monoesters formed from the reaction of a long-chain monohydroxy alcohol with a fatty acid molecule.* The block diagram for a wax is

| Fatty acid |———| Long-chain alcohol |

Many naturally occurring waxes are known. In most cases they function as a protective coatings. The leaf surfaces of many plants that grow in dry regions are coated with wax to minimize water loss. Human ears contain a wax, secreted by glands in the ear, that acts as a protective barrier against infection by capturing airborne particles. Beeswax, a secretion of this insect, is used as a structural

FIGURE 24.22
Waxes, esters of long-chain alcohols and fatty acids, are essential to the well-being of aquatic birds. Waxes coating feathers waterproof the bird and also thermally insulate it from the cold water. [William Staley from National Audubon Society/Photo Researchers, Inc.]

TABLE 24.5

Structures, Sources and Uses for Selected Important Waxes.

Wax	Structure of a Principal Ester Component	Source	Uses
Beeswax	$CH_3(CH_2)_{14}-\overset{\displaystyle O}{\overset{\|}{C}}-O-(CH_2)_{29}CH_3$ Myricyl palmitate	Bees	Candles, cosmetics, confections, medicines, art preservation
Carnauba wax	$(HO)_aCH_2(CH_2)_b-\overset{\displaystyle O}{\overset{\|}{C}}-O-(CH_2)_cCH_3$ $a = 0$ or 1; $b = 17-29$; $c = 31$ or 33	Carnauba palm trees	Coatings for perishable products; polishing candies and pills; auto and floor waxes
Rice bran wax	$CH_3(CH_2)_m-\overset{\displaystyle O}{\overset{\|}{C}}-O-(CH_2)_nCH_3$ $m = 20$ or 22; $n = 21-35$	Rice bran	Lipstick base, plastics processing aid
Jojaba wax	$CH_3(CH_2)_a-\overset{\displaystyle O}{\overset{\|}{C}}-O-CH_2(CH_2)_a-CH_3$ $a = 18$ or 20	Jojoba beans	Cosmetics and candles

material for the beehive. Commercially, waxes are marketed as protective coatings (automobile and floor waxes, and lotions) and for use in candles.

Waxes coat the hair of many animals and the feathers of birds. For aquatic birds, this wax coating is a waterproofing agent (see Figure 24.22). Removal of such wax from the feathers of aquatic birds has very serious consequences. The feathers become "wet," and the bird cannot maintain its buoyancy. Such wax loss does occur when birds come in contact with an oil spill. Oil, a mixture of nonpolar hydrocarbons (Section 19.3), becomes a solvent for the waxes. Loss of feather wax has another effect on aquatic birds. The insulative value of the feathers is lost, and some birds quickly die of exposure in cold water.

Table 24.5 gives structural formulas, sources, and uses for selected waxes. Natural waxes are always mixtures, like triglycerides, of numerous similarly structured esters. Sometimes natural waxes also contain other components besides esters. Such components include free alcohols, free carboxylic acids, and long-chain hydrocarbons. Paraffin wax (Section 19.4) is a completely non-ester material; it is a mixture of C_{20} and higher hydrocarbons. It is paraffin wax that is used in home canning to seal various types of jars.

24.13 Steroids

Steroids *are lipids with a fused-ring structure consisting of three six-membered rings and one five-membered ring.* The rings are customarily labeled by letters of the alphabet and each carbon atom by a number, as shown below.

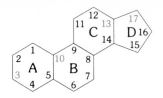

Steroid nucleus

Many different steroids have been isolated from plants, animals, and human beings. Location of double bonds within the fused ring system and the nature and location of substituents distinguish one steroid from another. Substituents located at carbons 3, 10, 13, and 17 are particularly common in biologically important steroids.

Steroids to be discussed in this section are cholesterol, bile salts, and steroidal hormones.

Cholesterol

Cholesterol is the most abundant steroid in the human body. The name has the -ol ending because cholesterol is an alcohol, with an —OH group on carbon 3 of the steroid nucleus. In addition, cholesterol has methyl groups bonded to carbon atoms 10 and 13 and a small branched hydrocarbon chain is present on carbon 17.

Cholesterol

Cholesterol is found in nerve and brain tissue and in the blood, and it is the main component of gallstones. Human blood plasma contains about 50 mg of free cholesterol per 100 mL and about 170 mg of cholesterol esterified with various fatty acids.

Cholesterol has received much recent publicity because of a possible correlation between blood cholesterol levels and the disease known as atherosclerosis (Section 22.5). Large amounts of cholesterol are present in the plaque buildup associated with this disease.

Most body cholesterol is synthesized in the liver from carbohydrates, proteins, and fats (our food). Thus, the elimination of cholesterol-containing foods from the diet will not necessarily lower cholesterol levels. The total caloric intake must also be reduced, and the types and amounts of fats ingested must be changed. It has been found that certain unsaturated fish and vegetable oils, when substituted for saturated fats, lower the blood cholesterol level. Plants do not synthesize cholesterol. Therefore food such as margarine, sunflower oil, and peanut butter contain no cholesterol. Animal fats do contain cholesterol, so that butter, milk, eggs, and meat are sources of cholesterol in the diet.

The problem of gall bladder "attacks" caused by gallstones is related to cholesterol. A large percentage of gallstones are almost pure cholesterol crystals that have precipitated from bile. Bile is a digestive juice excreted by the liver and stored by the gall bladder to aid in the solubilization of dietary fats within the small intestine.

Cholesterol is not totally "bad." It occurs, in amounts up to 25%, in cell membranes where it contributes to membrane rigidity. There is a direct relationship between cholesterol concentration and rigidity. Cholesterol also plays a vital biological role in chemical synthesis within the human body, being the starting material for the synthesis of numerous steroidal hormones, vitamin D, and bile salts.

Bile Salts

Bile salts *are emulsifying agents that make dietary lipids soluble in the aqueous environment of the digestive tract.* During digestion, bile salts are released from the gall bladder into the intestine, where they aid in the digestive processes by emulsifying (solubilizing) fats and oils. Their mode of action is much like that of soap in the washing process (Section 24.11); micelles are formed.

Cholesterol is the starting material for bile salt synthesis. The two principal bile salts are sodium glycocholate and sodium taurocholate.

Sodium glycocholate

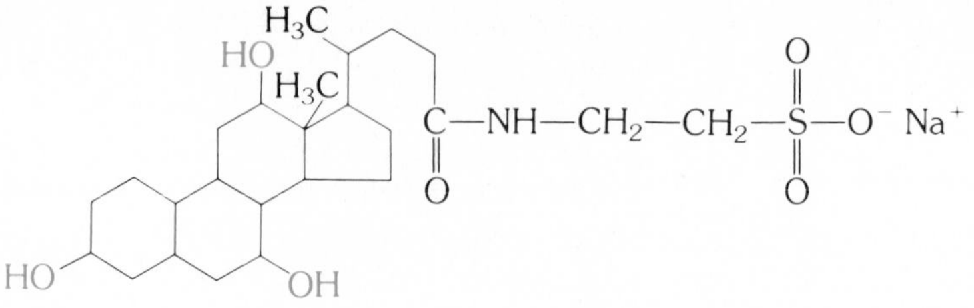

Sodium taurocholate

Note that both of these salts have three —OH groups (carbons 3, 7, and 12) compared to one —OH group in cholesterol. In addition, the carbon 17 side chain differs from that in cholesterol.

Fresh bile from the liver is golden yellow. On standing, bile darkens progressively from gold to green to blue and finally to brown, as the bile pigments present gradually oxidize. Total excretion of bile salts varies from 0.5 to 2.0 grams daily, and such excretion is responsible for the characteristic color of the feces.

Hormones

Hormones *are chemicals produced by ductless glands within the human body that provide a communication pathway between various tissues.* They are chemical messengers. Many, but not all, hormones in the human body are steroids. Such steroidal hormones are synthesized from cholesterol; hence they contain the four-ring fused system characteristic of that molecule.

Two areas of the human body are "centers" for steroidal hormone production. They are the adrenal glands and gonads (the testes in males and the ovaries in females). The adrenal glands are found at the top of each kidney. More than 30 different hormones are produced by the adrenal glands.

Three important adrenal hormones are cortisol, aldosterone, and cortisone.

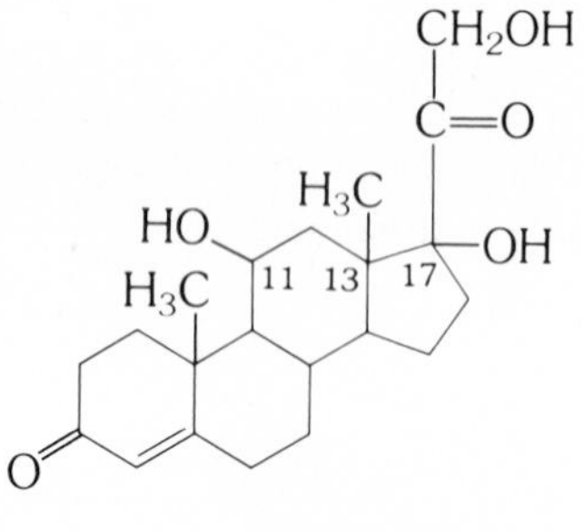

Cortisol

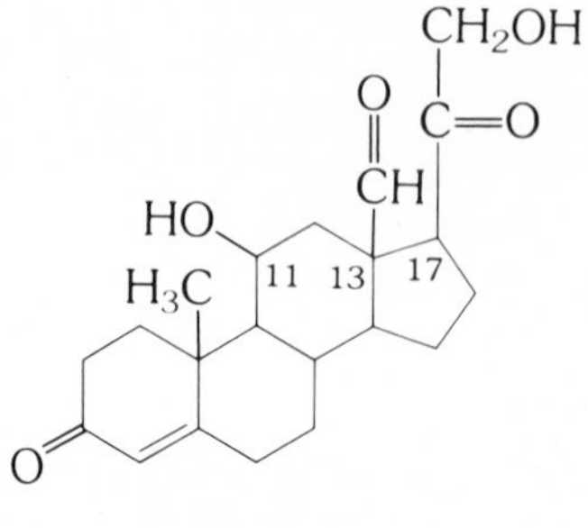

Aldosterone

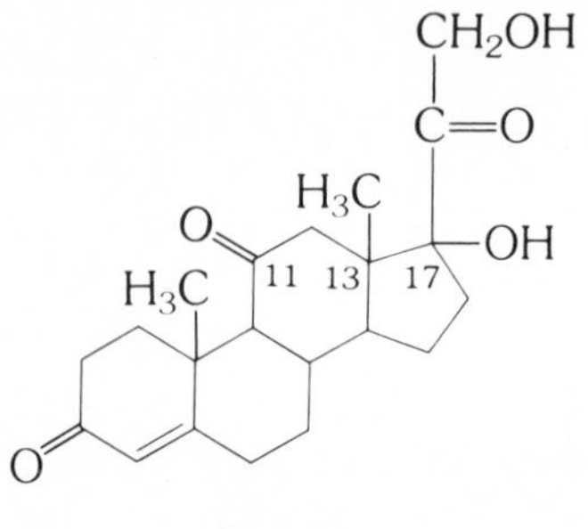

Cortisone

Notice the structural similarities of these three hormones. Cortisol and aldosterone differ only in the attachments at carbon 13 and the presence of an —OH group at carbon 17 in cortisol. Cortisol and cortisone differ only in the attachments at carbon 11.

Cortisol promotes formation of glycogen (Section 24.7) and stimulates degradation (digestion) of fats and proteins. Aldosterone regulates blood pressure by influencing the absorption of Na^+ and Cl^- ions in kidney tubules. Cortisone exerts powerful anti-inflammatory effects in the body.

CORTISONE AND ARTHRITIS

Synthetically produced cortisone, identical to the natural hormone, finds use as a prescription medication. When applied topically or injected into a diseased joint, it acts as an anti-inflammatory agent and is of great use in treating arthritis. Long-term use of cortisone in treating arthritis, however, has adverse side effects; it can weaken bones and make muscles atrophy. To combat this problem a synthetic hormone, prednisolone (with a cortisone-like structure) is now marketed. This synthetic homone has cortisone-like activity, but avoids the adverse side effects of long-term use of the natural hormone.

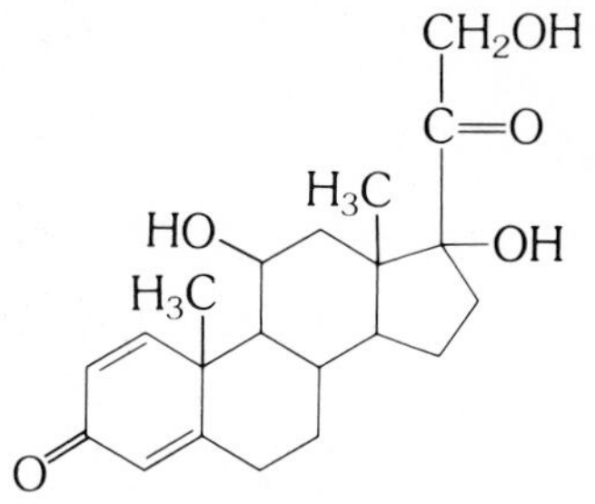

Prednisolone

Structurally related to cholesterol and the adrenal hormones previously discussed are the sex hormones produced by the testes of men and the ovaries of women.

Male sex hormones, called androgens, are responsible for the development of the male sex organs and secondary masculine sexual characteristics such as deep voice, greater muscular development, and facial hair. The most important androgen is testosterone.

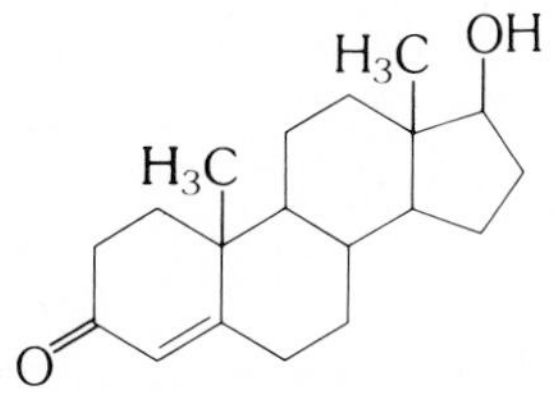

Testosterone

Since testosterone is involved in muscle development in men, athletes who compete in events where strength and muscle development are important (for example, weight lifting, shot put, and hammer throw) have sometimes resorted to using synthetic testosterone-like steroids in an attempt to improve performance. Such use of steroids has been officially banned by the world's amateur athletic organizations.

ANABOLIC STEROIDS AND ATHLETES

Synthetic steroids have been developed that have only masculinizing (androgenic) or muscle-building (anabolic) effects. It is the anabolic steroids that have been misused by athletics. Apart from athletics, anabolic steroids have medical purposes. These steroids are prescribed by physicians to correct hormonal imbalances or to prevent the withering of muscle in persons who are recovering from surgery or starvation. A commonly used (or misused) anabolic steroid is the compound norancholane.

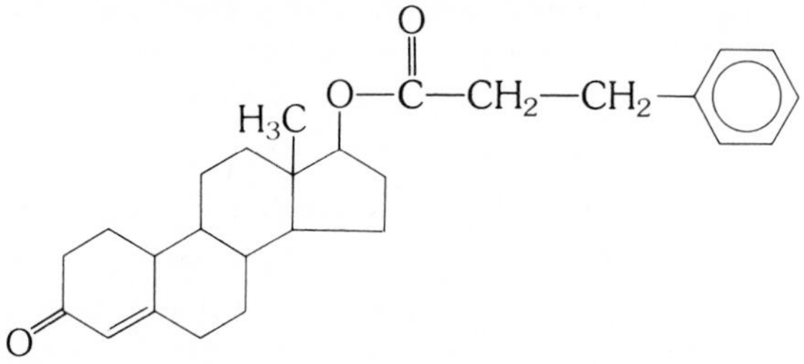

Norancholane

There are two important classes of female sex hormones—estrogens and progestins. The estrogens, produced primarily in the ovaries, control the female menstrual cycle, the development of breasts, and the other female sexual characteristics. The two most important estrogens are estrone and estradiol.

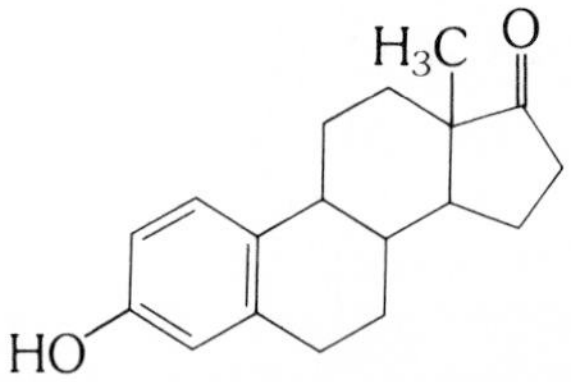

Estrone

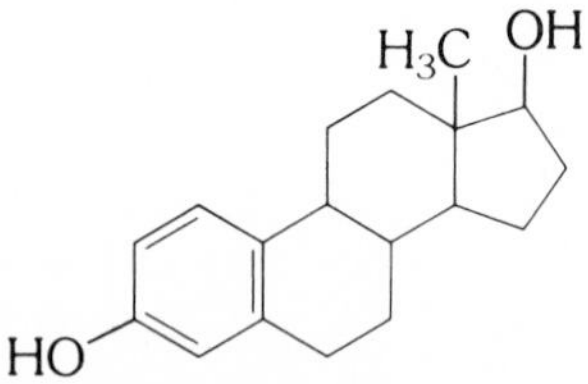

Estradiol

The estrogens differ from other steroids in that ring A of the steroid nucleus is an aromatic ring.

The most important progestin is the hormone progesterone.

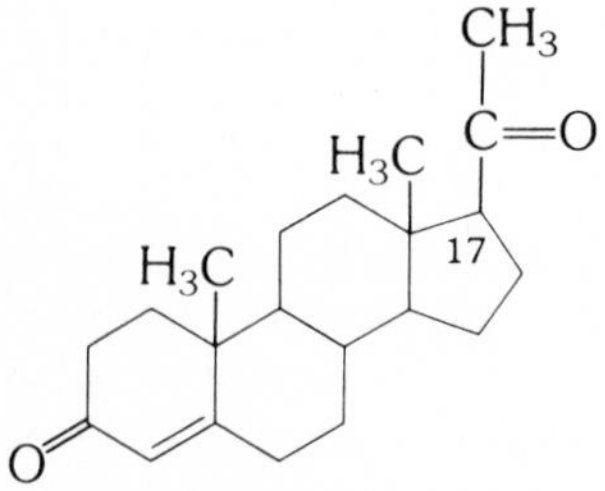

Progesterone

Progesterone causes changes in the uterine wall to prepare the uterus for pregnancy if the egg is fertilized and prevents the further release of eggs from the ovaries during a pregnancy. It is interesting to note the close similarity in structure between progesterone and the male hormone testosterone. They differ only in the attachments at carbon 17.

BIRTH CONTROL PILLS

Oral contraceptives (birth control pills) are a takeoff on the role of progesterone in preventing maturation and release of eggs during pregnancy. Oral contraceptives for women contain synthetic steroids that simulate the hormonal processes resulting from pregnancy and thus prevent ovulation (release of an egg). Progesterone itself is an ineffective oral contraceptive because a women's body breaks it down before it reaches the ovaries. Numerous progesterone-like derivatives are now available that overcome this problem and yet still mimic progesterone's hormonal action. "The pill" is the name used for drugs containing such compounds.

Most pills now marketed are combination contraceptives, containing both a progestational compound and an estrogenic compound. The widely prescribed Ortho-Novum tablet contains the progesterone-like norethindrone and estrogen-like mestranol. Table 24.6 gives the structures of these two compounds and the naturally occurring steroid hormones that they mimic. Note the similarities between the synthetic and natural hormones.

Pills currently marketed are much safer than early oral contraceptives, but there are still side effects. Adverse side effects include fluid imbalance, increased risk of blood clots, an increased appetite, and weight gain.

TABLE 24-6

A Comparison of the Structures of Synthetic and Natural Female Sex Hormones.

Mestranol (synthetic)

Estrone (natural)

Norethindrone (synthetic)

Progesterone (natural)

Practice Questions and Problems

TYPES OF BIOCHEMICAL SUBSTANCES

24-1 What are the two general groups of biochemical substances and what are the major types of compounds found in each group?

24-2 Discuss the composition of the human body in terms of major types of biochemical substances present.

CLASSIFICATION OF CARBOHYDRATES

24-3 Give a general definition for a carbohydrate.

24-4 What functional groups can be found in carbohydrates?

24-5 Distinguish between the following types of carbohydrates: a monosaccharide, a disaccharide, and a polysaccharide.

IMPORTANT MONOSACCHARIDES

24-6 Draw open-chain structural formulas for each of the following monosaccharides.
  a. Glucose   b. Galactose
  c. Fructose   d. Ribose

24-7 Explain the principal differences and similarities between the structures of the members of each of the following pairs of monosaccharides.
  a. Glucose and galactose
  b. Glucose and fructose
  c. Glucose and ribose

24-8 Identify the monosaccharide with each of the following characteristics.
  a. Most abundant monosaccharide
  b. Also known as dextrose
  c. Also known as levulose
  d. Also known as blood sugar
  e. Used in hospitals as an intravenous source of nourishment
  f. Found in brain and nerve tissue as part of very complex molecules
  g. Has the sweetest taste of all monosaccharides
  h. A component of RNA

24-9 The structure of glucose is sometimes written in an open-chain form and other times as a cyclic structure. Explain why either form is acceptable.

24-10 Identify each of the following structures as an α- or β-isomer.

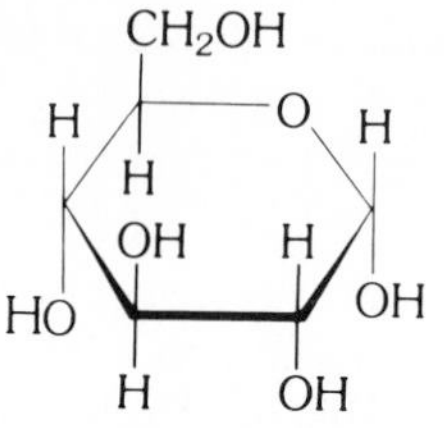

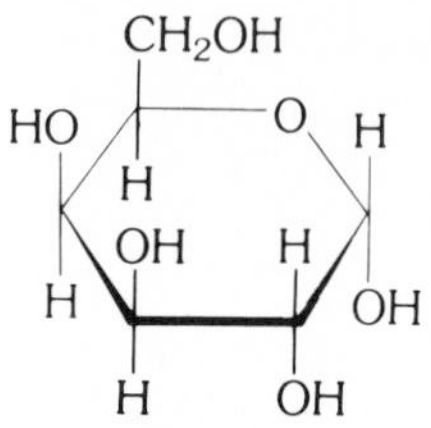

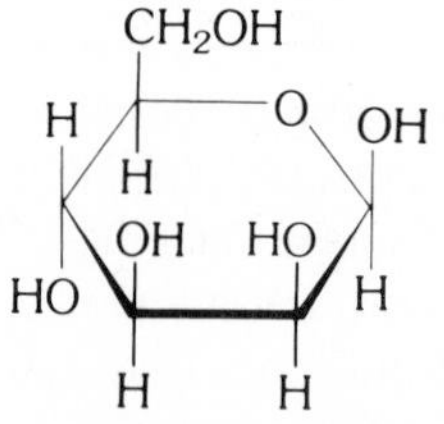

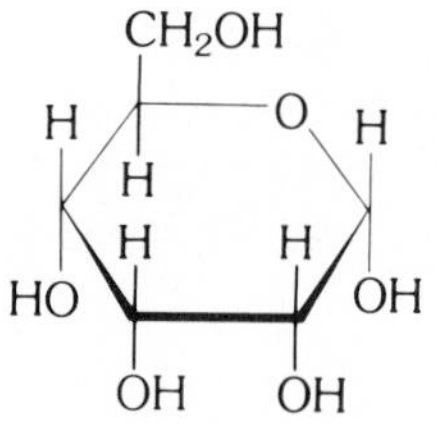

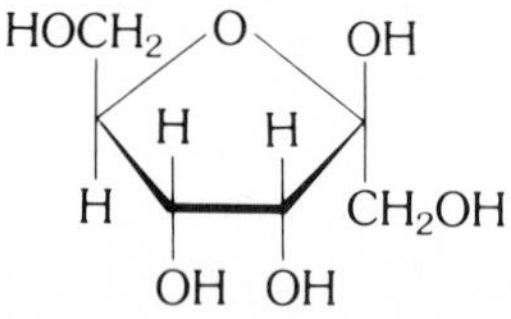

24-11 Fructose and glucose both contain six carbon atoms. Why do the cyclic forms of fructose have a five-membered ring instead of the six-membered ring that is found in the cyclic forms of glucose?

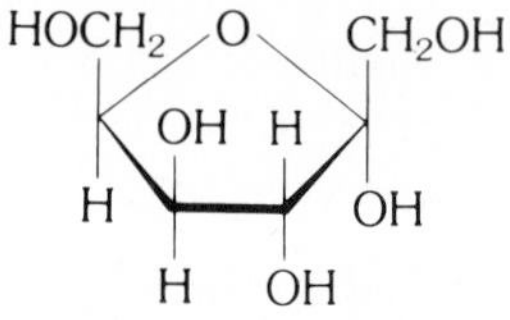

IMPORTANT DISACCHARIDES

24-12 Which monosaccharide is a component of all three of the disaccharides sucrose, maltose, and lactose.

24-13 Identify a disaccharide that fits each of the following descriptions.
  a. Household table sugar
  b. Converted to lactic acid when milk sours
  c. An ingredient in infant formulas used as substitutes for mother's milk

d. Formed in germinating grain

e. Hydrolyzes when cooked with acidic foods to give invert sugar

24-14 What monosaccharides are produced from the hydrolysis of each of the following disaccharides?
a. Sucrose   b. Maltose   c. Lactose

24-15 Describe two types of "allergies" to milk that account for the fact that milk is not a good food for everyone.

ARTIFICIAL SWEETENERS

24-16 Why is "invert sugar" sweeter than the original sucrose from which it was obtained?

24-17 Explain why aspartame is considered a low-calorie sweetener even though it has the same caloric content as sucrose (4 calories/gram).

POLYSACCHARIDES

24-18 Describe the differences and similarities in the structures of each of the following pairs of polysaccharides.
a. Glycogen and amylopectin
b. Amylose and glycogen
c. Amylose and cellulose
d. Amylose and amylopectin

24-19 Name a polysaccharide that fits each of the following descriptions.
a. The unbranched polysaccharide in starch
b. The most abundant polysaccharide in starch
c. A storage form of carbohydrates in animals and humans
d. A structural carbohydrate in plants

24-20 Explain why human beings cannot digest cellulose, but mammals such as deer, sheep, and cows live quite comfortably on a diet high in cellulose content.

24-21 Although no caloric value is obtained from cellulose, it is still considered important in the diet of human beings. Explain.

24-22 What is the biological role of glycogen?

24-23 What is the difference between the processes of glycogenesis and glycogenolysis?

CLASSIFICATIONS OF LIPIDS

24-24 Give a formal definition for the class of biochemicals called lipids.

24-25 In which of the following solvents would you expect lipids to be soluble?

a. H_2O b. $CH_3\text{-}CH_2\text{-}O\text{-}CH_2\text{-}CH_3$
c. $CH_3\text{-}CH_2\text{-}OH$ d. CCl_4

FATTY ACIDS

24-26 How does a fatty acid differ in structure from a simple carboxylic acid?

24-27 Name the saturated fatty acids containing 16, 18, and 20 carbon atoms.

24-28 Compare the structures of stearic acid and oleic acid. How are they different? What do they have in common?

24-29 Give the names of four naturally occurring fatty acids having 18 carbon atoms and indicate how many carbon–carbon double bonds are present in each acid.

24-30 Compare the melting points of arachidic acid and arachidonic acid. Suggest a reason for the dramatic difference between the melting points of these two acids.

FATS AND OILS

24-31 What is the difference between a simple triglyceride molecule and a mixed triglyceride molecule?

24-32 Define in terms of structural features the class of compounds called triglycerides.

24-33 Draw the structure of a triglyceride formed from glycerol and each of the following.
a. Three molecules of palmitic acid
b. Two stearic acid molecules and one myristic acid molecule.
c. One palmitic acid, one stearic acid, and one cerotic acid molecule.

24-34 How does a fat differ from an oil in terms of the following?
a. Melting point   b. Chemical structure

24-35 Explain the meaning of the term "polyunsaturated" when it is used in advertising for vegetable oils.

24-36 Using the information in Table 24.4, determine which member of each of the following pairs of naturally occurring triglyceride mixtures is more unsaturated.
a. Lard or beef tallow
b. Peanut oil or linseed oil
c. Human fat or corn oil

CHEMICAL REACTIONS OF FATS AND OILS

24-37 Give the general name of the products produced from each of the following.

a. Hydrolysis of a triglyceride

b. Saponification of a triglyceride

c. Complete hydrogenation of a triglyceride

24-38 Write two structural equations, one for the acid hydrolysis and one for the basic hydrolysis (saponification), of the following triglyceride.

$$CH_2-O-\overset{\overset{\textstyle O}{\|}}{C}-(CH_2)_{14}-CH_3$$
$$CH-O-\overset{\overset{\textstyle O}{\|}}{C}-(CH_2)_{14}-CH_3$$
$$CH_2-O-\overset{\overset{\textstyle O}{\|}}{C}-(CH_2)_7-CH=CH-(CH_2)_7-CH_3$$

24-39 Why can only unsaturated triglycerides undergo hydrogenation?

24-40 Describe how margarines are produced from vegetable oils.

24-41 Why do animal fats and vegetable oils become rancid when exposed to moist warm air?

24-42 Describe the differences between hydrolytic rancidity and oxidative rancidity.

24-43 Explain the molecular basis for the cleansing action of soap.

24-44 What are the structural differences between the active ingredient (cleansing agent) in soaps and detergents?

WAXES

24-45 Draw the general structure of a wax..

24-46 How does a wax differ from a triglyceride in (a) structure and (b) chemical properties?

24-47 Draw the structure of a wax formed from palmitic acid and cetyl alcohol $[CH_3-(CH_2)_{14}-CH_2-OH]$.

24-48 Draw the structure of the principal ester component of beeswax.

24-49 Describe a function that may be served by a wax in each of the following locations.

a. On the surface of a leaf

b. Within the human ear

c. On the feathers of an aquatic bird

STEROIDS

24-50 Draw the fused hydrocarbon ring system characteristic of steroids.

24-51 Draw the structure of cholesterol.

24-52 Explain why a diet rich in cholesterol is not necessarily the principal cause of a person's high blood cholesterol concentration.

24-53 What is the starting material for the synthesis of bile salts?

24-54 What role do bile salts play in the process of digestion in humans?

24-55 What are hormones and from what source are most steroidal hormones synthesized?

24-56 What are the biological effects of each of the following adrenal hormones?

a. Cortisol   b. Aldosterone

c. Cortisone

24-57 Prednisolone is preferred over cortisone as an antiinflammatory agent. Why?

24-58 Name the primary male sex hormone and the three principal female sex hormones.

24-59 What role do the estrogens and progesterone serve in preparation for pregnancy?

24-60 What role do steroid compounds have in the action of oral contraceptive pills?

24-61 Why is progesterone ineffective as an oral contraceptive?

24-62 Why would an athletic be tempted to use anabolic steroids?

24-63 List all structural differences in each of the following pairs of steroids.

a. Cholesterol and sodium glycocholate

b. Cortisol and cortisone

c. Testosterone and progesterone

d. Estrone and estradiol

25

Biochemistry: Proteins and Nucleic Acids

807

25.1 Occurrence and Functions of Proteins

Proteins are components of all living cells. They are, thus, found in all forms of life, from the simplest (bacteria, algae, and other microorganisms) to the most complex (human beings). Proteins are required in the diets of animals and humans in order for them to synthesize tissues, enzymes, certain hormones, and some blood components. In addition, proteins are used in the maintenance and repair of existing tissues and as a source of energy. The importance of proteins in life processes is recognized in the name. The word protein is derived from the Greek *proteios*, which means "of first importance."

Next to water, proteins are the most abundant substances in most cells—approximately 15% of the cell's mass (Section 24.1). All proteins contain the elements carbon, hydrogen, oxygen, nitrogen, and sulfur. Other elements such

TABLE 25.1

The Biological Functions of Selected Proteins

General Function	Example	Specific Function
Transportation	Hemoglobin	Carries oxygen in the blood to the cells
	Lipoprotein	Carries fatty acids in the blood
Structural	Collagen	Forms connective tissue and cartilage
	Keratin	Forms hair, wool, skin, nails, feathers, and horns
Defense	Gamma globulin	Helps fight infectious diseases
	Prothrombin	Causes blood clotting
Catalytic (enzymes)	Trypsin	Hydrolyzes proteins
	Sucrase	Hydrolyzes sucrose
Storage	Ferritin	Stores iron for reuse
	Egg albumin	Supplies amino acids to developing embryo
Regulatory	Insulin	Regulates glucose metabolism

as phosphorus, iodine, and iron are essential constituents of certain specialized proteins. Casein, the main protein of milk, contains phosphorus, an element very important in the diet of infants and children. Hemoglobin, the oxygen-transporting protein of blood, contains iron. The presence of the element nitrogen in proteins sets them apart from carbohydrates and lipids. The average nitrogen content of proteins is about 16%.

Proteins are one of the three major categories of foods needed by humans, the other two being carbohydrates and fats. Foods high in protein content include fish, beans, nuts, cheese, eggs, poultry, and meat. Such foods tend to be, in general, some of the more expensive types of food and, thus, are least available in underdeveloped countries of the world. Protein deficiency is a major problem among the world's undernourished peoples.

An "unbelievable" number of different types of protein molecules, each with a different function, exist in the human body. A typical human cell contains about 9000 different proteins. It is estimated that about 100,000 different proteins are present in humans. Table 25.1 is a partial list of the biological functions of proteins.

25.2 Polymeric Nature of Proteins

Proteins are extremely large molecules with formula masses that vary from about 6000 to several million amu. Their immense size (in the molecular sense) can be appreciated by comparing glucose with hemoglobin, a relatively small protein. Glucose has a formula mass of 180 amu, and the formula mass of hemoglobin is 65,000 amu. The molecular formula of glucose is $C_6H_{12}O_6$, and that of hemoglobin is $C_{2952}H_{4664}O_{832}N_{812}S_8Fe_4$.

Proteins, like polysaccharides, are polymers. In polysaccharides the monomeric units are monosaccharides (usually glucose); see Section 24.4). In proteins the monomeric units are amino acids, substances containing both an amino group (Section 17.9) and a carboxyl group (Section 17.7). **Proteins** *are polymers of amino acids in which the amino acids are linked together by amide linkages formed between the carboxyl group of one amino acid and the amino group of another amino acid.* The starting point for a discussion of proteins is, thus, amino acids, the building blocks for proteins.

25.3 Amino Acids—Building Blocks for Proteins

An **amino acid** *is an organic compound that contains both an amino (—NH_2) group and a carboxylic acid (—COOH) group.* In amino acids found in proteins, it is almost always true that the amino and carboxyl groups are separated from each other by a single carbon atom. The general formula for such amino acids is

The R group in the general formula, called the *side chain*, may be hydrogen, a alkyl group, an aromatic ring, or a heterocyclic ring. The nature of this R group is responsible for the differences among the various amino acids.

The simplest amino acid is the one with hydrogen as the R group.

The common name of this amino acid is *glycine*. Next in complexity is the amino acid *alanine*, in which R is a methyl group.

$$H_2N-CH-C\underset{OH}{\overset{O}{\diagup}}$$
$$\underset{CH_3}{|}$$

More than 30 different amino acids have been identified as constituents of protein. Only 20 of these are, however, commonly encountered. Table 25.2 lists the structures of these 20 most common amino acids. Because IUPAC names for these amino acids are quite cumbersome, common names, which are given in the table, are almost always used. The names of amino acids are often abbreviated by using the first three letters of the acid's common name. Such name abbreviations are also given in Table 25.2.

Within Table 25.2, the amino acids are grouped according to the polarity properties of their side chains. Note that there are three categories: (1) nonpolar side chain, (2) polar side chain, and (3) very polar side chain.

The properties of specific proteins are mostly determined by the properties of the side chains present on the amino acids. For example, water-insoluble structural proteins (such as those in hair, wool, and tendons) contain high percentages of amino acids with nonpolar side chains. Membrane proteins also have large numbers of nonpolar side chains. The more water-soluble proteins (such as albumin and hemoglobin) contain large amounts of amino acids with polar side chains.

25.4   Essential Amino Acids

All of the amino acids included in Table 25.2 are necessary constituents of human protein. Some but not all of them can be synthesized within the body from other materials. There is strong evidence that the human body is incapable of producing 10 of these 20 amino acids fast enough or in sufficient quantities to sustain normal growth. These 10 amino acids, which are called *essential amino acids*, must be obtained from our food. **Essential amino acids** *are amino acids that must be obtained from food because they cannot be biosynthesized in sufficient quantity*. An adequate human diet must include foods that contain these 10 amino acids (see Table 25.3).

A dietary protein that contains all ten of the essential amino acids is called a *complete protein*. Most animal proteins, including casein from milk and those found in meat, fish, and eggs, are complete proteins. Casein contains all 10 essential amino acids plus 9 of the 10 nonessential amino acids.

TABLE 25.2

The Common Amino Acids Found in Proteins

(R groups are in color, and the three-letter abbreviations are in parentheses.)

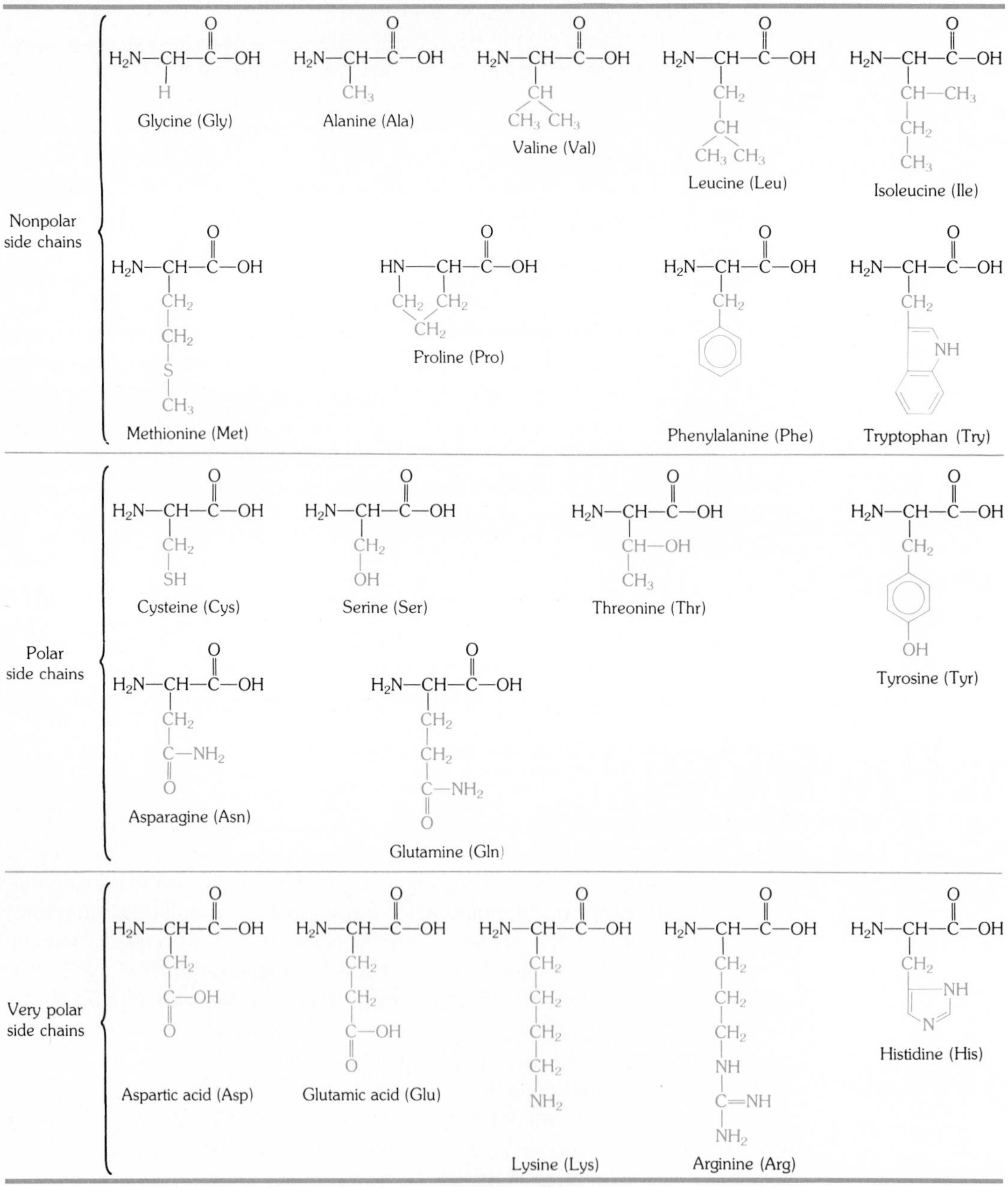

Nonpolar side chains

Glycine (Gly)   Alanine (Ala)   Valine (Val)   Leucine (Leu)   Isoleucine (Ile)

Methionine (Met)   Proline (Pro)   Phenylalanine (Phe)   Tryptophan (Try)

Polar side chains

Cysteine (Cys)   Serine (Ser)   Threonine (Thr)   Tyrosine (Tyr)

Asparagine (Asn)   Glutamine (Gln)

Very polar side chains

Aspartic acid (Asp)   Glutamic acid (Glu)   Lysine (Lys)   Arginine (Arg)   Histidine (His)

TABLE 25.3

The Ten Essential Amino Acids

Isoleucine	Threonine
Leucine	Tryptophan
Lysine	Valine
Methionine	Arginine*
Phenylalanine	Histidine*

* Arginine and histidine are essential for children, but may not be essential for adults.

LIQUID PROTEIN DIETS

Gelatin, a substance obtained from the hydrolysis of collagen (found in meat), is an important animal protein that is not complete. It lacks tryptophan and is low in several other amino acids. This is mentioned because many "liquid protein" diets are gelatin based. People on such diets, if this is their only protein source, will be lacking some essential amino acids.

VEGETARIAN DIETS

Most plant proteins are incomplete. Rice protein lacks the amino acids lysine and threonine; wheat protein lacks lysine; corn protein lacks lysine and tryptophan; and soy protein is very low in methionine. Vegetarians must, thus, eat a range of plant foods in order to obtain all of the essential amino acids.

Dietary proteins may be rated in terms of biological value on a percentage scale. Such percentage values for selected proteins are given in Table 25.4. Note that, in general, animal proteins have higher biological value than do plant proteins.

TABLE 25.4

Biological Value of Selected Dietary Proteins

Food	Biological Value (%)
Whole hen's egg	94
Whole cow's milk	84
Fish	83
Beef	73
Soybeans	73
White potato	67
Whole grain wheat	65
Whole grain corn	59
Dry beans	58

25.5 Peptide Formation

A carboxylic acid and an amine can react with each other to give an amide as a product. The general equation for such a reaction is

$$R-C\overset{O}{\underset{OH}{\big<}} \quad + \quad \overset{H}{\underset{H}{\big>}}N-R' \longrightarrow R-\overset{O}{\overset{\|}{C}}-\overset{H}{\underset{}{N}}-R' + H_2O$$

Carboxylic acid   Amine   Amide

Two amino acids can react with each other in a similar way. The carboxyl group of one amino acid reacts with the amino group of the other amino acid. The product is a molecule containing two amino acids linked by an amide bond.

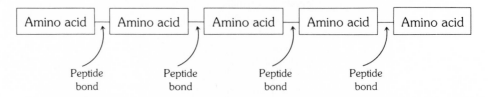

In amino acid chemistry, amide bonds that link amino acids together are given the special name of *peptide bond*. A **peptide bond** *is a bond between the carboxyl group of one amino acid and the amino group of another amino acid.*

Under proper conditions many amino acids can bond together to give chains of amino acids linked by numerous peptide bonds. For example, four peptide bonds are present in a short chain of five amino acids.

| Amino acid | Amino acid | Amino acid | Amino acid | Amino acid |

Peptide bond   Peptide bond   Peptide bond   Peptide bond

Short to medium-sized chains of amino acids are known as *peptides*. A **peptide** *is a sequence of amino acids, of up to 50 units, in which the amino* *acids are joined together through amide (peptide) bonds.* A compound containing two amino acids joined by a peptide bond is specifically called a *dipeptide*; three amino acids in a chain give a *tripeptide*, and so on. When the length of the chain becomes more than one wants to indicate in the name, it is merely called a *polypeptide*.

The order in which amino acids bond to each other is important. Consider the reaction between the amino acids alanine and glycine to give the following dipeptide.

Alanine   Glycine

There is an amino group at one end (at the left) of this dipeptide and a carboxyl group at the other end (at the right). The amino end is called the *N-terminal end* and the carboxyl end the *C-terminal end*. Using amino acid abbreviations (Table 25.2), we designate this dipeptide as

Ala-Gly

When using such notation, by convention the N-terminal end of the peptide is always written on the left.

A second dipeptide of glycine and alanine exists which is isomeric with the preceding one. It has the amino acid sequence

Gly-Ala

In this dipeptide, the glycine residue is the N-terminal end and the alanine residue is the C-terminal end.

Ala-Gly and Gly-Ala are different molecules (isomers). These isomers, which differ only in the order of attachment of the amino acids, have different chemical and physical properties. From this simple example we learn that sequence of amino acids in a chain is very important. Changing the sequence of amino acids creates a new polypeptide with different properties than the original one.

The number of isomeric peptides possible increases rapidly as the length of the peptide chain increases. Let us consider the tripeptide Ala-Ser-Cys as another example. In addition to this sequence, five other arrangements of these three components are possible, each one representing another isomeric tripeptide: Ala-Cys-Ser, Ser-Ala-Cys, Ser-Cys-Ala, Cys-Ala-Ser, and Cys-Ser-Ala.

The artificial sweetener aspartame (Section 24.6) has a structure involving the dipeptide Asp-Phe. It is the methyl ester of this dipeptide.

Aspartame

If the order of the amino acids is reversed, the dipeptide loses much of its sweet taste and cannot be used as an artificial sweetener.

One analogy often drawn is the similarity between polypeptide structure and word structure. When the 26 letters of the English alphabet are properly sequenced, words can be formed to convey information. Polypeptides that function biologically are formed from the 20 amino acids, each in its own proper sequence. The proper sequence of letters is necessary for a word to make sense, as is the amino acid sequence for a biologically active polypeptide. Furthermore, the letters that form a word, like the amino acids of a polypeptide are written from left to right. As any dictionary of the English language will document, a huge variety of words can be formed by different letter sequences. Imagine the possible number of sequences as longer polypeptides are considered. There are 1.55×10^{66} sequences possible for the 51 amino acids found in insulin! From these possibilities, the normal body reliably produces only *one*, illustrating the remarkable precision of life processes. From the simplest bacterium to the human brain cell, only those protein sequences needed by the cell are produced. The fascinating process of protein biosynthesis and the way in which genes in DNA direct this process are discussed later in this chapter.

25.6 Structural Levels for Proteins

Proteins *are polypeptides that contain more than fifty amino acid units.* The dividing line between a polypeptide and a protein is arbitrary. The important point is that proteins are polymers containing a large number of amino acid units linked together by peptide bonds. Polypeptides are "short" chains of amino acids. Some proteins have molecular masses in the millions. Some proteins also contain more than one polypeptide chain.

There are three levels of organization within the structure of a protein. These levels are referred to as *primary, secondary,* and *tertiary* structure. An understanding of all three levels is necessary for understanding protein function. The next three sections of text consider these three structural levels. Although they are considered separately, we need to remember that it is the total combination of all three levels of structure that control the function of a protein.

25.7 Primary Structure of Proteins

The **primary structure** *of a protein is the sequence of amino acids present in the peptide chain or chains of the protein.* Knowledge of primary structure tells us what amino acids are present, the number of each, and the length and number of polypeptide chains present. Every type of protein molecule present in a biological organism has a different sequence of amino acids, and it is this sequence that allows the protein to carry out its function, whatever it may be.

Insulin, the substance that regulates blood sugar level and a deficiency of which leads to diabetes, was the first protein for which primary structure was determined. This project, which took over 8 years, was completed in 1951. Today, several hundred proteins have been sequenced, that is, the order of amino acids within the polypeptide chain or chains has been determined. Sequencing times, using modern technology, are much shorter than that for insulin; it is now in terms of months rather than years.

Figure 25.1 shows the primary structure of human insulin. The amino acids present are parts of two polypeptide chains: chain A contains 21 amino acids and chain B contains 30 amino acids. The A and B chains are cross-linked by two disulfide bonds, and there is a third disulfide linkage between positions 6 and 11 of the A chain. Disulfide bonds are not part of the primary structure of a protein. They are, however, very important in overall structure of a protein and are discussed further in Section 25.9.

The primary structure of a specific protein is often similar among different species. For example, the primary structures of insulin in cows, pigs, sheep, and horses are similar to each other and also to human insulin. Table 25.5 shows the differences, which occur at only four amino acid positions, among these insulins.

ANIMAL INSULIN AND DIABETICS

Evidently the changes caused by the substitutions listed in Table 25.5 are not sufficient to alter significantly the biological function of "foreign" insulin compared to human insulin. Both pig and cow insulins, obtained from slaughterhouse animals, are used extensively in the treatment of human diabetics whose supplies of human insulin are inadequate. A few individuals have an allergic response to the animal insulin; however, the great majority of human diabetics can utilize the nonhuman insulin without complications. Within the next decade, human insulin will be available for all diabetics as a result of advances in the field of genetic engineering (Section 25.23). Genetically altered bacteria are now

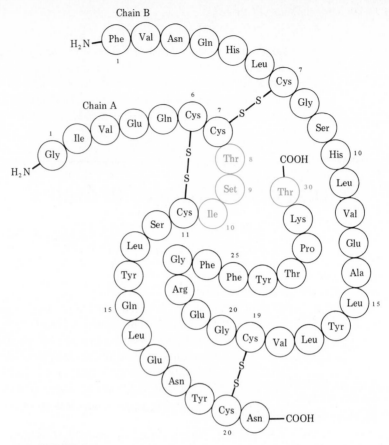

FIGURE 25.1

Diagram showing the primary structure of human insulin. The locations where variance occurs between human and certain animals are shown in color. (See Table 25.5 for comparisons of variance.)

TABLE 25.5

Variation of Amino Acids in Positions 8, 9, and 10 of Chain A and Position 30 of Chain B for Various Insulins

	Chain A Position			Chain B
Species	8	9	10	Position 30
Human	Thr	Ser	Ile	Thr
Pig	Thr	Ser	Ile	Ala
Cow	Ala	Ser	Val	Ala
Sheep	Ala	Gly	Val	Ala
Horse	Thr	Gly	Ile	Ala

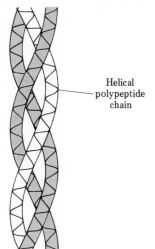

FIGURE 25.6

The "weaving" together of three helical polypeptide chains to form a triple helix.

Helical polypeptide chain

about a common axis, to give a rope-like arrangement of polypeptide chains (*see* Figure 25.6).

Some cross-linking between chains in the triple helix occurs through covalent bond formation. The extent of cross-linking in the collagen of meat is a function of the age of the animal from which the meat was obtained. The older the animal, the greater the cross-linking and the tougher the meat. The process of tanning, which coverts skin to leather, involves increasing the degree of cross-linking.

Because of the weak nature of hydrogen bonds (Section 8.16), secondary structural features of proteins are easily disrupted. More information concerning such disruption and its effects on protein function is given in Section 25.11.

25.9   Tertiary Structure of Proteins

*The **tertiary structure** of a protein is the overall three-dimensional shape that results from the attractive forces among amino acid side chains (R groups) that are widely separated from each other within the chain.* The "twists" and "folds" in a protein chain relate to tertiary structure.

The dividing line between secondary and tertiary structure can be somewhat "hazy". The secondary structure of any protein refers to the arrangement and interactions of amino acids in a relatively small portion of the protein, while tertiary structure describes the arrangement and interactions of amino acids for the entire molecule.

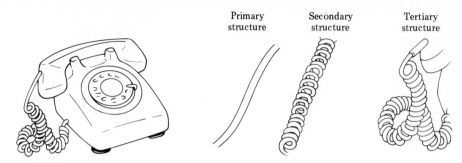

FIGURE 25.7

Three levels of structure of a telephone cord.

A good analogy for the relationships among the primary, secondary, and tertiary structures of a protein is that of a telephone cord (Figure 25.7). The primary structure is the long straight cord. The coiling of the cord into a helical arrangement gives the secondary structure. The supercoiling arrangement the cord adopts after one hangs up the receiver is the tertiary structure.

Proteins are classified into two general types based on tertiary structure considerations: (1) globular proteins and (2) fibrous proteins. A **globular protein** *has a roughly spherical or globular overall shape.* Globular proteins are usually distorted spheres, being more elongated, much like an ellipsoid or even an oblate spheroid (a squashed sphere). A **fibrous protein** *has a long, thin string-like shape.* Such proteins are composed of long strands or flat sheets.

Four types of attractive interactions have been found to contribute to the tertiary structure, either globular or fibrous, of a protein: (1) disulfide bonds, (2) electrostatic attractions (salt linkages), (3) hydrogen bonds, and (4) hydrophobic attractions. All four are interactions between amino acid R groups. This is a major distinction between tertiary structure interactions and secondary structure interactions. Tertiary structure interactions involve the R groups of the amino acids, whereas secondary structure interactions involve the functional groups associated with the peptide linkages between amino acid units.

Disulfide bonds form between the —SH groups of two cysteine amino acids. They are covalent bonds and differ in this respect from the other three types of interactions to be discussed.

Disulfide bonds may involve two cysteine units in the same chain, or the two cysteines may be in different chains. Referring back to Figure 25.1, we note the presence of both types of sulfide bonds in the structure of insulin.

Electrostatic interactions, sometimes called salt bridges, always involve amino acids with ionized side chains. Such amino acids are those with very polar side chains (Table 25.2), that is, the acidic amino acids and the basic amino acids. The two R groups, one acidic and one basic, are held together by simple ion–ion attractions.

Hydrogen bonds can occur between amino acids with polar R groups. A variety of polar side chains can be involved, especially those that possess the following functional groups.

Hydrogen bonds are relatively weak and are easily disrupted by changes in pH and temperature.

Hydrophobic attractions result when two nonpolar side chains are in close proximity to each other. In aqueous solution, globular proteins usually have their polar R groups directed outward, toward the aqueous solvent (which is also polar), and their nonpolar R groups inward (away from the polar water molecules). The nonpolar R groups interact with each other. Hydrophobic interactions are common between phenyl rings and alkyl side chains. Although this type of interaction is weaker than hydrogen bonding or electrostatic interactions, it is a significant force because of the large number of such interactions that occur.

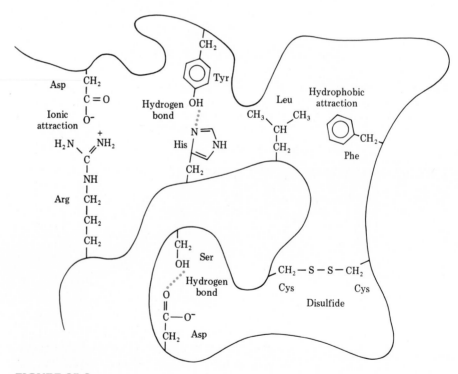

Figure 25.8 summarizes the types of interactions that contribute to the tertiary structure of a protein.

FIGURE 25.8

Types of interactions between amino acid R groups that lead to the tertiary structure of a protein.

25.10 Protein Hydrolysis

When a protein or polypeptide is heated to boiling in a solution of strong acid or strong base, the peptide bonds of the amino acid chain are hydrolyzed and free amino acids are produced. This hydrolysis reaction is the reverse of the formation reaction for a peptide bond (Section 25.5). Amine and carboxylic acid functional groups are regenerated.

Within the human body, hydrolysis of protein takes place without heating, at body temperature, with the help of enzymes that catalyze the hydrolysis reaction. Numerous different enzymes are involved. Some enzymes attack only N-terminal or C-terminal amino acids, whereas others break peptide bonds at specific internal amino acid locations.

Digestion of protein is simply enzyme-catalyzed hydrolysis of ingested protein. The free amino acids produced from this process are absorbed through the intestinal wall into the bloodstream and transported to the liver. Here they become the raw materials needed for the synthesis of new protein tissue. Such synthesis involves reassembling the amino acids into new and different combinations from that in which they were ingested. Excess amino acids are not stored in the body; hence, a continuous, balanced dietary intake of protein is needed. Such protein must contain all of the essential amino acids (Section 25.4).

25.11 Protein Denaturation

Denaturation *is the process in which the secondary and tertiary structures of a protein are disrupted, causing a loss in the protein's biological activity.* In Sections 25.8 and 25.9 we saw that the three-dimensional shape of a protein is determined by secondary and tertiary structures. Denaturation alters this shape, and biological activity is lost. In denaturation, the primary structure of a protein remains unchanged.

When denaturation occurs under mild conditions, the protein can often be restored to its original structure and effectiveness by carefully reversing the conditions of denaturation. Such restoration of the protein's original shape is called *renaturation* (see Figure 25.9). Denaturation under strong conditions is irreversible and usually such denatured proteins precipitate out of solution, a process called *coagulation*.

FIGURE 25.9

The processes of protein denaturation and protein renaturation.

Table 25.6 is a listing of several physical and chemical agents that cause denaturation. The effectiveness of a given denaturing agent depends on the type of protein upon which it is acting.

Perhaps the most dramatic example of protein denaturation occurs when egg white (a concentrated solution of the protein albumin) is poured onto a hot surface. The clear albumin solution immediately changes into a white solid with a jelly-like consistency (see Figure 25.10). A similar process occurs when hamburger juices encounter a hot surface. A brown jelly-like solid forms.

One of the reasons many foods are cooked is to denature the protein present so that it is more easily digested. It is easier for digestive enzymes to "work on"

TABLE 25.6

Selected Physical and Chemical Denaturing Agents

Denaturing Agent	Mode of Action
Heat	Disrupts hydrogen bonds by making molecules vibrate too violently; produces coagulation, as in the frying of an egg
Microwave radiation	Causes violent vibrations of molecules that disrupt hydrogen bonds
Ultraviolet radiation	Probably operates much as the action of heat (e.g., sunburning)
Violent whipping or shaking	Causes molecules in globular shapes to extend to longer lengths, which then entangle (e.g., beating egg white into meringue)
Detergents	Probably affect hydrogen bonds and salt bridges
Organic solvents (e.g., ethanol, 2-propanol, acetone)	May interfere with hydrogen bonds because these solvents also can form hydrogen bonds; quickly denature proteins in bacteria, killing them (e.g., the disinfectant action of 70% ethanol)
Strong acids and bases	Disrupt hydrogen bonds and salt bridges; prolonged action leads to actual hydrolysis of peptide bonds
Salts of heavy metals (e.g., salts of Hg^{2+}, Ag^+, Pb^{2+})	Metal ions combine with —SH groups and form precipitates (these salts are all poisons)
Reducing agents	Oxidize disulfide linkages to produce —SH groups

FIGURE 25.10
Protein denaturation is the major change that occurs when an egg is cooked ''over easy.''

denatured (unraveled) protein. Additionally, cooking foods kills microorganisms through protein denaturation. For example, ham and bacon harbor parasites that cause trichinosis. Cooking the ham or bacon denatures parasite protein.

In surgery, heat is often used to seal small blood vessels. This process is called *cauterization*. Also small wounds can be sealed using cauterization. Heat-induced denaturation is used to sterilize surgical instruments and in canning foods because bacteria are destroyed when the heat coagulates their protein.

The body temperature of a patient with fever may typically rise to 102, 103, or even 104°F. A temperature above 106°F (41°C) is extremely dangerous, for at this level the enzymes of the body begin to be inactivated. Enzymes, which function as catalysts for almost all body reactions, are protein. Inactivation of enzymes, through denaturation, can have lethal consequences on body chemistry.

The effect of ultraviolet radiation from the sun, a nonionizing radiation (Section 15.13) is similar to that of heat. Denatured skin proteins cause most of the problems associated with sunburn.

A curdy precipitate of casein, the principal protein in milk, is formed in the stomach when the hydrochloric acid of gastric juice reacts with milk. The

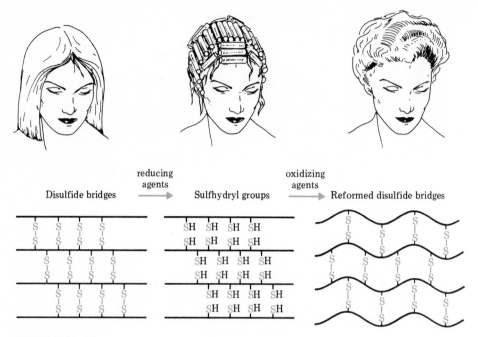

reducing
agents
oxidizing
agents

Disulfide bridges ⟶ Sulfhydryl groups ⟶ Reformed disulfide bridges

FIGURE 25.11
Permanent waves can be created in straight hair by breaking disulfide linkages in hair
protein and then reforming these linkages in different locations.

CHEESE AND YOGURT

curdling of milk that takes place when milk sours or cheese is made results from
the presence of lactic acid, a by-product of bacterial growth. Yogurt is prepared
by growing lactic acid-producing bacteria in skim milk. The coagulated denatured protein gives yogurt its semisolid consistency.

Serious eye damage can result from contact with acids or bases, often
resulting in a clouded cornea caused by irreversibly denatured and coagulated
protein. Hence the rule that students wear protective eyewear in the chemistry
laboratory.

ALCOHOL AS A
DISINFECTANT

Alcohols are an important type of denaturing agent. Denaturation of bacterial
protein takes place when isopropyl or ethyl alcohol is used as a disinfectant. This
accounts for the common practice of swabbing the skin with alcohol before
giving an injection. Interestingly, pure isopropyl or ethyl alcohol is less effective
than the commonly used 70% alcohol solution. Pure alcohol quickly denatures
and coagulates the bacterial surface, thereby forming an *effective* barrier to
further penetration by the alcohol. The 70% solution denatures more slowly and
allows complete penetration to be achieved before coagulation of the surface
proteins takes place.

The process used in waving hair — that is, in giving a person a permanent —
involves reversible denaturation. Hair is a protein with many disulfide

(—S—S—) linkages as part of its tertiary structure (Section 25.9). It is these disulfide linkages that give the overall shape to hair protein. In the permanent process, hair is first treated with a reducing agent (a waving lotion) that breaks the disulfide linkages in the hair, converting each linkage to two sulfhydryl (—SH) groups. The "reduced" hair, whose tertiary structure has been disrupted, is then wound on curlers to give it a new configuration. Finally, the reduced and rearranged hair is treated with an oxidizing agent (a neutralizer) to form disulfide linkages at new locations within the hair. The new shape and curl of the hair are maintained by the newly formed disulfide bonds and the resulting tertiary structure accompanying their formation. Of course, as new hair grows out, the "permanent" will have to be repeated. Figure 25.11 shows the steps involved in obtaining a permanent hair wave.

25.12 Nucleic Acids

Nucleic acids are responsible for the remarkable ability of cells to produce exact replicas of themselves. Such a process requires that certain types of information be passed unchanged from one cell generation to the next. This storage and transfer of information are accomplished by nucleic acids.

Like polysaccharides (Section 24.7) and proteins (Section 25.2), nucleic acids are polymers. The repeating unit for the nucleic acid polymeric structure is a nucleotide. **Nucleic acids** *are polymeric molecules in which the repeating unit is a nucleotide.* Prior to discussing nucleotide structure (Section 25.13), we consider where nucleic acids are found within the human body, and the major biological functions of these substances.

Two kinds of nucleic acids exist—deoxyribonucleic acids (DNA) and ribonucleic acids (RNA). (The structural differences between DNA and RNA will be considered in Section 25.13.) The majority of DNA is found in the nucleus of a cell. RNA is found both in the nucleus and outside the nucleus in the cytoplasm of the cell. *Cytoplasm* is the material of the cell, exclusive of the nucleus, as is shown in Figure 25.12. As a general rule, RNA molecules are shorter polymers and decompose much more readily than DNA. Within the nucleus, the DNA is bonded to proteins, in structures called *chromosomes* (Section 25.19), the carriers of genetic information.

Two major biological functions exist for nuclei acids.

1. They store genetic information, transfer it between cells, and transmit it from one generation to the next generation.
2. They are responsible for the control and direction of protein synthesis in cells.

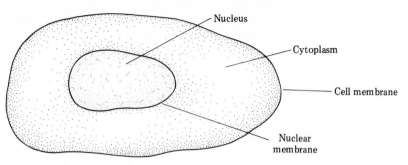

FIGURE 25.12

A simplified drawing of a cell showing its two major regions. DNA is found in the nucleus and RNA is found in both nucleus and cytoplasm.

In human cells, DNA is involved in the first of these major nucleic acid roles, and both DNA and RNA are involved in the second.

25.13 Nucleotides

The monomers for nucleic acid polymers, nucleotides, are more complex in structure than carbohydrate monomers (monosaccharides; see Section 24.4) or protein monomers (amino acids; see Section 25.3). Within each nucleotide monomer are three subunits. A **nucleotide** *is a molecule composed of a five-carbon sugar bonded to both a phosphate group and a nitrogeneous base.* Figure 25.13 shows the general structure of a nucleotide in terms of its component subunits, which we now examine in more detail.

One of two five-carbon monosaccharide molecules, ribose or deoxyribose, is always present in a nucleotide. Structurally, these two compounds are very similar. The only difference between them occurs at carbon 2 of the ring structure. In ribose, an —H and an —OH group are bonded to this carbon atom, while in deoxyribose two hydrogen atoms are present.

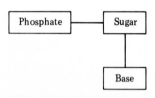

FIGURE 25.13

A nucleotide has three components: a nitrogen-containing heterocyclic base, a five-carbon sugar, and a phosphate group. Both the base and the phosphate are attached to the sugar.

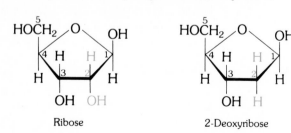

Ribose   2-Deoxyribose

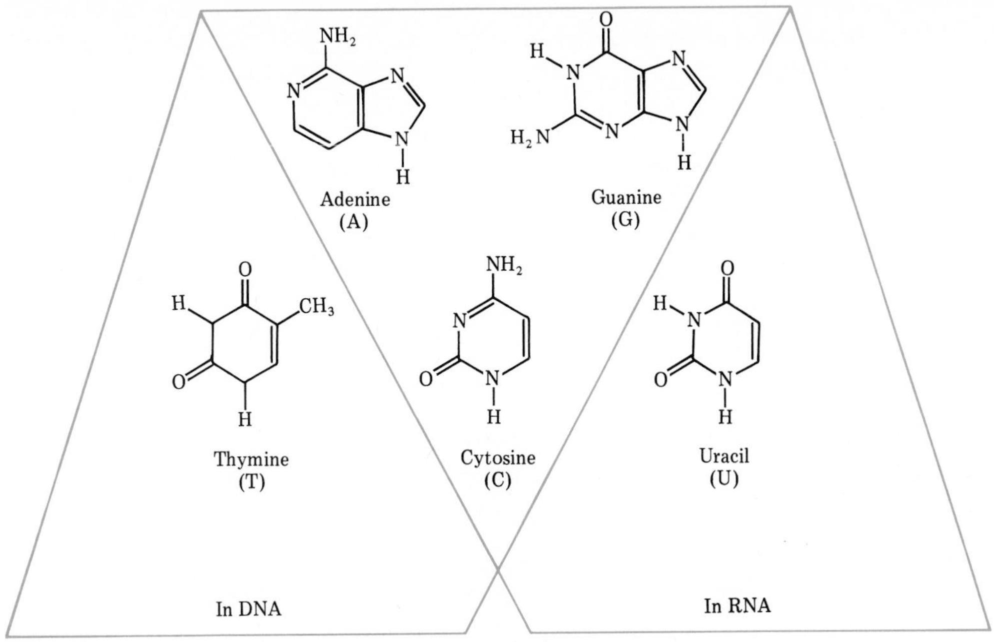

FIGURE 25.14
The five nitrogenous bases found in nucleic acid nucleotides.

The two general types of nucleic acids, RNAs and DNAs, can be distinguished from each other on the basis of the type of sugar unit present. Nucleotides present in RNA molecules always contain ribose—hence the R in RNA. Nucleotides present in DNA molecules always contain deoxyribose—hence the D in DNA.

Five different nitrogen-containing basic compounds—all heterocyclic amines (Section 17.9)—are found in nucleotides. They are of two types: single ring compounds (thymine, cytosine, and uracil) and double ring compounds (adenine and guanine) (see Figure 25.14). These compounds are considered bases because their nitrogen atoms tend to accept protons (Lewis base behavior; see Section 10.4). Adenine, guanine, and cytosine, abbreviated A, G, and C, are found in both DNA and RNA. Uracil (U) is normally found only in RNA, while thymine (T) occurs only in DNA.

The phosphate group, an inorganic component of the nucleotide, is derived from phosphoric acid.

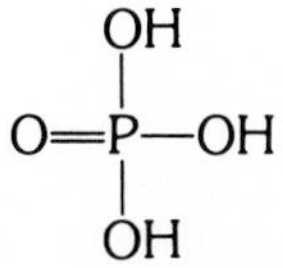

The formation of a nucleotide from these three types of components is represented in the following reaction.

Guanine

Phosphoric acid

Ribose

Nucleotide

Notice that water is removed during bond formation. Since ribose is involved in the preceding equation, the nucleotide formed must be one found in RNA. The base and phosphate group are always attached to the sugar at the positions shown in the reaction.

25.14 **Primary Nucleic Acid Structure**

The arrangement for nucleotide units in a nucleic acid polymer is shown in Figure 25.15. The sugar unit of one nucleotide is bonded to the phosphate group of the next nucleotide in the sequence.

The complete chemical structure for a short segment of a nucleic acid (a tetranucleotide) is shown in Figure 25.16. We know that this segment represents

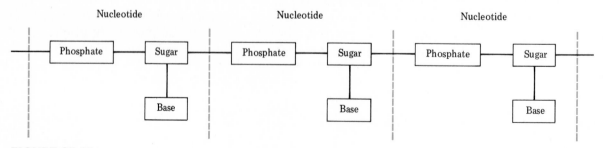

FIGURE 25.15

Diagram of the structure of a nucleic acid polymer showing the nucleotide repeating units.

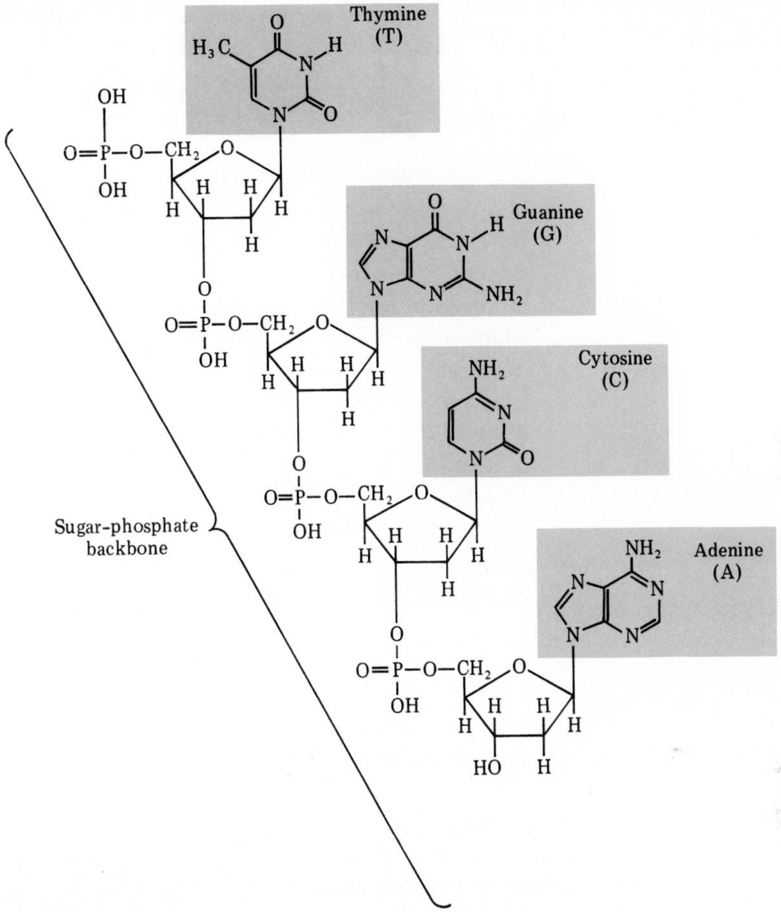

FIGURE 25.16

Chemical structure of a four-nucleotide-long segment of a DNA molecule.

a portion of a DNA molecule rather than an RNA molecule since the sugar present is deoxyribose rather than ribose.

Generalizations about the structures of nucleic acids, which can be drawn from figures like 25.15 and 25.16, include the following.

1. *The "backbone" of a nucleic acid molecule is a constantly alternating sequence of sugar and phosphate groups.* For DNA molecules, this alternation sequence involves deoxyribose and phosphate.

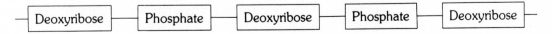

For RNA molecules, the alternation sequence involves ribose and phosphate.

| Ribose | Phosphate | Ribose | Phosphate | Ribose |

2. *The variable part of a nucleic acid molecule is the sequence of bases attached to the sugar units of the backbone.* It is the sequence of these base side chains that distinguishes various DNAs from each other and various RNAs from each other. Only four types of base side chains are involved in any given nucleic acid structure. This is a much simpler situation than that previously encountered for proteins where 20 side chain entities (amino acids) were available for use (Section 25.5).

3. *Each phosphoric acid molecule (phosphate group) involved in the back-bone of the nucleic acid still possesses one of its three original —OH groups.* (The other two have become involved in nucleotide formation) This remaining —OH group is free to exhibit acidic behavior; that is, it is free to lose its proton.

$$
\begin{array}{ccc}
\quad O & \quad O \\
\quad \| & \quad \| \\
O\!-\!\overset{}{P}\!-\!O & \longrightarrow & O\!-\!\overset{}{P}\!-\!O + H^{+} \\
\quad | & \quad | \\
\quad OH & \quad O^{-}
\end{array}
$$

It is such behavior by many phosphate groups along the backbone that gives nucleic acids their acidic properties.

A more chemically oriented definition of a nucleic acid than the one given in Section 25.12 is now in order. A **nucleic acid** is *a polymeric molecule containing an alternating phosphate–sugar "backbone" system with a nitrogeneous base (heterocyclic amine) side chain attached to each sugar unit.* Such a definition specifies the generalized primary structure of a nucleic acid.

Like proteins (Section 24.8), nucleic acids have secondary structure in addition to primary structure. Since secondary structural features for DNAs and RNAs are distinctly different, we will discuss these secondary features separately.

25.15 Double Helix Structure of DNA

Although the "backbone" and the "base sequence" in a *single strand* of DNA are important parts of its overall structure, they do not represent the complete structure of DNA in most cells. DNA has not only primary structure but also secondary structure, as do proteins (Section 25.8).

A key observation leading to the determination of DNA's complete structure involved measurement of the amount of the side chain bases A, T, G, and C present in DNA obtained from many different types of organisms. An interesting pattern emerged. The amount of A was found to always equal the amount of T, and the amounts of C and G were always equal.

The relative amounts of each of these "base pairs" in DNA varies depending upon the life form from which the DNA was obtained. (Each animal or plant has a unique base composition.) However, the relationships

$$\% \ A = \% \ T \qquad \text{and} \qquad \% \ C = \% \ G$$

always hold. For example, human DNA contains 30% adenine, 30% thymine, 20% guanine, and 20% cytosine.

An explanation for these percentage figures came in 1953, when the "complete" structure for DNA was first proposed. (Three scientists who contributed to this structural proposal, the English chemists Maurice Wilkins and Francis Crick and the American James Watson, later were awarded a Nobel Prize for their achievement.)

A DNA molecule contains two polynucleotide chains coiled around each other in a double helix. The bases (side chains) of each polynucleotide extend inward (within the interior of the double helix) toward the other polynucleotide, with each base in one chain hydrogen bonded to a base in the opposite chain. Often, the secondary structure of a DNA molecule is compared to a twisted ladder (see Figure 25.17). The rungs of the ladder correspond to the hydrogen-bonded

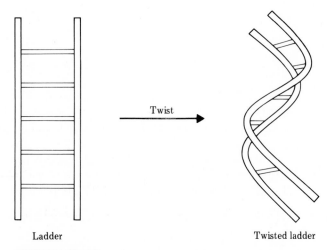

Twist

Ladder   Twisted ladder

FIGURE 25.17

If a flexible ladder is twisted, its shape resembles that of a double-stranded DNA molecule. The rungs of the ladder are analogous to the base pairs, and the supporting structure is analogous to the alternating deoxyribose and phosphate units.

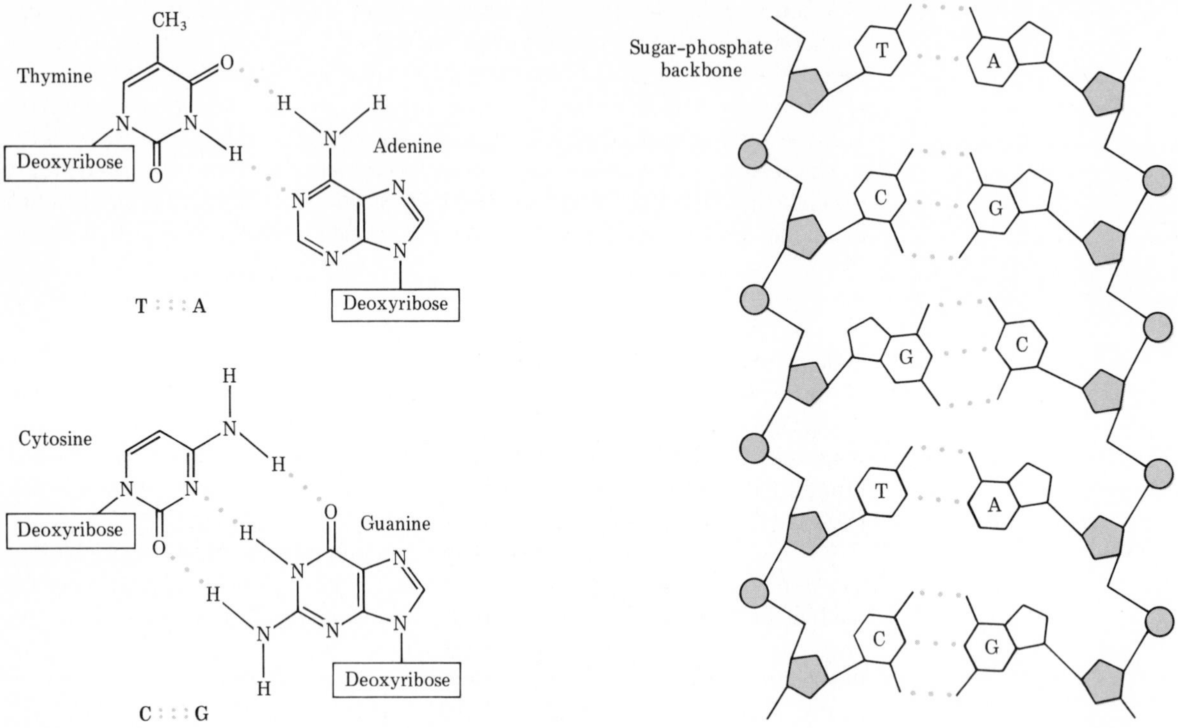

FIGURE 25.18

The complementary base pairing, through hydrogen-bonding interactions, between thymine and adenine and between cytosine and guanine.

bases, and the twisted vertical supports for the rungs correspond to the alternating sugar and phosphate "backbone."

The pairing of bases, through hydrogen bonding, is very selective. Because of size limitations within the interior of the double helix, only certain base combinations are found to occur. The base cytosine (C) is always found "opposite" guanine (G), and adenine (A) is always found "opposite" thymine (T). No other pairing combinations are found. The bases C and G and the bases A and T are said to be *complementary bases*, and their interaction with each other represents *complementary base pairing*. Complementary base pairing explains, very simply, the fact that the amounts of A and T, like the amounts of C and G, are always equal in any DNA molecule.

Complementary base pairing, through hydrogen bonding, also explains why two strands of DNA polymer are attracted to each other. Although hydrogen

2. The base thymine found in DNA is not found in RNA. In its place is the base uracil (Figure 25.14). Uracil can base pair (hydrogen bond) with adenine as could thymine.

3. RNA is a single-stranded molecule rather than double stranded (double helix) as is DNA. Consequently, base pairing is not involved, and there is no requirement for equal amounts of complementary bases as with DNA.

4. RNA molecules are much smaller than DNA molecules, ranging from as few as seventy-five nucleotides to a few thousand nucleotides.

We should note that the single-stranded nature of RNA (item 3 in the list) does not prevent a *portion* of an RNA molecule from folding back upon itself and forming a double-helical region. If the base sequences along two portions of a RNA strand are complementary, a "hairpin loop" structure such as that shown in Figure 25.21, results.

Through transcription (Section 25.19), DNA produces three types of RNA, which are distinguished on the basis of functionality. The three types are ribosomal RNA (rRNA), messenger RNA (mRNA), and transfer RNA (tRNA). Characteristics of these three types of RNA are as follows.

Ribosomal RNA *combines with a series of proteins to form complex structures, called ribosomes, that serve as the physical sites for protein synthesis.* Ribosomes have molecular masses on the order of 3 million. More than 80% of the RNA in a cell is rRNA. The rRNA present in ribosomes has no informational function.

Messenger RNA *carries genetic information (instructions for protein synthesis) from DNA to the ribosomes.* It is a template (pattern) made from DNA. The size (molecular mass) of mRNA varies with the length of the protein whose synthesis it will direct. Less than 2% of the total RNA in a cell is mRNA.

Transfer RNA *delivers specific individual amino acids to the ribosomes, the site of protein synthesis.* These RNAs are the smallest of the RNAs, possessing only 75 to 90 nucleotide units.

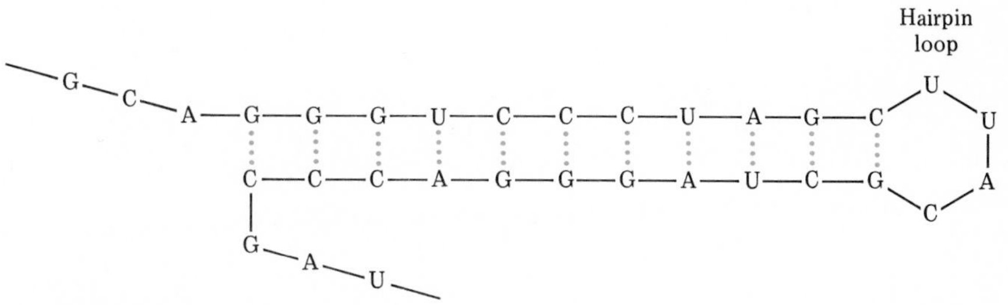

FIGURE 25.21

A hairpin loop in single-stranded RNA, caused by complementarily base pairing between bases at different sites along the strand.

25.19 Transcription: RNA Synthesis

The mechanics of transcription are in many ways similar to what happens during DNA replication (Section 25.16). Four steps are involved.

1. A *portion* of the DNA double helix unwinds so as to expose nucleotide bases. The unwinding process is governed by the enzyme RNA polymerase rather than DNA polymerase (replication enzyme).
2. Free *ribonucleotides* align along *one* of the exposed strands of bases of the DNA, forming new base pairs. In this process, U rather than T aligns with A in the base-pairing process. Since it is free ribonucleotides rather than free deoxyribonuclides that are involved in the base pairing, ribose, rather than deoxyribose becomes incorporated into the nucleic acid "backbone."
3. Linkage of the aligned ribonucleotides occurs.
4. Transcription ends when the RNA polymerase enzyme encounters a sequence of bases that is "read" as a stop signal. The newly formed RNA molecule and the RNA polymerase enzyme are released, and the DNA then rewinds to reform the original double helix.

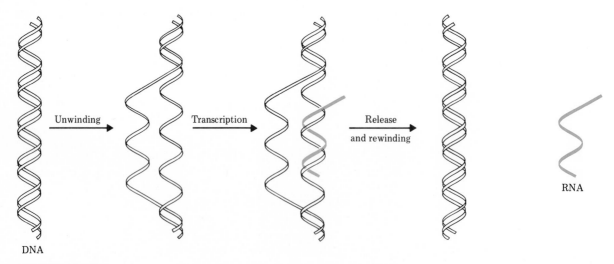

DNA

FIGURE 25.22
The process of transcription of DNA to form RNA involves the unwinding of a portion of the DNA double helix.

Figure 25.22 shows the overall process of transcription of DNA to form RNA.

Within a DNA chain are instructions for the synthesis of numerous mRNA molecules, which in turn will direct the synthesis of numerous specific proteins. During transcription, the DNA molecule unwinds, at the command of RNA polymerase, only at the particular spot where the appropriate base sequence is found for the mRNA (and protein) of concern. Such short segments of DNA, containing instructions for the formation of particular mRNAs, are called *genes*. A **gene** *is a segment of a DNA molecule that contains the base sequence for the production of a single, specific RNA molecule.*

"GENE CONTENT" OF A CELL

In humans, most genes are composed of 1000 to 3500 nucleotide units. Hundreds of genes can exist along a DNA molecule strand. The DNA found in one human cell contains an estimated 5.5 billion nucleotide base pairs, making up nearly 1 million genes.

The "gene content" of a human cell, approximately 1 million genes, is an extraordinary amount of information. To put the informational content of a human cell in perspective consider the following example. Let us assume you have "special abilities" such that you can see, with the naked eye, DNA molecules with their base pairs and that you decided to write down the characteristics (base sequence) of all genes in a human cell. You use one page of a lab notebook for each gene, and it takes 5 minutes (a very rapid speed) to record each gene's characteristics. When you finish the project, your notebook, 1 million pages long, will be equivalent to 1250 800-page textbooks. With a recording time of 5 minutes per gene, it would take you, working 8 hours a day 5 days a week, 40 years to complete your work. All this information is found within one human cell.

A **chromosome** *is an individual DNA molecule tightly coiled about a group of small proteins.* Different organisms have different numbers of chromosomes within a cell. The chromosome number is therefore a distinguishing characteristic of a species. A normal human has 46 chromosomes in each body cell, an onion 16, a pine tree 24, a corn plant 20, a fruit fly 8, a cat 38, and a mosquito only 6. Each chromosome contains an enormous, single molecule of DNA.

25.20 The Genetic Code

The nucleotide (base) sequence of a mRNA molecule is the "informational part" of such a molecule. The sequence in a given mRNA determines the amino acid sequence for the protein synthesized under that mRNA's direction.

How can the base sequence of a mRNA molecule (which involves only *four* different bases—A, C, G, and U) encode enough information to direct proper sequencing of *twenty* amino acids in proteins? Obviously, if a single base codes for a single amino acid, we do not have enough bases—four bases and twenty amino acids. A two-nucleotide code for amino acids will not suffice either. How many different "words" of two letters can be formed from an "alphabet" of four letters—the letters A, C, G, and U? The answer is 16. The "words" would be AA, AC, AG, AU, CA, CC CG, CU, GA, GC GG GU, UA, UC, UG, and UU. Sixteen words is not enough to code for 20 amino acids. A code involving three-letter words, however, will give us more than enough combinations to code for the 20 amino acids. Sixty-four different three-letter "words" are possible using our A, C, G, and U base "alphabet."

Research has now shown that three-letter "words"—that is, three-nucleotide base sequences—are used by mRNA molecules to specify amino acids needed in protein synthesis. Such three-nucleotide sequences are called *codons*. A **codon** *is a three-nucleotide sequence in a mRNA molecule that codes for a specific amino acid needed during the process of protein synthesis.*

Which amino acid is specified by which codon? (We have 64 codons to choose from.) Such codon–amino acid relationships were deciphered by researchers by adding different *synthetic* mRNA molecules (whose base se-

TABLE 25.7

The Genetic Code—Codons Specific for Each Amino Acid Found in Protein (*the amino acids are listed in alphabetical order*)

Amino Acid	Codons	Number of Codons
Alanine	GCA, GCC, GCG, GCU	4
Arginine	AGA, AGG, CGA, CGC, CGG, CGU	6
Asparagine	AAC, AAU	2
Aspartic acid	GAC, GAU	2
Cysteine	UGC, UGU	2
Glutamic acid	GAA, GAG	2
Glutamine	CAA, CAG	2
Glycine	GGA, GGC, GGG, GGU	4
Histidine	CAC, CAU	2
Isoleucine	AUA, AUC, AUU	3
Leucine	CUA, CUC, CUG, CUU, UUA, UUG	6
Lysine	AAA, AAG	2
Methionine	AUG	1
Phenylalanine	UUC, UUU	2
Proline	CCA, CCC, CCG, CCU	4
Serine	UCA, UCC, UCG, UCU, AGC, AGU	6
Threonine	ACA, ACC, ACG, ACU	4
Tryptophan	UGG	1
Tyrosine	UAC, UAU	2
Valine	GUA, GUC, GUG, GUU	4
Termination	UAA, UAG, UGA	3
Total number of codons		64

quences were known) to cell extracts and then determining the structure of any newly formed protein. After a great number of such experiments, researchers finally matched all 64 possible codons with aspects of protein synthesis. It was found that 61 of the 64 codons (formed by various combinations of the bases A, C, G, and U) related to specific amino acids and that the other 3 combinations were termination codons ("stop" signals) for protein synthesis. Collectively, these relationships between three-base sequences and amino acid identities are known as the *genetic code*. The **genetic code** *is a list of possible codons and the amino acid for which each codon stands* (see Table 25.7). The determination of this code is one of the most remarkable of twentieth century scientific achievements. The Nobel Prize in chemistry was awarded, in 1968, to Marshall Nirenberg and H. G. Khorane for their work in unraveling this essential code.

A casual glance at the information in Table 25.7 reveals that the code is *degenerate*; that is, there is more than one codon for most amino acids. Degeneracy does not imply chaos or confusion. It only means that there is more than one "code word" for certain amino acids. (By analogy, in the English language, two or more words often have the same meaning.) It is important to point out that the "reverse" of degeneracy is not part of the code. No codon specifies more than one amino acid. This leads to order. [Many can specify one (degeneracy), but one cannot specify many (disorder).]

The genetic code also has the feature of being *nonoverlapping*. This means that parts of two adjacent codons cannot function as a new codon. For example, in the base sequence GCA CUA (codons for alanine and leucine) we cannot take the last of the first codon, the A, and the first two parts of the second codon, CU, and combine them to form ACU (a codon for threonine).

Finally, studies of the genetic code in many organisms indicate that so far the code has been the same in almost all of them—that is, the code is almost *universal*. The same codon specifies a given amino acid whether the cell is a bacterial cell, a corn plant cell, or a human cell.

25.21 Anticodons

The codons in mRNA are read and translated into an amino acid sequence by tRNA molecules. The general shape of tRNA molecules is crucial to their "translation" function. Figure 25.23, a schematic representation of a typical tRNA molecule, shows that it is folded and twisted into regions of parallel strands

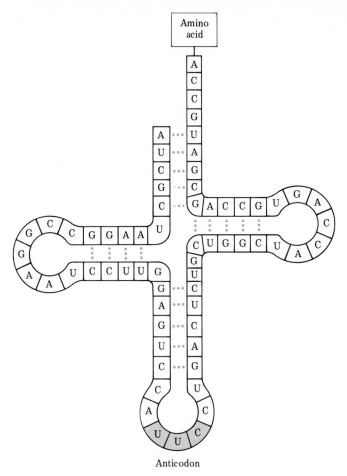

FIGURE 25.23

Cloverleaf structure of a tRNA molecule. The anticodon is located in the hairpin loop opposite the open end of the cloverleaf.

and regions of loops. The net result is a "cloverleaf" shape. Of particular importance in the cloverleaf structure is the hairpin loop at the bottom of the structure. It contains a sequence of three bases called an *anticodon*. An **anticodon** *is a three-base sequence in a tRNA molecule that serves as a complement, during protein synthesis, to a codon in a mRNA molecule.*

It is the interaction between anticodon (tRNA) and codon (mRNA) that leads to the proper placement of an amino acid into a growing peptide chain during protein synthesis. This interaction, which involves complementary base pairing, is shown in Figure 25.24.

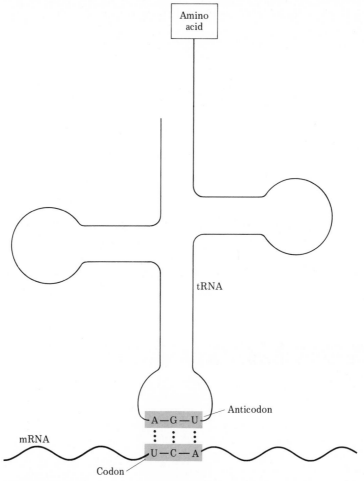

FIGURE 25.24

The interaction between anticodon (tRNA) and codon (mRNA), which involves complementary base pairing, governs the proper placement of amino acids in a protein.

25.22 Translation: Protein Synthesis

The substances needed for the translation part of protein synthesis are mRNA, rRNA, tRNA, amino acids, and a number of enzymes. The translation process can be conveniently discussed in terms of four steps.

1. A molecule of mRNA binds to a ribosome (rRNA and protein) at a specific site.
2. Molecules of tRNA carry amino acids to proper binding sites on the ribosome.
3. Amino acids positioned in proper sequence at ribosome binding sites, through codon–anticodon interactions, are joined together through peptide linkages.
4. Protein synthesis is terminated by a "stop" codon on the mRNA molecule.

Attachment of mRNA to Ribosome

A ribosome contains two subunits, a large subunit and a small subunit. The smaller unit is where the mRNA attaches itself and where the mRNA codons are "read." The complete ribosome participates in the synthesis of the protein chain. Ribosomes have two special sites for protein synthesis, a P site and an A site. Our notation for a ribosome and its synthesis sites is as follows.

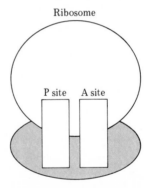

The messenger RNA attaches itself to the smaller ribosomal subunit so that its first codon is at the P site and the second codon is at the A site.

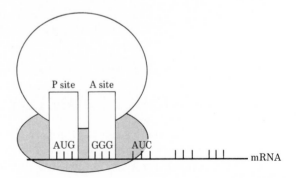

Amino Acid Transport by tRNA

The amino acids required for protein synthesis are brought to the ribosomal site by tRNA molecules. The carrier tRNAs have the amino acids chemically bonded to them. Specific tRNAs carry specific amino acids. Cells have at least one kind of tRNA that specifically binds each of the 20 common amino acid molecules. The process by which a given tRNA molecule recognizes the amino acid it is to carry is controlled by an enzyme that is simultaneously specific for the structure of the tRNA molecule and the structure of the amino acid molecule. The amino acid that bonds to a given tRNA molecule is the one whose identity is "consistent" with the anticodon present in the tRNA structure.

The "landing point" for an amino acid-carrying tRNA molecule is a mRNA codon that is complementary to its anticodon. All amino acids are brought into position at the ribosome by this principle: base pairing between a tRNA anticodon and a mRNA codon.

Initiation of protein synthesis occurs when tRNA molecules bring amino acids to the P and A synthesis sites.

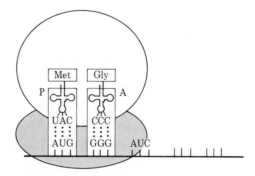

In our illustration methionine (Met) is the amino acid brought to the first site and glycine (Gly) is at the second site. The base pairing (between anticondon and codon) involves only two combinations—A with U and G with C (Section 25.15).

Peptide Bond Formation Between Amino Acids

With amino acids in place at both the P and A sites of the ribosome, an enzyme promotes the formation of a peptide bond between the two amino acids. In this process, the amino acid at the P site is detached from its tRNA and becomes attached to the amino acid at the A site.

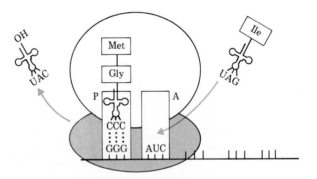

$$CH_3-S-CH_2-CH_2-CH_2-CH$$ (Met, NH_2, O=C, O)

Gly (NH_2, CH_2, O=C, O)

P — UAC — AUG

A — CCC — GGG — AUC

$$CH_2-S-CH_2-CH_2-CH$$ (NH_2, O=C, NH, CH_2, O=C, O)

Gly

"Empty" tRNA

P — OH — UAC — AUG

A — CCC — GGG — AUC

Ribosome moves in this direction.

The "empty" tRNA at the P site now leaves that site to return to the cytoplasm to pick up another molecule of its specific amino acid and be available for further use. Simultaneously with the tRNA release from the P site, the ribosome shifts along the mRNA. This "shift" puts the newly formed dipeptide at the P site, and the third codon of mRNA is now available, at site A, to accept a tRNA molecule whose anticodon complements it.

OH — UAC

Ile — UAG

Met
Gly

P — CCC — GGG

A — AUC

Now a repetitious process begins. The third codon, at the A site, accepts an incoming tRNA with its accompanying amino acid, and then the entire dipeptide at the P site is transferred and bonded to the A site amino acid to give a tripeptide. The "empty" tRNA at the P site is released, a ribosome shift along the mRNA occurs, and we are ready to begin formation of the tetrapeptide.

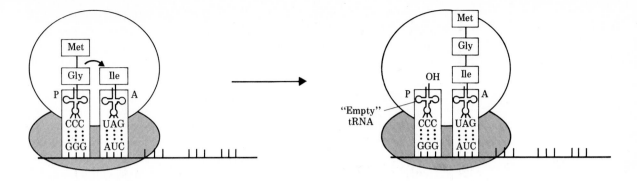

Protein Chain Termination

The polypeptide continues to grow by way of translation until all necessary amino acids are in place and bonded to each other. Appearance in the mRNA codon sequence of one of the three stop codons (UAA, UAG, or UGA; see Table 25.7) terminates the process, and the newly formed protein is released from the ribosome.

25.23 Recombinant DNA and Genetic Engineering

During the last decade, increased scientific knowledge concerning how DNA molecules behave under various chemical conditions has opened the door to a new and exciting field of research called *genetic engineering*. Human insulin, human growth hormone, and human interferons are now being manufactured in research laboratories (and commercially, in small amounts) using genetic engineering techniques.

HUMAN INSULIN FROM BACTERIA

Commercially, a number of companies are producing human insulin. About 5% of the population of the United States suffer from diabetes, many of whom depend on injections of insulin obtained from the pancreas of slaughterhouse animals. Genetically transformed bacteria now produce human insulin, which is available for individuals who react adversely to insulin from normal animal sources. Pituitary growth hormone (used in treating human dwarfism) and human interferon can now also be prepared by genetically transformed bacteria, making larger quantities of these important proteins available.

Genetic engineering procedures are centered around a type of DNA called recombinant DNA. **Recombinant DNA** *molecules are DNA molecules that* *have been synthesized by splicing a segment of DNA, usually a gene, from one organism into DNA from another organism.* In 1978, the first report appeared that the gene (from rats) for producing insulin had successfully been incorporated into a strain of *Escherichia coli* bacteria and that the bacteria had then begun to produce insulin. The exciting possibilities of the process were immediately evident to the scientific community, and genetic engineering become a rapidly developing field. Let us examine the theory and procedures used in obtaining, through genetic engineering, recombinant DNA.

The bacterium *E. coli*, a one-celled organism found in the intestinal tract of animals and humans, is the organism most used in experiments involving recombinant DNA. (Yeast cells are also now used, with increasing frequency, in such research.) Important reasons for their extensive use include

1. The genetics of this one-celled species is well known.
2. A number of strains of *E. coli* have been developed that cannot live outside of specially prepared culture media, so there is little danger that these special strains could grow and reproduce outside the laboratory.
3. The genetic code for *E. coli* is the same as that for human cells. (Recall, from Section 25.20, that the genetic code is almost universal.)

Structurally, relative to DNA, one difference exists between *E. coli* cells and human cells. Not all DNA within such bacteria is found in its single chromosome. Besides chromosomal DNA, *E. coli* also contain DNA in the form of small circular double-stranded molecules called *plasmids*, several of which exist within a cell. These plasmids, which carry only a few genes, replicate independently of the chromosome. They also have the characteristic of being relatively easily transferred from one cell to another. It is the plasmids of *E. coli* that are used in recombinant DNA work.

Let us consider "overall" steps in the process of obtaining *E. coli* cells that contain recombinant DNA. We will then consider details of specific steps.

1. Cells of a specific strain of *E. coli* are placed in a solution that dissolves cell membranes, releasing the contents of the cells.
2. The released cell components are separated into fractions, one of which contains the plasmids. The isolated plasmid fraction is the material used in further steps.
3. A special enzyme, called a restriction enzyme, is used to cleave—that is, open up—the double-stranded DNA from the circular plasmids. The result is linear (noncircular) DNA molecules.
4. The same restriction enzyme is then used to remove a desired gene from a chromosome in another organism.
5. The gene, from step 4, and the plasmid (linear DNA material), from step 3, are mixed in the presence of the enzyme DNA ligase, which splices the

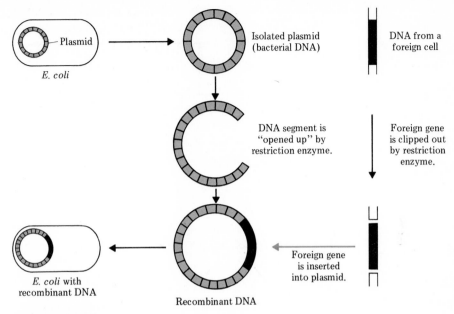

FIGURE 25.25

Recombinant DNA is made by inserting a gene obtained from a foreign cell into DNA obtained from another type of organism.

two together. This splicing reaction takes place at both ends of the gene, resulting in a new circular plasmid (recombinant DNA).

6. The newly synthesized plasmids (recombinant DNA) are placed in a live *E. coli* culture where they are taken up by the *E. coli* bacteria. The *E. coli* culture into which the plasmids are placed need not be identical to the strain from which the plasmids were originally obtained.

Figure 25.25 summarizes the steps involved in obtaining recombinant DNA and incorporating it into a living cell.

In step 3 of the preceding sequence, a restriction enzyme is mentioned. *Restriction enzymes* are enzymes that can recognize specific base sequences in DNA and then cleave this base sequence in a "staircase" manner. To understand how a restriction enzyme works, let us consider one that cleaves DNA between G and A bases in the sequence GAATTC. Such an enzyme cleaves the double-helix structure of a DNA molecule in the manner shown in Figure 25.26. Note that the double helix is not cut straight across; the individual strands are cut at different points, giving a "staircase" cut. This type of cut produces unpaired bases on each fragment resulting from the cut. Such unpaired base ends are called "sticky ends" because they are ready to "stick" (pair up) with a complementary section of DNA if they can find one.

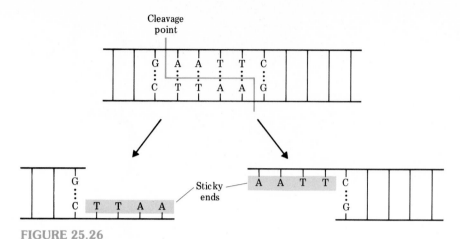

FIGURE 25.26

A restriction enzyme cleaves DNA in a staircase manner rather than cutting it straight across.

If the same restriction enzyme used to cut a plasmid is also used to cut a gene from another DNA molecule—it must be cut in two places, at both ends of the gene—the "sticky ends" of the gene will be complementary to those of the plasmid. Thus the plasmid and gene can readily combine to give a new modified plasmid molecule. This modified plasmid molecule is called recombinant DNA. Figure 25.27 shows diagrammatically the "match" between "sticky ends" that occurs when plasmid and gene combine.

Step 6 in the previously listed procedure involves placing the recombinant DNA (modified plasmids) into a live *E. coli* culture where they are taken up by the *E. coli* bacteria. This process is called transformation. **Transformation** *is the process of incorporating recombinant DNA into a host cell.* Usually only a few bacteria of the millions in a culture are transformed successfully. This makes the task of finding the transformed species difficult unless a specific label or marker is included in the recombinant DNA.

One powerful technique for selecting transformed bacteria is based upon antibiotic resistance. Not all bacteria are resistant to antibiotics such as penicillin. Those that are have DNA plasmids that contain penicillin-resistant genes. If a DNA plasmid containing a penicillin-resistant gene is used in the recombinant DNA, transformation of nonresistant bacteria will make them resistant to penicillin. Only the transformed cells will grow in media containing penicillin, allowing the selection of those containing the desired gene. The untransformed, nonresistant bacteria will not survive.

Once selection of transformed cells is accomplished, large numbers of identical cells called *clones* are obtained. **Clones** *are cells that have descended*

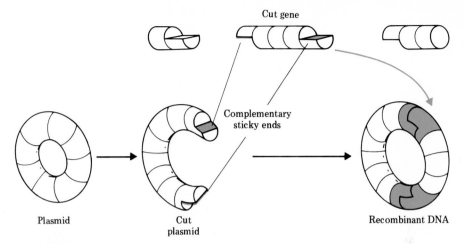

Cut gene

Complementary
sticky ends

Plasmid

Cut
plasmid

Recombinant DNA

FIGURE 25.27

The "sticky ends" of cut plasmid and cut gene that combine to form recombinant DNA are complementary.

from a single cell and have identical properties. Within a few hours, a single genetically altered cell can give rise to thousands of clones. Each clone has the capacity to synthesize the protein directed by the foreign gene it carries.

Many researchers who spend great effort and expense to develop new genetically altered strains of bacteria actually patent their "inventions." Anyone who wishes to use them may be required to pay royalties to the inventor. For further protection of their inventions, some genetic engineers also include their own "trademark" gene within their modified plasmids. Since bacteria are so easily passed around (or stolen), when future questions arise as to who developed a certain genetically altered strain, the engineer can show the presence of his or her own gene, which was included in the original transformation.

Researchers are not limited to selection of naturally occurring genes for transforming bacteria. Chemists have developed nonenzymatic methods of linking nucleotides together, such that they can construct artificial genes of any sequence they so desire. In fact, benchtop instruments are now available that can be programmed by a microprocessor to synthesize any DNA base sequence *automatically* (see Figure 25.28). The operator merely enters a sequence of desired bases, starts the instrument, and returns later to obtain the product. This synthesis requires only about 15 to 20 minutes per nucleotide. Such flexibility in manufacturing DNA has opened many doors, accelerated the pace of recombinant DNA research, and redefined the term "designer genes." It also solves the problem of trying to locate a particular gene (one out of thousands present) in an animal DNA molecule that is to be used to make recombinant DNA.

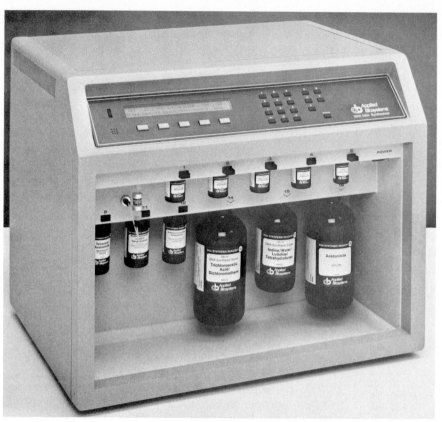

FIGURE 25.28
A fully automated benchtop DNA synthesizer. [Courtesy of Applied Biosystems.]

Practice Questions and Problems

AMINO ACIDS

25-1 What two functional groups are found in all amino acids?

25-2 What distinguishes one amino acid from another?

25-3 Draw the structural formula for each of the following amino acids.
a. Leucine   b. Serine   c. Lysine

25-4 Give the name and structure for each of the following amino acids.

a. Glu   b. Gly   c. Ala
d. Phe   e. Tyr   f. Cys

25-5 How is the structure of proline different from that of the other 19 common amino acids?

25-6 Give a definition for a protein in terms of amino acids.

ESSENTIAL AMINO ACIDS

25-7 Why are some amino acids called "essential," while others are not?

25-8 Explain why vegetarians should carefully watch their intake of plant proteins.

25-9 Why should a proper diet include protein from a variety of sources?

PEPTIDES

25-10 How are the terms amide bond and peptide bond related?

25-11 How many amino acids are present in a hexapeptide? How many peptide bonds are present in a hexapeptide?

25-12 Explain what is meant by the N-terminal end and the C-terminal end of a peptide.

25-13 Draw the complete structural formula for the tripeptide Try-Ala-Cys.

25-14 Draw structural formulas for two different dipeptides that can be made from one glycine and one cysteine molecule.

25-15 There are a total of six different sequences for a tripeptide containing one molecule each of serine, valine, and glycine. Using three-letter abbreviations for the amino acids, write the six possible sequences of amino acids.

25-16 How many different polypeptides containing two molecules of glycine and two molecules of leucine are possible?

PRIMARY STRUCTURE OF PROTEINS

25-17 Describe the difference in meaning between the terms polypeptide and protein.

25-18 Describe what is meant by the primary structure of a protein.

25-19 If you know the amino acid composition of a protein, can you predict its primary structure? Explain your answer.

25-20 Draw a structural representation, containing four units of the carbon–nitrogen "backbone" structure common to all polypeptides and proteins.

25-21 Would pig or cow insulin be the better alternate source for human diabetics, based solely upon primary structure?

SECONDARY STRUCTURE OF PROTEINS

25-22 What is meant by the secondary structure of a protein?

25-23 What are the three types of secondary structure commonly found in protein.

25-24 What type of intermolecular force holds the "turns" of an alpha-helix in position?

25-25 What type of intermolecular force holds parallel protein chains together in the beta-pleated sheet structure?

25-26 Draw a diagram illustrating the type of hydrogen bond present in the following.
a. An alpha-helix   b. A beta-pleated sheet

25-27 What secondary structural feature accounts for the great strength of the protein collagen?

25-28 Can more than one type of secondary structure be present in the same protein? Explain your answer.

25-29 What do we mean when we say that a section of a protein has a random coil arrangement?

TERTIARY STRUCTURE OF PROTEINS

25-30 What is meant by the tertiary structure of a protein?

25-31 Describe what is meant by the terms "globular protein" and "fibrous protein."

25-32 What are four types of attractive interactions that give rise to tertiary structure for proteins.

25-33 Identify the type of amino acid side chain involved in each of the following tertiary structure interactions.
a. Hydrophobic
b. Hydrogen bonding
c. Salt bridge

25-34 Distinguish between the origin of the attractive interactions in tertiary protein structure and the origin of those in secondary protein structure.

PROTEIN HYDROLYSIS

25-35 Identify the products from the complete hydrolysis of the tripeptide Val-Gly-Ala.

25-36 Will hydrolysis of the dipeptides Ala-Val and Val-Ala give the same products? Explain your answer.

PROTEIN DENATURATION

25-37 Which structural levels of a protein are affected by denaturation?

25-38 In what way is the protein in a cooked egg the same as in a raw egg?

25-39 Why is 70% isopropyl alcohol more effective as a disinfectant than pure isopropyl alcohol?

25-40 Explain why both reducing and oxidizing agents are used in the process of giving someone a hair "permanent."

NUCLEOTIDES

25-41 Describe the relationship of nucleotides to nucleic acids.

25-42 What are the three chemical "subunits" that make up a nucleotide?

25-43 What is the difference between the five-carbon sugars found in DNA and RNA?

25-44 Name the nitrogeneous bases found in DNA nucleotides.

25-45 Name the nitrogeneous bases found in RNA nucleotides.

25-46 Draw the structural formula for each of the following.
a. A nucleotide from DNA containing the base guanine
b. A nucleotide from RNA containing the base uracil

NUCLEIC ACID STRUCTURE

25-47 Explain what is meant by the "backbone" of a nucleic acid molecule.

25-48 How does the "backbone" of a DNA molecule differ from that of a RNA molecule?

25-49 What distinguishes various DNA molecules from each other?

DOUBLE-HELIX STRUCTURE OF DNA

25-50 Describe the secondary structural features of a DNA molecule.

25-51 What type of bonding holds a DNA double-helix molecule together?

25-52 Using the concept of complementary base pairing, write complementary strands for each of the following base sequences.
a. ACGTAT   b. TTACCGT   c. GCACTAAGTC

25-53 Draw diagrams that show the hydrogen bonds that stabilize the following types of base pairing in DNA.
a. T and A   b. G and C

25-54 The "base content" of a particular DNA is 36%

thymine. What is the percentage of each of the following bases in the molecule?
a. Adenine   b. Guanine   c. Cystosine

DNA REPLICATION

25-55 What is DNA replication?

25-56 What is the function of DNA polymerase in the replication process?

25-57 What principle insures that an exact duplicate copy of a DNA molecule results from replication?

25-58 Where are the DNA strands from a parent DNA molecule found within the daughter DNA molecules obtained from replication?

RNA MOLECULES

25-59 What are four major differences between RNA and DNA molecules?

25-60 What is the primary function of each of the following types of RNA?
a. tRNA   b. mRNA   c. rRNA

TRANSCRIPTION: RNA SYNTHESIS

25-61 What are the four steps in the process of transcription?

25-62 What will be the base sequence of the RNA formed from the following DNA template?

TACGGGTAACTCG

25-63 What is a gene?

25-64 What is the relationship between a gene and a chromosome?

THE GENETIC CODE

25-65 What is a codon?

25-66 In what type of molecule are codons found?

25-67 What interrelationship is defined by the genetic code?

25-68 Which amino acid is coded for by each of the following codons?
a. CUU   b. AAU   c. AGU
d. GGG   e. GUA   f. CAC

25-69 Explain why the base sequence ATC could not be a codon.

25-70 Explain what is meant by each of the following characterizations of the genetic code.

a. Degenerate   b. Nonoverlapping
c. Universal

25-71 Using the genetic code, predict the sequence of amino acids coded by the mRNA sequence

—AUG—AAA—GAA—GAC—CUA—UAC—CCC—

25-72 Why would two-base sequences in mRNA not make an appropriate code for amino acids?

ANTICODONS

25-73 What is an anticodon?

25-74 Describe the general shape of a tRNA molecule.

25-75 Draw a sketch of a tRNA molecule and label each of the following.
a. The site where an amino acid is attached during protein synthesis
b. The site where the anticodon is located

25-76 What principle governs the codon–anticodon interaction that occurs during protein synthesis?

25-77 What is the anticodon needed to interact with each of the following codons?
a. CCA   b. CGG   c. AUA   d. CAU

TRANSLATION: PROTEIN SYNTHESIS

25-78 What are the four major steps in the process of translation.

25-79 What role do ribosomes play in protein synthesis?

25-80 In the growth step of protein synthesis, at which site in the ribosome does new peptide-bond formation actually take place?

25-81 During protein synthesis, what two changes occur at a ribosome immediately after peptide bond formation has occurred?

RECOMBINANT DNA AND GENETIC ENGINEERING

25-82 How does recombinant DNA differ from "normal" DNA?

25-83 What are *E. coli* plasmids?

25-84 Describe what occurs when a particular restriction enzyme "operates" on a segment of double-stranded DNA.

25-85 What is transformation?

25-86 How are plasmids obtained from *E. coli* bacteria?

25-87 What are clones?

25-88 List two human proteins of commercial importance that can be produced by genetically altered bacteria.

Exponential Notation

In scientific work, we frequently encounter very large and very small numbers. As an illustration of this situation, consider the following information about human blood, a substance whose major component is water (92% by mass). In one drop of blood ($\frac{1}{20}$ mL) there are approximately

$$1,600,000,000,000,000,000,000$$

molecules of water. A molecule of water is the smallest possible unit of water (see Section 2.4). Obviously, molecules must be very small entities if there are that many of them in a single drop of blood. In fact, a single water molecule has a mass of

$$0.000000000000000000000030 \text{ gram}$$

Such large and small numbers are difficult to use. Recording them is not only a time-consuming task, but also one that is very prone to error; often too many or too few zeros are recorded. Also, these numbers are awkward to work with in calculations. Consider the problem of multiplying the preceding two numbers together; handling all of the zeros is a "mind-boggling" endeavor.

A method exists for expressing cumbersome multidigit numbers in compact form. This method is called exponential notation and it eliminates the need to write all the zeros. **Exponential notation** is a system in which an ordinary decimal number is expressed as a number between 1 and 10 multiplied by 10 *raised to a power.* The two previously cited numbers that deal with molecules of water are expressed in exponential notation as

$$1.6 \times 10^{21} \text{ molecules} \quad \text{and} \quad 3.0 \times 10^{-22} \text{ gram}$$

respectively. Note that in exponential form six digits serve for the previous 22 and 23.

Exponents

As the name implies, exponential notation involves the use of exponents. An **exponent** *is a number written as a superscript that follows another number and indicates how many times the first number is to be multiplied by itself.* The following examples illustrate the use of exponents.

$$4^3 = 4 \times 4 \times 4 = 64$$

$$2^4 = 2 \times 2 \times 2 \times 2 = 16$$

$$10^5 = 10 \times 10 \times 10 \times 10 \times 10 = 100,000$$

Exponents are also frequently referred to as *powers* of numbers. Thus, 4^3 may be read as "four to the third power," and 2^4 as "two to the fourth power." Raising a number to the second power is often called "squaring," and to the third power "cubing."

Negative as well as positive exponents exist. A negative sign in front of an exponent means that the number raised to the indicated power, is in the denominator of a fraction in which 1 is the numerator. The following examples illustrate this interpretation.

$$10^{-2} = \frac{1}{10^2} = \frac{1}{10 \times 10} = \frac{1}{100} = 0.01$$

$$10^{-4} = \frac{1}{10^4} = \frac{1}{10 \times 10 \times 10 \times 10} = \frac{1}{10,000} = 0.0001$$

Writing Numbers in Exponential Notation

A number written in exponential notation has two parts: (1) a *coefficient*, which is written first, and is a number between 1 and 10, and (2) an *exponential term*, which is 10 raised to a power. The coefficient part is always multiplied by the exponential term. Using the exponential form of the number 276 as an example, we have

Coefficient ⌐   ⌐ Exponent

$$2.76 \times 10^2$$

Multiplication sign ⌐   └ Exponential term

The rules for converting numbers from decimal to exponential notation are very simple.

1. The value of the exponent is determined by counting the number of places the original decimal point must be moved to give the coefficient.
2. If the decimal point is moved to the *left* to get the coefficient, the exponent is a *positive number*. If the decimal point is moved to the *right* to get the coefficient, the exponent is a *negative number*.

Example A.1 illustrates the use of these rules.

EXAMPLE A.1

Express the following numbers in exponential notation.
(a) 731   (b) 2230.7   (c) 0.000733

SOLUTION

(a) The coefficient, which must be a number between 1 and 10, is obtained by moving the decimal point in 731 two place to the left.

$$7\,31.$$

This gives us 7.31 as the coefficient. The value of the exponent in the exponential term is +2 because the decimal point was shifted two places to the left. A shift of the decimal point to the left will always result in a positive exponent. Multiplying the coefficient by the exponential term, 10^2, gives us the exponential notation form of the number.

$$7.31 \times 10^2$$

(b) The coefficient is obtained by moving the decimal point in 2230.7 three places to the left.

$$2\,230.7$$

The power of ten is +3, indicating that the decimal point was moved three places to the left to obtain the coefficient. Multiplying the coefficient and exponential terms together gives

$$2.2307 \times 10^3$$

(c) Because the number 0.000733 has a value less than 1, the decimal point must be moved to the right in order to obtain a coefficient with a value between 1 and 10.

$$0.000733$$

Moving the decimal point to the right means that the exponent will have a negative sign; in this case it is −4. Thus, in exponential notation we have

$$7.33 \times 10^{-4}$$

Converting from Exponential Notation to Decimal Notation

In order to convert a number in exponential notation, such as 6.02×10^{23}, into a regular decimal number, we start by examining the exponent. The value of the exponent tells how many places the decimal point must be moved. If the exponent is positive, the decimal point is moved to the right to give a number greater than 1; if it is negative, the decimal point is moved to the left to give a number less than 1. Zeros may have to be added to the number as the decimal point is moved.

EXAMPLE A.2

Express the following exponential numbers in ordinary decimal notation.
(a) 3.76×10^{-3} (b) 6.2×10^{5} (c) 5.3×10^{2}

SOLUTION
(a) The exponent -3 tells us the decimal point will be located three places to the left of where it is in 3.76. We add two zeros to accommodate the decimal point change.

Added zeros

$$0.003.76 = 0.00376$$

Decimal point shift

(b) The exponent $+5$ tells us the decimal point will be located five places to the right of where it is in 6.2. We add four zeros to mark the new decimal place correctly.

Added zeros

$$6.20000 = 620,000$$

Decimal point shift

(c) The decimal point will be moved two places to the right of where it is in 5.3. This gives the number 530 as the answer.

$$5.30 = 530$$

B

Exponential Notation and Mathematical Operations

A major advantage of writing numbers in exponential notation is that it greatly simplifies the mathematical operations of multiplication and division.

Multiplication in Exponential Notation

The multiplication of two or more numbers expressed in exponential notation involves two separate steps.

STEP 1   The coefficients (the decimal numbers between 1 and 10) are multiplied together in the usual manner.

STEP 2   The exponents of the powers of ten are *ADDED* algebraically.

EXAMPLE B.1

Carry out the following multiplications in exponential notation.
(a) $(2.33 \times 10^3) \times (1.55 \times 10^4)$
(b) $(1.13 \times 10^3) \times (5.81 \times 10^{-6})$
(c) $(1.25 \times 10^3) \times (1.85 \times 10^1) \times (2.05 \times 10^6)$

SOLUTION

(a) Multiplying the two coefficients together gives

$$2.33 \times 1.55 = 3.61$$

Multiplication of the two powers of 10 to give the exponential part of the answer requires that we add the exponents to give a new exponent.

$$10^3 \times 10^4 = 10^{(3+4)} = 10^7$$

Combining the new coefficient with the new exponential term gives the answer.

$$3.61 \times 10^7$$

(b) Multiplying the two coefficients together gives

$$1.13 \times 5.81 = 6.57$$

When multiplying the exponential terms, we will have to add exponents with different signs. To do this, we first determine the larger number without considering sign (6 is larger than 3) and then subtract the smaller number from it ($6 - 3 = 3$). The sign is always the sign of the larger number, which is minus in this case. Thus

$$10^3 \times 10^{-6} = 10^{3+(-6)} = 10^{-3}$$

Combining the coefficient and the exponential term gives

$$6.57 \times 10^{-3}$$

(c) Multiplying the three coefficients together gives

$$1.25 \times 1.85 \times 2.05 = 4.74$$

All of the exponential terms have the same sign—positive.
When adding numbers of the same sign, whether positive or negative, just add the numbers and place the common sign in front of the sum.

$$10^3 \times 10^1 \times 10^6 = 10^{(3)+(1)+(6)} = 10^{10}$$

Combining the coefficient and the exponential term gives

$$4.74 \times 10^{10}$$

Division in Exponential Notation

The division of numbers in exponential notation, like multiplication, involves two steps.

STEP 1   The coefficients are divided in the usual manner.

STEP 2   The exponents of the powers of 10 are *SUBTRACTED* algebraically. The exponent in the denominator (bottom) is always subtracted from the exponent in the numerator (top).

Note that in multiplication the exponents are *added*, and in division the exponents are *subtracted*.

EXAMPLE B.2

Carry out the following divisions in exponential notation.

(a) $\dfrac{8.42 \times 10^6}{3.02 \times 10^4}$ (b) $\dfrac{4.20 \times 10^{-3}}{1.19 \times 10^{-7}}$ (c) $\dfrac{9.44 \times 10^{-10}}{8.23 \times 10^6}$

SOLUTION

(a) Performing the indicated division of the coefficients gives

$$\frac{8.42}{3.02} = 2.79$$

The division of exponential terms involves the algebraic subtraction of exponents.

$$\frac{10^6}{10^4} = 10^{(+6)-(+4)} = 10^2$$

Algebraic subtraction involves changing the sign of the number to be subtracted and then following the rules for addition. In this problem the number to be subtracted $(+4)$ becomes (-4) upon changing the sign. Then we add $(+6)$ and (-4). The answer is $(+2)$, as shown. Combining the coefficient and the exponential term gives

$$2.79 \times 10^2$$

(b) The new coefficient is obtained by dividing 4.20 by 1.19.

$$\frac{4.20}{1.19} = 3.53$$

The exponential part of the answer is obtained by subtracting (-7) from (-3). Changing the sign of the number to be subtracted gives $(+7)$. Adding $(+7)$ and (-3) gives $(+4)$. Therefore,

$$\frac{10^{-3}}{10^{-7}} = 10^{(-3)-(-7)} = 10^4$$

Combining the coefficient and the exponential term gives

$$3.53 \times 10^4$$

(c) Dividing the number 9.44 by 8.23 will give the new coefficient.

$$\frac{9.44}{8.23} = 1.15$$

Division of the exponential terms involves subtracting $(+6)$ from (-10). Changing the sign of the number to be subtracted gives (-6). Adding (-6) and (-10) gives (-16).

$$\frac{10^{-10}}{10^{+6}} = 10^{(-10)-(+6)} = 10^{-16}$$

Combining the two parts of the problem yields an answer of

$$1.15 \times 10^{-16}$$

Selected Answers

Chapter 1

1–2 **a.** nonmatter **d.** matter
1–8 **a.** physical **d.** physical **g.** chemical
1–9 **a.** physical **d.** physical **g.** physical
1–11 **a.** chemical **d.** physical **g.** chemical
1–12 **a.** physical **d.** chemical **g.** chemical
1–14 **a.** law **d.** fact
1–19 **a.** false **d.** true **g.** false
1–20 **a.** hypothesis **d.** law

Chapter 2

2–1 **a.** heterogeneous mixture, homogeneous mixture
2–2 **a.** heterogeneous mixture **d.** pure substance
2–3 **a.** pure substance with one phase **d.** heterogeneous mixture with two phases **g.** heterogeneous mixture with three phases
2–4 **a.** chemically homogeneous and physically homogeneous **d.** chemically homogeneous and physically heterogeneous **g.** chemically heterogeneous and physically heterogeneous
2–9 **a.** substance A is a compound **d.** no classification is possible for substance H **g.** no classification is possible for substance K
2–10 **a.** true **d.** true **g.** true
2–14 **a.** compound **d.** mixture
2–15 **a.** false **d.** false **g.** true
2–16 **a.** 13 **d.** 16
2–17 **a.** hydrogen and helium **d.** nitrogen and oxygen **g.** hydrogen and oxygen
2–20 **a.** neon **d.** argon **g.** tin
2–21 **a.** Li **d.** B **g.** Ni
2–25 **a.** compound **d.** compound **g.** element
2–27 **a.** 4 **d.** 1 **g.** 3
2–28 **a.** one S and two O **d.** one Ca, one S, and four O **g.** one Be, two C, and two N
2–32 **a.** three **d.** SO_2, H_2O

Chapter 3

3–2 not consistent: c, e, f
3–5 **a.** 2.2 g
3–10 **a.** electron **d.** proton **g.** proton and neutron **j.** neutron
3–11 **a.** false **d.** true
3–16 **a.** the number of protons and also the number of electrons **d.** the total number of subatomic particles present in the atom

3–17 **a.** 24 protons, 29 neutrons, 24 electrons **d.** 16 protons, 18 neutrons, 16 electrons **h.** 1 proton, 2 neutrons, 1 electron
3–18 **a.** $^{58}_{28}Ni$ **d.** $^{18}_{8}O$ **g.** $^{197}_{79}Au$
3–21 **a.** $^{14}_{7}X$ and $^{15}_{7}X$
3–22 **a.** same number of protons, same number of electrons, and different numbers of neutrons **d.** same number of protons, same number of electrons, and different numbers of neutrons
3–25 same: b, c, e
3–26 **a.** $^{8}_{5}B$ **d.** $^{12}_{5}B$
3–30 28.09 amu

Chapter 4

4–3 **a.** Ga **d.** Cl **g.** Sc
4–4 **a.** group **d.** period **g.** period
4–6 **a.** orbital **d.** subshell **g.** subshell
4–7 **a.** 5 **d.** 50
4–8 **a.** 8 **d.** 2
4–9 **a.** spherical **d.** cloverleaf
4–11 **a.** 10 **d.** 32
4–14 **a.** 2s (lowest) and 4s (highest) **d.** 3d (lowest) and 5f (highest)
4–16 **a.** $1s^2 2s^2 2p^5$ **d.** $1s^2 2s^2 2p^6 3s^2 3p^6 4s^2$ **g.** $1s^2 2s^2 2p^6 3s^2 3p^6 4s^2 3d^{10} 4p^6 5s^2 4d^{10} 5p^6 6s^2 4f^{14}$-$5d^{10} 6p^6$
4–19 **a.** p area **d.** p area
4–20 **a.** p subshell **d.** d subshell
4–21 **a.** p^3 **d.** s^2
4–22 **a.** noble gas **d.** transition element **g.** representative element
4–23 **a.** nonmetal **d.** nonmetal **g.** nonmetal
4–24 **a.** H, O, C, and N **d.** Na, Mg, K, Ca, V, Cr, Mn, Fe, Co, Ni, Cu, Zn, Mn, and Sn **g.** H, C, N, O, F, Si, P, S, Cl, As, Se, and I **j.** V, Cr, Mn, Fe, Co, Ni, Cu, Zn, and Mn

Chapter 5

5–1 **a.** 1 **d.** 2
5–2 **a.** 2 **d.** 1
5–4 **a.** Mg **d.** S
5–8 **a.** +3 **d.** +1
5–10 **a.** 12 protons and 10 electrons **d.** 3 protons and 2 electrons
5–11 **a.** +2 **d.** +1
5–12 **a.** two lost **d.** one lost
5–15 **a.** KCl **d.** K_2S

871

5—16 **a.** MgO   **d.** Li$_2$O
5—17 **a.** Na$_2$S   **d.** AlP
5—21 **a.** potassium phosphide   **d.** calcium oxide
5—22 **a.** yes   **d.** no
5—23 **a.** +3   **d.** +4
5—24 **a.** copper (II) oxide and copper (I) oxide
   d. tin(II) bromide and tin(IV) bromide
5—25 **a.** gold(I) chloride   **d.** aluminum chloride
5—26 **a.** KBr   **d.** Be$_3$P$_2$
5—27 **a.** NiS   **d.** Au$_2$S$_3$
5—35 **a.** F   **d.** Mg
5—36 **a.** H—Cl (0.9), H—O (1.4), H—F (1.9)
   d. P—N (0.9), S—O (1.0), Be—F (2.5)
5—38 **a.** polar covalent   **d.** nonpolar covalent
5—40 **a.** polar bonds and a nonpolar molecule   **d.** polar
   bonds and a polar molecule
5—41 **a.** polar   **d.** polar
5—42 **a.** sulfur hexafluoride   **d.** chlorine dioxide
5—43 **a.** ICl   **d.** N$_2$O$_3$
5—44 **a.** NO$_3^-$ **d.** OH$^-$
5—45 **a.** none   **d.** hypochlorite
5—46 **a.** Ba(NO$_3$)$_2$ **d.** Au$_2$SO$_4$
5—47 **a.** sodium carbonate   **d.** calcium hydroxide
5—48 **a.** KHCO$_3$ **d.** Cu(OH)$_2$

Chapter 6

6—1 **a.** qualitative   **d.** quantitative
6—3 **a.** micro   **d.** milli
6—4 **a.** centigram   **d.** megameter
6—5 **a.** milligram, centigram, decigram   **d.** microliter,
   milliliter, megaliter
6—7 **a.** 4.832 g   **d.** 4.832 × 10^{-3} kg
6—8 **a.** 3 × 10^{-6} km   **d.** 2.4 × 10^{-11} km
6—10 2.5 L
6—12 1.1 × 10^5 drops
6—13 **a.** 1 inch   **d.** 1 pound
6—15 6.6 × 10^{-6} lb
6—17 2.1 × 10^5 mL
6—20 15.2 mL
6—22 112 lbs
6—24 10.40
6—26 41°C
6—28 274°C
6—30 308 K
6—33 **a.** light and heat energy   **d.** mechanical energy
6—35 **a.** potential   **d.** potential
6—37 **a.** 300 cal   **d.** 9.3 cal
6—38 **a.** 0.155 cal/g °C
6—40 6040 cal

Chapter 7

7—1 **a.** 84.01 amu   **d.** 98.96 amu
7—2 **a.** 158.12 amu   **d.** 176.14 amu
7—3 **a.** 5.94% H, 94.06% O   **d.** 54.47% C, 9.16% H,
   36.36% S
7—6 **a.** 1.81 ×10^{24} molecules SO$_2$ **d.** 6.02 ×10^{23}
   molecules NH$_3$
7—7 **a.** 4.52 × 10^{23} atoms Ni   **d.** 4.52 × 10^{23}
   molecules C$_9$H$_8$O$_4$
7—9 **a.** 58.44 g NaCl   **d.** 461.0 g PbI$_2$

7—10 **a.** 73.0 g N   **d.** 54.1 g H$_2$O
7—11 **a.** 0.0250 mole Br   **d.** 0.0375 mole B$_4$H$_{10}$
7—14 **a.** 6.00 moles N, 3.00 moles O   **d.** 4.00 moles O
7—15 **a.** 8.00 moles   **d.** 12.0 moles
7—16 **a.** 3.14 × 10^{23} atoms Be   **d.** 6.15 × 10^{23} atoms U
7—17 **a.** 63.55 g Cu   **d.** 1.06 × 10^{-22} g Cu
7—18 **a.** 0.185 mole Ag   **d.** 6.6 × 10^{-2} mole S
7—19 **a.** 6.14 × 10^{22} atoms P   **d.** 1.54 × 10^{23} atoms P
7—20 **a.** 5.3 × 10^{-16} g C$_{21}$H$_{36}$O$_2$ **d.** 1.5 × 10^{-17} g
   C$_{21}$H$_{36}$O$_2$
7—22 **a.** N$_2$ + O$_2$ → 2 NO   **d.** 2 Na + 2 H$_2$O → 2 NaOH
   + H$_2$
7—23 **a.** CH$_4$ + 2 O$_2$ → CO$_2$ + 2 H$_2$O   **d.** 2 C$_4$H$_{10}$ +
   13 O$_2$ → 8 CO$_2$ + 10 H$_2$O
7—24 **a.** 2 Fe$_2$O$_3$ +3 C → 4 Fe + 3 CO$_2$ **d.** 2 Al (NO$_3$)$_3$
   + 3 H$_2$SO$_4$ → Al$_2$(SO$_4$)$_3$ + 6 HNO$_3$
7—26 **a.** 2.00 moles Fe   **d.** 3.00 moles Fe$_2$O$_3$
7—27 **a.** 1.00 mole N$_2$ **d.** 1.00 mole N$_2$
7—28 **a.** 9.76 g C$_7$H$_{16}$ **d.** 158 g Fe$_3$O$_4$
7—30 **a.** 766 g KClO$_3$ **d.** 59.7 g KClO$_3$

Chapter 8

8—2 **a.** gaseous   **d.** gaseous
8—12 **a.** 0.954 atm   **d.** 0.816 atm
8—16 **a.** 2.10 L   **d.** 7.88 L
8—17 **a.** 2.4 atm   **d.** 0.33 atm
8—22 **a.** 5.42 mL   **d.** 16.0 mL
8—23 **a.** 71°C   **d.** 1744°C
8—27 **a.** 228°C   **d.** 1312°C
8—28 **a.** 1.80 atm   **d.** 5.23 atm
8—31 **a.** 0.53 atm
8—34 **a.** endothermic   **d.** exothermic
8—35 **a.** boiling point   **d.** decomposition   **g.** boiling point

Chapter 9

9—3 **a.** saturated   **d.** saturated
9—5 **a.** dilute   **d.** concentrated
9—8 **a.** very soluble   **d.** very soluble
9—9 **a.** soluble   **d.** soluble   **g.** soluble
9—10 **a.** 6.78%(m/m)
9—13 **a.** 4.00%(v/v)
9—15 **a.** 1.8%(m/v)
9—16 **a.** 6.0 M
9—18 **a.** 85.5 mL solution
9—21 **a.** an increase
9—26 **a.** osmosis   **d.** 3%(m/m) solution

Chapter 10

10—12 **a.** HI(s) → H$^+$(aq) + I$^-$(aq)   **d.** KOH(s) → K$^+$(aq)
   + OH$^-$ (aq)
10—13 **a.** HF (acid) and H$_2$O (base)   **d.** HCO$_3^-$ (acid) and
   H$_2$O (base)
10—14 **a.** HClO + H$_2$O → H$_3$O$^+$ + ClO$^-$ **d.** H$_3$O$^+$
   + OH$^-$ → H$_2$O + H$_2$O
10—16 **a.** strong   **d.** weak
10—17 **a.** strong   **d.** strong
10—20 **a.** monoprotic   **d.** monoprotic
10—25 **a.** NaOH + HCl → NaCl + H$_2$O   **d.** 3 Ba(OH)$_2$ +
   2 H$_3$PO$_4$ → Ba$_3$(PO$_4$)$_2$ + 6 H$_2$O
10—27 **a.** acid   **d.** salt   **g.** acid

10–31 **a.** 1.0×10^{-11} **d.** 5.0×10^{-9}
10–33 **a.** acidic **d.** basic
10–34 **a.** 7 **d.** 1
10–36 **a.** 1.0×10^{-3} **d.** 1.0×10^{-13}
10–41 **a.** oxidized **d.** loses
10–42 **a.** +4 **d.** +2 **g.** +5
10–43 **a.** +3 **d.** +7 **g.** +1
10–44 **a.** yes **d.** no
10–45 **a.** Fe_2O_3 (reduced, oxidizing agent) and A1 (oxidized, reducing agent)
  d. F_2 (reduced, oxidizing agent) and NaCl (oxidized, reducing agent)

Chapter 11

11–2 65% by mass
11–4 Approximately $-80°$ C
11–11 Na^+, Mg^{2+}, and Ca^{2+} (metallic) and Cl^-, SO_4^{2-}, and HCO_3^- (nonmetallic)
11–17 Ca^{2+}, Mg^{2+}, and Fe^{2+} ions
11–23 90% distillation
11–27 No phase change is needed

Chapter 12

12–1 H, S, O, N, and Ca
12–8 **a.** $Ca_3(PO_4)_2 + 3 H_2SO_4 \rightarrow 3 CaSO_4 + 2 H_3PO_4$
  d. $2 SO_2 + O_2 \rightarrow 2 SO_3$
12–11 lack of chemical reactivity; low boiling point
12–18 **a.** $2 NH_3 + CO_2 \rightarrow (NH_2)_2CO + H_2O$
  d. $2 NH_3 + H_3PO_4 \rightarrow (NH_4)_2HPO_4$
12–22 **a.** $N_2 + 3H_2 \rightarrow 2 NH_3$
12–28 **a.** approximately 129 mm Hg
12–32 **a.** CaO **d.** $Ca(OH)_2$
12–33 **a.** $CaCO_3 \rightarrow CaO + CO_2$
  d. $CaO + SiO_2 \rightarrow CaSiO_3$
12–36 $Ca(OH)_2 + CO_2 \rightarrow CaCO_3 + H_2O$

Chapter 13

13–6 white P, P_4 tetrahedra; red P, long chains of P_4 tetrahedra
13–10 **a.** $3 Ca_3(PO_4)_2 + 6 SiO_2 + 10 C \rightarrow 6 CaSiO_3 + 10 CO + P_4$
13–17 2000–2500°C temperature and 100,000–150,000 atm pressure
13–19 **a.** cutting, grinding, and polishing tools
  d. absorbant of unwanted materials
13–21 **a.** reactant; energy $+ 6 CO_2 + 6 H_2O \rightarrow C_6H_{12}O_6 + 6 O_2$
13–26 **a.** baking soda
  d. sodium aluminum sulfate
13–30 refrigeration and beverage carbonation
13–32 $SiO_2 + 2 C \rightarrow Si + 2 CO$
13–39 tetrahedral SiO_4^{4-} ion
13–41 **a.** beryl
13–45 $Be_3Al_2Si_6O_{18}$ with chromium present as an impurity
13–47 **a.** B_2O_3

Chapter 14

14–2 Al - 61,000, Ca - 26,000, Mg - 24,000, Fe - 22,000, Na - 21,000, K - 8,900, Ti - 3,700, Mn - 370, and Sr - 110

14–4 Al, Fe, and Mn
14–10 iron ore, limestone, and coke
14–13 CO, generated from the reaction of CO_2 and coke
14–14 **a.** $C + O_2 \rightarrow CO_2$
  $CO_2 + C \rightarrow 2 CO$
  d. $CaCO_3 \rightarrow CaO + CO_2$
14–17 carbon steel and alloy steel
14–26 Cu_2S
14–33 (1) electrical materials, (2) plumbing materials, and (3) manufacture of alloys
14–35 (1) jewelry and arts, (2) electronics and aerospace applications, and (3) dental applications
14–40 Ag_2S
14–41 **a.** 92.5% Ag
14–43 (1) photographic applications, and (2) electrical and electronic products
14–47 synthetic rubies and sapphires
14–54 (1) container and packaging industry, (2) the transportation sector, and (3) the building and construction industry

Chapter 15

15–1 **a.** $^{14}_{7}N$ **d.** sodium-24
15–2 **a.** $^{10}_{4}Be$ and beryllium-10 **d.** $^{256}_{99}Es$ and einsteinium-256
15–4 **a.** $^{200}_{84}Po \rightarrow ^{196}_{82}Pb + ^{4}_{2}\alpha$
  d. $^{244}_{96}Cm \rightarrow ^{240}_{94}Pu + ^{4}_{2}\alpha$
15–5 **a.** $^{25}_{11}Na \rightarrow ^{25}_{12}Mg + ^{0}_{-1}\beta$
  d. $^{104}_{45}Rh \rightarrow ^{104}_{46}Pd + ^{0}_{-1}\beta$
15–6 **a.** $^{8}_{5}B \rightarrow ^{8}_{4}Be + ^{0}_{1}\beta$
  d. $^{103}_{47}Ag \rightarrow ^{103}_{46}Pd + ^{0}_{1}\beta$
15–7 **a.** $^{59}_{28}Ni + ^{0}_{-1}e \rightarrow ^{59}_{27}Co$
  d. $^{161}_{67}Ho + ^{0}_{-1}e \rightarrow ^{161}_{66}Dy$
15–9 **a.** $^{0}_{-1}\beta$ **d.** $^{0}_{1}\beta$
15–10 **a.** $^{181}_{74}W + ^{0}_{-1}e \rightarrow ^{181}_{73}Ta$
  d. $^{59}_{29}Cu \rightarrow ^{59}_{28}Ni + ^{0}_{1}\beta$
15–11 **a.** $^{4}_{2}\alpha$ **d.** $^{17}_{8}O$
15–12 **a.** $^{9}_{4}Be + ^{4}_{2}\alpha \rightarrow ^{12}_{6}C + ^{1}_{0}n$
  d. $^{246}_{96}Cm + ^{12}_{6}C \rightarrow ^{254}_{102}No + 4 ^{1}_{0}n$
15–14 **a.** elements 43, 61, 84–109
  d. Sn
15–16 **a.** 14 to 10 before; 13 to 11 after; decrease
15–20 **a.** 1/4 **d.** 1/8
15–22 280 years
15–23 **a.** both processes **d.** fission
15–24 **a.** fusion **d.** neither
15–26 **a.** fissionable isotope produced from Th-232
  d. starting material for producing fissionable U-233
15–40 **a.** cigarette smoking **d.** air travel

Chapter 16

16–6 **a.** C_8H_{18} **d.** $C_{35}H_{72}$
16–8 **a.** 7 **d.** 7
16–9 **a.** 3-methylpentane **d.** 3-methylhexane
16–10 **a.** 1 **d.** 1
16–14 **a.** yes **d.** yes
16–16 **a.** 8 **d.** 14
16–17 **a.** methylcyclopropane **d.** 1,3-dimethylcyclopentane

16–20 **a.** unsaturated alkene **d.** unsaturated cycloalkene
16–21 **a.** 2-pentene **d.** 4-methyl-2-hexyne **g.** 1,3-pentadiene
16–25 **a.** saturated **d.** nonaromatic unsaturated
16–26 **a.** 12 **d.** 8
16–27 **a.** benzene **d.** 1,2,4-trimethylbenzene

Chapter 17

17–3 **a.** halide **d.** alcohol **g.** amide
17–7 **a.** 1,1-dibromoethane **d.** iodomethane
17–8 **a.** 2 **d.** 2
17–11 **a.** 2-butanol **d.** 2,4-dimethyl-2-pentanol
17–13 **a.** 1,3-propanediol
17–15 **a.** methanol **d.** methanol
17–19 **a.** m-propylphenol or 3-propylphenol **d.** 3,4-dichlorophenol
17–21 **a.** methyl propyl ether or 1-methoxypropane
17–22 **a.** mixed
17–32 **a.** yes **d.** no **g.** yes
17–35 **a.** 1
17–41 **a.** pentanoic acid **d.** ethanoic acid
17–47 **a.** methylethanoate
17–51 **a.** secondary **d.** primary
17–52 **a.** ethylmethylamine **d.** t-butylamine
17–53 **a.** yes **d.** yes

Chapter 18

18–3 addition, condensation, and rearrangement polymer
18–5 initiation, propagation, and termination
18–21 1/4 natural rubber and 3/4 synthetic rubber
18–25 **a.** diamine and diacid **d.** dialcohol and diacid
18–30 ester linkages are present between monomers
18–31 ethylene glycol and terephthalic acid
18–36 extensive cross-linking in both cases
18–44 dialcohol and diisocyanate
18–48 pigment, binder, and solvent
18–49 water-based latex and oil-based
18–52 60% water-based and 40% oil-based

Chapter 19

19–2 **a.** 15% **d.** 87% **g.** 92%
19–4 reservoir rock, cap rock, and an anticline trap
19–10 **a.** 10
19–17 60
19–18 **a.** butane
19–25 Kerogen is extensively cross-linked
19–26 mined, crushed, and heated
19–35 deep mining (40%) and strip mining (60%)
19–42 **a.** 20% in 1965, 36% in 1975, and 31% in 1985
19–48 approximately 10
19–50 approximately 30

Chapter 20

20–4 **a.** 780,800 ppm **d.** 9.340 ppm
20–7 **a.** carbon monoxide **d.** carbon monoxide
20–8 **a.** transportation **d.** stationary fuel combustion
20–10 **a.** -31% **d.** -26%
20–12 transportation
20–14 9 ppm (NAAQS) and 8 ppm (average)
20–19 NO and NO_2

20–22 $N_2 + O_2 \rightarrow 2\ NO$
$2\ NO + O_2 \rightarrow 2\ NO_2$
20–24 0.53 ppm (NAAQS) and 0.25 ppm (average)
20–37 0.12 ppm (NAAQS) and 0.13 ppm (average)
20–42 SO_2 and SO_3
20–47 0.03 ppm (NAAQS) and 0.01 ppm (average)
20–51 SO_x and NO_x
20–53 $CaCO_3 + H_2SO_4 \rightarrow CaSO_4 + CO_2 + H_2O$
20–56 almost equal in magnitude
20–59 75 $\mu g/m^3$ (NAAQS) and 50 $\mu g/m^3$ (average)

Chapter 21

21–4 **a.** 1 ppm **d.** 100–10,000 ppm
21–6 **a.** CO_2, NH_3 and HNO_3, and H_2SO_4
21–13 detergents
21–16 builders
21–23 osmotic process
21–26 $2\ FeS_2 + 7\ O_2 + 2\ H_2O \rightarrow 2\ FeSO_4 + 2\ H_2SO_4$
$4\ FeSO_4 + 10\ H_2O + O_2 \rightarrow 4\ Fe\ (OH)_3 + 4\ H_2SO_4$
21–35 **a.** Cd, Cr, Cu, Pb, and Hg
21–38 bioaccumulation in fish
21–41 screening, grit removal, and sediment removal

Chapter 22

22–2 aspirin, acetaminophen, and ibuprofen
22–8 **a.** ibuprofen
22–9 **a.** ibuprofen **d.** aspirin and acetaminophen
22–11 morphine and codeine
22–16 **a.** codeine **d.** propoxyphene
22–21 **a.** bulk-forming; seeds of psyllium plant **d.** irritant; bisacodyl
22–31 prevent synthesis of folic acid
22–37 simulate action of norepinephrine
22–40 coffee beans and tea leaves
22–42 **a.** percolated coffee **d.** baking chocolate
22–44 tobacco leaves
22–50 **a.** mild sedation and tranquility **d.** unconsciousness
22–54 complex phenol

Chapter 23

23–5 antibacterial agent
23–13 glycerin, propylene glycol, and sorbitol
23–15 a substituted ammonium salt
23–21 $Mg_3(Si_2O_5)_2(OH)_2$
23–36 a film-forming agent, a plasticizer, a resin, a solvent, and a mixture of perfumes
23–39 acetone, ethyl acetate, and methyl ethyl ketone
23–44 $Ca_{10}(PO_4)_6(OH)_2$
23–50 an abrasive and a detergent
23–52 **a.** helps prevent tartar formation **d.** helps remove plaque
23–54 SnF_2, Na_2PO_3F, and NaF

Chapter 24

24–4 aldehyde, ketone, and hydroxy groups
24–8 **a.** glucose **d.** glucose **g.** fructose
24–10 **a.** alpha **d.** alpha
24–12 glucose
24–13 **a.** sucrose **d.** maltose
24–14 **a.** glucose and fructose

24—19 a. amylose **d.** cellulose
24—25 a. no **d.** yes
24—36 a. lard
24—49 a. minimizes water loss
24—53 cholesterol

Chapter 25

25—4 a. glutamic acid **d.** phenylalanine
25—10 they are the same
25—16 six
25—23 alpha-helix, beta-pleated sheet, and triple helix
25—33 a. nonpolar
25—37 secondary and tertiary
25—38 same primary structure
25—44 adenine, guanine, cytosine, and thymine
25—51 hydrogen bonding
25—52 a. TGCATA
25—54 a. 36%
25—57 complementary base pairing
25—62 AUGCCCAUUGAGC
25—66 mRNA molecules
25—68 a. leucine **d.** glycine
25—76 complementary base pairing
25—77 a. GGU **d.** GUA
25—80 the A site
25—88 insulin, growth hormone

Index

Boldfaced terms are defined on the boldfaced page(s).

877